DOCUMENTING DOMESTICATION

DOCUMENTING DOMESTICATION

NEW GENETIC AND ARCHAEOLOGICAL PARADIGMS

Edited by

MELINDA A. ZEDER

DANIEL G. BRADLEY

EVE EMSHWILLER

BRUCE D. SMITH

UNIVERSITY OF CALIFORNIA PRESS
Berkeley   Los Angeles   London

University of California Press, one of the most distinguished
university presses in the United States, enriches lives around the
world by advancing scholarship in the humanities, social sciences,
and natural sciences. Its activities are supported by the UC Press
Foundation and by philanthropic contributions from individuals and
institutions. For more information, visit *www.ucpress.edu*.

University of California Press
Berkeley and Los Angeles, California

University of California Press, Ltd.
London, England

© 2006 by
The Regents of the University of California

Library of Congress Cataloging-in-Publication Data

Documenting domestication: new genetic and archaeological
paradigms / edited by Melinda A. Zeder ... [et al.].
  p. ; cm.
  Includes bibliographical references and index.
  ISBN-13: 978-0-520-24638-6 (alk. paper)
  1. Plants, Cultivated–Genetics.
  2. Plant remains (Archaeology).
  3. Domestic animals–Genetics.
  4. Animal remains (Archaeology).
  [DNLM: 1. Animals, Domestic–genetics. 2. Adaptation,
Biological–genetics. 3. Archaeology. 4. Crops, Agricultural–genetics.
5. Evolution. QH 432 D637 2006] I. Zeder, Melinda A.
  SB106.G46D63   2006
  631.5′233–dc22
                                              2005036362

10 09 08 07 06
10 9 8 7 6 5 4 3 2 1

The paper used in this publication meets the minimum requirements
of ANSI/NISO Z39.48-1992 (R 1997) *(Permanence of Paper)*.

CONTENTS

LIST OF CONTRIBUTORS vii

LIST OF TABLES ix

LIST OF FIGURES xi

1 Documenting Domestication: Bringing Together Plants, Animals, Archaeology, and Genetics 1
Melinda A. Zeder, Daniel G. Bradley, Eve Emshwiller, and Bruce D. Smith

SECTION ONE
Archaeological Documentation of Plant Domestication
Bruce D. Smith, section editor

2 Documenting Domesticated Plants in the Archaeological Record 15
Bruce D. Smith

3 Seed Size Increase as a Marker of Domestication in Squash (*Cucurbita pepo*) 25
Bruce D. Smith

4 A Morphological Approach to Documenting the Domestication of *Chenopodium* in the Andes 32
Maria C. Bruno

5 Identifying Manioc (*Manihot esculenta* Crantz) and Other Crops in Pre-Columbian Tropical America through Starch Grain Analysis: A Case Study from Central Panama 46
Dolores R. Piperno

6 Phytolith Evidence for the Early Presence of Domesticated Banana *(Musa)* in Africa 68
Ch. Mbida, E. De Langhe, L. Vrydaghs, H. Doutrelepont, Ro. Swennen, W. Van Neer, and P. de Maret

7 Documenting the Presence of Maize in Central and South America through Phytolith Analysis of Food Residues 82
Robert G. Thompson

SECTION TWO
Genetic Documentation of Plant Domestication
Eve Emshwiller, section editor

8 Genetic Data and Plant Domestication 99
Eve Emshwiller

9 DNA Sequence Data and Inferences on Cassava's Origin of Domestication 123
Kenneth M. Olsen and Barbara A. Schaal

10 Relationship between Chinese Chive *(Allium tuberosum)* and Its Putative Progenitor *A. ramosum* as Assessed by Random Amplified Polymorphic DNA (RAPD) 134
Frank R. Blattner and Nikolai Friesen

11 Using Multiple Types of Molecular Markers to Understand Olive Phylogeography 143
Catherine Breton, Guillaume Besnard, and André A. Bervillé

12 Origins of Polyploid Crops: The Example of the Octoploid Tuber Crop *Oxalis tuberosa* 153
Eve Emshwiller

SECTION THREE
Archaeological Documentation of Animal Domestication
Melinda A. Zeder, section editor

13 Archaeological Approaches to Documenting Animal Domestication 171
Melinda A. Zeder

14 A Critical Assessment of Markers of Initial Domestication in Goats *(Capra hircus)* 181
Melinda A. Zeder

15 The Domestication of the Pig *(Sus scrofa)*: New Challenges and Approaches 209
Umberto Albarella, Keith Dobney, and Peter Rowley-Conwy

16 The Domestication of South American Camelids: A View from the South-Central Andes 228
Guillermo L. Mengoni Goñalons and Hugo D. Yacobaccio

17 Early Horse Domestication on the Eurasian Steppe 245
Sandra L. Olsen

SECTION FOUR
Genetic Documentation of Animal Domestication
Dan Bradley, section editor

18 Documenting Domestication: Reading Animal Genetic Texts 273
Daniel G. Bradley

19 Genetic Analysis of Dog Domestication 279
Robert K. Wayne, Jennifer A. Leonard, and Carles Vilà

20 Origins and Diffusion of Domestic Goats Inferred from DNA Markers: Example Analyses of mtDNA, Y Chromosome, and Microsatellites 294
G. Luikart, H. Fernández, M. Mashkour, P. R. England, and P. Taberlet

21 Mitochondrial DNA Diversity in Modern Sheep: Implications for Domestication 306
Michael W. Bruford and Saffron J. Townsend

22 Genetics and Origins of Domestic Cattle 317
Daniel G. Bradley and David A. Magee

23 Genetic Analysis of the Origins of Domestic South American Camelids 329
Jane C. Wheeler, Lounès Chikhi, and Michael W. Bruford

24 Genetic Documentation of Horse and Donkey Domestication 342
Carles Vilà, Jennifer A. Leonard, and Albano Beja-Pereira

INDEX 355

LIST OF CONTRIBUTORS

UMBERTO ALBARELLA
Department of Archaeology, University of Sheffield, Sheffield, UK.

ALBANO BEJA-PEREIRA
Centro de Investigação em Biodiversidade e Recursos Genéticos Campus Agrário de Vairão, Universidade do Porto, Portugal; and CNRS/UMR 5553, Laboratoire de Biologie des Populations d'Altitude, Université Joseph Fourier, Grenoble, France.

ANDRÉ A. BERVILLÉ
INRA/UMR 1097-DGPC, Montpellier, France.

GUILLAUME BESNARD
UNIL,DEE, Bâtiment de Biologie, Lausanne, Switzerland.

FRANK R. BLATTNER
Taxonomy and Evolutionary Biology, Institute of Plant Genetics and Crop Plant Research (IPK), Gatersleben, Germany.

DANIEL G. BRADLEY
Smurfit Institute of Genetics, Trinity College. Dublin, Ireland.

CATHERINE BRETON
Ingénieur CIFRE AFIDOL, Aix-en-Provence, France.

MICHAEL W. BRUFORD
Cardiff School of Biosciences, Cardiff University, Cardiff, Wales, UK.

MARIA C. BRUNO
Department of Anthropology, Washington University, St. Louis, MO, USA.

LOUNÈS CHIKHI
UMR 5174, "Evolution et Diversité Biologique," Université Paul Sabatier, Toulouse, France.

E. DE LANGHE
Laboratory for Tropical Crop Improvement and INIBAP Transit Centre, KULeuven, Belgium.

KEITH DOBNEY
Department of Archaeology, University of Durham, Durham, UK.

H. DOUTRELEPONT
Archaeology Section, Royal Museum of Central Africa, Tervuren, Belgium.

EVE EMSHWILLER
Department of Botany, Field Museum of Natural History, Chicago, IL, USA; and Botany Department, University of Wisconsin-Madison, Madison, WI, USA.

PHILLIP R. ENGLAND
CSIRO Marine & Atmospheric Research, Wembley, Western Australia.

HELENA FERNÁNDEZ
CNRS/UMR 5553, Laboratoire de Biologie des Populations d'Altitude, Université Joseph Fourier, Grenoble, France.

NIKOLAI FRIESEN
Botanical Garden, University of Osnabrück, Osnabrück, Germany.

JENNIFER A. LEONARD
Department of Evolutionary Biology, Uppsala University, Uppsala, Sweden/Genetics Program, Department of Vertebrate Zoology; and National Museum of Natural History, Smithsonian Institution, Washington, DC, USA.

GORDON LUIKART
CNRS/UMR 5553, Laboratoire de Biologie des Populations d'Altitude, Université Joseph Fourier, Grenoble, France; CIBIO, Centro de Investigação em Biodiversidade e Recursos Genéticos, Campus Agrário de Vairão, Universidade do Porto, Vairão, Portugal; and Division of Biological Sciences, University of Montana, Missoula, MT, USA.

DAVID A. MAGEE
Biotrin International, Mount Merrion, Co. Dublin, Ireland.

P. DE MARET
Free University of Brussels. Brussels, Belgium.

MARJAN MASHKOUR
CNRS/ESA 8045, Laboratory of Comparative Anatomy, National Museum of Natural History, Paris, France.

CH. MBIDA
Director of the Cultural Patrimonies, Ministry of Culture, Yaounde, Cameroon.

GUILLERMO L. MENGONI GOÑALONS
Sección Arqueologia, Instituto de Ciencias Anthropologicas, Facultad de Filosofía y Letras, Universidad de Buenos Aires, Buenos Aires, Argentina.

KENNETH M. OLSEN
Department of Biology, Washington University, St. Louis, MO, USA.

SANDRA L. OLSEN
Section of Anthropology, O'Neil Research Center, Carnegie Museum of Natural History, Pittsburgh, PA, USA.

DOLORES R. PIPERNO
Archaeobiology Program, Department of Anthropology, National Museum of Natural History, Smithsonian Institution, Washington, DC, USA; and Smithsonian Tropical Research Institute, Panama.

PETER ROWLEY-CONWY
Department of Archaeology, University of Durham, Durham, UK.

BARBARA A. SCHAAL
Department of Biology, Washington University, St. Louis, MO, USA.

BRUCE D. SMITH
Archaeobiology Program, National Museum of Natural History, Smithsonian Institution, Washington, DC, USA.

RO. SWENNEN
Laboratory for Tropical Crop Improvement and INIBAP Transit Centre, KULeuven, Belgium.

PIERRE TABERLET
CNRS/UMR 5553, Laboratoire de Biologie des Populations d'Altitude, Université Joseph Fourier, Grenoble, France.

ROBERT G. THOMPSON
Archaeobiology Laboratory, Department of Anthropology. University of Minnesota, Minneapolis/St. Paul, MN, USA.

SAFFRON J. TOWNSEND
Institute of Zoology, Regent's Park, London, UK.

WIM VAN NEER
Royal Belgian Institute of Natural Sciences, Brussels, Belgium; and Katholieke Universiteit Leuven, Laboratory of Comparative Anatomy and Biodiversity, Leuven, Belgium.

CARLES VILÀ
Department of Evolutionary Biology, Uppsala University, Uppsala, Sweden.

L. VRYDAGHS
Royal Museum of Central Africa, Tervuren, Belgium; and Royal Belgian Institute of Natural Sciences, Brussels, Belgium.

ROBERT K. WAYNE
Department of Organismic Biology, Ecology and Evolution, University of California, Los Angeles, CA, USA.

JANE C. WHEELER
CONOPA, Lima, Peru.

HUGO D. YACCOBACCIO
Sección Arqueologia, Instituto de Ciencias Antropológicas, Facultad de Filosofía y Letras, Universidad de Buenos Aires, Buenos Aires, Argentina.

MELINDA A. ZEDER
Archaeobiology Program, National Museum of Natural History, Smithsonian Institution, Washington, DC, USA.

LIST OF TABLES

3.1 Direct Accelerator Mass Spectrometer radiocarbon dates on early *Cucurbita pepo* seeds in the Americas: Guilá Naquitz and Phillips Spring. 28

4.1 Andean *Chenopodium* taxa. 33
4.2 Summary of published data on archaeological *Chenopodium* seed size and testa thickness from early sites in the Andes and Eastern North America. 35
4.3 Modern *Chenopodium* samples. 38
4.4 Seed diameter (mm) for modern *Chenopodium* Taxa. 39
4.5 Summary data for modern *Chenopodium* specimens from the southern Lake Titicaca Basin, Bolivia. 40
4.6 Morphological characteristics of modern *Chenopodium* taxa. 41

5.1 A key for bell-shaped starch granules from manioc and other species. 54
5.2 Starch grain size in modern wild and domesticated *Manihot*. 57

6.1 Radiocarbon dates recovered from excavated pits at the Nkang site. 70
6.2 Plant taxa represented in charcoal recovered from the Nkang site. 71
6.3 Animal taxa represented at the Nkang site. 72
6.4 Modern source material for *Musa* and *Ensete* phytoliths. 73
6.5 Diagnostic morphological characters in phytoliths of *Musa* and *Ensete*, compared with the characters exhibited by the phytoliths recovered from the Nkang site. 77

7.1 Rondel phytolith taxonomy. 85
7.2 Squared chord distance results of blind test identifying cob types. 88
7.3 Squared chord distance results of replicability tests. 88
7.4 Central and South American maize and residue samples. 91

8.1 Species and subspecies of *Zea*. 104

9.1 Cassava samples used in analyses. 127
9.2 Summary of variation in the nuclear loci across taxa. 128
9.3 Allele sharing between cassava and *Manihot esculenta* ssp. *flabellifolia*, by locus and across all loci. 129

10.1 Accessions used in the comparison of *Allium ramosum* and *A. tuberosum*. 138

11.1 List of oleaster populations analyzed, their locations, the markers tested, and the mitotype results. 144
11.2 List of the cultivars studied. 146
11.3 Distribution of molecular markers in oleasters and olive cultivars. 147
11.4 Key for prediction of the wild vs. feral status of any oleaster tree. 151

12.1 Archaeological reports of oca tuber remains or representations in art. 155
12.2 Diagnostic differences among the three ncpGS sequence classes. 161

14.1 Revised fusion sequence and ages for goats. 193
14.2 AMS dates on bones from Zagros sites. 194
14.3 Long-bone fusion scores for goats and gazelles from archaeological sites. 199
14.4 Proportions of male and female goats in sites from the Zagros. 201
14.5 Proportions of unfused or fusing bones among male and female goats in sites from the Zagros. 201

16.1 Matrix of dental morphology on South American camelid incisors. 231
16.2 Archaeological sites in the South-Central Andes. 236

17.1 Mortality profiles for Bronze and Iron Age horses. 250

17.2	Mortality profile for Copper Age Botai culture horses.	250	
17.3	Selection of Early Horse Remains from Sites in the Near East and Europe.	252	
17.4	Comparison of proportions of adult female and male horses at Botai.	258	
17.5	Distribution of Botai horses by sex/age categories, based on mandible and maxilla MNIs.	258	
17.6	Comparison of proportions of juvenile and adult horses at Botai.	259	
17.7	Botai P2 bevel measurements taken by Olsen.	260	
17.8	Artifact raw material from horse elements at Botai.	262	
19.1	Mitochondrial DNA haplotypes found in dog breeds and their distribution.	284	
19.2	Samples of ancient Native American dogs studied by Leonard et al.	286	
20.1	Average heterozygosity and allele number for 22 microsatellite loci in each of nine goat breeds.	301	
20.2	Hierarchical distribution of mtDNA (HVI) diversity within and among populations (and continental groups of populations) for goats, cattle and humans.	304	
21.1	Samples analyzed in this study, broken down by numbers per breed and region.	309	
21.2	Frequency of *NsiI* restriction site occurrence for all three major restriction profiles.	313	
22.1	Intra-regional mtDNA genetic diversity.	324	
22.2	Intra-regional microsatellite genetic diversity.	324	
23.1	Three locus genotypes for samples where all three types of data are available.	334	
23.2	Pairwise genetic distances between the four SACs.	335	
24.1	Assignment test.	347	
24.2	Nucleotide diversity (and standard deviation) for each mtDNA donkey clade, for each continent and each hypothetical domestication center.	350	

LIST OF FIGURES

3.1 The size of *Cucurbita pepo* seeds from archaeological and paleontological sites in North America compared to three taxa of modern wild *C. pepo* gourds.   26

3.2 The size of *Cucurbita pepo* seeds from Guilá Naquitz compared to six taxa of modern wild *Cucurbita* gourds.   26

3.3 The location of Guilá Naquitz Cave in the Valley of Oaxaca.   27

3.4 *Cucurbita pepo* seeds from Guilá Naquitz Cave yielding AMS dates in the Early Archaic.   27

3.5 Increase in the size of peduncles of domesticated *Cucurbita pepo* in preceramic habitation zones of Guilá Naquitz Cave between ca. 6500 and 5800 BC.   29

3.6 The location of the Phillips Spring archaeological site in south-central Missouri.   29

4.1 *Chenopodium* seed morphology.   36

4.2 Map of the southern Lake Titicaca Basin.   37

4.3 Measuring testa thickness from a scanning electron micrograph.   38

4.4 Seed diameter (mm) for modern *Chenopodium* taxa.   39

4.5 Testa thickness (microns) in modern *Chenopodium* taxa.   39

4.6 Scatterplot of log testa thickness and log diameter illustrating the ratio of testa thickness to diameter for four *Chenopodium* taxa.   39

4.7 Scanning electron micrographs of experimentally charred modern *Chenopodium*, showing intact testa and pericarp morphology.   42

5.1 Map of Panama with location of archaeological sites and lakes discussed in the text.   47

5.2 Maps of Central and South America showing the distribution of wild *Manihot* species.   48

5.3 *Manihot* roots.   50

5.4 Starch grains from various crop plants of the Neotropics.   52

5.5 Compound starch grains from manioc still in aggregations as they originally formed in amyloplasts.   53

5.6 Starch grains from *Manihot esculenta* ssp. *flabellifolia* showing how they have eccentric hila and lack fissures.   56

5.7 Starch grains from *Manihot carthaginensis* showing how they have eccentric hila and lack fissures found in manioc.   56

5.8 Starch grain population from *Manihot aesculifolia* showing how it is dominated by simple, bell-shaped granules.   58

5.9 Some of the grinding stones examined from the Aguadulce Rock Shelter.   59

5.10 Starch grains from manioc from the Aguadulce Rock Shelter.   61

5.11 Starch grains from *Dioscorea* found on an edge-ground cobble at the Aguadulce Rock Shelter.   63

5.12 Starch grains from a tuber of *Dioscorea cymosula*, a wild species from Panama.   64

6.1 Location of the Nkang site in central Cameroon.   69

6.2 Cross-section of pit F9 at the Nkang site, which yielded *Musa* phytoliths.   70

6.3 Morphological characteristics of *Musa* and *Ensete* phytoliths.   73

6.4 Scanning electron micrograph of a modern *Musa* phytolith.   74

6.5 Scanning electron micrograph of a modern *Ensete* phytolith.   74

6.6 Light microscopy views of phytoliths recovered from the Nkang site.   75

6.7 Detailed side view of a phytolith recovered from the Nkang site.   75

6.8 Detailed top views of a phytolith from the Nkang site.   76

7.1 Entire decorated rondel in planar view and tilted rondel.   86

7.2 Indented rondel.   86

7.3 Four rondels in planar view.   86

7.4 Three maize ears showing *tga1* phenotypes.   87

7.5	Squared chord distance results of blind test identifying maize types.	88	11.3	Dendrogram based on the SSR dataset for oleasters.	148
7.6	Squared chord distance results of replicability tests.	89	11.4	Results from multiple correspondence analyses based on SSR data.	149
7.7	Location map of archaeological sites and maize collection areas mentioned in text.	90	11.5	Maps of possible routes followed by oleasters and cultivars in the Mediterranean Basin.	150
7.8	Squared chord distance results of modern maize comparative samples.	92	12.1	Diversity of tuber morphology and pigmentation in *Oxalis tuberosa* cultivated by a single household in the Campesino community of Viacha, Pisac District, Cusco Department, in southern Peru.	154
7.9	Squared chord distance results of La Emerenciana, Pirincay, Chancay, and Sierra Gorda vessels.	92	12.2	Distribution of wild species in the *O. tuberosa* alliance.	156
7.10	Squared chord distance results of Nicoya Peninsula vessels.	92	12.3	Simplified diagrams of the derivation of homologous and homeologous (partially homologous) chromosomes in an octoploid.	158
7.11	Squared chord distance results of Palmitopamba vessels.	93	12.4	The internal transcribed spacer region (ITS) of nuclear ribosomal DNA.	159
8.1	Cluster analysis of taxon means of Jaccard's Coefficient of Genetic Similarity values among infraspecific taxa of *Cucurbita pepo* and *C. argyrosperma*.	110	12.5	The amplified portion of chloroplast-expressed glutamine synthetase (ncpGS).	159
9.1	The cultivated cassava plant, *Manihot esculenta* ssp. *esculenta*.	124	12.6	One of 208 trees that resulted from analysis of ncpGS sequences from members of the *O. tuberosa* alliance, including cloned sequences from three plants of cultivated oca.	162
9.2	Diagrams of the three low-copy nuclear gene regions used for DNA sequence analysis.	125	12.7	One of the scenarios that was congruent with the ncpGS sequencing results.	162
9.3	Locations of *Manihot esculenta* ssp. *flabellifolia* populations and *M. pruinosa* populations sampled along the eastern and southern borders of the Amazon basin.	126	14.1	Map of the Fertile Crescent showing the distribution of wild goats *(Capra aegargrus)* and wild sheep *(Ovis orientalis)* and archaeological sites mentioned in the text.	182
9.4	The *G3pdh* haplotype genealogy (gene tree).	128	14.2	Map of Iran and Iraq showing locations of modern and archaeological samples.	190
9.5	Wild populations most closely related to cassava.	129	14.3	Modern goat breadth and depth measurements of selected long bones.	191
9.6	Maximum likelihood distance tree for wild populations and cassava accessions.	130	14.4	Modern goat length measurements of selected long bones.	191
10.1	Possible evolutionary relationships of crop species and their wild relatives.	136	14.5	Diachronic view of changes in the length of the second phalanx (GL) of goats from the Middle Paleolithic to the present day.	192
10.2	RAPD reaction of 29 *A. ramosum* and *A. tuberosum* accessions with Operon primer AB04, electrophoretically separated on a 1.5% agarose gel.	139	14.6	Diachronic view of changes in the depth of the distal humerus (Dd) of gazelle from the Epi- and Late Paleolithic to the present day.	196
10.3	Phenogram of a neighbor-joining analysis of 137 RAPD characters of accessions of wild *A. ramosum* and the crop plant *A. tuberosum*, together with the closely related *A. oreiprason* as outgroup taxon.	139	14.7	Breadth and depth measurements of goat long bones from Ganj Dareh arranged by age at fusion.	197
10.4	Unrooted neighbor-joining analysis of 127 RAPD characters of 29 accessions of wild *A. ramosum* and the crop *A. tuberosum*.	140	14.8	Breadth and depth measurements of goat long bones from Ali Kosh arranged by age at fusion.	198
11.1	Oleasters near Tamanar, Morocco.	144	14.9	Sex-specific harvest profiles of goats and gazelles from Zagros sites.	200
11.2	Map of oleaster and cultivar mitotypes in the Mediterranean Basin.	147	14.10	Sex-specific harvest profiles of goats from different levels at Ganj Dareh.	202

15.1	Astragalus length (GLl) for a variety of samples.	213
15.2	Map of the Alpine region and northern Italy showing the locations of sites mentioned in the text.	214
15.3	M3 measurements from Alpine Neolithic sites.	214
15.4	Variation in *Sus* tooth and bone measurements at the Mid Neolithic site of Rivoli (northern Italy).	215
15.5	Variation in *Sus* postcranial bone measurements at three Neolithic sites in northern Italy.	215
15.6	*Sus* mandibles from second millennium deposits at Chagar Bazar, northern Syria, showing abnormal wear and breakage of the tooth crowns.	220
15.7	*Sus* mandibles from second millennium deposits at Chagar Bazar, northern Syria, showing heavy dental calculus deposits.	220
16.1	Map of the Andean area showing localities discussed in the text.	229
16.2	Size variation in contemporary guanaco based on the proximal width and proximal depth of the first phalanx.	232
16.3	Size gradient in contemporary camelids using the Andean guanaco as standard.	233
16.4	Temporal trends in the use of camelids for the South-Central Andes 11,000–8500 BP.	235
16.5	Histogram showing the log difference between measurements of modern North Andean guanaco and archaeological specimens from several sites located in the South-Central Andes.	237
16.6	Bivariate plot of measurements of the distal metacarpal of selected large camelid specimens from northwestern Argentinean sites dated from 4100 to 2000 BP taken following Kent's protocols.	238
17.1	Map showing major sites in the Eurasian steppe.	253
17.2	Bit wear on lower second premolar from Malyan.	255
17.3	Botai hunting tools and traces.	259
17.4	Modern horse cranium in Mongolia with pole-axing wound.	259
17.5	Horse cranium with possible pole-axing wound from Botai.	259
17.6	Graph of Botai P2 bevel measurements taken by Olsen.	260
17.7	Lingual view of lower second premolar of *Equus lambei*.	261
17.8	Thong-smoother made on a horse mandible from the Botai culture settlement of Krasnyi Yar.	261
17.9	Scanning electron micrograph of the notch of a thong-smoother from Botai.	261
17.10	Horse cranium and cervical vertebrae in a pit outside a house at Krasnyi Yar.	263
17.11	Plan of the Botai culture settlement of Krasnyi Yar.	263
17.12	Plan of the Botai culture settlement of Vasilkovka.	264
18.1	Neighbor-joining networks linking mtDNA sequences from four domestic animal samples.	274
18.2	Synthetic map showing the first principal component resulting from the allele frequencies at six milk protein genes in 70 breeds across Europe and Turkey.	277
19.1	Allozyme and microsatellite electrophoresis.	280
19.2	Example of the use of DNA sequence for the construction of phylogenetic trees.	280
19.3	Difference between the species divergence time and the estimates obtained using genetic and morphological data.	281
19.4	Neighbor-joining relationship tree of wolf and dog mitochondrial DNA control region sequences.	282
19.5	Neighbor-joining phylogeny of modern domestic dogs from throughout the world and ancient domestic dogs from the Americas, with the coyote as outgroup.	287
19.6	Statistical parsimony network of Clade I modern dogs from throughout the world and ancient dogs from America.	288
20.1	Photos illustrating the morphological diversity and distinctiveness of two important goat breeds.	295
20.2	Phylogenetic tree of domestic and wild goat cytochrome *b* gene sequences.	296
20.3	Phylogenetic tree constructed from two Y-chromosome gene fragments.	297
20.4	Relationships among nine domestic goat breeds and two wild *Capra* species.	298
20.5	Phylogenetic trees from mitochondrial DNA (control region) sequences.	299
20.6	Map showing the distribution of maternal and paternal DNA lineages.	300
20.7	Mismatch distributions for mtDNA types from the major lineage of goat sequences (*C. hircus* A).	302

21.1	Neighbor-joining phenogram of control region sequences of domestic and wild sheep.	312	23.1	Minimum spanning network representing the relationships among cytochrome *b* mitochondrial haplotypes.	331
21.2	Restriction profiles of ovine control region showing banding patterns diagnostic for each haplogroup.	313	23.2	Allele size and frequency histograms for LCA 19 and YWLL 46.	332
21.3	Geographic distribution of mitochondrial CR region haplogroups in domestic sheep [HPG] A, B, and C.	314	23.3	Two-dimensional factorial correspondence plot for allele frequencies at four microsatellite loci in all SACs.	333
22.1	Scaled two-dimensional representation of genetic distances between cattle populations.	320	23.4	Admixture analysis using four microsatellite loci.	336
			23.5	Posterior density distributions for admixture contributions (p_1) calculated from a combined analysis of four loci.	337
22.2	Phylogenies of seven cattle mtDNA breed samples superimposed on approximate sample origins.	321			
22.3	Unrooted neighbor-joining phylogenies constructed using control region sequences from cattle, impala and Grant's gazelle, drawn to the same scale.	322	24.1	Phylogenetic tree of mtDNA control region sequences in horses.	345
			24.2	Genetic diversity in horse breeds.	346
			24.3	Pair-wise number of female migrants per generation ($N_f m_f$) and absolute number of migrants per generation *(Nm)* between horse breeds.	348
22.4	Neighbor-joining phylogeny of *Bos indicus*, *Bos taurus*, and extinct British *Bos primigenius* mtDNA control region haplotypes.	322			
22.5	A graph of the mean pairwise difference for each breed plotted by breed groupings.	323	24.4	Phylogenetic tree based on mtDNA control region sequences from donkeys (*Equus asinus*) and their kin.	349
22.6	Skeleton and reduced media networks for *Bos taurus* cattle haplotypes.	325	24.5	Unrooted tree of all domestic donkey mtDNA haplotypes	349

CHAPTER 1

Documenting Domestication
Bringing Together Plants, Animals, Archaeology, and Genetics

MELINDA A. ZEDER, DANIEL G. BRADLEY,
EVE EMSHWILLER, AND BRUCE D. SMITH

Introduction

Domesticates and the process of their domestication have been central, foundation areas of study in both biology and archaeology for more than 100 years. Although Charles Darwin's voyage on the *HMS Beagle* is often credited as the source of his theories about natural selection and biological evolution, Darwin opened *The Origin of Species* (Darwin 1859) with a chapter on human-induced variation under domestication, later following it with a two-volume work dedicated entirely to domesticated plants and animals (Darwin 1868). While Darwin was developing his revolutionary theories, Gregor Mendel was conducting experiments in cross-breeding varieties of garden peas at an Augustinian monastery in what is now the Czech Republic. Mendel's experiments yielded key insights into the inheritance of traits (Mendel 1865), providing Darwinian evolution with its driving mechanisms and creating the field of genetics.

Domestication was also the focus of the first scientifically oriented archaeological investigations, with Raphael Pumpelly's turn-of-the-century excavations, at the site of Anau in Turkmenistan, that tested his theories about the role of climate change in the emergence of agriculture (Pumpelly 1905). Pumpelly's work on agricultural origins was to have a profound influence on the later work of V. Gordon Childe, who argued that agricultural origins represented one of the two great transforming revolutions in human history (Childe 1951).

The interdisciplinary investigations of Robert Braidwood in the Fertile Crescent (Braidwood and Howe 1960; Braidwood et al. 1983) and Richard MacNeish in central Mexico (MacNeish 1967) in the 1950s and 1960s built on this legacy, and they set the precedent of bringing together researchers from diverse disciplines of botany, zoology, geology, and archaeology to jointly explore fundamental questions of when, where, how, and why humans made the transition from hunting and gathering to herding and farming. The origin of plant and animal domestication has remained among the "big questions" of archaeological inquiry ever since.

Advances in molecular and archaeobiological techniques over the past two decades have resulted in a virtual explosion of studies exploring the origins of plant and animal domestication. New genetic techniques have allowed plant and animal geneticists to identify the wild progenitors of key domestic species and the likely geographic context of initial domestication, to explore the diffusion of domestic crops and livestock, and even to begin to decode the specific genetic shifts that structured the transformation of a wild species into a domestic one. At the same time archaeobiologists have developed increasingly specialized approaches to the study of plant and animal domestication. The use of high-power scanning electron microscopy (SEM) and the development of advanced techniques for the recovery and identification of plant macro- and microfossils have resulted in sophisticated approaches for distinguishing wild from domestic plants and animals in archaeological assemblages. Above all, the ability to precisely and directly date plant and animal remains from archaeological sites has had a profound impact on calibrating the chronology of plant and animal domestication.

One of the consequences of these technological advances has been the development of increasingly more specialized approaches to the study of plant and animal domestication. The complexity of the topic demands this kind of focus, but the degree of specialization required to adequately explore questions about plant and animal domestication has also served to create a number of disciplinary silos that too often inhibit productive communication among the diverse range of researchers engaged in documenting domestication. Archaeobotanists working on the domestication of plants in the Old World, for example, do not regularly reference the work of those focusing on New World domesticates, and those who specialize in the analysis of macrofossil remains (e.g., seeds and fruit fragments) often do not fully appreciate the contributions of those who focus on the analysis of microfossils (e.g., pollen, phytoliths, and starch grains). Archaeozoologists working on livestock domestication rarely collaborate with archaeobotanists working on the domestication of crop plants, even when these species were domesticated at the same time by the same people. Molecular biologists interested in the origins of domestication of various plant or animal species often have little familiarity with recent archaeological research on the same species. Biologists working on the evolutionary genetics of crop plants do not regularly communicate with those working on similar problems with domestic livestock.

With all the focused attention on documenting the domestication of individual plant and animal species using highly specialized analytical approaches, it is easy to lose sight of the fact that there are fundamental common features that underlie the process of domestication, regardless of the species involved or the geographic or temporal context of its domestication. Every instance of domestication grows out of a mutualistic relationship between a plant or animal species and a human population that has strong selective advantages for both. For the target plant or animal species, human agency in their propagation and care provides a distinct competitive advantage, allowing the domesticate to expand far outside the environmental parameters that define the geographic range of its wild progenitor. And although breeding livestock and cultivating crops may not, at least initially, have provided humans a more bountiful or nutritious resource base than the procurement of wild resources, domesticates provided an important measure of security and predictability in human subsistence economies that fueled human population growth and expansion into new and challenging environments. It is this synergistic process, this coming together of a plant or animal population with a human population in an increasingly dependent mutualism, that all researchers interested in domestic origins seek to understand.

The course of this evolving relationship is profoundly shaped by both the biology of a target species and the cultural context of its human partners. There are different traits or attributes that make certain plant and animal species, and certain individuals within these species, more likely candidates for domestication than others. There are a variety of environmental and social variables that will make some human groups more likely to enter into this relationship with certain plant and animal species than others. The course of the domestication process of a plant or an animal will be shaped by the nature of its relationship with humans—the degree of active human intervention into its life cycle, the nature of the resource of interest to humans, the extent to which it becomes isolated from populations not involved in this partnership with humans. The degree of genetic plasticity, the flexibility of the species to adapt behaviorally and physiologically to new selective pressures introduced by human control, plays a central role in determining the course of the domestication process. Studying this process, then, requires understanding the central unifying principles that undergird all instances of domestication, as well as the unique factors that shape the course of domestication for individual species in different cultural and environmental settings.

Documenting domestication requires identifying markers that can be directly linked to this evolving relationship. These markers may take the form of morphological change in the target species, changes in its genetic structure, a restructuring of its population biology, or the transformation of its ecological context. Markers of domestication may also be found in the tools, settlement patterns, or even the ideology of the human partners in the domestication process. Detecting these markers and understanding their relationship to domestication require the expertise and analytical skills of a diverse range of researchers schooled in a variety of disciplines. But addressing overarching questions of how and why this domestication happened in a wide array of plant

TABLE 1.1
The Four-Celled Matrix for This Volume

	Archaeology	Genetics
Plants	Archaeological Documentation of Plant Domestication	Genetic Documentation of Plant Domestication
Animals	Archaeological Documentation of Animal Domestication	Genetic Documentation of Animal Domestication

and animal species involving human groups around the world ultimately requires being able to draw these different specialized perspectives together to build a context for cross-illumination.

This volume marks a first attempt to bring together a wide range of researchers working on plant and animal domestication from the perspectives of archaeology and genetics. All of the contributors to this volume are pursuing the common goal of documenting domestication. This book began with a symposium hosted at the annual meetings of the American Association for the Advancement of Science in 2001, during which four researchers working on the domestication of plants and animals from archaeological and genetic perspectives provided an overview of the accomplishments in their area of study and the challenges and opportunities for future research: Bruce Smith spoke about the archaeology of plant domestication; John Doebley about the genetics of plant domestication; Melinda Zeder, the archaeology of animal domestication; and Dan Bradley, the genetics of animal domestication. Following the same four-celled matrix that structured the 2001 AAAS symposium (see Table 1.1), this volume is divided into four sections, with each of the four volume editors (Dan Bradley, Eve Emshwiller, Bruce Smith, and Melinda Zeder) responsible for one of the sections.

Each section editor was asked to compile a collection of case studies that provided outstanding examples of the documentation of a plant or an animal species from the perspectives of either archaeology or genetics. Case study authors were instructed to explore the linkage between the process of domestication in the species they study and the marker or markers they use to document its domestication. Thus the section on the archaeological documentation of plant domestication, edited by Bruce Smith, includes chapters that use both macro-morphological markers (e.g., seed size, testa thickness) and micro-morphological markers (e.g., phytoliths and starch grains) to document the archaeological presence of a range of domesticates: pepo squash, quinoa, manioc, bananas, and maize. The section on the genetic documentation of plant domestication, edited by Eve Emshwiller (with a substantial initial contribution by Deena Decker-Walters, who selected the case study chapters and completed their initial editing), features the use of a wide range of genetic analytical techniques (e.g., RAPDs, RFLPs, DNA sequence data) applied to different plant genomes (mitochondrial, nuclear, chloroplast) to document the domestication and diffusion of cassava (manioc), the Chinese chive, olives, and oca. Melinda Zeder's section, on archaeological approaches to the study of animal domestication, features case studies that use various combinations of morphological and nonmorphological markers to document domestication in sheep and goats, pigs, South American camelids, and horses. Probably because the number of animal domesticates is more limited than the number of domesticated plants, there is more overlap between the archaeological and genetic animal case study chapters, with the section on the genetic documentation of animal domestication, edited by Dan Bradley, featuring the use of mitochondrial, chromosomal, and microsatellite DNA (both modern and ancient) to trace the ancestry of domestic dogs, goats, sheep, cattle, South American camelids, horses, and donkeys.

In addition, editors were asked to provide an overview chapter for their section in which they consider overarching issues in the documentation of domestication in their area of expertise, tying together themes raised in the case study chapters and bringing in additional examples from the work of other researchers on species not featured in their sections. The intent of these overview chapters is to provide a more general discussion of the accomplishments, as well as some of the shortcomings, of work in each general area and to outline the directions for future work.

No attempt is made in this volume at worldwide coverage of all key domestic species, nor does this volume seek to provide an encyclopedic rendering of the global story of the when and where, or even the why, of plant and animal domestication. There are a number of books available that attempt to do this already, either for plants (e.g., Cowan and Watson 1992), or animals (e.g., Zeuner 1963; Clutton-Brock 1999), or both plants and animals (e.g., Smith 1998). Instead, the volume features a wide range of approaches growing out of archaeology and genetics that highlight new and emerging paradigms for documenting domestication and point the way for future work in this quickly expanding research frontier. More than simply producing a "how-to" manual for documenting domestication, we hoped that by bringing these different approaches together, by featuring plants, animals, archaeology, and genetics in a single volume, we would be able to point the way to a deeper understanding of domestication as a general biological and cultural process. In this introductory chapter, we highlight some of the cross-cutting themes that have emerged from the four sections of the book as a means of moving toward this more synthetic understanding of domestication and the complex challenges that confront researchers seeking to document it.

Archaeological Approaches to Documenting Plant and Animal Domestication

Perhaps the most significant difference in the archaeological documentation of plants and animals is that in plants the new selective pressures introduced under domestication operate directly on morphological traits, whereas in animals they operate on behavioral attributes. This basic, though often overlooked, difference in the process of plant and animal domestication has a profound effect on the markers and methods archaeologists use to document plant and animal domestication.

There are a number of widely recognized traits that make some plants more attractive candidates for domestication than others. Primary among these are the generalized ability to colonize and adapt to the open, disturbed habitats created by human activities, particularly around settlements (Smith 1992). In these new anthropogenically created ecological frontiers, humans and colonizing plants came together to begin the mutualistic partnership that led to domestication. Once humans began to store and deliberately plant the harvested seed stock of certain of the more promising plant species (e.g., those with more palatable, less difficult to process seeds or fruits), a whole suite of adaptive responses was triggered that resulted in clear-cut morphological changes in target species. Sometimes referred to as the "adaptive syndrome of domestication" (Chapter 2), these morphological changes include increases in seed size and the thinning of seed coats, which allowed the candidate domesticate to sprout more quickly, crowd out competitors, and ensure that its seeds would be harvested and replanted the next season. Also included in this suite of adaptive morphological responses to domestication are the changes in or loss of seed dispersal mechanisms. In cereal grasses, for example, domestication is marked by a change in dispersal mechanisms from the brittle, shattering seed heads that ensure the propagation of wild plants, to the tough rachis that binds seeds to the grain head and ensures that harvested grain is transported back to human settlements and included in next season's seed stock. Although humans may not have deliberately selected for these specific morphological changes, these largely automatic responses on the part of the plant were, in fact, induced by deliberate human attempts to intercede in the life cycle of the plant in order to control its productivity and are thus considered leading-edge markers of domestication in annual seed plants. Deliberate human selection for such attributes as larger fruit size or changes in starch composition are usually considered later developments in the unfolding process of plant domestication.

The challenge faced in documenting domestication in a plant species, then, includes identifying those morphological markers that are the direct result of the genetic changes in plants caused by the new selective pressures arising from the evolving domestic partnership. Thus, Smith's use of seed size in squash in Chapter 3 and Bruno's use of testa thickness in quinoa (Chapter 4) both capture the leading edge of this adaptive syndrome of domestication of these two crop plants. Changes in fruit size of squash detected through analysis of peduncle morphology (Smith, Chapter 3) or in the size and form of starch grains of manioc (Piperno, Chapter 5) are both indicative of deliberate human selection for desirable traits in cultivated crop plants, as is Thompson's phytolith evidence for the distinguishing of soft and hard glume varieties of corn (Chapter 7). In all of these case study examples, then, there are clear causal links between the process of domestication of the various plant species studied and the genetically controlled morphological markers used to document the process. The challenge met by the various authors in this section of the book has been to discover the morphological markers of domestication in a variety of different plant species, to analyze how these markers relate to the process of the species' domestication, and then to develop clear-cut, replicable methods for detecting them in the archaeological record.

There are also a number of characteristics that make certain animal species more attractive candidates for domestication than others, but these attributes are primarily behavioral rather than morphological—attributes such as a social structure based on a dominance hierarchy, tolerance of penning, sexual precocity, and, above all, the possession of a less wary, less aggressive nature (Clutton-Brock 1999; Chapter 13). This last attribute is probably the single most important preadaptive factor that makes an animal species, and individuals within that species, attractive candidates for domestication. In a process similar to the invasion of anthropogenically disturbed habitats by colonizing plant species, the domestication of some animal species was probably set into motion when less wary individuals approached human settlements to feed off human refuse and stored food stocks. This is especially likely in the domestication of the dog and the pig, both species capable of omnivory with diets that overlap with those of humans, although it may also apply to such species as sheep and goats that may have been drawn to the same cereal stands utilized by humans. Certainly more concentrated selection for reduced aggression came into play once humans began to consciously tend, breed, and move herds of animals.

As with plants, different morphological changes in animals may come about at different points of the domestication process. For example, there is a suite of behavioral, physiological, and morphological traits that are hypothesized to be genetically linked to a reduction of aggressive behavior (e.g., greater playfulness, early onset of sexual maturity, changes in brain size and organization, changes in cranial morphology; Kruska 1988; Hemmer 1990) that should be seen early on in the domestication process. In certain species such as pigs and dogs (and perhaps other commensals such as sparrows, mice, and rats) that were initially drawn to human settlements to feed off refuse and food stock, some of these changes may have been set into motion even before there was any deliberate human intervention in their life cycle. Other morphological changes may have come about largely as side effects of the human intervention in the selection of breeding partners (e.g., changes in male horn morphology in bovid domesticates and perhaps in the body size of males), while a suite of other changes may have ensued once managed animals were moved out of their natural habitat and introduced into new environments (either through directed adaptations to new environmental conditions or through a more random founder effect). Finally, and probably even later than in plants, other morphological changes may have come about through deliberate attempts to breed animals for specific desirable traits (e.g., fat content, fiber quality, milk yield, strength, and speed). But because all but these latter morphological traits were tangential, and often delayed, side effects of the more direct selection for behavioral attributes in early animal domesticates, and because other factors may contribute to similar morphological changes in animals entirely outside of the context of domestication (e.g., change in body size as a response to climatic changes), researchers have tended to turn to a combination of nonmorphological markers, as well as a number of non–genetically controlled morphological markers to document domestication in animals.

Thus, while Albarella et al. (Chapter 15) and Mengoni Goñalons and Yacobaccio (Chapter 16) employ genetically controlled morphological markers that come into play at different points in the domestication process of pigs and South American camelids, they also use a suite of non–genetically controlled plastic responses to domestication (e.g., changes in health and diet), as well as data on population structure and evidence of human control, to document domestication in these species. Zeder (Chapter 14) questions the value of traditional morphological markers used to document domestication in sheep and goats, arguing that changes in harvest strategies reflecting human management can be detected in goats hundreds of years before the manifestation of any genetically controlled morphological responses to domestication. Olsen (Chapter 17) bases her case for horse domestication in Eurasia on a broad suite of nonmorphological markers including biogeographic and abundance data, population structure, skeletal part distribution, and a range of archaeological attributes from settlement patterns, lithic technology, and ritual behavior.

Not all studies of plant domestication, however, rely on documenting the morphological changes resulting from domestication. Indeed, as shown in Mbida et al.'s chapter on bananas in Africa (Chapter 6), being able to demonstrate the introduction of a non-native plant far outside its natural range can be used as a marker of the diffusion of a domestic crop outside the site of its initial domestication. Other nonmorphologically based approaches to documenting plant domestication that are coming into increasing use involve the reconstruction of plant communities or evidence of burning to detect human activities of land clearing and cultivation (Colledge 1998; Piperno et al. 1991; Smith, Chapter 2). Moreover, the extraction of plant residues (starches and phytoliths) from processing tools (as featured in the chapters by Piperno and Thompson) provide special insight into human use of domesticates, similar to the insights into animal exploitation gained by looking for evidence of milk residues or horse tackle (see Olsen, Chapter 17). New evidence for the use of wild cereals seen in both macrofossils and starch grains embedded in grinding stones recovered from the 20,000-year-old site of Ohalo on the shores of the Sea of Galilee in Israel (Weiss et al. 2004; Piperno et al. 2004b) suggests that intensive use of cereals may have preceded any evidence of morphological change in these plants by several millennia. If so, it is possible that nonmorphological markers that signal a change in the relationship between humans and wild plant resources will play an increasingly important role in tracking the leading edge of the process of plant domestication, just as they have with animal domestication.

There are a number of methodological similarities that run through the chapters on the archaeological documentation of plant and animal domestication that should be highlighted here. The first is the use of modern collections. The case study chapters on plants all make use of modern collections to provide critical reference class comparative data needed to develop effective methods for distinguishing domesticates in the archaeological record, as do the chapters on the archaeological documentation of sheep and goats, pigs, and South American camelids. All the archaeology chapters also explicitly detail the methods used to document domestication in plant and animal species in a way that allows other researchers to apply them to different data sets. Early efforts tended to treat the archaeological documentation of plant and animal domestication as the purview of a limited number of practitioners skilled in the art of distinguishing domestic from wild in the archaeological record. Thankfully, this era is long gone, as the chapters in this part of the volume amply demonstrate. Above all, the archaeological documentation of domestication must rest on secure, empirically grounded methods that can be applied by different researchers to different archaeological assemblages. The chapters on the archaeological documentation of plant and animal domestication in this volume set important standards for how this is to be done. Finally, all of the archaeology chapters emphasize the ability to directly date archaeobiological remains using small sample radiocarbon dating techniques (i.e., atomic mass spectrometer, or AMS, dating) to reconstruct the temporal sequence of domestication. Coupled with the development of methods capable of identifying domesticates in the archaeological record, direct dating of the remains of early domesticates provides an unprecedented degree of temporal resolution to the documentation of domestication.

Genetic Documentation of Plant and Animal Domestication

The genetic documentation of domestication focuses on the changes in genetic structure of plants and animal populations that come about through the domestic partnership with humans. These changes include those that control the morphological changes used by both archaeobotanists and archaeozoologists to mark plant and animal domestication in archaeological assemblages, as well as the physiological and behavioral responses to domestication that are harder to trace in the archaeological record. They also include the largely neutral, noncoding loci with their variable rates of mutation that are widely used in tracing the evolutionary ancestry of domesticates.

One of the major differences in the genetic response of plants and animals to the selective factors introduced by domestication stems from differences in generation time that affects both the pace of the domestication process and the rate of accumulation of change in DNA sequences. Generation time varies in both plants and animals, correlating very roughly with the "body size" of the organism. Although there are some long-lived tree crops and some short-lived domesticated animals, there are among the important domesticates a large number of annual plants with yearly generation times and many larger animals with relatively longer generation times. Truly annual plants (those that complete their entire life cycle within a single year) are likely to have a more rapid rate of DNA sequence evolution than longer-lived organisms. Their short life cycle also allows them to respond more quickly to the effects of selection, so that they may be domesticated more rapidly than long-lived taxa. This means that the genetic responses to domestication will happen more quickly in most crop plants than they will in most animal domesticates. The more rapid cycling of genetic material in plants (as well as the fact that the selective pressures introduced by domestication operate more directly on morphological features of plants) is another reason why one might expect to see morphological changes more quickly in plants undergoing domestication than in animals.

Even though their above-ground parts might die back each year, many plants cultivated on an annual cycle are not true annuals at all. These plants do not complete their life cycle from seed to seed in a single year, but rather persist and are propagated vegetatively by means of tubers, bulbs, cuttings, and the like (see Olsen and Schaal, Chapter 9, for manioc; Blattner and Friesen, Chapter 10, for Chinese chive; and Emshwiller, Chapter 12, for oca). Thus, the generation time, from seed to offspring seed, may be extremely long in vegetatively propagated crops. Whereas the breakthroughs in techniques for animal "cloning" have made the headlines in recent years, cloning plants is very old news indeed, as people have been cloning plants for many millennia. Although the early history of domesticated root crops is just beginning to emerge from the archaeological record, it is likely that they exhibit the same temporal range of domestication as seed-propagated crops (Sauer 1952; Harris 1969; Lathrap et al. 1977). The implications of clonal propagation versus sexual reproduction for domestication and evolution of crop plants are enormous. Harlan (1992) wrote that "[a]mong vegetatively propagated crops, selection is absolute and the effects immediate." Here he was referring to the initial domestication of crops from a wild ancestor, whereby favored individuals could be propagated and form new cultivars immediately. Nevertheless, the "absolute" effects of selection continue to affect crop evolution long after initial domestication. Cloning provides domesticators with the advantage that favored types can be propagated indefinitely with neither the need to prevent outcrossing nor the problems of inbreeding. Without variation there is no evolution, however. So, unless there is some sexual reproduction to reshuffle genes, evolution proceeds very slowly, depending entirely on somatic mutations. Although such mutations may indeed produce morphological changes that can be selected by farmers, these mutations may or may not be detectable with molecular markers. Thus clonal crops may have low variability, and the variability that exists today may be difficult to study. The paucity of variable markers may provide insufficient power to resolve phylogenetic and/or genealogical relationships of these clonal crops, and intraspecific morphological variation may correlate poorly with molecular variation.

Another important difference in the biology of plants and animals with special relevance to the genetic study of domestication is the prevalence of polyploidy in plants. Polyploidy is extremely rare in animals, and all domesticated animals are diploid, with two sets of chromosomes, one full set contributed by each parent. In contrast, a majority of crop plants are polyploid, with anywhere from three to eight or more different sets of chromosomes representing multiple ancestral genomes. Moreover, multiple genomes of a polyploid may have been contributed either by ancestors from within a single species or, through interspecific hybridization, by a number of different species. While interspecific hybridization is a factor in the evolution of numerous crop plants, it is rare in domesticated animals and usually results in sterile offspring and is thus without any evolutionary impact. Although polyploidy is argued to have a basically neutral effect on the ability of a species to be domesticated (Hilu 1993), polyploid evolution is clearly an important, and complicating, part of the story of crop evolution that is technically very challenging to study (see Emshwiller, Chapter 12).

Another promising emergent area in the genetic analysis of domesticates concerns the study of the genes and gene complexes that are specifically selected for (or against) by domestication. Here plants have the advantage over animals. Many plants produce an abundance of seed, so that a large number of offspring from any particular cross can provide large datasets for mapping quantitative trait loci (QTL), multiple genes that all affect a particular phenotypic feature to some extent. This technique is primarily used in plant breeding, but crop domestication researchers can draw on QLT mapping data to better understand the genetic basis for domestication (Rieseberg 1998; see Emshwiller, Chapter 8). In addition, the dense chromosome maps now available for several important crop plants can aid in the choice of appropriate markers for crop evolution studies. For example, the intensive work on corn genetics has identified several key "domestication genes" responsible for such characteristics as branching and glume architecture, and studies of rice have successfully isolated genes that control starch composition. Identification of domestication genes in animals is not as well along as in plants, but is certainly on the horizon for such species as cattle, pigs, and dogs (see Bradley's overview, Chapter 18). There is clear promise here for identifying not only the genes responsible for later morphological changes in domestic animals, but also those for the changes in behavior that constitute the front line of selection in animals undergoing domestication.

Many genetic studies of plant and animal domesticates do not focus on the particular genes responsible for the changes in morphology, behavior, and physiology that distinguish domesticates from their wild progenitors. Instead, these studies generally concentrate on variation in neutral genes or in noncoding genetic regions that can be used to trace the evolutionary history of domesticates and their wild progenitors. Here again, there are important differences between the biology of plants and the biology of animals that differentially affect the genetic study of plant and animal domesticates. One of the most significant of these differences concerns the rate of DNA sequence evolution in the organellar and nuclear genomes of plants and animals. Compared to the loci in the nuclear genome (chromosomes), loci on the organellar genomes are technically easier to work with in tracing phylogenetic lineages because they are usually inherited from only one parent and they occur in many identical copies in each cell. The large number of identical copies means that organellar loci are much easier to amplify from degraded DNA samples, which is particularly useful when working with ancient DNA. The fact that they are identical means it is not necessary to separate different copies (alleles) of the same gene on different chromosomes. For a locus on the nuclear chromosomes, in contrast, if an individual carries different alleles for that locus, the DNA sequences might not be readable unless the alleles are separated by the cumbersome process of molecular cloning.

However, the rates of evolution among the nuclear, mitochondrial, and chloroplast genomes are not equal. Moreover, there are significant differences in the relative rates of evolution in these genomes in plants and animals. In animals, mitochondrial DNA (mtDNA) is highly polymorphic, with a rate of molecular evolution (accumulation of mutations) that is 5 to 10 times higher than in the nuclear genome. The high degree of variation resulting from the rapid evolution of this genome makes mtDNA ideal for studying the divergence between wild and domestic populations under the relatively short time scale over which domestication operated.

This is why studies of animal domestication tend to use sequences of mtDNA loci more frequently than the more slowly evolving nuclear genes (nDNA). Indeed, all of the chapters on the genetic documentation of animal domestication rely primarily on mtDNA for tracking the maternal lineage of animal domesticates. Moreover, as also demonstrated in the chapters in this volume, since the effective population size of mtDNA is one-quarter that of nuclear DNA, it has proven particularly useful in the study of population dynamics because it is capable of detecting population bottlenecks that are likely to occur in domesticates, but which have less impact on nuclear DNA. Ancient DNA analysis holds special promise in the study of animal domestication because the copy number of mtDNA makes it relatively easy to recover from archaeological animal bones and its variability makes it powerful for tracing the evolutionary history of domesticated animals.

Although nuclear DNA is less variable than mtDNA in animals, and therefore generally less useful in phylogenetic studies of relatively shallow time depth, Y-chromosome nuclear DNA provides important information on the paternal line, which is of particular relevance to understanding the history of stock breeding (Chapter 20, for goats; Chapter 23, for camelids; and Chapter 24, for horses). Noncoding nuclear microsatellite DNA, contributed by both parents, has also proven useful in the study of animal domesticates, especially in detection of the very shallow time-depth variation as that expected between domestic breeds of animals (see Chapter 20; Parker et al. 2004).

Plants, however, do not have any genome compartment that evolves as quickly as animal mtDNA. The mitochondrial genomes of plants, unlike those of animals, are large in size and have frequent rearrangements that make them unsuitable for restriction site analysis, yet their DNA sequences evolve very slowly, because of a slower mutation rate. Sequences on the chloroplast genome of plants (cpDNA) evolve somewhat faster than plant mtDNA. Even though cpDNA has been particularly useful in phylogenetic studies at deep evolutionary levels, however, it still lacks sufficient variation to address some questions at the shorter time scale involved in domestication. Although not approaching the rate of evolution in animal mtDNA, loci on the nuclear genome in plants evolve at a rate similar to nDNA of mammals, that is, about 4 times faster than cpDNA and 12 times faster than plant mtDNA (Wolfe et al. 1987, 1989; Gaut 1998). And while it lacks the non-recombining nature and uniparental inheritance of organelle genomes, nuclear DNA can provide a great deal of information about the evolutionary history of domestic plants (and animals). This is true because, even if sequences of individual loci may have little variation, it is possible to use nDNA in various kinds of fragment analysis, in which small bits of the DNA from throughout the genome are analyzed, providing greater variability. Thus, while there are examples of the use of cpDNA in crop evolution studies, often it is only nuclear loci that have enough intraspecific variation to be useful in the documentation of plant domestication. Nuclear DNA is also necessary for the study of polyploids to provide evidence of all ancestral genomes. The reliance on nuclear DNA results in the use of an array of different methods such as microsatellites (SSRs), AFLPs, RAPDs, and other highly polymorphic markers that provide the variation that crop researchers need to study plant domestication (discussed in Chapter 8).

Among case studies of crop domestication using genetic data in this volume, only Breton et al. use all three genome compartments to study olive (see Chapter 11), including microsatellites on the chloroplast genome and RFLP-based haplotypes of the mitochondrial genome, as well as random amplified polymorphic DNA (RAPD) data, anonymous markers that are presumed to be primarily located on the nuclear chromosomes. RAPDs are also employed in Chinese chive (Chapter 10). Nuclear DNA is used by Olsen and Schaal in cassava (both microsatellites and DNA sequences of three loci in Chapter 9) and Emshwiller in oca (DNA sequences of two loci in Chapter 12).

Finally, mindful that at least half of the audience for this volume have little or no background in genetics, all authors of the genetics chapters are careful to explain clearly the choice of genetic markers used and the techniques employed in both extracting the genetic data and analyzing them. Reading these chapters will not qualify a nongeneticist to conduct genetic analysis, but the chapters will at the very least serve to demystify genetic analysis for the nonspecialist. The thoughtful attention to clear explication of the choice of markers and methods used in these analyses also underscores one of the central messages of this volume. Just as the markers used in the archaeological documentation of plant and animal domestication must be clearly linked to the process of domestication of the species at hand, the methods used in genetic analysis of domesticates must be chosen with full understanding of the biology of the species, its likely response to the process of domestication, and the potential of various techniques for illuminating different aspects of this process. There is a bewildering (at least to most archaeologists) array of different techniques available for DNA sequence and fragment analysis. As Emshwiller points out in her overview chapter, it is very important to avoid applying analytical techniques without understanding what these techniques do and how the data they yield relate to the process of domestication.

Contributions of Archaeology and Genetics to Documenting Domestication

It is critical that the genetic analysis of this history of domestication be conducted with not only an appreciation for the biology of plant and animal domesticates, but also an understanding of the cultural context of the human partners in the process. And this is where genetics and archaeology come together to provide a richly detailed understanding of domestication. This volume highlights several promising areas for the intersection of archaeology and genetics (see also Zeder et al. 2006).

Identification of Wild Progenitors of Plant and Animal Domesticates

One of the primary aims of the use of phylogenetic analysis of domesticates has been to identify the progenitor of domestic crop and livestock species, and this is certainly a major goal of the genetic chapters in this volume. In some cases these studies confirm, or at least support, the identification of progenitor species that had been based on morphological, geographic, or cytological studies (e.g., Chapter 20 for goats and Chapter 21 for sheep). Other studies are able to sort out long-standing controversies about the complicated parentage of other domesticates, more or less definitively identifying the sole progenitor species and ruling out a variety of other possible scenarios of relatedness that involved various hybridization events, back crossing, and feralizations of escaped domesticates (e.g., Chapter 9 for cassava, Chapter 11 for olives, and Chapter 23 for South American camelids). In other studies long-accepted progenitor species are shown instead to be closely related sister species (Chapter 10 for Chinese chives), while the complex hybrid parentage of the octoploid tuber oca is confirmed and the search for the parental species narrowed in another (Chapter 12). In some cases where the wild progenitor of a domesticate is either largely or entirely extinct, ancient DNA has been able to provide important clues on the ancestry of key domestic species (Chapter 22 for cattle and Chapter 24 for horses).

Correct identification of the wild progenitor of a domestic species is key to the development of archaeological markers capable of distinguishing between the remains of wild and domestic individuals in the archaeological record. Early allozyme research identifying the wild progenitor of the Andean crop plant quinoa, for example, contributed important direction to Bruno's development of morphological markers capable of distinguishing domestic quinoa from its progenitor in the archaeological record (Chapter 4). The resolution of the long and complicated debate over the ancestry of other prominent Andean domesticates—alpaca and llama (Chapter 23)—has been instrumental in the development of metric approaches to distinguishing between these domesticates and their wild progenitors discussed in Mengoni and Yacobaccio (Chapter 16).

Documenting the Number and Location of Domestication Events

Another obvious area of intersection between genetics and archaeology is the documentation of the number and location of domestication events. Genetic analysis identifying the progenitor and the likely geographic center of initial domestication of maize (reviewed in Chapter 8), for example, has provided critical direction to ongoing archaeological research in the Balsas River Valley of West Central Mexico that seeks to use a wide range of macro- and micro-morphological markers to track the initial domestication of corn (Piperno et al. 2004a). The combination of Olsen and Schaal's (Chapter 9) discovery that the likely geographic locus of manioc domestication was in the southern Amazon basin, with Piperno's breakthrough work on distinguishing wild and domestic manioc on the basis of starch grain analysis (Chapter 6), holds great promise for being able to trace the initial domestication and subsequent diffusion of this important crop plant despite the poor preservation of archaeobotanical remains in tropical soils. Bradley's genetic evidence that cattle underwent a possible third independent domestication event in northern Africa (Chapter 22) will certainly hearten archaeologists who have been arguing for North African cattle domestication on the basis of biogeographic evidence (Close and Wendorf 1992).

This last case example highlights one of the more exciting outcomes of bringing together genetics and archaeology, and that is the discovery that many plants and animal species were domesticated multiple times in multiple places. Earlier approaches to documenting domestication tended to follow a Vavilovian model that saw domestic crops, and livestock, as dispersing from a handful of single centers of origin. Recent molecular and archaeobiological research, however, suggests that multiple domestication events may be more common than originally thought, especially in animals. In fact, all of the animal domesticates (dogs, goats, sheep, cattle, pigs, South American camelids, and horses) discussed in this volume appear to have experienced multiple genetically independent domestication events.

Merging molecular and archaeological data holds special promise for documenting these multiple domestication events. Rejecting earlier models that saw domestic squash as having diffused into eastern North America out of a single center of initial domestication in Mexico, for example, Smith and colleagues used archaeobotanical and modern biogeographic data to build a case for an independent domestication of squash in eastern North America (Chapters 2 and 3), with recent genetic analysis confirming two independent domestication events (Sanjur et al. 2004). Genetic evidence of three domestic lineages in both domestic sheep and goats (Chapters 20 and 21) is in line with mounting archaeobiological evidence of multiple domestication events for these species, with at least two of these events having taken place at different parts of the broad arc of the Fertile Crescent in Southwest Asia. Improved techniques for detecting the leading edge of caprine domestication through the construction of sex-specific harvest profiles hold real promise of being able to pinpoint the location and timing of these events (Chapter 14). Albano and colleagues' discovery of two independent domestication events in the donkey, one centering on the Nubian ass in the Nile valley region and the other on the Somali ass in the Horn of Africa (Chapter 24), can be placed within the context of expanding overland and ocean trade routes both out of and into Africa. Understanding the role of the donkey in the emergence of international trade networks in the ancient world provides the cultural context for the domestication of this species, as well as adding to our understanding of the mechanics of these trading systems. Genetic detection of the introduction of Zebu cattle that likely entered the continent through the Horn of Africa sheds even more light on the same ocean trading network (Chapter 22), as does tracking the diffusion of the banana from the Southeast Asian site of initial domestication across the African continent (Chapter 6; and Zeder et al. 2006).

It is important to remember, however, that genetically independent domestication events are not necessarily culturally, or even entirely biologically, independent.

Knowledge of domestication can move between peoples and be applied to local wild plant and animal resources. It is also possible that many of the apparently independent domestication events in animals arose when either domestic males or females (or perhaps both) were moved into an area and served as a kind of seed stock, breeding with local wild populations. Depending on the sex of the domestic seed stock and the lineage traced by genetic markers (maternal or paternal), this level of contact could well go undetected. Archaeological investigations hold the promise of providing more definitive evidence of the degree of cultural contact between different centers of origin and the timing of these domestication events.

Tracing the Dispersal of Domesticates

The application of population genetic techniques to the study of domesticates allows primary domestication events to be distinguished from later or more restricted separate domestications (Chapter 19; Chapter 20; Chapter 22). These models also provide special insight into the dispersal of domesticates out of a center or centers of initial domestication. This kind of information will prove invaluable to archaeologists tracing the origin and spread of domesticates. Sheep and goats, for example, both seem to have expanded quite quickly across a wide region, while the dispersal of taurine cattle out of the Near East into Europe seems to have been a more directional, gradual process. Tracking the movement of dogs across Eurasia and into the New World (Chapter 19), and seeing the genetic stamp of this movement on modern dog breeds (Parker et al. 2004), provides a molecular map of the human migration with special relevance to resolving current controversies about the route of humans into the New World (see Straus 2000). The study of ancient and modern mtDNA of the Pacific rat has been instrumental in testing various proposed models for human migration in Oceania (Matisoo-Smith and Robins 2004).

Application of different genetic techniques for tracking maternal and paternal lineages provides special insight into how humans and their domestic animals moved into new areas and how people later set about improving domestic stock through hybridization and selective breeding. While hybridization is often seen as a confounding obstacle in the identification of the progenitors of domesticates (discussed in virtually every one of the genetics chapters in this volume), it is also an important part of the evolutionary history and cultural legacy of domesticates and one that leaves its stamp on the genetic structure of modern domesticates. Genetic data that implicate males as playing the primary role in breed differentiation in horses (Chapter 24), and that give witness to the complicated crosses between males and females in alpacas and llamas that have been used to improve both the yield and quality of fiber production (Chapter 23), provide insights into the history of stock breeding beyond those archaeologists could ever hope to gain from the archaeological record.

Archaeologists, in turn, offer important ground-truthing opportunities for testing and refining these genetically based dispersal models. The development of archaeological markers capable of detecting different phases of the domestication process has great potential for tracking the course of the dispersal of domesticates and their subsequent modification. The archaeological chapters on horse domestication (Chapter 17) and the domestication of South American camelids (Chapter 16), for example, offer a number of morphological and nonmorphological markers that can be used to trace the process of domestication and dispersal of these domesticates and that directly complement the insights gained from genetic analysis. Piperno's work, which tracks the movement of manioc out of the Amazon Basin and into Central America (Chapter 6), is an example of the archaeological advances in the study of crop dispersals that complement genetic analysis, as is her and others' work on tracking the diffusion of corn out of central Mexico, both southward into South America (Pearsall et al. 2003; Freitas et al. 2003), and northward into the southwestern United States (Vierra 2005), and even later into eastern North America (Riley et al. 1994). Thompson's discovery of phytolith markers capable of distinguishing different varieties of hard and soft glume corn (Chapter 7) holds the promise of allowing archaeologists to identify the temporal and geographic context of the shifts in selection on genes that are central to the evolutionary history of corn.

Documenting the Temporal Sequence of Domestication

Documenting the temporal context of domestication is a special area of interest to both archaeologists and molecular biologists. The use of an assumed regular rate of DNA base changes to estimate the divergence time between a wild progenitor and its domestic descendant (the so-called molecular clock) has been a source of considerable, highly publicized controversy. In particular there has been a long and lively debate among molecular biologists and paleoanthropologists over estimates of the timing of the origin and dispersal of humans based on molecular evidence and estimates derived from fossil evidence (see Stringer 2003 vs. Wolpoff 1989). The documentation of domestication has not been without similar discordances between molecular and fossil evidence. The chapter on the genetic documentation of dog domestication (Chapter 19) in this volume revisits one of the best known of these debates. Early work by Wayne and his colleagues (Vilà et al. 1997) used molecular data to estimate that the divergence of the wolf and dog lineages took place at about 135,000 years ago, more than 100,000 years earlier than the first morphological evidence of dog domestication is found in the archaeological record (seen first in both Europe and Asia at about 14–15,000 years ago). In Chapter 19 here, the team revises this date upward somewhat, but still endorses a molecular-clock estimate that puts the divergence of wolves and dogs considerably earlier (by at least several millennia) than the 15,000-year date for dog domestication suggested by fossil evidence. Wayne et al. argue that the discordance between these dates may be attributed to the fact that the genetic divergence between dogs and wolves, operating primarily on behavioral attributes having to do with reduced aggression, preceded the expression of morphological change in dogs. As discussed above, it is true that the expression of morphological change is often delayed in animal domesticates. However, the changes in the skull morphology that archaeologists use to discriminate dogs from wolves in the fossil record should find early expression in wolves undergoing domestication, since this trait is likely directly linked to the genetic selection for reduced aggression that is central to the domestication of these animals. A delay of thousands of years in the expression of this feature is hard to imagine. Evidence for the divergence of several lineages of domestic dog prior to human expansion into the New World, and the elimination of North American wolves as the source of an additional domestication event in dogs, on the other hand, suggests that the divergence of Old World dogs and wolves happened considerably earlier than the peopling of the New World, thought to have taken place over the course of several successive waves of immigration beginning about 14,000 years ago. Clearly, more work on both the genetic and archaeological side of the issue is needed.

For the most part, however, genetic studies featured here do not take a molecular-clock approach to fixing the temporal context of domestication. In his overview chapter (Chapter 18), Bradley recommends merging phylogeographic evidence for the origin and dispersal of domestic lineages gained from genetic analysis with directly dated fossil evidence for domestication provided by archaeologists. A similar approach is advocated by those working on the genetic study of plant domesticates, especially since the nuclear DNA sequences and microsatellites generally used in these studies violate the basic assumption of a regular rate of mutation on which the molecular clock rests. In fact, the molecular-clock hypothesis is controversial in general, for any timescale, taxon, or genome type. Even if one ignores the evidence for unequal mutation rates among different groups and different genomes, the timescale of domestication is much too short to be appropriate for molecular clocks, which are better calibrated for species that diverged millions or tens of millions of years ago, not populations (such as domesticates) diverging thousands of years ago (see Ho et al. 2005; Ho and Larson 2006; Dobney and Larsen in press). Thus, rather than an area of divergence, tying inferences derived from population genetics to archaeological data—in effect "anchoring genetic data to the archaeological narrative," as Bradley puts it—seems one of the more promising areas for cross-illumination of molecular and archaeological approaches.

The Promise of Ancient DNA

The development of techniques for the extraction and analysis of ancient DNA opens a whole new arena for cross-disciplinary work on documenting domestication. Despite difficulties in its extraction and replication, ancient DNA provides a tremendous opportunity for the integration of archaeology and biology in the study of plant and animal domestication. Thousands of years of selective breeding, hybridization, and introgression between wild and domestic populations makes it difficult to directly apply genetic data on modern domesticates to understanding the origin and early dispersal of domesticates. Ancient DNA (aDNA) offers a much more direct window on the process, potentially allowing for the definitive identification of the who, what, where, and when of domestication. Because aDNA is more likely to be preserved in animal bone than in charred plant remains, most of the application of aDNA to the study of domestication has centered on animal domesticates. The effectiveness of high-copy mtDNA in tracking shallow-time-depth evolutionary change in animals is another factor that contributes to the effectiveness of aDNA in the study of animal domestication. In fact, four of the six chapters on the genetic documentation of animal domestication in this volume make use of aDNA in their studies (Chapter 19 on dogs, Chapter 20 on goats, Chapter 22 on cattle, and Chapter 24 on horses). Moreover, the chapter on pig domestication in the section on the archaeological documentation of animal domestication (Chapter 15) also incorporates recent aDNA information on material from Japan that has been key in tracking the dispersal of domestic pig in East Asia. Like this latter example, the other uses of aDNA featured here generally aid in tracking the dispersal of animal domesticates. As yet it has not been possible to extract and replicate aDNA from the bones of animals dating to the earliest phases of the domestication process. But given the staggering advances of the last decade in genetic analysis, it seems just a matter of time before this is possible. The promise of being able to track the timing of genetic change in animal domesticates along with that of the other non-morphological and morphological indicators of animal domestication is exciting indeed. The potential of being able to detect the genetic basis for the behavioral shifts that are the leading edge of the domestication process in animals would transform our understanding of animal domestication. Work that has been accomplished on the study of ancient DNA from dried plant remains suggests that this goal is not too far-fetched. Combined analysis of morphology and ancient DNA of directly dated archaeological corn cobs from sites in Mexico and the southwestern United States by Jaenicke-Després and colleagues (2003) has been successful in demonstrating the results of human selection on three corn domestication genes as early at 4,400 years ago. This and similar studies (E. G. Erickson et al. 2005) underscore the future promise of combining archaeology and genetics.

From Documenting to Explaining Domestication

Attempts to explain why domestication occurred have tended to focus on universal, single forcing mechanisms that could be uniformly applied to all cases of domestication. Climate change, population pressure, resource optimization, and social tensions are just some of the proposed prime movers of plant and animal domestication (Childe 1951; Cohen 1977; Richardson et al. 2001; Hayden 1992). In part this one-size-fits-all approach to explaining domestication grows out of earlier views that domestication was limited to a small number of centers from which domesticates dispersed around the globe. These universalist approaches to explaining domestication were also framed in a relative vacuum of empirical information on the domestication of plants and animals, either because this information did not yet exist, or because the information that did exist did not agree with these overarching explanatory frameworks.

The explosive development of archaeological and genetic approaches to documenting domestication has radically transformed this situation. We know now that the people on every continent were actively engaged in the domestication of a wide range of plant and animal species, that some species were domesticated multiple times in multiple places, that people engaged in various forms of low-level food production that never resulted in full domestication, that the dispersal of domesticates and their adoption and alteration in new settings are all part of a complex and endlessly varied story of domestication. Clearly there are some universal principles that come into play in all these stories, both cultural and biological. Climate, community, optimization, adaptation, co-evolution, and selection all serve to shape the process of domestication wherever it occurred. But there are other highly localized factors that play important roles in each instance. Understanding what these factors are and how they shaped the unfolding process of domestication are equally, perhaps even more, important in explaining domestication.

We have made great strides in our ability to track this process in a richer, more detailed way than would have been thought possible just a decade ago. The decade ahead holds even greater promise for building on this level of achievement. But realizing this potential requires integrating the different insights gained from the full range of archaeological and genetic approaches to documenting domestication in plants and animals. The individual chapters in this book provide new paradigms for documenting domestication in plants and animals. The approaches they describe will certainly yield new breakthroughs in the study of domestication in the years to come. But perhaps the greatest contribution of a volume like this is that it creates a forum for conversation between those engaged in the varied and technically difficult pursuit of documenting domestication that can only result in a more profound understanding of domestication and its enduring impact on both the natural and cultural world.

Acknowledgments

The editors would like to thank Marcia Bakry (Senior Illustrator, Department of Anthropology, National Museum of Natural History, Smithsonian Institution) for her tireless efforts in revising art work throughout this volume.

References

Braidwood, R. J. and Howe, B. 1960. *Prehistoric investigations in Iraqi Kurdistan*. The Oriental Institute of the University of Chicago Studies in Ancient Oriental Civilization, No. 31. Chicago: University of Chicago Press.

Braidwood L., R. J. Braidwood, B. Howe, C. A. Reed, and P. J. Watson. 1983. *Prehistoric archaeology among the Zagros flanks*. Oriental Institute Publications, No. 105. Chicago: The Oriental Institute.

Childe, V. G. 1951. *Man makes himself*. New York: The New American Library of World Literature.

Close, A. and F. Wendorf. 1992. The beginnings of food production in the eastern Sahara. In *Transitions to agriculture in prehistory*, A.-B. Gebauer and T. D. Price (eds.), pp. 63–72. Madison, WI: Prehistory Press.

Clutton-Brock, J. 1999. *Domesticated animals*, 2nd edition. London: British Museum of Natural History.

Cohen, M. 1977. *The food crisis in prehistory: Overpopulation and the origins of agriculture*. New Haven: Yale University Press.

Colledge, S. 1998. Identifying pre-domestication cultivation using multivariate analysis. In *The origins of agriculture and plant domestication*, A. Damania, J. Valkoun, G. Willcox, and C. Qualset (eds.), pp. 121–131. Alleppo, Syria: International Center for Agricultural Research in the Dry Areas (ICARDA).

Cowan, C. W. and P. J. Watson. 1992. *The origins of agriculture*. Washington, D.C.: Smithsonian Institution Press.

Darwin, C. R. 1859. *On the origin of species by means of natural selection, or the preservation of favoured races in the struggle for life*. London: John Murray.

———. 1868. *The variation of animals and plants under domestication*. 2 vols. London: John Murray.

Dobney, K. and G. Larson. 2006. DNA and animal domestication: More windows on an elusive process. *Journal of Zoology*, in press.

Erickson, D. L., B. D. Smith, A. C. Clark, D. H. Sanweiss, and N. Tuross. 2005. An Asian origin for a 10,000 year old domesticated plant in the Americas. *Proceedings of the National Academy of Sciences USA* 102: 18315–18320.

Freitas, F. O., G. Bendel, R. G. Allaby, and T. A. Brown. 2003. DNA from primitive maize landraces and archaeological remains: Implications for the domestication of maize and its expansion into South America. *Journal of Archaeological Science* 30: 901–908.

Gaut, B. S. 1998. Molecular clocks and nucleotide substitution rates in higher plants. In *Evolutionary Biology*, Volume 30, M. K. Hecht, R. J. Macintyre, and M. T. Clegg (eds.), pp. 93–120. New York and London: Plenum Press.

Harlan, J. R. 1992. *Crops and Man*, 2nd edition. Madison, WI: American Society of Agronomy, Crop Science Society of America.

Harris, D. R. 1969. Agricultural systems, ecosystems and the origins of agriculture. In *The Domestication and Exploitation of Plants and Animals*, P. J. Ucko and G. W. Dimbleby (eds.), pp. 3–15. London: Duckworth.

Hayden, B. 1992. Models of domestication. In *Transition to agriculture in prehistory*, A. B. Gebauer and T. D. Price (eds.), pp. XX–YY. Madison, WI: Prehistory Press.

Hemmer, H. 1990. *Domestication: The decline of environmental appreciation*. Cambridge: Cambridge University Press.

Hilu, K. W. 1993. Polyploidy and the evolution of domesticated plants. *American Journal of Botany* 80: 1494–1499.

Ho, S. Y. W. and G. Larson. 2006. Molecular clocks: When times are a-changin'. *Trends in Genetics*, 22: 79–83.

Ho, S. Y. W., M. J. Phillips, A. Cooper, and A. J. Drummond. 2005. Time dependency of molecular rate estimates and systematic overestimation of recent divergence times. *Molecular Biology and Evolution* 22: 1561–1568.

Jaenicke-Després, V., E. S. Buckler, B. D. Smith, M. T. P. Gilbert, A. Cooper, J. Doebley, and S. Pääbo. 2003. Early allelic selection in maize as revealed by ancient DNA. *Science* 302: 1206–1208.

Kruska, D. 1988. Mammalian domestication and its effect on brain structure and behavior. In *Intelligence and evolutionary biology*, H. J. Jerison and I. Jerison (eds.), pp. 211–250. New York: Springer-Verlag.

Lathrap, D. W., D. Collier, and H. Chandra. 1977. Our father the cayman, our mother the gourd: Spinden revisited, or a unitary model for the emergence of agriculture in the New World. In *Origins of Agriculture*, C. A. Reed (ed.), pp. 713–752. The Hague: Mouton.

MacNeish, R. 1967. A summary of subsistence. In *The prehistory of Tehuacan Valley*, Volume 1, D. Byers (ed.), pp. 3–13. Austin: University of Texas Press.

Matisoo-Smith, E. and J. H. Robbins. 2004. Origins and dispersal of the Pacific peoples: Evidence from mtDNA phylogenies of the Pacific Rat. *Proceedings of the National Academy of Sciences, USA.* 101: 9167–9172.

Mendel, G. 1865. Versuche über Pflanzen-Hybriden. *Vorgelegt in den Sitzungen.* 8.

Parker, H. G., L. V. Kim, N. B. Sutter, S. Carlson, T. D. Lorentzen, T. B. Malek, G. S. Johnson, H. B. DeFrance, E. A. Ostrander, and L. Kruglyak. 2004. Genetic structure of the purebred domestic dog. *Science* 304: 1160–1164.

Pearsall, D. M., K. Chandler-Ezell, and A. Chandler-Ezell. 2003. Identifying maize in neotropical sediments and soils using cob phytoliths. *Journal of Archaeological Science* 30: 611–627.

Piperno, D. R., M. B. Bush, and P. A. Colinvaux. 1991. Paleoecological perspectives on human adaptation in Central Panama. II. The Holocene. *Geoarchaeology* 6: 227–250.

Piperno, D. R., A. J. Ranere, E. Moreno, J. Iriare, M. Lachniet, J. Jones, I. Holst, R. Dickau, and M. Pohl. 2004a. Environmental and agricultural history in the central Balsas watershed, Mexico: Results of preliminary research. Paper read at the Society for American Archaeology meeting, Montreal.

Piperno, D. R., E. Weiss, I. Holst, and D. Nabel. 2004b. Processing of wild cereal grains in the Upper Paleolithic revealed by starch grain analysis. *Nature* 407: 894–897.

Pumpelly, R. 1905. *Explorations in Turkestan with an account of the Basin of Eastern Persia and Sistan; Expedition of 1903 under Raphael Pumpelly.* Washington, D.C.: Carnegie Institute of Washington.

Richardson, P. J., R. Boyd, and R. L. Bettinger. 2001. Was agriculture impossible during the Pleistocene but mandatory during the Holocene? A climate change hypothesis. *American Antiquity* 66: 387–412.

Rieseberg, L. H. 1998. Genetic mapping as a tool for studying speciation. In *Molecular Systematics of Plants II*, D. E. Soltis, P. S. Soltis, and J. J. Doyle (eds.), pp. 459–487. Norwell, MA: Kluwer Academic Publishers.

Riley, T., G. Walz, C. Bareis, A. Fortier, and K. Parker. 1994. Accelerator mass spectrometry (AMS) dates confirm early *Zea mays* in the Mississippi Valley. *American Antiquity* 59: 490–498.

Sanjur, O., D. Piperno, T. Andres, and L. Wessel-Beaver. 2004. Using molecular markers to study plant domestication: The case of *Cucurbita*. In *Biomolecular archaeology: Genetic approaches to the past*, D. M. Reed (ed.), pp. 128–150. Occasional Paper No. 32 of the Center for Archaeological Investigations, Southern Illinois University, Carbondale. Carbondale, IL: CAI, Southern Illinois University, Carbondale.

Sauer, C. O. 1952 *Agricultural origins and dispersals.* New York: American Geographical Society.

Smith, B. D. 1992. The floodplain weed theory of plant domestication in Eastern North America. In *Rivers of change*, B. D. Smith (ed.), pp. 19–34. Washington, D.C.: Smithsonian Institution Press.

———. 1998. *The emergence of agriculture.* New York: W. H. Freeman.

Straus, L. 2000. Solutrean settlement in North America? A review of reality. *American Antiquity* 65: 219–226.

Stringer, C. 2003. Human evolution: Out of Ethiopia. *Nature* 423: 692–695

Vierra, B. (ed.). 2005. *Current perspectives on the Late Archaic across the borderlands.* Austin: University of Texas Press.

Vilà, C., P. Savolainen, J. E. Maldonado, I. R. Amorim, J. E. Rice, R. L. Honeycutt, K. A. Crandall, J. Lundeberg, and R. K. Wayne. 1997. Multiple and ancient origins of the domestic dog. *Science* 276: 1687–1689.

Walpoff, M. H. 1989. Multiregional evolution: The fossil alternative to Eden. In *The human revolution: Behavioural and biological perspectives on the origins of modern humans*, P. Mellars and C. B. Stringer (eds.), pp. 62–108. Edinburgh: Edinburgh University Press.

Weiss, E., W. Wetterstrom, D. Nadel, and O. Bar-Yosef. 2004. The broad spectrum revolution revisited: Evidence from plant remains. *Proceedings of the National Academy of Sciences, USA* 101: 9551–9555.

Wolfe, K. H., W.-H. Li, and P. M. Sharp. 1987. Rates of nucleotide substitution vary greatly among plant mitochondrial, chloroplast, and nuclear DNAs. *Proceedings of the National Academy of Sciences, USA.* 84: 9054–9058.

Wolfe, K. H., P. M. Sharp, and W.-H. Li. 1989. Rates of synonymous substitution in plant nuclear genes. *Journal of Molecular Evolution* 29: 208–211.

Zeder, M. A., E. Emshwiller, B. D. Smith, and D. G. Bradley. 2006. Documenting domestication: The intersection of genetics and archaeology. *Trends in Genetics*, in press.

Zeuner, F. 1963. *A history of domesticated animals.* London: Hutchinson.

SECTION ONE

ARCHAEOLOGICAL DOCUMENTATION OF PLANT DOMESTICATION

Bruce D. Smith, Section Editor

CHAPTER 2

Documenting Domesticated Plants in the Archaeological Record

BRUCE D. SMITH

Introduction

The transition from a hunting-and-gathering way of life, involving exclusive dependence on wild plants and animals, to economies that included a reliance on domesticated species, has long been recognized as a major turning point in human history. This transition occurred many times, in many different regions of the world, as human societies either domesticated plants and animals for the first time, or more frequently, adopted already domesticated species introduced from other regions into their local economies. Since this major transformation eventually encompassed most of the inhabitable portions of the earth, there exists in the archaeological record a rich variety of individual, regional-scale, historical-developmental histories that offer ever-increasing opportunities for comparative analysis and interpretation. As the chapters in this volume demonstrate, evidence for this transition has been steadily accumulating from different parts of the world over the past half century, with technological and methodological advances since the 1980s rapidly accelerating access both to previously poorly known regions, and to previously unrecognized data classes.

Along with the dramatic increase over the past two decades in the quality and quantity of archaeological information available regarding the timing and nature of the transition to food production in an expanding number of different world areas, and the remarkable improvements in available technology and methodological approaches, there has also been a growing appreciation for the developmental complexity and degree of variability that existed from region to region as human societies initially incorporated domesticates into their local economies. While the challenge of recognizing and illuminating common causal variables involved in the transition to food production and compressing them into developmental frameworks that might have broad-ranging explanatory power continues to encourage the formulation of general theories of agricultural origin, such universal explanations are also increasingly being viewed as largely peripheral to the much more intellectually central and challenging questions involved in reaching a better understanding of the specifics of each regional developmental trajectory. This is not to say that explanatory frameworks with a broad, even global, reach are not worth attempting; rather, given the growing recognition that the transition was often long and complex, and that it exhibited substantial variability between and within regions, and in light of the powerful new technologies and approaches that have been developed to investigate the shift over to food production, there are much more immediate, rewarding, and basic questions to be addressed. Many of these basic questions surround the identification of domesticates in the archaeological record. How can the presence of domesticates, and more specifically domesticated plants, be recognized in archaeological contexts?

Archaeological Indications of Domesticated Plants

Interestingly, at the same time that it is becoming increasingly evident that the transition to food production is a long, complex, and richly variable process from region to region, so, too, are researchers grappling with the recognition that "domestication" is not as simple and straightforward a concept as most definitions would suggest. *Domestication* is most often defined in terms of two salient characteristics: first, that the newly created "species" is observably distinct from its wild relatives; and second, that without continued human protection, it would cease to exist (see Smith 2001). Given this definition, searching for domesticated plants in the archaeological record has often focused on simply finding evidence of a number of well-described and distinctive morphological changes, such as an increase in seed size (Chapter 3) or a reduction in seed-coat thickness (Chapter 4). While this search for morphological change continues to be a very successful and rapidly expanding approach to identifying plant domesticates in the archaeological record, particularly as an increasing variety of morphological markers at smaller and smaller scales is considered (Chapters 5–7), researchers are also looking much more closely at the cause and effect linkage that exists between domestication and morphological change. In order to consider this linkage, it helps to recognize that morphological change falls at the tail end of what usefully can be characterized as a three-part process.

At the front end of this process, domestication is initiated by the creation of a new pattern of behavior, a new relationship, which develops and is sustained between a human society and a target species. This new relationship of domestication is qualitatively different from the patterns of interaction that had previously existed between the humans and the wild populations of the plant or animal species in question. The new relationship alters in some significant way the set of selective pressures that defined success and survival for wild populations of the target species; over time, the newly established selective pressures can result in changes in the genetic profile of those populations of the target species that are drawn into this new pattern of interaction with humans. The new genetic profile of the domesticate subset of the target species may also, in time, be reflected in observable morphological changes in the target species. When characterized in this admittedly simplistic manner, this basic causal chain relationship of *behavioral change* → *genetic change* → *morphological change* brings a number of interesting issues and questions into clearer focus.

It clarifies, for example, the three general categories of evidence for domestication that can be sought in the archaeological record. There are examples of direct, first-hand evidence for *behavioral change* in the archaeological record: the presence of a relationship of domestication existing between a human population and a target plant species could include indications of landscape modification associated with the cultivation of crop plants, including pollen and phytolith profiles, other indications of forest clearance for field systems, and the increased presence of likely domesticates and their weedy companions (Delcourt et al. 1986; Piperno et al. 1991; Piperno 1993; Colledge 1998); water management projects such as canals or check dams (e.g., Denham et al. 2003); preserved field systems (Toll 1995); increased and improved plant food storage facilities (Cowan 1987: 26; Fritz 1984); and the appearance or improvement of technology associated with field preparation (e.g., silica sheen on hoes) or plant processing. Such categories of direct evidence for the existence of relationships of domestication, of course, often do not occur until long after reliance on crop plants has been established, limiting their value as markers of either initial domestication or the initial introduction of a domesticate into a regional economy. These direct indications of human management of crop plants also vary considerably in terms of their relative strength as evidence of domesticated plants, in that potential alternative causes for many of them have to be taken into consideration (DeBoer 1975). Colledge, for example (1998), has shown that a distinctive suite of weedy species that is commonly present in modern agricultural fields in the Near East also appears in archaeobotanical assemblages from sites in the region that predate any compelling morphological markers of domesticated crop plants. She offers a strong argument that the appearance of this group of invasive weeds provides good evidence of very early human preparation of cultivated fields, and of their subsequent incidental harvesting along with cultivated crop plants. Miller, however (2002), has alternatively suggested that the appearance of such distinctive weedy species in archaeobotanical assemblages from the Near East might reflect the use of dung fuel, and that the carbonized (i.e., burned) weed seeds recovered from archaeological sites could in some cases reflect the diet of local grazing animals, rather than human cultivators.

Moving along the causal chain of domestication to *genetic changes*, it is only recently, and relatively rarely, that the genetic profile of plant specimens recovered from archaeological contexts has been successfully accessed in a way that provides information on particular traits or sets of traits that serve to differentiate between wild and domesticated taxa of a species, and which can be employed to establish the presence of a domesticated crop. Even though this ability to directly consider specific genes and alleles linked to domestication in archaeological plant material has only recently been demonstrated (e.g., Jaenicke-Després et al. 2003), and is of course dependent on the identification of the genes that control particular characteristics by geneticists working with modern crop plants (e.g., Dorweiler and Doebley 1997), it holds the promise of allowing the individual trajectories of crop evolution across time and space to be traced with much greater precision.

Finally, *morphological change* provides the largest and most productive category for documenting the presence of domesticated plants in the archaeological record. Unlike the relatively circumstantial and sometimes late appearing categories of evidence briefly described above as being potentially indicative of behavioral change associated with domestication, morphological change can often be strongly and directly linked to human societies creating new sets of selective pressures associated with crop management. A number of such archaeologically visible morphological markers of domestication have been identified for different plant species, particularly seed plants, and will be discussed later in this chapter.

When the process of domestication is broken down and presented as a simple three-link causal chain in which human-initiated behavioral changes produce genetic change in target species, which in turn can result in morphological changes, it also raises the key question of what the criteria are for identifying "domestication" in each of the three categories: behavioral change, genetic change, and morphological change. What kinds of changes in the relationship between a human society and a target species are necessary, for example, in order for the newly created pattern of interaction to qualify as "domestication"?

Over the long history of our species, humans have developed a wide range of different relationships with the plant and animal components of their biotic communities that have involved varying kinds and degrees of intervention and encouragement. A number of different approaches have been proposed for sorting these various interaction pairs into levels of increasing human intervention in the life cycle of target species leading up to "full" domestication (Ford 1985; Harris 1989; Rindos 1984; Smith 2001). These alternative frameworks for classifying different sustained patterns of interaction all underscore both the range of different ways in which humans encourage and improve the selective fitness of target species, and the absence of a clear, crisp, and universally applicable demarcation line for what constitutes relationships of "domestication" between human groups and target species of plants and animals. Similarly, as genetic approaches to comparative analysis of modern and ancient domesticates and their wild relatives become more common and more comprehensive, related discussion will ensue regarding the number and nature of genetic changes necessary in order for an archaeological specimen to be designated as a domesticate. Moving along the causal chain from genetic change to resultant morphological change, the same questions can be raised regarding what kinds of morphological changes constitute evidence of domestication. Rather than focusing exclusively on exactly what the behavioral, genetic, and morphological changes are that determine whether a relationship of domestication has been established, researchers are becoming increasingly interested in more broadly exploring and more accurately describing the full range of human–target species relationships that appear to cluster in the area of domestication, and might be described as encompassing key elements of domestication.

This causal chain characterization of the process of domestication also helps to underscore a number of other fairly obvious related issues. The establishment of such a new and sustained pattern of interaction, a new human–target species relationship of domestication, for example, is clearly the independent variable or component in the causal chain—the behavioral relationship *is* domestication. Genetic and morphological changes related to these relationships of domestication are the result of, and indicators of, the interaction pair that has been created. If the relationship is not sustained, then the resultant selective pressures, genetic changes, and morphological changes will fade, along with the relative fitness of the domesticate subset of the species, and it will cease to exist. This does not mean that the target species as a whole will become extinct, just that the particular genetic profile selected for, as humans intervene in the life cycle of the target species, will disappear. As the anthropogenic habitat and human interaction are withdrawn, and the associated selective pressures vanish, the selective forces extant outside the human environment will again dominate. The archaeological and historical records provide numerous examples of domesticated crop plants that once flourished and have since disappeared, even though their wild relatives continue to thrive in areas where their natural habitat still survives. In eastern North America, for example, wild stands of marshelder *(Iva annua)* and chenopod *(Chenopodium berlandieri)* can still be found along river floodplains and as successful invaders of anthropogenic disturbed soil settings, even though domesticated crop plant subsets of these species once grown by human societies for thousands of years have long vanished from the landscape (Smith 2002).

Similarly, when the causal and temporal directionality inherent in the process of domestication is clarified, the possibility of relationships of domestication existing in the absence of any morphological change in the target species is also clearly underscored. To be domesticated, a crop plant does not, of necessity, have to exhibit easily observable morphological change. Although new sets of selective pressures and adaptive opportunities created by an initially established relationship of domestication should result in some degree of sustained genetic change in the subset of the target plant species in question, it does not necessarily mean that any obvious (and archaeologically observable) changes in morphology will result. The fact that distinctive and dramatic morphological changes often do take place, and relatively quickly, in annual seed plant species has tended to obscure the parallel absence of such rapid and easily observed changes in perennial root crops (Turner and Peacock 2005; Smith 2005).

There is, of course, another obvious and very significant reason why morphological evidence for many domesticated crop plant species is rare to nonexistent in the archaeological record. Low archaeological visibility for either wild or domesticated taxa of a plant species can be the simple result of the plant in question not leaving behind much in the way of organic residue, and thus not providing an easily identifiable archaeological signature in terms of any recognizable and diagnostic remains. At the same time, many of the crop plants that have such low-visibility profiles (e.g., root crops such as manioc and yams, and some tree crops such as bananas) often have a range of cultivation that is largely limited to regions where preservation of organics in archaeological contexts is very poor (e.g., high-rainfall tropical forest environments). For such low-visibility plant species and such poor preservation regions, morphological change has had only very limited utility as an indicator of domesticated plants in the archaeological record, and a much greater reliance has been placed on evidence of forest clearing and technological change by researchers tracing the transition to food production in the archaeological record of the tropics. This has changed dramatically in recent years, however, as the interpretive value of new categories of morphological markers of domestication (i.e., phytoliths and starch grains; see Chapters 5, 6, and 7) is being demonstrated in a growing number of regions of the world. These new categories of evidence, and the expanding search for them worldwide, is beginning, finally, to fill in the record of transition to food production in the lower latitudes, and to redress the imbalance of emphasis on, and evidence for, the mid-latitude domestication of annual seed plants.

The Adaptive Syndrome of Domestication and Morphological Change in Annual Seed Plants

As early as the 1930s, researchers were looking for and finding morphological changes in archaeological plant remains that were considered to be evidence of domestication in seed plants (e.g., Gilmore 1931; Jones 1936). These morphological markers of domestication recognized more than seventy years ago, including an increase in seed size and a concentration of seeds in terminal seed heads (e.g., Gilmore 1931 for larger seed heads in sunflowers), were initially recognized as reflecting deliberate selection on the part of humans. Such deliberate selection for desirable traits would have entailed direct hands-on inspection of plants during the growing season, as well as harvested seeds or even prepared food dishes, in search of any number of preferred attributes (color, increased yield, flavor, texture, resistance to disease, etc.).

In a series of articles in the last half of the twentieth century, however, a number of scholars (Darlington 1956, 1969; Harlan and de Wet 1965; Zohary 1969, 1984; Harlan, de Wet, and Price 1973; de Wet and Harlan 1975; Heiser 1988) identified and described in detail a range of distinct morphological changes in seed plants that could be expected to result not from deliberate intentional selection for desirable traits on the part of humans, but as the result of *unconscious selection*: "selection resulting from human activities not involving a deliberate attempt to change the organism" (Heiser 1988: 78). These unintentional changes would result from a particular, clearly defined new set of human activities involving target species of plants—specifically, the sustained harvesting and planting of stored seed stock. The articles by Darlington, Heiser, Harlan, and others were of substantial theoretical importance because they demonstrated a clear and compelling cause and effect relationship along the causal chain of domestication, beginning with the development of a new relationship between human societies and the target species of plants.

This new relationship involved humans establishing garden or field plots of target species that were spatially or environmentally partitioned from wild stands. Each year, at the end of the growing season, human interveners would set aside, in a protected storage context, some portion of the seeds they harvested from their managed plots, and this seed stock would then be sowed in prepared plots at the beginning of the next growing season. This particular new form of human intervention in the life cycle of the target plant populations would have established an entirely new set of rules or selective pressures regarding the likelihood of any individual plant's seeds being included in next year's seed stock, and these new rules were distinctly different from those that applied to wild populations of the species still extant in natural stands.

In wild stands, for example, patterns of production, dispersal, and subsequent survival of seeds on the ground largely dictate which plants will have offspring represented in the next generation. In order to minimize the impact of on-the-plant seed predation from birds and other foragers, a staggered schedule of seed production, along with seeds that are shed as soon as they mature, increases the probability of seeds surviving to reach the ground. Once on the ground, a number of attributes, including the protection of a thick seed coat, reduces the likelihood of loss resulting from predation or physical damage.

In a managed plot, in contrast, where a one-time human harvest dictates which plants will contribute seed stock to the next generation, the rules for evolutionary success shift dramatically. Plants with staggered seed maturation schedules and rapid seed dispersal upon maturation will contribute fewer seeds to the human harvest than those plants that hold on to their seeds, and which have a higher percentage of mature seeds at harvest time. Thus the simple act of harvesting the managed plot at a scheduled time each year will strongly favor those plants that have all of their seed maturing simultaneously (just before harvest), and that do not disperse or shed their seed, but hold the seed tightly on the plant. In addition to uniform maturation and seed retention, human harvest also unconsciously selected for other morphological changes, including the presentation of seed to the human harvesters in compact clusters at the top of the main stem of the plant, where they were more likely to be seen and thus more likely to contribute to next year's seed stock.

In addition to these human harvesting selective pressures associated with the new relationship established between humans and target populations of plants (the sustained planting of stored seed stock), there were also considerable changes in the rules regarding seedling success in prepared seed beds during the early months of the growing season, relative to selective pressures that shaped growing season reproductive success in wild stands. Once humans took over responsibility for both the storage and protection of seed stock, and determined the correct time to plant at the beginning of the growing season, several attributes that had been of critical selective advantage to seeds exposed on the ground from the end of one growing season to the beginning of the next, would cease to be advantageous, and would in fact actually diminish a seed's probability of successfully developing into a plant that would survive and contribute to the next harvest. In the wild, for example, where loss of seeds to predation and to mistimed germination is an annual gamble, a good general, long-term strategy is to hedge bets by producing numerous small seeds with thick seed coats that will germinate at different times at the beginning of the next growing season. Some of the seeds will be missed by predators, and, of those, some will germinate at the correct time. Once humans began to store seeds in a protected environment, and to monitor weather and soil conditions at the beginning of the growing season to attempt a successful sowing, however, small seed size and any constraint on germination, such as a thick seed coat, were no longer an advantage but a hindrance. It is the seeds that can germinate and grow the most quickly after sowing that have the best chance of capturing the available sunlight, shading out the competing seedlings around them, and surviving to be harvested. In a human-managed growing environment, then, larger seeds, with more stored start-up food reserves and with thinner seed coats that will sprout quickly, have the selective advantage in terms of contributing to next year's seed stock.

Taken together, the five unconsciously generated morphological changes briefly described above constitute in large measure what is commonly termed *The Adaptive Syndrome of Domestication* (Hammer's 1984 *Domestikationssyndrom*). When humans began to harvest, store, and plant seeds over a sustained period, they unconsciously, inadvertently, created a new and distinctive selective environment to which the target plant populations under their management adapted through genetic and morphological change. Of the adaptive responses by target populations of seed plants to the new human-created environment, the most important (and most likely to be visible in the archaeological record, Wilson 1981) are (1) simultaneous ripening of seeds; (2) compaction of seeds in highly visible terminal stalk/branch "packages"; (3) seed retention (loss of natural seed dispersal mechanisms); (4) increase in seed size; and (5) simultaneous and rapid seed germination (loss of germination dormancy, reduction in seed-coat thickness).

It is interesting that all five of these morphological markers of domestication in seed plants, which are often mistakenly considered to actually comprise domestication itself, in fact not only fall at the tail end of the causal chain of domestication, but also can result from, and indicate the development of, a single, specific new pattern of human intervention in the life cycle of the target populations—the deliberate and sustained planting of stored seed stock. In many respects, then, archaeological evidence for the domestication of seed plants as reflected in these morphological changes is evidence for a specific and clearly defined type of new relationship between humans and target populations of selected plant species (human planting of stored seed stock) that can be classified under the broader general heading of domestication.

Following the recognition and description of this particular causal chain of domestication, involving the adaptive syndrome of domestication in seed plants, by Darlington, Harlan, Heiser, and others, archaeologists have focused considerable attention on improving standards for identifying the relevant morphological markers in the archaeological record. This has involved efforts to increase sample size and improve characterization and measurement of key attributes both in archaeobotanical assemblages and in modern and ancient reference class materials used as a baseline of comparison. All the chapters in this section of the volume underscore the importance of being able to convincingly demonstrate that archaeological specimens exhibit morphology that clearly reflects domestication, and is clearly distinguishable from that produced in wild populations of the species in question.

Sample size is often important in this regard, since morphology associated with domestication will sometimes occur in wild populations. In *Chenopodium*, for example, where a reduction in seed-coat thickness (loss of predation defense and germination dormancy) is the primary morphological marker of the adaptive syndrome of domestication looked for in archaeological contexts (Chapter 4), analysis of modern wild populations of the same species has shown that a small percentage of seeds (1–3%) will have thin seed coats that might be mistaken for evidence of domesticated crop plants in an archaeological context (Smith 2002: 149). This problem of low-frequency production of thin-testa seeds in wild populations, however, is substantially reduced as the number of thin-testa seeds recovered from an archaeological context is increased. If thousands of thin-testa *Chenopodium* seeds are found in an ancient storage context, for example, questions that could be raised regarding a single thin-testa seed recovered from an archaeological deposit are no longer relevant.

M. Kislev carried out a similar study of a different morphological marker of domestication in his analysis of barley remains from the early Neolithic site of Netiv Hagdud in Israel (Bar-Yosef and Gofer 1997) He was looking for evidence of seed retention (loss of natural seed dispersal), and found that 4 percent of the 3,200 seed attachment (rachis) fragments recovered from the site did not indicate easy separation (non-brittle rachis structure), which indicated seeds that were retained on the plant rather than dispersed—a clear marker of the adaptive syndrome of domestication, or so he initially thought. His subsequent consideration of modern wild populations of barley, however, showed that strong (non-brittle) rachis morphology indicating seed retention in fact occurred in the wild at a frequency high enough (10%) to make it difficult to make a case for the Netiv Hagdud barley as being domesticated (Kislev 1989). In a similar manner, Smith in Chapter 3 and Piperno in Chapter 5 in this section underscore efforts increasingly being made to survey a wide range of modern and ancient (when available) wild baseline reference class populations, to ensure that the morphology observed in archaeological specimens does in fact fall outside the range of variation that exists across broad environmental and taxonomic categories.

Interestingly, most of the morphological changes associated with the adaptive syndrome of domestication in annual seed plants, which resulted from unconscious selection on the part of human societies, were nonetheless attributes that were advantageous to the humans responsible for harvesting and preparing the seeds for consumption. Larger, uniformly mature seeds with thinner seed coats that were tightly packaged and tightly held on plants represented a significant qualitative improvement in terms of human harvesting, processing time, and energy costs. These changes yielded an unintentional yet nonetheless very substantial economic benefit for humans, and were all attributes that annual seed-crop farming societies would subsequently expand and improve upon through intentional, premeditated, deliberate or "methodical" selection (Heiser 1988). As a result, it is sometimes difficult to establish a basis on which to recognize when methodical human selection of annual seed crops was added onto the preceding morphological changes resulting from unintentional selection, or whether particular morphological changes observed in archaeological specimens are indicative of unintentional or methodical selection. The central point to be kept in mind in the frequent discussions that arise regarding whether a particular morphological change in seed plants resulted from unintentional or methodical selection is that those changes associated with the adaptive syndrome of domestication, while certainly subject to deliberate methodical human manipulation, could also have resulted from unintentional selection resulting from sustained harvesting and planting of stored seed stock.

No such discussions of unintentional versus deliberate or methodical selection surround the other major class of plant domesticates—perennial root crops—since for them, deliberate human selection formed the core of the behavioral relationship of domestication.

Methodical Selection through Cloning and Morphological Change in Perennial Root Crops

Until relatively recently, perennial plants with starch-rich underground organs, primarily propagated vegetatively through the planting of clonal fragments of parent plants, have had very low visibility in the archaeological record. As a group, such *root crops* do not produce any relatively decay-resistant distinctively diagnostic plant parts such as phytoliths, pollen, or seeds. In addition, they are often grown in regions with high rainfall and resultant poor preservation of archaeobotanical remains. Because of such problems of preservation, archaeologists long questioned the possibility of recovering direct archaeological evidence of a number of important present-day, lower-latitude crop plants, including manioc (*Manihot esculenta*), sweet potato (*Ipomea batatas*), yam (*Dioscorea trifida*), and arrowroot (*Maranta arundinacea*).

Given the difficulties of finding evidence of morphological change associated with domestication in apparently nonexistent archaeobotanical remains, researchers seeking evidence for the initial domestication of these perennial root crops have focused on direct, first-hand evidence of human behavioral change in the archaeological record indicative of the establishment of a relationship of domestication. Some of the strongest classes of evidence of this type involve clear and consistent pollen and phytolith profile indications of widespread modification of forest environments, including burning—evidence of forest clearance for field plots (Piperno 1993; Piperno et al. 1991). Distinctive tool forms, including manioc graters and griddles (DeBoer 1975; Roosevelt 1980), also provide evidence of human behavioral change that may have been linked to the processing of domesticated root crops.

The archaeological record for domestication of perennial root crops also differs from that of annual seed plants in that any unintentional selection during the initial stages of human intervention in the life cycle and management of root crops, if it occurred, could be much more difficult to isolate or identify in the archaeological record. The creation of a new relationship of domestication between human societies and target species of root crops did not involve storage and planting of stored seed stock, but rather the immediate replanting of fragments of harvested roots to form the crop for the next growing season. If planting plots were partitioned from wild plants, and new unintentional selective pressures were established in the human-modified environment, it is certainly possible that managed plots could produce plants that were unintentionally distinct in terms of genetic profile and morphology. To date, however, there has been little discussion of how the adaptive syndrome of domestication might be manifested in root crops (see Chapter 5).

Since no clear unintentional cause and effect linkage has yet been established between the new pattern of human behavior associated with the initial domestication of root crops (sustained vegetative propagation involving the resowing of harvested root fragments, Piperno and Pearsall 1998) and any resultant adaptive response by the target populations that produced a set of distinctive (and archaeologically visible) morphological changes, most researchers have centered their discussions of early root-crop management on evidence for deliberate or methodical selection.

In plants propagated vegetatively, "selection is absolute and its effects immediate" (Harlan 1975). Thus, from the very beginning, we might well expect that there may have been conscious selection in that preference may have been given to certain clones over others. The future variation, of course, would not be limited to the cultivated clones, for the primitive vegetatively propagated crops were also capable of reproducing sexually and starts from spontaneous seedlings might also be incorporated in futureplantings (Hawkes 1969; Heiser 1988: 80).

One of the most obvious ways in which present-day domesticated root crops differ from their wild progenitors and relatives is in terms of the larger size of their subterranean organs, and changes in the starch composition of these organs. Early methodical or deliberate selection on the part of humans during the periodic process of resowing appears to have focused on these two attributes. Selection for certain forms of starch, from the range that would have been present in the populations of plants that were targeted early in the process of human intervention and sustained replanting, would have focused on starch types that were best suited for the methods of processing and preparation being employed (Chapter 5).

Fortunately for researchers interested in identifying early domesticated root crops in the archaeological record, the individual starch grains making up the underground organs produced by present-day domesticated crop plants have been found not only to differ in terms of morphology, and often in size, from starch grains produced by related wild taxa, but also to exhibit morphology that is diagnostic at the species level (Chapter 5). As a result, it is now possible to identify starch grains in archaeological contexts that provide direct archaeological evidence of morphological changes associated with domestication for a number of different important root crops. In several landmark articles, Ugent and his colleagues (Ugent et al. 1982, 1984, 1986) described and discussed starch-grain structure in remarkably well preserved tubers of several different domesticates recovered from archaeological deposits along the dry desert coast of Peru. The question remained, however: how likely was it that starch grains could be recovered from less than ideal contexts of archaeological preservation beyond the dry coast of Peru?

The answer, surprisingly enough, as discussed in Chapter 5, is that given the right archaeological context, and careful field recovery and laboratory protocols, starch grains that have been preserved intact on the surface and in the cracks of grinding tools over long periods of time (Piperno et al. 2004) can provide clear, unequivocal documentation of the processing of a range of different wild and domesticated plant foods, including domesticated root crops grown in previously poorly documented low-latitude regions of the world.

Phytoliths (opal silica bodies), provide the other critically important class of direct micro-morphological indicators of plant usage in low-latitude zones of poor plant preservation. These small pieces of "glass," formed in the living cells of plants during their life span for structural support and to deter predation, exhibit considerable morphological variation, are resistant to decay, and can be recovered in substantial numbers from a range of archaeological contexts, including ancient fields (Pearsall 1987), storage and refuse pits, house floors, and other features (Chapter 6), as well as from food processing contexts such as cooking vessel residues (Chapter 7). As is the case with starch grains, recovery of phytoliths from food processing contexts, particularly phytolith forms closely associated with the edible portions of plants (e.g., corn cobs, seed coats, squash rinds), provides a strong case for the dietary role of the species represented.

Along with considering the archaeological context of where phytoliths are found (e.g., imbedded in ceramic-vessel cooking residues or scattered in general midden deposits), researchers looking for evidence of domesticated crop plants in phytolith assemblages also weigh the relative diagnostic value of the phytoliths in question. Considerable effort in this regard has been invested over the decades in building increasingly comprehensive modern reference class collections of phytolith forms produced by different taxonomic groups across most regions of the world (Piperno 1988, 2006). At the same time, application of standardized nomenclatural systems (e.g., SACDBT 1962a, 1962b), now allow for the accurate characterization of the wide range of phytolith forms that occur in plants, and for tracking their taxonomic patterns of occurrence (Chapter 6).

Distinctive phytolith morphology sometimes allows researchers not only to distinguish between different species (Piperno et al. 2002), but also to also differentiate between phytolith forms produced by domesticated crop plants and their wild progenitors, and thus to identify when domesticates first appear in the archaeological record. Phytoliths occur by the thousands, for example, in the fruit wall or rind of more than a dozen species of wild *Cucurbita* gourds that grow in different areas of the Americas, quite likely as a deterrent to predation (Piperno et al. 2000, 2002). These *Cucurbita* rind phytoliths are very distinctive in form and can be easily identified in archaeological phytolith assemblages. A half dozen of these species of wild *Cucurbita* gourds were brought under domestication at different times in North and South America, and there is a clear archaeological record of methodical human selection over time for larger fruit size. Comparative studies of modern squash have shown that as fruit size increases, the morphologically distinct phytoliths imbedded in the rind also increase in size. As a result, it has been possible to establish when larger fruit forms of domesticated *Cucurbita equadorensis* first appeared in human settlements along the northwest Pacific coast of South America, based on an observed increase in the size of rind phytoliths (Piperno and Stothert 2003).

Such fine-grain differentiation of domesticated crop plants and their wild relatives on the basis of changes in phytolith size or shape, however, is not essential when researchers are tracking the diffusion of crop plants outside the geographical range of their wild ancestor and any other wild relative that might produce similarly shaped phytolith forms. Since domesticated bananas belonging to the Southeast Asian genus *Musa*, for example, produce a phytolith form that can be confidently distinguished from those produced by African bananas of the genus *Ensete*, the appearance and diffusion of domesticated *Musa* across Africa, far from any related wild taxa, can be traced through analysis of archaeological phytolith assemblages (Chapter 6).

As the distance separating the geographical range of wild relatives and the archaeological site yielding the evidence of a proposed domesticate, including phytoliths, decreases, however, the case for human intervention and exportation of a crop plant outside its natural range is correspondingly weakened. In eastern North America, for example, the abundant recovery of maygrass seeds (*Phalaris caroliniana*) from archaeological sites outside the documented modern range of its wild relatives, and in association with other clearly domesticated seed crops, has been proposed as evidence of human management and active cultivation. In the absence of any morphological change associated with domestication, however, researchers have been hesitant to label these archaeological maygrass specimens as representing a domesticated crop plant, instead placing it in a gray-area category of "quasi-cultivation" (Smith 2002: 106–108). One of the obvious complicating factors that come increasingly into play as distance from the documented natural range of wild relatives is reduced is the possibility that the natural range of a species has changed through time. As a result of such complicating factors, human exportation arguments have to be individually considered within the context of other available lines of evidence. Take, for example, the recovery of distinctive phytoliths of *Musa* at the Kuk Swamp site in New Guinea, at an elevation several thousand feet above the habitat range of wild *Musa* populations. Direct evidence of human behavior associated with crop management and domestication in the form of considerable landscape modification for drainage has also been documented at Kuk Swamp (Denham et al. 2003), and together with the phytoliths of *Musa* has been used as evidence of human cultivation of this important tree crop. As is the case with maygrass, the available evidence (*Musa* phytoliths and landscape modification) is employed to argue for the cultivation of bananas, rather than for the presence of domesticated bananas. The long and complex history of human landscape modification associated with crop management at Kuk Swamp further complicates easy interpretation of the timing and nature of human cultivation of bananas in this high elevation setting, far above the present-day habitat of wild *Musa*.

Similarly, in some archaeological contexts, the profile or composition of a phytolith assemblage, rather than the presence of a particular distinctive phytolith form, can document the presence of a domesticated crop plant.

In Chapter 7, for example, Thompson shows that when compared to modern reference, the profiles of phytolith assemblages recovered from residues scraped from the interior of cooking vessels can document not only the presence of maize, but whether a hard- or soft-glumed variety had been grown and cooked.

As all of these recent research initiatives show, from New Guinea and West Africa through Panama and South America, starch grain and phytolith analysis is witnessing remarkable advances in data recovery and analysis, and is opening up vast new geographical and taxonomic regions for consideration, throughout the tropics and beyond. Such micromorphological approaches to tracking the domestication and diffusion of root crops, along with parallel efforts to identify and document annual seed plant domestication, clearly hold the promise of continuing to expand and improve, particularly in concert with associated archaeological approaches to documenting animal domestication, and genetic approaches to domestication of both plants and animals.

Discussion: Future Areas of Inquiry

Characterized above as a causal chain, the process of domestication begins with the initiation by humans of new patterns of interaction and intervention in the life cycle of plants that establishes a new and qualitatively distinct relationship between the intervening humans and the target populations of the species in question. The new sets of selective pressures, and the newly defined rules for success among target populations that are inherent in this new relationship of domestication, will often lead to altered genomes or genetic profiles within target populations, which in turn may be manifested in physical morphological change. These morphological changes, when they appear, can reflect both the adaptive syndrome of domestication, resulting from unconscious selection on the part of humans, and deliberate or methodical selection. In either case, the continued existence of these genetic and morphological changes depends entirely on humans actively maintaining their established patterns of intervention. If the humans withdraw their management efforts, the altered genomes and morphology of target populations will largely trend back within the range of variation exhibited in "wild" populations of the species, and the domesticates will disappear.

For both annual seed plants and perennial root-crop species with starch-rich underground organs, the particular patterns of sustained human intervention at the beginning of the causal chain that qualify for inclusion under the general heading of domestication have been well described and documented. For annual seed plants, this qualitatively significant new relationship is based on sustained human planting of stored seed stock harvested each year from a managed garden or field plot. For perennial root crops, human intervention involves sustained replanting of clonal fragments of parent plants soon after harvest. These two well-defined behavior sets have formed the historical foundation for the vast majority of the world's food crop economies. At the same time that the clear description of the cause and effect relationship between deliberate and sustained planting of seed plants and root crops and the resultant morphological changes associated with both unconscious and methodical selection has resulted in improved recognition and documentation of those changes in archaeological assemblages, it has also served to focus attention on a number of productive pathways for future inquiry.

First among these, of course, is the challenge and promise of uncovering new datasets and new approaches relevant to identifying plant domesticates in the archaeological record. The remarkable success of phytolith and starch grain research in recent years (Chapters 5, 6, 7) in documenting the diffusion of crop plants, particularly in low-latitude tropical environments, provides clear indications that archaeological research focusing on morphological evidence for the domestication of plants will continue to expand rapidly in terms of geographical, taxonomic, and technological reach.

Second, as genetic research on modern and archaeologically recovered domesticates and their wild progenitors continues to develop as a field of inquiry (Chapters 8–12), there will be increasing opportunities to compare and integrate relevant information from all three links in the causal chain of domestication, involving consideration of evidence of the establishment of causal behavior sets, genetic change, and morphological change.

Finally, the closer focus on defining, understanding, and circumscribing the concept of domestication has also increased interest and attention on other forms of human intervention in the life cycle of plants that resemble in important ways the two specific behavior sets of domestication discussed above. While the deliberate and sustained planting of stored seed stock and the replanting of clonal parent plant fragments have understandably been acknowledged as representing a sufficient level and form of intervention in the life cycle of target populations to qualify as "domestication," such a designation also raises the question of what other categories of human intervention in the life cycle of plant populations look similar to, and might also appropriately be grouped with or in proximity to, domestication.

The value of such searches for similar behavior sets is not in establishing a clear and concise boundary definition for what does and does not constitute domestication, but rather in beginning to identify and accurately characterize additional forms of human intervention in the life cycle of plants that exist, or may have existed in the past, beyond the two major behavioral relationships discussed above. Just as the either-or dichotomy of hunting and gathering versus agriculture has long hindered consideration of human societies and economic solutions that occupied the vast middle ground of low-level food production (Smith 2001), so, too, has consideration of the full range of qualitatively different forms of human intervention in the life cycle of plants been constrained by an emphasis on sorting plants into either wild or domesticated, with domestication being based on evidence of morphological change. While the search for morphological markers of domestication in the archaeological record will continue to be critically important, it will be interesting also to look for evidence of behavioral relationships of significant human intervention and manipulation that did not result in obvious physical changes that are easily observed in specimens of target plant populations preserved in the archaeological record.

Take, for example, the recent reassessment of the patterns of management of a number of perennial root-crop species by Indian groups of the Northwest Coast of North America long characterized as "hunter-gatherers." The small, starch-rich underground organs of a number of "wild" species have long been recognized as playing some role in the diet of Northwest Coast societies (e.g., Pacific silverweed, rice-root, springbank clover, Nootka lupine). Far from simply being collected in the wild, however, these species are now documented as having been the subject of sustained and substantial long-term human management (Deur and Turner 2005). Ethnohistorical records and recent surveys document that the spatial extents of natural stands of these species were artificially expanded with considerable investment of human labor, including the replenishment of soil, construction of retaining walls, seasonal weeding and cultivation, transplanting, and selective harvesting. Ownership of sections of maintained "gardens" was marked and respected. While there is no evidence to date of any genetic or morphological changes in target populations resulting from this human management of root crops along the Northwest Coast, the relationship of intervention that existed would seem generally comparable to that documented for cultivation of root crops in other regions of the world in terms of the type and amount of human investment of time and labor. The relative dietary importance of the Northwest Coast root crops is at this point difficult to ascertain accurately, but the question of their status as possible "domesticates" opens up the broader and largely unexplored general set of questions regarding what other forms of substantial human management of plant populations may have existed in the past, and what record of them might exist unrecognized in the archaeological record.

References

Bar-Yosef, O. and A. Gofer (eds.). 1997. *An early Neolithic village in the Jordan Valley part 1: The archaeology of Netiv Hagdud.* American School of Prehistoric Research Bulletin 43. Cambridge, MA: Peabody Museum of Archaeology and Ethnology, Harvard University.

Colledge, S. 1998. Identifying pre-domestication cultivation using multivariate analysis. In *The origins of agriculture and plant domestication*, A. Damania, J. Valkoun, G. Willcox, and C. Qualset (eds.), pp. 121–131. Alleppo, Syria: International Center for Agricultural Research in the Dry Areas (ICARDA).

Cowan, C. W. 1987. *First farmers of the Middle Ohio Valley.* Cincinnati, OH: Cincinnati Museum of Natural History.

Darlington, C. 1956. *Chromosome botany.* London: Allen and Unwin.

———. 1969. *The evolution of man and society.* London: Allen and Unwin.

DeBoer, W. 1975. The archaeological evidence for manioc cultivation: A cautionary note. *American Antiquity* 40: 419–433.

Delcourt, P., H. Delcourt, P. Cridlebaugh, and J. Chapman. 1986. Holocene ethnobotanical and paleoecological record of human impact on vegetation in the Little Tennessee River Valley, Tennessee. *Quaternary Research* 25: 330–349.

Denham, T. P., S. G. Haberle, C. Lentfer, R. Fullagar, J. Field, M. Therin, N. Porch, and B. Winsborough. 2003. Origins of agriculture at Kuk Swamp in the highlands of New Guinea. *Science* 301: 189–193.

Deur, D. and N. Turner (eds.) 2005. *Traditions of plant use and cultivation on the Northwest Coast.* Seattle: University of Washington Press.

de Wet, J. and J. Harlan. 1975. Weeds and domesticates. Evolution in the man-made habitat. *Economic Botany* 29: 99–107.

Dorweiler, J. E. and J. Doebley. 1997. Developmental analysis of Teosinte Glume Architecture1: A key locus in the evolution of maize (*Poaceae*). *American Journal of Botany* 84: 1313–1322.

Ford, R. 1985. The process of plant food production in prehistoric North America. In *Prehistoric food production in North America*, R. Ford (ed.), pp. 1–18. Anthropological Papers, No. 75. Ann Arbor, MI: Museum of Anthropology, University of Michigan.

Fritz, G. 1984. Identification of cultigen amaranth and *Chenopodium* from rock shelters in northwest Arkansas. *American Antiquity* 49: 558–572.

Gilmore, M. 1931. Vegetal remains of the Ozark bluff-dweller culture. *Papers of the Michigan Academy of Science, Arts and Letters* 14: 83–102.

Hammer, K. 1984. Das Domestikationssyndrom. *Kulturpflanze* 32: 11–34.

Harlan, J. 1975. *Crops and Man.* Madison, WI: American Society of Agronomy.

Harlan, J. and J. de Wet. 1965. Some thoughts about weeds. *Economic Botany* 19: 16–24.

Harlan J., J. de Wet, and E. Price. 1973. Comparative evolution of cereals. *Evolution* 27: 311–325.

Harris, D. 1989. An evolutionary continuum of people-plant interaction. In *Foraging and farming: The evolution of plant exploitation*, D. Harris and G. Hillman (eds.), pp. 11–26. London: Unwin Hyman.

Hawkes, J. 1969. The ecological background of plant domestication. In *The domestication and exploitation of plants and animals*, P. Ucko and G. Dimbleby (eds.), pp. 17–29. Chicago: Duckworth.

Heiser, C. 1988. Aspects of unconscious selection and the evolution of domesticated plants. *Euphytica* 37: 77–85.

Jaenicke-Despré, V., E. Buckler, B. D. Smith, M. Gilbert, A. Cooper, J. Doebley, and S. Pääbo. 2003. Early allelic selection in maize as revealed by ancient DNA. *Science* 302: 1206–1208.

Jones, V. 1936. The vegetal remains of Newt Kash Hollow Shelter. In *Rock shelters in Menifee County, Kentucky*, W. Webb and W. Funkhouser, pp. 147–167. University of Kentucky Reports in Archaeology and Anthropology 3(4). Lexington, KY: University of Lexington.

Kislev, M. 1989. Pre-domesticated cereals in the pre-pottery Neolithic A period. In *People and culture in change*, I. Hershkovitz (ed.), pp. 147–151. British Archaeological Reports, International Series 508(i). Oxford: British Archaeological Reports.

Miller, N. F. 2002. Tracing the development of the agropastoral economy in southeastern Anatolia and northern Syria. In *The dawn of farming in the Near East*, R. T. J. Cappers and S. Bottema (eds.), pp. 85–94. Studies in Early Near Eastern Production, Subsistence, and Environment 6. Berlin: Ex Oriente.

Pearsall, D. M. 1987. Evidence for prehistoric maize cultivation on raised fields at Peñón del Río, Guayas, Ecuador. In *Pre-hispanic agricultural fields in the Andean region*, W. M. Denevan, K. Mathewson, and G. Knapp (eds.), vol. 359(ii), pp. 279–295. British Archaeological reports International. Oxford: British Archaeological Reports.

Piperno, D. R. 1988. *Phytolith analysis: An archaeological and geological perspective*. San Diego: Academic Press.

———. 1993. Phytolith and charcoal records from deep lake cores in the American tropics. In *Current research in phytolith analysis: Applications in archaeology and paleoecology*, D. M. Pearsall and D. R. Piperno (eds.), pp. 58–71. MASCA Research Papers in Science and Archaeology, Vol. 10. Philadelphia, PA: The University Museum of Archaeology and Anthropology, University of Pennsylvania.

———. 2006. *Phytolith analysis in archaeology and environmental history*. Walnut Creek, CA: AltaMira Press.

Piperno, D. R. and D. M. Pearsall. 1998. *The origins of agriculture in the lowland Neotropics*. New York: Academic Press.

Piperno, D. R. and K. E. Stothert. 2003. Phytolith evidence for Early Holocene *Cucurbita*. Domestication in Southwest Ecuador. *Science* 299: 1054–1057.

Piperno, D. R., M. B. Bush, and P. A. Colinvaux. 1991. Paleoecological perspectives on human adaptation in Central Panama. II. The Holocene. *Geoarchaeology* 6: 227–250.

Piperno, D. R., I. Holst, T. C. Andres, and K. E. Stothert. 2000. Phytoliths in *Cucurbita* and other neotropical *Cucurbitaceae* and their occurrence in early archaeological sites from the lowland American tropics. *Journal of Archaeological Science* 27: 193–208.

Piperno, D. R., I. Holst; L. Wessel-Beaver; and T. C. Andres. 2002. Evidence for the control of phytolith formation in *Cucurbita* fruits by the hard rind (Hr) genetic locus: Archaeological and ecological implications. *Proceedings of the National Academy of Sciences, USA* 99: 10923–10928.

Piperno, D. R., E. Weiss, I. Holst, and D. Nabel. 2004. Processing of wild cereal grains in the Upper Paleolithic revealed by starch grain analysis. *Nature* 407: 894–897.

Rindos, D. 1984. *The origins of agriculture*. Orlando, FL: Academic Press.

Roosevelt, A. 1980. *Parmana: Prehistoric maize and manioc subsistence along the Amazon and Orinoco*. New York: Academic Press.

SACDBT (Systematics Association Committee for Descriptive Biological Terminology), 1962a. Terminology of simple symmetrical plane shapes. Chart 1a. *Taxon* 11: 145–156.

———. 1962b. Terminology of simple symmetrical plane shapes. Chart 1a. *Taxon* 11: 245–247.

Smith, B. D. 2001. Low level food production. *Journal of Archaeological Research* 9: 1–43.

———. 2002. *Rivers of change*. (Paperback edition). Washington, D.C.: Smithsonian Institution Press.

———. 2005. Low level food production and the Northwest Coast. In *Traditions of plant use and cultivation on the Northwest Coast*, D. Deur and N. Turner (eds.), pp. 37–66. Seattle: University of Washington Press.

Toll, M. (ed.) 1995. *Soil, water, biology, and belief in prehistoric and traditional Southwestern agriculture*. New Mexico Archaeological Council Special Publication No. 2. Albuquerque: New Mexico Archaeological Council.

Turner, N, and S. Peacock. 2005. Solving the perennial paradox. In *Traditions of plant use and cultivation on the Northwest Coast*, D. Deur and N. Turner (eds.), pp. 101–150. Seattle: University of Washington Press.

Ugent, D., S. Pozorski, and T. Pozorski. 1982. Archaeological potato tuber remains from the Casma Valley of Peru. *Economic Botany* 36: 182–192.

Ugent, D., S. Pozorski, and T. Pozorski. 1984. New evidence for ancient civilization of *Canna edulis* in Peru. *Economic Botany* 38: 417–432.

Ugent, D., S. Pozorski, and T. Pozorski. 1986. Archaeological manioc (*Manihot*) from Coastal Peru. Economic Botany 40: 78–102.

Wilson, H. 1981. Domesticated *Chenopodium* of the Ozark bluff dwellers. *Economic Botany*. 35: 233–239.

Zohary, D. 1969. The progenitors of wheat and barley in relation to domestication and agricultural dispersal in the Old World. In *The domestication and exploitation of plants and animals*. P. Ucko and G. Dimbleby (eds.), pp. 47–66. Chicago: Duckworth.

———. 1984. Modes of evolution in plants under domestication. In *Plant biosystematics*, W. Grant, pp. 579–596. Toronto: Academic Press.

CHAPTER 3

Seed Size Increase as a Marker of Domestication in Squash (*Cucurbita pepo*)

BRUCE D. SMITH

Introduction

Among the most basic questions surrounding the origin of any plant or animal domesticate are the identity of its wild ancestor and the temporal, spatial, and cultural context of its initial domestication. Inherent in these questions of where and when a domesticate first emerged is the issue of whether the domestication was a single isolated development, or if multiple independent domestications of a species occurred at different times in different places. The geographical extent and degree of internal partitioning of the range of a wild ancestor obviously is an important factor in whether single or multiple domestications take place. A single origin, for example, has long been proposed for a number of the Near Eastern cereals whose wild ancestors had quite restricted and internally unpartitioned geographical ranges (Zohary 1996). Contrary to Lentz's recent proposition (Lentz et al. 2001: 375), such single origins for some cereal domesticates should not be considered as evidence that plant domestication in general involves single domestication events followed by diffusion. Multiple domestications are always a possibility where wild plant or animal progenitor populations extend across a broad and varied geographical range.

In the Americas, *Cucurbita pepo* squash (pumpkins, acorn, and summer varieties), is, along with *Chenopodium* (Chapter 4), one of the best-documented species to have been domesticated more than once from wild ancestor populations having a broad geographical range. More than 10 different taxa of wild *Cucurbita* gourds have overlapping geographical distributions that today extend from northern South America up into the mid-latitude eastern United States (Nee 1990). A half-dozen domesticated squashes were developed out of these wild *Cucurbita* taxa (i.e., *C. maxima, C. moschata, C. ficafolia, C. argyrosperma, C. equadorensis, C. pepo*: Robinson and Decker-Walters 1997; Piperno and Stothert 2003; Sanjur et al. 2002, 2004). Among these, *Cucurbita pepo* was independently domesticated twice—first in south central Mexico about 10,000 years ago and then again, on the basis of currently available archaeological evidence, in the mid-latitude eastern woodlands of North America by about 5,000 years ago (Smith 1997, 2000, 2002). In both Mexico and the eastern United States, this independent domestication of *C. pepo* squash is indicated by clear changes in the morphology of *C. pepo* recovered from human habitation layers in archaeological sites. These morphological changes distinguish the early domesticated crop plants from all of their related wild taxa, including their direct wild ancestors, and they have also been linked to a specific set of causative human behaviors—the deliberate and sustained planting of stored seed stock.

Morphological Markers of Domestication

In the 1960s and 1970s, Harlan, de Wet, and Price published a series of articles that outlined a clear set of morphological criteria for distinguishing between wild and domesticated seed plants based on changes in reproductive propagules resulting from the storage, planting, and harvesting of seed stock over a number of years (Harlan and de Wet 1965; Harlan et al. 1973; de Wet and Harlan 1975). These morphological changes, which reflect responses by the now managed plants to new sets of selective pressures, both in the seedbed and at harvest, are automatic and independent of any deliberate or directed selection on the part of humans; they have collectively been termed "the adaptive syndrome of domestication" (Chapter 1, 2; Smith 1998, 2002). Those plants with seeds tightly clustered and tightly held at the terminal ends of their stalks, for example, would have a selective advantage at harvest in that they would be more likely to be seen and successfully collected and thus to contribute to next year's planting. Similarly, seeds that sprout quickly because of reduced germination dormancy (thinner seed coat—see Chapter 4) and that grow rapidly (greater stored start-up energy, larger seed) in seedbeds would have a selective advantage in terms of competing for sunlight and surviving to be harvested.

Unfortunately, most of the morphological changes in seed plants identified by Harlan and others as resulting from deliberate human planting and harvesting are difficult to document in archaeobotanical assemblages. One of the most obvious, and most archaeologically visible, of these changes that mark initial domestication is an increase in seed size in response to seedbed selective pressures.

In order to use seed size as a criterion for differentiating between the seeds of wild and domesticated *C. pepo* squash, a comprehensive standard, or baseline, of comparison of seed measurements for wild *Cucurbita* taxa is needed. Figures 3.1

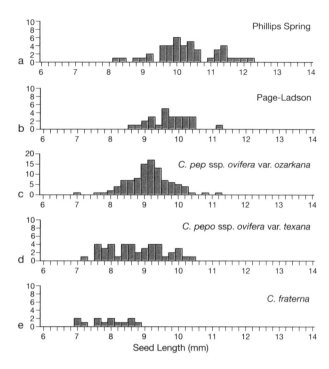

FIGURE 3.1 The size of *Cucurbita pepo* seeds from archaeological and paleontological sites in North America compared to three taxa of modern wild *C. pepo* gourds. (a) Phillip Spring (n = 45, mean = 10.5); (b) Page-Ladson (n = 35, mean = 9.9); (c) *C. pepo* ssp. *ovifera* var. *ozarkana* (n = 100, mean = 9.2); (d) *C. pepo* ssp. *ovifera* var. *texana* (n = 42, mean = 8.9); (e) *C. pepo* ssp. *ovifera* var. *fraterna* (n = 13, mean = 7.9).

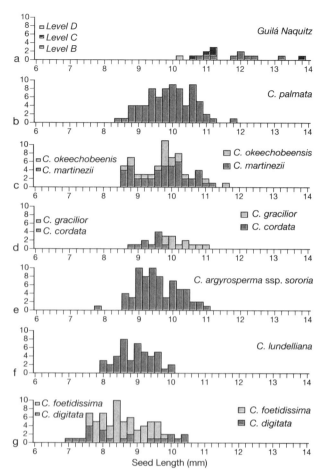

FIGURE 3.2 The size of *Cucurbita pepo* seeds from Guilá Naquitz compared to six taxa of modern wild *Cucurbita* gourds. (a) Guilá Naquitz, Levels B, C, and D (n = 15); (b) *C. palmata* (n = 75, mean = 10.2); (c) *C. okeechobeensis* (n = 20, mean = 9.9) and *C. martinezii* (n = 46, mean = 9.9); (d) *C. gracilior* (n = 10, mean = 10.7) and *C. cordata* (n = 10, mean = 9.5); (e) *C. argyrosperma* ssp. *sororia* (n = 75, mean = 9.7); (f) *C. lundelliana* (n = 50, mean = 9.0); (g) *C. foetidissima* (n = 55, mean = 8.7) and *C. digitata* (n = 20, mean = 8.8).

and 3.2 provide such a general wild *Cucurbita* gourd seed profile. More than 700 seeds from 63 separate accessions, representing 12 modern species, subspecies, or varieties of wild *Cucurbita* gourds were measured. With the exception of *C. pedatifolia* and *C. galeottii*, all known North American (Mexican and U.S.) wild *Cucurbita* species are represented in Figures 3.1 and 3.2 (see Nee 1990: 58). In addition to the 12 modern taxa included in Figures 3.1 and 3.2, seed length information is also presented for the Page-Ladson seed assemblage (Figure 3.1b), a collection of late Pleistocene seeds of a wild *Cucurbita* gourd recovered from American Mastodon (*Mammut americanum*) dung deposits in north Florida (Newsom et al. 1993). Of the more than 700 modern seeds measured, the smallest was 7.0 mm long (Figures 3.1c and e, Figure 3.2g), while the largest was 11.9 mm in length (Figure 3.2b). The largest seeds in 10 of the 13 different taxa exceeded 10 mm in length, but none were larger than 12 mm long (only 6 of the 717 seeds were longer than 11 mm and only 1 measured more than 11.5 mm in length). These maximum-length values provide a strong basis for defining the upper size-range ceiling for seeds of North American wild *Cucurbita* gourds at 11–12 mm in length.

At the species level of comparison, the three *C. pepo* taxa listed in Figure 3.1c–e (*C. pepo* ssp. *ovifera* var. *ozarkana*, *C. pepo* ssp. *ovifera* var. *texana*, *C. pepo* ssp. *fraterna*) provide a smaller, taxonomically tighter, reference class, with their upper limit for seed length (ca. 11 mm) representing a more closely drawn wild/domesticated boundary line than the 12-mm seed length ceiling derived from the 13-taxa reference class. These three modern taxa of wild, or "free-living," *Cucurbita* gourds actually represent an even tighter, more focused reference class in that recent genetic research (Sanjur et al. 2002, 2004) has identified them as not only closely related to each other, but also as a group representing the present-day descendants of the wild progenitor populations of *Cucurbita* gourds that gave rise to the "eastern" lineage of domesticated squash (*C. pepo*. ssp. *ovifera*) about 5,000 years ago. A previous study (Decker-Walters et al. 1993) had identified one of these three—the wild Ozark gourd (*C. pepo* ssp. *ovifera* var. *ozarkana*)—as the progenitor of the *C. pepo* ssp. *ovifera* domesticate

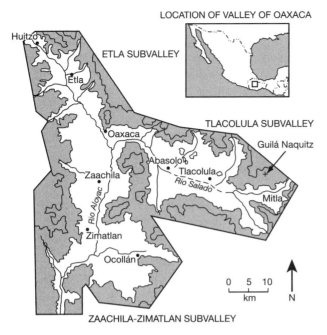

FIGURE 3.3 The location of Guilá Naquitz Cave in the Valley of Oaxaca (after Flannery 1986: Figure 3.1).

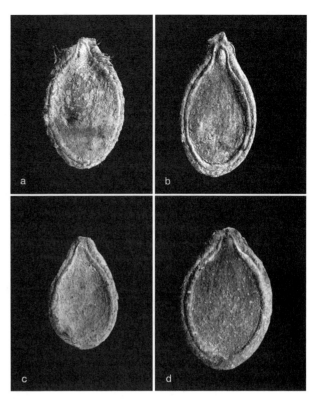

FIGURE 3.4 *Cucurbita pepo* seeds from Guilá Naquitz Cave yielding AMS dates in the Early Archaic. (a) seed from Square C11, Zone B2 (INAH 5) measuring 12.5 mm in length and yielding an AMS radiocarbon date of 6475 cal. BC; (b) seed from Square E9, Zone C (INAH 14) measuring 13.8 mm in length and yielding an AMS radiocarbon date of 7975 cal BC; (c) seed from Square E11, Zone B1 (INAH 23) measuring 13.2 mm in length and yielding an AMS radiocarbon date of 8025 cal. BC; (d) seed from Square C11, Zone B1, measuring 12.1 mm in length and yielding an AMS radiocarbon date of 6265 cal. BC (Table 3.1).

lineage, but the progenitor pool and the geographical range within which domestication occurred has now been expanded. Unfortunately, present-day populations of the wild ancestor of the Mexican domesticated *C. pepo* squash lineage have yet to be located. Since the seeds of modern cultivars of the "eastern" lineage of *C. pepo* are smaller than those of the Mexican lineage, this size difference could also have existed, some scholars have suggested, in the wild ancestor populations (Robinson and Decker-Walters 1997: 7). As a result, although 11 mm represents a good minimum length boundary for identifying domesticated *C. pepo* seeds in eastern United States archaeological contexts, the larger reference class value of 12 mm provides a more prudent minimum seed-length boundary for the designation of domesticated C. pepo in Mexico, at least until the wild progenitor of the Mexican C. pepo lineage has been identified.

Given these minimum seed-length values of 11 mm and 12 mm for establishing the initial appearance of domesticated *C. pepo* squash in the eastern United States and Mexico, we can turn to consideration of the archaeological sites and human habitation layers that have yielded the oldest seeds exceeding these seed-length thresholds.

The *Cucurbita pepo* Seed and Peduncle Assemblage of Guilá Naquitz Cave

Situated in a small tributary canyon of the Rio Mitla in the Valley of Oaxaca (Figure 3.3), Guilá Naquitz Cave was excavated by Kent Flannery in 1966 (Flannery 1986). The small 8- by 10-m cave opening at the base of the canyon wall was found to contain a 20-cm-thick habitation layer (Zone A) dating to AD 620–740, and below this, a number of much earlier habitation layers (Zones B, C, D, E) resulting from a series of intermittent short-term seasonal occupations by small family groups that dated between about 8,000 and 10,500 years ago.

These early occupation layers at Guilá Naquitz Cave (Zones B1, B2, B3, C, D) yielded more than 250 *Cucurbita* rind fragments, a half-dozen fruit stems or peduncles, and 17 seeds that were assigned to the genus *Cucurbita* by Thomas Whitaker and Hugh Cutler (Whitaker and Cutler 1971, 1986). In their initial analysis of the Guilá Naquitz *Cucurbita* assemblage, Whitaker and Cutler did not have the benefit of being able to determine the age of seeds directly through accelerator mass spectrometer (AMS) small sample radiocarbon 14 dating; in addition, clear standards for establishing the domesticated status of archaeological squash specimens had not yet been fully formulated.

The Guilá Naquitz *Cucurbita* seeds reanalyzed in 1996–1997 were all identified as *C. pepo*, based on a number of distinctive morphological characteristics (e.g., uniform color, prominent marginal ridge, short margin hairs, naked margin [see Figure 3.4]; Smith 1997, 2000). Nine of these seeds fell below

TABLE 3.1
Direct Accelerator Mass Spectrometer Radiocarbon Dates on Early *Cucurbita pepo* Seeds in the Americas:
Guilá Naquitz and Phillips Spring

Seed Length (mm)	Laboratory Sample	Site and Location	Conventional Age[a]	Dendrocalibrated Age[b]
10.8	ß47293	Phillips Spring Unit K2 (ISM#1502)	4440 ± 75	5025 ± 75
11.4	ß91404	Guilá Naquitz C11/B1	7720 ± 60	8430 ± 60
12.0	ß91406	Guilá Naquitz C11/B1	7610 ± 60	8375 ± 60
12.1	ß91405	Guilá Naquitz C11/B1	7690 ± 50	8415 ± 50
12.5	ß100763	Guilá Naquitz C11/B2	7710 ± 50	8425 ± 50
13.2	ß100766	Guilá Naquitz E11/B1	8990 ± 60	9975 ± 60
13.8	ß100764	Guilá Naquitz E9/C	8910 ± 50	9925 ± 50

[a] In radiocarbon years before present.
[b] Intercept in calendar years before present.

the 12-mm seed-length boundary line for distinguishing between the seeds of wild and domesticated *C. pepo* (compare Figure 3.2a with Figure 3.1c–e and Figure 3.2b–g), indicating the continuing use of wild *Cucurbita* gourds throughout the early occupations of the cave. In contrast, six seeds fell at or above the 12-mm dividing line, indicating the cultivation of domesticated *C. pepo* plants by the early inhabitants of Guilá Naquitz. The fruit stems, or peduncles, and fruit end rind fragments having peduncle scars, recovered from the early occupation levels at Guilá Naquitz provide further evidence that the inhabitants of this cave both harvested wild *Cucurbita* gourds and cultivated domesticated *C. pepo* squash. Ten of the peduncle scars and peduncles conformed to the wild gourd morphotype in terms of form and size (maximum basal diameter 13 mm and roughly circular outline). A second group of seven peduncles and peduncle scars, in contrast, were considerably larger than the wild morphotype profile (maximum basal diameter 14.7–23.6 mm), and all had a distinctive angular pentagonal outline at the point of attachment to the fruit. A characteristic 10-ridge lobing pattern also identifies them as *C. pepo* (Smith 1997).

While the recovery of *C. pepo* seeds that are substantially larger than those produced by wild *Cucurbita* gourds in the early habitation layers of the cave provides good evidence for the adaptive syndrome of domestication—an automatic adaptive response by squash plants to new seed bed selective pressures associated with deliberate and sustained planting—the size increase in peduncles reflects deliberate human selection for larger fruits. Such evidence of intentional selection for desired characteristics follows more than a thousand years after an increase in seed size indicates deliberate planting and initial domestication.

A total of 40 radiocarbon dates have been obtained on plant remains from Guilá Naquitz Cave (Smith 2000; Piperno and Flannery 2001), including four direct AMS dates on *C. pepo* peduncles and six direct AMS dates on *C. pepo* seeds (Table 3.1). The five directly dated *C. pepo* seeds that fall above the 12-mm maximum length for the wild morphotype reference class, and which can be considered as representing domesticated plants, are all more than 8,400 years old, with the two oldest of the dated seeds indicating that the family-sized groups that were seasonally occupying Guilá Naquitz were cultivating a large-seeded *C. pepo* squash as early as 10,000 years ago. In contrast, the first peduncles to fall above the 13-mm upper boundary for wild morphotype *Cucurbita* gourds do not appear until 8,400 years ago, and they show a subsequent steady size increase over time (Figure 3.5).

The *C. pepo* Seed and Peduncle Assemblage from the Phillips Spring Site

Located adjacent to an artesian spring on a terrace of the Pomme de Terre River in west central Missouri (Figure 3.6), the water-saturated habitation layers of the Phillips Spring site yielded abundant and well-preserved plant remains when excavated in the 1970s (Kay et al. 1980), including uncarbonized rind fragments, seeds, and fruit stems assignable to the genus *Cucurbita*. The earliest habitation zone at Phillips Spring, Unit K2, sealed beneath a large rock-lined hearth, yielded three standard radiocarbon dates (4310 ± 70, 4222 ± 57, 4240 ± 80 BP radiocarbon years), along with approximately 125 whole seeds and fragments identified as *C. pepo* (Kay et al. 1980: 814; King 1980: 218, 1985: 81). When initially analyzed, 62 of the *C. pepo* seeds from Unit K2 (the "squash and gourd zone") provided both length and width measurements. They ranged in length from 8.3 to 12.2 mm, with a mean length of 10.5 mm (King 1985: 82). Fifteen years later, reanalysis of the Phillips Spring K2 *Cucurbita* assemblage produced the same results, even though length and width measurements could be obtained on only 45 seeds (Figure 3.1a; Smith 2000: 54). Assignment of the Unit K2 *Cucurbita* seeds to *C. pepo* was also confirmed (based on marginal ridge and hair morphology), as was the temporal placement of the "squash and gourd zone." A direct AMS

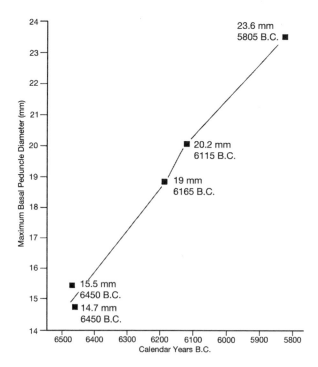

FIGURE 3.5 Increase in the size of peduncles of domesticated *Cucurbita pepo* in preceramic habitation zones of Guilá Naquitz Cave between ca. 6500 and 5800 BC (see Smith 2000: 40).

radiocarbon date of 4440 ± 75 BP radiocarbon years (5025 calibrated calendar years BP) was obtained on one of the Unit K2 seeds (Table 3.1; Smith 2000: 58).

Generally accepted as representing an early stage of domesticated *C. pepo* in the eastern United States, the Unit K2 Phillips Spring seed assemblage has a mean length value larger than all but one of the 13 wild *Cucurbita* gourd reference taxa listed in Figures 3.1 and 3.2. Significantly, although only two Phillips Spring seeds met or exceeded the 12.0-mm domestication boundary employed for the Guilá Naquitz assemblage, 12 of the 45 Unit K2 *C. pepo* seeds measured for the present study exceeded the 11.0-mm wild seed length ceiling proposed for the eastern United States, providing strong evidence for deliberate planting, an adaptive response to seedbed selective pressure, and the independent domestication of *C. pepo* ssp. *ovifera* in the eastern United States.

The four fruit-end fragment peduncle scars in the Unit K2 *Cucurbita* assemblage, in contrast, show no indication of morphological changes beyond the parameters of the wild morphotype gourd profile. They compare closely to the wild Ozark gourd (*C. pepo* ssp. *ovifera* var. *ozarkana*) in both diameter and circular outline, and they appear to be from small globular fruits. As is the case in the developmental sequence for the Mexican lineage of domesticated *C. pepo*, a wide range of changes in fruit morphology indicative of deliberate and sustained human selection also appears in the 2,000-year span following the 5000 BP Unit K2 Phillips Spring assemblage and its evidence for initial domestication of *C. pepo* in the eastern United States. These changes

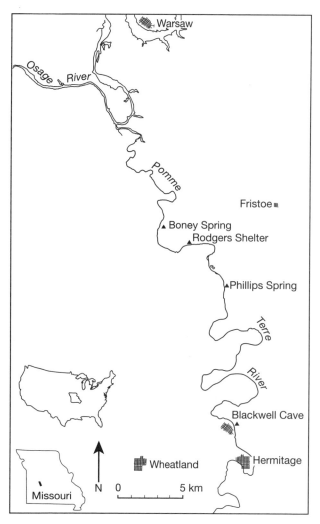

FIGURE 3.6 The location of the Phillips Spring archaeological site in south central Missouri.

include a diversification of fruit forms, surface lobing and warting, and an increase in both fruit size and rind thickness (Cowan 1997: 73).

Discussion—Future Research Directions

This chapter marshals archaeological evidence in support of the proposition that *pepo* squash (*Cucurbita pepo*) was independently domesticated twice in the Americas: the pumpkin lineage (*Cucurbita pepo* ssp. *pepo* in Mexico by 10,000 years ago, and the acorn–summer squash lineage (*Cucurbita pepo* ssp. *ovifera*) in the eastern United States by 5,000 years ago. For each of these separate domestication processes, the supporting arguments rest on an observed morphological change—an increase in the size of *pepo* squash seeds recovered from an archaeological context beyond what has been documented in modern reference class taxa of wild *Cucurbita* gourds. It is important to keep in mind that, while the evidence compiled to date and presented here constitutes a good beginning—an initial general spatial and temporal outline of domestication in this species—there is still considerable room for strengthening the supporting arguments, and for expanding the context and scope of consideration of *C. pepo* domestication.

The most obvious way in which these supporting arguments can be strengthened is through enlarging the modern reference class of seed size measurements, thereby ensuring that the full range of taxonomic and geographical variation in seed size for all of the present-day wild taxa of *Cucurbita* gourds is encompassed. This is particularly true for the three modern wild taxa that have the greatest genetic similarity to the eastern acorn–summer squash lineage. While two of the three taxa that compose the *C. pepo* ssp. *ovifera* wild progenitor gourd complex—the Ozark and Texas wild gourds (*C. pepo* ssp. *ovifera* varieties *texana* and *ozarkana*)—have been studied to some extent throughout their geographical range, and have yielded reasonably large and representative samples of seeds for modern comparison collections, the third (*C. pepo* ssp. *ovifera* var. *fraterna*) is still relatively poorly documented in terms of both seed size and natural habitat. Based on the apparent strong selective pressures controlling seed and fruit size in all of the present-day taxa of wild morphotype *Cucurbita* gourds, however, any expansion of modern reference class seed size values beyond that already documented would not be expected.

Of even more importance, in terms of strengthening our modern reference class of wild *Cucurbita* taxa, will be to identify in Mexico, if they are still extant, modern descendant populations of the wild progenitor of the Mexican pumpkin lineage of domesticated *pepo* squash. If such modern populations of the wild gourd that gave rise to *Cucurbita pepo* ssp. *pepo* more than 10,000 years ago still survive somewhere in Mexico, they have not yet been located and documented, in spite of fairly substantial field survey coverage for wild *Cucurbita* taxa.

At the same time, it will be just as important to expand the reference class of ancient collections of wild *Cucurbita* taxa recovered from human contexts and from natural deposits (e.g., Page-Ladson, Figure 3.1b) over broad spatial and temporal ranges. It is important to be able to document that the same strong selective pressures operating to control seed size in modern wild *Cucurbita* gourds were also present in the past, and that wild gourds in the early Holocene did not produce seeds that were larger than their present-day descendant populations. This ancient reference class of *Cucurbita* represents an important aspect of any supporting argument for domestication. A claim of domesticated status for squash seeds recovered from an archaeological context based on seed size increase is considerably strengthened if the candidate for early domesticate seeds can be shown not only to be larger than those produced by modern wild taxa, but also to be larger than seeds produced by presumably wild gourd taxa of the past. The ideal supporting argument in this regard would be a rich and well-documented deep-time-depth regional archaeobotanical record of human use of a small-seeded wild *Cucurbita* gourd, leading up to the appearance of a larger seeded (domesticated) form in the archaeological sequence. Supporting evidence of this kind does exist to some extent in both Mexico (Figure 3.2) and eastern North America (Figure 3.1) in the form of small squash seeds, likely from wild taxa, occasionally being recovered from scattered archaeological contexts that predate the appearance of large seeds from early domesticates. This reference class, however, needs to be substantially improved in terms of both overall seed count and temporal and spatial coverage.

Another obvious area of future research regarding the early history of domesticated *pepo* squash in the Americas will involve tracing the timing of introduction of the Mexican pumpkin lineage of *pepo* squash into eastern North America from the Southwest, and the subsequent manner in which it was variously added into the existing indigenous food production economies of different regions of the East. There has been considerable recent research on the initial introduction of other crop plants from Mexico, including other species of squash (e.g., Fritz 1994), and it should be possible to distinguish in the archaeological record between the indigenous acorn–summer squash lineage (*C. pepo* ssp. *ovifera*) and the introduced Mexican pumpkin lineage (*C. pepo* ssp. *pepo*) on the basis of genetic profiles as well as seed, fruit, and peduncle morphology.

Finally, at the same time that there is considerable room for strengthening and expanding the supporting arguments for when and where domestication of this important crop plant took place, there is also a great deal to be learned regarding the role of this species in the low-level food production economies of the human societies that first domesticated it in Mexico and the eastern United States so long ago, and the extent to which it represented a significant change, or if it initially at least was just a modest supplement to an otherwise stable and relatively secure way of life.

References

Cowan, C. W. 1997. Evolutionary changes associated with the domestication of *Cucurbita pepo*: Evidence from eastern Kentucky. In *People, plants, and landscapes*, K. Gremillion (ed.), pp. 63–85. Tuscaloosa: University of Alabama Press.

Decker-Walters, D., T. Walters, C. W. Cowan, and B. D. Smith. 1993. Isozyme characterization of wild populations of *Cucurbita pepo*. *Journal of Ethnobiology* 13: 55–72.

de Wet, J. M. J. and J. R. Harlan. 1975. Weeds and domesticates: Evolution in the man-made habitat. *Economic Botany* 29: 99–107.

Flannery, K. 1986. *Guilá Naquitz. Archaic foraging and early agriculture in Oaxaca, Mexico*. New York: Academic Press.

Fritz, G. 1994. Precolumbian *Cucurbita argyrosperma* ssp. *argyrosperma* in the western woodlands of eastern North America. *Economic Botany* 48: 280–292.

Harlan, J. and J. M. J. deWet. 1965. Some thoughts on weeds. *Economic Botany* 18: 16–22.

Harlan, J. J., M. J. deWet, and E. G. Price. 1973. Comparative evolution of cereals. *Evolution* 27: 311–325.

Kay, M., F. B. King, and C. K. Robinson. 1980. Cucurbits from Phillips Spring: New evidence and interpretation. *American Antiquity* 45: 806–822.

King, F. B. 1980. Plant remains from Phillips Spring, a multicomponent site in the western Ozark highland of Missouri. *Plains Anthropologist* 25: 217–227.

———. 1985. Early cultivated Cucurbits in eastern North America. In *Prehistoric Food Production in North America*, R. I. Ford (ed.), pp. 73–98. Ann Arbor, Mich.: Museum of Anthropology, University of Michigan, Anthropological Papers No. 75.

Lentz, D. W., M. E. D. Pohl, K. O. Pope, and A. R. Wyatt. 2001. Prehistoric sunflower (*Helianthus annuus* L.) domestication in Mexico. *Economic Botany* 55: 370–376.

Nee, M. 1990. The domestication of *Cucurbita* (Cucurbitaceae). *Economic Botany* 44 (supplement): 56–69.

Newsom, L., S. D. Webb, and J. S. Dunbar. 1993. History and geographic distribution of *Cucurbita pepo* gourds in Florida. *Journal of Ethnobiology* 13: 75–79.

Piperno, D. and K. Flannery. 2001. The earliest archaeological maize (*Zea mays* L.) from highland Mexico: New accelerator mass spectrometry dates and their implications. *Proceedings of the National Academy of Sciences* 98: 1324–1326.

Piperno, D. and K. Stothert. 2003. Phytolith evidence for Early Holocene *Cucurbita* domestication in southwest Ecuador. *Science* 299: 1054–1057.

Robinson, R. W. and D. Decker-Walters. 1997. *Cucurbits*. New York: CAB International.

Sanjur, O., D. Piperno, T. Andres, and L. Wessel-Beaver. 2002. Phylogenetic relationships among domesticated and wild species of *Cucurbita* inferred from a mitochondrial gene: Implications for crop plant evolution and areas of origin. *Proceedings of the National Academy of Sciences, USA* 99: 535–540.

———. 2004. Using molecular markers to study plant domestication: The case of *Cucurbita*. In *Biomolecular archaeology: Genetic approaches to the past*. D. M. Reed (ed.), pp. 128–150. Occasional Paper No. 32 of the Center for Archaeological Investigations, Southern Illinois University, Carbondale. Carbondale, Ill.: CAI, Southern Illinois University, Carbondale.

Smith, B. D. 1997. The initial domestication of *Cucurbita pepo* in the Americas 10,000 years ago. *Science* 276: 932–934.

———. 1998. *The emergence of agriculture*. New York: W.H. Freeman.

———. 2000. Guilá Naquitz revisited. In *Cultural evolution: Contemporary viewpoints*, G. Feinman and L. Manzanilla (eds.), pp. 15–59. New York: Kluwer Academic/Plenum Publishers.

———. 2002. *Rivers of change* (paperback edition). Washington, D.C.: Smithsonian Institution Press.

Whitaker, T. and H. Cutler. 1971. Prehistoric cucurbits from the Valley of Oaxaca. *Economic Botany* 25: 123–127.

———. 1986. Cucurbits from preceramic levels at Guilá Naquitz. In *Guilá Naquitz: Archaic foraging and early agriculture in Oaxaca, Mexico*, K. Flannery (ed.), pp. 275–279. New York: Academic Press.

Zohary, D. 1996. The founder crops of southwest Asian agriculture. In *The origins and spread of agriculture and pastoralism in Eurasia*, D. Harris (ed.), pp. 142–158. Washington, D.C.: Smithsonian Institution Press.

CHAPTER 4

A Morphological Approach to Documenting the Domestication of *Chenopodium* in the Andes

MARIA C. BRUNO

Introduction

Chenopods were grown as domesticated crops in many regions of the Americas prior to European contact, and they are known to have been independently domesticated in the Andes of South America, in Mexico, and in the eastern United States. Today, however, while they remain an important component of the Andean diet, chenopods survive only as a minor crop in Mexico and are no longer commercially grown north of Mexico. Ironically, less is known about the domestication of the Andean chenopods, *Chenopodium quinoa* and *Chenopodium pallidicaule*, than about the domestication of their extinct eastern North American counterpart, *C. berlandieri* ssp. *jonesianum*. In the Andes, the genetic and archaeobotanical research necessary to identify wild progenitor populations and to document where and when initial chenopod domestication occurred is just beginning. Narrowing down the time and place of Andean *Chenopodium* domestication is not only essential to advancing our general knowledge of agricultural origins in South America, but also necessary for evaluating alternative developmental models regarding the transition from Archaic (ca. 8000–1500 BC) to Formative (1500 BC–AD 500) period socioeconomic systems in the south central Andean highlands.

In this case study, I employ approaches developed by researchers working on eastern North American chenopods to examine the seed morphology of four modern *Chenopodium* taxa encountered in the southern Lake Titicaca Basin, Bolivia. A set of morphological characteristics is identified that permits the identification of these species in the archaeological record, and which can be employed in distinguishing the seeds of wild chenopods from those of domesticated crop plants.

In the eastern United States, investigators have shown that seeds of the domesticate *C. berlandieri* ssp. *jonesianum* have a significantly thinner seed coat (testa) than those of its wild counterpart *C. berlandieri* ssp. *berlandieri* (Wilson 1981; Smith 1984, 1985a, 1985b). Accurate measurement of testa thickness with scanning electron microscopy (SEM), along with direct accelerator mass spectrometer (AMS) radiocarbon dating, of *Chenopodium* seeds from archaeological sites throughout the eastern United States has provided rather complete temporal and spatial evidence for the domestication of this crop plant in this region (Smith 1984, 1985a, 1985b, 1988; Smith and Cowan 1987; Fritz and Smith 1988; Gremillion 1993a, 1993b).

Changes in testa thickness have also been shown to have occurred during the domestication of the Andean chenopods (Nordstrom 1990; Eisentraut 1998; Bruno 2001). In the Andes, however, distinguishing between wild and domesticated chenopods based on testa thickness is complicated by the variety of different *Chenopodium* species that were utilized.

Chenopods of the Andes

Humans interact with a range of different chenopod taxa in the Andes, including two domesticates (*C. quinoa* Willd. and *C. pallidicaule* Aellen), several varieties of agricultural weed (*C. quinoa* ssp. *milleanum* Aellen, also referred to as *C. quinoa* var. *melanospermum* Hunziker), and wild species such as *C. ambrosioides* Aellen, and *C. hircinum* Schrader (see Table 4.1).[1] *C. quinoa*, or quinoa, is the most widely recognized Andean chenopod and is still grown by indigenous farmers throughout the Andes. The other, less well-known, domesticated chenopod is *C. pallidicaule*, or kañawa. It is categorized as a "rustic" domesticate because it maintains several wild characteristics such as self-seeding, differential maturation, and easily shattering seeds (Gade 1970). These characteristics, however, make it a reliable crop in the high, dry regions of the Andean *altiplano*. *C. quinoa* ssp. *milleanum*, *C. quinoa* var. *melanospermum*, and *C. hircinum* are all spontaneous, weedy chenopods that are recognized by their black seeds and are commonly referred to as *quinoa negra* or *ajara* (Hunziker 1943, 1952; Nelson 1968; Wilson 1988b, 1990). The first two are agricultural weeds, conspecific with *quinoa*, that grow throughout the Andes. *C. hircinum* is a wild chenopod common to the eastern Andean slopes and the plains of Argentina, Uruguay, and Paraguay (Wilson 1988a, 1988c, 1990). Though not specifically an agricultural weed, *C. hircinum* often occupies habitats created by human disturbance. *C. ambrosioides*, or *paiko*, is a wild chenopod encountered throughout the Andes and is used as a vermifuge (La Barre 1948, 1959; Bastien 1987; Franquemont et al. 1990).

Identifying *Chenopodium* Domestication: Taxonomic and Genetic Approaches

Chenopods are weedy, colonizing plants that are adapted to disturbed habitats (Baker 1972; Holzner and Numata 1982). As a result, theories of weed domestication proposed by

TABLE 4.1
Andean *Chenopodium* Taxa

	C. quinoa	*C. pallidicaule*	*C. quinoa* ssp. *milleanum*/ *C. quinoa* var. *melanospermum*	*C. hircinum*	*C. ambrosioides*
Section	Cellulata	Leiosperma	Cellulata	Cellulata	Ambrina
Common Names	Quinoa, quinua, jupa, quingua	Kañawa, cañihua	Quinoa negra, ajara, ayara	Quinoa negra, ajara, ayara	Paiko
Relation to Humans	Domesticate	Domesticate	Agricultural weed	Thrives in disturbed habitats, including agricultural fields	Thrives in disturbed habitats, collected for medicine
Distribution	Southern Colombia to northern Chile and western Argentina	Peruvian and Bolivian altiplano	Southern Colombia to northern Chile and western Argentina	Eastern slopes of the Andes; valleys of Paraguay, Uruguay, and Argentina	Tropical and subtropical South America
Physical Characteristics	Tall and little branched with an exserted leafy panicle	Short, bushy and branched with many small inflorescences concealed in foliage	Tall with little to some branching and diffuse to compact inflorescence	Tall with little to some branching and diffuse to compact inflorescence	Short, bushy, and branched with ebracteate inflorescence

SOURCE: Based on Aellen and Just 1943; Hunziker 1943, 1952; Simmonds 1965; Gandarillas 1974; Gade 1970; Tapia et al. 1979; Wilson 1980, 1981, 1990; National Research Council 1989.

Anderson, Harlan, de Wet, and others are clearly applicable to *C. quinoa* and *C. pallidicaule* (Anderson 1952; Harlan and de Wet 1965; de Wet 1973; Harlan et al. 1973; de Wet and Harlan 1975; Kuznar 1993; Pickersgill 1977; Pearsall 1989, 1993; Smith 1992). Most regional specialists believe that the Andean domesticates arose from wild populations that colonized anthropogenic habitats (Sauer 1952; Pearsall 1980, 1989, 1992; Kuznar 1993). Researchers have also suggested that the process of domestication resulted in crop/weed complexes. Genetic research has focused on understanding the phylogenetic relationships of the crop/weed complexes, as well as on identifying their wild progenitors.

The *quinoa* crop/weed complex, consisting of the domesticate *C. quinoa* and the black-seeded types *C. quinoa* var. *melanospermum* and *C. quinoa* ssp. *milleanum*, has been defined by several botanists (Nelson 1968; Heiser and Nelson 1974; Wilson and Heiser 1979; Heiser 1976; Pickersgill 1977; Wilson 1988b, 1990). Wilson's allozyme studies suggest that the crop and its weedy relatives (all of which he refers to as *ajara*), "have moved together from a temporal and geographic point of origin along the same geographic and phylogenetic path. It is therefore difficult to place *ajara* in a basal, intermediate, or derived phyletic position relative to *quinoa*, i.e., its phyletic distance from the progenitor taxon is equivalent to that of *quinoa*" (Wilson 1990: 99). Analysis of random amplified polymorphic DNA (RAPD) of *C. quinoa* and *C. quinoa* var. *melanospermum* by Ruas et al. supports Wilson's assessment that the crop/weed complex is a "monophyletic, possibly co-evolving unit" (Ruas et al. 1999: 26).

Identifying the wild progenitor of the *quinoa/ajara* crop/weed complexes has been more difficult. Wilson (1990) hypothesizes that a wild North American tetraploid progenitor (*C. berlandieri*) traveled to Mexico and South America via human migrations or by long-distance bird dispersals. The populations in each region were subsequently domesticated independently: *C. berlandieri* var. *jonesianum* in eastern North America, *C. berlandieri* ssp. *nuttalliae* in Mexico, and *C. quinoa* in the Andes. Since the greatest modern diversity of *quinoa* occurs in the south central Andean highlands, researchers following a Vavilovian model have posited that domestication occurred in this region (Gandarillas 1974; Wilson 1988a,

1990). Wilson's allozyme research, however, has indicated that an earlier form of *C. hircinum* is the strongest candidate for having been the wild progenitor of *C. quinoa*. Electrophoretic variation places *C. hircinum* in a basal position relative to the other Andean tetraploid chenopods, making the present-day geographical range of *C. hircinum* (eastern Andean slopes and plains) a likely location for the initial domestication of *C. quinoa* (Wilson 1988c, 1990).

In contrast to *quinoa*, very little is known about *kañawa* domestication. Based on its weedy character, a "dump heap" scenario of domestication is plausible. Heiser (1976) suggests that the wild form of *kañawa* could have been a weed in *quinoa* fields. Early farmers probably recognized its tolerance to drought and freezing temperatures and began managing it. Taxonomic studies have yet to verify whether a weed evolved alongside the crop. Hunziker (1943, 1952) classified a dark-seeded form of *kañawa* as *C. pallidicaule* var. *melanospermum*. Leon (1964) described a "wild," dark-seeded form of *kañawa* growing in the Lake Titicaca region, which local people called *ayara* or *quitacawa*. While the "wild" form described by Leon may represent a sympatric weed, no subsequent studies have investigated this possibility, making it difficult to determine whether a companion weed evolved along with *kañawa*. Simmonds (1965) indicates that *kañawa* is predominantly inbreeding and that there is only limited evidence for introgression with wild or weedy taxa. To date there have not been any genetic studies focusing on the domestication of *kañawa*; however, the RAPD analysis by Ruas et al. (1999) shows that *C. pallidicaule* is a distinct group from the crop/weed complex of *C. quinoa*.

Chenopodium Domestication: A Morphological Approach

Researchers interested in finding evidence for the initial domestication of *Chenopodium* in the archaeological record of the Andes have looked primarily for the kinds of changes in seed morphology that might be expected. The first morphological analyses of *Chenopodium* domestication in the Andes focused on seed size (seed diameter) because an increase in seed size is a common change associated with the domestication of seed plants (Harlan et al. 1973; Harlan 1975). Pearsall's (1980, 1989) analysis of chenopod seeds from Pachamachay and Panaulauca Caves, Peru, and Browman's (1986) similar study of archaeological chenopod seeds from Chiripa, Bolivia, both recognized an increase in seed diameter over time, and a clear bimodal distribution of seed size (Browman 1986: 145; Pearsall 1989: 322). Both Pearsall and Browman propose that the documented increase in seed size over time reflected selection for larger seeds under domestication, but both researchers also acknowledge difficulties with this conclusion. None of the archaeological seeds in the assemblages they studied were as large as seeds of modern domesticated populations, and the range of seed diameter documented for the modern domesticate, *C. pallidicaule* (1.0–1.5 mm) overlaps other wild chenopods (0.8–1.2 mm). (See Table 4.2.)

Clearly a simple increase in seed size does not represent an adequate morphological marker of domestication for *Chenopodium* in the Andes. Browman and Pearsall also suggest that the bimodal pattern observed in seed diameter measurements might reflect two *Chenopodium* taxa, perhaps *C. quinoa* and *C. pallidicaule*, but no follow-up studies to test this hypothesis further have been carried out.

In the same time frame, researchers interested in chenopod domestication in the eastern United States also found negligible differences in seed size between wild and domestic chenopods (Asch and Asch 1977; Smith 1985a). In an effort to find other possible morphological means of distinguishing between wild and domesticated chenopods, researchers in eastern North America turned to scanning electron microscopy to examine the micro-morphological features of *Chenopodium* seeds more precisely (Smith 1988).

The *Chenopodium* fruit has a round, starchy perisperm with an embryo that wraps around the perimeter. A membranous layer called the testa, or seed coat, encapsulates the perisperm and embryo and controls germination dormancy (Figure 4.1). In 1981, Hugh Wilson used scanning electron microscopy (SEM) to compare the testa thickness of modern Mexican domesticates *huauzontle* and *chia* (varieties of *C. berlandieri* ssp. *nuttalliae* (Safford) Wilson and Heiser), a modern wild eastern North American chenopod (*C. bushianum*), and pale-seeded archaeological chenopods from the Holman rock shelters in northwestern Arkansas. The SEM micrographs revealed that both the Arkansas and Mexican chenopods had thin testas, whereas the wild seeds had thick testas. Wilson suggested that humans selected seeds with thin testas because they had reduced germination dormancy, another trait desirable in seed crops (Harlan 1975). Analysis of seeds from other eastern North American sites clearly established that low testa thickness (less than about 20 microns) was the most reliable indicator of *Chenopodium* domestication in the region (Smith 1984, 1985a, 1985b, 1988; Fritz and Smith 1988).

Other morphological differences were noted between wild and domesticated chenopods, many of which are correlated with a reduction in testa thickness. Many domesticated chenopods are lighter in color than are their wild counterparts. This is attributed to the thin, nearly transparent testa, which leaves the white perisperm visible (Wilson 1981). Because some dark-colored seeds have thin testas (Fritz and Smith 1988) and archaeological seeds are usually charred, this characteristic is of limited usefulness in the analysis of seeds recovered from an archaeological context. Margin configuration also differs between wild and domesticated chenopods: domesticated seeds are truncate, whereas wild seeds are round to biconvex (Asch and Asch 1977; Wilson 1981; Smith 1985a; Figure 4.1). Wilson (1981, 1988a) suggests that the reduced testa allows "fruit morphology to be influenced by structural features of the embryo" (1988a: 487). The flat (in cross-section) cotyledons of the embryo create a truncate margin in the fruit. Finally, seed coat texture varies: domesticates tend to have smooth seed coats

TABLE 4.2

Summary of Published Data on Archaeological *Chenopodium* Seed Size and Testa Thickness from Early Sites in the Andes and Eastern North America

Source	Site/Phase	Seed Size (mm) Range	Mean	Testa Thickness (microns) Range or Mean
Andean Highlands				
Pearsall 1980	Pachamachay, Peru			
	Ceramic	0.75–1.80	1.24	nd
	Preceramic/Ceramic	0.75–1.55	1.10	nd
	Preceramic	0.75–1.00	0.90	nd
Browman 1986	Chiripa, Bolivia			
	300 BC–AD 50	0.40–1.90	1.00	nd
	600–300 BC	0.50–1.90	1.00	nd
	850–600 BC	0.40–2.00	0.90	nd
	1350–1000 BC	0.40–1.30	0.70	nd
Nordstrom 1990	Panaulaca Cave, Peru			
	Pancan, Peru (combined)			
	AD 365–1300	0.65–2.50	nd	1–60
	1620–700 BC	0.65–1.65	nd	5–40
	3000–1620 BC	0.50–1.85	nd	5–30
	3000 BC	0.85	nd	12
Eisentraut 1998	Quelcatani			
	WXX: Middle Formative	0.50–1.50	0.83	7–23.7
	WXXI: Early Formative	0.50–1.50	0.91	4.3–19.7
	WXXII: Early Formative	0.50–1.15	0.87	8.8–16.0
	WXXIII: Early Formative/			
	Late Archaic	0.50–2.00	0.92	10.4–23.4
	WXXVII: Archaic[a]	0.50–1.15	0.86	8.2–12.4
Eastern North America				
Smith and Fritz 1998	ca. 2000 BP			
	Russell Cave, AL	nd	nd	7–16
	Ash Cave, OH	nd	nd	11–21
	Edens Bluff, AR	nd	nd	12–15
	White Bluff, AR	nd	nd	12–16
Gremillion 1993a, 1993b	Late Archaic (1500–1000 BC)			
	Newt Kash (16420)	nd	nd	15.8
	Newt Kash (16446)	nd	nd	20.7
	Early Woodland (1000–300 BC)			
	Big Bone, TN cultigens	nd	nd	21.2
	Big Bone, TN seeds	nd	nd	31.1
	Salts Cave, KY (vestibule) cultigens	nd	nd	10.5
	Salts Cave, KY (vestibule) weeds	nd	nd	35.1

[a] Stratigraphically level WXXVII is thought to be Archaic because it falls below a stratum dated to 3600 BP. However, *Chenopodium* seeds from this level were AMS dated and yield a date of 2740 ± 50 BP.

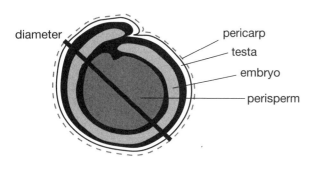

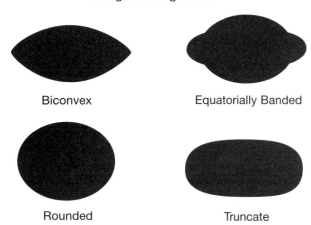

FIGURE 4.1 *Chenopodium* seed morphology (after Bruno and Whitehead 2003).

as opposed to the reticulate seed coats of the wild and weedy chenopods. Employing these features, researchers have been able to determine how, when, and where North American chenopods were domesticated (Smith 1985a, 1988, 1990, 1995; Smith and Cowan 1987; Fritz and Smith 1988; Gremillion 1993a, 1993b).

A similar examination of Andean chenopods in search of differences in testa thickness and associated morphological changes was first carried out by Nordstrom (1990), who considered both archaeological and modern samples from central highland Peru. Her study demonstrated that there are distinct differences in testa thickness of modern Andean *Chenopodium* taxa. Domesticated populations were found to have relatively thin seed coats (*C. quinoa* 0–20 microns and *C. pallidicaule* 5–25 microns), while wild taxa had relatively thick testas (*C. salinium* 25–80 microns and *C. ambrosioides* 15–80 microns).[2] Acknowledging some overlap in the wild and domesticate populations, Nordstrom suggests that this might be due to introgression. In her analysis of chenopod assemblages from two early Peruvian sites, Panaulaca and Pancan (3000 BC–AD 1300), she found that 89 percent of specimens had seed coats between 5 and 25 microns thick. The earliest of the thin-coated seeds came from a level at Panaulauca Cave dating to approximately 3000 BC.

Eisentraut (1998) carried out the second SEM analysis of Andean chenopods on archaeological specimens from Quelcatani Cave and Camata in Peru. Occupation of these two sites spanned the transition between the Archaic Period (7000–3500 BP) and Formative Period (3500–2000 BP). She found seeds with testa thickness values of less than 25 microns in each level she examined. She described the thin-testa seeds as generally truncate and the thick-testa seeds as biconvex, rounded, or equatorially banded. The surfaces of the thin-testa seeds were smooth or smooth and dimpled, while the thick-testa seeds were either smooth and wavy or reticulate-aveolate in surface appearance. Seed size was quite consistent over time (range of means 0.83–0.92 mm with a standard deviation of only 0.90 mm). An AMS radiocarbon date of 2740 ± 50 BP on a thin-testa seed from Quelcatani Cave was the earliest date obtained in her study.

These landmark studies demonstrated that the methods used to identify domesticated chenopods in the eastern United States could also be employed in documenting the initial domestication of *Chenopodium* in the Andes. Although these studies have clearly shown that testa thickness is a diagnostic indicator of chenopod domestication, they do not address the potential problem of interspecific morphological variation in Andean chenopods. Given the wide range of morphological variation documented for different chenopod taxa in the Andes, the current study of chenopods from the southern Lake Titicaca Basin, Bolivia considers various potential morphological indicators of domestication, such as testa thickness, within a larger context of variability across taxa.

Case Study: Chenopods of the Southern Lake Titicaca Basin, Bolivia

The southern Lake Titicaca Basin is located in the southern portion of the Andean *altiplano*, a high (2500–4000 m) plateau between the eastern and western Andean cordilleras that extends from Peru to southern Bolivia. The southern Lake Titicaca region is dry (750 mm annual precipitation) and cold (9° C annual temperature) because of a combination of elevation and inverse weather patterns (Boulangé and Aquize Jaén 1981; Binford and Kolata 1996).

Modern *Chenopodium* species of the southern Lake Titicaca Basin provided a baseline of comparison for my analysis of archaeological chenopods recovered by the Taraco Archaeological Project (TAP) at Chiripa, Bolivia (1500 BC– AD 100) (Figure 4.2). Chiripa is one of the most important Formative Period sites in the southern Andean highlands (Browman 1981; Mohr-Chavez 1988; Hastorf 1999, 2003), and it exhibits many of the key cultural developments associated with this period, including sedentary villages, long-distance trade networks, small civic-ceremonial centers, and agriculture.

TAP recovered large quantities of charred *Chenopodium* seeds from their excavations (Whitehead 1999). Although

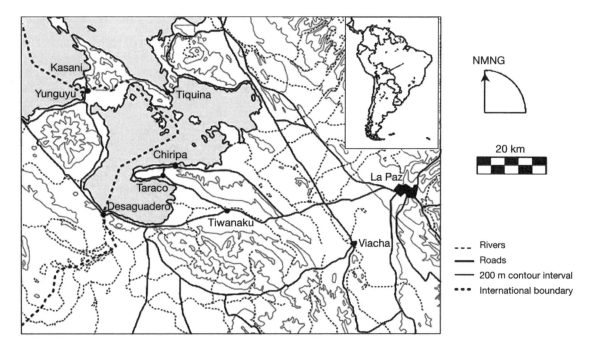

FIGURE 4.2 Map of the southern Lake Titicaca Basin (from Hastorf 1999).

scholars had long assumed that the people at Chiripa were agriculturalists, this conclusion was based primarily on indirect data (Whitehead 2000). Morphological examination of chenopod seeds recovered from Chiripa provided the first direct test for domesticated chenopods and a food production economy being present at Chiripa during the Formative Period (Bruno 2001; Bruno and Whitehead 2003).

Materials

The modern *Chenopodium* taxa examined were *C. quinoa*, *C. quinoa* var. *melanospermum*, *C. pallidicaule*, and *C. ambrosioides*. Dr. Christine Hastorf and her students collected most of the samples in this study during previous field seasons at Chiripa and Tiwanaku. I also collected specimens from Chiripa, Tiwanaku, and Achuta Grande (a community 5 km east of Tiwanaku) in July 2000. Although most of the wild and weed seed samples were taken from whole plants, many were contributions from the harvests of local farmers and from collections they had made of the weeds that grow in and around their fields. Therefore, each sample represents seeds from various plants within the same species (Table 4.3).

All the modern *Chenopodium* samples were charred in a muffle furnace for comparison with the charred archaeological specimens. Smith (1985a) and Nordstrom (1990) both recorded a 5 percent decrease in seed diameter with charring. Smith (1985a) also found that charring did not affect seed-coat thickness. Because these values had been established previously, I did not quantify changes in my own charring trials. I did, however, measure seed diameter and testa thickness for one uncharred specimen of each taxon for eneral comparison. With the possible exception of *C. ambrosioides*, all taxa conform to the patterns observed by Smith and Nordstrom. In the case of *paiko*, the uncharred testa thickness measurement (23.5 microns) is much larger than average charred measurement (range 11–14.5 microns). The possible significance of this result requires investigation beyond the scope of the project reported here.

Morphological Analysis of *Chenopodium* Seeds

Measuring Seed Diameter

I obtained seed diameter data from two different sets of specimens using two different measuring techniques. The two sets of specimens were (1) several large samples of modern charred seeds ($N = 2605$) used to test the utility of seed diameter in distinguishing wild from domesticated populations, and (2) smaller samples of modern charred and uncharred seeds ($N = 46$) for subsequent examination of testa thickness using SEM. The digital imaging system of the Washington University Ethnobotany Laboratory was used to determine seed diameter measurements for the first set of specimens.[3] The second set of seeds was measured using a Wild M3 dissecting microscope with an ocular micrometer.

In his analysis of North American assemblages, Smith found that the "beak diameter" (the tip of the beak to the opposite edge) most often represented the maximum fruit diameter, as opposed to the perpendicular measurement, which is the line perpendicular to the beak (Smith 1985a: 120; see Figure 4.1). Because most of the seeds I analyzed tended to be more oval than circular and the "beak diameter" was not always the maximum seed diameter, the location of my "maximum diameter" measurements varied.

TABLE 4.3
Modern *Chenopodium* Samples

Species, Local Common Name	Location
C. quinoa quinoa blanca	Achuta Grande
C. quinoa quinoa blanca	Chiripa
C. quinoa quinoa blanca	Tiwanaku[a]
C. quinoa quinoa blanca	La Paz Market[a]
C. quinoa quinoa amarillo	Achuta Grande
C. quinoa jupa amarillo	Chiripa[a]
C. quinoa jupa amarco	Chiripa[a]
C. quinoa var. *melanospermum* ajara	Chiripa[a]
C. quinoa var. *melanospermum* jupa ajara	Chiripa[a]
C. quinoa var. *melanospermum* ajara negra	Chiripa[a]
C. quinoa var. *melanospermum* jupa ajaru	Chiripa[a]
C. pallidicaule kañawa	Tiwanaku
C. pallidicaule kañawa	LaPaz, Ingavi, Pirkuta[a]
C. ambrosioides paiko	Chiripa

[a] Indicates samples collected by Hastorf.

Measuring Testa Thickness

A Hitachi S-450 Scanning Electron Microscope was employed to measure seed coat thickness (Figure 4.3).[4] For each taxon, the metric measurements included maximum seed diameter and testa thickness. Ratios of testa thickness to seed diameter were also calculated in order to provide a relative measure of testa thickness that takes seed size into account. Nonmetric traits recorded (when present) included pericarp patterning, seed-coat texture, and margin configuration. These traits were best observed with the SEM, but could also be seen under a dissecting microscope at magnifications of 10–50×.

Results

Distinct patterns were observed among the four modern *Chenopodium* taxa examined, especially when all of the morphological attributes were considered together.

Seed Size

A total of 150 to 200 charred seeds were measured from each of the 15 samples ($N = 2605$) to determine the range of seed size for the four modern taxa. Of the four different taxa, *C. quinoa*[5] is the largest, and in order of decreasing size are *quinoa negra*, *kañawa*, and *paiko* (Table 4.4, Figure 4.4). These differences are statistically significant ($F = 2337.18$, df = 3, 2601, $p < 0.001$), but the size ranges overlap. Therefore, identification of species or domestication based on seed size alone is difficult.

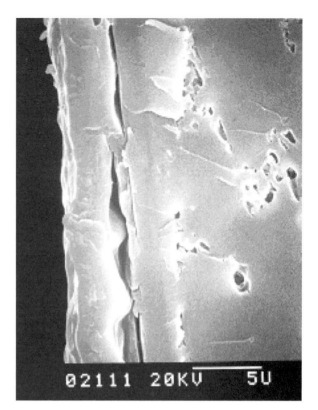

FIGURE 4.3 Measuring testa thickness from a scanning electron micrograph. The testa is the layer of tissue on the left of the micrograph. The average width is measured with a ruler and recorded in centimeters. The measurement in microns is obtained by multiplying the centimeter value by 10,000 and then dividing this value by the degree of magnification.

Testa Thickness

A very clear pattern emerges for testa thickness. The cultigens *C. quinoa* and *C. pallidicaule* have thin testas, 1.25–3.75 microns ($n = 11$; $x = 2.39$ microns; SD = 0.719) and 4.25–7.50 microns ($n = 12$; $x = 5.95$ microns; SD = 1.11), respectively (Table 4.5, Figure 4.5). Conversely, the weed *C. quinoa* var. *melanospermum* has a thicker seed coat of 22–51 microns ($n = 16$; $x = 39.50$ microns; SD = 8.74). *C. ambrosioides* is problematic because, according to the charred seed coat measurements, it has a relatively small testa thickness range of 11–14.5 microns ($n = 6$; $x = 12.58$; SD = 1.59), yet the one uncharred specimen is 23.5 microns. As mentioned above, the discrepancy between charred and uncharred testa thickness requires further inquiry. The differences in testa thickness among taxa are statistically significant ($F = 139.23$, df = 3, 41, $p < 0.001$).

Seed Size and Testa Thickness

To assess the relationship between seed size and testa thickness, I calculated a ratio of testa thickness over seed diameter (Table 4.5). Significant differences were recognized among the taxa ($F = 353.325$, df = 3, 41, $p < 0.001$). When diagrammed, this ratio provides a clear visual indication of the obvious differences between taxa: the two domesticates are

TABLE 4.4
Seed Diameter (mm) for Modern *Chenopodium* Taxa

Taxon	N	Mean	Minimum	Maximum	Standard Deviation
Quinoa	1256	1.9159	1.05	2.95	0.2736
Quinoa negra	600	1.4969	1.06	1.93	0.1293
Kañawa	450	1.1681	0.84	1.44	0.1046
Paiko	299	1.0678	0.79	1.38	0.092

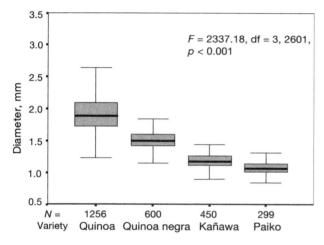

FIGURE 4.4 Seed diameter (mm) of modern *Chenopodium* taxa. In these box plots and those in other figures, the vertical lines represent the range, the dark horizontal bars represent the mean, and the rectangles represent one standard deviation.

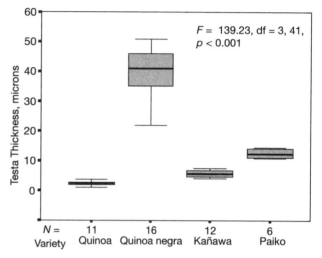

FIGURE 4.5 Testa thickness (microns) in modern *Chenopodium* taxa.

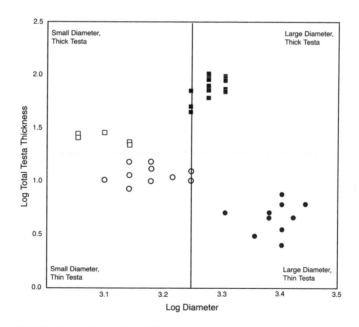

FIGURE 4.6 Scatterplot of log testa thickness and log diameter illustrating the ratio of testa thickness to diameter for four *Chenopodium* taxa. Kañawa (*C. pallidicaule*) (empty circle) and quinoa (*C. quinoa*) (black circle) are the domesticated varieties. Paiko (*C. ambrosioides*) (empty square) and quinoa negra (*C. quinoa* var. *melanospermum*) (black square) are the wild/weedy varieties.

TABLE 4.5
Summary Data for Modern *Chenopodium* Specimens from the Southern Lake Titicaca Basin, Bolivia

Testa Thickness (microns)	Total Testa Thickness (microns)[a]	Log[b] Total Testa Thickness	Seed Diameter (mm)	Log Seed Diameter	Ratio Testa/ Diameter[c]	Log Ratio Testa/ Diameter	Texture	Margin Configuration
Modern Quinoa								
3.00	6.00	.78	2.20	3.34	.003	−2.56	Smooth	Truncate
1.25	2.50	.40	2.00	3.30	.001	−2.90	Smooth	Truncate
2.25	4.50	.65	2.10	3.32	.002	−2.67	Smooth	Truncate
2.50	5.00	.70	1.60	3.20	.003	−2.51	Smooth	Truncate
2.50	5.00	.70	1.90	3.28	.003	−2.58	Smooth	Truncate
3.75	7.50	.88	2.00	3.30	.004	−2.43	Smooth	Truncate
2.50	5.00	.70	1.90	3.28	.003	−2.58	Smooth	Truncate
2.25	4.50	.65	1.90	3.28	.002	−2.63	Smooth	Truncate
3.00	6.00	.78	2.00	3.30	.003	−2.52	Smooth	Truncate
1.75	3.50	.54	2.00	3.30	.002	−2.76	Smooth	Truncate
1.50	3.00	.48	1.80	3.26	.002	−2.78	Smooth	Truncate
Modern Quinoa negra								
35.00	70.00	1.85	1.60	3.20	.044	−1.36	Reticulate	Rounded
47.00	94.00	1.97	1.60	3.20	.059	−1.23	Reticulate	Biconvex
51.00	102.00	2.01	1.50	3.18	.068	−1.17	Reticulate	Biconvex
39.00	78.00	1.89	1.50	3.18	.052	−1.28	Reticulate	Biconvex
48.00	96.00	1.98	1.50	3.18	.064	−1.19	Reticulate	Biconvex
45.00	90.00	1.95	1.50	3.18	.060	−1.22	Reticulate	Biconvex
45.00	90.00	1.95	1.50	3.18	.060	−1.22	Reticulate	Biconvex
22.00	44.00	1.64	1.40	3.15	.031	−1.50	Reticulate	Rounded
25.00	50.00	1.70	1.40	3.15	.036	−1.45	Reticulate	Biconvex
36.00	72.00	1.86	1.60	3.20	.045	−1.35	Reticulate	Biconvex
50.00	100.00	2.00	1.50	3.18	.067	−1.18	Reticulate	Biconvex
43.00	86.00	1.93	1.60	3.20	.054	−1.27	Reticulate	Biconvex
30.00	60.00	1.78	1.50	3.18	.040	−1.40	Reticulate	Biconvex
35.00	70.00	1.85	1.40	3.15	.050	−1.30	Reticulate	Biconvex
45.00	90.00	1.95	1.60	3.20	.056	−1.25	Reticulate	Biconvex
36.00	72.00	1.86	1.50	3.18	.048	−1.32	Reticulate	Biconvex
Modern Kañawa								
5.00	10.00	1.00	1.40	3.15	.007	−2.15	Canaliculate	Rounded
7.50	15.00	1.18	1.20	3.08	.012	−1.90	Canaliculate	Rounded
5.70	11.40	1.06	1.10	3.04	.010	−1.98	Canaliculate	Rounded
5.00	10.00	1.00	1.20	3.08	.010	−2.08	Canaliculate	Rounded
5.75	11.50	1.06	1.10	3.04	.008	−1.98	Canaliculate	Rounded
5.00	10.00	1.00	1.00	3.00	.010	−2.00	Canaliculate	Rounded
4.25	8.50	0.93	1.10	3.04	.010	−2.11	Canaliculate	Rounded
6.50	13.00	1.11	1.20	3.08	.007	−1.97	Canaliculate	Rounded
5.50	11.00	1.04	1.30	3.11	.010	−2.07	Canaliculate	Rounded
6.25	12.50	1.10	1.40	3.15	.008	−2.05	Canaliculate	Rounded
7.50	15.00	1.18	1.10	3.04	.008	−1.87	Canaliculate	Rounded
7.50	15.00	1.18	1.10	3.04	.008	−1.87	Canaliculate	Rounded

TABLE 4.5 (continued)

Testa Thickness (microns)	Total Testa Thickness (microns)[a]	Log[b] Total Testa Thickness	Seed Diameter (mm)	Log Seed Diameter	Ratio Testa/ Diameter[c]	Log Ratio Testa/ Diameter	Texture	Margin Configuration
Modern Paiko								
14.00	28.00	1.45	.90	2.95	.031	−1.51	Punctate	Rounded
13.50	27.00	1.43	.90	2.95	.030	−1.52	Punctate	Rounded
11.00	22.00	1.34	1.10	3.04	.020	−1.70	Punctate	Rounded
11.00	22.00	1.34	1.10	3.04	.020	−1.70	Punctate	Rounded
11.50	23.00	1.36	1.10	3.04	.020	−1.68	Punctate	Rounded
14.50	29.00	1.46	1.00	3.00	.029	−1.54	Punctate	Rounded

[a] Because the testa thickness is measured on only one side of the seed, I double the testa thickness value in order to account for the entire area represented by the testa when calculating the testa/diameter ratio.
[b] A natural log was applied to transform the values.
[c] Diameter originally measured in mm and then converted to microns before calculating the ratio. To calculate the ratio, I divide total testa thickness (microns) by diameter (microns).

TABLE 4.6
Morphological Characteristics of Modern *Chenopodium* Taxa

	C. pallidicaule Aellen	*C. quinoa* Willd.	*C. quinoa* var. *melanospermum* Hunziker	*C. ambrosioides* L.
Common Names	Kañawa, cañihua	Quinoa, jupa	Quinoa negra, ajara	Paiko
Status	Domesticate	Domesticate	Weed	Wild
Pericarp	Irregularly punctate	Reticulate-aveolate	Reticulate-aveolate	Smooth
Seed Coat Texture	Canaliculate	Smooth	Reticulate	Foveate
Margin Configuration	Rounded to truncate	Truncate	Biconvex to equatorially banded	Rounded
Seed Diameter (charred)	1.0–1.4 mm	1.6–2.2 mm	1.4–1.6 mm	0.9–1.1 mm
Testa Thickness (charred)	4.25–7.5 microns	1.25–3.75 microns	22–55 microns	11–14.5 microns

relatively large with thin seed coats, whereas the weedy and wild varieties are relatively small with thick seed coats (Figure 4.6).

Seed Coat Texture

Each taxon has a distinct seed coat texture (Figure 4.7). Like its domesticated cousins in Mexico and the eastern United States, *C. quinoa* has a very smooth to slightly undulating seed coat surface (Figure 4.7a). *C. quinoa* var. *melanospermum* is reticulate-aveolate, again similar to the North American weedy forms (Figure 4.7b, Smith 1985a). *C. pallidicaule* has a canaliculate pattern that is most distinct near the beak and along the margins (Figure 4.7c). Finally, *C. ambrosioides* has a distinctive foveate or pitted surface (Figure 4.7d).

Pericarp Patterning

When the pericarp is preserved archaeologically, it too allows distinguishing among *Chenopodium* taxa. Both *C. quinoa* and *C. quinoa* var. *melanospermum* have a reticulate pericarp (Aellen and Just 1943; Wilson 1980) (Figure 4.7a–b). *C. pallidicaule*, however, is characterized by a smooth to ribbed pericarp (Figure 4.7c). Finally, *C. ambrosioides* does not appear to exhibit a distinctive pattern, but is rather smooth (Figure 4.7d).

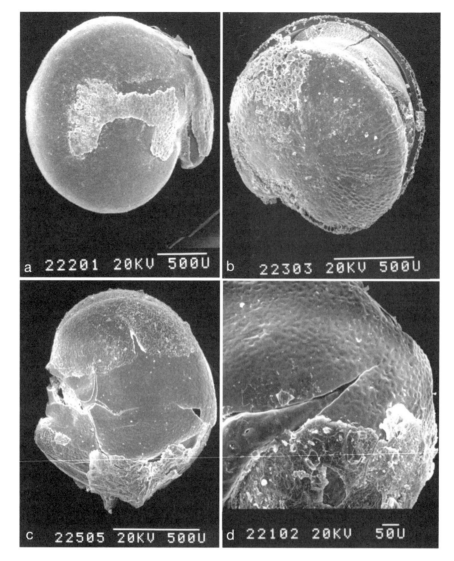

FIGURE 4.7 Scanning electron micrographs of experimentally charred modern *Chenopodium*, showing intact testa and pericarp morphology: (a) *C. quinoa*: the thin testa is intact, and a patch of the pericarp remains in the center; (b) *C. quinoa* var. *melanospermum*: the thick testa is split, and some pericarp remains; (c) *C. pallidicaule*: the testa is intact, and pericarp remains on the margins; (d) *C. ambrosioides*: the testa is intact, and some of the pericarp remains at the bottom of the image.

Margin Configuration

Margin configuration is discernible on seeds that have not been distorted by puffing. The variation between Andean wild and domesticated taxa is similar to what has been observed in eastern North America (Asch and Asch 1977; Smith 1985a) and Mexico (Wilson 1980). *C. quinoa* seeds are truncate, whereas *C. pallidicaule* is generally truncate to round. *C. quinoa* var. *melanospermum* can be equatorially-banded or biconvex. *C. ambrosioides* has a very round, almost spherical configuration.

Discussion

This case study presents a set of morphological characteristics that aid in the identification of individual Andean *Chenopodium* taxa, and in differentiating wild, weedy, and domesticated specimens from each other (Table 4.6). By first assessing the natural variation present in each taxon, one can more confidently identify morphological variation caused by human selection.

In the Andes it is not possible to identify a domesticated chenopod seed based on testa thickness alone. *C. ambrosioides*, a wild plant, has a charred testa thickness of less than 20 microns. Although this may be an effect of charring, most archaeological *Chenopodium* assemblages in the Andes are charred. As a result, it is necessary to identify the taxon first using nonmetric features. Seed coat texture, pericarp patterning, and to a lesser extent margin configuration are diagnostic traits for distinguishing among *C. quinoa*, *C. quinoa* var. *melanospermum*, *C. pallidicaule*, and *C. ambrosioides*.

Within a given species, testa thickness and the testa thickness/seed size ratio are useful for separating domestic varieties from both weedy and wild varieties. The differences in testa thickness between *C. quinoa* and *C. quinoa* var. *melanospermum* can be attributed to human selection for reduced germination dormancy. Selection for a larger seed size may also have occurred, because the modern *C. quinoa* seeds are significantly larger than are their black-seeded counterparts.

Conclusions and Directions of Future Research

Comparative analysis of modern chenopod taxa and archaeological specimens resulted in the identification of a quinoa-like form and a quinoa negra–like form in the archaeological samples from Chiripa, indicating that domesticated chenopods were under cultivation at Chiripa as early as 1500 BC (ca. 3500 BP)[6] (Bruno 2001; Bruno and Whitehead 2003). This provides the earliest evidence to date of domesticated chenopods in the Andes based on direct AMS dates from charred chenopods.

Eisentraut (1998) obtained a direct AMS date of 2740 ± 50 BP calibrated on a thin-testa seed identified from Quelcatani Cave, Peru. Thin-testa seeds come from levels at Panaulauca Cave, Peru, dating (by conventional radiocarbon dating) to as early as 3000 BC (ca. 5000 BP) (Nordstrom 1990). Based on these independent studies, ca. 5000 BP would appear to be a liberal estimate of when *Chenopodium* was first brought under domestication in the central Andean highlands. A more conservative estimate of around 3500 BP could also be proposed for the south central Andean highlands. *Chenopodium* remains from associated contexts in highland Peru and coastal Chile date almost 2000 years earlier (Pearsall 1992; Kuznar 1993; Aldenderfer 1999), but their domesticated status has not been verified. While the domesticated chenopods from Chiripa and Quelcatani are probably forms of *C. quinoa*, it is not clear whether the other early examples are *quinoa* or *kañawa*. Clearly, more research, both genetic and morphological, is needed on the Andean *Chenopodium* taxa.

I recommend that future morphological studies of chenopod domestication in the Andes follow the multiple attribute comparative approach outlined here, which takes into account the taxonomic diversity of chenopods in the Andes. Direct AMS dating of identified taxa from different regions and time periods will greatly advance our understanding of *Chenopodium* domestication in the Andes.

Acknowledgments

I would like to thank the farmers of Chiripa, Tiwanaku, and Achuta Grande for sharing their knowledge and chenopods for this study. I also thank Christine Hastorf, William Whitehead, and the Taraco Archaeological Project for giving me the opportunity to carry out this project. William Whitehead and Eduardo Machicado helped draft several of the figures. David Browman, Gayle Fritz, Patty Jo Watson, Lisa Hildebrand, Christine Hastorf, and Jason Kaufman provided guidance in research and useful comments on this article. Michael Veith piloted the SEM in the biology department at Washington University. Finally, I would like to thank Bruce Smith and Melinda Zeder for the invitation to contribute to this volume. The research was funded by the Washington University in St. Louis School of Arts and Science's Pre-Doctoral Summer Research Grant (2000), Summer Writing Grant (2001), and Faculty Research Grant (2001 awarded to David Browman).

Notes

1. This is not an exhaustive list of Andean chenopods, but it does represent the most commonly cited species found in anthropogenic habitats.
2. *C. salinium* is a dark-seeded species.
3. The system includes a Sony CCD black-and-white video camera attached to an Olympus SZ-PT microscope that feeds into a Gateway PC. A live image is captured from the microscope and displayed on the computer monitor using the image analysis software Scion Image. The microscope and imaging software are calibrated to a millimeter scale, and measurements can be taken on the screen using the software's "analysis" function. Each measurement is automatically recorded in Scion Image. I saved these data as a text file and transported them to a statistical software package, SPSS 8.0, which I used for all statistical analyses.
4. Seed samples were mounted on aluminum SEM specimen mounts using 12-mm diameter Carbon Adhesive Tabs (Electron Microscopy Sciences, Ft. Washington, PA 19034; cat. #77825-12). After measuring the diameter under the light microscope, specimens were cut in half using a razor blade, and placed on a stub with the newly exposed surface facing up. Mounted samples were coated with about 50 nm of gold in a Polaron E5000 Sputter Coater. Micrographs for all specimens were taken at magnifications of 1000× to 4000×. Images of the desired area were captured using Polaroid 55 P/N (positive/negative) film. Because the SEM employed did not have on-screen measuring capabilities, testa thickness measurements were taken directly from the micrographs (Figure 4.3).
5. Several varieties of quinoa are available today, two of which were included here: *quinoa blanca* and *quinoa amarilla*. Although there was some variation in seed size between these populations, differences between the testa measurements were not statistically significant. Therefore, I have included them all under the single category "quinoa" for this discussion.

6. This AMS date was obtained on five charred *Chenopodium* seeds larger than 1.2 mm from a single context. In the seeds that remained in this sample, I identified domesticated specimens (see Whitehead 1999 for details on radiocarbon dates and Bruno and Whitehead 2003 for details on the analysis of the sample).

References

Aellen, P. and T. Just. 1943. Key and synopsis of the American species of the genus Chenopodium L. *American Midland Naturalist* 30: 47–76.

Aldenderfer, M. S. 1999. The Late Preceramic–Early Formative transition on the South-Central Andean Littoral. In *Pacific Latin America in prehistory: The evolution of archaic and formative cultures*, Michael Blake (ed.), pp. 213–222. Pullman, Wash.: WSU Press.

Anderson, E. 1952. *Plants, man, and life*. Boston: Little Brown.

Asch, D. L. and N. B. Asch. 1977. Chenopod as cultigen: A re-evaluation of some prehistoric collections from eastern North America. *Midcontinental Journal of Archaeology* 2: 3–45.

Baker, H. G. 1972. The evolution of weeds. *Annual Review of Ecology and Systematics.* 5: 1–24.

Bastien, J. W. 1987. *Healers of the Andes: Kallawaya herbalists and their medicinal plants*. Salt Lake City: University of Utah Press.

Binford, M. W. and A. L. Kolata. 1996. The natural and human setting. In *Tiwanaku and its hinterland: Archaeology and paleoecology of an Andean civilization*, A. L. Kolata (ed.), Vol. 1, pp. 57–88. Washington, D.C.: Smithsonian Press.

Boulangé, B. and E. Aquize Jaén. 1981. Morphologie, hydrographie et climatologie du Lac Titicaca et de son Basin Versant. *Revue d'Hydrobiologie Tropicale* 14: 269–287.

Browman, D. L. 1981. New light on Andean Tiwanaku. *American Scientist* 69: 408–419.

———. 1986. Chenopod cultivation, lacustrine resources, and fuel use at Chiripa, Bolivia. In *New world paleoethnobotany: Collected papers in honor of Leonard W. Blake*, E. E. Voigt and D. M. Pearsall (eds.), pp. 137–172. *The Missouri Archaeologist* Vol. 47. Columbia: Missouri Archaeological Society.

Bruno, M. C. 2001. *Formative agriculture? The status of* Chenopodium *domestication and intensification at Chiripa, Bolivia (1500 B.C.–100 B.C.)*. Master's Thesis, Department of Anthropology, Washington University in St. Louis.

Bruno, M. C. and W. T. Whitehead. 2003 *Chenopodium* cultivation and Formative period agriculture at Chiripa, Bolivia. *Latin American Antiquity* 14: 339–355.

de Wet, J. M. J. 1973. Evolutionary dynamics of cereal domestication. *Bulletin of the Torrey Botanical Club.* 102: 307–312.

de Wet, J. M. J. and J. R. Harlan. 1975. Weeds and domesticates: Evolution in the man-made habitat. *Economic Botany* 29: 99–107.

Eisentraut, P. J. 1998. Macrobotanical remains from southern Peru: A comparison of Late Archaic–Early Formative period sites from the Puna and Suni zones of the Western Titicaca basin. Ph.D. dissertation, Department of Anthropology, University of California, Santa Barbara. Ann Arbor, MI: University of Michigan Microfilms.

Franquemont, C., T. Plowman, E. Franquemont, S. King, C. Niezgoda, W. Davis, and C. Sperling. 1990. The ethnobotany of Chinchero, an Andean community in southern Peru. *Fieldiana, Botany* No. 24. Chicago: Field Museum of Natural History.

Fritz, G. J. and B. D. Smith. 1988. Old collections and new technology: Documenting the domestication of *Chenopodium* in eastern North America. *Midcontinental Journal of Archaeology* 13: 3–27.

Gade, D. W. 1970, Ethnobotany of Cañihua (*Chenopodium pallidicaule*), rustic seed crop of the Altiplano. *Economic Botany* 24: 55–61.

Gandarillas, H. 1974. *Genetica y origin de la quinia*. La Paz, Bolivia: Instituto Nacional del Trigo, Departmento de Estudios Economicos, Estadisticas y Comercializacion.

Gremillion, K. J. 1993a. Crop and weed in prehistoric eastern North America: The *Chenopodium* example. *American Antiquity* 58: 496–509.

———. 1993b. The evolution of seed morphology in domesticated *Chenopodium*: An archaeological case study. *Journal of Ethnobiology* 13: 149–169.

Harlan, J. R. 1975. *Crops and man*. Madison, Wisc.: American Society of Agronomy.

Harlan, J. R. and J. M. J. de Wet. 1965. Some thoughts on weeds. *Economic Botany* 19: 16–24.

Harlan, J. R., J.M.J. de Wet, and E. G. Price. 1973. Comparative evolution of cereals. *Evolution* 27: 311–325.

Hastorf, C. A. 1999. Introduction to Chiripa and the site area. In *Early settlement at Chiripa, Bolivia: Research of the Taraco Archaeological Project*, C. A. Hastorf (ed.), pp. 1–6. Contributions of the Archaeological Research Facility No. 57. Berkeley: Archaeological Research Facility, University of California at Berkeley.

Hastorf, C. A. 2003. Community with the ancestors: Ceremonies and memory in the Middle Formative at Chiripa, Bolivia. *Journal of Anthropological Anthropology* 31: 305–332.

Heiser, C. B. 1976. Origins of some cultivated new world plants. *Annual Review of Ecology and Systematics* 10: 309–326.

Heiser, C. B. and D. C. Nelson. 1974. On the origin of cultivated chenopods (*Chenopodium*). *Genetics* 78: 503–505.

Holzner, W. and M. Numata. 1982. *Biology and ecology of weeds*. The Hague: Junk.

Hunziker, A. 1943. Las especies alimenticias de *Amarantus* y *Chenopodium* cultivadas por Los Indios de América. *Revista Argentina Agronomica.* 10: 146–154.

———. 1952. *Los pseudocereales de la agricultura indígena de América*. Buenos Aires: Acme Agency.

Kuznar, L. A. 1993. Mutualism between *Chenopodium*, herd animals, and herders in the south central Andes. *Mountain Research and Development* 13: 257–265.

La Barre, W. 1948. The Aymara Indians of the Lake Titicaca Plateau, Bolivia. Memoir Series of the American Anthropological Association No. 68. New York: Kraus.

———. 1959. Materia medica of the aymara Lake Titicaca Plateau, Bolivia. *Webbia* 15(1): 47–49.

Leon, J. 1964. *Plantas alimenticias Andinas*. Lima, Peru: Instituto Interamericano de Ciencias Agricolas Zona Andina.

Mohr-Chavéz, K. L. 1988. The significance of Chiripa in Lake Titicaca Basin developments. *Expedition* 30: 2, 17–26.

National Research Council. 1989. *Lost crops of the Incas: Little-known plants of the Andes with promise for worldwide cultivation*. Washington, D.C.: National Academy Press.

Nelson, D. C. 1968. *Taxonomy and origins of* Chenopodium quinoa *and* Chenopodium nuttalliae. PhD Dissertation, Indiana University. Ann Arbor, MI: University of Michigan Microfilms.

Nordstrom, C. 1990. Evidence for the domestication of *Chenopodium* in the Andes. Report to the National Science Foundation. University of California, Berkeley Paleoethnobotany Laboratory Reports #19.

Pearsall, D. M. 1980. Ethnobotanical report: Plant utilization at a hunting base camp. In *Prehistoric hunters of the high Andes*, John Rick (ed.), pp. 191–231. New York: Academic Press.

———. 1989. Adaptation of prehistoric hunter-gatherers in the High Andes: The changing role of plant resources. In *Foraging and farming*, D. Harris and G. Hillman (eds.), pp. 318–332. London: Unwin Hyman.

———. 1992. Origins of plant cultivation in South America. In *The origins of agriculture*, C. Cowan and P. J. Watson (eds.), pp. 173–205. Washington, D.C.: Smithsonian Institution Press.

———. 1993. Domestication and agriculture in the New World. In *The last hunters–first farmers*, T. D. Price and A. B. Gebauer (eds.), pp. 157–192. Santa Fe: School of American Research Press.

Pickersgill, B. 1977. Biosystematics of crop-weed complexes. *Kulturpflanze* 29: 377–388.

Ruas, P. M., A. Bonifacio, C. F. Ruas, D. J. Fairbanks, and W. R. Andersen. 1999. Genetic relationship among 19 accessions of six species of *Chenopodium* L., by Random amplified polymorphic DNA fragments (RAPD). *Euphytica* 105: 25–32.

Sauer, C. O. 1952. *Agricultural origins and dispersals*. New York: American Geographical Society.

Simmonds, N. W. 1965. The grain chenopods of the tropical American highlands. *Economic Botany* 19: 223–235.

Smith, B. D. 1984. *Chenopodium* as a prehistoric domesticate in eastern North America: Evidence from Russell Cave, Alabama. *Science* 226: 165–167.

———. 1985a. The role of *Chenopodium* as a domesticate in pre-maize garden systems of the eastern United States. *Southeastern Archaeology* 4: 51–72.

———. 1985b. *Chenopodium berlandieri* ssp. *jonesianum*: Evidence for a Hopewellian domesticate from Ash Cave, Ohio. *Southeastern Archaeology* 4: 107–133.

———. 1988. SEM and the identification of micro-morphological indicators of domestication in seed plants. In *Scanning electron microscopy in archaeology*, S. L. Olsen (ed.), pp. 203–214. BAR International Series 452. Oxford: BAR.

———. 1990. Origins of agriculture in eastern North America. *Science* 246: 1566–1571.

———. 1992. The floodplain weed theory of plant domestication in eastern North America. In *Rivers of change*, by B. D. Smith, pp. 19–33. Washington, D.C.: Smithsonian Institution Press.

———. 1995. Seed plant domestication in eastern North America. In *Last hunters–first farmers*, T. D. Price and A. B. Gebauer (eds.), pp. 193–213. Santa Fe: School of American Research Press.

Smith, B. D. and C. W. Cowan. 1987. Domesticated *Chenopodium* in prehistoric eastern North America: New accelerator dates from eastern Kentucky. *American Antiquity* 52: 355–357.

Tapia, M., H. Gandarillas, S. Alandia, A. Cardozo, A. Mujica, R. Ortiz, V. Otazu, J. Rea, and B. Salas. 1979. Quinia y kaniwa cultivos Andinos. *Centro Internacional de Investigaciones para El Desarrolo (CIID)*. Instituto Interamericano de Ciencias Agricolas (IICA). Editorial IICA.

Whitehead, W. T. 1999. Paleoethnobotanical evidence. In *Early settlement at Chiripa, Bolivia: Research of the Taraco Archaeological Project*, C. A. Hastorf (ed.), pp. 95–104. Contributions of the Archaeological Research Facility No. 57. Berkeley: Archaeological Research Facility, University of California at Berkeley.

———. 2000. Perspectives on agriculture from Formative and Tiwanaku sites in the Bolivian Altiplano. Paper presented at the 65th Annual Meeting of the Society of American Archaeology, Philadelphia, PA, April 7.

Wilson, H. D. 1980. Artificial hybridization among species of *Chenopodium* sect. *Chenopodium*. *Systematic Botany* 5: 253–263.

———. 1981. Domesticated *Chenopodium* of the Ozark Bluff Dwellers. *Economic Botany* 35: 233–239.

———. 1988a. Allozyme variation and morphological relationships of *Chenopodium hircinum*. *Systematic Botany* 13: 215–228.

———. 1988b. Quinua biosystematics: Domesticated populations. *Economic Botany* 42: 461–477.

———. 1988c. Quinua biosystematics II: Free-living populations. *Economic Botany* 42: 478–494.

———. 1990. Quinua and relatives (*Chenopodium* sect. *Chenopodium* subsect. *Cellulata*). *Economic Botany* 44 (Suppl.): 92–110.

Wilson, H. D. and C. B. Heiser. 1979. The origin and evolutionary relationships of 'Huauzontle' (*Chenopodium nuttalliae* Safford), domesticated chenopod of Mexico. *American Journal of Botany* 66: 198–206.

CHAPTER 5

Identifying Manioc (*Manihot esculenta* Crantz) and Other Crops in Pre-Columbian Tropical America through Starch Grain Analysis
A Case Study from Central Panama

DOLORES R. PIPERNO

Introduction

In both the high elevation Andes and the lowland tropical forests of the Americas, different plant species were brought under domestication, not for their seeds or fruits but for their starch-rich underground organs (Sauer 1950; Harlan 1992). The major root crop species domesticated in the Neotropics include manioc (*Manihot esculenta* Crantz), sweet potato (*Ipomoea batatas* (L.) Lam.), yams (*Dioscorea trifida* L.f.), yautia (*Xanthosoma sagittifolium* (L.) Schott & Endl.), arrowroot (*Maranta arundinacea* L.), and lirén (*Calathea allouia* (Aubl.) Lindl.) (Sauer 1950; Harlan 1992; Piperno and Pearsall 1998). Until recently, our understanding of where and when these major tropical forest crop plants were initially domesticated, and their subsequent history of dispersal, has been severely hampered by their limited documented presence in the archaeological record. Except for in the most arid environmental settings, soft, starchy plant structures are rarely preserved as archaeobotanical remains. At the same time, the major root crops such as manioc, sweet potato, and yams contribute little to the pollen and phytolith records because of either limited production or lack of taxonomic specificity (Piperno 1998).

Fortunately, however, it is now clear that human preparation and use of root crops do, in fact, often leave behind clear signatures in the archaeological record in the form of diagnostic strain grain residues preserved intact on stone processing tools. This chapter presents a case study starch grain analysis of plant grinding stones recovered from the Aguadulce Rock Shelter in Panama, and documents the arrival of manioc into southern Central America from its cradle of origin in southern Amazonia. Criteria are presented for identifying manioc starch grains and distinguishing them from those of closely related wild taxa and other plants in the Neotropical flora. Starch grain evidence for other domesticated and economic root and seed crops found at the Aguadulce shelter is discussed more briefly.

Although no food staple known to us, including manioc, is presently thought to have been taken under cultivation and domesticated in Panama, a convergence of other factors makes this country a productive area of research into early agricultural history in the Neotropics. Panama's location at the intersection of two continents made it an obligatory passageway for crop-plant dispersal. Archaeological and ethnohistoric records both indicate that maize, avocado (*Persea americana* Mill.), and some chili peppers (*Capsicum* spp.) arrived from the north well in advance of the earliest Spanish, while southern introductions, including manioc, sweet potato, squash (*Cucurbita moschata* Duchesne), beans (*Phaseolus* spp.), pineapple (*Ananas comosus* (L.) Merrill), and peanuts (*Arachis hypogaea* L.), had been dispersed across the Isthmian region prior to European contact. Moreover, the central Pacific watershed of Panama, the region under discussion here, has been a focus of archaeological and paleo-ecological research over the last 25 years (Figure 5.1) (e.g., Cooke and Ranere 1984, 1992a, 1992b; Cooke, 1998; Piperno 1995; Piperno et al. 1991a, 1991b; Piperno and Pearsall 1998; Ranere 1992; Ranere and Cooke 1996, 2003). A continuous 11,000-year-long sequence of human occupation and plant exploitation has been recovered from excavation of more than 15 rock shelters and open-air sites. The lithic, faunal, ceramic, shell, and plant records are well studied and dated, and arguably are among the most complete and best understood sets of artifacts and ecofacts in the humid tropics. Systematic archaeological foot surveys and analyses of the lithics and ceramics recovered from surface finds also provide robust insights into major regional population trends through time.

The flora of Panama has been intensively studied for over 50 years (D'Arcy 1987; Correa et al. 2004). Long-term political stability plus the presence of the Smithsonian Tropical Research Institute, a bureau of the Smithsonian Institution dedicated to tropical biology and anthropology, continue to make the country a focus for collecting efforts. Hence, efforts to study historical aspects of plant use are not handicapped by a serious lack of knowledge concerning modern species representation and distribution. Not surprisingly, given that the flora often reflects a combination of disparate elements whose centers of origination were far to the north or south, congeneric wild relatives of crop plant species are often uncommon or not present. This is reflected in the large genus *Manihot*, for which only three wild species, *M. carthagenesis*, *M. aesculifolia* (formerly *M. gualanensis*), and *M. brachyloba*, have been described from the Isthmus.

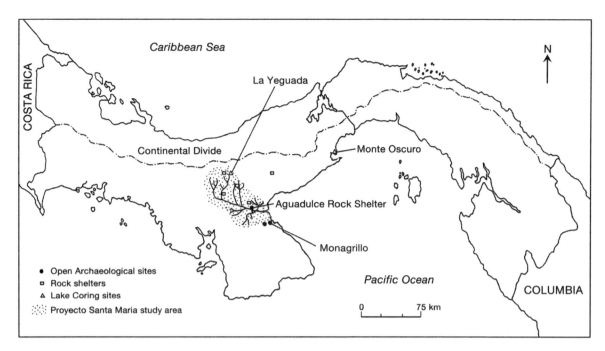

FIGURE 5.1 Map of Panama with location of archaeological sites and lakes discussed in the text.

Prior Research on Manioc Origins and Dispersals

Manioc (also "yuca" from the island Arawak or "mandioca" from the Tupí-Guaraní) is the most important indigenous root crop in traditional swidden (slash and burn) cultivation systems today in the Neotropics. At the time of European contact, manioc was grown from Argentina to Mexico and was consumed on a regular basis in many parts of its range (Sauer 1950). Early accounts by chroniclers also make it clear that in the Antilles and eastern South America, manioc and other root crops such as sweet potato were more important food staples than were seed crops such as maize (Sauer 1950).

Manihot is a very large genus, containing nearly 100 wild species distributed from the southern United States to Argentina (Rogers 1963) (Figures 5.2a and 5.2b). Arguing mainly from morphological data, researchers have suggested that manioc was domesticated independently more than once, that it was possibly of Mexican origin, or that it was a compilospecies with genes pooled from several different wild relatives from different regions of the tropics (Rogers 1963; Rogers and Appan 1973; Jennings 1995). Recent molecular studies by Olsen and Schall using single-copy nuclear genes and microsatellites, however, appear to show conclusively that manioc was domesticated only once, in southern Amazonia, from a single wild progenitor species: *M. flabellifolia* (Pohl) Ciferri (hereafter *M. e.* ssp. *flabellifolia*) (Olsen and Schaal 1999; Chapter 9, this volume).

The ecological preferences of the modern populations of *M. e.* ssp. *flabellifolia* that are genetically closest to manioc, and thus likely to be the location of its direct ancestor, are the seasonal tropical forests of southwestern Brazil, or what Olsen and Schaal (1999) refer to as the transitional area between the *cerrado* (savanna scrub) and the lowland Amazonian rain forest. Tropical forests are classified as seasonal when they have a prolonged, annual dry period lasting from four to seven months, during which little to no rain falls and many trees drop their leaves. Soils of these forests, which once occupied large parts of the lowland tropical landscape before they were cut to make room for agriculture, are typically much more fertile than those of ever-wet or aseasonal tropical forest because they are not as highly leached (Piperno and Pearsall 1998). Such kinds of arboreal formations, which are neither too wet nor too dry (e.g., annual precipitation between 1,200 and 2,000 mm) and contain easily combustible biomass that further enhances soil fertility at the end of the dry season when the seed and root beds are prepared, appear to have supported the domestication and early spread of numerous other crop plants of the lowland tropical forest (Piperno and Pearsall 1998; Piperno 2006a). Thus we might expect that some of manioc's earliest movements after domestication would have been into regions with similar ecological characteristics.

Archaeologists have expended considerable effort in attempting to document the history of manioc domestication and use, but have been hampered by the poor preservation of botanical remains in humid climates. As is the case for many root crops of South American derivation, the earliest undisputed, macrobotanical remains of manioc come from the arid desert coast of Peru, where they are dated to about 4000 BP (all dates are in uncalibrated radiocarbon years) (Towle 1961; Ugent et al. 1986). In contrast, the history of manioc use and spread in the more humid

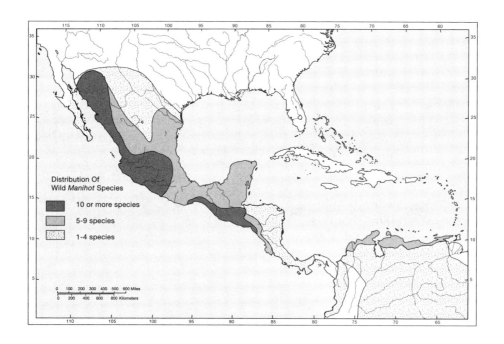

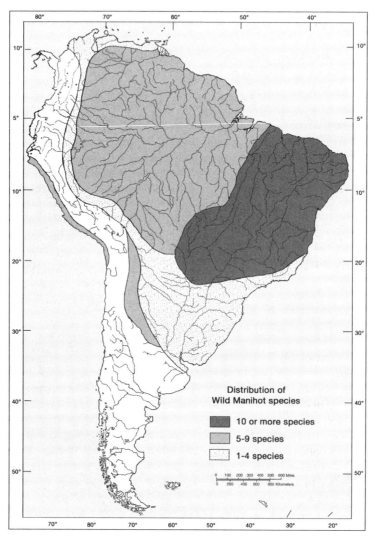

FIGURE 5.2 Maps of Central and South America showing the distribution of wild *Manihot* species (adapted from Rogers 1963).

areas of South America, where poisonous or "bitter" forms high in cyanogenic glucosides were widely grown at the time of European contact, has been based primarily on identifying artifacts associated with the processing of manioc tubers (i.e., small, longitudinal chips of stone called grater chips and large ceramic platters called *budares* or griddles). These objects are well described ethnographically as used for grating the roots and then toasting the sieved, pressed, and dried flour (e.g., Carneiro 1983; Roosevelt 1980: 130–135). Similar kinds of stone and ceramics recovered from archaeological sites located in the humid lowlands of Colombia, eastern Venezuela, and Brazil dating from about the fourth millennium BP onward have been used to argue for the cultivation of bitter manioc (see Roosevelt 1980: 9–11 for an overview). As various researchers have noted, however, it might prove difficult to confidently identify all small chipped stone tools as "grater flakes," and that even if chipped stone functioned in this manner, plants other than manioc may have been involved. Similarly, arguments by analogy employed to infer the function of ceramic platters may not be secure in all archaeological situations (DeBoer 1975; Pickersgill and Heiser 1977; Roosevelt 1980). Ceramic griddles, for example, were used in the Antilles during the sixteenth century for cooking *Zamia* roots and conceivably could work equally well for toasting maize cakes (Sturtevant 1969; DeBoer 1975).

Recent starch grain studies carried out by Linda Perry (2001, 2002) on quartz grater flakes from the Middle Orinoco region in Venezuela dated to a few centuries after the time of Christ add considerable empirical weight to these concerns since a wide variety of starches, but none from manioc, were isolated from these artifacts. In contrast to bitter manioc, sweet or nonpoisonous varieties of the plant, which were being grown west of the Andes and throughout Central America and Mesoamerica when Europeans arrived, are usually simply peeled and boiled, baked, or fried, often contributing to soups. These varieties would leave no inferential evidence for their presence archaeologically in terms of specific types of stone tools or ceramics.

In Mesoamerica, scattered pollen and other evidence for manioc has been reported. Remains of tiny bits of starchy plant tissue, found in coprolites recovered from dry caves in the Tehuacán Valley and Tamaulipas and estimated to date between 2900 and 2200 BP, have tentatively been identified as manioc (Callen 1967). Whether these specimens derive from *M. esculenta* or from a locally occurring wild species is not clear at this time. Very large pollen grains of *Manihot* spp. recovered from sediment cores from the Gulf Coast of Mexico and Belize, where they are dated to ca. 5800 and 4500 BP, respectively, are likely to be from *M. esculenta*, based on their size (> 200 μm), which exceeds that documented for wild *Manihot* specimens studied, and the context of recovery—former agricultural fields (Pohl et al. 1996; Pope et al. 2001).

Identifying the Domestication of Root Crops

Most researchers interested in the origin of crop plants recognize sustained deliberate planting—the cycle of repeated planting and harvesting in prescribed and prepared planting beds (Piperno and Pearsall 1998: 6–8)—as representing a key threshold in plant domestication. Plant domestication is thus most easily identified archaeologically by changes in the phenotypic attributes of plants wrought under that human selection pressure (e.g., non-shattering rachises; variation in fruit color and pattern; or increased seed, phytolith, or pollen size over that documented for extant wild relatives. See Chapters 2, 3, and 4, this volume). In the tropics, however, documenting domestication is not always so easy. More than a few examples can be found of plants that are routinely planted and harvested in gardens without any obvious and dramatic effect on the phenotype (Piperno and Pearsall 1998: 6–7). These are practices of plant food production, but identifying them archaeologically is obviously a far more difficult task than in those situations where phenotypic markers of deliberate planning are evident.

Modern-day roots of *Manihot esculenta* cannot be confused with any produced by their wild congenerics. They are larger, and unlike those of their wild relatives, they uniformly possess starch-rich, non-woody root stock in their older life stages. Sweet varieties have also lost the toxic substances that protect wild species from their natural predators (Figure 5.3). These differences in overall root size or chemical composition, however, offer little help in distinguishing between wild and domesticated manioc roots in archaeological contexts since, as already noted, preservation is so poor in humid climates. Differences in size and morphology of the starch grains of wild versus domesticated roots could provide a means for differentiating between wild and domesticated manioc if the starches, in fact, are preserved and can be retrieved for study in a largely unaltered state.

As will be shown below, starches of domesticated manioc do differ substantially in morphology, and often in size, from wild species, including *M. e.* ssp. *flabellifolia*, and can be distinguished confidently from these and other taxa. How this is achieved is described in detail below, but little can be offered at this time in a way of explanation as to the cause(s) for the changes in starch grain morphology that occurred under systematic cultivation. It is likely, however, that some of the differences in starch grain morphology between wild and domesticated *Manihot* have to do with deliberate human selection for those specific types of starches that are most useful for culinary practices from all those that were available in wild manioc roots as the domestication process proceeded (see Ugent et al. 1987 for a similar view on why starches from domesticated potatoes can be differentiated from wild relatives). These issues will be explored in future work.

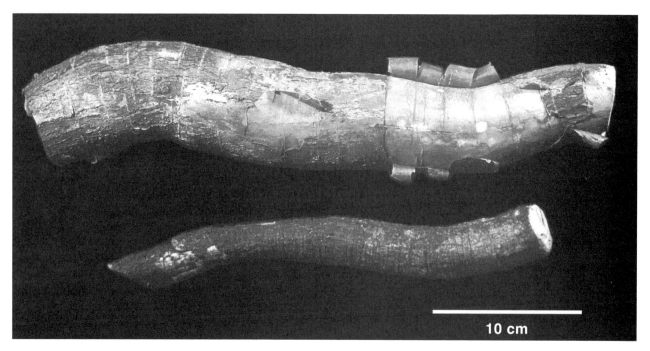

FIGURE 5.3 *Manihot* roots: (top) *M. escutenta* from Panama; (bottom) a wild species of *Manihot* (*M. aesculifolia*).

Starch Grain Analysis: Research Designs and Data Sets

Basic Issues of Starch Grain Research

Starch grains, which are found in large quantities in most higher plants, are the major form in which plants store their carbohydrates, or energy. Starch grain molecules are composed of amylose and amylopectin. The proportion of these two polymers varies depending upon the plant species. It is well understood that plants produce two major types of starch grains: *reserve starch*, which is formed in tiny cellular organelles known as amyloplasts, and *transitory starch*, formed largely in chloroplasts (e.g., Reichert 1913; Shannon and Garwood 1984; Zeeman et al. 2004). The former type, as its name implies, is found in plant organs that have a major storage function, such as roots, rhizomes, corms, tubers, and seeds. Transitory starch granules are formed primarily in leaves and other vegetative structures such as stems. Unlike reserve starches, transitory starches are converted by plants into sugars during photosynthesis soon after they are manufactured, often within the same 24-hour period, and thus are a much shorter-lived part of a plant's life cycle than are reserve starches.

Moreover, all investigators who have studied starches are agreed that reserve starches are made by plants in a highly diverse array of forms that may be genus- and even species-specific. In contrast, transitory granules are documented to be mostly of the same basic types, and therefore are of much more limited utility in identification. Additional research is needed to determine if they routinely survive for long periods of time in archaeological contexts, and might therefore be of value in documenting the remains of vegetative structures (Haslam 2004). It is clear, however, that the reserve starch grains made by seeds and subterranean organs of plants are by far the most useful for identifying plants (e.g., Haslam 2004; Loy 1994, et al. 1992; Perry 2001, 2002; Piperno et al. 2000b, 2004; Reichert 1913; Torrence et al. 2004).

There is a large literature on starch grain properties and morphology to which researchers interested in archaeological applications can refer (e.g., Reichert 1913; Whistler et al. 1984; Sivak and Preiss 1998; Loy 1994; Ugent and Verdun 1982; Cortella and Pochettino 1994; Ugent et al. 1981, 1982, 1984, 1986, 1987; Therin et al. 1999). A large number of atlases and keys of starch grains containing descriptions, measurements, comparisons, and photographs of starches from wild and domesticated plant species have been compiled as a result of older (e.g., Reichert 1913; Seidemann 1966) and newer investigations (e.g., Holst and Piperno, in preparation; Loy 1994, et al. 1992; Perry 2001, 2002; Pearsall et al. 2004; Piperno et al. 2000b, 2004; Torrence et al. 2004). Such information is also routinely published in the dedicated starch journal, *Die Stärke*.

Basic Aspects of Starch Grain Identification

Botanists and now the group of paleo-ethnobotanists who are actively carrying out starch grain research are well agreed that the morphology of starch granules can be specific to, and diagnostic of, a particular individual genus or species. The morphological features that allow for identification can be observed with a compound light microscope. The use of cross-polarized light is valuable for securely identifying an archaeological granule as starch, particularly when one is just starting out in starch grain research, but polarized light obscures many of the diagnostic features one uses to identify the grains. These criteria include (1) overall granule shape in three dimensions, contour, and surface features (e.g., single or compound; spherical, ellipsoidal, ovoid, or bell-shaped; rough or smooth); (2) form and position of the hilum (centric or eccentric), which is the botanical center of the granule; (3) number and characteristics of pressure facets; (4) presence and characteristics of fissures—internal cracks emanating from the hila of some starch granules formed when the granule begins to grow outward from the hilum and quite literally cracks; (5) presence/absence of demonstrable (well-defined) lamellae—growth bands formed during the development of the granules that are readily visible on the surface of some kinds of starches under ordinary light at powers up to 400×; and (6) size and morphology of the extinction cross (e.g., relative lengths and angle of the arms) (Figure 5.4a–h). Polarized light is necessary for studying criterion 6 but, as noted above, other traits are best identified when cross-polarized light is *not* used.

Reference Collections and the Identification of Manioc

The author's modern starch grain reference collection encompasses more than 400 species of plants from 40 different families, including all domesticated species known or thought to have been grown in pre-Columbian Panama, plus their wild congeneric and familial relatives, noncultivated species presently used as foodstuffs (e.g., Fabaceae, Marantaceae, and various tree fruits), and plants commonly used for medicinal, craft, and ritualistic purposes in local indigenous economies today (e.g., Bennett 1962; Duke 1968, 1975; Hazlett 1986). Comparative analysis of the species comprising this starch reference collection has confirmed that for a range of domesticated crop plants (e.g., maize, *Phaseolus* and *Canavalia* beans, manioc, sweet potato, yams, arrowroot, yautia, jícama (Pachyrhizus), peanuts [*Arachis hypogaea* L.], squashes [*Cucurbita* spp.], and palms), the diagnostic attributes previously described by other investigators do in fact allow different domesticates to be distinguished on a morphological basis from each other as well as from other plants in this reference collection (Piperno and Holst 1998; Piperno et al. 2000b; Holst and Piperno, in preparation; Figure 5.4a–h).

With the more widespread use of starch grain analysis by paleo-ethnobotanists, including its applications to regions where crop plants may originally have been brought under cultivation and domesticated, there has been an increased emphasis on comparing starches in domesticated species with starches in their closely related wild relatives (e.g., Loy 1994; Loy et al. 1992; Perry 2001, 2002; Piperno et al. 2000b, 2004). For obvious reasons, this aspect of starch research will continue to strongly engage the interest of archaeobotanical investigators both in the Old World and the New. What attributes, then, distinguish the starch grains of domesticated manioc roots from those of wild relatives?

Three wild species of *Manihot* have been reported in Panama: *M. carthaginensis*, *M. aesculifolia*, and *M. brachyloba*. We studied the first two but were unable to find the latter, which seems to be confined to the remote, wetter forests of the eastern-most part of the country (D'Arcy 1987). Roots from a total of 29 other species of wild *Manihot* from Brazil, Mexico, and Venezuela, including *M. e.* ssp. *flabellifolia*, were also analyzed. For three of these species, seeds were included in the analysis. Most of these plants were sampled by the author from the large collections of *Manihot* made by David Rogers and other botanists during the last 40 years, which are housed at the herbarium of the New York Botanical Garden. The specimen of *M. e.* ssp. *flabellifolia* included in the study was kindly provided by Kenneth Olsen and Barbara Schaal from their greenhouse collections at Washington University in St. Louis.

Roots, stems, and seeds from other herbaceous and arboreal genera of the manioc family Euphorbiaceae that commonly occur in Panama, such as *Jatropha* spp. and *Cnidoscabus* spp. were also studied (although these are of little or no economic value today). Published starch keys and atlases were also consulted. Starches from the wild *Manihot* samples were compared with those from roots of 12 different populations of modern domesticated manioc from Panama. Morphological and size determinations for *Manihot* and other taxa incorporated sampling of multiple reference points within a single specimen in order to account for within-sample variability. Size determinations are based on a sample size of at least 50 starch grains. Upper and lower size-range boundaries and confirmation of single-grain species diagnostics were achieved by scanning thousands of additional starch grains from each species after the sum was tabulated.

Our particular focus was on a class of starch grains called bell-shaped or hemispherical granules that form dominant components of starch assemblages from manioc and closely related plants (Reichert 1913; Ugent et al. 1986). They often have inflated proximal ends and from one to three distinct and rounded pressure facets located on the distal or basal part of the grain (Figures 5.4a and 5.5). Hemispherical granules are easily distinguished from other types of compound grains, including those that dominate the starch assemblages of other cultivars like sweet potato and yautia (*Xanthosoma*), which originally form in larger aggregates in amyloplasts, and thus are angular and have more numerous pressure facets (compare Figures 5.4a and 5.4b). Figure 5.5 illustrates how bell-shaped compound granules are formed in manioc: more than one grain originally occupies an amyloplast and forms aggregates with others. When modern samples are prepared for microscopic study, or when root materials are deposited into sedimentary environments, the great majority of these aggregates break apart. The points at which they were joined together, called the pressure facets, can be readily identified, however, and the determination that one is dealing with compound grains is easily made.

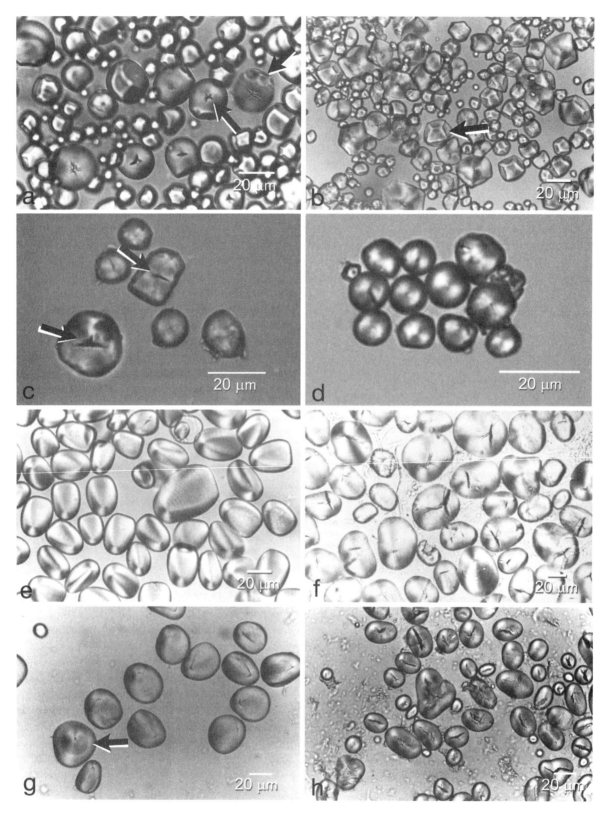

FIGURE 5.4 Starch grains from various crop plants of the Neotropics. The black arrow on the right in Figure 5.4a indicates the single type of bell-shaped grain diagnostic of manioc. Notice that starch grains from hard and soft endosperm maize varieties can be distinguished (Figures 5.4c and 5.4d): (a) manioc; (b) sweet potato; (c) maize, Race Jala; (d) maize, Race Harinoso de Ocho (eight-rowed flour corn); (e) yam (*D. trifida*); (f) arrowroot; (g) lirén; (h) common beans (*Phaseolus vulgaris*). Arrows point to various features of starch grains useful in identification (e.g., centric hilum in 5.4a (the black arrow on the left), angled compound grain with many pressure facets in 5.4b, long fissures in 5.4c). (After Piperno 2004: Figure 1.)

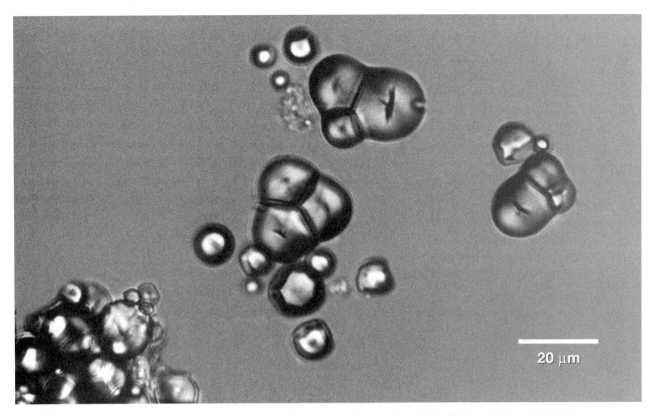

FIGURE 5.5 Compound starch grains from manioc still in aggregations as they originally formed in amyloplasts.

The results reported here, together with previously published reports concerning Neotropical plants (Perry 2001, 2002; Piperno and Holst 1998, Piperno et al. 2000b), indicate that bell-shaped grains are relatively uncommon in tropical species and are mainly confined to subterranean organs of the plants. These bell-shaped grains are particularly characteristic of two genera in the reference collection used in this study—*Manihot* and *Pachyrhizus*—and may also occur rarely in the Marantaceae and Dioscoraceae. The seeds of wild *Manihot* and other Euphorbeaceae that were studied produce oils, not starches.

In manioc and other taxa, bell-shaped granules do not form completely homogeneous populations when the important traits used for taxon identification described above are considered. For example, some granules of a single plant may have well-developed fissures, while others may not. Many granules may possess centric fissures, while a much smaller proportion of the starch granule population will have eccentric fissures. Types of grains that appear to be unique to individual species of plants do occur in manioc (below) and other plants, and are obviously of great importance. Such species-diagnostic grains, however, may account for only a relatively small percentage of all the granules present, and might be missed in archaeological assemblages where sample sizes are not particularly large. It is clear that attribute combinations used in a multiple grain analysis represents an effective, and perhaps the most conservative, means to distinguish species (Reichert 1913; Ugent et al. 1982, 1987; Cortella and Pochettino 1994).

Such an approach employs all the morphological characteristics that account for a population of starches in a single species (rather like having a complete skeleton of an early hominid rather than just a piece of a skull or a pelvis) and takes into account intra and interspecific variation in grain attributes.

Table 5.1 contains population signatures for bell-shaped granules in manioc, wild *Manihot* species, and some of the few other plants in the reference collections that were found to produce bell-shaped starches. (Table 5.2 on page 57 shows the mean size and ranges values for these starch granules.) The key in Table 5.1 was constructed through a stepwise determination of the key features of starch granules observed when scanning and counting starch preparations mounted on slides, starting with overall grain shape and contours and ending with the presence or absence of lamellae. Bell-shaped granules in roots of domesticated manioc have smooth and highly transparent surfaces, one to five unbanded pressure facets with predominantly rounded edges, hila that are almost always centric and open, fissures that are combinations of wing-, crossed-, y-shaped, and deeply stellated types, and no demonstrable lamellae (Figures 5.4a and 5.5). In contrast, starch granules in manioc's putative wild ancestor, *M. e.* ssp. *flabellifoli*, have much narrower pressure facets than those in manioc, and lack or have unremarkable fissures, none of which are stellated. Their hila are mostly eccentric (Figure 5.6). The grains are also much smaller and appear less smooth and less transparent than in manioc. Thirty-one other wild

TABLE 5.1
A Key for Bell-Shaped Starch Granules from Manioc and Other Species

Species (Source)	Pressure Facet Edges			Hilum			Hilum		Fissure			Lamellae
	Rounded	Straight	Banded	Open	Closed		Centric	Eccentric	Present	Absent	Stellate	
Manihot esculenta Crantz (Panama)	97%	3%	0%	97%	3%		91%	9%	72%	28%	30%	0%
Manihot esculenta ssp *flabellifolia* (Pohl) Ciferri (Brazil, K. Olsen and B. Schaal, St. Louis collection)	90%	10%	0%	51%	49%		2%	98%	27%	73%	0%	10%
Manihot carthaginensis (Jacquin) Muell. Arg. (Venezuela, D. J. Rogers No. 499)	65%	35%	59%	52%	48%		44%	56%	34%	66%	0%	4%
Manihot aesculifolia (H.B.K.) Pohl (Panama, C. Galdames)	85%	15%	71%	66%	34%		56%	44%	8%	92%	0%	6%
Manihot fruticulosa Rogers @ Appan (Goiás, Brazil, H.S. Irwin et al. 11/18/95)	92%	8%	50%	74%	26%		48%	52%	34%	66%	24%	20%
Manihot pringlei Watson (Tamaulipas, Mexico, D. J. Rogers No. 528)	82%	18%	29%	56%	44%		58%	42%	22%	78%	18%	44%
Manihot oaxacana Rogers @ Appan (Oaxaca, Mexico, D. J. Rogers No. 507)	100%	0%	0%	25%	72%		32%	68%	4%	96%	0%	50%
Manihot filamentosa Pittier (Guárico, Venezuela, D. J. Rogers No. 495)	100%	0%	0%	26%	74%		42%	58%	10%	90%	0%	6%
Manihot divergens Pohl (Goiás, Brazil, H. S. Irwin et al. 10/22/65)	100%	0%	0%	24%	76%		14%	86%	12%	88%	0%	2%
Manihot chloristicta Standley @ Goldman (Colima, Mexico, D. J. Rogers No. 512)	100%	0%	0%	8%	92%		22%	78%	8%	92%	0%	0%
Pachyrhizus panamensis (Panama, wild, C. Galdamez)	10%	90%	90%	15%	85%		20%	80%	12%	88%	0%	5%

TABLE 5.1 (continued)

Species (Source)	Pressure Facet Edges			Hilum		Hilum		Fissure		
	Rounded	Straight	Banded	Open	Closed	Centric	Eccentric	Present	Absent	Stellate	Lamellae
Pachyrhizus erosus L. (Guerrero, Mexico, cultivated, D. R. Piperno and I. Holst)	18%	82%	68%	4%	96%	86%	14%	50%	50%	10%	38%
Dioscorea convolvulaceae (Panama, wild, C. Galdamez)	100%	0%	0%	89%-	11%	13%	87%	94%	6%	9%	87%

NOTE: Percentages of stellated fissures for each species are of all fissure types present in the species' assemblage. Percentages of banded pressure facets are of all straight facets present, since rounded pressure facets are never banded. Collectors and collection localities are listed under the taxon.

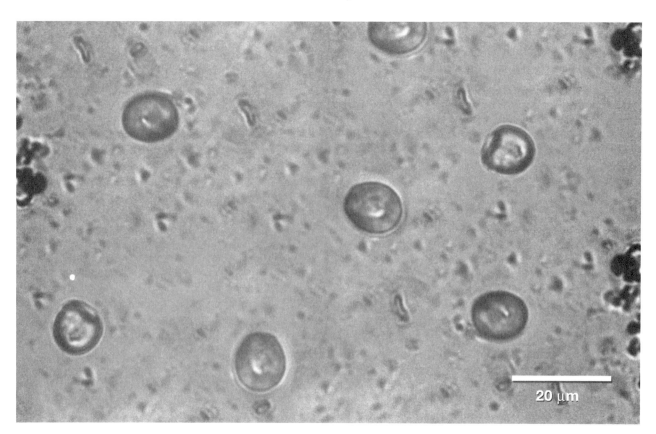

FIGURE 5.6 Starch grains from *Manihot esculenta* ssp. *flabellifolia* showing how they have eccentric hila and lack fissures.

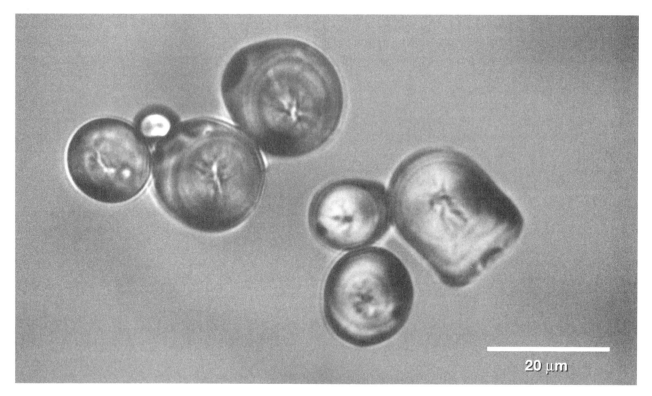

FIGURE 5.7 Starch grains from *Manihot carthaginensis* showing how they have eccentric hila and lack fissures found in manioc.

TABLE 5.2
Starch Grain Size in Modern Wild and Domesticated *Manihot*

Species	Mean Length (μm)	Range (μm)
Manihot esculenta	13–16	6–28
Manihot esculenta ssp. *flabellifolia*	8	4–12
Manihot carthaginensis	15	8–24
Manihot aesculifolia	10	6–16
Manihot fruticulosa	13	6–30
Manihot pringlei	13	6–28
Manihot oaxacana	6	4–8
Manihot filamentosa	9	4–14
Manihot divergens	9	4–16
Manihot chloristicta	8	4–10
Manihot angustiloba (Torrey) Muell. Arg. (Sinaloa, Mexico, D. J. Rogers No. 519)	11	6–16
Manihot attenuata Muell. Arg. (Goías, Brazil, H. S. Irwin et al. No. 9421)	11	8–20
Manihot pentaphylla (Goías, Brazil, H. S. Irwin et al. No. 11887)	8	4–12
Manihot grahami Hooker (Tallahassee, Florida, D. J. Rogers No. 499, escape from cultivation)	6	4–10
Manihot gracilis ssp. *gracilis* (Rogers @ Appan)	9	4–12
Manihot caerulescens Pohl (Rogers @ Appan) (Mato Grasso, Brazil, H. S. Irwin No. 16210)	12	6–19

species of *Manihot* native to Panama and other regions of the Neotropics can similarly be distinguished from manioc using these same criteria. Table 5.1 presents a representative survey of these starch grains in roots from 8 of the 31 wild species studied (Figure 5.7).

A type of granule was also identified in manioc that was not present in other reference class species, and which has not been previously described. It appears to be unique to *M. esculenta* (Figure 5.4a, white arrow). This newly identified diagnostic granule has five to eight pressure facets located on the distal end in the form of a corona, along with a stellated fissure.

Overall, there is a strong tendency for wild *Manihot* to have closed and eccentric hila. Also distinct from manioc, most wild species do not have many grains that develop fissures, and very few of the fissures that do develop are of the stellate variety. In addition, when fissures are present, they occur on grains with eccentric hila. Some wild species also have significant percentages of grains with lamellae. As a result of the process of domestication, then, the following changes to starch morphology occurred in manioc: (1) the hila moved from an eccentric to a centric position (notice how available wild and domesticated specimens of *Pachyrhizus* also differ in this respect; Table 5.1), and the hila became much more likely to be open rather than closed; (2) fissures became more frequent, varied, and conspicuous; and (3) grains became much larger.

The criteria just described also easily discriminate manioc from taxa of other genera that contain bell-shaped or hemispherical granules. For example, in tubers of *Pachyrhizus* spp. (the yam bean), another genus commonly used and cultivated in some regions of the Americas, many grains have acutely angled and banded pressure facets in addition to closed hila and possess simple or no fissures. These grains appear to be genus specific. The small proportion (about 20 to 30%) of starch grains in sweet potato that are bell-shaped have eccentric hila and lamellae and, interestingly, sometimes have a distinctive brown to brownish-red tinge. The extinction crosses of sweet potato and manioc also differ.

In summary, comparative analysis of the starch grain reference class yielded no species of plant that produces a population of bell-shaped starch granules comparable to that of manioc, nor were any identified in a comprehensive survey of available published keys and indices. Indeed, based on this analysis, any starch grain recovered from an archaeological site in Panama that exhibits a centric and open hilum, a y-, wing-shaped, or stellated fissure, without lamellae, and has rounded and unbanded pressure facet edges, is likely to be from manioc.

A recent independent study by Linda Perry of a large, modern reference collection of sweet and bitter forms of manioc and other economic plants from a variety of families from the Middle Orinoco Basin, Venezuela, is also

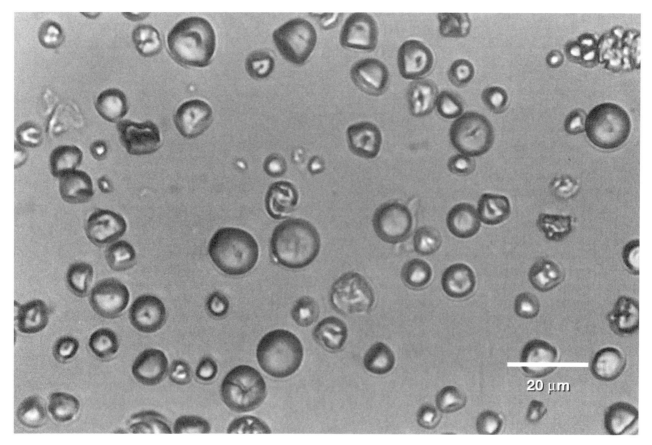

FIGURE 5.8 Starch grain population from *Manihot aesculifolia* showing how it is dominated by simple, bell-shaped granules. Single grains or populations of grains such as these found archaeologically cannot identify the presence of manioc.

important with regard to establishing standard starch markers for manioc (Perry 2001, 2002). Her research is concordant with ours in indicating that manioc can be identified using the criteria described herein, and she notes that the idiosyncratic type of starch granule with a stellate fissure and multiple basal pressure facets that we isolated only from manioc discriminates manioc in her study region as well. Interestingly, her data suggest that starches from sweet and bitter forms of manioc may not be distinguishable.

The extent to which keys such as ours can be employed in identifying manioc in other regions of the Neotropics awaits further study. Available information presented here on wild species of *Manihot* from Brazil, Venezuela, and Mexico suggests promise in this regard. Some areas where caution is required should also be pointed out. For example, starch grains designated here as "simple bell-shaped granules," that is, those with centric hila but without prominent fissures or distinctive kinds of pressure facets, cannot, in and of themselves, be employed to identify manioc. Other taxa, including many wild *Manihot* species, also produce such simple, bell-shaped forms, sometimes in high frequencies in grain assemblages (Figure 5.8), whereas in manioc they constitute a smaller relative percentage of the total starch grain assemblage. Therefore, an archaeological population consisting only of simple, bell-shaped grains could easily indicate that manioc is not represented. Another note of caution in archaeological identification involves the relative size of starch grains. While starch granules in wild *Manihot* are often considerably smaller than those in domesticated varieties, size alone does not universally separate wild from domesticated *Manihot* (Table 5.2).

Archaeological Evidence for Manioc and Other Roots from the Aguadulce Rock Shelter

The Archaeological Context

The Aguadulce Rock Shelter is located on the Pacific coastal plain of central Panama approximately 17 km from the sea (Ranere and Hansell, 1978; Piperno et al. 2000b; Figure 5.1). Today, the region receives 1,600 mm of precipitation annually, distributed on a highly seasonal basis, and has a mean annual temperature of 26°C. In the absence of human interference, the vegetation would be a deciduous tropical forest, one that in its overall floristic composition, structure, and other characteristics would be substantially similar to the forests where *M. e.* ssp. *flabellifolia* grows naturally today.

The Aguadulce shelter is one of a large number of prehistoric cultural occupations located within or near the watershed of the Río Santa Maria (Figure 5.1). As described above, intensive archaeological and paleo-ecological research during the last 25 years documents a cultural sequence beginning with the arrival of people using Clovis-like tools at about 11,000 BP (Cooke 1998; Ranere and Cooke 2003). Systematic archaeological foot surveys carried out by the Proyecto Santa Maria (Cooke and Ranere 1984) indicate that human populations grew substantially between 11,000 and 7000 BP. Pollen, phytolith, and charcoal studies of lake sediments indicate that at ca. 7000 BP slash and burn cultivation was initiated in the seasonally dry Pacific-watershed forests of the Río Santa Maria watershed and nearby areas (Piperno et al. 1991b; Bush et al. 1992; Piperno and Jones 2003).

By 8000 to 7000 BP, specialized stone implements used to grind and process plant materials were widely distributed in central Panama, and a variety of plants were exploited (Ranere 1992; Ranere and Cooke 1996; Piperno and Pearsall 1998; Piperno et al. 2000a; Piperno 2006a). Some of these, including arrowroot and squashes of the genus *Cucurbita*, both of which were likely introduced from northern South America, were cultivated. The manufacture of ceramics was under way in central Panama by ca. 5000 BP (Cooke 1995). Sedentary villages appear by ca. 2400 BP. In contrast to earlier and smaller settlements of the region, they were located to exploit major zones of alluvium and were numerous by the time of the Common Era.

Excavations at the Aguadulce shelter carried out in 1973–1975 and 1997 (Piperno et al. 2000b; Ranere and Hansell 1978, 2000) uncovered the remains of three distinct human occupations. Zone D, the oldest, is a yellow (inside the dripline) or red (beyond the dripline) silty clay atop weathered bedrock. Zone D represents a sequence of occupations dating between 11,000 and 7000 BP, based on the presence of characteristic Paleoindian (11,000–10,000 BP) and Early Preceramic (10,000–7000 BP) bifacial reduction lithic technologies (Ranere 1992; Ranere and Cooke 1996) and a series of Accelerator Mass Spectrometer (AMS) 14C dates obtained on sediment phytolith samples from Zone D (10,725 ± 80 BP, 10,529 ± 184 BP, 8423 ± 79 BP, 7061 ± 81 BP and 5560 ± 80 BP: Wilding 1967; Kelly et al. 1991; Mulholland and Prior 1993; Piperno et al. 2000b). Four of the five dates fall comfortably within the ca. 11,000 to 7000 BP time frame of occupation indicated by the Zone D lithic assemblages. The single outlier date (5560 ± 80 BP) was obtained on a very small (130 microgram) sample that was recovered in a soil column directly underneath a larger Zone C sample dating to 6910 ± 60 BP). No milling stones were present in Zone D.

Zone C, a dark brown, clayey silt with angular rock, yielded two AMS 14C phytolith dates: 6910 ± 60 BP, cited above, at its base and 6207 ± 60 BP stratigraphically higher. A terminal date for Zone C is estimated to be ca. 5000 BP. The initiation of Zone C marks a dramatic increase both in

FIGURE 5.9 Some of the grinding stones examined from the Aguadulce Rock Shelter: (a) and (b), edge-ground cobbles from Zone C; (c) the boulder milling-stone base from Zone C. This tool contained manioc. (After Piperno and Holst 1998: Figures 11–13.)

the intensity of human occupation and the rate of sediment accumulation. As in other sites from the region dated to this period, bifacial reduction disappears from the chipped-stone industry, which is now characterized by expedient core reduction and bipolar reduction (Ranere and Cooke 1996). A characteristic form of groundstone tool termed "edge-ground cobble" (the grinding facet is located on the narrow edge of the stone) first appears near the base of this zone, together with boulder milling-stone bases (Figures 5.9a–c). Use-wear and replicative studies carried out by Ranere suggested that this distinctive tropical forest tool complex was probably used to process roots and tubers (Ranere 1980). Faunal remains, including animal bone and shell, are encountered for the first time in Zone C and are abundant and well preserved in sediments underneath the overhang. Ceramics are still absent.

Zone B is a tan silt with much less angular rock than Zone C. It yielded a single AMS phytolith date of 4250 ± 60 BP, and is marked by the introduction of "Monagrillo" pottery, the earliest found in Central America (Cooke 1995). The Zone B occupations of Aguadulce shelter have been dated between ca. 5000 to 3000 BP, based on 14C dates from the Aguadulce shelter and from the nearby Monagrillo shell midden type site (Figure 5.1) as well as from other sites containing this distinctive ceramic tradition in central Panama (Cooke 1995). The edge-ground cobbles from this zone tend to be more heavily worn than those in Zone C.

Recovery of Archaeological Stone Tools and Sediments, and Their Laboratory Analysis

The analysis of plant microfossils such as starch grains and phytoliths that are found still adherent to archaeological stone tools is an important means of documenting plant use and subsistence. The microfossils present in such tool-use residues can be linked to specific classes of artifacts whose use in processing plant foods has already been established, and the age of the tools can often be confidently determined through radiocarbon dating of associated cultural material.

Preliminary analysis of milling stones recovered from the first set of excavations at Aguadulce shelter in 1973–1975 indicated that they contained starch grains from a variety of plants, including possibly manioc (Piperno and Holst 1998). However, the sample size of tools and starches was small. Furthermore, archaeological sediments that had been closely associated with the grinding stones that were analyzed, and that could potentially provide additional information on the sources of the starches isolated from the tools (e.g., whether some of the starches had become attached to tools not through plant processing but from contact with the sediments) were unavailable for analysis. Our modern reference collection at that time was also considerably smaller than it is now, and did not encompass wild species of Manihot. We returned to the site in 1997 in order to increase the sample of milling stones and starch grains. A protocol for the laboratory analysis of archaeological starch grains was also developed in order to maximize their recovery from plant grinding implements. Many aspects of our field and laboratory methodology, presented below, are applicable to any class of stone artifacts. As described in Piperno et al. (2000b), the following protocols were used during field work and then in the laboratory.

Artifacts and sediments for starch analysis retrieved during excavations were immediately put into plastic bags, using starch-free plastic gloves. After artifacts were removed from the ground, sediments from directly beneath and around their periphery to a distance of 5–10 cm were sampled. After excavations were completed, column samples were taken every 5 cm from profiled walls of the squares from which tools had been recovered.

Once in the laboratory, grinding stones were examined under a stereoscopic microscope at a power of 100×. In order to investigate closely the distribution of starch grains on the stone tools, all of the tool surfaces, including those with no detectable use wear as well as those where obvious grinding facets and pounding edges could be identified, were sampled by inserting the point of a fine needle into visible cracks and crevices. Between 10 and 13 locations covering the entire tool were sampled in this manner. The residue loosened and removed was then mounted in water on a microscope slide and examined with polarized and unpolarized light at a power of 400×. Groundstone tools were subsequently washed with a brush underneath running water, and starch was removed from the recovered sediment using techniques developed and described in detail elsewhere (Therin et al. 1999; e.g., addition of a heavy metal liquid at a density of 1.8 made from a solution of cesium chloride, CsCl). Washed tools were shaken at moderate speed in an ultrasound for 5 to 10 minutes to ensure that most of the strongly adherent sediment and starches were removed from them, and then starch was isolated from this suspension using CsCl as before. Sediment samples from surrounding deposits were also studied using the same methods. Ten to 20 cubic centimeters of sediment sampled from beneath and around the peripheries of the grinding stones were analyzed for starch grains, along with 24 sediment samples from two stratigraphic columns of excavation squares from which grinding stones had been recovered. In order to provide a control comparison, four nonartifactual, angular rocks that were a natural component of the site, along with the sediment closely associated with them, were also sampled and analyzed in the same manner.

A total of 18 edge-ground cobbles and one milling stone base was analyzed. Of these, 12 contained starch residue from a variety of plants, including yams, maize, arrowroot, and legumes (Piperno and Holst 1998; Piperno et al. 2000b). Thirteen bell-shaped granules with attributes found only in manioc roots were recovered from two of these tools, an edge-ground cobble (catalogue number 42) and the boulder milling-stone base (catalogue number 26a) (Figure 5.10c). As described in more detail below, both of these were securely stratified in Zone C, which lacks ceramics and dates from about 7000 B.P. to 5000 B.P. The starch grains, all of which possessed extinction crosses like those in modern manioc when examined under cross-polarized light, had rounded pressure facets and centric, open hila (Figure. 5.10a–e). They exhibited no demonstrable lamellae, and were smooth and transparent. Eight of them had crossed, y- to wing-shaped (Figure 5.10a), and stellated fissures; two were of the latter type (Figure 5.10c, 5.10d). As described above, when any starch grain with these fissures, centric and open hila, and without lamellae is recovered from a Panamanian site, it is likely to represent manioc.

One of the grains from tool #42, with pressure facets in the form of a corona on the distal end and a stellated fissure, is the type seen only in manioc (Figure 5.10c). The remaining four bell-shaped grains recovered from tool #42 either lacked or had simple fissures (Figure 5.10b, 5.10e). These also occur in lower frequencies than do grains with more conspicuous types of fissures in modern manioc.

In summary, the assemblages of bell-shaped starch grains recovered from the Zone C tools and sediments closely match the starch grain profile characteristic of modern

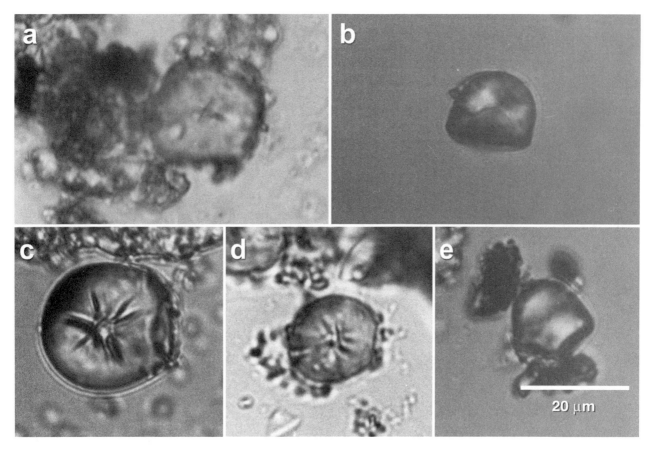

FIGURE 5.10 Starch grains from manioc from the Aguadulce Rock Shelter that exhibit rounded pressure facets, centric open hila, and other characteristics of manioc: (a) starch granule exhibiting crossed-y to wing-shaped fissure; (b) bell-shaped starch grain with simple fissure; (c) starch grain type seen only in manioc, with corona form pressure facet on distal end and a stellated fissure; (d) another bell-shaped starch grain with a stellated fissure; (e) bell-shaped starch grain with simple fissure.

manioc, and only manioc. Along with having comparable frequency values for diagnostic features, the mean size (14–18 microns) and ranges (10–24 microns) of the archaeological grains are also completely compatible with modern varieties of manioc (Table 5.2). A single, simple, bell-shaped starch grain (without a fissure or other identifying features) was recovered from another Zone C edge-ground cobble. As discussed above, although these types of grains are found in manioc, they cannot unequivocally identify its presence.

The laboratory protocol with its series of steps used to analyze and remove starches from the tools and sediments yielded the following results (Piperno et al. 2000b). The systematic needle probes of all the surfaces of the stones resulted in starch recovery only from the grinding and/or pounding facets. In the case of the tools from the 1997 excavations, where closely associated sediment was analyzed, every starch grain save one (from Zone B and, therefore, not associated with the preceramic tools containing manioc) was recovered from sediments that were firmly adhered to the artifacts when we lifted them from the ground, and that we subsequently removed by washing with a hard brush and using sonification. Sediments sampled from the immediate proximity of the tools, in contrast, were nearly devoid of starch (with the exception of one starch grain of maize recovered from a Zone B sample). The analysis of two dozen different sediment samples from two stratigraphic columns yielded only a single starch grain, further indicating that starch preservation in the sediments is very poor. Finally, no starch was recovered from the analysis of four nonartifactual stones and the closely associated sediment. Given the known properties of starch grains, their survival in tropical sediments for time periods longer than a few days to a week is highly improbable (Haslam 2004). In any case, if some of the starches on the tools in this study were a result of secondary deposition from the site's sediments shortly after the tools had been discarded, starch recovery from unused tool facets and the nonartifactual stones would also be expected.

The edge-ground cobble and milling-stone base on which the manioc starches occurred are dated to the late preceramic period (7000 to 5000 BP), based on radiocarbon dates from the Aguadulce shelter listed above, other lithics characteristic of the late preceramic period with which they were associated (e.g., a chipped stone assemblage dominated by flakes from irregular and bipolar cores), and an absence of ceramics in Zone C (Piperno et al. 2000b; Ranere 1992; Ranere and Cooke 1996). No visible disturbances or intrusive cultural materials such as small ceramic fragments were present in the contexts that yielded the tools. The edge-ground cobble containing manioc starch grains from near the base of Zone C (tool #42) is one of the oldest plant grinders from the site (it also demonstrates the less intensively worn edges characteristic of the earliest grinding stones from the region) and should date to nearly 7000 BP, an age supported by a phytolith date of 6910 ± 60 BP obtained from a sediment sample that was taken from immediately below it. The milling-stone base was also well stratified in the middle of Zone C. The preceramic age of these tools and the manioc they contain appears secure.

No manioc starches were recovered from edge-ground cobbles from the Zone B occupation of Aguadulce (Piperno et al. 2000b). The absence of manioc and other root starches (below) on these later tools may indicate an increasing importance of seed crops like maize in the ceramic-phase subsistence economy. Alternatively, it may simply point to changes in food preparation techniques associated with the availability of ceramics that altered how manioc was processed and made it less visible through study of the types of artifacts considered here (e.g., roots may have been peeled with chipped stone tools or modified shells, as they are today, and then boiled). Paleo-ecological evidence from La Yeguada, a lake located near Aguadulce (Figure 5.1), indicates that forest clearing as a result of slash and burn cultivation that had started at about 7000 BP intensified at ca. 4000 BP, a pattern consistent with an increased reliance on seed cropping, which is more demanding on tropical soils than is root crop cultivation (Piperno et al. 1991b). Thus it is quite possible that both changes in processing techniques and an increased emphasis on maize and other crops may explain the absence of manioc starch grains in Zone B. There is no indication that manioc starches would have been preserved differentially once these residues accumulated on the grinding stones.

Starches from Other Domesticated and Wild Plants at the Aguadulce Rock Shelter

In addition to manioc, starch grains from yams (*Dioscorea* spp.), arrowroot (*Maranta arundinacea*), and maize were recovered from Zone C grinding stones, with maize and some of the yams occurring on the edge-ground cobbles that contained manioc. Yam grains, which starch analysts agree are genus specific, are easily identified by their ovoid shapes, cuneiform-shaped depressions, and lamellae (Figure 5.4e). Arrowroot produces diagnostic ovoid grains with eccentric hila and softly defined, proximal fissures (Reichert 1913; Piperno and Holst 1998; Perry 2001; Figure 5.4f). As was the case with manioc, yam grains were not present on the grinding stones studied from Zone B.

The yam starch grains recovered from the two stratigraphically earliest edge ground cobbles from Zone C have shapes characteristic of wild species. In contrast, those removed from an edge-ground cobble found near the top of Zone C, which also possesses the heavy use wear on three sides that is characteristic of later tools (Piperno et al. 2000b), are identical in morphology and size to those of *Dioscorea trifida*, the only species of yam that was domesticated in the Americas, probably in northeastern South America (compare Figures 5.4e and 5.11). We believe these *Dioscorea* grains are very probably from *D. trifida*. Comparisons of starches from a large sample of 15 different wild *Dioscorea* species from Panama collected during the past four years by Carmen Galdamez, a botanist at the Smithsonian Tropical Research Institute, with those of numerous specimens of *D. trifida*, show that grain morphology in wild species is distinct from *D. trifida*, and, in addition, exhibits a high degree of within-tuber heterogeneity (Figure 5.12). The yam grains isolated in substantial numbers from Zone B tool # 350 (n = 16) are uniform in shape (compare Figures 5.4e, 5.11, and 5.12). Grain uniformity is another possible result of selection under cultivation in yams. The modern yam discussed in Piperno et al. (2000b) whose starches presented the only possible confusion with *D. trifida*, has been positively identified as a variety of *Dioscorea alata* (Carmen Galdamez, personal communication to D. Piperno, August 2004), an Asian yam that was introduced following European contact and is still commonly grown in Panama.

An important root crop conspicuous by its absence in grinding-stone residues is *yautia* (*Xanthosoma sagittifolium*), a member of the Aroid family that took the place of the Old World taro (*Colocasia*) in pre-Columbian economies, and which is still commonly cultivated in Neotropical forest settings. There is a straightforward, likely explanation for its absence. Unlike arrowroot and many species of yams, which require pounding and maceration before they can be cooked and eaten, yautia is simply peeled and boiled, and so it would not be likely to be found on grinding implements even if it had been consumed at the site.

Maize starch grains, which can be confidently identified in the Neotropics on the basis of morphology and size (see Haslam 2003; Perry 2001, 2002; Pearsall et al. 2004; Piperno and Holst 1998), are present in high numbers along with the root and tuber remains on Zone C grinding stones, sometimes constituting the most common starch type present (Piperno et al. 2000b). Along with maize starch grains, phytoliths from maize cobs, derived primarily from the chaff adhered to the kernels when they were processed, have also been identified on Zone C grinding stones. Analyses by a number of researchers of large, modern reference collections of wild and domesticated *Zea* and non-*Zea* grasses have established that cob phytoliths (in part called "rondels"), which are of different types than those found in leaves, can be used throughout the Americas to identify the presence of maize cobs in archaeological contexts (Bozarth 1993; Mulholland 1993; Irarte et al. 2004; Pearsall 2000; Pearsall et al. 2003, Piperno, 2006b, Piperno and Pearsall 1993, Piperno et al. 2001, Chapter 7, this volume).

It is also well established that the formation and morphology of the cob phytoliths are under the control of an important genetic locus in *Zea* called *tga1*, which, in

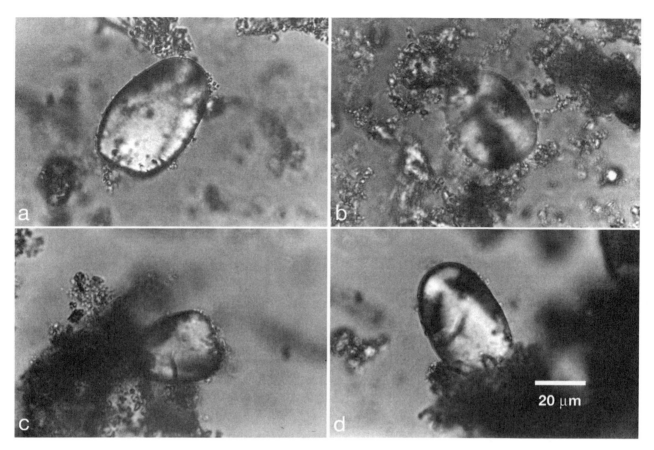

FIGURE 5.11 Starch grains from *Dioscorea* found on an edge-ground cobble at the Aguadulce Rock Shelter. They are identical in size and shape to those found in domesticated *Dioscorea trifida*. This type of starch grain does not occur in wild species of Dioscorea from Panama studied by the author.

addition to phytolith characteristics, controls the hardness (lignification) and size of the outer glume subtending the kernel in wild and domesticated maize (e.g., Dorweiler and Doebley 1997; Pearsall et al. 2003; Piperno and Flannery 2001; Piperno et al. 2001; Chapter 7, this volume). Because of the actions of *tga1*, hard- and soft-glumed varieties of *Zea* produce distinguishable phytolith assemblages (discussed below) (e.g., Dorweiler and Doebley 1997; Pearsall et al. 2003; Piperno and Pearsall 1998; Piperno et al. 2001).

The starch grain and phytolith record from the Aguadulce shelter stone tools thus adds to the coherent and multi-proxy corpus of data from central Panama indicating that late preceramic economies (7000 BP to 5000 BP) were mixed root- and seed-crop systems, and that maize was introduced into the Isthmian region by early in the seventh millennium BP (see Piperno and Pearsall 1998 and Piperno 2006a and b for an extended discussion). Numerous maize starch grains along with cob phytoliths have also been retrieved recently from over 20 manos and metates recovered from intact house floors dating to the Valdivia III period (ca. 4300 BP) at the site of Real Alto in southwest Ecuador (Pearsall et al. 2004). These findings add to the considerable evidence accumulated by Pearsall and others previously that maize was well established in northern South America by ca. 5000 BP (Pearsall et al. 2003; Piperno and Pearsall 1998).

Endosperm type exerts a considerable influence on starch morphology in maize (e.g., Cortella and Pochettino 1994; Perry 2001, 2002; Piperno et al. 2000b; Reichert 1913). Because of the way they are packed into amyloplasts when they are formed, starches of maize races with hard endosperms (e.g., popcorns and flint corns) are angular and four-to five-sided, with a rough, grooved surface. In contrast, starches from flour corns are not densely packed together in amyloplasts and predominantly have spherical shapes and smooth contours (compare Figures 5.4c and 5.4d). Maize starches from late preceramic and early ceramic stone tools from Aguadulce were all of the hard endosperm type, consistent with the presence primarily of a popcorn or other similar kernel variety before ca. 3000 BP. In addition, maize cob phytoliths from the earliest Zone C stones, and from associated sediments, have characteristics indicating that they were derived from a hard-glumed type of corn (see Piperno and Pearsall 1998: 220–226 and Piperno et al. 2001 for illustrations and detailed discussions). Using starch grain and phytolith analyses of residues recovered from stone tools in tandem, therefore, can provide information regarding important phenotypic characteristics of prehistoric maize: endosperm type and glume hardness.

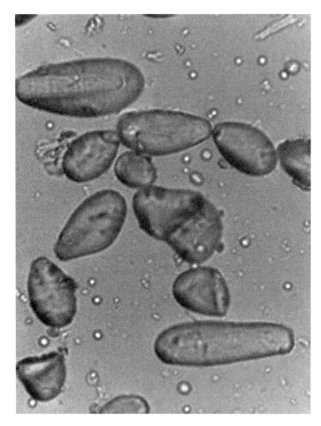

FIGURE 5.12 Starch grains from a tuber of *Dioscorea cymosula*, a wild species from Panama. Like grains from all wild yam species studied, they are distinct in morphology when compared with *D. trifida*, and also highly irregular and variable within a single tuber.

Discussion and Future Directions of Research

Manioc was an extremely important crop in the lowland tropical forest during pre-Columbian times. Our studies provide the earliest direct evidence for its cultivation in the Americas, and support hypotheses formulated many years ago (Sauer 1952; Lathrap 1970) that the lowland Neotropical forest was an early and independent center of food production in the New World. Moreover, the Aguadulce starch grain record establishes what appears to have been a fairly routine exploitation of yams in addition to the cultivation of maize and arrowroot during the late preceramic period in southern Central America, and suggests that *Dioscorea trifida* was introduced into the Isthmian region from northern South America by ca. 5000 BP. Edge-ground cobbles and their associated boulder milling-stone bases, an early and important plant-processing stone tool complex of tropical forest peoples from Ecuador to Panama (Ranere 1992), were apparently used to turn a variety of root and seed crops into foods. Because a range of as yet unidentified types of starch grains has also been recovered from these tools, it is likely that these tools were also used to process nonfood plants, for example, those with medicinal value or from which poisons of various kinds used to stun game were extracted.

Our studies of starch grains in wild *Manihot* species from outside Panama suggest that considerable potential exists for the identification of manioc in other regions of the Neotropics, and importantly, it appears that it will be possible to distinguish manioc starches from those of its wild progenitor, *M. e.* ssp. *flabellifolia*. Nevertheless, it needs to be remembered that *Manihot* is a large genus and that regional reference collections must be established, particularly in its centers of diversity, before archaeological identifications are attempted. Manioc was, in all probability, a cultivar of southern Amazonia (Olsen and Schaal 1999; Chapter 9, this volume), and thus we should expect early evidence for the crop in South America. Indications of this nature are starting to come to light, as manioc-type starches have been recovered from grinding stones of preceramic age (ca. 8000 BP to 6000 BP) from sites in the Middle Porce Valley of Colombia (Juan-Tresseras 1999).

Although this study focused on grinding-stone residues, it should be emphasized that chipped stone tools hold considerable promise of yielding equally valuable results. Loy et al. (1992), for example, demonstrated that starch grains can survive for very long periods of time on flaked stone artifacts in the Old World tropics. Analysis of chipped tool residues could prove especially useful in identifying varieties of manioc that did not require much processing before consumption. Flaked stone knives and shells, still routinely used to peel and scrape skins of manioc and other roots today, might be particularly informative in terms of yielding evidence of root processing in archaeological contexts. And finally, the importance of Linda Perry's studies should be underscored, as they showed how "grater flakes" from the Orinoco Basin (which were associated with ceramic griddles and which have long been assumed to be indirect evidence for the cultivation of bitter varieties of manioc), while yielding several hundred starch grains from a wide range of plants, including maize, arrowroot, and yams, *did not produce any from manioc* (Perry 2001, 2002). Maize starches were among the most common types present. This, of course, is not to say that no prehistoric period ceramic griddles or grater chips were used for preparing manioc, but in the two sites from the Middle Orinoco region studied by Perry, other factors appear to have been at work. Now that starch grain analysis has put into very clear focus how ethnographic analogies and indirect forms of evidence might be misleading us about prehistoric plant use, it may provide us with that long hoped-for avenue for empirically documenting the history of the root-crop complex of tropical America.

Acknowledgments

I thank Irene Holst for her considerable insight into starch grain analysis, and for her excellent technical support. Carmen Galdames, botanist at the Smithsonian Tropical Research Institute (STRI), collected and identified many of the wild plants in the modern starch grain reference collection. Special thanks to Ira Rubinoff, director of the STRI, who unfailingly supported the author's research through many years. This research was supported by the STRI, a Scholarly Studies Grant from the Smithsonian Institution, and a grant to the STRI from the Andrew W. Mellon Foundation.

References

Bennett, C. 1962. The Bayano Cuna Indians, Panama: An ecological study of livelihood and diet. *Annual Association of American Geography* 52: 32–50.

Bozarth, S. R. 1993. Maize (*Zea mays*) cob phytoliths from a central Kansas Great Bend Aspect archaeological site. *Plains Anthropologist* 38: 279–286.

Bush, M. B., D. R. Piperno, P. A. Colinvaux, P. E. DeOliveira, L. A. Krissek, M. C. Miller, and W. E. Rowe. 1992. A 14,300-year paleoecological profile of a lowland tropical lake in Panama. *Ecology* 62: 251–275.

Callen, E .O. 1967. Analysis of the Tehuacán coprolites. In *The prehistory of the Tehuacán valley, environment, and subsistence*, Volume 1, D. S. Byers (ed.), pp. 261–289. Austin: University of Texas Press.

Carneiro, R. L. 1983. The cultivation of Manioc among the Kuikuru of the upper Xingú. In *Adaptive responses of native Amazonians*, R. B. Hames and W. T. Vickers (eds.), pp. 65–108. New York: Academic Press.

Cooke, R. G. 1995. Monagrillo, Panama's first pottery: Summary of research, with new interpretations. In *The emergence of pottery: Technology and innovation in ancient societies*, W. K. Barnett and J. W. Hoopes (eds.), pp. 169–184. Washington, D.C.: Smithsonian Institution Press.

———. 1998. Human settlement of Central America and northern South America. *Quaternary International* 49/50: 177–190.

Cooke, R. G. and A. J. Ranere. 1984. The "Proyecto Santa Maria": A multidisciplinary analysis of prehistoric adaptations to a tropical watershed in Panama. In *Recent developments in isthmian archaeology*, F. W. Lange (ed.), pp. 3–30. British Archaeological Reports International Series 212. Oxford: BAR.

Cooke. R. G. and A. J. Ranere. 1992a. Prehistoric human adaptations to the seasonally dry forests of Panama. *World Archaeology* 24: 114–133.

Cooke, R. G. and A. J. Ranere. 1992b. The origins of wealth and hierarchy in the central region of Panama, with observations on its relevance to the phylogeny of Chibchan-speaking polities in Panama and elsewhere. In *Wealth and hierarchy in the intermediate area*, F. W. Lange (ed.), pp 243–316. Washington, D.C.: Dumbarton Oaks.

Correa, M. D. A., C. Caldames, and M. Stapf (eds.). 2004. *Catálogo de las Plantas Vasculares de Panamá*. Bogotá, Colombia: Editora Nove Art.

Cortella, A. R. and M. L. Pochettino. 1994. Starch grain analysis as a microscopic diagnostic feature in the identification of plant material. *Economic Botany* 48: 171–181.

D'Arcy, W. G. 1987. *Flora of Panama: Index and checklist*. St. Louis: Missouri Botanical Garden.

DeBoer, W. 1975. The archaeological evidence for manioc cultivation: A cautionary note. *American Antiquity* 40: 419–433.

Dorweiler, J. E. and J. Doebley. 1997. Developmental analysis of *Teosinte Glume Architecture 1*: A key locus in the evolution of maize (Poaceae). *American Journal of Botany* 84: 1313–1322.

Duke, J. 1968. *Darien ethnobotanical dictionary*. Columbus, OH: Battelle Memorial Institute.

———. 1975. Ethnobotanical considerations on the Cuna Indians. *Economic Botany* 29: 278–293.

Harlan, J. R. 1992. *Crops and man*, 2nd edition. Madison, WI: American Society of Agronomy and Crop Science Society of America.

Haslam, M. 2003. Evidence for maize processing on 2000-year-old obsidian artifacts from Copán, Honduras. In *Phytolith and starch research in the Australian-Pacific-Asian regions: The state of the art*, D. M. Hart and L. A. Wallis (eds.), pp. 153–161. Canberra, Australia: Pandanus Books, The Australian National University.

———. 2004. The deposition of starch grains in soils: Implications for archaeological residue analyses. *Journal of Archaeological Science*, 31: 1715–1734.

Hazlett, D. L. 1986. Ethnobotanical observations from Cabecar and Guaymi settlements in Central America. *Economic Botany* 40: 339–352.

Holst, I. and D. R. Piperno, in preparation. *A starch grain identification manual for tropical plants*. Lanham, MD. AltaMira Press.

Iriarte, J., I. Holst, O. Marozzi, C. Listopad, E. Alonso, A. Rinderknecht, and J. Montaña. 2004. Evidence for cultivar adoption and emerging complexity during the Mid-Holocene in the La Plata Basin, Uruguay. *Nature* 432: 614–617.

Jennings, D. L. 1995. Cassava: Manihot esculenta (Euphorbiaceae). In *Evolution of crop plants*, J. Smartt and N. W. Simmonds (eds.), pp. 128–132. New York: Wiley.

Juan-Tresseras, J. 1999. Análisis arqueobotánico (fitolotos y almidones) del yacimiento de "El Morro" Valle Medio del Porce, Dpto. Antioquia, Colombia. Unpublished manuscript in possession of the author.

Kelly, E. F., R. G. Amundson, B. D. Marino, and M. J. Deniro. 1991. Stable isotope ratios of carbon in phytoliths as a quantitative method of monitoring vegetation and climate change. *Quaternary Research* 35: 222–233.

Lathrap, D. W. 1970. *The Upper Amazon*. New York: Praeger.

Loy, T. H. 1994. Methods in the analysis of starch residues on stone tools. In *Tropical archaeobotany: Applications and new developments*, J. G. Hather and J. G. London (eds.), pp. 86–114. London: Routledge.

Loy, T. H., S. Wickler, and M. Spriggs. 1992. Direct evidence for human use of plants 28,000 years ago: Starch residues on stone artifacts from the northern Solomon islands. *Antiquity* 66: 898–912.

Mulholland, S. C. 1993 A test of phytolith analysis at Big Hidatsa, North Dakota. In *Current research in phytolith analysis: Applications in archaeology and paleoecology*. D. M. Pearsall and D. R. Piperno (eds.), pp. 131–145. Philadelphia: MASCA, The University Museum of Archaeology and Anthropology.

Mulholland, S. C. and C. Prior. 1993. AMS radiocarbon dating of phytoliths. In *Current research in phytolith analysis: Applications in archaeology and paleoecology*. D. M. Pearsall and D. R. Piperno (eds.), pp. 21–23. Philadelphia: MASCA, The University Museum of Archaeology and Anthropology.

Olsen, K. M. and B. A. Schaal. 1999 Evidence on the origin of cassava: Phylogeography of *Manihot esculenta*. *Proceedings of the National Academy of Science, USA*. 96: 5586–5591.

Pearsall, D. M. 2000. *Paleoethnobotany: A handbook of procedures*. 2nd edition. San Diego: Academic Press.

Pearsall, D. M., K. Chandler-Ezell, and A. Chandler-Ezell. 2003. Identifying maize in Neotropical sediments and soils using cob phytoliths. *Journal of Archaeological Science* 30: 611–627.

Pearsall, D. M., K. Chandler-Ezell, and J. A. Zeidler. 2004. Maize in ancient Ecuador: Results of residue analysis of stone tools from the Real Alto site. *Journal of Archaeological Science* 31: 423–432.

Perry, L. 2001. *Prehispanic subsistence in the Middle Orinoco Basin, Venezuela: Starch analyses yield new evidence*. Unpublished PhD dissertation, Southern Illinois University, Carbondale. Ann Arbor: University of Michigan Microfilms.

———. 2002. Starch analyses reveal multiple functions of quartz "manioc" grater flakes from the Orinoco basin, Venezuela. *Intercienca* 27: 635–639.

Pickersgill, B. and C. B. Heiser Jr. 1977. Origins and distribution of plants domesticated in the New World tropics. In *Origins of agriculture*, C. A. Reed (ed.), pp. 803–835. The Hague: Mouton Publishers.

Piperno, D. R. 1995. Plant microfossils and their application in the New World tropics. In *Archaeology in the lowland American tropics: Current analytical methods and recent applications*, P. W. Stahl (ed.), pp. 130–153. Cambridge: Cambridge University Press.

———. 1998. Paleoethnobotany in the Neotropics through microfossils: New insights into ancient plant use and agricultural origins in the tropical forest. *Journal of World Prehistory* 12: 393–449.

———. 2006a. A behavioral ecological approach to the origins of plant cultivation and domestication in the seasonal tropical forests of the New World. In *Foraging theory and the transition to agriculture*, D. Kennett and B. Winterhalder (eds.), pp. 137–166. Berkeley: University of California Press.

———. 2006b. *Phytoliths: A comprehensive guide for archaeologists and paleoecologists*. Lanham Md: AltaMira Press.

Piperno, D. R. and K. Flannery. 2001. The earliest archaeological maize (*Zea mays* L.) from highland Mexico: New AMS dates and their implications. *Proceedings of the National Academy of Sciences, USA* 98: 2101–2103.

Piperno, D. R. and I. Holst. 1998. The presence of starch grains on prehistoric stone tools from the lowland Neotropics: Indications of early tuber use and agriculture in Panama. *Journal of Archaeological Science* 25: 765–776.

Piperno, D. R. and I. Holst. 2004. Crop domestication in the American tropics: Starch grain analysis. In *Encyclopedia of plant and crop science*, R. M. Goodman (ed.), pp. 330–332. New York: Marcel Dekker.

Piperno, D. R. and J. Jones, 2003. Paleoecological and archaeological implications of a late Pleistocene/early Holocene record of vegetation and climate from the Pacific coastal plain of Panama. *Quaternary Research*: 59: 79–87.

Piperno, D. R. and D. M. Pearsall. 1993. Phytoliths in the reproductive structures of maize and teosinte: Implications for the study of maize evolution. *Journal of Archaeological Science* 20: 337–362.

Piperno, D. R. and D. M. Pearsall. 1998. *The origins of agriculture in the lowland Neotropics*. San Diego: Academic Press.

Piperno, D. R., M. B. Bush, and P. A. Colinvaux. 1991a. Paleoecological perspectives on human adaptation in central Panama. I. The Pleistocene. *Geoarchaeology* 6: 210–226.

Piperno, D. R., M. B. Bush, and P. A. Colinvaux. 1991b. Paleoecological perspectives on human adaptation in central Panama. II. The Holocene. *Geoarchaeology* 6: 227–250.

Piperno, D. R., I. Holst, T. C. Andres, and K. E. Stothert. 2000a. Phytoliths in Cucurbita and other Neotropical Cucurbitaceae and their occurrence in early archaeological sites from the lowland American tropics. *Journal of Archaeological Science* 27: 193–208.

Piperno, D. R., A. J. Ranere, I. Holst, and P. Hansell. 2000b. Starch grains reveal early root crop horticulture in the Panamanian tropical forest. *Nature* 407: 894–897.

Piperno, D. R., I. Holst, A. J. Ranere, P. Hansell, and K. E. Stothert. 2001. The occurrence of genetically-controlled phytoliths from maize cobs and starch grains from maize kernels on archaeological stone tools and human teeth, and in archaeological sediments from southern Central America and northern South America. *The Phytolitharien*. 13: 1–7.

Pohl, M. D., K. O. Pope, J. G. Jones, J. S. Jacob, D. R. Piperno, S. de France, D. L. Lentz, J. A. Gifford, F. Valdez Jr., M. E. Danforth, and J. K. Josserand. 1996. Early agriculture in the Maya lowlands. *Latin American Antiquity* 7: 355–372.

Pope, K. O., M. E. D. Pohl, J. G. Jones, D. L. Lentz, C. von Nagy, F. J. Vega, and I. R. Quitmyer. 2001. Origin and environmental setting of ancient agriculture in the lowlands of Mesoamerica. *Science* 292: 1370–1373.

Ranere, A. J. 1980. Stone tools and their interpretation. In *Adaptive radiations in prehistoric Panama*, O. F. Linares and A. J. Ranere (eds.), pp. 118–138. Cambridge, MA: Peabody Museum of Archaeology and Ethnology, Harvard University.

———. 1992. Implements of change in the Holocene environment of Panama. In *Archaeology and environment in Latin America*, O. R. Ortiz-Troncoso and T. Van der Hammen (eds.), pp. 25–44. Amsterdam: Universiteit van Amsterdam.

Ranere, A. J. and R. Cooke. 1996. Stone tools and cultural boundaries in prehistoric Panama: An initial assessment. In *Paths to Central American prehistory*, F. W. Lange (ed.), pp. 49–77. Niwot, Col.: University Press of Colorado.

Ranere, A. J. and R. G. Cooke. 2003. Late glacial and early Holocene occupation of Central American tropical forests. In *Under the canopy: The archaeology of tropical rain forests*, J. Mercader (ed.), pp. 211–248. New Brunswick, N.J.: Rutgers University Press.

Ranere, A. J. and P. Hansell. 1978. Early subsistence patterns along the Pacific coast of central Panama. In *Prehistoric Coastal Adaptations*, B. L. Stark and B. Voorhies (eds.), pp. 43–59. New York: Academic Press.

Ranere, A. J. and P. Hansell. 2000. The Aguadulce shelter revisited: Early and middle Holocene occupation in the llanos of the Gran Cocle. Paper read at the 65th annual meeting of the Society for American Archaeology, April 2002, New Orleans.

Reichert, E. T. 1913. *The differentiation and specificity of starches in relation to genera, species, etc*. Washington, D.C.: Carnegie Institution of Washington.

Rogers, D. J. 1963. Studies of *Manihot esculenta* Crantz and related species. *Economic Botany* 90: 43–54.

———. 1965. Some botanical and ethnological considerations of *Manihot esculenta*. *Economic Botany* 19: 369–337.

Rogers, D. J. and S. G Appan. 1973. *Flora Neotropica*. Monograph no. 13, *Manihot Manihotoides*. New York: Hafner Press.

Roosevelt, A. C. 1980. *Parmana: Prehistoric maize and manioc subsistence along the Amazon and Orinoco*. New York: Academic Press.

Sauer, C. O. 1950. Cultivated plants of South and Central America. In *Handbook of South American Indians*, J. Steward (ed.), pp. 487–543. Bureau of American Ethnology Bulletin, no. 143, vol. 6, Washington, D.C.: U.S. Government Printing Office.

———. 1952. *Agricultural origins and dispersals*. New York: American Geographical Society.

Seidemann, J. 1966. *Stärke-atlas*. Berlin: Paul Parey.

Shannon, J. C. and Garwood, D. L. 1984. Genetics and physiology of starch development. In *Starch: Chemistry and technology*, R. L. Whistler, J. N. Bemiller, and E. F. Paschall (eds.), pp. 26–85. Orlando: Academic.

Sivak, M. N. and Preiss, J. 1998. *Starch: Basic science to biotechnology*. San Diego: Academic Press.

Sturtevant, W. C. 1969. History and ethnography of some West Indian starches. In *The domestication and exploitation of plants and animals*, P. J. Ucko and G. W. Dimbleby (eds.), pp. 177–199. Chicago: Aldine.

Therin, M., R. Fullagar, and R. Torrence. 1999. Starch in sediments: A new approach to the study of subsistence and land use in Papua New Guinea. In *The prehistory of food*, C. Gosden and J. Hather (eds.), pp. 438–462. London: Routledge.

Towle, M. A. 1961. *The ethnobotany of Pre-Columbian Peru*. Chicago Viking Fund Publications in Anthropology, no. 30. Chicago: Viking.

Ugent, D. and M. Verdun. 1982. Starch grains of the wild and cultivated Mexican species of *Solanum*, subsection potato. *Phytologia* 53: 351–363.

Ugent, D., S. Pozorski, and T. Pozorski. 1981. Prehistoric remains from the sweet potato from the Casma Valley of Peru. *Phytologia* 49: 401–415.

Ugent, D., S. Pozorski, and T. Pozorski. 1982. Archaeological potato tuber remains from the Casma Valley of Peru. *Economic Botany* 36: 182–192.

Ugent, D., S. Pozorski, and T. Pozorski. 1984. New evidence for ancient civilization of Canna edulis in Peru. *Economic Botany* 38: 417–432.

Ugent, D., S. Pozorski, and T. Pozorski. 1986. Archaeological manioc (Manihot) from coastal Peru. *Economic Botany* 40: 78–102.

Ugent, D., T. Dillehay, and C. Ramire. 1987. Potato remains from a Late Pleistocene settlement in south central Chile. *Economic Botany* 41: 17–27.

Whistler, R. L., J. N. Bemiller, and E. F. Paschall. 1984. *Starch: Chemistry and technology*. Orlando, FL: Academic Press.

Wilding, L. P. 1967. Radiocarbon dating of biogenetic opal. *Science* 156: 66–67.

Zeeman, S. C., S. M. Smith, and A. M. Smith. 2004. The breakdown of starch in leaves. *New Phytologist* 163: 1469–1537.

CHAPTER 6

Phytolith Evidence for the Early Presence of Domesticated Banana *(Musa)* in Africa

CH. MBIDA, E. DE LANGHE, L. VRYDAGHS, H. DOUTRELEPONT,
RO. SWENNEN, W. VAN NEER, AND P. DE MARET

Introduction: The African Musaceae

The banana was domesticated from two wild Southeast Asian progenitor species: *Musa acuminata* and *M. balbisiana* (Simmonds and Shepard 1955). The process of banana domestication was complex, involving intra- and interspecies hybridization, polyploidization, seed sterility, parthenocarpy, and somatic mutations. Over time, humans spread this important domesticated crop plant throughout the tropics of Asia, Africa, and the Americas.

In Africa, four categories of bananas are found today: those of recent introduction (RI), the Indian Ocean Complex (IOC), the Eastern African AAA highland group (EA-AAA), and the African Plantain group (AP-AAB) (De Langhe et al. 1996). These four banana groups likely reached Africa in successive waves. The RI group, about 10 cultivars grown today near towns and townships for their quality as a dessert fruit, was introduced during postcolonial times. The Indian Ocean Complex is mainly distributed along the coastal areas of the Indian Ocean (Madagascar, East Africa, South India, Sri Lanka, Malaysia, and western Indonesia). In Africa, the IOC includes an almost complete spectrum of banana genomes (edible-AA, AB, AAA, AAB, ABB), but only a limited number of cultivars have been morphologically identified for each of these genome types (Simmonds 1959; De Langhe et al. 1996). They have never become a staple food in Africa. Representatives of the hardy AB and ABB genomes of the IOC have diffused inland in Africa in recent times. The East African AAA Highland group is made up of 60 cultivars (Karamura 1999; Sebasigari 1990) that have not been found elsewhere in the world. They are grown as a basic source of food and for beer-making, and dominate the landscape throughout an area stretching from Uganda southward to Mozambique and southern Africa. Other genomic combinations (AB, ABB, and AAB) are rarely represented in this group (De Langhe et al. 1996). In the rainforest zones of East Africa, the African Plantains (a subgroup of the AAB genome) are a staple food and play a dominant role in village economies. Cultivar diversity of plantains in this area is both substantial (Swennen 1990; Swennen and Vuylsteke 1987) and clearly of African origin (De Langhe et al. 1996; De Langhe and de Maret 1999).

Because the banana does not produce durable seed, pollen, or wood, the initial introduction of this important crop plant into Africa has been difficult to document archaeologically. As a result, theories regarding the likely timing of the arrival and diffusion of banana complexes in Africa were, until 1996, based solely on linguistic, ethnobotanical, and genetic information. In 1990, for example, Vansina (1990: 64) proposed that the banana did not arrive in East Africa before about AD 1–200, and that its diffusion throughout Equatorial Africa was complete by around AD 500. Based on an exhaustive comparative cultural lexicon study, however, Berchem (1989–1990) concluded that the banana was probably an earlier introduction. He argued that the Malagasy were not the first to introduce the banana into Africa, and that the Bantu must have known the plant before they came into contact with the Malagasy. In a broader biological context, De Langhe and collaborators considered the present-day geographical distribution of banana complexes in Africa along with estimated mutation rates, and proposed that the plantain had reached the East African coast and was introduced into the forest regions of East Africa by 3000 BP (De Langhe et al. 1996).

From the 1990s on, it was recognized that direct evidence for the initial introduction and subsequent diffusion of the banana in Africa could come from archaeological records. A range of plant species, including bananas (Tomlinson 1959), produce very small phytoliths of diagnostic value. Phytoliths are hard, opal silica bodies that are often well preserved in the archaeological record. Thus, banana phytoliths could potentially provide a clear record of the early history of this crop plant in Africa and other tropical areas of the world. In 1985 banana phytoliths were identified in early sediments at Kuk, a Papua New Guinean site in the Upper Wahgi Valley near Mt. Hagen (Wilson 1985). Since Wilson's landmark study, research on modern reference collections has further documented the distinctive morphology of banana phytoliths (Piperno 1988; Runge 1996, 1999; Runge and Runge 1997; Pearsall 1998; Kealhofer and Piperno 1998). In recent years, banana phytoliths have been reported in a steadily increasing number of archaeological sites in Africa (Doutrelepont et al. 1996; Leju et al. 2006), Papua New Guinea (Bowdery 1999; Lentfer 2001; Denham et al. 2003), Laos, and Peninsular Malaysia (Bowdery 1999), Easter Island (Scott-Cummings 1998; Vrydaghs et al. 2004), and Pakistan (Fuller and Madella 2001; Madella 2003). Phytolith analysis thus offers a way of obtaining direct evidence for the early history of the banana in Africa, and testing the theory that the banana supported the Bantu expansion and the substantial increase in village density in the forests of East Africa around 3000 BP.

An Archaeobotanical Case Study: Nkang, South Cameroon

Archaeological and Paleo-environmental Context

Archaeological investigations along the Atlantic front of central Africa identify three periods: the Late Stone Age (before 4000 BP), the Stone to Metal Period (4000–2000 BP), and the Iron Age (2000–1800 BP) (de Maret 1996; Oslisly and White 2000). The Late Stone Age is characterized by microlithic tools recovered over a broad area reaching from the Lope (Gabon) to Monte Alen (Equatorial Guinea) and Nguti and Shum Laka (Cameroon). During the Stone to Metal Period, occupation sites characterized by large storage pits (ca. 15 cubic meters) and abundant polished stone tools and ceramics are typically located on the top of small hills and mounts close to rivers. This period corresponds more or less with a drier climatic phase (Kibangian B) recorded for all of central Africa (Maley 1992, 1993; Schwartz 1992). Representative of this period are the sites of Obogogo, Nkometou, Ndindan, and Okolo in Cameroon (Atangana 1988; Claes 1985; Mbida 1992), and Rivière Denis, Okala, and Lope in Gabon (Clist 1990; Oslisly and White 2000; de Maret 1992, 1996). Subsequent large Iron Age settlements are also situated on hilltop settings, typically encircled by large refuse pits spaced five to eight meters apart. Archaeological assemblages of Iron Age settlements typically include abundant ceramics, polished stone tools, iron slag, charcoal, and charred nuts. The sites of Obogogo, Okolo, and Oliga in Cameroon have occupational components representative of this period (Elouga 2000; Oslisly et al. 2001). At the Okolo site, a posthole pattern indicating a two-roomed rectangular structure was documented in association with several pits (Atangana 1988; de Maret 1992). Along the coast of Gabon, a similar but slightly younger sequence has been documented, with Stone to Metal Period occupations dating from 3500 to 2500 BP and Iron Age settlements from about 2500 to 1400 BP (Clist 1990; Oslisly and White 2000). The temporal placement of these sites is based on radiocarbon dates obtained from charcoal recovered from refuse pits. Although a number of plant species have been identified at Iron Age sites in southern Cameroon and Gabon, including oil palm (*Elais guineensis*), aiele (*Canarium schweifurthii*) and angongui or anzabili (*Antocaryon klaineanum*) and *Coula edulis* (Clist 1990; de Maret 1983; Mbida 1996; Oslisly and White 2000), until recently no clear evidence of cultivated crops or domestic animals had been recovered from settlements along the Atlantic coast of central Africa. The first such evidence for cultivation of crop plants was recovered from the site of Nkang in southern Cameroon.

Analysis of phytolith assemblages recovered from a series of refuse pits at the Nkang site offered an excellent opportunity to look for direct evidence of the early presence of domesticated banana in a location more than 1,000 km to the west of where *Musa* is thought to have first arrived in Africa. Located at the present-day settlement of Nkang, in south

FIGURE 6.1 Location of the Nkang site in central Cameroon (from Figure 1, Mbida et al. 2001).

Cameroon, the archaeological site of Nkang can be confidently placed within the larger regional context of cultural development in the region. It is the most northern of a group of large Stone to Metal Period settlements recorded in the Yaounde area. The present-day rural settlement of Nkang (11°19′E; 4°16′N) is located about 10 km east of Monatelé, the administrative town of the Lekié division, and 70 km to the northwest of Yaounde, the capital of Cameroon (Figure 6.1).

Average annual precipitation in this degraded rainforest region is high (1,360 mm), occurring mostly during March to June and September to October. The present-day inhabitants of the area, the Eton, are farmers who grow cacao, plantain, groundnut, cassava, and maize, and raise poultry, sheep, and goats. Hunting, fishing, and collecting are secondary activities.

Excavations at Nkang exposed 11 pits spaced at about eight-meter intervals and apparently unrelated to any other archaeological feature. These pits were bottle-like, ovoid or cylindrical in cross-section (Figure 6.2), and their volume ranged from 3–15 cubic meters. When the primary functions of these pits were abandoned, they served as refuse pits and were filled with local soil, broken ceramics and stone tools, iron slag, charcoal, and faunal remains. Excavation of these secondary refuse pits involved bisection, with both halves of the pit fill then removed in strata ranging in thickness between 5 and 50 cm, depending on the thickness of the depositional layers. All visible ceramic, lithic, and faunal materials were retrieved. Charcoal fragments larger than two millimeters were systematically recovered by hand, and soil samples were taken from each depositional layer for laboratory analysis.

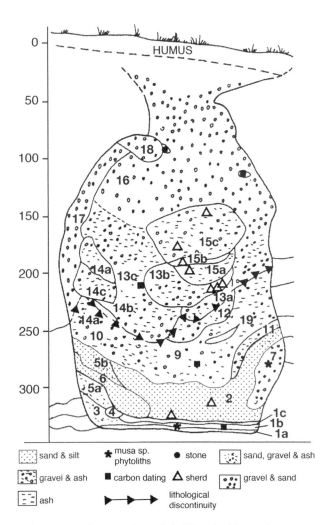

FIGURE 6.2 Cross-section of pit F9 at the Nkang site, which yielded *Musa* phytoliths (from Figure 2, Mbida et al. 2001).

TABLE 6.1
Radiocarbon Dates Recovered from Excavated Pits at the Nkang Site

Structure	Depth (cm)	Lab. Code	C-14 Dates (years BP)	cal-BC 2 Sigma
F 6	?	Lv-1940	2580 ± 110	850–410
F 9	300–350	Lv-1944	2490 ± 110	840–370
F 9	250–300	Lv-1943	2490 ± 80	790–400
F 9	200–250	Lv-1942	2400 ± 60	770–350
F 3	?	Lv-1939	2420 ± 70	770–350
F 7 bis	?	Lv-1941	2340 ± 70	800–150
F 13	140–170	Lv-1945	2310 ± 90	800–100
F 14	190–210	Lv-1943	2170 ± 80	390–AD 1

SOURCE: From Table 1, Mbida et al. 2000.

Ceramic assemblages recovered from the excavated pits included spherical, ellipsoid, and ovaloid vessel forms, often with collars or necks, similar to those recovered from other sites of the Obogogo group: Obogogo (Claes 1985, 1992); Nidindan (Mbida 1992); and Okolo (Atangana 1988). Lithics recovered from the Nkang refuse pits consisted mainly of grinding stones, hammerstones, whetstones, and polished axes. About one kilogram of iron slag was also recovered from two pits (F7NF, F7bis), indicating that ironworking was also carried out at the site.

The chronology of the Nkang site is based on a series of eight radiocarbon dates on charcoal samples (Lv-1939–1946) recovered from six of the excavated pits. Three of these samples (Lv-1942, 1943, 1944) recovered from pit F9 yielded dates between 2490 and 2400 radiocarbon years BP (calibrated 840–350 BC), with a 95 percent level of confidence inter-date overlap (Mbida 1996: 638–639; Table 6.1), placing the Nkang site in the Stone to Metal Age period (de Maret 1996).

Reconstruction of the environmental context of the Nkang site 2,500 years ago is based on archaeobotanical and archaeofaunal data. Identification of the archaeobotanical material (charcoal, seeds, and phytoliths) was carried out with the aid of the modern references collections of the Laboratory of Wood Biology and Xylarium (Royal Museum of Central Africa, Belgium). In addition, several monographs and articles with illustrated material and identification keys were consulted (Metcalfe and Chalk 1950, 1983; Metcalfe 1971; Twiss et al. 1969; Piperno 1988, 1989; Rapp and Mulholland 1992; Runge 1996, 1999; Vrydaghs 2003). The faunal remains were identified through comparison with the reference collection of the Royal Museum of Central Africa (Belgium) and facilitated by the use of literature on the postcranial osteomorphology and osteomorphometry of African mammals (Walker 1985; Peters 1988, 1989; Van Neer 1989).

Species present in the wood charcoal assemblages of the Nkang refuse pits represented a number of different forest facies, including semi-evergreen guineo-congolean rainforest with mean rainfall above 1,200 mm per year, as well as pioneer and old secondary forest. Some swamp or riparian wood species were also identified (Table 6.2). The taxonomic diversity of wood species varied from pit to pit, with F14 and F9 having high species counts, and the other six pits containing only a limited number of taxa. Phytoliths deriving from Palmaceae and Poaceae and identifiable as Panicoid grasses (Twiss et al. 1969) are mostly indicative of open spaces, while Zingiberaceae phytoliths reflect secondary forest species. Phytolith evidence for a wet environment (Cyperaceae) was also recovered. Opal grains derived from woody tissues were present (up to 2 percent) (Bullock et al. 1985). Such a low level of abundance is typical for a forested environment (Twiss et al. 1969; Runge 1999).

Although less informative than the plant remains, faunal data recovered from the Nkang pits indicated species typical of both closed forest canopy and open environments. Some of the species of wooded habitats include land snails (*Achatina*), forest buffalo (*Syncerus caffer nanus*), forest duikers (*Cephalophus* sp.) and bushbuck (*Tragelaphus scriptus*), while

TABLE 6.2
Plant Taxa Represented in Charcoal Recovered from the Nkang Site

Plant Taxa	F1	F3	F6	F7	F7 NF	F9	F13	F14
Anacardiaceae								
Antrocaryon micraster	—	—	—	—	—	x	—	—
Lannea antiscorbutica	—	—	—	—	—	x	—	—
L. welwitschii	—	—	—	—	—	—	—	x
Apocynaceae								
Strophantus intermedius	—	—	—	—	—	—	—	x
Palmae (Arecaceae)								
Elaeis guineensis	x	x	x	x	x	x	x	x
Burseraceae								
Canarium schweihnfurthii	x	x	x	x	x	x	x	x
Euphorbiaceae								
Anthrostemma senegalensis	—	—	—	—	—	—	—	x
Spondianthus preussii	—	—	—	—	—	—	—	x
Uapaca sp.	—	—	—	—	—	—	—	—
Flacourtaceae								
Caloncoba welwitschii	—	—	—	—	—	—	—	x
Meliaceae								
Carapa procera	—	—	—	—	—	—	—	x
Entandrophragma sp.	—	—	—	—	—	x	—	—
Trichilia prieuriana	—	—	—	—	—	—	—	x
Mimosaceae								
Albizia ealaensis	—	—	—	—	—	x	—	—
A. ferruginea	—	—	—	—	—	—	—	x
Newtonia sp.	—	—	—	—	—	x	—	—
Piptadeniastrum africanum	—	—	—	—	—	x	—	—
Ochnaceae								
Ochna multiflora	—	—	x	—	—	—	—	—
Sapindaceae								
Chytranthus macrobotrys	—	—	—	—	—	—	—	x
Sapotaceae								
Chrysopltyllum pruniformis	—	—	—	—	—	—	—	x

SOURCE: From Table 2, Mbida et al. 2000.

waterbuck *(Kobus ellipsiprymnus)* inhabits woodlands and clearings close to water (Table 6.3).

In summary, paleo-environmental markers are dominated by species typical of forests, although indicators of open habitats are also present. The overall assemblage pattern reflects a mix of primary, pioneer, and old secondary forests. The identification of two species of catfish, as well as the presence of riparian species in the charcoal assemblage, indicates the presence of nearby streams or rivers. Other species represented in the charcoal assemblage provide indications of nearby habitat settings. *Canarium*, for example, is typical of old secondary forest, while *Chytranthus* indicates waterlogged sites. To produce fruit, Elais requires sunny, wet, and warm places (i.e., clearings). Faunal remains representing species typical of savanna settings were also identified. Hence, a forested environment made up of different facies including clearings within walking distances from the site is suggested. The various landscapes reflected by the identified botanical and faunal remains are not inconsistent with a regional-scale climatic episode. A central African climatic downturn (climatic phase Kibangian B) documented as occurring at about 2700–3000 BP (Maley 1992, 1993; Schwartz 1992) is particularly well documented in southern Cameroon at Lake Ossa (Reynaud-Ferrera et al. 1996) and Barombi Mbo (Giresse et al. 1994). This event is likely related to a change in the sea surface temperature upwellings of the Guinean Gulf (Maley

TABLE 6.3
Animal Taxa Represented at the Nkang Site

Animal Taxa	F1	F3	F5	F6	F7	F7 Bis	F7 NF	F9	F13	Total
Freshwater gastrapods										
Lanistes libycus	—	—	—	2	—	—	—	—	—	2
Potadoma cf. *freethii*	—	—	—	1	—	—	1	3	—	5
Terrestrial gastropods										
Achatina sp.	1	—	—	—	2	—	—	7	—	10
Limicolaria sp.	—	—	—	—	—	2	1	2	—	5
Marine gastropod										
Trochidae indet.	—	—	—	—	—	—	—	1	—	1
Bivalves										
Aspatharia sp.	—	—	—	1	—	—	—	2	—	3
Crustaceans										
Freshwater crab (Decapoda indet)	—	—	—	—	—	1	—	—	—	1
Fish										
Catfish 1 (*Chrysichthys* sp.)	—	—	—	—	—	1	—	—	—	1
Catfish 2 (Clariidae)	—	—	—	1	—	—	—	—	—	1
Nile perch (*Lates nilticus*)	—	—	—	—	—	1	—	—	—	1
Wild mammals										
Small rodents	—	4	—	—	—	—	—	—	3	7
Cane Rat (*Thryonomys* sp.)	—	—	—	1	1	—	—	1	—	3
Hippopotamus (*Hippopotamus amphibious*)	—	—	—	—	—	1	—	—	—	1
Bushbuck (*Tragelaphus scriptus*)	—	—	—	—	1	—	—	—	—	1
Waterbuck (*Kobus ellipsiprymnus*)	—	—	—	—	—	7	1	—	—	8
Kob (*Kobus kob*)	—	—	—	—	—	5	—	—	—	5
Medium-sized duikers (*Cephalophus* sp.)	—	—	1	2	—	7	—	—	—	10
Forest buffalo (*Syncerus caffer nanus*)	—	—	—	1	—	2	—	1	—	4
Domestic Mammals										
Goat (*Capra hircus*)	—	—	—	1	—	—	—	—	—	1
Sheep (*Ovis aries*)	—	2	—	—	—	—	—	—	—	2
Sheep or goat	—	1	—	—	—	1	—	—	—	2
Total identified	1	7	1	10	4	28	3	17	3	74
Unidentified gastropods	—	—	—	—	—	1	1	3	—	5
Unidentified mammals	—	—	1	8	—	15	—	1	—	25

SOURCE: From Table 4, Mbida et al. 2000.

et al. 2001), which would have caused a retraction and fragmentation of forested habitats. The occupational period of Nkang falls within this period, and the proposed paleoenvironmental reconstruction for Nkang conforms to such a climatic downturn.

Musaceae Phytoliths from Nkang

DIFFERENTIATING *MUSA* FROM *ENSETE* PHYTOLITHS

In Africa, the Musaceae family is represented by two genera: *Musa* and *Ensete*. Wild species of *Musa* have never been reported on the African continent, while the genus *Ensete* is represented by two endemic and widespread species: *Ensete ventricosum* and *E. giletti*. As a result, if any phytoliths recovered from pits at the Nkang site could be unequivocally assigned to the genus *Musa*, they would directly point to the introduction and cultivation of domesticated edible bananas in central Africa 2,500 years ago. Therefore, phytoliths produced in the genus *Musa* should be distinguishable from those produced by species of the genus *Ensete*.

The first description of Musaceae phytoliths, published by Tomlinson (1969), does not offer any such clear distinction,

TABLE 6.4
Modern Source Material for *Musa* and *Ensete* Phytoliths

	Accession	Name of Cultivar or Species
Musa	BS 001	"Pisang lilin" edible AA
	BS 048	"Valery" AAA–Cavendish
	BS 694	"Pisang Jambe" AAA or AAAA
	BS 754	"Corne I" AAB–Horn Plantain
	ITC 0767	"Dole" ABB–Bluggoe
	ITC 1123	"Yangambi km 5" AAA
	ITC 0346	"Giant Cavendish" AAA–Cavendish
	ITC 0002	"Dwarf Cavendish" AAA–Cavendish
Ensete	ITC 1389	*Ensete gilletii*
	ITC 1387	*E. ventricosum*

SOURCE: From Table 1, Mbida et al. 2001.

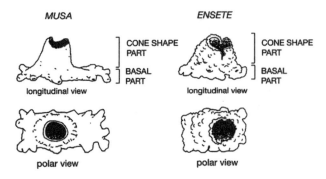

FIGURE 6.3 Morphological characteristics of *Musa* and *Ensete* phytoliths (from Figure 3, Mbida et al. 2001).

however, and is rather confusing because he uses a number of different, poorly defined terms: "silica-bodies," "trough-shaped silica bodies," and "stegmata." The trough-shaped bodies described are characterized by a shallow depression and a more-or-less rectangular form appearing in longitudinal rows adjacent to the fibrous part of the bundle sheets. Tomlinson also describes some spherical bodies associated with the trough-like bodies. In *Ensete*, he notes that the trough-shaped silica bodies are absent, while in *E. arnoldianum*, silica bodies intermediating between the stegmata and the spherical bodies were seen. Similarly, Wilson's description of the silica bodies of *Musa* and *Ensete* is unclear and does not provide sufficient detail for differentiation between the two genera. While the reference collection includes some *Ensete* material (*E. glaucum* (Roxb.) Cheesman syn. *Musa calosperma* F. Muell) (Simmonds 1962), Wilson (1985) does not provide any distinctive morphological criteria for distinguishing between the two taxa.

In general, phytolith keys for plants of the New World tropics such as those provided by Piperno (1988: 65) and the University of Missouri online phytolith database (Pearsall 1998) include descriptions of the *Heliconia* (class V.D.1, Piperno 1988: 65) and *Musa* phytolith (class V.D.2, Piperno 1988: 65). The *Heliconia* phytolith is described as a small solid plug with a deep trough located in the center. *Musa* phytoliths are described as possessing a shallow trough that often is not centrally located and often is positioned laterally rather than on the elevated part of the phytolith. For both, a size range of 5–20 microns is given. No phytolith descriptions are presented for *Ensete*. Similarly, no description of *Ensete* phytoliths is included in the South Asian collection described by Kealhofer and Piperno (1998) or in the African collection studied by Runge (1996, 1999). The limited nature of the morphological descriptions provided for Musaceae phytoliths, as well as the absence of robust qualitative (shape, surface ornamentation) and quantitative (size, frequency) information, underscores the need for a comprehensive and focused comparative study designed to document morphological differences in the phytolith assemblages of *Ensete* and *Musa*.

For comparative purposes, modern plant material representing each of the four banana complexes introduced into Africa, as well as the two endemic *Ensete* species, were studied (Table 6.4). These materials were obtained from the Royal Botanical Garden of Meise, Belgium, and from the Laboratory of Tropical Crop Improvement, Katholieke Universiteit Leuven, Belgium. The latter institution hosts the *in vitro* world collection of Musaceae, where all germplasm is maintained under slow-growing conditions (Van den Houwe et al. 1995; Van den Houwe and Swennen 1998; Van den Houwe and Panis 2000). The *Ensete* material also included one specimen collected in Cameroon at an altitude of 1,600 meters in a mixed savanna and gallery forest setting (Mbu Baforchu-Bamenda), and belonging to *E. gilletii*, the only *Ensete* species in western Africa. In order to allow archaeological comparisons, all reference samples were studied by optical microscopy in transmitted light at magnification x400 (OT) as well as by scanning electron microscopy (SEM) (Mbida et al. 2001).

Phytolith assemblages extracted from both *Musa* and *Ensete* genera included a volcaniform phytolith (Figures 6.3–6.5) composed of a base and a raised part. In *Musa* phytoliths, the raised part is a truncated cone or subcylinder. The truncation follows a saddle plane and presents a crater with a thin and continuous rim (Figures 6.4a and 6.4b). A single indentation may be observed in its lower part. In other cases, the rim is crenated, or the crater is partially covered by a tongue of the rim. The basal part is generally roughly parallelepiped to subcubical but some specimens exhibit a boat-shaped basal part. Processes are distributed all along the sides of the base (Figure 6.5) and in some specimens along the lower face (Figure 6.4a).[1] They have morphological characters similar to those of the major cone. X-ray analysis indicates a typical silicium curve. When viewed under the scanning electron microscope, the phytolith sculpture appears smooth to irregularly verrucate. Optical microscopy showed the surface of both the elevated and basal part as psilate. No striking differences in phytolith morphology could be observed among the eight *Musa* accessions studied. According to ICPN 1.0

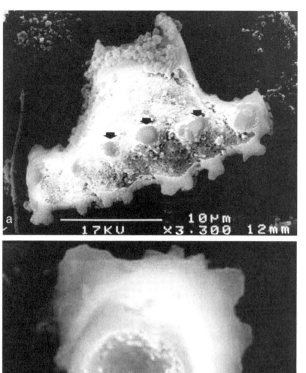

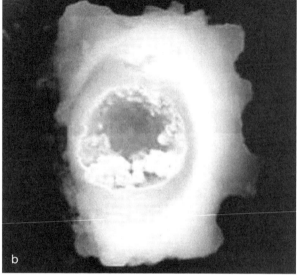

FIGURE 6.4 Scanning electron micrograph of a modern *Musa* phytolith: (a) side view—note the protuberances on the bottom as well as the side of the base (from Plate 1, Mbida et al. 2001); (b) top view.

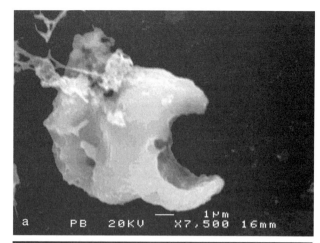

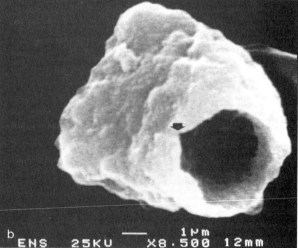

FIGURE 6.5 Scanning electron micrograph of a modern *Ensete* phytolith: (a) side view showing the marked convex slope of the elevation; (b) top view showing the rough and the thick rim of the crater and the indentation (arrow) (from Plate 1, Mbida et al. 2001).

protocol (Madella et al. 2005), the *Musa* phytolith can be described as being a volcaniform concave psilate phytolith.

Ensete phytoliths, by contrast, have a raised part that is a truncated paraboloid (Figure 6.5a). The truncation follows a horizontal to saddle plane. The crater presents a rough and thick rim, commonly having one to three crenations (Figure 6.5b). Some craters exhibit a small tongue as with *Musa*. The basal part is roughly parallelepiped to subcubical. No processes can be noticed on the lateral and lower face of the phytoliths, but small tubular diverticula are observed on the base of the paraboloid (Figure 6.5a–b). No differences in phytolith morphology could be detected between the two *Ensete* accessions that were studied. Again, X-ray analysis indicates a typical silicium curve. In SEM and light microscopy, the surface of both the elevated and the basal parts of the *Ensete* phytolith are densely and heavily verrucate. Following ICPN 1.0 protocol (Madella et al. 2005), the *Ensete* phytolith can be described as being a volcaniform convex verrucate phytolith.

ANALYSIS OF ARCHAEOLOGICAL PHYTOLITHS FROM NKANG

How do the phytoliths recovered from the Nkang site pits and assigned to the family Musaceae compare to these reference class standards (Figure 6.6a and 6.6b)? When viewed through a light microscope, the phytoliths recovered from the Nkang pits exhibit a truncated concave cone or subcylinder (Figure 6.7a), a continuous rim without deep indentations (Figure 6.8a), parallelepiped to cubical basal part (Figures 6.6a and 6.8b), psilate surface (Figure 6.7a), and processes (Figures 6.7b and 6.8b). This combination of features indicates that they can be classed as being a volcaniform psilate concave phytolith, and are therefore clearly representative of the genus *Musa* and are not derived from the genus *Ensete*. Table 6.5 provides a synopsis of the salient morphological features of *Musa* and *Ensete* phytoliths, along with those features exhibited by the Nkang site phytoliths.

This assignment of the Nkang site Musaceae phytoliths to the genus *Musa* is also supported by published descriptions of *Musa* and *Ensete* phytolith morphology. Previous research

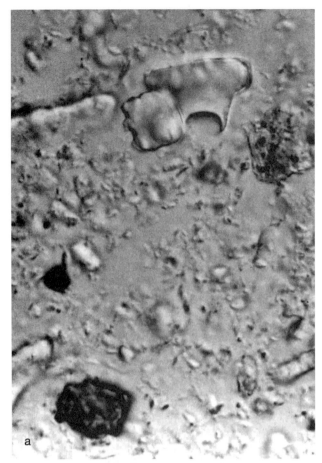

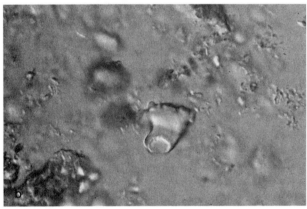

FIGURE 6.6 Light microscopy views of phytoliths recovered from the Nkang site: (a) nearly top view (left piece) and side view (right piece) of volcaniform phytoliths observed in Hor. 1b of pit F9 (×400); (b) nearly side view of another volcaniform phytolith observed in Hor. 1b of pit F9 (×400).

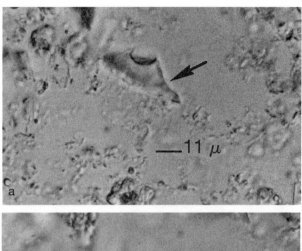

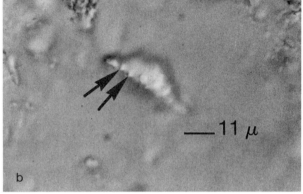

FIGURE 6.7 Detailed side view of a phytolith recovered from the Nkang site (from Plate 1, Mbida et al. 2001): (a) the concave slope of the elevation (arrow); (b) the base showing several protuberances (arrows).

considered a number of different portions of the *Musa* phytolith (Piperno 1988; Wilson 1985). Wilson focuses on the elevated part, which he describes as having "an oval depression on one surface. The rim of this depression appears to be robustly built, and is a most characteristic feature" (Wilson 1985: 91). Piperno notes variations of the base, which may vary from "roughly rectangular to both ends pointed" (Piperno 1988: 65). These brief characterizations correlate with the more detailed description presented here, providing further proof that *Musa* phytoliths are mineral bodies with regular morphological characteristics that can clearly and concisely be differentiated from *Ensete* phytoliths on a number of grounds.

In a side view, *Musa* phytoliths present a conical or subcylindrical elevated part with concave sides. The summit of the elevation is truncated by a wavy plane. This combination of observations clarifies the elevation as a truncated cone or subcylinder. In a top view, the shape of the basal part is parallel sided, mostly square[1] to transversely oblong. The SEM side view shows that the base is not a thin sheet, but rather a subparellepiped or subcube.

In contrast, the lateral view of the *Ensete* elevated part presents an ovoidal to spherical outline with convex sides. This description corresponds to a paraboloid (Peele and Church 1948). The top view indicates the base to be parallel sided with an outline varying from squarish to broadly oblong and oblong.

Do other taxa produce phytoliths similar to those described for *Musa*? Piperno, for example, describes a somewhat similar morphotype form as present in the epiderm of Burseraceae fruits (Piperno 1988: 105, 243, Plates 84, 85; Piperno, conversation with L. Vrydaghs, 2 August 2001). Since one Burseraceae

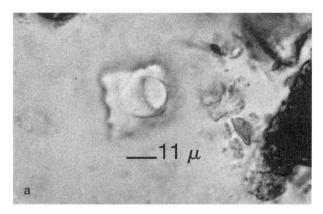

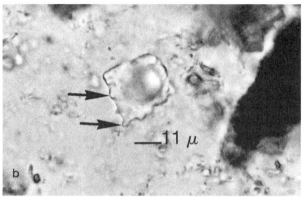

FIGURE 6.8 Detailed top views of a phytolith from the Nkang site (from Plate 1, Mbida et al. 2001): (a) the cone, showing the thin continuous rim; (b) protuberances along the base (arrows).

species (*Canarium schweinfurthii* Engl.) was identified in the wood charcoal assemblage from pit F9 at Nkang, we conducted phytolith research on modern seed material of this species. Only mineralized vessel morphotypes were extracted from the studied material (Vrydaghs and Doutrelepont 2000). As the atlas of leaf phytolith (Vrydaghs 2003) does not provide further evidence of Burseraceae for the deposits of pit F9, this possibility was ruled out.

Roughly similar general morphologies composed of a base and an elevated part do occur in other taxa. This elevation may be conical or subcylindrical, truncated or untruncated. Based on published descriptions of reference material, however, such elevated features never include a crater with a rim (Brown 1984; Deleebeeck et al. 2000; Fredlund and Tiezen 1994; Piperno and Pearsall 1999; Runge 1999). In other cases, the elevated part is hemispherical or combines a lower hemisphere with a cone. It may also be diabolo shaped. Bases may be transversely elliptic to transversely narrowly elliptic with limited thickness and without sculpture. To date, these morphologies are to be found only in Monocotyledons, mostly Gramineae. Phytoliths with an untruncated conical elevation can easily be distinguished by one or several of the following features as well. The same phytolith may present several cone-shaped parts or a single cone. The base may be circular or sinuous, or else irregularly polygonal with at least five sides (Ollendorf 1992). The surface ornamentation may be verrucate with an apical, interapical, or peripherical distribution of small ribs on the lateral sides of the basal part. The latter morphologies are found in Cyperaceae or Orchidaceae. Hence, the *Musa* morphotype, with its combination of distinct features, cannot be confused with of any other known phytoliths on the African continent.

The dates obtained for Pit F9 indicate a period between 2490 and 2400 BP (Table 6.1). No radiocarbon dates, however, were obtained on the material from the Pit F7NF shaft. Fortunately, the chronological placement of the Pit F7NF sample can be inferred from F7, a later pit, partially dug into F7NF sediments. Many ceramic refits occurred between F7 and neighboring features like F7bis, which yielded a date of 2340 ± 70 BP (cal. 800–150 BC, Lv-1941), and F6, which dated to 2580 ± 70 BP (cal. 850–410 BC Lv-1940). The refitting of the shards and the style of pottery in all the excavated pits led to the conclusion that these features were contemporary. Features F9 and F7NF also can be assigned on the basis of ceramic assemblages, both typologically and contextually, to the early phase of Nkang occupation (Mbida 1996). Based on the high degree of cultural association of the material recovered from well-sealed structures like refuse pits, the close agreement of the radiocarbon dates available for the Nkang site, and their close correlation with other related Stone to Metal Period sites in this area of Cameroon, one can conclude that the phytoliths under discussion are securely dated to the middle of the third millennium BP.

The Early Food-Producing Complex in Southern Cameroon

Since wild bananas have never been reported on the African continent, the occurrence of *Musa* phytoliths at the Nkang site unequivocally points to the cultivation of introduced domesticated bananas during the third millennium BP. This conclusion is consistent with other evidence of food production recovered from the Nkang pit features. Five bones of domesticated sheep and goat were found in pits F3, F6, and F7bis, indicating that stock breeding was part of the Nkang economy (Mbida et al. 2000). These caprine bones are from deposits that have yielded radiocarbon dates between 2580 ± 110 and 2420 ± 70 BP (Table 6.1), and can be assigned to the same occupation as the banana phytoliths. Similar indications of small livestock breeding have been reported for Nigeria for the second half of the first millennium BC at the site of Daima (Connah 1976, 1981), and from levels securely dated between 1800 and 1500 calibrated BC at the site of Gajiganna (Breunig 1995; Breunig et al. 1996), suggesting that food production economies were well established over a broad area during the first millennium BC. But, beyond village settlements, were there other major changes in the landscape caused by this change in subsistence economy? Specifically, what was the impact of the banana cultivation on fields or village garden plots?

Plantains are edaphically much more demanding than most other bananas. They develop poorly ramified rooting

TABLE 6.5
Diagnostic Morphological Characters in Phytoliths of *Musa* and *Ensete*, Compared with the Characters Exhibited by the Phytoliths Recovered from the Nkang Site

	Musa	*Ensete*	*Archaeological*
Cone-Shaped Part			
Morphology, slope	Concave	Convex	Concave
Morphology, rim	Smooth or crenate and verrucose	Thick, rough with indentations	Smooth and no indentation
Apex Truncation	Saddle-like	Flat	Saddle-like
Surface	Smooth	Verrucate	Smooth
Basal Part			
Morphology	Rectangular to squarish, or boat shaped	Rectangular to squarish	Rectangular to squarish
Surfaces	Protuberances	No protuberance	Protuberances

SOURCE: From Mbida et al. 2001, Table 2.

systems with relatively few root hairs and thus require easy access to nutrients. The roots, therefore, manifest a high trophism toward organic material (Swennen 1984, Chapter 5; Swennen et al. 1986, 1988; Wilson et al. 1985). On the poor ferralsols of the African forest, where material is concentrated on the shallow uppermost horizon, the rooting system will be superficial and very vulnerable. Even slightly more fertile soils are not suitable for sustainable plantain production unless there is a large amount of organic material in the upper horizons. The Nkang soils are ultisoils with two to four percent of organic matter in the upper horizon (Hor A) and a pH between 4 and 6. Hence, they are not entirely suitable for banana cultivation. The pits fill of pit F9 was made of local sediments partly deposited by runoff erosion. In the past, the edaphic conditions seem to have been comparable to the modern one. These conditions, combined with the consequences of the climatic downturn, including a possible lengthening of the dry season (Maley 1992; Schwartz 1992), suggest that despite the availability of clearings in the forests, there were considerable challenges to sustainable plantain banana cultivation.

Compared to the natural soil today, the older deposits of pit F9 (Hor 1 to 12) show a secondary enrichment in carbonate and phosphate, the latter related to the presence of organic matter (bones and plant tissues) in the soil. These are the kinds of soil conditions expected with refuse disposal, as well as with the use of these pits as latrines and wells. Usually, refuse dumps are located in the backyards of houses and in refuse pits close to the houses. While at Nkang there were no archaeological features associated with the pits, at Okolo pits are associated with postholes delimiting a rectangular surface divided into two rooms (Atangana 1988; de Maret 1991). Thus, the archaeological context of the phytoliths along with the edaphic requirements of bananas and the likely paleoenvironmental conditions of the time seem to point to the greater likelihood that plantain cultivation took place in backyards close to settlements instead of at the border of the forest clearing. The measurements of low magnetic susceptibility for most of the F9 horizons support this conclusion since, according to Le Borgne (1955, 1960), the most significant origin for high magnetic susceptibility might be the consequence of slash and burn.

A center of secondary diversity for AAB plantains exists in the humid lowlands of western and central Africa (Swennen 1990). Both the diversity of cultivars and their socio-economic importance suggest that they have been cultivated in the region for millennia. Botanical evidence and historic-linguistic studies led to the hypothesis that plantains arrived in East Africa prior to 3000 BP (De Langhe et al. 1996). The evidence for cultivation of domesticated *Musa* at Nkang supports this hypothesis but implies that the crop diffused across the continent quite rapidly over a period of only a few hundred years. Since plantain is an ideal staple crop for agriculture in the rain forest (Vansina 1990), its early presence in Cameroon could explain the increase in village density in the forest environment during the Stone to Metal Period, and could improve our understanding of the early stages of Bantu expansion (de Maret 1996).

It could further be suggested that the crop complex package of yam (*Dioscorea* sp.), taro (*Colocasia* sp.) and banana arrived at the same time in Africa. The propagated yam (*Dioscorea alata* L) is originally from Southeast Asia, but several endemic *Discorea* species exist on the African continent, so that phytoliths can not be used as a direct archaeological marker for the introduction of non-African domesticated *Dioscorea* species. Taro (*Colocasia esculenta* (L.) Schott) unfortunately, does not produce phytoliths. Starch grain analysis (see Piperno, Chapter 5 this volume), which presents a promising avenue for investigation of the early history both of yams and taro, has not yet been conducted for the African continent. Consequently, Musaceae phytoliths provide archaeologists and archaeobotanists with the only direct evidence of early cultivation of introduced domesticated crop plants in humid tropical Africa. Interestingly, the recovery and analysis of phytolith assemblages could enable researchers to test suggestions (e.g., Clark 1976) that cultivation of *Ensete* may have preceded the introduction of domesticated *Musa* into Africa.

Diagnostic phytoliths of *Musa* are being recovered from an ever-increasing number of archaeological sites (Bowdery 1999; Scott-Cummings 1998; Madella 2003; Vrydaghs et al. 2004), indicating the clear potential for tracking the diffusion of this crop plant over a broad range of tropical zones throughout the world outside the range of its wild progenitors. Tracking the course of banana domestication within this natural habitat zone where there are wild, semidomesticated, and domesticated varieties of bananas requires another set of analytical approaches (Denham et al. 2003; Vrydaghs and De Langhe 2003). In the eastern Pacific, the Austronesian migration is thought to have played a major role in the dispersal of the banana, while slave trade was most probably instrumental in bringing the banana to the New World. Before pursuing the early history of the banana in the Americas, however, it will be essential to rule out possible confusion between *Musa* and *Heliconia* phytoliths. The *Heliconia* phytolith is elongate, elliptic to blocky, and lacking the volcano shape of *Musa* phytoliths; however, it has a deep crater almost cutting the phytolith into two parts. Specimens with and without projections have been described (Pearsall 1998). *Heliconia* phytoliths have been recovered from a number of prehistoric and colonial period contexts in the Neotropics (Jones 1993; Piperno 1988, 1991; Pearsall 1994; L. Vrydaghs and H. Doutrelepont, personal observations on archaeological material from Peru transmitted by L. Scott Cummings, November 1996). In contrast, *Musa* phytoliths have yet to be reported for any archaeological context in the Americas. Given the broad spatial and temporal spans represented by the sites that have yielded negative results (e.g., western Ecuador (1550–1700 BC; AD 250–450); central Pacific Panama (6610, 4750, 2000 BP), it seems highly unlikely that the African plantain reached the New World in pre-Colombian times.

In conclusion, the *Musa* phytolith appears to be a quite promising tool in the search for direct evidence of early tropical agriculture. The identification of *Musa* phytoliths at the Nkang site demonstrates that tropical plant cultivation began much earlier than previously assumed in humid Africa, and phytolith analysis holds the promise of shedding further light on the early history of banana cultivation in Africa and other tropical landscapes worldwide.

Notes

1. The descriptive lexicon for the base refers to the terminology of simple symmetrical plane shapes published in 1962 by the Systematic Association Committee for Descriptive Biological Terminology (1962 a and b). As to the general phytolith descriptive terminology, the definitions can be found at http://users.skynet.be/fa100812/index.html (Vrydaghs 2004).

Acknowledgments

This research was made possible by grants from the Directorate General for International Cooperation, Belgium; the National Lottery; and the Belgian National Fund for Scientific Research. The archaeozoological contribution represents results from the Belgian program on Interuniversity Poles of Attraction initiated by the Belgian State Prime Minister's Office, Federal Services. The authors also wish to thank Dr. J. Crowther of the University of Wales for his analysis, Dr. B. D. Smith for his editing work, and Miss Y. Baele and Mr. D. Van Aubel for their assistance with the illustrations. The authors gratefully acknowledge the editors who invited us to publish our results.

References

Atangana, C. 1988. *Archéologie au Cameroun méridional: Étude du site d'Okolo*. Thèse de doctorat de troisième cycle en Archéologie. Université de Paris I: Panthéon-Sorbonne.

Berchem, J., 1989–1990. Sprachbeziehungen im Bereich des Kulturwortschatzes zwischen den Bantusprachen und dem Malagasy. *Sprache und Geschichte in Afrika* 10/11: 9–169.

Bowdery, D. 1999. Phytoliths from tropical sediments: Reports from Southeast Asia and Papua New Guinea. *IPPA Bulletin* 18: 159–168.

Breunig, P. 1995. Gajiganna und Konduga. Zur frühen besiedlung des Tschadbeckens in Nigeria. *Beiträge zur allgemeinen und verleichenden archäologie* 15: 3–48.

Breunig, P., K. Neumann, and W. Van Neer. 1996. New research on the Holocene settlement and environment of the Chad basin in Nigeria. *The African Archaeological Review* 13: 111–145.

Brown, D. A. 1984. Prospects and limits of a phytolith key for grasses in the central United States. *Journal of Archaeological Sciences* 11: 345–368.

Bullock, P., N. Fedoroff, A. Jongerius, G. Stoops, and T. Tursina. 1985. *Handbook for soil thin section description*. Mount Pleasant, Wolverhampton: Waine Research.

Claes, P. 1985. Contribution à l'étude des céramiques anciennes des environs de Yaoundé. Mémoire de licence, Université Libre de Bruxelles. Unpublished thesis.

———. 1992. A propos des céramiques de Mimboman et d'Okolo: Premières analyses. In *L'Archéologie au Cameron*, J. M. Essomba (ed.), pp. 215–227. Paris: Kartala.

Clark, J. D. 1976. The domestication process in sub-Saharan Africa with special reference to Ethiopia. *Origine de l'Elevage et de la Domestication*. Colloque XX: 83–105. IXe Congrès Union Internationale des Sciences Préhistoriques et Protohistoriques, Nice.

Clist, B. 1990. Des derniers chasseurs aux premiers métallurgistes: Sédentarisation et débuts de la métallurgie du fer (Cameroun, Gabon, Guinée équatoriale). In *Paysages quaternaires de l'Afrique centrale atlantique*, R. Lanfranchi and D. Schwartz (eds.), pp. 458–479. Paris: ORSTOM editions. Collection Didactiques.

Connah, G. 1976. The Daima sequence and the prehistoric chronology of the Lake Chad region of Nigeria. *Journal of African History* 17: 321–352.

———. 1981. Man and a lake. In *Le Sol, la Parole et l'Ecrit*. Melanges en homage à Mauny, Collectif, pp. 161–178. Paris: Société Française d'Histoire d'Outre-Mer.

De Langhe, E. and P. de Maret. 1999. Tracking the banana: Significance to early agriculture. In *The prehistory of food. Appetites for change*, C. Gosden and J. Hather (eds.), pp. 377–396. London and New York: Routledge.

De Langhe, E., R. Swennen, and D. Vuylsteke. 1996. Plantain in the early Bantu world. In *The growth of farming communities in Africa from the Equator southwards*, J. E. G. Sutton (ed.), *Azania* 29–30: pp. 147–160. Nairobi: The British Institute in Eastern Africa.

Deleebeeck, N., C. Cocquyt, and P. Goetghebeur. 2000. Phytolith study of reed (*Phragmites australis* (Cav.) Steudel (Poaceae]) along Schelde and Durme (Belgium). In *Man and the (palaeo) environment. The phytolith evidence. Abstracts (Lectures, Posters and Phytolith Systematic Workshop)*, Vrydaghs, L. and Degraeve, A. (eds.), 3rd international meeting on phytolith research (IMPR): 13. Tervuren, Belgium, 21–25 August 5.

Denham, T. P., S. G. Haberle, C. Lentfer, R. Fullagar, J. Field, M. Therin, N. Porch, and B. Winsborough. 2003. Origins of agriculture at Kuk Swamp in the highlands of New Guinea. *Science* 301: 189–193

Doutrelepont, H., L. Vrydaghs, E. De Langhe, B. Janssens, R. Swennen, C. Mbida, P. de Maret. 1996. Banana in Africa. In *The state of-the-art of phytoliths in soils and plants*, A. Pinilla Navarro, M. J. Machado, and J. Juan-Tresserras (eds.), 27. First European Meeting on Phytolith Research. 23–26 September: 27. Madrid, Spain.

Elouga, M., 2000. Carte archéologique du nord de la Sanaga. Paysages des sites et mise en évidence de la transgression forestière sur la savane (centre du Cameroun). In *Dynamique à long terme des écosystèmes forestiers intertropicaux*, M. Servant and S. Servant-Vildarhy (eds.), pp. 133–137. IRD: UNESCO.

Fredlund, G. G. and L. T. Tieszen. 1994. Modern phytolith assemblages from the North American Great Plains. *Journal of Biogeography* 21: 321–335.

Fuller, D. and M. Madella. 2001. Issues in Harappan archaeobotany: Retrospect and prospect. In *Indian archaeology in retrospect*, Volume II, S. Settar and R. Korisettar (eds.), pp. 317–390. Protohistory. Archaeology of the Harappan civilization. MANOHAR in association with the Indian Council of Historical Research.

Giresse, P., J. Maley, and P. Brenac. 1994. Late quaternary palaeoenvironments in the Lake Brombi Mbo (West Cameroon) deduced from pollen and carbon isotopes of organic matter. *Palaeogeography, Palaeoclimatology, Palaeoecology* 107: 65–78.

Jones, J. G. 1993. Analysis of pollen and phytoliths in residue from a colonial period ceramic vessel. In *Current research in phytolith analysis: Applications in archaeology and palaeoecology*, D. M. Pearsall and D. R. Piperno (eds.), pp. 31–36. Philadelphia: The University Museum of Archaeology and Anthropology.

Karamura, D. A. 1999. Numerical taxonomy studies of the East-African highland Bananas (*Musa* AAA-East-Africa) in Uganda. PhD dissertation, The University of Reading. Published by International Network for the Improvement of Banana and Plantain (INIBAP). Rome: International Plant Genetic Resources Institute (IPGRI).

Kealhofer, L. and D. Piperno. 1998. *Opal phytoliths in Southeast Asian flora*. Smithsonian Contribution to Botany 88. Washington, D.C.: Smithsonian Press.

Le Borgne, E. 1955. Susceptibilité magnétique anormale du sol superficiel. *Annales de Géophysique* 11: 399–419.

———. 1960. Influence du feu sur les propriétés magnétiques du sol et du granite. *Annales de Géophysique* 16: 159–195.

Leju, B. J., R. Robertshaw, D. Taylor. 2006. Africa earliest Bananas? *Journal of Archaeological Science* 33: 102–113.

Lentfer, C. 2001. Musaceae phytoliths in the Kundil's section at Kuk, Papua New Guinea. A paper presented at the symposium, The state of the art in phytolith and starch research in the Australian-Pacific-Asian regions. A conference hosted by the Center for Archaeological Research at the Australian National University, 1–3 August 2001, 19.

Madella, M. 2003. Investigating agriculture and environment in South Asia: present and future contributions from opal phytoliths. In *Indus Ethnobiology. New perspectives from the field*, St. A. Weber and W. R. Belcher (eds.), pp. 199–249. Lexington Books.

Madella, M., A. Alexandre, and T. Ball. 2003. The international code of phytolith nomenclature 1.0. *Phytolitarien* 15: 7–16.

Madella, M., A. Alexandre, and T. Ball. 2005. International Code for Phytolith Nomenclature 1.0. *Annals of Botany* 96: 253–260.

Maley, J. 1992. *Mise en évidence d'une péjoration climatique entre ca 2500 et 2000 and BP en Afrique tropicale humide*. Bulletin de la Société Geologique de France 163: 363–365.

———. 1993. The climatic and vegetational history of the equatorial regions of Africa during the upper Quaternary. In *The archaeology of Africa. Food, metals and towns*, T. Shaw, P. Sinclair, A. Bassey and A. Okpoko (eds.), pp. 43–52. One World Archaeology 20. London and New York: Routledge.

Maley, J., P. Brenac, S. Bigot, and V. Moron. 2001. Variations de la végétation et des paléoenvironnements en forêt dense africaine au cours de l'Holocène. Impact de la variation de températures marines. In *Dynamique à long terme des écosystèmes forestiers intertropicaux*, M. Servant and S. Servant-Vildarhy (eds.), pp. 205–220. IRD: UNESCO.

de Maret, P. 1983. Nouvelles données sur la fin de l'âge de la pierre et les débuts de l'âge du fer dans la moitié méridionale du Cameroun. In *The Proceedings of the Ninth Congress of the Pan-African Association of Prehistory and Related Studies*, B. Andah, P. de Maret, and R. Soper (eds.), pp. 198–202. Jos: Rex Charles Publications.

———. 1991. La recherche achéologique au Cameroun. In *La recherche en sciences humaines au Cameroun*, P. Salmon and J. S. Symoens (eds.), pp. 37–51. Bruxelles: Académie Royale des Sciences d'Outre-mer.

———. 1992. Sédentarisation, agriculture et métallurgie du sud Cameroun. Synthèse des recherches depuis 1978. In *L'archéologie au Cameroun*, J. M. Essomba (ed.), pp. 247–262. Paris: Karthala.

———. 1996. Pits, pots and the far west stream. *Azania* 29–30: 318–323.

Mbida, C. 1992. Etude préliminaire du site de Ndindan et datation d'une première série de fosses. In *L'archéologie au Cameroun*, J. M. Essomba (ed.), pp. 263–284. Paris: Karthala.

———. 1996. *L'émergence de communautées villageoises au Cameroun meridional. Etude archéologique des sites de Nkang et Ndindan*. Unpublished PhD. Facultée de Philosophie et Lettre. ULB.

Mbida, C., W. Van Neer, H. Doutrelepont, and L. Vrydaghs. 2000. Evidence for banana cultivation and animal husbandry during the first millennium BC in the forest of southern Cameroon. *Journal of Archaeological Science* 27: 151–162.

Mbida, C., H. Doutrelepont, L. Vrydaghs, R. Swennen, H. Beeckman, E. De Langhe, and P. de Maret. 2001. First archaeological evidence of banana cultivation in central Africa during the third millennium before present. *Vegetation History and Archaeobotany* 10: 1–6.

Metcalfe, C. R. 1971. *Anatomy of the Monocotyledons. v. Cyperaceae.* Oxford: Clarendon Press.

Metcalfe, C. R. and L. Chalk. 1950. *Anatomy of the Dicotyledons.* Volumes 1 and 2. Oxford: Clarendon Press.

Metcalfe, C. R. and L. Chalk. 1983. *Anatomy of the Dicotyledons.* 2nd ed. Volume 2. Oxford: Clarendon Press.

Ollendorf, A. L. 1992. Toward a classification scheme of Sedge (Cyperaceae) phytoliths. In *Phytolith systematics. Emerging issues, advances in archaeological and museum science*, vol. 1, G. Rapp Jr. and S. C. Mulholland (eds.), pp. 91–111. New York and London: Plenum Press.

Oslisly, R. and L. White. 2000. La relation homme/milieu dans la réserve de la Lope (Gabon) au cours de l'Holocène: Les implications sur l'environnement. In *Dynamique à long terme des écosystèmes forestiers intertropicaux*, M. Servant and S. Servant-Vildarhy (eds.), pp. 241–250. IRD: UNESCO.

Oslisly, R., C. Mbida, and P. Kinyock. 2001. Premiers résultats de la recherche archéologique sur le littoral du Cameroun entre Kribi et Kampo. A paper presented at the XIV Congrès de l'Union Internationale des Sciences Préhistoriques et Protohistoriques. 2–8 September 2001. Liège: Université de Liège, 343.

Pearsall, D. M. 1994. Investigating New World tropical agriculture: Contributions from phytolith analysis. In *Tropical Archaeobotany. Applications and new developments*, J. Hatter (ed.), pp. 155–138. World Archaeology 22, London and New York: Routledge.

———. 1998. Phytoliths in the flora of Ecuador: The University of Missouri online phytolith database [http://www.missouri.edu/~phyto]. With contributions by A. Biddle, K. Cghandler-Ezell, S. Collins, S. Stewart, C. Vientimilla, Dr Zhijun Zhao, and N. A. Duncan, page designer and editor.

Pele, R. and J. A. Church. 1948. *Mining engineers' handbook.* Volume 2. New York: John Wiley & Sons.

Peters, J. 1988. Osteomorphological features of the appendicular skeleton of African buffalo, Syncerus caffer (Sparrman, 1779) and of domestic cattle, *Bos primigenius* f. *taurus* Bojanus, 1827. *Zietschrift für Säugetierkunde* 53: 108–123.

———. 1989. Osteomorphological features of the appendicular skeleton of gazelles, genus *Gazella* Blainville 1816, bohor reedbuck, redunca redunca (Pallas, 1767) and bushbuck, Tragelaphus scriptus (Pallas, 1766). *Anatomia, Histologia, Emvryologia* 18: 97–113.

Piperno, D. R. 1988. *Phytolith analysis. An archaeological and geological perspective.* New York: Academic Press.

———. 1989. The occurrence of phytoliths in the reproductive structures of selected tropical angiosperms and their significance in tropical paleoecology, paleoethnobotany and systematics. *Review of Palaeobotany and Palynology* 61: 147–173.

———. 1991. Phytolith analysis in the American tropics. *Journal of World Prehistory* 5: 155–191.

Piperno, D. R. and D. Pearsall. 1999. *The silica bodies of tropical American grasses: Morphology, taxonomy, and implications for grass systematics and fossil phytolith identification.* Smithsonian Contributions to Botany 85. Washington, D.C.: Smithsonian Press.

Rapp, G., Jr. and S. C. Mulholland (eds.). 1992. Phytolith systematics: Emerging issues. In *Advances in archaeological and museum science.* Volume 1. New York and London: Plenum Press.

Reynaud-Farrera, I., J. Maley, and D. Wirrman. 1996. Vegetation et climat dans les forêts du sud-ouest Cameroun depuis 4770 BP: Analyse pollinique des sédiments du Lac Ossa. *Comptes Rendus de l'Académie des Sciences de Paris* 322, IIa, pp. 749–755.

Runge, F. 1996. Opal phtolithe in Pflanzen aus den humiden und semi-ariden Osten und ihre bedeutung für die klima- und vegetationsgschichte. *Botanisch Jahrbook Systematik* 118: 303–363.

———. 1999. The opal phytolith inventory of soils in central Africa: Quantities, shapes, classification, and spectra. *Review of Palaeobotany and Palynology* 107: 23–53.

Runge, F. and J. Runge. 1997. Opal phytoliths in East African plants and soils. In *The state-of-the-art of phytoliths in soils and plants*, A. Pinilla, J. Juan-Treserras. and M. J. Machado (eds.), pp. 71–81. Monografias del centro de Ciencas Medioambientales 4, Consejo Superior de Investigaciones Cientificas, Madrid.

Schwartz, D. 1992. *Assèchement climatique vers 3000 BP et expansion Bantu en Afrique centrale atlantique: Quelques réflexions.* Bulletin de la Société Géologique de France 163: 353–361.

Scott-Cummings, L. 1998: A review of recent pollen and phytoliths studies from various contexts on Easter Island. In *Easter Island in Pacific context South Seas symposium*, C. M. Stevenson (ed.), pp. 100–106. The Easter Island Foundation occasional paper 4. Los Osos, CA: Bearsville and Cloud Mountain Press.

Sebasigari, K. 1990. Principaux caractères de détermination dans la caractérisation morphologique des bananiers triploids acuminate de l'Afrique de l'Est. In *Identification of genetic diversity in the genus Musa*, R. L. Jarret (ed.), Montpellier: International Network for the Improvement of Banana and Plantain (INIBAP).

Simmonds, N. W. 1959. *Bananas.* Tropical Agriculture Series. London: Longmans.

———. 1962. *The evolution of the bananas.* London: Longmans.

Simmonds N. W. and K. Shepherd, 1955. The taxonomy and origins of the cultivated bananas. *Journal of the Linnaean Society (Botany)* 55: 302–312.

Swennen, R. 1984. *A physiological study of the suckering behaviour in Plantain (Musa cv. AAB).* PhD thesis n 132. Dissertationes de Agricultura. Katholieke Universiteit te Leuven. Belgium.

———. 1990. Limits of morphotaxonomy: Names and synonyms of plantains in Africa and elsewhere. In *The identification of genetic diversity in the genus Musa*, R. L. Jarret (ed.). Proceedings of an International Workshop, Los Banos, Philippines, pp. 172–210. 5–10 September 1988. Montpellier, France: International Network for the Improvement of Banana and Plantain (INIBAP).

Swennen, R. and D. Vuylsteke. 1987. Morphological taxonomy of plantain (*Musa* cultivars AAB) in West Africa. In *Banana and Plantain Breeding Strategies*, G. Persley, E. De Langhe (eds.), pp. 165–171. Proceedings of an international workshop. ACIAR Proceedings No. 21. Cairns, Australia, 13–17 October 1986.

Swennen, R., E. De Langhe, J. Janssen, and D. Decoene. 1986. Study of the root development of some *Musa* cultivars in hydroponics. *Fruits* 41: 515–524.

Swennen, R., G. F. Wilson, and D. Decoene. 1988. Priorities for future research on the root system and corm in plantains and bananas in relation with nematodes and the banana weevil. Nematodes and the borer weevil in bananas: Present status of research and outlook. Proceedings of a workshop, Bujumbura, Burundi, 7–11 December 1987, pp. 91–96. Montpellier, France: International Network for the Improvement of Banana and Plantain (INIBAP).

Systematics Association Committee for Descriptive Biological Terminology. 1962a. Terminology of simple symmetrical plane shapes. (Chart 1). *Taxon* 11: 145–156.

———. 1962b. Terminology of simple symmetrical plane shapes. (Chart 1a). *Taxon* 11: 245–248.

Tomlinson, P. C. 1959. An anatomical approach to the classification of the Musaceae. *Journal of the Linnaean Society (Botany)* 55: 779–809.

———. 1969. *Anatomy of the Monoctyledons III. Commelinales-Zingiberales*. Oxford: Clarendon Press.

Twiss, P. C., E Suess, and R. M. Smith. 1969. Morphological classification of grass phytoliths. *Soil Sciences Society of America Proceedings*. 33: 109–115.

Van den Houwe, I. and B. Pannis. 2000. In vitro conservation of banana: Medium term storage and prospects for cryopreservation. In *Conservation of plant genetic resources in vitro*, Volume 2, M. K. Razdan and E. Cocking E. (eds), pp. 225–257. M/S Science Publishers, U.S.A

Van den Houwe, I. and R. Swennen. 1998. La collection mondiale du bananier (*Musa* spp.) au Centre de Transit de l'INIBAP à la K.U. Leuven: Strategies de conservation et mode d'opération. *Biotechnologie, Agronomie Société et Environnement* 2: 36–45.

Van den Houwe, I., K. De Smet, H. Tezenas du Montcel, and R. Swennen. 1995. Variability in storage potential of banana shoot culture under medium term storage conditions. *Plant Cell, Tissue and Organ Culture* 42: 269–274.

Van Neer, W. 1989. *Contribution to the archaeozoology of Central Africa*. Tervuren: Annales du Musée Royal de l'Afrique Centrale, Sciences Zoologiques 259.

Vansina, J. 1990. *Paths in the rainforests*. Madison: University of Wisconsin Press.

Vrydaghs, L. 2003. Studies in opal phytolith. Material and identification criteria. Volume 2, Part 2: Atlas. PhD dissertation, Ghent University. Unpublished.

———. 2004. Phytolith Glossary. [http://users.skynet.be/fa100812/index.html]. Page designer B. Kojic.

Vrydaghs, L. and E. De Langhe. 2003. The *Musa* phytolith: An opportunity to rewrite history. *INIBAB Annual Report* 2002: 14–17

Vrydaghs, L. and H. Doutrelepont. 2000. Analyses phytolitariennes: Acquis et prespectives. In *Dynamique à long terme des éosystèmes forestiers intertropicaux*, M. Servant and S. Servant-Vildary (eds.), pp. 389–399. IRD: UNESCO.

Vrydaghs, L., C. Cocquyt, T. Van de Vijver, and P. Goetghebeur. 2004. Phytolithic evidence of the introduction of Schoenoplectus californicus subsp. totora at Easter Island. *Rapa Nui Journal* 18: 95–106.

Walker, R. 1985. *A guide to postcranial bones of East-African animals*. Norwich, U.K.: Hylochoerus Press.

Wilson, S. M. 1985. Phytolith analysis at Kuk, an early agricultural site in Papua New Guinea. *Archaeology in Oceania* 20: 90–96.

Wilson, G. F., R. Swennen, and E. De Langhe. 1985. Effects of mulch and fertilizer on yield and longevity of a medium and giant plantain and a banana cultivar. *Proceedings of the 3rd meeting, Abidjan, Côte d'Ivorie, 27–31. May 1985*. International Association for Research on Plantain and Bananas, 109–111.

CHAPTER 7

Documenting the Presence of Maize in Central and South America through Phytolith Analysis of Food Residues

ROBERT G. THOMPSON

Introduction

Documenting the rates and routes of diffusion of maize *(Zea mays)* throughout the Americas, and its changing economic role, remains a central challenge for scholars interested in understanding the complex socioeconomic patterns of pre-Columbian development in the New World. Approaches that combine consideration of both biological and anthropological evidence clearly hold great promise in addressing this problem (Smith 2001). To date, most archaeobotanical studies of maize in the Americas have relied on the archaeological excavation of plant macrofossils in the form of charred seeds and other plant parts, or on the recovery of microfossils in the form of pollen and phytoliths (opal silica bodies contained in plant tissue) from archaeological sediments. This chapter also employs phytoliths to document the presence of maize in the archaeological record, but in a manner that relies on clear evidence of its role as a food source from culturally and temporally secure contexts—the analysis of opal phytolith assemblages present in directly datable (by AMS, accelerator mass spectrometer radiocarbon dating) food residues recovered intact from ceramic vessels. The biological foundation for the study of phytolith assemblages from maize cob chaff and food residues is first outlined, and the method employed in looking for evidence of maize in residue deposits in ceramic vessels from different regions of Central and South America is described.

Food Residue Phytolith Assemblages

Opal phytolith assemblages contained in food residues (e.g., the charred encrustations on the interior of ceramic cooking vessels, human coprolites, calculus deposits removed from teeth) provide more direct, firsthand evidence of human diets than those recovered from archaeological sediments and used for paleo-ecological research or dietary reconstruction (Thompson and Mulholland 1994). In an assemblage of phytoliths obtained from the sediments of an archaeological excavation, any silica-producing past or present-day plant in the vicinity of the site represents a potential contributing source for a portion of the recovered phytolith assemblage. Food residues recovered from cooking vessels, in contrast, can reasonably be expected to contain only those phytoliths that were contained in the plants actually cooked in the pots.

In addition, since different plant parts produce different assemblages of phytoliths (Mulholland 1989), food residue phytolith analysis can focus on the recognition and analysis of phytolith profile signatures for the portions of plants actually cooked and consumed, rather than having to consider total plant phytolith assemblages. As a result, the study of phytoliths recovered from food residues involves the identification, description, and comparative analysis of fewer sets of phytolith forms, and allows more robust statistical comparison of food residue phytolith assemblages with potential contributing food plants.

Food residues remaining on the interior walls of ceramic cooking vessels provide direct evidence of past meals. Rather than being scraped away, such food residues were often allowed to accumulate over time in order to seal the walls of the vessel and reduce permeability. Many types of analysis have been attempted on these cooking pot residues, with varying degrees of success (Heron et al. 1991). Interestingly, one of the earliest analyses of ceramic cooking residues focused on phytoliths. Edman and Soderberg (1929) were able to demonstrate that opal phytoliths recovered from the food residues of a five-thousand-year-old Chinese cooking vessel provided evidence for early rice use in China. Despite this early work, the full potential of food residue phytolith analysis has yet to be realized. The presence of opal phytoliths in food residues can often be difficult to establish without removal of organic components. Hastorf and DeNiro (1985), for example, examined carbon isotope ratios of food residues from ceramic vessels recovered from sites in the Mantaro River valley of Peru in an effort to recognize the presence and dietary importance of maize over a period of several hundred years. Despite microscopic examination of the residues studied, they reported that "analysis of the encrusted organic matter by light and scanning electron microscopy (SEM) indicates that all morphological characters useful for identification had been destroyed during burning" (Hastorf and DeNiro 1985: 490). After dissolution and removal of the organic component of the residues, however, assemblages of silica bodies were observed as being present in the Mantaro residue samples, providing a complementary line of evidence (Thompson 1993).

A number of studies in the early 1990s showed that different component parts of maize plants and maize cobs contain various kinds and amounts of phytoliths (Bozarth 1993;

Piperno and Pearsall 1993; Thompson 1993; Thompson and Mulholland 1994). Although maize kernels do not themselves contain phytoliths, adjacent structures—glumes—contain abundant opal silica bodies. Since the stripping of maize kernels from cobs in preparation for cooking does not involve the perfect separation of kernels from adherent glumes, substantial amounts of glume phytoliths accompany kernels into cooking vessels. It is these glume phytoliths that are deposited in the food residues of ceramic vessels used to cook maize. Opal phytoliths can withstand the heat of cooking. An early method of extracting phytolith assemblages from modern reference plants, in fact, involved "dry ashing"—high-temperature incineration of the portion of the plant from which the phytoliths were to be obtained (Rovner1983).

Looking Beyond Alternative Timetables for the Initial Introduction of Maize into South America

There is a general consensus that the wild ancestor of maize is teosinte *(Zea mays ssp. parviglumis)*, an annual wild grass, and that domestication occurred only once, in the Balsas River Drainage of southwestern Mexico, by 7,000 to 9,000 years ago (Bennetzen et al. 2001; Matsuoka et al. 2002). By the time of European contact, maize cultivation had reached as far north as Manitoba and Ontario, and as far south as Chile and Argentina. Between the initial domestication of maize in the Balsas, and the eventual arrival of maize farming at its northern and southern limits of production, there is still much to be learned regarding the timing and nature of its complex history of diffusion and adoption, and its changing role and relative dietary importance.

Pearsall (1978) and Piperno (1984) pioneered the use of plant microfossils to trace the southward dispersal of maize out of Mexico. Based on the recovery from archaeological sediments of large, cross-shaped phytoliths considered to be unique to *Zea mays*, they proposed that maize was introduced into lowland Central America and northwestern South America very soon after it was initially domesticated. They outline a time frame of diffusion in which maize was dispersed rapidly from its origin in southwestern Mexico into and through the lowlands of Central America, and was well established in coastal Ecuador by the Valdivia III period at ca. 5000 BP (Pearsall and Piperno 1990). Alternatively, a much later introduction of maize into South America has been proposed: at around 3500 BP, based on the initial appearance of macrofossil maize remains in archaeological contexts (Fritz 1994a, 1994b; Russ and Rovner 1989; Smith 1998; Staller 2003). Genetic analysis of modern distribution patterns of *Zea mays* varieties also appears to support a more recent dispersal of maize southward into low-elevation environments:

> Among archaeologists, there have been two models for the early diversification of maize. According to one, because the oldest directly dated fossil maize comes from the Mexican highlands, the early diversification of maize occurred in the highlands with maize spreading to the lowlands at a later date. The second model interprets maize phytoliths from the lowlands as the oldest maize, and accordingly places the early diversification of maize in the lowlands. Our data suggest that maize diversified in the highlands before it spread to the lowlands (Matsuoka et al. 2002: 6083).

Maize cobs from the Guilá Naquitz site, which have been directly dated by the AMS radiocarbon method and described in detail by Piperno and Flannery (2001) and by Benz (2001), indicate that a very primitive form of maize was being grown in Oaxaca in the southern highlands of Mexico by 6250 BP (calibrated). Benz argues that the Guilá Naquitz maize does not reflect much diversification, and that it is statistically indistinguishable from the highland Mexican maize populations of 700 years later (Benz 2001: 2106). He argues that, judging from the early Guilá Naquitz cobs, several fundamental characteristics of domesticated maize were not yet fixed in the populations represented. Benz's description of the early Guilá Naquitz maize, and its lack of diversification, would appear to support the shorter time frame proposed for the dispersal of maize into South America, based on the initial appearance of maize cob fragments.

Microfossil, macrofossil, and genetic classes of evidence thus appear to provide support for two quite different time frames for the introduction of maize cultivation into northern South America, and the debate regarding which of the two is correct has gained considerable attention in recent years. The current case study analysis of food residues from cooking vessels does not directly address this debate, for obvious reasons: cooking vessels were not a necessary prerequisite for maize processing. Corn could have been grown and subsequently processed for consumption as either an alcoholic beverage (Smalley and Blake 2003) or as a food crop in ways that did not involve cooking pots, and therefore could have been present as a cultivated plant in areas of Central and South America both before and after the development of ceramic vessels, without leaving any residue signatures.

Ceramic vessel residue analysis does offer, however, the opportunity to move beyond the ongoing debate regarding the exact timetable and patterns of initial diffusion of maize southward out of the Balsas, and to address an expanded and very interesting set of related research questions. Throughout the Americas, when maize was initially introduced into new areas, it often did not immediately and abruptly reshape the diets, economies, and ways of life of pre-Columbian societies. It often would play a relatively small dietary role for long periods of time before shifting to center stage as a major crop plant in economies shifting over more dramatically to a reliance on food production. Residue analysis of cooking vessels (or other processing technologies; see Piperno, Chapter 5, this volume), while not always capable of addressing questions regarding the presence or absence of different domesticates, does provide clear evidence regarding the changing economic, social, and political roles of a variety of crop plants, including maize.

Methodology

Recovery of Residue and Maize Chaff Samples

Samples of maize cob chaff (bits of hard and soft glume) were recovered from modern cobs by simple hand scraping. This manual removal of chaff would have been similar to the process used to remove kernels from ears on a large scale (as could be expected during the preparation of either food or *chicha*—maize beer). The chaff was then treated with heated nitric acid, which dissolved the organic matter, leaving the opal phytoliths. Following nitric acid removal of organics, the solutions of nitric acid, which contained the remaining phytoliths, were placed in centrifuge tubes and centrifuged at 3000 rpm for 15 minutes at a time, concentrating the phytoliths in the bottom of the test tubes. The supernatant nitric acid was then pipetted off and replaced with distilled water. After five repetitions of centrifuging, pipetting off the supernatant liquid, and replacing it with distilled water, the procedure was duplicated, replacing the distilled water with ethanol. Slides were made by pipetting phytoliths onto slides, allowing the alcohol to dry, and sealing the phytoliths under a cover slip with permount.

The procedure for sampling residues is similar. First, the soil contact layer is removed from the residue by scraping it with a razor blade. The blade or a dental pick is used to remove the residue from the pottery wall, taking care not to sample the vessel. The food residues are then placed in nitric acid, and from there the procedure is much the same as for maize cobs. Food residues generally take much longer for the dissolution of the organic material than do maize cob chaff samples.

Phytolith Taxonomy

Comparing assemblages of phytoliths recovered from modern lineages of maize, from archaeological cobs, and from food residues, requires the use of a phytolith taxonomy that is both flexible enough and inclusive enough to allow description of the types of phytoliths recovered from a number of genetic and environmental backgrounds. Several taxonomies of phytoliths have been developed based on morphological features, plant type, and tissue of origin (e.g. Piperno and Pearsall 1993; Pearsall et al. 2003). In an effort to develop a taxonomic scheme useful for classifying disaggregated assemblages of phytoliths, Mulholland and Rapp (1992) classified phytoliths recovered from Graminae based solely on their morphological characteristics. This classification scheme proved effective in describing phytoliths recovered from both sediments and plants, allowing the statistical comparison of assemblages, and it is used in modified form here. During the current study, the description of literally thousands of maize cob phytoliths led to the revision of the Mulholland and Rapp taxonomy: fine-tuning an already functional classification scheme through the incorporation of more of the observed, three-dimensional variation. This expanded taxonomy of rondel phytoliths, employed in the present study, is presented as Table 7.1.

Examination of a rondel phytolith in planar view (planar view is defined as the short cylinder resting in an upright position) with a light microscope results in the appearance of a thin face (the larger face) and a thick face (the smaller face). Each of these faces may be entire (either a complete circle or oval) or indented (with a concavity or concavities in the outline). In addition, either face may bear small projections, described as decorations.

Using this system, phytoliths are classified according to the planar view of their faces. A-2-B, for instance, represents a phytolith with an entire outline of the thin face, a thick face of different size as the thin face, and an entire thick face with small projections. Figure 7.1 illustrates this type of phytolith. Note that one side of the B face is flattened, but not indented. Figure 7.2 illustrates a similar phytolith, tilted on its side. The constriction in the middle of the phytolith is clearly shown, as are the projections from one face. Figure 7.3 has four rondels in planar view. Rondel A in Figure 7.3 has two indentations in each face, directly across from each other, and the thick face has multiple small projections. The classification of this form would be C-3-D-3. Rondel B has entire faces of approximately the same size, and the thick face has multiple projections. The form of this rondel is described as A-1-B. Phytolith C has entire faces (one side of each face is flattened but not indented). No decorations are present on this rondel. This phytolith is described as A-1-A. Phytolith D has two entire faces of similar size, each of them decorated, and would be described as B-1-B.

Statistical Analysis of Assemblage Data

A sample of 100 rondel phytoliths in planar view was selected from each residue and maize chaff assemblage and described morphometrically. The data on the 100 forms from each sample were then entered in a database. In addition to the detailed examination of the phytoliths, a statistical technique new to phytolith analysis was applied, allowing a more sophisticated comparison of the differences between assemblages.[1] The statistical comparison in question is known as *squared chord distances*, and is described as: $d_{ij} = {}_k(p_{ik}{}^5 - p_{jk}{}^5)^2$.

In this equation, d_{ij} is the squared chord distance between sample i and sample j, p_{ik} is the proportion of phytolith type k in sample i, and p_{jk} is the proportion of phytolith type k in sample j (Prentice 1980; Overpeck et al. 1985). This technique is employed by palynologists interested in comparing modern and fossil pollen assemblages in an effort to find present-day analogs for ancient assemblages. It is an elaboration of summing squared differences, adjusting for the signal to noise ratio created by the presence of highly and sparsely represented phytolith categories. Using this technique, modern pollen assemblages are codified, and the resulting data

TABLE 7.1
Rondel Phytolith Taxonomy

PHYTOLITHS WITH AN ENTIRE THIN FACE

   A. Thin face is a complete circle or oval outline, without decorations.

   B. Thin face is a complete circular or oval outline, with decorations present.

RELATION OF THIN FACE TO THICK FACE

   A-1. Thin face is approximately the same size as the thick face.

   B-1. Thin face is approximately the same size as the thick face, with decorations present.

   A-2. Thin face is substantially different in size from the thick face.

   B-2. Thin face is substantially different in size from the thick face, with decorations present.

PHYTOLITHS WITH AN INDENTED THIN FACE

   C-(1, 2, 3, etc.) thin face has one to several indentations.
      1. Thin face has one indentation.
      2. Thin face has two indentations.
      3. Thin face has two indentations directly across from each other.
      4. Thin face has three indentations.

   D-(1, 2, 3, etc.) thin face has one to several indentations, and decorations are present.
      1. Thin face has one indentation.
      2. Thin face has two indentations.
      3. Thin face has two indentations directly across from each other.
      4. Thin face has three indentations.

PHYTOLITHS WITH AN ENTIRE THICK FACE

   A. Thick face is a complete circle or oval outline, without decorations.

   B. Thick face is a complete circular or oval outline, with decorations present.

PHYTOLITHS WITH AN INDENTED THICK FACE

   C-(1, 2, 3, etc.) Thick face has one to several indentations.
      1. Thick face has one indentation.
      2. Thick face has two indentations.
      3. Thick face has two indentations directly across from each other.
      4. Thick face has three indentations.

   D-(1, 2, 3, etc.) thick face has one to several indentations, and decorations are present.
      1. Thick face has one indentation.
      2. Thick face has two indentations.
      3. Thick face has two indentations directly across from each other.
      4. Thick face has three indentations.

RONDEL PHYTOLITH FACE SHAPES

   1. Rondel face is circular.
   2. Rondel face is shaped like a square with rounded corners.
   3. Rondel face is shaped like an oval.
   4. Rondel face is shaped like an oval with squared corners on the ends.
   5. Rondel face is a rounded triangular form.
   6. Rondel face is a triangular form.
   7. Rondel face is irregular and does not fit in a category above.

are then compared to the codified data for fossil assemblages. In a similar manner, squared chord distances are employed in the current study to establish the relative dissimilarity of each prehistoric residue sample and each modern plant sample.

Genetic Control of Silica Body Assemblages: Distinguishing Silica Body Assemblages from Different Maize Types in the Archaeological Record

Variation in phytolith assemblages has long provided an important diagnostic tool for taxonomists. A number of studies have looked at teosinte fruitcase and maize cob phytolith profiles and documented variation in phytolith assemblages between different types of maize (Pearsall et al. 2003; Piperno and Pearsall 1993; Piperno et al. 2001; Thompson 1993; Thompson and Mulholland 1994). Similarly, Ball et al. (1999) noted differences in the silica body assemblages of different species of *Triticum*, and Terrrel and Wergin (1981) documented that silica deposition in *Zizania* differed among the species within the genera. Dorweiler and Doebley (1997), Dorweiler (1996), and Dorweiler et al. (1993) have identified a single locus in maize: *teosinte glume architecture* (*tga*1), which controls several aspects of glume morphology, both macroscopic and microscopic, including the deposition of silica in the form of phytoliths and silicified epidermal cells (Dorweiler and Doebley 1997: 1314). The team's work revealed that *tga*1 controls the induration, orientation, length, and shape of the glume (Dorweiler 1996: 20). The pleiotropic effects suggest that *tga*1 may represent a regulatory locus:

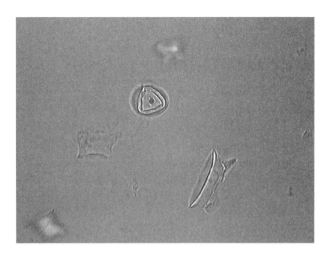

FIGURE 7.1 Entire decorated rondel in planar view and tilted rondel (1000× magnification).

> We have investigated several features of glume development to understand how *tga*1 controls glume induration (hardening). We compared the effects of the maize and teosinte alleles in the maize inbred W22. In this background, increased induration of the glumes in teosinte homozygotes *(tga*1 + *teosinte/tga*1 + *teosinte)* is attributable to a thicker abaxial mesoderm of lignified cells. Silica deposition in the abaxial epidermal cells of the glumes is also affected. The standard W22 line *(Tga*1 + *maize/Tga*1 + *maize)* has high concentrations of silica in the short cells of the epidermis of the glume, but the long cells have virtually no silica. In contrast, teosinte allele homozygotes deposit silica in both the short and long cells of the glume epidermis. Silica deposition also appears to be affected by genetic background. The teosinte background modifies the phenotype of *tga*1 plants toward a more uniform distribution of silica, whereas the maize background modifies the phenotype of *tga*1 plants toward concentration of silica in the short cells (Dorweiler 1996: 20).

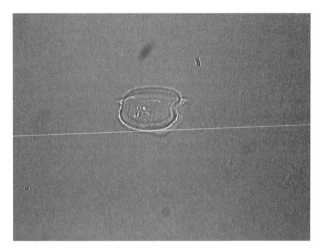

FIGURE 7.2 Indented rondel (1000× magnification).

Dorweiler found that, overall, the homozygous samples examined for each population displayed quite dramatic differences in morphology and silica distribution based on the effect of a single genetic locus: "measurements [morphological] of the heterozygote generally fall between the two homozygotes, though semi-dominance of the maize allele results in measurements of the heterozygote which more closely resemble values in the maize allele homozygotes" (Dorweiler 1996: 59). This was congruent with results obtained by studying the deposition of silica in the glumes of the three types:

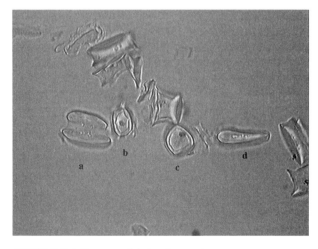

FIGURE 7.3 Four rondels in planar view (1000× magnification).

> The effects of *tga*1 on silica deposition have some archaeological relevance. Long after most plant material has decomposed, the insoluble silica crystals from within

FIGURE 7.4 Three maize ears showing *tga*1 phenotypes.

the cells, called phytoliths, remain. Phytolith size and three-dimensional structure can be analyzed to determine the species and relative proportions of plants that were growing in a particular area at a given time (Piperno 1984).

There is even some evidence that suggests maize and teosinte may be distinguishable by the relative proportions of each phytolith type (Piperno 1984). This difference was determined using leaf tissue. Our results showing the effects of *tga*1 and genetic background on the deposition of silica in the glume indicate that it will be important to analyze glume phytoliths in an archaeological context (Dorweiler and Doebley 1997: 1320–1321).

Ears from each of the three maize types used in the Dorweiler and Doebley (1997) study were made available for the present analysis (Figure 7.4). The original maize variety used in this study is W22. This is one of the inbred lines of maize, homozygous at all genetic locations. At the locus examined, each member of the population represented the *Tga*1 + *maize*/*Tga*1 + *maize* genotype. A second population consisted of plants with the *tga*1 + *teosinte*/*tga*1 + *teosinte* genetic background. The source of the teosinte allele in W22-TGA was *Z. mays* ssp. *mexicana* Wilkes 48703. The third population was heterozygous, with the expression of *Tga*1 + *maize*/*tga*1 + *teosinte*. Individual ears of each type of maize studied by Dorweiler (1996) were numbered and cataloged as members of each group: WM (maize inbred line W22), HM *(Tga*1 + *maize*/*tga*1 + *teosinte)*, or TM *(tga*1 + *teosinte*/*tga*1 + *teosinte)*.

A blind test was then conducted, based on a detailed examination of a sample of 100 rondel phytoliths recovered from cobs from each of the three groups (WM, HM, and TM). A total of 11 samples were described in detail, and a data table including all samples was created. The 11 blind phytolith samples were then compared with a second set of phytolith samples extracted from 12 ears of maize (4 from each group) used to establish the criteria by which each group could be recognized (Table 7.2). The numbers in Table 7.2 represent the distance between each blind sample and the representative ears of maize. The lower the number, the greater the similarity between the blind rondel phytolith sample and the rondel phytolith samples extracted from representative ears of the three groups. The shaded areas represent the groups of samples to which the blind samples were most similar. The blind samples that represent the W22-TGA homozygous population are readily identified. Blind samples 1, 3, 6, and 15 match well to the TM group, with generally about half the distance between themselves and the heterozygous populations as between themselves and the W22 homozygous populations (Table 7.2). The W22 homozygous blind samples (2, 5, 11, and 13) also separate clearly. Each of these samples is roughly half as distant from the heterozygotes as from the W22-TGA homozygotes. The blind samples representing phytolith assemblages from heterozygous ears are less readily distinguished. Samples 4, 8, and 14 represent heterozygous populations, and have morphology more similar to W22 than to W22-TGA homozygotes. This is reflected in the dissimilarity indices (Table 7.2) and presented visually in Figure 7.5.

Replicability

Characterization of the phytolith assemblages of each type of maize allowed the recognition of blind samples based on a replicable, verifiable set of data. To test the replicability of this study, a research assistant was trained in both the description of rondel phytolith forms and the image analysis techniques used to conduct length and width measurements. She then built a database of morphological descriptions and measurements of rondels from the same cobs used to create the initial database.[2] To further test the potential for chaff phytolith assemblages to reflect different types of maize, and the possible variation in chaff phytolith profiles within single cobs, the assistant analyzed blind samples of chaff from whole cobs, as well as chaff taken from the proximal, medial, and distal thirds of cobs.

Table 7.3 and Figure 7.6 show the replicability test results using the squared chord distance technique to compare blind samples with phytolith samples from cobs representing the three maize groups. In all cases, it was possible to match the blind sample with the correct group. A single sample (TM 7C) provided a problematic result. Although TM 7C is correctly classified as TM when average distances are compared, the two closest matches are between this sample and heterozygous database populations.

These results show that the comparison of phytolith production from the chaff of maize cobs allows blind samples of maize to be matched to their source group. As suggested by Dorweiler and Doebley (1997), genetically based changes in the morphology of maize cupules and glumes linked to the *tga*1 locus are reflected in the assemblages of phytoliths produced. As a result, the squared chord distance technique outlined above would appear to hold great promise in characterizing the expression of the *tga*1 locus. *Tga*1 acts as a regulatory gene during plant development. Both direct and pleiotropic effects of *tga*1 result in differing patterns of

TABLE 7.2
Squared Chord Distance Results of Blind Test Identifying Cob Types

	TM 6	TM 12	TM 16	TM 20	HM 2	HM 3	HM 7	HM 8	WM 3	WM 4	WM 5	WM 11
Blind Sample 1	85	99	109	75	159	191	182	193	249	330	280	331
Blind Sample 2	239	250	217	223	174	212	246	214	104	137	87	99
Blind Sample 3	72	93	79	88	141	201	163	175	250	308	261	328
Blind Sample 4	149	149	148	139	50	78	87	90	118	160	132	168
Blind Sample 5	246	265	233	241	146	187	202	173	85	89	73	87
Blind Sample 6	70	71	74	62	163	186	145	161	256	322	295	340
Blind Sample 8	184	164	142	169	97	80	98	85	101	122	100	133
Blind Sample 11	264	286	243	251	137	174	204	168	70	71	64	72
Blind Sample 13	263	267	238	250	126	148	163	143	69	79	77	83
Blind Sample 14	141	166	145	146	73	94	112	89	119	169	114	180
Blind Sample 15	70	96	85	62	142	196	168	171	232	302	244	306

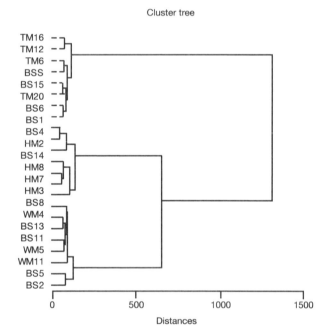

FIGURE 7.5 Squared chord distance results of blind test identifying maize types.

silica deposition in the glumes of different varieties of maize, providing distinctive maize rondel phytolith assemblage profiles or "fingerprints."

In addition, this technique can be used to trace one of the most important changes in the domestication of maize from teosinte: the loss of the indurate fruitcase (Dorweiler and Doebley 1997). Well-chewed quids, coprolites, and the residues remaining on any grinding equipment could provide phytolith assemblages from chaff. These assemblages, in turn, can indicate selection for softer glumes that do not encapsulate kernels, and trace the ubiquity of the use of new maize varieties across communities. Beyond the development of softer glumes, *tga*1, a regulatory locus, has been shown to interact with other genes during the development of the maize cob. The resulting rondel assemblages provide a tool for identification more powerful than simply reflecting the hardness of glumes. Hart et al. (2003), for example, showed that maize glume phytolith assemblage fingerprints can be employed to identify different lineages within the Northern Flint Maize Complex, which is considered to be a genetically distinct and relatively homogenous group (Doebley et al. 1986). This ability to distinguish between different varieties of maize within the homogenous Northern Flint Complex was substantially improved once phytolith size was factored out of the analysis. Size is an aspect of Graminae phytoliths demonstrated to be most affected by environmental, rather than genetic, factors. When length and width were removed from consideration in the analysis of the Northern Flint Complex, varietal identification of maize became more statistically robust, as did comparisons of the maize samples and residues.

Application of the Approach: A South American Case Study

Materials and Methods

Staller and Thompson (2002) presented a multidisciplinary analysis of data recovered during excavation of the La Emerenciana site in Ecuador, which indicated that maize was present in South America by late Valdivian times but did not seem to be a significant part of the diet. Phytolith assemblages recovered from three Valdivian vessels as part of that study indicated the presence of maize. In the current study, one of the La Emerenciana vessels was re-examined, along with two pottery vessels from Pirincay, a highland Formative Period site in Ecuador. Two vessels from the site of Palmitopamba, a multicomponent Yumbo Period site with a later Inca component, located in the rain forest region of western

TABLE 7.3
Squared Chord Distance Results of Replicability Tests

		WM	WM	WM	WM	HM	HM	HM	HM	TM	TM	TM	TM
A	HM14 B	192	142	132	125	51	67	52	64	88	75	72	91
B	WM20	95	70	81	93	158	178	149	138	245	215	191	239
C	WM6 B	79	57	74	60	121	142	124	118	188	172	140	170
D	TM15	266	188	198	177	77	76	105	96	64	70	83	61
E	TM7 B	230	164	168	144	71	64	88	80	59	70	73	61
F	HM6	182	111	104	126	60	45	48	34	84	67	73	83
G	TM2	200	140	162	129	67	70	103	87	61	78	63	67
H	WM6 A	82	58	63	64	162	174	159	150	243	221	181	234
I	HM5	113	69	68	73	65	79	70	74	118	116	99	125
J	WM6 C	70	51	57	59	145	170	149	151	228	217	178	212
K	TM7	246	186	206	173	92	84	114	109	75	95	104	78
L	WM14	64	50	63	67	154	175	147	151	229	220	179	221
M	TM7 C	152	103	110	116	79	79	98	94	81	84	59	98
N	HM14 C	133	100	91	99	78	94	75	85	148	136	115	154
O	HM14 A	172	115	98	109	54	52	36	51	112	98	71	109
P	TM7 A	244	189	196	167	77	85	113	99	65	82	86	47
Q	HM 14	126	82	81	70	56	66	52	56	95	90	70	95
R	WM6	74	67	68	85	186	207	177	168	271	242	206	244

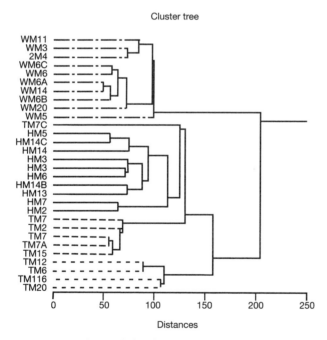

FIGURE 7.6 Squared chord distance results of replicability tests.

Pinchincha province, Ecuador, have also been examined. Pottery from Ancon and Sierra Gorda, Peru, representative of the Early Horizon, also yielded phytolith assemblages, as did ceramic vessels recovered from the Nicoya Peninsula of Costa Rica, demonstrating the potential of this technique for identifying maize in food residues in the lowlands of Central America (Figure 7.7).

In order to provide a set reference baseline of comparison for analysis of the food residues samples recovered from these cooking vessels, phytolith assemblages were extracted from maize chaff samples taken from modern cobs collected across lowland Central America and northwestern South America (Figure 7.8, Table 7.4). Comparison of the phytolith assemblages recovered from cooking vessel food residues with those extracted from the cob chaff specimens of the modern reference collection was carried out using the same methods as were employed in blind tests to establish the utility of the technique.

Results

The phytolith assemblages composing the cob chaff reference set were first compared to each other in order to establish the extent to which samples could be correctly separated. As shown in the Ward's cluster analysis of the squared chord distance data on the maize samples (Figure 7.8), the 45 samples of cob chaff sorted into three main groups. The first

FIGURE 7.7 Location of archaeological sites and maize collection areas mentioned in text: (1) Nicoya Peninsula; (2) Boruca; (3) Talamanca Reserve; (4) Bri Bri; (5) Palmitopamba and Mitad del Mundo; (6) Banos and Sarayacu; (7) Pirncay and La Emerenciana; (8) Ancon and Sierra Gorda; and (9) Jauja and Huancayo.

group includes *canguil* and *pisankalla*, maize varieties cultivated today in widely separated areas of South America (Ecuador and Argentina). Despite their geographic separation, genetic profiles of these two groups indicate that they are closely related.[3] Also within this group are *confite punteaguado* from Jauja, Peru; chulpi from Huancayo, Peru; and a cob of a *morado* variety from Chancay, Peru. The second cluster contains most of the remaining South American lowland and highland samples, and a yellow flint type collected from Bri Bri in Costa Rica (Figure 7.8). The third cluster includes most of the Central American samples, along with a sample from the Amazon basin region of Ecuador. The presence of the Ecuadorian sample in the third cluster may represent actual affinities with the Central American types, or it may simply be an artifact of the relatively small size of the reference class database.

Clear patterns were also evident when archaeological food residue phytolith assemblages were compared with the cob chaff reference collection groupings. The phytolith assemblage recovered from cooking vessel food residues at the La Emerenciana site in Ecuador, dated to 4100 BP, for example, can be seen to cluster very closely with phytolith assemblages recovered from cooking vessels of the nearby but much later occupied highland site of Pirincay (ca. 2700–2200 BP), as well as with phytolith assemblages extracted from modern varieties of maize (the *canguil, pisankalla* group) (Figure 7.9). The observed similarities between pre-Columbian cooking vessel food residue phytolith assemblages and those extracted from the cob chaff of modern indigenous varieties of maize does not, of course, indicate an exact one-to-one correlation. Maize evolution is a continuous process. What is clear, however, is that the maize being grown at La Emerenciana 4,100 years ago and at Pirincay 2,700 years ago, as well as several modern Ecuadorian varieties, are quite comparable in terms of their relative expression of *tga*1. Food residues recovered from cooking vessels at two Early Horizon settlements farther south along the Peruvian Coast—Ancon and Sierra Gorda—again yielded very similar phytolith assemblages that clustered closely together within the *canguil, pisankalla* group of maizes (Figure 7.9).

TABLE 7.4
Central and South American Maize and Residue Samples

Abbreviation	Sample	Location	Comments
Central and South American Maize Samples			
Matpur	Dark purple maize from Matambu 10 row	Matambu, Costa Rica	Collected by Thompson 1996
Mat Y	Tropical Flint 12 row	Matambu, Costa Rica	Collected by Thompson 1996
Mor	Morocho	Palmitopamba, Ecuador Quito. Ecuador	Quito sample obtained from UWM Herbarium
BrBrY	Tropical Flint12 row	Bri Bri, Costa Rica	Collected by Thompson 1996
Talamanca	Tropical Flint	Talamanca Reserve, Costa Rica	Collected by Thompson 2000
Bor Br	Tropical Flint	Boruca, Costa Rica	Collected by Thompson 1996
SAR8R A	8-row flint	Sarayacu, Ecuador Eastern Amazon Lowlands	Collected by Thompson 2003
SAR8R B	8-row flour	Sarayacu, Ecuador Eastern Amazon Lowlands	Collected by Thompson 2003
Morpp	Morocho 16 rows	Palmitopamba, Ecuador	Collected by Thompson 2003
Moramsar	Morocho Amarillo 16 rows	Sarayacu, Ecuador	Collected by Thompson 2003
Morsar	Morocho 16 row	Sarayacu, Ecuador	Collected by Thompson 2003
Cyl8r	8-row flour maize	Mitad del Mundo	Collected by Thompson 2003
Morrojpp	Morocho roja 16 row	Palmitopamba, Ecudaor	Collected by Thompson 2003
Ambanos	Amarillo 12 row	Banos	Collected by Thompson 2003
Morampp	Morocho amarillo	Palmitopamba, Ecuador	Collected by Thompson 2003
Amar	Amarillo	Quito, Ecuador	Collected by Hugh Iltis
Amarmdm	Amarillo	Mitad del Mundo, Ecuador	Collected by Thompson 2003
Canguil	Popcorn	Quito, Ecuador	Collected by Hugh Iltis
Ch mor	Morado	Chancay, Peru	Collected by Thompson 2000
Mcomm	Mais Commun	Jauja, Peru	Collected by Hastorf
Confpunt	Confite punteaguado	Jauja, Peru	Collected by Hastorf
ChulpH	Chulpi	Huancayo, Peru	Collected by Hastorf
Pisky	Pisankalla popcorn	Yutopian, Argentina	Collected by Thompson 1996
Piskg	Pisankalla popcorn	University of Minnesota Greenhouse	Grown in 1998
Food Residue Samples			
Sierra Gorda	Early Horizon	Peru	
Ancon	Early Horizon	Peru	
A086	Formative	Pirincay, Ecuador	
A080	Formative	Pirincay, Ecuador	
LE 5623	Formative	La Emerenciana, Ecuador	Directly AMS dated
PP Yumbo	800–1500 AD	Palmitopamba, Ecuador	4 samples
PP Inca	Ca. 1500 AD	Palmitopamba, Ecuador	
CR 128-7	Unknown	Nicoya Peninsula, Costa Rica	
CR 128-34	Unknown	Nicoya Peninsula, Costa Rica	
CR 128-42	Unknown	Nicoya Peninsula, Costa Rica	

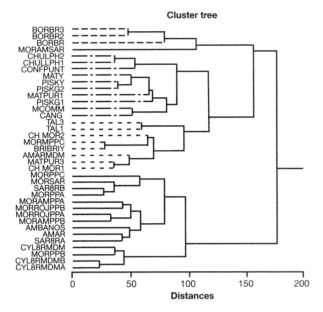

FIGURE 7.8 Squared chord distance results of modern maize comparative samples.

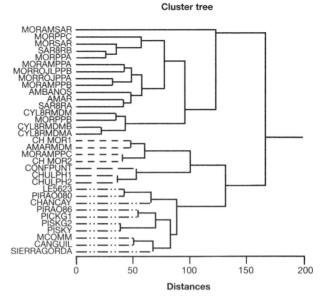

FIGURE 7.9 Squared chord distance results of La Emerenciana, Pirincay, Chancay, and Sierra Gorda vessels.

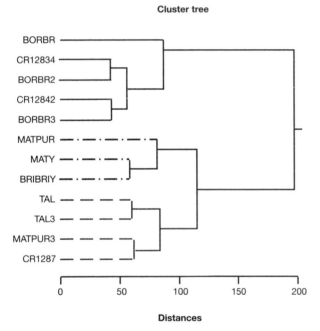

FIGURE 7.10 Squared chord distance results of Nicoya Peninsula vessels.

A third clear example of temporal continuity in maize cob glume and cupule architecture within a region can be seen in a comparison of food residue phytolith assemblages from lowland Central America with present-day maize varieties of the region. Ceramic vessels collected from the Nicoya peninsula of Costa Rica in 1877, currently in the collections of the Anthropology Department of the University of Minnesota, yielded food residue phytolith assemblages that were very similar to those found in one of a larger group of different, modern Costa Rican varieties of maize exhibiting very distinctive morphologies. Figure 7.10 presents a cluster analysis of the results of the comparison of the Costa Rican food residue and cob chaff assemblages. Each of the modern maize variety morphotypes collected produced a distinctive and unique assemblage of phytoliths. Both of the residue samples produced similar assemblages to those recovered from a brown-kerneled tropical flint maize collected at Boruca. Stone (1949) noted that 12 varieties of maize were being grown at Boruca. That a variety of maize currently being grown at Boruca would yield a cob chaff phytolith assemblage so similar to that recovered from ca. 1877 ceramic vessels of the Nicoya peninsula is not surprising, since the geographical and temporal distance between them is not great.

The site of Palmitopamba, in western subtropical Ecuador, provides an example of using residue analysis to study the social implications of maize use. The occupation of the site spanned both the Yumbo and Inca periods. The Yumbo were tropical forest people who occupied the western flank of the Andes in northern Ecuador from about 800 AD into the late 1700s. The Inca conquest of the northern Ecuadorian highlands is believed to have begun in the 1470s, and Inca material believed to date until the mid 1500s has been recovered from the site. Palmitopamba is a steep hill that was modified into a series of platforms by the Yumbo. These platforms are apparently not agriculturally related, but rather are civic and or religious in nature. The Inca subsequently utilized this terraced hillside as a fort, and archaeological research on the nature of potential Inca modifications of the hillside is ongoing (Lippi 2004).

The terraces at Palmitopamba have yielded both Yumbo and Inca pottery. Much of the pottery recovered retained food organic encrustations on the interior surfaces. These encrustations represent the residue of material cooked in the

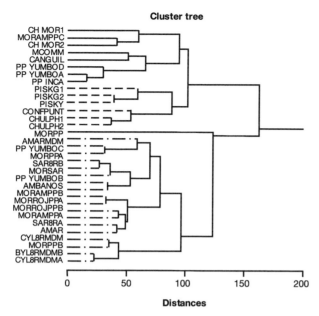

FIGURE 7.11 Squared chord distance results of Palmitopamba vessels.

vessels, presumably for food. So far, five vessels have been sampled for phytoliths. The Inca and Yumbo vessels analyzed yielded phytolith assemblages reflecting continuity in terms of expression of *tga1* (Figure 7.11). Yumbo vessels yielded phytolith evidence of canguil, and another type of assemblage similar to the morocho and morochilla types grown in the region today.

Discussion

The case study presented in this chapter draws from and builds upon the work of a number of other researchers. More than 75 years ago, Edman and Soderberg (1929) first showed that the phytolith signature of a domesticated plant (rice) could be recovered from the residue encrusted on an ancient cooking vessel. Twenty-five years ago, Pearsall (1978) and Piperno (1984) identified a distinctive, cross-shaped phytolith form they considered in size and shape to be unique to and diagnostic for the most important domesticated plant of the Americas—*Zea mays*. A decade later, a number of researchers demonstrated the value of considering total phytolith assemblages, rather than focusing on individual "diagnostic" phytoliths, and the significance of distinctively different phytolith assemblages being produced by different component parts of the maize plant (Mulholland 1987, 1989; Piperno 1988; Thompson 1993; Thompson and Mulholland 1994). In addition, Mulholland (1989, 1993) pioneered the development of statistical approaches to identifying maize cob phytolith assemblage signatures in archaeological sediments. In the mid 1990s, Dorweiler, Doebley, and others (Dorweiler 1996; Dorweiler and Doebley 1997; Dorweiler et al. 1993) documented the genetic basis for variation in the cupule and glume architecture (and phytolith signature) of maize cobs.

This study demonstrates that it is not only possible to identify the presence of maize cob chaff in the organic residue recovered from ceramic cooking vessels (or a range of other possible processing and depositional contexts; i.e., coprolites, quids, tooth calculus, grinding slabs), but it is also possible to determine with considerable accuracy the relative effect of *tga1* on silica bodies in the maize being prepared for consumption. In effect, it is possible to provide information regarding which varieties of maize were being grown, processed, and consumed.

The research potential of being able to identify maize cob phytolith assemblages at the subspecific level is shown in the two Formative Period South American examples presented here. A comparison of phytolith assemblages recovered from ceramic vessels from two Formative Period sites in Ecuador that were separated by more than 1500 years (La Emerenciana, 4100 BP; Pirincay, 2700–2200 BP) demonstrates the continued cultivation of a type of maize with similar glume silica deposition characteristics, and also the appearance at Pirincay of a second type of maize with different glume silica deposition. So while the debate involving the timing of the diffusion of maize into South America cannot be resolved through ceramic food residue analysis, the presence of a distinct maize lineage in Ecuador by the late Formative can be firmly established. Another South American example shows a similar close continuity between maize types recovered from Early Horizon period vessels (ca. 2000 BP) at the sites of Ancon and Sierra Gorda, along the south coast of Peru. The movement of popcorn lineages across the continent of South America appears to be a very ancient phenomenon. Anderson (1947) described the presence of pointed popcorn in burial contexts from early pottery bearing sites throughout South America.

A third case study example, which compared phytolith assemblages from ceramic vessels collected more than 125 years ago in Costa Rica with contemporary maize varieties from the region, again indicates remarkable continuity of glume architecture through time, and underscores the significant research potential of a wide range of materials currently curated in museum collections. Residues from newly excavated or long archived ceramics, whether pre-Columbian or more recent, along with groundstone tools, coprolites, or quids can provide AMS dates and phytolith assemblages. Clearly, a wealth of information regarding the initial diffusion and subsequent roles of maize in societies across the Americas remains to be recovered from museum collections.

A later prehistoric site in Ecuador, Palmitopamba, provides a demonstration of the potential for this research to aid in understanding the social implications of maize use. The Inca presence in the northwestern edge of their territory is not well understood, and their relations with the Yumbo may not have been as conquerors. The Inca are well known for introducing their own varieties of maize as they moved into and subsumed various polities (Hastorf 1994). If further ongoing analyses support this initial result, it would appear that the Inca presence in this region did not include the introduction of new maize types, but rather the adoption of local types. This is consistent with the ethnohistoric data on the region (Lippi 2004), which suggest that the Inca never subjugated the Yumbo.

It is important to recognize that phytolith assemblages have a good probability of being preserved intact and can be extracted from a wide range of long-archived materials, even when they have been subjected to a variety of undescribed cleaning techniques. Organic residues, after all, are simply a matrix containing silica cells. The fundamental requirement for the analysis of long-archived ceramic specimens is the same as for freshly excavated ones, which is that the surface contact layer be mechanically removed. The remaining residue should contain only phytoliths resulting from deposition during the use of the pot. While the processing of food residues from ceramic sherds is similar to that for maize cob chaff samples, more time is usually required for the nitric acid treatment in removal of the organic component.

The approach outlined here for the extraction, analysis, and characterization of maize cob phytolith assemblages from ancient ceramic vessels also holds great promise for recognizing the presence of other domesticated grasses in the archaeological record. Patterson et al. (2001) outline a range of similar genetic changes that occur in the domestication of grasses. Maize, rice, and wheat have all undergone convergent genetic changes, underscoring that the process of domestication is similar across the grasses. In a recent analysis of residues in storage vessels from the site of Tel Kedesh, both the approach outlined here and a similar technique reported by Ball et al. (1998) were employed to search for evidence of wheat. Both approaches pointed to the presence of *Triticum aestievum*, or breadwheat (Berlin et al. 2003), indicating that the approach presented here can be successfully employed to characterize and identify assemblages of phytoliths from the chaff of any of the domesticated grasses. Given the ubiquity of domesticated grasses, and the central economic and social role afforded them, this approach holds the promise of providing biologically firm identifications and culturally significant insights worldwide.

Notes

1. Margaret Davis, Randy Calcote, and Shinya Sugita recommended the squared chord distance technique as an appropriate way to compare phytolith assemblages.
2. Because the assistant had a limited time (only three weeks) to learn the techniques before conducting her own study, it was decided to compare her blind samples to her own database, rather than to Thompson's initial database. This was done to mitigate the effects of incomplete learning. As a result, even if her description of a rondel type differed from Thompson's, it would remain consistent throughout the tests.
3. Two of the *pisankalla* samples were grown in a greenhouse at the University of Minnesota, in different soils and under different growing conditions than the parent cob, which was collected in Catamarca province, Argentina. The phytolith assemblages recovered from these cobs are more similar to each other than to samples from other lineages, supporting the concept that genetics controls silica deposition in different lineages.

References

Anderson, E. 1947. Popcorn. *Natural History* 56: 227–230.

Ball, T., J. Gardner, and J. Brotherson. 1998. Identifying phytoliths produced by the inflorescence bracts of three species of wheat (*Triticum monococcum* L., *T. dicoccon* Schrank, and *T. Aestivum* L.) using computer-assisted image and statistical analysis. *Journal of Archaeological Science* 23: 619–632.

Ball, T., J. Gardner, and N. Anderson. 1999. Identifying inflorescence phytoliths from selected species of wheat (*Triticum monococcum* L., *T. dicoccon* Schrank, and *T. Aestivum* L.) and barley (*Hordeum vulgare* and *H. spontaneum*). *American Journal of Botany* 86: 1615–1623.

Bennetzen, J., E. Buckler, V. Chandler, J. Doebley, J. Dorweiler, B. Gaut, M. Freeling, S. Hake, E. Kellogg, R. S. Peothig, V. Walbot, and S. Wessler. 2001. Genetic evidence and the origin of maize. *Latin American Antiquity* 12: 84–86.

Benz, B. 2001. Archaeological evidence of teosinte domestication from Guilá Naquitz, Oaxaca. *Proceedings of the National Academy of Science, USA* 98: 2104–2106.

Berlin, A., T. Ball, R. Thompson, and S. Herbert. 2003. Ptolemaic agriculture, "Syrian Wheat," and *Triticum aestivum*. *Journal of Archaeological Science* 30: 115–127.

Bozarth, S. R. 1993. Maize *(Zea mays)* cob phytoliths from a central Kansas Great Bend Aspect archaeological site. *Plains Anthropologist* 38: 279–286.

Doebley, J., M. Goodman, and C. Stuber. 1986. Exceptional genetic divergence of northern flint corn. *American Journal of Botany* 73: 64–69.

Dorweiler, J. 1996. Genetic and evolutionary analysis of glume development in maize and teosinte. PhD thesis, University of Minnesota. Ann Arbor: University of Michigan Microfilms.

Dorweiler, J. and J. Doebley. 1997. Developmental analysis of teosinte glume architecture1: A key locus in the evolution of maize (Poaceae). *American Journal of Botany* 84: 1313–1322.

Dorweiler, J., A. Stec, J. Kermicle, and J. Doebley. 1993. Teosinte glume architecture1: A genetic locus controlling a key step in maize evolution. *Science* 262: 233–235.

Edman, G. and E. Soderberg. 1929. *Auffindung von reiss in einer tonscherbe aus einer etwas funftausend jahrigen Chinesischen siedlung.* Bulletin of the Geological Society of China, 8: 363–365.

Fritz, G. 1994a. Are the first American farmers getting younger? *Current Anthropology* 35: 305–309.

———. 1994b. Reply to Piperno. *Current Anthropology* 35: 639–643.

Hart, J., R. Thompson, and H. J. Brumbach. 2003. Phytolith evidence for early maize in the northern Finger Lakes region of New York. *American Antiquity* 68: 619–640.

Hastorf, C. 1994. The changing approaches to maize research. In *History of Latin American archaeology*, Augusto Oyuela-Caydedo (ed.), pp. 139–154. Aldershot: Brookfield.

Hastorf, C. and M. DeNiro. 1985. New isotopic method used to reconstruct prehistoric plant production and cooking processes. *Nature* 315: 489–491.

Heron, C., R. Evershed, and L. Goad. 1991. Effects of migration of soil lipids on organic residues associated with buried potsherds. *Journal of Archaeological Science* 18: 641–659.

Lippi, R. 2004. *Tropical forest archaeology in western Pinchincha, Ecuador.* Case Studies in Archaeology, J. Quilter (ed.). Belmont, Calif.: Thomson Wadsworth.

Matsuoka, Y., Y. Vigoroux, M. Goodman, J. Sanchez, E. Buckler, and J. Doebley. 2002. A single domestication for maize shown by multilocus microsatellite genotyping. *Proceedings of the National Academy of Science, USA.* 99: 6080–6084.

Mulholland, S. 1987. *Phytolith studies at Big Hidatsa, North Dakota.* PhD thesis, University of Minnesota. Ann Arbor: University of Michigan Microfilms.

———. 1989. Phytolith shape frequencies in North Dakota grasses: A comparison to general patterns. *Journal of Archaeological Science* 16: 489–511.

———. 1993. A test of phytolith analysis at Big Hidatsa, North Dakota. In *Current research in phytolith analysis: Applications in archaeology and paleoecology*, D. Pearsall and D. R. Piperno (eds.), pp. 131–145. Philadelphia: MASCA, The University Museum of Archaeology and Anthropology.

Mulholland, S. and G. Rapp. 1992. A morphological classification of grass silica bodies. In *Phytolith systematics: Emerging issues*, G. Rapp, Jr. and S. Mulholland (eds.), pp. 65–89. Advances in Archaeological and Museum Science, Volume 1. New York: Plenum Press.

Overpeck, J., T. Webb III, and I. Prentice. 1985. Quantitative interpretation of fossil pollen spectra: Dissimilarity coefficients and the method of modern analogs. *Quaternary Research* 23: 87–108.

Patterson, A. and Y. R. Lin. 1995. Convergent domestication of cereal crops by independent mutations at corresponding genetic loci. *Science* 269: 1714–1719.

Pearsall, D. M. 1978. Phytolith analysis of archaeological soils: Evidence for maize cultivation in Formative Ecuador. *Science* 199: 177–178.

Pearsall, D. M. and D. R. Piperno. 1990. Antiquity of maize cultivation in Ecuador: Summary and reevaluation of the evidence. *American Antiquity* 55: 324–337.

Pearsall, D. M., K. Chandler-Ezell, and A. Chandler-Ezell. 2003. Identifying maize in Neotropical sediments and soils using cob phytoliths. *Journal of Archaeological Science* 30: 611–627.

Piperno, D. R. 1984. A comparison and differentiation of phytoliths from maize and wild grasses: Use of morphological criteria. *American Antiquity* 49: 361–383.

———. 1988. *Phytolith analysis: An archaeological and geological perspective.* San Diego: Academic Press.

Piperno, D. R. and K. Flannery. 2001. The earliest archaeological maize (*Zea mays* L.) from highland Mexico: New accelerator mass spectrometry dates and their implications. *Proceedings of the National Academy of Science, USA* 98: 2101–2103.

Piperno, D. R. and D. M. Pearsall. 1993. Phytoliths in the reproductive structures of maize and teosinte: Implications for the study of maize evolution. *Journal of Archaeological Science* 20: 337–362.

Piperno, D. R. and D. M. Pearsall. 1998. *The Origins of Agriculture in the Lowland Neotropics.* San Diego: Academic Press.

Piperno, D. R., I. Holst, A. J. Ranere, P. Hansell, K. E. Stothert. 2001. The occurrence of genetically controlled phytoliths from maize cobs and starch grains from maize kernels on archaeological stone tools and human teeth, and in archaeological sediments from southern Central America and northern South America. *The Phytolitharien* 13: 1–7.

Prentice, I. 1980. Multidimensional scaling as a research tool in quaternary palynology: A review of theory and methods. *Review of Palaeobotany and Palynology* 31: 71–104.

Rovner, I. 1983. Plant opal phytolith analysis: Major advances in archaeobotanical research. In *Advances in Archaeological Method and Theory*, Volume 6, M. Schiffer (ed.), pp. 225–266. New York: Academic Press.

Russ, J. and I. Rovner. 1989. Stereological identification of opal phytolith populations from wild and cultivated *Zea*. *American Antiquity* 54: 784–792.

Smalley, J. and M. Blake. 2003. Sweet beginnings. *Current Anthropology* 44(5): 675–703.

Smith, B. D. 1998. *The emergence of agriculture*, 2nd edition. New York: Scientific American Library.

———. 2001. Documenting plant domestication: The consilience of biological and archaeological approaches. *Proceedings of the National Academy of Sciences, USA.* 98: 1324–1326.

Staller, J. 2003 An examination of the paleobotanical and chronological evidence for an early introduction of maize *(Zea mays)* into South America. *Journal of Archaeological Science* 29: 33–50.

Staller, J. and R. Thompson. 2002. A multidisciplinary approach to understanding the initial introduction of maize into coastal Ecuador. *Journal of Archaeological Science* 29: 33–50.

Stone, D. 1949. The Boruca of Costa Rica. Papers of the Peabody Museum Volume 26 (2). Cambridge, MA: Harvard University.

Terrell, E. and W. Wergin. 1981. Epidermal features and silica deposition in lemmas and awns of *Zizania* (Graminae). *American Journal of Botany* 68: 697–707.

Thompson, R. 1993. Opal phytolith evidence complements isotope studies of food residues from the Upper Mantaro Valley, Peru. Paper presented at the Northeast Conference of Andean Archaeology and Ethnohistory. Boston, Mass.

Thompson, R. and S. Mulholland. 1994. The identification of corn in food residues on utilized ceramics at the Shea Site (32CS101). *Phytolitherian Newsletter* 8: 7–11.

SECTION TWO

GENETIC DOCUMENTATION OF PLANT DOMESTICATION

Eve Emshwiller, Section Editor

CHAPTER 8

Genetic Data and Plant Domestication

EVE EMSHWILLER

Chapter Overview

This is an extremely exciting time for the study of plant domestication. Genetic techniques are providing answers to previously insoluble questions about crop origins. Although interest in understanding these origins has persisted for many years, the recent availability of new methods, especially those based on the polymerase chain reaction (PCR), has led to an explosive increase in the number of studies of crop domestication and evolution. Previously unasked questions are being investigated, such as identifying the genes responsible for the changes in plant form brought about by domestication and assessing the effects of domestication on variation in the genome. In addition, some studies with different goals, like characterizing the genetic diversity of crops and their wild relatives for conservation or breeding purposes, also provide information about crop domestication as an ancillary benefit.

A comprehensive review of the enormous number of recent papers on this topic is beyond the scope of this chapter. My goal here is to discuss the different manners in which genetic data have been used to address questions of crop domestication. I begin with a brief enumeration of the kinds of questions about plant domestication that can be studied with molecular data, followed by a description of the kinds of genetic data and their advantages and pitfalls. To illustrate how these data have been applied to each of the kinds of questions, I use examples from a few well-studied crops. I make special reference to the wealth of studies of maize, because genetic data have been applied to many different aspects of the domestication of this important crop. Finally, I will introduce the four case study chapters that appear in this book and conclude with a brief discussion of future prospects.

Aspects of Domestication Studied with Genetic Data

Crop domestication is a process of evolutionary change in the genetics of plant populations, brought about by human influence. That influence is wrought through a combination of deliberate selection for favorable traits (conscious selection) and automatic selection (unconscious selection) as a response to human modification of the environment and human control of plant propagation (see Harlan et al. 1973; de Wet and Harlan 1975; Zohary 1984; Heiser 1988). The term *domestication* is sometimes used more broadly. It may refer to symbiosis, mutualism, or "co-evolution" among species—with or without the involvement of humans—or it may refer to various kinds of resource management considered to be domestication of landscapes or the environment (e.g., Rindos 1989; Yen 1989; Terrell, et al. 2003; Smith 2005). But the discussion herein is limited to the effects of human influence on the genetics of individual (plant) taxa.

Different crops that are used and harvested in the same way (e.g., grain crops) tend to undergo similar changes in form when domesticated, due to similar changes in selective pressures. These common phenotypic changes constitute an "adaptive syndrome of domestication" (de Wet and Harlan 1975; Hammer 1984 cited by Heiser 1988; Harlan 1992; Smith 1998). Yet those crops that are used, harvested, and propagated in different ways will have different modifications (Zohary 1984). The transformations that are observable in the archaeological record are morphological, such as an increase in seed size or the loss of fragility of the rachis in grain crops. But domestication can involve changes that are less easily observed, such as changes in chemical composition (e.g., glycoalkaloids, Johns 1989; Johns and Alonso 1990), in day length sensitivity, or in phenology (e.g., timing or uniformity of fruit ripening). It is obvious that the genes that control these particular traits are affected by domestication. Less obvious is that the evolutionary selection on these genes can reduce molecular variation in other parts of the genome as well, especially in chromosomal regions that are linked to the selected genes. This selection leads to effects dubbed *genetic bottlenecks* when the reduction in diversity affects the entire genome, or the *footprint* or *signature* of selection when it affects only a small zone linked to the gene.

Genetic data are being used to investigate all of these domestication effects. However, to investigate how domestication has been manifested genetically in any particular crop, that crop must be compared to the wild species from which it was domesticated, so its origins must be studied first. In the following summary of the kinds of questions about plant domestication that may be studied with genetic data, I begin with the applications of genetic data to questions of a crop's origins, followed by applications to the study of evolution under domestication.

Crop Origins

The first set of questions relates to the basic "What? Where? When?" of crop domestication. Usually the first step is to identify the wild progenitor of the crop, that is, the species that was brought under domestication. In some cases this question has already been well established on the basis of other data (see Zohary and Hopf 1994), but in other cases, even today, the wild progenitor of a crop is unknown, and genetic data are used to help identify the ancestral species. Some crops have not diverged much in morphology from wild relatives, in which case genetic data can test prior morphology-based hypotheses about the progenitor taxon. The morphologically similar wild taxon may be supported by the genetic data as a probable progenitor, or alternatively, the wild plants may be revealed as escaped from cultivation, as hybrid derivatives, or as a closely related taxon that was nevertheless not the progenitor (see Chapter 10). Once the progenitor taxon is identified, the crop and the progenitor are usually considered to belong to a single species, because they are not only capable of continued interbreeding, but usually are indeed continuing to exchange genes (see Zohary and Hopf 1994). Exceptions include cases in which interspecific hybridization and/or polyploidy are involved in the crop's history.

Many crops are polyploids; that is, they have more than two chromosome sets. The study of the origins of polyploid crops constitutes a special case, because it involves not only identifying the wild species that was domesticated, but also finding the diploid taxa from which the genomes of the polyploid are derived. Polyploids may be derived from a single diploid species, or they may arise through hybridization of several species or well-differentiated populations. Wild polyploid taxa may be domesticated, or domestication may precede polyploidization. I address studies of these crops further in the case study chapter on *Oxalis tuberosa* (Chapter 12).

Whether or not the crop is polyploid, it is important to remember that the origins of the crop occurred in the past, and thus both the crop and the extant populations of the progenitor taxon are *descendants* of the original progenitor. Although the time since domestication is relatively short for the accumulation of mutations (except in rapidly evolving markers such as microsatellites, discussed later in this chapter), there can potentially be changes, nevertheless, in both the crop and progenitor populations due to selection and genetic drift. This is especially true in polyploids, in which the coming together of divergent genomes can produce both rapid and gradual changes that are only beginning to be understood (see reviews of this active area of study by Soltis and Soltis 2000; Wendel 2000; Rieseberg 2001; Levy and Feldman 2002; Kellogg 2003).

Once the progenitor taxon is known, genetic variation across the range of the wild populations can be used to pinpoint the geographic area in which the crop arose. A related question is whether the plant was domesticated only once, or in several places independently. The question of independent domestications has been the subject of reviews by Blumler (1992) and Zohary (1996, 1999). Recent debates (e.g., about barley, discussed below) underscore the idea that different interpretations of the same data can lead to different conclusions. In some cases there is insufficient variation in the wild populations to allow this question to be addressed. Contrary to the "centers of origin" concept (Vavilov 1992) is the idea that the crop may have been domesticated over a widespread, diffuse area—with continual geneflow between the incipient crop and its wild relatives—rather than in a single locality (see Harlan 1971). However, the evidence for multiple domestications within a particular region has been debated (Blumler 1992).

The possibility of free-living (spontaneous) populations that are either recent or ancient escapes from cultivation has long been recognized to be a complicating factor in studies of crop origins (de Candolle 1884). These may be merely *adventive* (not persisting), or they may have become naturalized after their introduction to a new area. Genetic data are complementary to morphological and archaeological data in providing additional criteria for distinguishing such escapes from relict progenitor populations. In addition to the expectation that a crop will have a subset of the alleles found in its progenitor (Doebley 1989), other criteria based on gene trees may be applied (Chapter 9).

In addition to questions of what taxon was domesticated and where, genetic data can sometimes help resolve questions about the crop's dispersal from its initial area of domestication. Potentially, genetic data may help to explore the effects of constraints on the dispersal of particular crops (Diamond 1997).

It is sometimes hoped that genetic data would also address questions of the temporal scale of domestication. However, these questions are usually better addressed using archaeological data, with genetic data taking a subordinate and complementary role. As mentioned earlier, the time scales concerned mean that there would be few mutations accumulated. Other problems with the use of a molecular clock assumption are discussed in the case study chapters on cassava and oca (Chapter 9; Chapter 13).

Evolution under Domestication

Whereas the questions of the origins of crops have been studied for a very long time, some other aspects of crop domestication have only recently been addressed. These aspects include the evolutionary effects of domestication on the genetics of crop populations, the extent of hybridization and geneflow between domesticated and wild populations, and the genetic basis of the morphological and other changes that occur during domestication. These new areas of inquiry have been made possible only by the molecular tools that are now applied to systematics, population genetics, and genetic mapping.

An important effect of domestication on crop population genetics is through a *domestication bottleneck*, or founder effect of domestication. The term "population bottleneck" refers to the concept that when a population experiences a small population size at some time in its history, only a small proportion of the genetic diversity is retained, because of the effects of random sampling and genetic drift. The founder effect is a special case of this phenomenon, referring to situations in which a small number of individuals colonize a new habitat (e.g., an island). In the case of crops undergoing domestication only a small number of individuals might be brought under cultivation, leading to a "founder effect of domestication" (Ladizinsky 1985). However, if domestication occurred slowly, over a wide area, or in multiple events, the founder effect may be lessened (Zohary 1996). The extent of the bottleneck is studied by assessing the amount of variation (differences among alleles) at a particular locus (genetically defined site; pl. *loci*).

Other studies have focused on assessing the extent and effects of geneflow between crops and their wild or weedy relatives. Introgressive hybridization was defined by Anderson and Hubricht (1938, p. 396) as the case in which "through repeated back-crossing of the hybrids to the parental species there is an infiltration of the germplasm of one species into that of another." Rieseberg and Wendel (1993) expand the concept to include hybridization within species, thus defining introgression as "gene exchange between species, subspecies, races, or any other set of differentiated population systems." The extent, importance, and evolutionary effects of this process have been discussed for many years (e.g., Heiser 1949, 1973; Rieseberg and Ellstrand 1993; Rieseberg and Wendel 1993; Ellstrand et al. 1999). Some researchers assert that introgression has increased the genetic diversity of crops because new alleles are introduced from wild relatives, whereas others contend that most geneflow is in the direction from crops into wild and weedy populations, rather than the reverse. In addition to being the focus of research, the possibility of past and continuing introgression between crops and wild species also complicates the interpretation of molecular data, with respect to resolving the questions about crop origins discussed above.

The final area of study discussed here has only begun to be addressed since the early 1990s: the dissection of the genetic basis of the changes that occurred under domestication. This area of research combines the evolutionary methods of molecular systematics and population genetics with the genetic techniques of plant breeding, and it is new enough that it was included as a primary part of the discussion of future prospects in crop evolution studies in the review by Doebley (1992b). Because this research depends on comparing the traits and the genetics of the crop with those of its wild progenitor, it obviously can begin only after the progenitor has been identified. These studies are aimed at determining how many and which traits were the targets of selection, the number of genes involved with each trait, and the extent of the contribution of each gene. These studies were also aimed at distinguishing whether selection acted upon the protein-coding genes themselves, or instead upon the promoter regions of the genes or other regulatory genetic regions. The mapping of quantitative trait loci (QTLs, discussed later in this chapter) has now been applied to the study of "domestication genes" in several crops. The level of molecular variation in and near these genes is another active area of research that can confirm whether these loci were indeed the targets of selection, and it may even be used to identify additional genes that were involved with domestication.

The information gained from addressing this entire range of questions about crop domestication has relevance to many other studies. For example, it informs questions about the geographical and temporal setting of the origins of agriculture—information which in turn can influence the debates about why the transition to agriculture occurred. Understanding the genetic changes that underlie the accelerated evolution of crops under human influence is relevant for "understanding the genetic basis of fundamental evolutionary processes" (Burke et al. 2002b). Conservation of crop genetic diversity requires knowledge of the closest relatives of crop plants and the distribution of genetic variation among crop and wild populations. Finally, there can be implications for practical crop improvement, especially given new understandings that one cannot predict from a wild plant's phenotype as to what sort of useful genes it may have that can be applied to crop improvement (Tanksley and McCouch 1997; Bernacchi et al. 1998; Jiang et al. 1998).

Kinds of Genetic Data Used in the Study of Crop Domestication

Non-molecular Data

Although this chapter concentrates on contributions to the study of crop domestication by molecular techniques that have become available only in recent decades, genetic data also include non-molecular traits that have a much longer history of research. As early as Darwin (1883) and de Candolle (1884)—before the genetic basis of inheritance was understood—crop domestication was studied by observation of morphological variation in crops and their wild relatives. In his seminal studies of crop origins, Vavilov's (1992) inferences on the "centers of origin" of crops were based on variation in morphology, phenology, disease resistance, fertility of hybrids, and other traits with a genetic basis. Before the molecular era, biosystematists used crossing studies to assess the fertility and viability of offspring of hybrids in order to infer relationships among species, and these tools, as well as morphometrics and numerical taxonomy, have each contributed to identifying crop origins.

Cytogenetic studies have also played an important role in revealing crop origins (see, e.g., the review by Pickersgill 1989). Classical cytology includes not only chromosome counts, determinations of ploidy level (the number of chromosome sets), and the study of chromosome morphology, but also genome analysis, in which the relationships of species are inferred based on the extent of chromosome pairing in their hybrids. Genome analysis permitted some of the likely progenitors of polyploid crops to be identified many decades ago. Although the pairing behavior of chromosomes in hybrids is more complex than originally thought (see, e.g., Benavente and Orellana 1989, 1991), the genome groups established by these methods have stood up under testing with newer methods in many cases (see Wendel and Albert 1992; Mason-Gamer 2001). More recently, the tools of cytogenetics have expanded to include not only traditional chromosome banding with various stains, but also techniques of "chromosome painting" using fluorescent *in situ* hybridization (FISH) (e.g., P. Zhang et al. 2002) and genomic *in situ* hybridization (GISH) (see review by Stace and Bailey 1999). Although these cytogenetic studies have also contributed to identification of crop progenitors, especially of polyploids, they are outside the scope of this chapter.

Molecular Data

Although techniques based on the polymerase chain reaction (PCR) and DNA sequencing predominate today, earlier contributions of macromolecular data to the understanding of the domestication of several crops used variation in proteins and in DNA restriction sites (*restriction fragment length polymorphism*, or RFLP). These contributions were the subject of earlier reviews by Doebley (1989; 1992b) and Gepts (1990b; 1993). Although they are becoming overshadowed by other techniques, these methods are still useful, and they have certain advantages not shared by some of the newer techniques, as discussed later in this chapter. There are now, however, many more different kinds of molecular techniques. Each technique has its own advantages and pitfalls, and none is a panacea for all research applications. Yet they all share the advantage that variation in DNA and proteins is not subject to the effects of environmental factors, so this variability provides a more direct estimate of genetic diversity than non-molecular data can (Powell et al. 1995).

PROTEIN DATA

As is the case for nearly all molecular techniques, the study of protein polymorphism depends on the separation of molecules (different sizes and/or charges) by electrophoresis, due to their unequal rates of mobility when an electrical current is passed through a buffer solution in a gel. Proteins studied in this way include isozymes and storage proteins.

Isozymes (or allozymes) are slightly different forms of a particular enzyme. Once they are separated in a starch gel they can be detected by providing them with their substrate. The advantages of isozymes are (1) they are co-dominant, so it is possible to detect both alleles in a diploid heterozygote (with the exception of "null," i.e., nonfunctional, alleles); (2) homology is relatively certain, with the possible exception of recently duplicated loci or polyploidy; and (3) they demand less time and money than most other methods. On the other hand, they usually are not as variable as many DNA markers.

Variation in seed or tuber storage proteins has also been used in crop domestication studies. Rather than having enzymatic activity, these proteins are accumulated in seed or tubers as a readily available source of nitrogen for rapid growth during germination or sprouting. Variation in such proteins, such as the phaseolin seed proteins of *Phaseolus* beans, is assessed by various kinds of electrophoresis, most commonly SDS (sodium dodecyl sulfate) polyacrylamide gel electrophoresis (SDS-PAGE) (Gepts 1990b). Seed storage proteins are highly variable, but they usually reflect a single locus and therefore have low genomic coverage, and like isozymes they do not reveal silent substitutions or differences in flanking regions and introns (Gepts 1990b).

DNA DATA

DNA data comprise either DNA sequences of a particular genetic locus or various kinds of "fragment analysis," that is, comparisons of relatively small stretches of DNA that are polymorphic (i.e., present or absent, or of differing lengths) in different individuals. The polymorphic "fragments" may result either from cutting up DNA with restriction enzymes, or from amplifying DNA segments using the polymerase chain reaction (PCR). The genetic loci analyzed, whether by sequencing or by fragment analysis, are located on one of the three genomes of plants: nuclear, mitochondrial, or chloroplast (plastid). These genomes differ in their rates of evolutionary change (Wolfe et al. 1987) and in whether they derive from both parents, as is the case for chromosomes that make up the nuclear genome, or only from the maternal ("seed") parent, as is usually the case for the chloroplast and mitochondrial genomes.

DNA sequence data differ from the polymorphic fragment markers in that they can be analyzed phylogenetically to obtain a gene tree (see the discussion in Chapter 9). However, DNA sequencing samples a smaller portion of the genome, often at greater cost, than most of those highly polymorphic markers. In general, the level of sequence variation is higher for loci on the nuclear genome than on the slower-evolving chloroplast and mitochondrial genomes (Wolfe et al. 1987), but even the nuclear genome may have insufficient variation to provide enough characters to resolve evolutionary relationships. Knowledge of the precise DNA sequence reduces the problems of homology assessment presented by markers that are assessed on the basis of length alone, but does not eliminate them entirely (see the discussion by Doyle and Doyle 1999).

Gene duplication introduces complications to the interpretation of gene trees because homologous genes (those descended from a common ancestral gene) may have diverged in two ways: through speciation (orthologous) or by gene duplication (paralogous). Even among single-copy genes, different genetic loci can have different histories, so caution is necessary in inferring relationships among taxa based on the gene tree (see Doyle 1992; Doyle and Gaut 2000; Gaut et al. 2000). It is also important to understand the pitfalls of assuming a constant and equal rate of change, or "molecular clock" (Gaut and Weir 1994; Gaut 1998), especially with respect to the increase in uncertainty that inevitably occurs when such calculations are several steps removed from any valid calibration point (see Graur and Martin 2004).

Methods for analysis of polymorphic fragments make use of restriction enzymes, the polymerase chain reaction (PCR), or a combination of these techniques. Restriction enzymes recognize a particular short (usually 4–6 bases) DNA sequence, called the *restriction site*, and cut the DNA in or near the sequence. The cut-up DNA fragments of different lengths are separated by electrophoresis and visualized, usually by annealing the fragments to a radioactively labeled probe. The variation among the fragments from each sample may be evaluated simply in terms of the shared presence or absence of fragments of similar size, known as RFLP (restriction fragment length polymorphism). When possible, it is preferable to infer the location of the restriction sites themselves and to score each individual for the presence or absence of the sites, rather than treating the data as "anonymous bands." Mapped restriction sites of the chloroplast genome or of ribosomal genes have been important in the study of several crops (Doyle and Beachy 1985; Doebley et al. 1987b). RFLP analysis is described further in the case study chapter on olive (Chapter 11).

The development of PCR amplification has led to an explosion of different techniques that take advantage of this method to amplify DNA segments, including not only the current methods of DNA sequencing, but also various kinds of PCR-based polymorphic markers (reviewed in Wolfe and Liston 1998). Primers of random sequence that amplify polymorphic regions of the genome are used in similar techniques known as random amplified polymorphic DNA (RAPD) and arbitrarily primed PCR (AP-PCR). Analyses of RAPD data are discussed further in the case study chapters on *Allium* and olive (Chapter 10; Chapter 11). Restriction enzymes and PCR methods are combined in the analysis of amplified fragment length polymorphism (AFLP), in which PCR is used to amplify a subset of the fragments cut by restriction enzymes (Vos et al. 1995).

The polymerase chain reaction is also used in analyses of several kinds of repetitive DNA, known as VNTR markers (variable number of tandem repeats). Segments of DNA sequence, from long to very short, are repeated throughout all known genomes. Longer repetitive sequences are known as *minisatellites*, whereas very short repeats (usually only 1–4 bases long) are known as *microsatellites* or simple sequence repeats (SSRs) (Morgante and Olivieri 1993; Tautz and Schlötterer 1994; Powell et al. 1995). Repetitive DNA can be assessed either as specific single loci, or as multilocus "fingerprints" (e.g., M13 minisatellite fingerprinting, Rogstad et al. 1988; Nybom et al. 1990; Sonnante et al. 1994; Stockton and Gepts 1994). Sequence tagged SSRs (ST-SSR) are microsatellites that are specific to a particular locus. Microsatellites are somatically stable, co-dominant markers that are highly polymorphic, with variation that is thought to result primarily, but not exclusively, from slippage during DNA replication (Morgante and Olivieri 1993; Tautz and Schlötterer 1994). Because they are co-dominant, their allele frequencies can be used in population genetic studies. However, they are usually specific to a particular species or closely related group of species, so they must be developed anew for each taxon of interest. Microsatellite markers are discussed further in the cassava and olive case study chapters (Chapter 9; Chapter 11). A related technique, Inter-SSRs (ISSRs), uses a single microsatellite primer to amplify a fragment between two anonymous SSR loci (Zietkiewicz et al. 1994; Sylvester 2003).

Mobile genetic elements called *retrotransposons*, which transpose themselves by means of reverse transcription of an RNA intermediate, are now being used as molecular markers as well (reviewed in Kumar and Hirochika 2001). Primers amplify the stretch of DNA either between two retrotransposons or between a microsatellite and a retrotransposon in various methods. These methods are known as Inter-Retrotransposon Amplified Polymorphism (IRAP) and REtrotransposon-Microsatellite Amplified Polymorphism (REMAP); Copia-SSR; and Inter-MITE polymorphisms (IMP) and Retrotransposon-based insertion polymorphisms (RBIP), all of which are similar multilocus PCR-based methods for detecting retrotransposon integration events in the genome (Flavell et al. 1998; Kalendar et al. 1999; Provan et al. 1999b; Chang et al. 2001; Baumel et al. 2002). Compared to somatically stable markers (e.g., microsatellites), retrotransposon-based markers may be more useful in vegetatively propagated crops, in which they show higher levels of polymorphism (Bretó et al. 2001) presumably because they can transpose in the absence of meiosis.

Several newer methods of assessing polymorphism combine several techniques, such as using both restriction enzymes and PCR amplification, or combining different kinds of PCR primers. One example is selectively amplified microsatellite polymorphic loci (SAMPL), which combines aspects of AFLP and SSR analyses (Witsenboer et al. 1997; Palacios et al. 2002). Another is sequence-related amplified polymorphism (SRAP), a method similar to AFLP, but less technically demanding (Li and Quirós 2001). Other methods have been infrequently used in studies of crop domestication to date, although they have proven useful in other contexts (see the review by Wolfe and Liston 1998). These methods include random amplified microsatellite polymorphisms (RAMP), which uses an ISSR primer and a RAPD primer; methods such as CAPS and PCR-MRSP, which use restriction enzymes to cut a PCR-amplified fragment; and various methods that modify the conditions of electrophoresis to separate slightly different fragments, such as single-strand conformation polymorphism (SSCP) and denaturing gradient gel electrophoresis (DGGE).

Each of these methods has advantages and disadvantages, and it is important to understand their limitations and appropriate applications. These are discussed in the case study chapters and well-reviewed elsewhere (e.g., Wolfe and Liston 1998; Zhang and Hewitt 2003), and are discussed only briefly here. Co-dominant markers (e.g., microsatellites) have advantages over those that are predominantly dominant (e.g., RAPD, ISSR) in many kinds of analyses, because they may provide allelic information. The predominantly dominant markers may include an unknown proportion of co-dominant markers, which can create problems because the assumptions of the analytical methods are violated (e.g., independence of characters). Some methods, such as RAPDs, are well known to be very sensitive to PCR conditions, so they have problems of repeatability (Jones et al. 1997). Because markers may be derived not only from the nuclear genome, but also from the plastid or mitochondrial genomes, assumptions of

Mendelian inheritance may also be violated. Whereas microsatellites (ST-SSRs) are co-dominant and have the highest number of alleles per locus, they require a greater investment in labor to develop them for each new taxon, so they are more expensive than methods that can be applied to any taxon without a need for prior sequence information (e.g., AFLP, RAPD, ISSR).

The variability and informative nature of different DNA fingerprinting techniques (RAPD, ISSR, AFLP and SSR) have been compared in soybean (Powell et al. 1996), potato (Spooner et al. 1996; Milbourne et al. 1997; McGregor et al. 2000), hop (Patzak 2001), sorghum (Uptmoor et al. 2003), and rice (Parsons et al. 1997; Virk et al. 2000), among other crops. Microsatellites usually have more alleles per locus than the other markers, whereas AFLP reveals the most profiles with a single primer pair. These methods differ not only in their power to discriminate, but more importantly, some of the comparisons found that there was little agreement in the results of phenetic analyses of the same accessions using different marker systems, which suggests caution when drawing conclusions with a single marker type. All of these methods may be problematic in polyploids, especially the dominant markers, because gene dosage cannot be assessed and peaks cannot easily be attributed to individual loci (Milbourne et al. 1997; McGregor et al. 2000).

Many of the polymorphic markers have problems with homology assessment. They generally rely on simple comparisons of fragment length, so they depend on the assumption that fragments of the same size from different plants are from the same (orthologous) locus. This assumption may be violated, however, especially in interspecific comparisons. Examples of such "size homoplasy" have been reported for RAPDs (Rieseberg 1996), minisatellites (Olsen 1999), and microsatellites (Doyle et al. 1998; Peakall et al. 1998). Methods such as sequence characterized amplified region (SCAR), which use markers for which the DNA sequence has been determined, can be used to improve homology assessment (Freyre et al. 1996). Another problem is that translocations and other genomic rearrangements can affect analyses if many markers (e.g., AFLP) are affected by a single translocation (Warburton et al. 2002).

Regardless of the kind of molecular data used, the inferences depend on how they are analyzed. Each of the different methods of phenetic and phylogenetic analysis has different assumptions, philosophical bases, and optimality criteria, and the choice of method is very controversial. Discussion of these methods is beyond the scope of this chapter, but practitioners should not allow themselves to practice "point and click" analyses (see Grant et al. 2003) without understanding the differences among analytical methods (an excellent place to begin is with Doyle and Gaut 2000).

How can these various kinds of data provide responses to the different questions about crop domestication? To illustrate, I begin with the ongoing saga of the investigation of several aspects of domestication of a single crop: maize, or corn. After relating the story of this one crop, I include briefer examples from other crops that illustrate a sampling of the breadth of approaches and questions regarding crop domestication.

TABLE 8.1
Species and Subspecies of *Zea* Following Doebley (1990)

Zea mays L. ssp. *mays*
Zea mays L. ssp. *parviglumis* Iltis & Doebley
Zea mays L. ssp. *mexicana* (Schrader) Iltis
Zea mays L. ssp. *huehuetenangensis* (Iltis & Doebley) Doebley

Zea luxurians (Durieu & Ascherson) Bird
Zea perennis (Hitchcock) Reeves & Mangelsdorf
Zea diploperennis Iltis, Doebley & Guzmán

Study of Maize Domestication with Genetic Data

No crop provides a richer example for genetic studies of its domestication than maize, *Zea mays* ssp. *mays*. The study of maize domestication with genetic methods has involved many researchers, many decades, and much controversy. However, because there has been so much research on this important crop, the study of maize domestication has gone beyond the questions of its origins to address other issues. Attention is now focused on the particular genes that were involved in the domestication of the crop, as well as the population genetic effects of domestication on the plant's genome.

Origins of Maize

The question of the origins of this crop has been an especially challenging one. Maize looks so unlike any wild grass that it was initially difficult to conceive that any known wild species could have given rise to the crop (Beadle 1972). The large wild grasses known as teosintes, which are now known to be the closest relatives of maize, were originally classified as species in the genus *Euchlaena*, rather than in the genus *Zea*. Eventually, the teosintes were brought into the same genus, and some of them were recognized as being the same biological species as maize (see Table 8.1).

Before the advent of biochemical and molecular genetic techniques, maize was studied with morphological, cytological, and classical genetic techniques. Detailed morphological study of traditional cultivars from throughout the Americas led to their classification into races (e.g., Cutler 1946). Cytological research on maize and its relatives described the numbers and morphology of chromosomes, and the position and inheritance of chromosome knobs (e.g., McClintock 1929). Two competing hypotheses about the origin of maize were put forward and were subsequently the center of decades of debate (Beadle 1939; Mangelsdorf and Reeves 1939). These hypotheses are each associated primarily with their major proponents, George Beadle and Paul Mangelsdorf, although both ideas were based on work by several researchers (reviewed in Doebley 2001).

Beadle was convinced that maize was domesticated from an annual teosinte, based on his earlier work with Rollins Emerson on the pairing behavior of chromosomes in fertile hybrids between maize and various teosintes (Beadle 1939). The competing hypotheses of Mangelsdorf, known as the tripartite theory, suggested that maize and teosinte diverged before domestication, and that maize was domesticated from an extinct wild maize (Mangelsdorf and Reeves 1939). The diversity of maize was hypothesized to have arisen due to "infection" (i.e., introgressive hybridization) of genes from the related genus *Tripsacum*. Mangelsdorf's tripartite theory held sway for many decades, but it began to lose supporters in the 1970s and 1980s as the evidence supporting teosinte as the progenitor of maize began to accumulate. In the early 1970s, Beadle broke his long silence on the issue to describe the results of examining 50,000 F_2 (second generation) descendants from a maize and teosinte (ssp. *mexicana*) cross, among which he found 1 in 500 that resembled one or the other of the parents (Beadle 1972). From these numbers he calculated that as few as four or five chromosomal regions may have been responsible for the major differences between maize and teosinte, a conclusion that is startlingly congruent with the recent results of QTL mapping (discussed below).

Molecular evidence that an annual teosinte is the closest relative of maize has accumulated, beginning in the 1980s. The first molecular data applied to understanding the evolution of maize were isozyme variation and mapped restriction sites of the chloroplast genome (reviewed in Doebley 1990a). Phenetic analyses of isozyme data of maize and several teosintes showed that the domesticate was most similar to teosinte populations of subspecies *Z. mays* ssp. *parviglumis*, specifically those from the central Balsas River valley of southwestern Mexico (Doebley et al. 1984; Doebley et al. 1987a). Thus, these data pointed to modern-day teosinte populations from the Balsas watershed as the ones that are most likely to be descended from the progenitor populations. The phylogenetic tree based on chloroplast DNA (cpDNA) was consistent with the idea that the cultigen, *Z. mays* ssp. *mays*, belongs to the same species as three wild teosinte subspecies (i.e., *Z. mays* sspp. *parviglumis*, *mexicana*, and *huehuetenangensis*) because their chloroplast haplotypes were intermingled together in the same clade (Doebley et al. 1987b; Doebley 1990a). Unfortunately, although it confirmed the close relationship of these taxa, this mixing of haplotypes meant that the cpDNA data did not have the power to distinguish which subspecies was the progenitor of maize.

Although the original tripartite hypothesis of Mangelsdorf has been discredited, a modified version, invoking hybridization of *Tripsacum* with a perennial teosinte to give rise to maize, still has supporters (MacNeish and Eubanks 2000; Eubanks 2001a; 2001b; 2001c). As part of the evidence for this hypothesis, Eubanks (1995, 1997) reports successful artificial hybridization of *T. dactyloides* and *Z. diploperennis*, and she cites similarities in ear and kernel morphology of these hybrids with early archaeological maize cobs. Other researchers have expressed doubt that these plants are truly *Tripsacum-Z. diploperennis* hybrids (Bennetzen et al. 2001), because their chromosome number, fertility, and RFLP profiles do not fit expectations for such a hybrid. *Tripsacum*-teosinte hybrids would potentially have great value as a bridge for introducing desirable alleles from *Tripsacum* into maize (Eubanks 1995, 1997; Throne and Eubanks 2002). For that reason it will be important to resolve the question of the parentage of the putative hybrid plants by sequencing some of the alleles that are thought to be derived from each of the parents.

Regardless of whether these taxa can be successfully hybridized with human help, however, the currently available DNA sequence data do not show evidence of *Tripsacum* ancestry in maize. *Tripsacum* and *Zea* are generally recognized as sister genera, that is, they are more closely related to each other than to any other grasses. However, analyses of DNA sequences consistently show that *Zea* alleles are closer to each other than any are to *Tripsacum* alleles. It is true, as pointed out by Eubanks (2001a), that many analyses of *Zea* sequences include only a few *Tripsacum* samples, if any. Even fewer are rooted with an outgroup from outside these two genera (but see Doebley et al. 1987b; Buckler and Holtsford 1996; Lukens and Doebley 2001). However, although not designed to explicitly test the relationship of *Zea* to *Tripsacum*, a few *Tripsacum* sequences were included in studies of sequence variation of several loci in *Zea*, such as *adh2* (Goloubinoff et al. 1993), ITS (Buckler and Holtsford 1996), *adh1* (Eyre-Walker et al. 1998; Tiffin and Gaut 2001), *glb* (Hilton and Gaut 1998; Tiffin and Gaut 2001), *te1* (White and Doebley 1999), *tb1* (Lukens and Doebley 2001), and *c1* and *waxy* (Tiffin and Gaut 2001). In all cases the *Zea* sequences grouped together and were divergent from those of *Tripsacum*, and none of the domesticated maize sequences of these loci were found to cluster with those from *Tripsacum*. Thus, data from these loci agree with the earlier studies of restriction sites of chloroplast DNA (Doebley et al. 1987b) in the separation of *Zea* and *Tripsacum*.

These different loci do not necessarily agree in their depictions of relationships among *Zea* species, but discordant patterns of gene trees among closely related species are not unusual. The intermingling of sequences from different *Zea* taxa may represent polymorphisms that were retained from ancestral species and/or introgression among species (see discussions in Hanson et al. 1996; Gaut et al. 2000; Tiffin and Gaut 2001). To date none of the phylogenies based on DNA sequences of nuclear loci show indications in maize of introgression from *Tripsacum*. Even if there were such introgression in the history of maize, however, such evidence might be difficult to find, because introgression by definition implies that only a few alleles from one species have "infiltrated" into the genome of the other species (Anderson and Hubricht 1938; Rieseberg and Wendel 1993). Additional loci will undoubtedly continue to be sampled, but caution in their interpretation will be necessary because of gene duplication through polyploidization. Most evidence indicates that the ancestral genome of maize was doubled, but there is disagreement about whether this occurred after the divergence of *Zea* and *Tripsacum* (Wilson et al. 1999) or in their common ancestor, that is, before the divergence into the two genera that exist today (Gaut and Doebley 1997; Gaut et al. 2000). In duplicated genes, sequences of the "correct" (orthologous) copies must be compared, or relationships may be misinterpreted (for a discussion of the challenges of interpreting duplicated loci, see Doyle and Gaut 2000).

A Single Domestication of Maize and Its Subsequent Dispersal

Knowledge of the progenitor permits other questions to be addressed. One of these is to determine whether maize was domesticated once, or in multiple independent events. This question has been the subject of abundant, but contradictory, speculation. On the one hand, the great morphological, cytological, and genetic diversity of maize suggests that maize may have been domesticated more than once, whereas the complexity of the morphological changes needed to transform teosinte into maize argues that this transformation is unlikely to have occurred more than once (reviewed in Matsuoka et al. 2002).

In a distance analysis of data from 99 microsatellite loci, Matsuoka et al. (2002) found that all of the 193 domesticated maize plants sampled from throughout the pre-Colombian range in North and South America clustered together, separately from 71 teosinte plants of three *Z. mays* subspecies. This single maize cluster is consistent with a single origin of domestication and agrees with earlier results based on isozymes and chloroplast DNA, neither of which gave any support for multiple domestications of maize (Doebley 1990a). The microsatellite data also confirmed earlier results indicating that ssp. *parviglumis* accessions from the central Balsas River valley were the teosinte populations closest to domesticated maize. However, as pointed out by Matsuoka et al. (2002), the support for Balsas teosinte must be tempered with the realization that undiscovered or extinct populations may be even closer to maize, and that species distributions may have changed since the time of the domestication.

A related task is to specify the geographic area of domestication and the subsequent paths of dispersal of the crop. Matsuoka et al. (2002) interpreted their microsatellite data as indicating that the most ancient of the surviving maize types sampled were those in the highlands of Mexico, from where maize eventually spread to lowland Mesoamerica and then into North and South America. Maize diversified into two major groups as the crop diffused north and south. One group eventually reached north into the eastern and central United States and southern Canada, with the other group dispersing through lowland South America and eventually up into the Andes (Matsuoka et al. 2002).

Geneflow between Maize and Teosinte

The microsatellite data of Matsuoka et al. (2002) also helped reevaluate geneflow from teosinte to maize. In certain areas of the highlands of Mexico, ssp. *mexicana* teosinte grows as a weed in cornfields. Maize accessions from these areas have an average of 2.3% ssp. *mexicana* germplasm, with a few highland accessions reaching 9–12%. Mexican maize from lowland areas where ssp. *mexicana* does not grow has only 0.4% ssp. *mexicana* alleles (range 0.2–2%). This confirms earlier evidence of allozyme alleles that occurred in ssp. *mexicana* and in sympatric maize populations, but not in maize from other areas, suggesting that these alleles had probably introgressed into maize from ssp. *mexicana* (Doebley et al. 1987a; Doebley 1990a). Introgression between ssp. *parviglumis* and maize is harder to assess, because there are no alleles that are diagnostic of one or the other. As with other examples of crop-to-wild geneflow, this case sounds a cautionary note concerning the potential release of genes from genetically engineered crops to their wild relatives (Doebley 1990b).

This geneflow also underscores a pitfall of attempting to pinpoint the progenitor populations within a species. It is not easy to assess the historical levels of geneflow among *Zea mays* populations. Hanson et al. (1996) discuss the possibility that substantial introgression with a secondary population (e.g., between ssp. *mexicana* and ssp. *mays*) could result in gene trees that disguise the origins of the crop from within a different population (e.g., ssp. *parviglumis*). A similar caveat regards distinguishing single versus multiple domestications. Even if there were multiple origins of domestication, they might be obscured if there was subsequently substantial mixing among the domesticated populations and little mixing with the wild populations, especially if much of the genome experienced the effects of selection during domestication. In this case there might be more alleles shared among the domesticated populations than with the separate progenitors, making the domesticated populations appear more similar to each other than to their respective progenitor populations in a distance analysis.

Maize Domestication Genes

One area in which maize researchers have been breaking new ground is in the study of *domestication genes*. These are genes that control the traits involved in the "adaptive syndrome of domestication." Several new methods make it possible to investigate whether a particular gene was involved in the domestication process, that is, whether it was one of the genes that directly experienced the effects of selection. One of these methods is the mapping of quantitative trait loci, or QTLs (see Tanksley 1993). A quantitative trait is one that shows continuous variation in some measurable quantity, indicating that it is probably controlled by several different genetic loci. In contrast, a trait controlled by single genetic locus (or very few) will show discrete variation in which Mendelian genetic segregation can be observed.

To identify chromosomal regions where genes controlling the traits are likely to be found, one must first have a "saturated" genetic map. There must be sufficient molecular markers scattered at known locations throughout the genome to detect associations among markers and traits. Parental plants that differ in the traits of interest are hybridized. To study domestication genes, the parents that are selected for crossing are a domesticated and a wild parent that differ in many domestication traits. A large number of descendants of the original cross (usually either F_2 or F_3 generation) are scored for the traits and also for the molecular markers. Thus, each individual plant must be measured for the traits of interest, and its genotype for the molecular markers must be determined (i.e., whether they are heterozygous or are homozygous for one or the other of the parental alleles). To find markers that are near the QTLs, several statistical procedures can be used to select those markers that are correlated with the traits of interest. For instance, a marker's LOD score (log of the odds) expresses in logarithmic form the ratio of the likelihood that there is such a gene near the marker to the likelihood that there is not. QTL mapping can identify chromosomal regions that may contain a gene affecting the trait, estimate how much of the trait's variation is controlled by the gene, and determine whether the gene operates in a dominant, additive, or over-dominant fashion (Doebley 1992a).

Initial QTL studies in *Zea* used two separate mapping populations, descendants of crosses between maize and ssp. *mexicana*, and maize with ssp. *parviglumis* (Doebley and Stec 1991, 1993; Doebley 1992a). In agreement with the initial estimates of Beadle (discussed earlier), these mapping studies found that five chromosomal regions control most of the differences between maize and teosinte (Doebley 1992a; Doebley and Stec 1993). Some of these traits appeared to be controlled by few genes of large effect, as Beadle had predicted, but other traits seemed to be under multigenic control (Doebley 1992a). Contrary to the prediction of the Catastrophic Sexual Transmutation Theory of Iltis (1983), now abandoned (Iltis 2000), the domestication traits segregated independently.

However, QTL mapping identifies only broad chromosomal regions, in each of which at least one gene that influences the trait may be found. It does not identify the genes themselves. A "candidate gene" approach can be used, in the case of well-studied crops like maize, to try to identify the individual gene in a particular QTL region. If there is a known gene that maps to the same chromosomal location (usually because of a mutation in that gene), that candidate locus can be further tested to determine whether it is the same functional gene as the QTL. The first maize "domestication gene" to be studied extensively is *teosinte branched1*, or *tb1*. This locus affects apical dominance in maize, such that the maize mutant has teosinte-like branching instead of a single dominant stalk. Both of the *Zea* QTL mapping populations consistently showed a strong effect on plant architecture in the same chromosomal region in which *tb1* was located.

Genetic complementation tests can determine whether two mutations that have the same phenotype are actually in the same functional gene. Complementation tests and fine-scale mapping did indeed show that *tb1* was the locus that had been identified by QTL mapping (Doebley et al. 1995a; Doebley and Wang 1997).

Finally, "transposon tagging" was used to clone *tb1* and determine its DNA sequence (Doebley and Wang 1997; Wang et al. 1999). In this technique a transposable element ("jumping gene") is inserted into the gene of interest, thus causing a new mutation. Then a genomic library is probed with the transposon and with molecular markers known to be located near the gene, in order to identify cloned restriction fragments carrying the gene itself. Once the sequence of *tb1* was determined in this way, PCR primers were designed to amplify and sequence larger numbers of individuals and determine the variability of the gene (see below and Wang et al. 1999).

Other loci that have been investigated in maize as possible domestication genes include *teosinte glume architecture1*, or *tga1* (Dorweiler et al. 1993; Doebley 1995; Dorweiler and Doebley 1997), which controls the formation of the teosinte cupulate fruitcase that encases the kernel; *terminal ear1*, or *te1* (Doebley 1995; White and Doebley 1999); and several loci affecting kernel weight (Doebley et al. 1994). However, one domestication gene candidate, *suppressor of sessile spikelets1* (*sos1*) (Doebley et al. 1995b), was shown by mapping and developmental studies not to have been involved in the evolution of maize, even though its phenotype of single, rather than paired, spikelets appears similar to teosinte (Doebley et al. 1995b).

Many of the proposed domestication genes are thought to be either regulatory loci (rather than protein-coding genes), or those in which the selection on the gene is thought to have affected the regulatory (promoter) regions of genes. This hypothesis is currently an active area of investigation, not only in maize (see below and Hanson et al. 1996; Dorweiler and Doebley 1997; Selinger and Chandler 1999; Wang et al. 1999; White and Doebley 1999), but also in other crops such as tomato (e.g., Cong et al. 2002).

Bottleneck of Domestication

Artificial selection, whether deliberate or not, is expected to result in a reduction in the genetic diversity in the crop in two related ways. The so-called "bottleneck of domestication" is an example of the "founder effect" phenomenon: When a small number of individuals reach a new habitat and found a large population, the diversity in that population is reduced because the small number of founders provided a reduced amount of diversity to their descendants. In the case of domesticates, only a small number of individuals will carry the mutations that are selected for in the process of domestication. The extent of the genetic bottleneck depends both on the number of individuals in the founding population and on the duration of the bottleneck. The extent of the bottleneck in the domestication of maize was first explicitly studied in two neutral loci, *adh1* (Eyre-Walker et al. 1998) and *glb1* (Hilton and Gaut 1998). These studies found that maize had 83% and 60%, respectively, of the diversity found at these loci in ssp. *parviglumis*. As discussed by Hilton and Gaut (1998), the data from these other loci are consistent with the hypothesis of ssp. *parviglumis* as progenitor of maize, although they are not an explicit test. The data indicate that (1) there are no fixed differences between these two subspecies, (2) the maize sequences are intermixed with a subset of *parviglumis* sequences on the neighbor-joining (NJ) distance tree, and (3) maize has reduced diversity compared to ssp. *parviglumis*.

This reduction in diversity was not as great as had been expected, however. The extreme morphological divergence between maize and teosinte would suggest that the genetic base should have been narrowed even more. Similar proportions of the sequence diversity of the wild progenitor have been found in pearl millet (67%, Gaut and Clegg 1993), and have since been found in other neutral loci in maize (e.g., White and Doebley 1999; Tenaillon et al. 2001; Whitt et al. 2002). Although the extent of the reduction in diversity can be used to estimate the population size during the domestication bottleneck, these estimates depend on assumptions of how long the bottleneck lasted, which is unknown (Eyre-Walker et al. 1998; Hilton and Gaut 1998). Whether these results from maize indicate that crops in general have retained more of the diversity of their ancestors than had been expected remains to be seen (see Hilton and Gaut 1998). Maize is a wind-pollinated outbreeder that suffers from inbreeding depression if self-pollinated. This breeding system is likely to have played a role in its high diversity. In addition, the extent of introgression from wild populations into the crop after the period of domestication cannot easily be determined, and this aspect complicates the interpretation of the results.

Testing the "Signature" or "Footprint" of Selection

An issue related to the domestication bottleneck is the further reduction in genetic diversity at loci that are the direct targets of selection, and in genetic regions linked to such loci. If selection favors a particular allele strongly enough that it is quickly brought to fixation in the population, other linked loci may also be brought to fixation. This reduction in diversity is known as a "selective sweep" or "genetic hitchhiking," and it depends on the amount of recombination in the genetic region involved. If a particular locus was involved in the domestication of a crop, such a reduction in diversity in the gene—as a "signature" or "footprint" of selection—can be expected.

The study by Wang et al. (1999) of variation in the *tb1* locus has uncovered the strongest evidence to date of the effects of selection on a domestication gene. However, the greatest reduction in genetic diversity was not in the protein-coding part of the gene (transcriptional unit), in which maize has 39% of the diversity of teosinte. Instead, the greatest reduction of diversity was in the regulatory region of the gene, "upstream" in the 5' nontranscribed region, where maize has only 2.9% of the diversity of ssp. *parviglumis*. The fact that the protein-coding part of the gene in maize still retains much of the variation found in teosinte implies that there was plenty of recombination between this part of the gene and the regulatory region, and thus a low "hitchhiking effect" on the transcriptional unit. This recombination leads Wang et al. (1999; 2001) to suggest that it probably took several hundred years for the mutation in the regulatory region of *tb1* to come to fixation. Similar analysis of the putative domestication gene *te1* (White and Doebley 1999) found reduction in diversity similar to that in neutral genes due to the domestication bottleneck, but found no evidence of direct selection on the gene. However, since the promoter region of this gene was not sequenced and the HKA test used is fairly conservative, the possibility remains that this gene was a target of selection during domestication. Other tests of disease resistance genes (L. Zhang et al. 2002) and genes involved in starch synthesis (Whitt et al. 2002) found some candidates that appear to have been targets of selection, whereas other candidates appear to have evolved in a neutral fashion.

Finally, Vigouroux et al. (2002) employed an ingenious reverse strategy to study the genetic effects of domestication. They looked for the signature of a "selective sweep" to identify genes that have experienced the effects of strong selective pressure, even in cases where the function of the gene was not yet known. Variation in microsatellite loci was assessed in maize and teosinte, and was compared to neutral expectations using the Ewens-Watterson test. Neutral theory predicts a certain ratio between the heterozygosity at a microsatellite locus compared to the number of alleles at that locus. A reduction in this ratio suggests that the microsatellite may be linked to a gene that has experienced selection. Several microsatellites were identified by Vigouroux et al. (2002) that fit this pattern, suggesting that they might be linked to genes involved in maize domestication. Because the microsatellites studied were from expressed sequence tags (ESTs), they were already known to be linked to particular genes, and the candidate genes identified in this study can be investigated further to assess their possible involvement in domestication (Vigouroux et al. 2002).

Research on the domestication of maize has involved many aspects, including (1) identifying the progenitor; (2) determining the number of domestication "events"; (3) tracing the geography of domestication and dispersal; (4) assessing geneflow between the crop and its progenitors; (5) mapping, identifying, and sequencing domestication genes; (6) assessing the extent of the domestication bottleneck; and (7) looking for the localized reduction of diversity that indicates the footprint of selection on particular genes. Investigations of other crops have involved investigations of many of the same questions, but have not necessarily followed the same pattern or found results similar to those of maize. The following sections provide briefer examples of other investigations that have differed in finding evidence (or not) of multiple domestications, in achieving varying success at pinpointing the geographic areas of origin of crops, in assessments of crop-to-wild geneflow, and in mapping and identification of other domestication genes.

Additional Examples from Studies of Domestication of Other Crops

Testing Hypotheses of Multiple Domestication with Genetic Data—the Example of *Cucurbita pepo*

Several hypotheses of multiple independent domestications of particular crops have been proposed, based on analyses of morphological or archaeological data. Genetic data are particularly well suited to complement these other data sources by providing an independent rigorous test of such hypotheses. Squashes provide a good example. Because domesticated squashes and ornamental gourds comprise several different species within the genus *Cucurbita* (five fully domesticated species and one "partially domesticated," Nee 1990), it is obvious that these separate species represent separate domestications. However, the possibility that there were independent domestications within one of those species, *C. pepo*, a hypothesis that was originally based upon morphological and archaeological data (Whitaker and Carter 1946; Heiser 1985), has now been confirmed by the accumulation of complementary molecular data from many sources.

Cucurbita pepo includes both free-living populations and cultivated squashes and gourds with a great diversity of forms, including acorn, zucchini, scallop, crookneck, fordhook, and pumpkin cultivars. A phenetic analysis of seed morphology traits (Decker and Wilson 1986) revealed variation in size and shape of seeds and distinguished among some wild and cultivated forms, but the variation in seed morphology among cultivars did not produce distinct groupings. On the other hand, molecular data have consistently supported the separation of *C. pepo* into two divergent groups. This separation was first revealed by analysis of variation in isozyme data (Decker 1985, 1988; Kirkpatrick et al. 1985; Decker and Wilson 1987), and it has been supported in subsequent analyses of additional samples using variation in isozymes (Decker-Walters et al. 1993), mapped restriction sites (Wilson et al. 1992), SSR and ISSR loci (Katzir et al. 2000), DNA sequences of the mitochondrial locus *nad1* (Sanjur et al. 2002), and RAPDs (Decker-Walters et al. 2002). As evidence has accumulated, it is clear that domesticated *C. pepo* squashes comprise two molecularly divergent groups, which were selected from populations that had already differentiated through geographical isolation long before humans domesticated them (see the discussion in Decker-Walters et al. 1993). These two divergent groups are now classified as two subspecies: *C. pepo* L. ssp. *pepo* and *C. pepo* ssp. *ovifera* (L.) Decker. For the intraspecific taxonomy of *C. pepo* see Decker-Walters et al. (2002).

Cucurbita pepo ssp. *pepo* includes pumpkins, zucchini and other marrow squashes, Mexican landraces, and a few ornamental gourds, and is known only from cultivated forms. No wild populations have yet been identified as good candidates to be the progenitor of this group (Decker-Walters et al. 2002). The progenitor may be undiscovered or extinct (Decker-Walters et al. 1993), but archaeological samples from southern Mexico (Smith 1997), as well as the inclusion of most landraces from Mexico in this group, suggest that *C. pepo* ssp. *pepo* may have been domesticated in southern Mexico (Decker-Walters et al. 2002).

Cucurbita pepo ssp. *ovifera* comprises both domesticated and free-living populations. This subspecies is further divided into taxonomic varieties. *Cucurbita pepo* ssp. *ovifera* var. *ovifera* includes some cultivated squash cultivars (e.g., acorn, crookneck, scallop) and most ornamental gourds. Isozyme and RAPD data have supported the separation of free-living populations in the United States into two distinct molecularly divergent groupings. Wild populations in Texas are classified as *C. pepo* ssp. *ovifera* var. *texana* (Scheele) Decker, whereas populations outside Texas to the east and north, in the Mississippi valley and the Ozarks, belong to *C. pepo* ssp. *ovifera* var. *ozarkana* Decker-Walters (Decker 1988; Decker-Walters et al. 1993, 2002).

A third subspecies—*C. pepo* ssp. *fraterna* (L. H. Bailey) Andres—is known only from free-living populations in northeastern Mexico. All known wild populations of *C. pepo* (i.e., var. *ozarkana*, var. *texana*, and ssp. *fraterna*) are more genetically similar to var. *ovifera* cultigens than to those of ssp. *pepo* (Wilson et al. 1992; Decker-Walters et al. 1993, 2002; Katzir et al. 2000; Sanjur et al. 2002). Whereas none of the free-living populations appear to be progenitors of ssp. *pepo*, molecular evidence supports var. *ozarkana* as the best candidate to be the progenitor of the cultigen var. *ovifera*. Isozyme alleles of var. *ovifera* cultivars are a subset of those in var. *ozarkana*, whereas var. *texana* and ssp. *fraterna* lack common alleles of the cultigen, and var. *texana* is also nearly fixed for an allele that is not found in any cultivars (Decker-Walters et al. 1993). The mitochondrial locus *nad1* was invariant, or nearly so, among accessions of ssp. *fraterna* and varieties *ovifera*, *texana*, and *ozarkana* of ssp. *ovifera* (Sanjur et al. 2002). This lack of resolution reflects the known slow evolutionary rate of mitochondrial genes (Wolfe et al. 1987). The results of Sanjur et al. (2002) are congruent with the prior isozyme-based hypotheses of Decker-Walter et al. (1993), not in conflict, as Sanjur et al. (2002) claim. However, the low resolution indicates that the *nad1* locus used by Sanjur et al. (2002) did not have the power to distinguish among these taxa. Phenetic analysis of RAPD data exclude ssp. *fraterna* as being outside (less similar to) the cluster that includes all cultivated and wild varieties of ssp. *ovifera* (see Figure 8.1 and Decker-Walters et al. 2002). These RAPD data confirm the separation of the intraspecific taxa that had been suggested based on prior isozyme data (e.g., Decker-Walters et al. 1993), and genetic differentiation of most of the free-living populations from the cultivars confirms their status as indigenous ("truly wild"), rather than escaped from cultivation (Decker-Walters et al. 2002). On the other hand, the possibilities of geneflow and/or a limited number of feral populations were suggested by

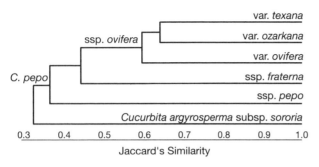

FIGURE 8.1 Cluster analysis of taxon means of Jaccard's Coefficient of Genetic Similarity values among infraspecific taxa of *Cucurbita pepo* and *C. argyrosperma*. Figure contributed by Deena Decker-Walters, based on data from Fig. 3 in Decker-Walters et al. 2002.

isozyme alleles that appeared to have introgressed into certain wild populations from cultivars (Decker-Walters et al. 1993), and by the clustering of those wild populations with cultivars in the results of phenetic analyses of RAPD data (Decker-Walters et al. 2002).

Thus, molecular genetic data have shown that *C. pepo* was domesticated independently from two divergent populations, and that *C. pepo* ssp. *ovifera* var. *ozarkana* is most likely descended from one of those progenitor populations. A separate domestication of *C. pepo* in North America, independent from domestication in Mexico, is also supported by archaeological evidence. *Cucurbita pepo* was domesticated much earlier in Mexico (Smith 1997), but *Cucurbita* remains in eastern North America were morphologically similar to wild gourds of the region (i.e., thin rinds, small seeds) prior to 5000 years ago (Smith 2002; Smith and Decker-Walters 2002; Chapter 3). Subsequent to that time, morphological changes indicating domestication are found in remains of eastern *C. pepo*, during the same period in which other indigenous crops of the region were domesticated (Smith 1998, 2002). Thus, as discussed earlier by Heiser (1985) such independent domestication of *Cucurbita* helps to support the hypothesis of an independent origin of agriculture north of Mexico.

Additional Studies of the Question of Multiple Domestication

The common bean, *Phaseolus vulgaris* L., provides another example in which the case for multiple domestications is strong. Evidence that domesticated common beans were domesticated at least twice began with phaseolin seed protein (Gepts and Bliss 1986; Gepts et al. 1986; Gepts 1990a), but additional evidence is now available from other data sources: allozymes (Koenig and Gepts 1989; Singh et al. 1991; Debouck et al. 1993), M13 DNA fingerprinting (Sonnante et al. 1994), RFLP (Becerra Velasquez and Gepts 1994), AFLP (Tohme et al. 1996), and RAPD and SCAR (Freyre et al. 1996), as well as additional seed protein data and agronomic and morphological traits (Islam et al. 2002; see additional references in review by Gepts 1996). Although *P. vulgaris* is generally considered a single species, wild populations had already diverged before domestication, and there are some barriers to interbreeding (i.e., hybrid weakness) between Mesoamerican and Andean populations, in wild as well as in cultivated beans (Gepts and Bliss 1985; Koinange and Gepts 1992). Initial studies found two gene pools, one extending from Mexico south through Central America into Colombia, and the other occurring in the Andes of southern Peru, Bolivia, and Argentina. Sampling in the northern Andes turned up additional complexity, however. These studies indicated that there may have been at least one other center of domestication in Ecuador and northern Peru, and perhaps an additional one in Colombia, in addition to the two primary centers in Mesoamerica and the central Andes (Debouck et al. 1993; Freyre et al. 1996; Tohme et al. 1996; Islam et al. 2002).

In the two examples just discussed, those of squash and beans, multiple domestications could be demonstrated because two requirements were met: (1) The wild populations had differentiated before domestication, and (2) the markers used were able to detect differences among the cultivars of the domesticate that correlated with differences among the wild populations. If these requirements are not met, it may be impossible to determine whether or not there were multiple domestications. For example, although linguistic and utilization data suggested separate origins in African yam bean *(Sphenostylis stenocarpa)* between strains cultivated for seed vs. those cultivated for tubers, and this division was reflected in differences in chloroplast DNA, corresponding cpDNA types were not found among wild populations, making it impossible from the available data to distinguish whether or not there were multiple domestications (Potter and Doyle 1992).

However, even with substantial genetic differentiation in cultivated and wild populations, new kinds of genetic data do not solve all problems, as continuing debate about the multiple origins of barley demonstrates. Suggestions of multiple domestications of barley are based on several sources of evidence (see reviews in Blumler 1992; Zohary 1999), including the two non-allelic loci that control the fragile rachis and the results of theoretical simulations of outbreeding rates in this primarily self-pollinating crop (e.g., Ladizinsky and Genizi 2001). However, recent studies of barley domestication using different molecular markers do not agree with each other. The differences in inference may result in part from different data types used—either RFLP (Molina-Cano et al. 1999); RAPD (Blattner and Badani Mendez 2001) or AFLP (Badr et al. 2000). As discussed earlier, different marker types frequently lead to different results. However, the different inferences are more likely due to different interpretations based on comparisons with morphological and archaeological data, analytical methods, or on theoretical grounds. Sampling is a particularly important issue. The claim that Morocco was a second area of origin of barley (Molina-Cano et al. 1999) is difficult to interpret without data from local domesticated populations for comparison, because it is not possible to exclude the possibility that the free-living

Moroccan plants sampled had escaped from cultivation and reverted to a non-shattering rachis (see Abbo et al. 2001; Blattner and Badani Mendez 2001). Other studies excluded from consideration as progenitors any plants that were believed—on morphological and ecological grounds (e.g., awn length or occurrence in secondary habitats)—to be escaped from cultivation or hybrid weedy races (Badr et al. 2000; Blattner and Badani Mendez 2001). In counterpoint to this tactic, the case study chapter on cassava (Olsen and Schaal, Chapter 9 of this volume) describes criteria by which molecular data can be used to distinguish whether such populations are feral escapes or probable progenitors.

A more serious caveat regards inferences of single domestications based on "monophyly" of cultivated material in "intraspecific phylogenies" (see Nixon and Wheeler 1990 for why applying these concepts within species is problematic). Several recent studies have based such conclusions about the origins of several founder crops of southwest Asia, such as wheats and barley, on distance-based analyses (specifically neighbor-joining, NJ) of AFLP data or other polymorphic markers (Heun et al. 1997; Badr et al. 2000; Ozkan et al. 2002). Recent simulations by Allaby and Brown (2003), however, have cast doubt on these methods to infer crop origins. Depending on the parameters of population biology modeled in the simulations, in some cases separately domesticated populations were merged in the results of NJ analyses, giving the appearance of a single origin. Although not suggesting that these founder crops necessarily had multiple origins, these simulations indicate that alternatives to strict single origins for these crops still remain as valid hypotheses, and they suggest caution in drawing strong conclusions on the results of such analyses alone.

In the case of the polyploid wheats, in particular, molecular evidence is accumulating to support earlier presumptions (see Feldman and Millet 2001) that hexaploid wheat originated more than once. Evidence is accumulating of at least two origins of *T. aestivum* (hexaploid bread wheat) from diverse groups within tetraploid wheats (emmer and hard wheats) (Ishii et al. 2001; von Buren 2001), as well as from multiple populations of the D-genome donor *Ae. taushii* (Talbert et al. 1998).

Polyploid Crops

The domestication of polyploid crops is discussed in detail in the chapter on *Oxalis tuberosa* (Emshwiller, Chapter 12 of this volume). However, it is worth mentioning here that evidence of all genomes of an allotetraploid or higher-level polyploid is best sought with the use of nuclear loci, particularly single-copy genes, in a phylogenetic context, with loci on organelle genomes as a complement. Some recent studies of polyploid origins include the use of *polygalactonurase* sequences in hexaploid kiwifruit, *Actinidia deliciosa* (Atkinson et al. 1997); nuclear *waxy* and plastid *rps16* sequences in the grain crop tef, *Eragrostis tef* (Ingram and Doyle 2003); *stearoyl-ACP desaturase* and *oleoyl-PC desaturase* sequences in peanut, *Arachis hypogaea* L. (Jung et al. 2003); and a comparison of cpDNA and ITS in coffee, *Coffea arabica* (Cros et al. 1998). Care must be taken in making inferences from polymorphic markers ("fragment analysis") using distance-based clustering algorithms that combine diploid and polyploid taxa in the same analysis (e.g., Hilu and Stalker 1995), because it is difficult to reject the possibility that these similarity-based analyses may group polyploids together as an artifact of shared ploidy level.

Crop-Wild-Weed Geneflow

Molecular data have advantages for testing hypotheses of introgressive hybridization. These data provide many independent, highly variable markers, most of which are neutral or nearly neutral (Rieseberg and Wendel 1993). They are not affected by the environment, and many types can provide evidence of both parents of a hybrid. In contrast, morphological traits in hybrids, while often thought of as being intermediate between the parental forms, are not necessarily so. In some cases they may be similar to one or the other parent or unlike either parent (i.e., showing novel or more extreme forms than either parent, see Rieseberg and Ellstrand 1993). Thus, hypotheses of introgression based on other data, such as intermediate morphology, may be tested with molecular data. Seiler and Rieseberg (1997) review tests by Rieseberg and colleagues of prior hypotheses proposed by Heiser regarding hybridization between wild and domesticated sunflower taxa (*Helianthus* spp.). The majority of the investigations using molecular data supported Heiser's hypotheses (e.g., Rieseberg et al. 1991), with the exception of introgression between cultivated sunflower and *H. bolanderi* (Rieseberg et al. 1988).

Even when hybridization is possible between two taxa, the extent of introgression is variable. Levels of geneflow may vary depending on the breeding system, phenology, and population sizes, and the consequences depend also on whether introduced alleles are neutral, detrimental, or beneficial (reviewed in Ellstrand et al. 1999). All parts of all genomes do not introgress from one species to another equally. Introgression of the chloroplast genome appears to occur more frequently than nuclear genes in sunflower (e.g., Rieseberg et al. 1991) and several other taxa sampled to date (Rieseberg and Wendel 1993). Within the nuclear genome, markers on different chromosomal segments introgress more often than others (Rieseberg et al. 1999a, 1999b).

The extent to which introgression between domesticates and free-living populations operates in both directions or is primarily unidirectional has long been an area of discussion (reviewed in de Wet and Harlan 1975; Pickersgill 1981; Doebley 1989; Rieseberg and Wendel 1993). The example from maize described earlier in this chapter indicates that introgression can indeed introduce alleles from wild populations into domesticates, and such introgression has also been demonstrated in *Gossypium* (cotton, Wendel et al. 1989; Percy and Wendel 1990). Geneflow in the other direction, from crops to free-living populations, was demonstrated in squash (*Cucurbita pepo*) with isozymes and RAPD data (Decker-Walters et al. 1993, 2002). In the case of common beans *(Phaseolus vulgaris)*, introgression was three to four times higher from domesticated beans to wild populations than the reverse in an assessment of AFLP markers (Papa and Gepts 2003). However, molecular data can provide evidence to demonstrate gene flow only if diagnostic crop or wild alleles can be identified; if they cannot be identified, introgression may go undetected. Therefore, although introgression may be demonstrated in individual cases, the issue of whether geneflow occurs more frequently from crops to wild or weedy populations than the reverse, as a general phenomenon, may not be resolvable.

The consequences of geneflow from domesticates into wild or weedy relatives is currently of particular interest because of concerns about the effects of the escape of genes from genetically modified organisms. The review by Ellstrand, et al. (1999) demonstrates that the possibility that such escape might make weeds more aggressive is "far from hypothetical." Indeed, several studies have indicated that weedy populations, rather than being derived directly either from plants escaped from cultivation or from selective forces acting on purely wild populations (see Pickersgill 1981; Barrett 1983), have derived from hybridization between crops and wild relatives (e.g., de Wet and Harlan 1975; Papa and Gepts 2003). Crops such as sunflower have wild populations with which they are capable of exchanging genes throughout their North American range (Burke et al. 2002a). Alleles introduced to wild populations through introgression from crops can persist for several generations (Whitton et al. 1997). Wild-weed-crop complexes have been observed in many crops, and may act over time to mitigate the effects of the bottleneck of domestication (Beebe et al. 1997). However, geneflow can also obscure crop origins, precluding the localization of the area of domestication (e.g., Coulibaly et al. 2002). Although molecular data can sometimes help to distinguish between introgression and the alternative explanations described by Heiser (1973), these challenges cannot always be resolved. The possibility that crop-wild geneflow continued subsequent to domestication may make it difficult to interpret patterns in the genetic data.

Genetic Basis of Domestication

Some patterns are emerging from studies of the genetic basis of domestication in various crops. Like the domestication traits in maize that were reviewed earlier in this chapter, many other genes selected during domestication appear to involve regulatory functions. The tight inflorescence phenotype of cauliflower and broccoli cultivars of *Brassica oleracea* is produced by modifications in a regulatory gene, *cauliflower (BoCAL)* (Purugganan et al. 2000). Map-based cloning in tomato has identified *fw2.2*, an important domestication gene that affects fruit weight (Frary et al. 2000), in which the mutations that were selected in domestication are probably in the promoter region rather than in the protein-coding region of the gene (Nesbitt and Tanksley 2002). Selected alleles of *fw2.2* regulate the timing of gene expression, causing the gene to become active about a week earlier in development (Cong et al. 2002). Not all changes in domestication traits affect gene regulation, however. The glutinous rice favored in East and Southeast Asia has starch composed of little or no amylose, due to a mutation in the *waxy* locus that prevents splicing of an intron from the gene (Olsen and Purugganan 2002).

Now that several crops have been studied with the techniques of QTL mapping, an emerging area of study is to compare the genetics of plant domestication among closely (or more distantly) related species (see reviews by Buckler et al. 2001; Paterson 2002; Theissen 2002; Frary and Doganlar 2003). Some themes are common among several crops, but further sampling is finding that there are few universals. For example, many domestication traits seem to be controlled by relatively few genes with large phenotypic effects (Koinange et al. 1996; Poncet et al. 2000, 2002; Doganlar et al. 2002; Weeden et al. 2002; Frary and Doganlar 2003). There is considerable variation, however, and a trait controlled by a single gene in one species may be affected by several loci in another domesticate. For instance, the shattering rachis, a classical domestication trait, is controlled by a single locus in sorghum and three loci in rice, but it is affected by ten loci in maize (Paterson et al. 1995). Unlike most other crops studied to date, domestication traits in sunflower are influenced by a large number of loci of moderate or slight effect (Burke et al. 2002b). A surprising proportion of the sunflower QTL alleles were in the "wrong" direction, that is, alleles from wild parents gave a cultivated-like phenotype, and vice versa. A similar unexpected finding in cotton found that more variation in fiber traits had been contributed to this polyploid crop by the genome from an ancestral group that does not produce spinnable fibers, than by the genome from a group that does produce fiber (Jiang et al. 1998). These findings underscore the idea that the potential of wild germplasm to contribute to future crop improvement cannot necessarily be determined by the phenotype of the wild plants (Tanksley and McCouch 1997).

Another pattern arising from QTL mapping studies is that domestication genes in some, but not all, crops seem to occur in clusters, in which loci controlling several different domestication traits are found linked together. The generality of this phenomenon is not yet known, but it is an active area of both theoretical and empirical research (e.g., Koinange et al. 1996; Le Thierry d'Ennequin et al. 1999; Xiong et al. 1999; Cai and Morishima 2000, 2002; Burke et al. 2002b; Doganlar et al. 2002; Frary and Doganlar 2003). Such clustering might result from selection that favors linkage in "co-adapted complexes," but it may also be due to pleiotropic effects (in which one gene controls several seemingly unrelated traits). For example, the *fin* locus in *Phaseolus vulgaris* affects the shift to reproductive growth, but this in turn affects the number of nodes on main stem, number of pods, and days to flowering, so these traits are not truly independent (Koinange et al. 1996).

Comparative mapping studies are also assessing synteny among taxa, that is, the extent to which genes are colinear in different taxa. The observation that the same genes or markers are often found in roughly the same order on the same chromosomes in taxa that diverged millions of years ago is being held up for closer examination by comparisons across varying taxonomic levels and by both larger and finer scales along chromosomes (Brubaker et al. 1999; Wilson et al. 1999; Buckler et al. 2001; Gaut 2001). Here crop domestication studies are informing general evolutionary questions. These studies have brought up the question of the extent of "convergent domestication," the idea that the same (orthologous) genetic loci may have been selected in different crops as they were domesticated (Paterson et al. 1995; Buckler et al. 2001; Paterson 2002). Mapping of QTLs in different grain crops, for instance, has found that genes controlling the same "domestication trait" may map to similar chromosomal regions in different crops, suggesting that some of them (not necessarily all) might be orthologous loci (Paterson et al. 1995). This is a particularly promising area of study that will not only have practical implications, but when extended to comparisons with wild taxa, will improve our understanding of morphological evolution in general.

Domestication Bottleneck—Founder Effect of Domestication

A loss of genetic diversity accompanying domestication has been demonstrated in several crops, including, among others, sunflower (Tang and Knapp 2003), cotton (Small et al. 1999; Iqbal et al. 2001), adzuki bean (Xu et al. 2000), potato (Provan et al. 1999a; Ortiz and Huaman 2001), and common bean and lima bean (Gepts 1996; Papa and Gepts 2003). An important observation is that even among studies of the same crop, the extent of the reduction in diversity depends on the marker system used (Gepts 1996; Provan et al. 1999a; Anthony et al. 2002). Additionally, the bottleneck resulting from dispersal of the crop from its homeland can be narrower than the domestication bottleneck itself. In some cases, as in potato and coffee, propagules derived from a very few individuals were disseminated around the world, resulting in a severe founder effect (Ortiz and Huaman 2001; Anthony et al. 2002). Modern breeding can decrease genetic diversity in advanced cultivars (Sonnante et al. 1994), but in other cases there has been deliberate introgression of alleles by breeders more recently (Ortiz and Huaman 2001).

The extent of the domestication bottleneck depends on many factors, including the level of polymorphism in the domesticated population, whether the domestication occurred in a narrow or broad area, whether there were single or multiple domestications, and the extent of introgression from wild relatives after domestication. It also depends on the reproductive system of the species, that is, whether it is outcrossing or self-compatible, whether pollination is by wind or insects, or whether propagation of the crop is primarily by seed or by vegetative means. Although the effects of these breeding systems on the diversity of wild species have been reviewed (Ellstrand and Roose 1987; Hamrick and Godt 1990, 1996), there has been little comparative study of the extent of the domestication bottleneck in crop taxa with different breeding systems.

Case Study Chapters Illustrating the Use of Genetic Data to Study Crop Domestication

The case study chapters in this volume provide examples of studies of crop origins using different kinds of DNA data. The questions under investigation involve identifying the wild progenitor of the crop under study, pinpointing the geographical area of its domestication, and/or tracing the paths of diffusion of the crop. The data used include DNA sequences (cassava and oca), RAPDs (*Allium* and olive), microsatellites (cassava and olive), and RFLPs of the mitochondrial genome (olive).

Kenneth M. Olsen and Barbara A. Schaal (Chapter 9 of this volume) discuss the use of DNA sequence data to investigate the origin of domestication of one of the most important tropical staple food crops, *Manihot esculenta* Crantz ssp. *esculenta*. This root crop, known alternatively as cassava, tapioca, manioc, mandioca, and yuca, is the world's sixth most important crop, yet it is under-investigated, perhaps because it is cultivated primarily for subsistence. Olsen and Schaal explain the advantages of using haplotype networks of DNA sequence data for historical studies, and they describe their inferences about the origin of cassava based on data from three low-copy nuclear genes: glyceraldehyde 3-phosphate dehydrogenase (*G3pdh*), β-glucosidase (*BglA*), and α-hydroxynitrile lyase (*Hnl*). They compare sequences of these loci from cultivated cassava accessions to those of wild populations of *M. esculenta* ssp. *flabellifolia* (Pohl) Ciferri and *M. pruinosa* Pohl, both of which were among wild taxa that have been proposed as progenitors of cassava, either individually or as putatively hybridizing taxa. They also compared the inferences based on DNA sequences to data from microsatellites from the same populations. *Manihot esculenta* ssp. *flabellifolia* was supported by their analyses of both DNA sequences and microsatellites as the progenitor of cassava, and there was no indication of a contribution from *M. pruinosa*. This study illustrates methods and criteria for distinguishing whether free-living populations are escaped from cultivation or are truly wild progenitors. In this case, the contention that ssp. *flabellifolia* might represent escaped cassava populations rather than the wild progenitor was not supported. Cultivated cassava had a subset of the ssp. *flabellifolia* alleles, as expected if ssp. *flabellifolia* were the progenitor. If the latter were merely escaped cassava, the reverse pattern would be expected, that is, greater diversity in cultivated ssp. *esculenta* than in ssp. *flabellifolia*. The combination of DNA sequence data and microsatellite frequencies permitted Olsen and Schaal to infer that cassava was domesticated in the transitional vegetation area along the southern edge of the Amazonian basin.

The chapter by Frank R. Blattner and Nikolai Friesen (Chapter 10) provides an example of the use of random amplified polymorphic DNA (RAPD) data in crop domestication research. They first explain how the technique works and describe its advantages and pitfalls. Then they report on the use of RAPD data to investigate the origin of *Allium tuberosum* Rottl. ex Spreng, commonly known as "Chinese chive" or "garlic sprouts." In Eastern Asia, this crop is the second-most economically important species of *Allium* (the genus that includes onion, garlic, leek, etc.). Blattner and Friesen hoped to determine whether the Chinese chive was domesticated in a single or in multiple regions, and to define the geographic area in which it was domesticated. The results were unexpected. Although the wild species *A. ramosum* L. resembles *A. tuberosum* so closely that some taxonomists have considered them to be the same species, the RAPD data rejected the hypothesis that *A. tuberosum* is a domesticated form of *A. ramosum*. Analyses of the RAPD data by phenetic (UPGMA, neighbor-joining, and principal coordinate analysis) and parsimony-based phylogenetic methods (using accessions of *A. oreiprason* Schrenk as outgroups) found that *A. ramosum* and *A. tuberosum* formed two separate groups. *Allium tuberosum* was not nested within *A. ramosum* in the RAPD-based trees, nor did it have a subset of the alleles of the latter species. Blattner and Friesen interpret these results as indicating that these taxa are sister species, rather than having a progenitor-descendant relationship. The basal branches in the *A. tuberosum* group, including the only diploid sampled in that group, were all plants from northern China and Korea. Because northern China and Mongolia are not well studied botanically, it is possible that progenitor populations may yet be found in that region.

Catherine Breton, Guillaume Besnard, and André Bervillé (Chapter 11) studied the domestication and dispersal of olive, *Olea europaea* L. ssp. *europaea*, using several different kinds of molecular markers, including not only RAPDs and microsatellites (SSRs), which are discussed in the other chapters, but also mitochondrial RFLPs (restriction fragment length polymorphisms). Wild relatives of olive, known as "oleasters" are considered to be the same subspecies as the cultivated olive, but a different taxonomic variety. Domesticated olives are designated as *O. europaea* ssp. *europaea* var. *europaea*, and oleasters are designated as *O. europaea* ssp. *europaea* var. *sylvestris* (Mill.) Lehr. (These oleasters should not be confused with *Elaeagnus angustifolia*, a species that is also known as "oleaster," "wild olive," or "Russian olive," but is not related to true olive.) Oleasters have sometimes been considered olives escaped from cultivation, as they are difficult to distinguish morphologically. However, this is refuted by the finding that they are more diverse in RAPDs and chloroplast SSRs than domesticated olive. Contrary to commonly held prior assumptions, Breton, Besnard and Bervillé found greater molecular diversity in oleasters of the western areas of the Mediterranean than in those of the eastern areas. The RFLP patterns distinguished four mitochondrial haplotypes, or "mitotypes," among olives, all but one of which were also found in oleasters. One mitotype predominates in both olives and oleasters in the eastern Mediterranean, whereas other mitotypes occur in both wild and cultivated populations in the western Mediterranean. The distributions of mitotypes in olives and oleasters around the Mediterranean basin lead Breton et al. to infer that in addition to at least one domestication of olives in the eastern Mediterranean, there was probably also a separate domestication, from progenitor oleasters with other mitotypes, in western areas as well.

The final case study chapter (Chapter 12) is an example of the use of DNA sequence data to study the origins of the genomes of a polyploid crop. *Oxalis tuberosa*, commonly known as "oca," is a tuber crop cultivated in the Andean highlands of South America that has eight sets of chromosomes (i.e., it is octoploid). Prior information on oca's progenitors was scant, but the crop clearly originated in the Andes, and cytologists had found a group of *Oxalis* species dubbed the "*Oxalis tuberosa* alliance" that all shared the same base chromosome number as oca. In Chapter 12 I summarize investigations of the origins of oca's genomes based on DNA sequences of two nuclear loci: (1) the frequently used ITS region, the internal transcribed spacers of nuclear ribosomal DNA, and (2) ncpGS, a nuclear gene that encodes a form of glutamine synthetase that is expressed in the chloroplast. Data from both loci confirmed the origins of oca from within the *O. tuberosa* alliance, but only one of the loci had the power to identify the probable progenitors. Data from sequences of ncpGS showed multiple sequence types within individual plants of oca, providing information on the multiple genomes of this species. The ncpGS results indicated that the octoploid crop is clearly of hybrid origin (i.e., it is allopolyploid). Comparison of the oca sequences with those of wild *Oxalis* species collected in the Andes of Peru and Bolivia identified two wild tuber-bearing taxa as possible progenitors of the crop, so oca might have formed through hybridization of these two progenitor candidates.

The investigations described in this volume on four crops—*Allium tuberosum*, *Manihot esculenta*, *Olea europaea*, and *Oxalis tuberosa*—all establish the basic information about the origins, progenitors, and diffusion of these domesticates that will form the basis for future research. Subsequent study of different aspects of the domestication of each of these crops will undoubtedly build upon these foundations. Interestingly, these are all vegetatively propagated crops, so it will be interesting to compare studies of their domestication to those of the many seed-propagated crops under study. Many aspects of their evolution are likely to differ from seed-propagated crops. The extent of the domestication bottleneck in particular will be interesting to compare, but there is also little known about how geneflow, domestication genes, and the footprint of selection are manifested in these crops that primarily are clonally propagated Thus, continuing study of each of these crops will contribute to our understanding of evolution under domestication.

Conclusions and Future Prospects

Genetic data have provided great insight into plant domestication in recent years, and the rate of new studies is accelerating. Some crops have been very well studied, going beyond the identification of wild progenitors to examine other questions, such as whether or not there were multiple independent domestication events; the extent of introgressive hybridization between the crop and its wild relatives; the genetic basis of domestication; and how much the process of domestication reduced variation, either overall (bottleneck effects) or at chromosomal areas tightly linked to the selected sites (footprint of selection). Whereas a few crops have been thoroughly studied, however, others have been neglected. It is natural to focus on crops with great worldwide economic importance, but to understand the generality of the conclusions drawn about domestication we need to know more about these other "orphan" crops. Studies of such crops will likely be just as informative about the evolutionary processes and history of domestication—perhaps more so, because they may be less confounded by modification from recent breeding efforts. The case studies in this book demonstrate that there are still many crops for which the current distributions, or even the identities, of the progenitor populations are as yet unknown.

As a greater foundation of studies in various crops accumulates, the possibilities for making comparisons among them will build. Some possibilities already have been mentioned above, such as the need for studies of the effects of domestication on vegetatively propagated (clonal) crops compared with seed-propagated (sexual) crops, or the need to test the hypothesis of "convergent domestication" through selection upon the same genes in different crops of the same plant family. Crops differ in many ways, including life history traits, breeding system, plant parts used, geographical origin, ecology, and whether they were primary (founder) crops or secondary domesticates. All of these differences may influence the process of domestication, but we do not yet have the data with which to compare them. In addition, as more crops progress from basic studies of their origins toward QTL studies of the genetics of their domestication, we will be better able to assess the generality of patterns such as domestication traits being controlled by a few genes of large effects. The extent of "hidden" beneficial alleles in the wild relatives of crops should also be explored further, as well as the basis for the apparent clustering of domestication genes on chromosomes.

One final area of study that still has great unrealized potential is the use of ancient DNA. The possibilities for the use of this technique with existing and future collections have only begun to be explored. Nevertheless, expectations about its promise have been scaled back from initial unrealistic hopes. DNA becomes degraded over time, and as it breaks up into small fragments it becomes more prone to artifacts such as mosaic sequences caused by PCR recombination (Goloubinoff et al. 1994; see also Cronn et al. 2002). Some alleles may be missed entirely, although amplification of smaller fragments may improve success (Sefc et al. 2003). The greatest challenge is the potential for contamination by DNA from modern sources, so rigorous precautions must be taken to ensure that the amplified DNA is truly ancient DNA (Goloubinoff et al. 1994; Brown 1999). Although preserved DNA has been recovered from unexpected contexts, only a small proportion of such samples yields usable DNA (Brown 1999; Manen et al. 2003).

Despite these technical limitations, ancient DNA can help resolve questions regarding crop domestication that might be insoluble otherwise. With the use of ancient DNA, it is possible to go beyond observable morphological traits in archaeological plant remains. Traits such as culinary quality can be inferred from ancient DNA, as in the detection of genetic variants that influence protein and starch composition of wheat and maize, which in turn affects the quality for making bread or tortillas, respectively (Brown 1999; Jaenicke-Després et al. 2003). Ancient DNA can elucidate the time and place of appearance of crop variations and their subsequent spread, as in the cases reviewed by Brown (1999), in which ancient DNA from wheat demonstrated earlier-than-expected dates for the presence of hexaploid wheat in Switzerland, or of high-quality glutenin alleles in ancient Greece. DNA from ancient maize cobs from Mexico and New Mexico provided temporal benchmarks showing that there had been selection on three domestication genes before maize was dispersed throughout the Americas (Jaenicke-Després et al. 2003). These particular examples were possible because of the quantity of information already available about the genetics and evolution of wheat and maize, so that appropriate comparisons to modern material could be made. Nevertheless, it is remarkable that inferences can be made about the evolution of plant parts that are not preserved, as when early selection on plant architecture could be shown, based on sequences of *tb1* in ancient maize cobs (Jaenicke-Després et al. 2003).

Although DNA from modern sources is much more complete than ancient DNA, inferences made from modern DNA about the history of domestication often depend on assumptions and models, and patterns of modern diversity may have alternative explanations. It is possible that ancient DNA will help resolve ongoing debates, such as those about multiple crop origins (see Jones and Brown 2000). Ancient DNA can provide data on the presence, if not the frequency, of alleles in the past, thus providing information that does not depend on molecular clock assumptions. In fact, these data can be used to test whether such assumptions are valid (Goloubinoff et al. 1993; Jones and Brown 2000). Because populations of a crop's wild progenitors may have shifted or reduced their geographic range, lost genetic diversity, or even become extinct, ancient DNA might provide a more complete picture than modern sources (Jones and Brown 2000). However, although ancient DNA may reveal the presence of alleles in the past that are not found in modern material, one must be cautious about asserting the opposite. Because ancient DNA often comprises very small samples, it is risky both to assume that alleles not sampled from a particular time in the past were not present at that time, and to base evolutionary rate calculations on that assumption (e.g., Freitas et al. 2003).

It is important to improve communication between researchers of domestication in the disciplines of molecular genetics and archaeology, as this volume attempts to do. Ancient DNA has the potential to provide a true bridge between these disciplines. Collaborating in studies of ancient DNA may finally bring together multidisciplinary teams that will not only share data, but also combine strengths to provide new insights into domestication. Through this interchange, future studies using these different approaches may be better able to inform and complement each other.

Acknowledgments

I thank Deena Decker-Walters for providing the summary phenogram of RAPD data from *Cucurbita pepo* (Figure 8.1), and for the initial invitation to contribute to this book; Bill Burger, Edna Davion, John Doebley, Bruce Smith, and Melinda Zeder for taking the time to read earlier versions of this manuscript; and Jeff Doyle for discussion. This work was supported by The Field Museum of Natural History.

References

Abbo, S., S. Lev-Yadun, and G. Ladizinsky. 2001. Tracing the wild genetic stocks of crop plants. *Genome* 44: 309–310.

Allaby, R. G. and T. A. Brown. 2003. AFLP data and the origins of domesticated crops. *Genome* 46: 448–453.

Anderson, E. and L. Hubricht. 1938. Hybridization in *Tradescantia*. III. The evidence for introgressive hybridization. *American Journal of Botany* 25: 396–402.

Anthony, F., M. C. Combes, C. Astorga, B. Bertrand, G. Graziosi, and P. Lashermes. 2002. The origin of cultivated *Coffea arabica* L. varieties revealed by AFLP and SSR markers. *Theoretical and Applied Genetics* 104: 894–900.

Atkinson, R. G., G. Cipriani, D. J. Whittaker, and R. C. Gardner. 1997. The allopolyploid origin of kiwifruit, *Actinidia deliciosa* (Actinidiaceae). *Plant Systematics and Evolution* 205: 111–124.

Badr, A., K. Muller, R. Schafer-Pregl, H. El Rabey, S. Effgen, H. H. Ibrahim, C. Pozzi, W. Rohde, and F. Salamini. 2000. On the origin and domestication history of barley (*Hordeum vulgare*). *Molecular Biology & Evolution* 17: 499–510.

Barrett, S. C. H. 1983. Crop mimicry in weeds. *Economic Botany* 37: 255–282.

Baumel, A., M. Ainouche, R. Kalendar, and A. H. Schulman. 2002. Retrotransposons and genomic stability in populations of the young allopolyploid species *Spartina anglica* C.E. Hubbard (Poaceae). *Molecular Biology and Evolution* 19: 1218–1227.

Beadle, G. W. 1939. Teosinte and the origin of maize. *Journal of Heredity* 30: 245–247.

———. 1972. The mystery of maize. *Field Museum of Natural History Bulletin* 43:2–11.

Becerra Velasquez, V. L. and P. Gepts. 1994. RFLP diversity of common bean (*Phaseolus vulgaris*) in its centers of origin. *Genome* 37: 256–263.

Beebe, S., O. Toro, A. V. Gonzalez, M. I. Chacon, and D. G. Debouck. 1997. Wild-weed-crop complexes of common bean (*Phaseolus vulgaris* L, Fabaceae) in the Andes of Peru and Colombia, and their implications for conservation and breeding. *Genetic Resources and Crop Evolution* 44: 73–91.

Benavente, E. and J. Orellana. 1989. Pairing competition between identical and homologous chromosomes in autotetraploid rye heterozygous for interstitial C-Bands. *Chromosoma* 98: 225–232.

———. 1991. Chromosome differentiation and pairing behavior of polyploids—An assessment on preferential metaphase-I associations in colchicine-induced autotetraploid hybrids within the genus *Secale*. *Genetics* 128: 433–442.

Bennetzen, J., E. Buckler, V. Chandler, J. Doebley, J. Dorweiler, B. Gaut, M. Freeling, S. Hake, E. Kellogg, R. S. Poethig, V. Walbot, and S. Wessler. 2001. Genetic evidence and the origin of maize. *Latin American Antiquity* 12: 84–86.

Bernacchi, D., T. Beck Bunn, Y. Eshed, Y. Lopez, V. Petiard, J. Uhlig, D. Zamir, and S. Tanksley. 1998. Advanced backcross QTL analysis in tomato. I. Identification of QTLs for traits of agronomic importance from *Lycopersicon hirsutum*. *Theoretical and Applied Genetics* 97: 381–397.

Blattner, F. R. and A. G. Badani Mendez. 2001. RAPD data do not support a second centre of barley domestication in Morocco. *Genetic Resources and Crop Evolution* 48. 13–19.

Blumler, M. A. 1992. Independent inventionism and recent genetic evidence on plant domestication. *Economic Botany* 46: 98–111.

Bretó, M. P., C. Ruiz, J. A. Pina, and M. J. Asíns. 2001. The diversification of *Citrus clementina* Hort. ex Tan., a vegetatively propagated crop species. *Molecular Phylogenetics and Evolution* 21: 285–293.

Brown, T. A. 1999. How ancient DNA may help in understanding the origin and spread of agriculture. *Philosophical Transactions of the Royal Society of London Series B—Biological Sciences* 354: 89–97.

Brubaker, C. L., A. H. Paterson, and J. F. Wendel. 1999. Comparative genetic mapping of allotetraploid cotton and its diploid progenitors. *Genome* 42: 184–203.

Buckler, E. S. IV and T. P. Holtsford. 1996. *Zea* systematics: ribosomal ITS evidence. *Molecular Biology and Evolution* 13: 612–622.

Buckler, E. S. IV, J. M. Thornsberry, and S. Kresovich. 2001. Molecular diversity, structure and domestication of grasses. *Genetical Research* 77: 213–218.

Burke, J. M., K. A. Gardner, and L. H. Rieseberg. 2002a. The potential for gene flow between cultivated and wild sunflower (*Helianthus annuus*) in the United States. *American Journal of Botany* 89: 1550–1552.

———. 2002b. Genetic analysis of sunflower domestication. *Genetics* 161: 1257–1267.

Cai, H. W. and H. Morishima. 2000. Genomic regions affecting seed shattering and seed dormancy in rice. *Theoretical and Applied Genetics* 100: 840–846.

———. 2002. QTL clusters reflect character associations in wild and cultivated rice. *Theoretical and Applied Genetics* 104: 1217–1228.

Chang, R. Y., L. S. O'Donoughue, and T. E. Bureau. 2001. Inter-MITE polymorphisms (IMP): a high throughput transposon-based genome mapping and fingerprinting approach. *Theoretical and Applied Genetics* 102: 773–781.

Cong, B., J. Liu, and S. D. Tanksley. 2002. Natural alleles at a tomato fruit size quantitative trait locus differ by heterochronic regulatory mutations. *Proceedings of the National Academy of Sciences of the United States of America* 99: 13606–13611.

Coulibaly, S., R. S. Pasquet, R. Papa, and P. Gepts. 2002. AFLP analysis of the phenetic organization and genetic diversity of *Vigna unguiculata* L. Walp. reveals extensive gene flow between wild and domesticated types. *Theoretical and Applied Genetics* 104: 358–366.

Cronn, R., M. Cedroni, T. Haselkorn, C. Grover, and J. F. Wendel. 2002. PCR-mediated recombination in amplification products derived from polyploid cotton. *Theoretical and Applied Genetics* 104: 482–489.

Cros, J., M. C. Combes, P. Trouslot, F. Anthony, S. Hamon, A. Charrier, and P. Lashermes. 1998. Phylogenetic analysis of chloroplast DNA variation in *Coffea* L. *Molecular Phylogenetics and Evolution* 9: 109–117.

Cutler, H. C. 1946. *Races of maize in South America.* Cambridge, MA: Botanical Museum, Harvard University.

Darwin, C. 1883. *The variation of animals and plants under domestication*, 2nd Edition. New York: D. Appleton.

Debouck, D. G., O. Toro, O. M. Paredes, W. C. Johnson, and P. Gepts. 1993. Genetic diversity and ecological distribution of *Phaseolus vulgaris* (Fabaceae) in northwestern South America. *Economic Botany* 47: 408–423.

de Candolle, A. 1884. *Origin of cultivated plants.* London, UK: Trench.

Decker, D. S. 1985. Numerical analysis of allozyme variation in *Cucurbita pepo*. *Economic Botany* 39: 300–309.

———. 1988. Origin(s), evolution, and systematics of *Cucurbita pepo* (Cucurbitaceae). *Economic Botany* 42: 4–15.

Decker, D. S. and H. D. Wilson. 1986. Numerical analysis of seed morphology in *Cucurbita pepo*. *Systematic Botany* 11: 595–607.

———. 1987. Allozyme variation in the *Cucurbita pepo* complex *Cucurbita pepo* var. *ovifera* vs. *Cucurbita texana*. *Systematic Botany* 12: 263–273.

Decker-Walters, D. S., T. W. Walters, C. W. Cowan, and B. D. Smith. 1993. Isozymic characterization of wild populations of *Cucurbita pepo*. *Journal of Ethnobiology* 13: 55–72.

Decker-Walters, D. S., J. E. Staub, S. M. Chung, E. Nakata, and H. D. Quemada. 2002. Diversity in free-living populations of *Cucurbita pepo* (Cucurbitaceae) as assessed by random amplified polymorphic DNA. *Systematic Botany* 27: 19–28.

de Wet, J. M. J. and J. R. Harlan. 1975. Weeds and domesticates: Evolution in the man-made habitat. *Economic Botany* 29: 99–109.

Diamond, J. 1997. *Guns, germs, and steel: The fates of human societies*. New York: Norton.

Doebley, J. 1989. Isozymic evidence and the evolution of crop plants. In *Isozymes in plant biology*, D. E. Soltis and P. S. Soltis (eds.), pp. 165–191. Portland, Or.: Dioscorides Press.

———. 1990a. Molecular evidence and the evolution of maize. *Economic Botany* 44: 6–27.

———. 1990b. Molecular evidence for gene flow among *Zea* species—Genes transformed into maize through genetic engineering could be transferred to its wild relatives, the teosintes. *Bioscience* 40: 443–448.

———. 1992a. Mapping the genes that made maize. *Trends in Genetics* 8: 302–307.

———. 1992b. Molecular systematics and crop evolution. In *Molecular Systematics of Plants*, P. S. Soltis, D. E. Soltis, and J. J. Doyle (eds.), pp. 202–222. New York: Chapman and Hall.

———. 1995. Genetic dissection of the morphological evolution of maize. *Aliso* 14: 297–304.

———. 2001. George Beadle's other hypothesis: One-gene, one-trait. *Genetics* 158. 487–493.

Doebley, J. and A. Stec. 1991. Genetic analysis of the morphological differences between maize and teosinte. *Genetics* 129: 285–295.

———. 1993. Inheritance of the morphological differences between maize and teosinte: Comparison of results for two F-2 populations. *Genetics* 134: 559–570.

Doebley, J. and R. L. Wang. 1997. Genetics and the evolution of plant form: An example from maize. *Cold Spring Harbor Symposia on Quantitative Biology* 62: 361–367.

Doebley, J. F., M. M. Goodman, and C. W. Stuber. 1984. Isoenzymatic variation in *Zea* (Gramineae). *Systematic Botany* 9: 203–218.

Doebley, J., M. M. Goodman, and C. W. Stuber. 1987a. Patterns of isozyme variation between maize and Mexican annual teosinte. *Economic Botany* 41: 234–246.

Doebley, J., W. Renfroe, and A. Blanton. 1987b. Restriction site variation in the *Zea* chloroplast genome. *Genetics* 117: 139–148.

Doebley, J., A. Bacigalupo, and A. Stec. 1994. Inheritance of kernel weight in two maize-teosinte hybrid populations: Implications for crop evolution. *Journal of Heredity* 85: 191–195.

Doebley, J., A. Stec, and C. Gustus. 1995a. *Teosinte branched1* and the origin of maize: Evidence for epistasis and the evolution of dominance. *Genetics* 141: 333–346.

Doebley, J., A. Stec, and B. Kent. 1995b. *Suppressor of sessile spikelets1 (Sos1)*: A dominant mutant affecting inflorescence development in maize. *American Journal of Botany* 82: 571–577.

Doganlar, S., A. Frary, M.-C. Daunay, R. N. Lester, and S. D. Tanksley. 2002. Conservation of gene function in the Solanaceae as revealed by comparative mapping of domestication traits in eggplant. *Genetics* 161: 1713–1726.

Dorweiler, J. E. and J. Doebley. 1997. Developmental analysis of *teosinte glume architecture1*: A key locus in the evolution of maize (Poaceae). *American Journal of Botany* 84: 1313–1322.

Dorweiler, J., A. Stec, J. Kermicle, and J. Doebley. 1993. *Teosinte glume architecture1*: A genetic locus controlling a key step in maize evolution. *Science* 262: 233–235.

Doyle, J. J. 1992. Gene trees and species trees: Molecular systematics as one-character taxonomy. *Systematic Botany* 17: 144–163.

Doyle, J. J. and R. N. Beachy. 1985. Ribosomal gene variation in soybean *(Glycine)* and its relatives. *Theoretical and Applied Genetics* 70: 369–376.

Doyle, J. J. and J. L. Doyle. 1999. Nuclear protein-coding genes in phylogeny reconstruction and homology assessment: Some examples from Leguminosae. In *Molecular systematics and plant evolution*, P. M. Hollingsworth, R. M. Bateman and R. J. Gornall (eds.), pp. 229–254. London: Taylor & Francis.

Doyle, J. J. and B. S. Gaut. 2000. Evolution of genes and taxa: A primer. *Plant Molecular Biology* 42: 1–23.

Doyle, J. J., M. Morgante, S. V. Tingey, and W. Powell. 1998. Size homoplasy in chloroplast microsatellites of wild perennial relatives of soybean *(Glycine* subgenus *Glycine)*. *Molecular Biology and Evolution* 15: 215–218.

Ellstrand, N. C. and M. L. Roose. 1987. Patterns of genotypic diversity in clonal plant species. *American Journal of Botany* 74: 123–131.

Ellstrand, N. C., H. C. Prentice, and J. F. Hancock. 1999. Gene flow and introgression from domesticated plants into their wild relatives. *Annual Review of Ecology and Systematics* 30: 539–563.

Eubanks, M. W. 1995. A cross between 2 maize relatives— *Tripsacum dactyloides* and *Zea diploperennis* (Poaceae). *Economic Botany* 49: 172–182.

———. 1997. Molecular analysis of crosses between *Tripsacum dactyloides* and *Zea diploperennis* (Poaceae). *Theoretical and Applied Genetics* 94: 707–712.

———. 2001a. An interdisciplinary perspective on the origin of maize. *Latin American Antiquity* 12: 91–98.

———. 2001b. The mysterious origin of maize. *Economic Botany* 55: 492–514.

———. 2001c. The origin of maize: Evidence for *Tripsacum* ancestry. *Plant Breeding Reviews* 20: 15–66.

Eyre-Walker, A., R. L. Gaut, H. Hilton, D. L. Feldman, and B. S. Gaut. 1998. Investigation of the bottleneck leading to the domestication of maize. *Proceedings of the National Academy of Sciences of the United States of America* 95: 4441–4446.

Feldman, M. and E. Millet. 2001. The contribution of the discovery of wild emmer to an understanding of wheat evolution and domestication and to wheat improvement. *Israel Journal of Plant Sciences* 49: S25–S35.

Flavell, A. J., M. R. Knox, S. R. Pearce, and T. H. N. Ellis. 1998. Retrotransposon-based insertion polymorphisms (RBIP) for high throughput marker analysis. *Plant Journal* 16: 643–650.

Frary, A. and S. Doganlar. 2003. Comparative genetics of crop plant domestication and evolution. *Turkish Journal of Agriculture and Forestry* 27: 59–69.

Frary, A., T. C. Nesbitt, A. Frary, S. Grandillo, E. van der Knaap, B. Cong, J. Liu, J. Meller, R. Elber, K. B. Alpert, and S. Tanksley. 2000. fw2.2: A quantitative trait locus key to the evolution of tomato fruit size. *Science* 289: 85–88.

Freitas, F. O., G. Bendel, R. G. Allaby, and T. A. Brown. 2003. DNA from primitive maize landraces and archaeological remains: Implications for the domestication of maize and its expansion into South America. *Journal of Archaeological Science* 30: 901–908.

Freyre, R., R. Rios, L. Guzman, D. G. Debouck, and P. Gepts. 1996. Ecogeographic distribution of *Phaseolus* spp (Fabaceae) in Bolivia. *Economic Botany* 50: 195–215.

Gaut, B. S. 1998. Molecular clocks and nucleotide substitution rates in higher plants. In *Evolutionary biology*, Volume 30, M. K. Hecht, R. J. Macintyre, and M. T. Clegg (eds.), pp. 93–120. New York and London: Plenum Press.

———. 2001 Patterns of chromosomal duplication in maize and their implications for comparative maps of the grasses. *Genome Research* 11: 55–66.

Gaut, B. S. and M. T. Clegg. 1993. Nucleotide polymorphism in the *Adh1* locus of pearl millet *(Pennisetum glaucum)* (Poaceae). *Genetics* 135: 1091–1097.

Gaut, B. S. and J. F. Doebley. 1997. DNA sequence evidence for the segmental allotetraploid origin of maize. *Proceedings of the National Academy of Sciences of the United States of America* 94: 6809–6814.

Gaut, B. S. and B. S. Weir. 1994. Detecting substitution-rate heterogeneity among regions of a nucleotide sequence. *Molecular Biology and Evolution* 11: 620–629.

Gaut, B. S., M. Le Thierry d'Ennequin, A. S. Peek, and M. C. Sawkins. 2000. Maize as a model for the evolution of plant nuclear genomes. *Proceedings of the National Academy of Sciences of the United States of America* 97: 7008–7015.

Gepts, P. 1990a. Biochemical evidence bearing on the domestication of *Phaseolus* (Fabaceae) beans. *Economic Botany* 44: 28–38.

———. 1990b. Genetic diversity of seed storage proteins in plants. In *Plant population genetics, breeding, and genetic resources*, A. H. D. Brown, M. T. Clegg, A. L. Kahler, and B. S. Weir (eds.), pp. 64–82. Sunderland, Mass.: Sinauer Associates Inc.

———. 1993 The use of molecular and biochemical markers in crop evolution studies. *Evolutionary Biology* 27: 51–94.

———. 1996. Origin and evolution of cultivated *Phaseolus* species. In *Advances in legume systematics* (Part 8): *Legumes of economic importance*, B. Pickersgill and J. M. Lock (eds.), pp. 65–74. Kew: Royal Botanic Gardens.

Gepts, P. and F. A. Bliss. 1985. F1 hybrid weakness in the common bean—Differential geographic origin suggests two gene pools in cultivated bean germplasm. *Journal of Heredity* 76: 447–450.

———. 1986. Phaseolin variability among wild and cultivated common beans *(Phaseolus vulgaris)* from Colombia. *Economic Botany* 40: 469–478.

Gepts, P., T. C. Osborn, K. Rashka, and F. A. Bliss. 1986. Phaseolin protein variability in wild forms and landraces of the common bean *(Phaseolus vulgaris)*—Evidence for multiple centers of domestication. *Economic Botany* 40: 451–468.

Goloubinoff, P., S. Pääbo, and A. C. Wilson. 1993. Evolution of maize inferred from sequence diversity of an *Adh2* gene segment from archaeological specimens. *Proceedings of the National Academy of Sciences of the United States of America* 90: 1997–2001.

———. 1994. Molecular characterization of ancient maize: Potentials and pitfalls. In *Corn and culture in the prehistoric New World*, S. Johannessen and C. A. Hastorf (eds.), pp. 113–125. Boulder, Colo.: Westview Press.

Grant, T., J. Faivovicha, and D. Pola. 2003. The perils of "point-and-click" systematics. *Cladistics* 19: 276–285.

Graur, D. and W. Martin. 2004. Reading the entrails of chickens: Molecular timescales of evolution and the illusion of precision. *Trends in Genetics* 20: 80–86.

Hammer, K. 1984. Das Domestikationssyndrom. *Kulturpflanze* 32: 11–34.

Hamrick, J. L. and M. J. W. Godt. 1990. Allozyme diversity in plant species. In *Plant population genetics, breeding, and genetic resources*, A. H. D. Brown, M. T. Clegg, A. L. Kahler, and B. S. Weir (eds.), pp. 43–63. Sunderland, Mass.: Sinauer Associates Inc.

———. 1996. Effects of life history traits on genetic diversity in plant species. *Philosophical Transactions of the Royal Society of London Series B—Biological Sciences* 351: 1291–1298.

Hanson, M. A., B. S. Gaut, A. O. Stec, S. I. Fuerstenberg, M. M. Goodman, E. H. Coe, and J. F. Doebley. 1996. Evolution of anthocyanin biosynthesis in maize kernels: The role of regulatory and enzymatic loci. *Genetics* 143: 1395–1407.

Harlan, J. R. 1971. Agricultural origins: Centers and noncenters. *Science* 174: 468–473.

———. 1992. *Crops and man*, 2nd edition. Madison, Wisc.: American Society of Agronomy, Crop Science Society of America.

Harlan, J. R., J. M. J. de Wet, and E. G. Price. 1973. Comparative evolution of cereals. *Evolution* 27: 311–325.

Heiser, C. B., Jr. 1949. Natural hybridization with particular reference to introgression. *Botanical Review* 49: 645–687.

———. 1973. Introgression reexamined. *Botanical Review* 39: 347–366.

———. 1988. Aspects of unconscious selection and the evolution of domesticated plants. *Euphytica* 37: 77–81.

———. 1985. Some botanical considerations on the early domesticated plants north of Mexico. In *Prehistoric food production in North America*, R. I. Ford (ed.), pp. 57–72. Ann Arbor, Mich.: Museum of Anthropology, University of Michigan.

Heun, M., R. SchaferPregl, R. Klawan, R. Castagna, M. Accerbi, B. Borghi, and F. Salamini. 1997. Site of einkorn wheat domestication identified by DNA fingerprinting. *Science* 278: 1312–1314.

Hilton, H. and B. S. Gaut. 1998. Speciation and domestication in maize and its wild relatives: Evidence from the *globulin-1* gene. *Genetics* 150: 863–872.

Hilu, K. W. and H. T. Stalker. 1995. Genetic relationships between peanut and wild species of *Arachis* sect *Arachis* (Fabaceae): Evidence from RAPDs. *Plant Systematics and Evolution* 198: 167–178.

Iltis, H. H. 1983. From teosinte to maize—the catastrophic sexual transmutation. *Science* 222: 886–894.

———. 2000. Homeotic sexual translocations and the origin of maize (*Zea mays*, Poaceae): A new look at an old problem. *Economic Botany* 54: 7–42.

Ingram, A. L. and J. J. Doyle. 2003. The origin and evolution of *Eragrostis tef* (Poaceae) and related polyploids: Evidence from nuclear *waxy* and plastid *rps16*. *American Journal of Botany* 90:116–122.

Iqbal, M. J., O. U. K. Reddy, K. M. El-Zik, and A. E. Pepper. 2001. A genetic bottleneck in the "evolution under domestication" of upland cotton *Gossypium hirsutum* L. examined using DNA fingerprinting. *Theoretical and Applied Genetics* 103: 547–554.

Ishii, T., N. Mori, and Y. Ogihara. 2001. Evaluation of allelic diversity at chloroplast microsatellite loci among common wheat and its ancestral species. *Theoretical and Applied Genetics* 103: 896–904.

Islam, F. M. A., K. E. Basford, R. J. Redden, A. V. Gonzalez, P. M. Kroonenberg, and S. Beebe. 2002. Genetic variability in cultivated common bean beyond the two major gene pools. *Genetic Resources and Crop Evolution* 49: 271–283.

Jaenicke-Després, V., E. S. Buckler, B. D. Smith, M. T. P. Gilbert, A. Cooper, J. Doebley, and S. Pääbo. 2003. Early Allelic selection in maize as revealed by ancient DNA. *Science* 302:1206–1208.

Jiang, C. X., R. J. Wright, K. M. El-Zik, and A. H. Paterson. 1998. Polyploid formation created unique avenues for response to selection in *Gossypium* (cotton). *Proceedings of the National Academy of Sciences*, USA 95: 4419–4424.

Johns, T. 1989. A chemical-ecological model of root and tuber domestication in the Andes. In *Foraging and farming—The evolution of plant exploitation*, D. R. Harris and G. C. Hillman (eds.), pp. 504–519. London: Unwin Hyman.

Johns, T. and J. G. Alonso. 1990. Glycoalkaloid change during the domestication of the potato, *Solanum* section *Petota*. *Euphytica* 50: 203–210.

Jones, C. J., K. J. Edwards, S. Castaglione, M. O. Winfield, F. Sala, C. vandeWiel, G. Bredemeijer, B. Vosman, M. Matthes, A. Daly, R. Brettschneider, P. Bettini, M. Buiatti, E. Maestri, A. Malcevschi, N. Marmiroli, R. Aert, G. Volckaert, J. Rueda, R. Linacero, A. Vazquez, and A. Karp. 1997. Reproducibility testing of RAPD, AFLP and SSR markers in plants by a network of European laboratories. *Molecular Breeding* 3: 381–390.

Jones, M. and T. Brown. 2000. Agricultural origins: The evidence of modern and ancient DNA. *Holocene* 10: 769–776.

Jung, S., P. L. Tate, R. Horn, G. Kochert, K. Moore, and A. G. Abbott. 2003. The phylogenetic relationship of possible progenitors of the cultivated peanut. *Journal of Heredity* 94: 334–340.

Kalendar, R., T. Grob, M. Regina, A. Suoniemi, and A. Schulman. 1999. IRAP and REMAP: Two new retrotransposon-based DNA fingerprinting techniques. *Theoretical and Applied Genetics* 98: 704–711.

Katzir, N., Y. Tadmor, G. Tzuri, E. Leshzeshen, N. Mozes-Daube, Y. Danin-Poleg, and H. S. Paris. 2000. Further ISSR and preliminary analysis of relationships among accessions of *Cucurbita pepo*. *Acta Horticulturae* 510: 433–439.

Kellogg, E. A. 2003. What happens to genes in duplicated genomes. *Proceedings of the National Academy of Sciences*, USA 100: 4369–4371.

Kirkpatrick, K. J., D. S. Decker, and H. D. Wilson. 1985. Allozyme differentiation in the *Cucurbita pepo* complex *Cucurbita pepo* var. *medullosa* vs. *Cucurbita texana*. *Economic Botany* 39: 289–299.

Koenig, R. and P. Gepts. 1989. Segregation and linkage of genes for seed proteins, isozymes, and morphological traits in common bean *(Phaseolus vulgaris)*. *Journal of Heredity* 80: 455–459.

Koinange, E. M. K. and P. Gepts. 1992. Hybrid weakness in wild *Phaseolus vulgaris* L. *Journal of Heredity* 83: 135–139.

Koinange, E. M. K., S. P. Singh, and P. Gepts. 1996. Genetic control of the domestication syndrome in common bean. *Crop Science* 36: 1037–1045.

Kumar, A. and H. Hirochika. 2001. Applications of retrotransposons as genetic tools in plant biology. *Trends in Plant Science* 6: 127–134.

Ladizinsky, G. 1985. Founder effect in crop-plant evolution. *Economic Botany* 39: 191–199.

Ladizinsky, G. and A. Genizi. 2001. Could early gene flow have created similar allozyme-gene frequencies in cultivated and wild barley? *Genetic Resources and Crop Evolution* 48.101–104.

Le Thierry d'Ennequin, M., B. Toupance, T. Robert, B. Godelle, and P. Gouyon. 1999. Plant domestication: A model for studying the selection of linkage. *Journal of Evolutionary Biology* 12: 1138–1147.

Levy, A. A. and M. Feldman. 2002. The impact of polyploidy on grass genome evolution. *Plant Physiology* 130: 1587–1593.

Li, G. and C. F. Quirós. 2001. Sequence-related amplified polymorphism (SRAP), a new marker system based on a simple PCR reaction: Its application to mapping and gene tagging in *Brassica*. *Theoretical and Applied Genetics* 103: 455–461.

Lukens, L. and J. Doebley. 2001. Molecular evolution of the teosinte branched gene among maize and related grasses. *Molecular Biology and Evolution* 18: 627–638.

MacNeish, R. S. and M. W. Eubanks. 2000. Comparative analysis of the Río Balsas and Tehuacán models for the origin of maize. *Latin American Antiquity* 11: 3–20.

Manen, J. F., L. Bouby, O. Dalnoki, P. Marinval, M. Turgay, and A. Schlumbaum. 2003. Microsatellites from archaeological *Vitis vinifera* seeds allow a tentative assignment of the geographical origin of ancient cultivars. *Journal of Archaeological Science* 30: 721–729.

Mangelsdorf, P. C. and R. G. Reeves. 1939. The origin of Indian corn and its relatives. *Texas Agricultural Experiment Station Bulletin* 574: 1–315.

Mason-Gamer, R. J. 2001. Origin of North American *Elymus* (Poaceae: Triticeae) allotetraploids based on granule-bound starch synthase gene sequences. *Systematic Botany* 26: 757–768.

Matsuoka, Y., Y. Vigouroux, M. M. Goodman, G. J. Sanchez, E. Buckler, and J. Doebley. 2002. A single domestication for maize shown by multilocus microsatellite genotyping. *Proceedings of the National Academy of Sciences*, USA 99: 6080–6084.

McClintock, B. 1929. Chromosome morphology in *Zea mays*. *Science* 69: 629.

McGregor, C. E., C. A. Lambert, M. M. Greyling, J. H. Louw, and L. Warnich. 2000. A comparative assessment of DNA fingerprinting techniques (RAPD, ISSR, AFLP and SSR) in tetraploid potato (*Solanum tuberosum* L.) germplasm. *Euphytica* 113: 135–144.

Milbourne, D., R. Meyer, J. E. Bradshaw, E. Baird, N. Bonar, J. Provan, W. Powell, and R. Waugh. 1997. Comparison of PCR-based marker systems for the analysis of genetic relationships in cultivated potato. *Molecular Breeding* 3: 127–136.

Molina-Cano, J. L., M. Moralejo, E. Igartua, and I. Romagosa. 1999. Further evidence supporting Morocco as a centre of origin of barley. *Theoretical and Applied Genetics* 98: 913–918.

Morgante, M. and A. M. Olivieri. 1993. PCR-amplified microsatellites as markers in plant genetics. *The Plant Journal* 3: 175–182.

Nee, M. 1990. The domestication of *Cucurbita* (Cucurbitaceae). *Economic Botany* 44: 56–68.

Nesbitt, T. C. and S. D. Tanksley. 2002. Comparative sequencing in the genus *Lycopersicon*: Implications for the evolution of fruit size in the domestication of cultivated tomatoes. *Genetics* 162: 365–379.

Nixon, K. C. and Q. D. Wheeler. 1990. An amplification of the phylogenetic species concept. *Cladistics* 6: 211–223.

Nybom, H., S. H. Rogstad, and B. A. Schaal. 1990. Genetic variation detected by use of the M13 DNA fingerprint probe in *Malus*, *Prunus* and *Rubus* Rosaceae. *Theoretical and Applied Genetics* 79: 153–156.

Olsen, K. M. 1999. Minisatellite variation in a single-copy nuclear gene: Phylogenetic assessment of repeat length homoplasy and mutational mechanism. *Molecular Biology & Evolution* 16:1406–1409.

Olsen, K. M. and M. D. Purugganan. 2002. Molecular evidence on the origin and evolution of glutinous rice. *Genetics* 162: 941–950.

Ortiz, R. and Z. Huamán. 2001. Allozyme polymorphisms in tetraploid potato gene pools and the effect on human selection. *Theoretical and Applied Genetics* 103: 792–796.

Ozkan, H., A. Brandolini, R. Schafer-Pregl, and F. Salamini. 2002. AFLP analysis of a collection of tetraploid wheats indicates the origin of emmer and hard wheat domestication in southeast Turkey. *Molecular Biology and Evolution* 19: 1797–1801.

Palacios, G., S. Bustamante, C. Molina, P. Winter, and G. Kahl. 2002. Electrophoretic identification of new genomic profiles with a modified selective amplification of microsatellite polymorphic loci technique based on AT/AAT polymorphic repeats. *Electrophoresis* 23: 3341–3345.

Papa, R. and P. Gepts. 2003. Asymmetry of gene flow and differential geographical structure of molecular diversity in wild and domesticated common bean (*Phaseolus vulgaris* L.) from Mesoamerica. *Theoretical and Applied Genetics* 106: 239–250.

Parsons, B. J., H. J. Newbury, M. T. Jackson, and B. V. FordLloyd. 1997. Contrasting genetic diversity relationships are revealed in rice (*Oryza sativa* L) using different marker types. *Molecular Breeding* 3: 115–125.

Paterson, A. H. 2002. What has QTL mapping taught us about plant domestication? *New Phytologist* 154: 591–608.

Paterson, A. H., Y. R. Lin, Z. K. Li, K. F. Schertz, J. F. Doebley, S. R. M. Pinson, S. C. Liu, J. W. Stansel, and J. E. Irvine. 1995. Convergent domestication of cereal crops by independent mutations at corresponding genetic loci. *Science* 269: 1714–1718.

Patzak, J. 2001. Comparison of RAPD, STS, ISSR and AFLP molecular methods used for assessment of genetic diversity in hop (*Humulus lupulus* L.). *Euphytica* 121: 9–18.

Peakall, R., S. Gilmore, W. Keys, M. Morgante, and A. Rafalski. 1998. Cross-species amplification of soybean (*Glycine max*) simple sequence repeats (SSRs) within the genus and other legume genera: Implications for the transferability of SSRs in plants. *Molecular Biology and Evolution* 15: 1275–1287.

Percy, R. G. and J. F. Wendel. 1990. Allozyme evidence for the origin and diversification of *Gossypium barbadense* L. *Theoretical and Applied Genetics* 79: 529–542.

Pickersgill, B. 1981. Biosystematics of crop-weed complexes. *Kulturpflanze* 29: 377–388.

———. 1989. Cytological and genetical evidence of the domestication and diffusion of crops within the Americas. In *Foraging and farming—The evolution of plant exploitation*, D. R. Harris and G. C. Hillman (eds.), pp. 426–439. London: Unwin Hyman.

Poncet, V., F. Lamy, K. M. Devos, M. D. Gale, A. Sarr, and T. Robert. 2000. Genetic control of domestication traits in pearl millet (*Pennisetum glaucum* L., Poaceae). *Theoretical and Applied Genetics* 100: 147–159.

Poncet, V., E. Martel, S. Allouis, K. M. Devos, F. Lamy, A. Sarr, and T. Robert. 2002. Comparative analysis of QTLs affecting domestication traits between two domesticated × wild pearl millet (*Pennisetum glaucum* L., Poaceae) crosses. *Theoretical and Applied Genetics* 104: 965–975.

Potter, D. and J. J. Doyle. 1992. Origins of the African yam bean (*Sphenostylis stenocarpa*, Leguminosae)—Evidence from morphology, isozymes, chloroplast DNA, and linguistics. *Economic Botany* 46: 276–292.

Powell, W., M. Morgante, C. Andre, J. W. McNicol, G. C. Machray, J. J. Doyle, S. V. Tingey, and J. A. Rafalski. 1995. Hypervariable microsatellites provide a general source of polymorphic DNA markers for the chloroplast genome. *Current Biology* 5: 1023–1029.

Powell, W., M. Morgante, C. Andre, M. Hanafey, J. Vogel, S. Tingey, and A. Rafalski. 1996. The comparison of RFLP, RAPD, AFLP and SSR (microsatellite) markers for germplasm analysis. *Molecular Breeding* 2: 225–238.

Provan, J., W. Powell, H. Dewar, G. Bryan, G. C. Machray, and R. Waugh. 1999a. An extreme cytoplasmic bottleneck in the modern European cultivated potato *(Solanum tuberosum)* is not reflected in decreased levels of nuclear diversity. *Proceedings of the Royal Society of London Series B—Biological Sciences* 266: 633–639.

Provan, J., W. T. B. Thomas, B. P. Forster, and W. Powell. 1999b. Copia-SSR: A simple marker technique which can be used on total genomic DNA. *Genome* 42: 363–366.

Purugganan, M. D., A. L. Boyles, and J. I. Suddith. 2000. Variation and selection at the CAULIFLOWER floral homeotic gene accompanying the evolution of domesticated *Brassica oleracea*. *Genetics* 155: 855–862.

Rieseberg, L. H. 1996. Homology among RAPD fragments in interspecific comparisons. *Molecular Ecology* 5: 99–105.

———. 2001. Polyploid evolution: Keeping the peace at genomic reunions. *Current Biology* 11: R925–R928.

Rieseberg, L. H. and N. C. Ellstrand. 1993. What can molecular and morphological markers tell us about plant hybridization? *Critical Reviews in Plant Sciences* 12: 213–241.

Rieseberg, L. H. and J. F. Wendel. 1993. Introgression and its consequences in plants. In *Hybrid zones and the evolutionary process*, R. G. Harrison (ed.), pp. 70–109. New York: Oxford.

Rieseberg, L. H., D. E. Soltis, and J. D. Palmer. 1988. A Molecular reexamination of introgression between *Helianthus annuus* and *Helianthus bolanderi* (Compositae). *Evolution* 42: 227–238.

Rieseberg, L. H., H. C. Choi, and D. Ham. 1991. Differential cytoplasmic versus nuclear introgression in *Helianthus*. *Journal of Heredity* 82: 489–493.

Rieseberg, L. H., M. J. Kim, and G. J. Seiler. 1999a. Introgression between the cultivated sunflower and a sympatric wild relative, *Helianthus petiolaris* (Asteraceae). *International Journal of Plant Sciences* 160: 102–108.

Rieseberg, L. H., J. Whitton, and K. Gardner. 1999b. Hybrid zones and the genetic architecture of a barrier to gene flow between two sunflower species. *Genetics* 152: 713–727.

Rindos, D. 1989. Darwinism and its role in the explanation of domestication. In *Foraging and farming: The evolution of plant exploitation*, D. R. Harris and G. C. Hillman (eds.), pp. 27–41. London: Unwin Hyman.

Rogstad, S. H., J. C. Patton, II, and B. A. Schaal. 1988. M13 repeat probe detects DNA minisatellite-like sequences in gymnosperms and angiosperms. *Proceedings of the National Academy of Sciences, USA* 85: 9176–9178.

Sanjur, O. I., D. R. Piperno, T. C. Andres, and L. Wessel-Beaver. 2002. Phylogenetic relationships among domesticated and wild species of *Cucurbita* (Cucurbitaceae) inferred from a mitochondrial gene: Implications for crop plant evolution and areas of origin. *Proceedings of the National Academy of Sciences, USA* 99: 535–540.

Sefc, K. M., R. B. Payne, and M. D. Sorenson. 2003. Microsatellite amplification from museum feather samples: Effects of fragment size and template concentration on genotyping errors. Auk 120: 982–989.

Seiler, G. J. and L. H. Rieseberg. 1997. Systematics, origin, and germplasm resources of the wild and domesticated sunflower. In *Sunflower technology and production*, Agronomy Monograph no. 35, pp. 21–65. Madison, Wisc.: American Society of Agronomy, Crop Science Society of America, Soil Science Society of America.

Selinger, D. A. and V. L. Chandler. 1999. Major recent and independent changes in levels and patterns of expression have occurred at the *b* gene, a regulatory locus in maize. *Proceedings of the National Academy of Sciences, USA* 96: 1507–1512.

Singh, S. P., R. Nodari, and P. Gepts. 1991. Genetic diversity in cultivated common bean: I. Allozymes. *Crop Science* 31:19–23.

Small, R. L., J. A. Ryburn, and J. F. Wendel. 1999. Low levels of nucleotide diversity at homoeologous *Adh* loci in allotetraploid cotton (*Gossypium* L.). *Molecular Biology and Evolution* 16: 491–501.

Smith, B. D. 1997. The initial domestication of *Cucurbita pepo* in the Americas 10,000 years ago. *Science* 276: 932–934.

———. 1998. *The Emergence of Agriculture*. New York: Scientific American Library, W. H. Freeman (paperback edition).

———. 2002. *Rivers of Change - Essays on Early Agriculture in Eastern North America*. Washington, D. C.: Smithsonian Institution Press (paperback edition).

———. 2005. Low-level food production and the northwest coast. In *Keeping it living: Traditions of plant use and cultivation on the northwest coast*, D. Deur and N. J. Turner (eds.), pp. 37–66. Vancouver, B.C./Seattle, Wash.: University of British Columbia Press/ University of Washington Press.

Smith, B. D. and D. Decker-Walters. 2002. *Cucurbita pepo* in eastern North America: The early history of a domesticate. Paper presented in the Symposium "Issues Concerning Early North American Agriculture, East and Southwest," Annual Meeting of the Society for American Archaeology, March 20–24, Denver, Colorado.

Soltis, P. S. and D. E. Soltis. 2000. The role of genetic and genomic attributes in the success of polyploids. *Proceedings of the National Academy of Sciences, USA* 97: 7051–7057.

Sonnante, G., T. Stockton, R. O. Nodari, V. L. B. Velasquez, and P. Gepts. 1994. Evolution of genetic diversity during the domestication of common bean (*Phaseolus vulgaris* L.). *Theoretical and Applied Genetics* 89: 629–635.

Spooner, D. M., J. Tivang, J. Nienhuis, J. T. Miller, D. S. Douches, and A. Contreras. 1996. Comparison of four molecular markers in measuring relationships among the wild potato relatives *Solanum* section *Etuberosum* (subgenus *Potatoe*). *Theoretical and Applied Genetics* 92: 532–540.

Stace, C. A. and J. P. Bailey. 1999. The value of genomic *in situ* hybridization (GISH) in plant taxonomic and evolutionary studies. In *Molecular systematics and plant evolution*, P. M. Hollingsworth, R. M. Bateman, and R. J. Gornall (eds.), pp. 199–210. London: Taylor & Francis.

Stockton, T. and P. Gepts. 1994. Identification of DNA probes that reveal polymorphisms among closely related *Phaseolus vulgaris* lines. *Euphytica* 76: 177–183.

Sylvester, H. A. 2003. Inter-simple sequence repeat restriction fragment length polymorphisms for DNA fingerprinting. *Biotechniques* 34: 942–944.

Talbert, L. E., L. Y. Smith, and M. K. Blake. 1998. More than one origin of hexaploid wheat is indicated by sequence comparison of low-copy DNA. *Genome* 41: 402–407.

Tang, S. and S. J. Knapp. 2003. Microsatellites uncover extraordinary diversity in native American land races and wild populations of cultivated sunflower. *Theoretical and Applied Genetics* 106: 990–1003.

Tanksley, S. D. 1993. Mapping polygenes. *Annual Review of Genetics* 27: 205–233.

Tanksley, S. D. and S. R. McCouch. 1997. Seed banks and molecular maps: Unlocking genetic potential from the wild. *Science* 277: 1063–1066.

Tautz, D. and C. Schlötterer. 1994. Simple sequences. *Current Biology* 4: 832–837.

Tenaillon, M. I., M. C. Sawkins, A. D. Long, R. L. Gaut, J. F. Doebley, and B. S. Gaut. 2001. Patterns of DNA sequence polymorphism along chromosome 1 of maize (*Zea mays* ssp *mays* L.). *Proceedings of the National Academy of Sciences, USA* 98: 9161–9166.

Terrell, J. E., J. P. Hart, S. Barut, N. Cellinese, A. Curet, T. Denham, C. M. Kusimba, K. Latinis, R. Oka, J. Palka, M. E. D. Pohl, K. O. Pope, P. R. Williams, H. Haines, and J. E. Staller. 2003. Domesticated landscapes: The subsistence ecology of plant and animal domestication. *Journal of Archaeological Method and Theory* 10: 323–368.

Theissen, G. 2002. Key genes of crop domestication and breeding: Molecular analyses. In *Progress in Botany*, K. Esser, U. Lüttge, W. Beyschlag, and F. Hellwig (eds.), pp. 189–203. Berlin: Springer.

Throne, J. E. and M. W. Eubanks. 2002. Resistance of tripsacorn to *Sitophilus zeamais* and *Oryzaephilus surinamensis*. *Journal of Stored Products Research* 38: 239–245.

Tiffin, P. and B. S. Gaut. 2001. Sequence diversity in the tetraploid *Zea perennis* and the closely related diploid *Z. diploperennis*: Insights from four nuclear loci. *Genetics* 158: 401–412.

Tohme, J., D. O. Gonzalez, S. Beebe, and M. C. Duque. 1996. AFLP analysis of gene pools of a wild bean core collection. *Crop Science* 36: 1375–1384.

Uptmoor, R., W. Wenzel, W. Friedt, G. Donaldson, K. Ayisi, and F. Ordon. 2003. Comparative analysis on the genetic relatedness of *Sorghum bicolor* accessions from Southern Africa by

RAPDs, AFLPs and SSRs. *Theoretical and Applied Genetics* 106: 1316–1325.

Vavilov, N. I. 1992. *Origin and geography of cultivated plants*. Cambridge, UK: Cambridge University Press [translation by D. Löve of Russian text originally published during 1920–1940].

Vigouroux, Y., M. McMullen, C. T. Hittinger, K. Houchins, L. Schulz, S. Kresovich, Y. Matsuoka, and J. Doebley. 2002. Identifying genes of agronomic importance in maize by screening microsatellites for evidence of selection during domestication. *Proceedings of the National Academy of Sciences, USA* 99: 9650–9655.

Virk, P. S., J. Zhu, H. J. Newbury, G. J. Bryan, M. T. Jackson, and B. V. Ford-Lloyd. 2000. Effectiveness of different classes of molecular marker for classifying and revealing variation in rice *(Oryza sativa)* germplasm. *Euphytica* 112: 275–284.

von Buren, M. 2001. Polymorphisms in two homeologous *g-gliadin* genes and the evolution of cultivated wheat. *Genetic Resources and Crop Evolution* 48. 205–220.

Vos, P., R. Hogers, M. Bleeker, M. Reijans, T. van de Lee, M. Hornes, A. Frijters, J. Pot, J. Peleman, M. Kuiper, and M. Zabeau. 1995. AFLP: A new technique for DNA fingerprinting. *Nucleic Acids Research* 23: 4407–4414.

Wang, R. L., A. Stec, J. Hey, L. Lukens, and J. Doebley. 1999. The limits of selection during maize domestication. *Nature* 398: 236–239.

———. 2001. Correction: The limits of selection during maize domestication (vol. 398, p. 236, 1999). *Nature* 410: 718.

Warburton, M. L., B. Skovmand, and A. Mujeeb-Kazi. 2002. The molecular genetic characterization of the "Bobwhite" bread wheat family using AFLPs and the effect of the T1BL.1RS translocation. *Theoretical and Applied Genetics* 104: 868–873.

Weeden, N. F., S. Brauner, and J. A. Przyborowski. 2002. Genetic analysis of pod dehiscence in pea *(Pisum sativum* L.). *Cellular & Molecular Biology Letters* 7: 657–663.

Wendel, J. F. 2000. Genome evolution in polyploids. *Plant Molecular Biology* 42: 225–249.

Wendel, J. F. and V. A. Albert. 1992. Phylogenetics of the cotton genus *(Gossypium)*: Character-state weighted parsimony analysis of chloroplast DNA restriction site data and its systematic and biogeographic implications. *Systematic Botany* 17: 115–143.

Wendel, J. F., P. D. Olson, and J. M. Stewart. 1989. Genetic diversity, introgression, and independent domestication of old world cultivated cottons. *American Journal of Botany* 76: 1795–1806.

Whitaker, T. W. and G. F. Carter. 1946. Critical notes on the origin and domestication of the cultivated species of *Cucurbita*. *American Journal of Botany* 33: 10–15.

White, S. E. and J. F. Doebley. 1999. The molecular evolution of *terminal ear1*, a regulatory gene in the genus *Zea*. *Genetics* 153: 1455–1462.

Whitt, S. R., L. M. Wilson, M. I. Tenaillon, B. S. Gaut, and E. S. Buckler IV. 2002. Genetic diversity and selection in the maize starch pathway. *Proceedings of the National Academy of Sciences, USA* 99: 12959–12962.

Whitton, J., D. E. Wolf, D. M. Arias, A. A. Snow, and L. H. Rieseberg. 1997. The persistence of cultivar alleles in wild populations of sunflowers five generations after hybridization. *Theoretical and Applied Genetics* 95: 33–40.

Wilson, H. D., J. Doebley, and M. Duvall. 1992. Chloroplast DNA diversity among wild and cultivated members of *Cucurbita* (Cucurbitaceae). *Theoretical and Applied Genetics* 84: 859–865.

Wilson, W. A., S. E. Harrington, W. L. Woodman, M. Lee, M. E. Sorrells, and S. R. McCouch. 1999. Inferences on the genome structure of progenitor maize through comparative analysis of rice, maize and the domesticated panicoids. *Genetics* 153: 453–473.

Witsenboer, H., J. Vogel, and R. W. Michelmore. 1997. Identification, genetic localization, and allelic diversity of selectively amplified microsatellite polymorphic loci in lettuce and wild relatives *(Lactuca* spp.). *Genome* 40: 923–936.

Wolfe, A. D. and A. Liston. 1998. Contributions of PCR-based methods to plant systematics and evolutionary biology. In *Molecular Systematics of Plants* II, D. E. Soltis, P. S. Soltis, and J. J. Doyle (eds.), pp. 43–86. Norwell, Mass.: Kluwer Academic Publishers.

Wolfe, K. H., W.-H. Li, and P. M. Sharp. 1987. Rates of nucleotide substitution vary greatly among plant mitochondrial, chloroplast, and nuclear DNAs. *Proceedings of the National Academy of Sciences, USA* 84: 9054–9058.

Xiong, L. X., K. D. Liu, X. K. Dai, C. G. Xu, and Q. F. Zhang. 1999. Identification of genetic factors controlling domestication-related traits of rice using an F-2 population of a cross between *Oryza sativa* and *O. rufipogon*. *Theoretical and Applied Genetics* 98: 243–251.

Xu, R. Q., N. Tomooka, D. A. Vaughan, and K. Doi. 2000. The *Vigna angularis* complex: Genetic variation and relationships revealed by RAPD analysis, and their implications for in situ conservation and domestication. *Genetic Resources and Crop Evolution* 47: 123–134.

Yen, D. E. 1989. The domestication of environment. In *Foraging and farming—The evolution of plant exploitation*, D. R. Harris and G. C. Hillman (eds.), pp. 55–75. London: Unwin Hyman.

Zhang, D. X. and G. M. Hewitt. 2003. Nuclear DNA analyses in genetic studies of populations: Practice, problems and prospects. *Molecular Ecology* 12: 563–584.

Zhang, L., A. S. Peek, D. Dunams, and B. S. Gaut. 2002. Population genetics of duplicated disease defense genes, *hm1* and *hm2*, in maize *(Zea mays* ssp. *mays* L.) and its wild ancestor *(Zea mays* ssp. *parviglumis)*. *Genetics* 162: 851–860.

Zhang, P., B. Friebe, and B. S. Gill. 2002. Variation in the distribution of a genome-specific DNA sequence on chromosomes reveals evolutionary relationships in the *Triticum* and *Aegilops* complex. *Plant Systematics and Evolution* 235: 169–179.

Zietkiewicz, E., A. Rafalski, and D. Labuda. 1994. Genome fingerprinting by simple sequence repeat (SSR)-anchored polymerase chain-reaction amplification. *Genomics* 20: 176–183.

Zohary, D. 1984. Modes of evolution of plants under domestication. In *Plant biosystematics*, W. F. Grant (ed.), pp. 579–586. Toronto: Academic Press.

———. 1996. The mode of domestication of the founder crops of southwest Asian agriculture. In *The origins and spread of agriculture and pastoralism in Eurasia*, D. R. Harris (ed.), pp. 142–158. London: University College London Press.

———. 1999. Monophyletic vs. polyphyletic origin of the crops on which agriculture was founded in the Near East. *Genetic Resources and Crop Evolution* 46: 133–142.

Zohary, D. and M. Hopf. 1994. *Domestication of plants in the old world*, Oxford Science Publications. Oxford: Oxford University Press.

CHAPTER 9

DNA Sequence Data and Inferences on Cassava's Origin of Domestication

KENNETH M. OLSEN AND BARBARA A. SCHAAL

Chapter Overview

Cassava *(Manihot esculenta* Crantz ssp. *esculenta)* is a major staple food of the humid tropics. Its starchy root, also known as tapioca, manioc, mandioca, and yuca, is the primary calorie source for over 500 million people (Best and Henry 1992). Cassava is cultivated primarily by subsistence farmers, and the ability of the cassava shrub to thrive on marginal land has ensured its increasing importance to tropical agriculture (FAO 2000). Although a native of the New World, cassava has become the major source of carbohydrates in sub-Saharan Africa (Cock 1985), and the sixth most important crop worldwide (Mann 1997).

Despite its global importance as a food source, cassava has traditionally received less attention by researchers than have temperate cereal crops, and for this reason it has sometimes been referred to as an "orphan" crop. One of the most basic research questions that have remained unresolved until recently concerns cassava's evolutionary and geographical origins. In this chapter we describe our investigation of cassava's origin of domestication using DNA sequence data from three low-copy nuclear genes. We then compare inferences drawn from these data with conclusions based on five microsatellite markers. Our research focuses on two long-standing questions concerning cassava's domestication: (1) From which wild species was cassava domesticated? and (2) Where in the New World did the domestication occur?

Prior Research

Background

Cassava (Figure 9.1) is a member of the genus *Manihot* (Euphorbiaceae), which comprises roughly 98 species distributed throughout the New World tropics. Most of these species occur in northern South America (~80 spp.); there is a secondary center of species diversity in Central America and Mexico (~17 spp.). Prior to the development of molecular markers, determination of cassava's relationship to the wild *Manihot* species was extremely problematic (Renvoise 1972; Rogers and Appan 1973). *Manihot* possesses few morphological traits that are reliable for species delimitation, and many species show tremendous intraspecific morphological variability. These patterns led to speculation (much of it as yet untested) that hybridization is extensive among co-occurring (i.e., sympatric) species in nature (e.g., Rogers and Appan 1973). Like the wild *Manihot* species, cultivated cassava also shows substantial variation in morphological traits. This variability was traditionally cited as evidence that the crop arose through hybridization from one or more interbreeding species complexes, either in Central America, South America, or potentially both of these regions (Rogers 1963, 1965; Rogers and Appan 1973; Ugent et al. 1986; Sauer 1993).

The most recent monographers of the genus (Rogers and Appan 1973) were strong proponents of this "compilospecies" hypothesis, arguing that the cassava probably had multiple origins in the New World. On the basis of morphology, they proposed that Central American species (particularly *M. aesculifolia* (Kunth) Pohl) were probably among the closest wild relatives of the crop. Although populations were observed in South America that bear a closer physical resemblance to cassava than the Central American species (see below), these were considered to be feral or escaped cassava plants (Rogers and Appan 1973).

Anthropological data proved scarcely more definitive than taxonomic studies in identifying cassava's origin. As a crop of humid tropical lowlands, cassava yields few diagnostic archaeobotanical remains. Those remains that are recovered tend to come from arid sites outside the probable area of initial cultivation (e.g., Ugent et al. 1986). Tracing cassava's origin through ethnological studies is further complicated by the crop's long-term, geographically widespread use. Cassava cultivation was probably already widespread throughout the New World tropics by 2000 years ago (Renvoise 1972; Pearsall 1992; Sauer 1993; Piperno et al. 2000), which has confounded efforts to identify a region of earliest cultivation.

Challenges to the "Compilospecies"

Beginning in the 1980s, two lines of evidence began to emerge that suggested that the hybrid origin domestication hypothesis was in need of reevaluation. First, the availability of molecular markers led to reassessments of cassava's genetic relationship to South American vs. Central American species. Specifically, analysis of RFLPs (restriction fragment length polymorphisms) (Bertram 1993; Fregene et al. 1994), AFLPs (amplified fragment length polymorphisms) (Roa et al. 1997), and DNA sequences (Schaal et al. in press) indicated that South American and Central American species form two distinct evolutionary lineages, and that cassava is genetically most

FIGURE 9.1 The cultivated cassava plant, *Manihot esculenta* ssp. *esculenta*.

similar to the South American lineage. These findings thus called into question cassava's putative affinity to the Central American *Manihot* species complex.

At the same time that these genetic data were beginning to accumulate, expanded collecting of *Manihot* populations in South America (Allem 1987, 1992, 1994) indicated the existence of naturally occurring populations that are morphologically similar enough to cassava to be considered the same species. If truly wild, these South American *M. esculenta* populations (referred to below as *M. esculenta* ssp. *flabellifolia* (Pohl) Ciferri; see Allem 1994; Roa et al. 1997) would be the most obvious candidates to be cassava's wild progenitor. Moreover, because they resemble the crop so much more closely than any other wild species, their presence would call into question the need to invoke interspecific hybridization in explaining the crop's origin. Rather, the more parsimonious explanation would be that the crop was descended directly from this wild taxon. These developments led to our own research in tracing cassava's evolutionary and geographical origins of domestication. In the following discussion we describe the molecular methods we have used in addressing these questions and the findings of our research.

Using Molecular Data to Address Questions of Domestication

Hybrid Origin vs. Single Progenitor?

Our approach in addressing cassava's origin has focused on determining the genetic relationship between the crop and the candidate progenitor, *M. esculenta* ssp. *flabellifolia*.

Specifically, we sought to determine whether *M. e. flabellifolia* is the wild progenitor of cassava (Allem 1994), or merely a feral derivative of the crop (Rogers and Appan 1973). Population genetics theory offers several predictions about the genetic variation in a crop and its wild ancestor that are useful in addressing this question.

First, a crop's overall genetic diversity is expected to be lower than that of its wild ancestor. The reason for this is that crop domestication involves population genetic processes (e.g., selection, founder events, population bottlenecks) that reduce the overall genetic diversity of the crop lineage relative to the progenitor. Second, the genetic variation found in the crop is expected to be largely shared with the progenitor. This is true because, on an evolutionary time scale, the crop and its progenitor have had very little time to diverge genetically (except at the few genes directly selected upon during domestication). Any substantial genetic differences between the crop and its putative progenitor would most likely indicate the contribution of additional, hybridizing species in the crop's origin. Third, the geographical distribution of genetic variation in the putative progenitor species can provide insight into the crop's geographical origin of domestication. Specifically, close genetic resemblance between the crop and a particular geographical subset of wild populations would strongly suggest that the crop was domesticated in the region of those populations.

On the basis of these predictions, three scenarios for cassava's origin may be envisioned. If the genetic diversity observed in cassava were found to be a subset of that in *M. e. flabellifolia*, this would suggest that the crop is in fact derived from *M. e. flabellifolia*. In this case, closer examination of *M. e. flabellifolia* populations could lead to insights into the geographical origin of cultivation. If the genetic variation of *M. e. flabellifolia* were instead a subset of that in the crop, this would strongly suggest that *M. e. flabellifolia* represents feral escapes from cultivation; in that case the true crop progenitor(s) would need to be sought elsewhere. Finally, if the crop's genetic diversity were partly, but not entirely shared with *M. e. flabellifolia*, this pattern might suggest an origin from both *M. e. flabellifolia* and additional hybridizing species. To test this third hypothesis, our sampling of wild *Manihot* populations included not only *M. e. flabellifolia* but also a close relative, *M. pruinosa* Pohl, which has a distribution that partially overlaps that of *M. e. flabellifolia* and which has been proposed to hybridize with this species (Allem 1992, 1999).

Choosing Appropriate Molecular Markers

Our analyses relied on two types of molecular markers: DNA sequences and microsatellites (also known as simple sequence repeats—SSRs). For both of these markers, the observed genetic variation is neutral in the sense that it does not encode any differences in obviously selectable morphological or physiological traits. Thus, our criteria for identifying origins of domestication are based solely on inferred genetic relationships. For both DNA sequences and microsatellites, our choice of specific genes depended largely on finding loci

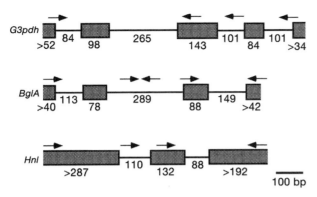

FIGURE 9.2 Diagrams of the three low-copy nuclear gene regions used for DNA sequence analysis. Arrows indicate positions of primers used in PCR amplification and sequencing. Boxes indicate exons, and lines indicate intervening introns. Numbers indicate length in base pairs (bp) of the intron and exon regions sequenced. The figure is modified from Fig. 2 of Olsen and Schaal (1999) and Figs. 7–1 and 7–4 of Olsen (2000).

that showed sufficient genetic variation within our study system to be useful for resolving genetic relationships.

DNA SEQUENCES

The DNA sequences come from portions of three enzyme-encoding genes found within the nuclear genome of *Manihot*: *G3pdh*, which encodes glyceraldehyde 3-phosphate dehydrogenase; *BglA*, which encodes a β-glucosidase; and *Hnl*, which encodes an α-hydroxynitrile lyase (Figure 9.2). All three of these genes are low-copy-number genes, meaning that they do not comprise regions of multiple, tandemly repeated gene copies (such as the internal transcribed spacer [ITS] region of nuclear ribosomal DNA, which is commonly used in molecular systematic studies). *G3pdh* was initially selected because it had previously been shown to be variable in other plant species (Strand et al. 1997), and this gene has been the major focus of our sequence-based analysis (Olsen and Schaal 1999; Olsen 2002). *BglA* and *Hnl* were chosen because both genes had previously been cloned from cassava and sequenced (Hughes et al. 1998; Liddle et al. 1998), which greatly facilitated further analysis of them in our evolutionary studies of *Manihot*.

For DNA sequence data, genetic variation is observed in the form of differences in the nucleotides (i.e., bases) that constitute the DNA. These differences occur in two forms: substitutions, where one base is substituted for another (e.g., GATC becomes GAAC); and insertions/deletions, where bases are added or deleted (e.g., GATC becomes GATTC). Most of these polymorphisms are observed at nucleotide positions where a mutation has no direct effect on the protein encoded by the gene. Typically, these are either within introns (regions of non-coding DNA separating the protein-encoding exons; see Figure 9.2); or at synonymous sites (i.e., nucleotide positions within exons where more than one base can encode the same amino acid). Because these polymorphisms have no functional consequences, they can accumulate at much faster evolutionary rates than mutations causing amino acid changes (which are often disadvantageous and selected against).

Across a gene, nucleotides are physically linked and are inherited as a unit, barring rare intragenic recombination events. Two copies of this DNA sequence that differ at one or more nucleotide sites are considered different alleles, or *haplotypes*, of that gene. Nucleotide differences between haplotypes presumably have arisen by random mutations occurring gradually over time. Therefore, a greater number of nucleotide differences indicates greater genetic divergence between two haplotypes. Because these genetic differences can be directly observed and quantified by DNA sequencing, the genealogical relationships among haplotypes can be inferred. When represented graphically, haplotype genealogies are often referred to as *gene trees*.

The genealogical information inherent in DNA sequences makes them a valuable tool for looking at genetic change within a species over time. Specifically, haplotype genealogies may be analyzed within the framework of coalescent theory, which provides a theoretical context for studying how population genetic forces influence patterns of relationship among haplotypes. Haplotype genealogies can enable powerful insights into historical population genetic forces and their lasting effects on the genetic structure of a species (reviewed by Schaal and Olsen 2000). Analysis of haplotype genealogies has thus allowed population genetics to be studied as dynamic processes, changing over time within a species lineage. This explicit genealogical perspective sets DNA sequence data apart from many of the more commonly used molecular genetic markers (e.g., microsatellites, AFLPs, RAPDs). Indeed, the genealogical analysis of DNA sequences arguably represents one of the major advances of population genetics within the past 50 years (Schaal and Olsen 2000).

MICROSATELLITES

The second type of molecular marker we analyzed was microsatellites, which are regions of short, tandemly repeated nucleotides (e.g., GAGAGA...). Insertion/deletion mutations at these loci result in variation in the number of tandem repeats, which in turn affects the overall size of the repeat region (e.g., GAGAGA vs. GAGAGAGAGA). These repeat-length differences can be detected via differential migration of the DNA segments in porous gels subjected to electrophoresis; length variants can therefore be identified and treated as distinct alleles of the same gene. As with DNA sequence haplotypes, most microsatellite repeat variation does not affect protein function.

Unlike DNA sequences, genealogical relationships of microsatellite alleles are not easily inferred, and this limits their use in phylogenetic studies. On the other hand, microsatellites often show abundant, easily scored variation, and data from multiple loci are more easily obtained than with DNA sequencing. Thus, microsatellites can provide

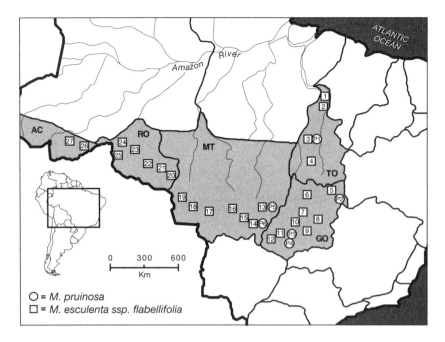

FIGURE 9.3 Locations of *Manihot esculenta* ssp. *flabellifolia* populations (squares) and *M. pruinosa* populations (circles) sampled along the eastern and southern borders of the Amazon basin. Shaded regions indicate the five Brazilian states in which populations were observed and sampled. Abbreviations of Brazilian states: AC, Acre; GO, Goiás; MT, Mato Grosso; RO, Rondônia; TO, Tocantins. Population numbers correspond to the list of population names in Table 1 of Olsen (2002); the figure is modified from Fig. 1 of Olsen and Schaal (1999).

multiple, independent assessments of genetic relationships for a study system. In this respect, they are similar to AFLPs. Microsatellites offer an important advantage over AFLPs, however, in that there is little ambiguity in the assignment of specific fragments to individual loci (i.e., they can be treated as co-dominant). This allows information on heterozygosity and other aspects of an individual's genotype to be analyzed (see, e.g., Olsen and Schaal 2001). In our study, we examined genetic variation at five microsatellite loci to provide a complementary analysis to the DNA sequence data.

Materials and Methods

Research Design

STUDY SYSTEM

Manihot esculenta ssp. *flabellifolia* occurs in the belt of transitional vegetation separating the lowland rainforest of the Amazon basin and the seasonally dry *cerrado* (savanna) of the Brazilian Shield plateau. It grows as a clambering understory shrub in moist forest patches (often in ravines), along the eastern and southern borders of the Amazon basin (Figure 9.3). Although it has been collected primarily in central and western Brazil and eastern Peru (Allem 1992, 1994), the distribution of this species likely extends south into lowland Bolivia and Paraguay. In our work, we refer to all wild *M. esculenta* populations as *M. e. flabellifolia*, including those previously segregated as "*M. e. peruviana*" (see Allem 1994 and his subsequent reassessment in Roa et al. 1997).

Manihot esculenta ssp. *flabellifolia* differs from cultivated cassava primarily in root and stem traits. Besides greatly increased root tuberization, the crop also tends to have thickened, knobby stems with short internodes, as well as swollen leaf scars and stipule bases (see Fig. 9.1). This increased resource allocation to the stem is thought to reflect artificial selection for ease of vegetative propagation; cassava is propagated primarily by stem cuttings.

The other wild species included in the study, *M. pruinosa*, is a shrub that grows in dense *cerrado* to the southeast of the Amazon basin, where its distribution partially overlaps that of *M. e. flabellifolia*. Although *M. e. flabellifolia* and *M. pruinosa* specialize on different habitats, the patchiness of vegetation within the rainforest-cerrado transition zone (see Prance 1987) allows the two species to grow in close proximity. Based on its morphological similarity to cassava, *M. pruinosa* has been grouped into cassava's "secondary gene pool" of potentially interfertile species (Allem 1992); it is the only such species that is sympatric with *M. e. flabellifolia*. For these reasons, *M. pruinosa* is arguably the most likely candidate to be hybridizing with *M. e. flabellifolia* and/or cassava. It was included in our study specifically to test whether cassava's origins extend beyond the candidate progenitor, *M. e. flabellifolia*.

TABLE 9.1
Cassava Samples Used in Analyses

CIAT Accession	Country of Origin	Common Name
M ARG 11	Argentina	*Duro do Valle 30*
M BOL 3	Bolivia	*Rasada de Bolivia*
M BRA 12	Brazil	—
M BRA 931	Brazil	*Enganha Ladrão*
M COL 1505	Colombia	*UCV 2096*
M COL 1684	Colombia	*Charay*
M COL 2215	Colombia	*Venezolana 1*
M CR 32	Costa Rica	*Yuca Mangi*
M CUB 74	Cuba	*Señorita*
M ECU 82	Ecuador	*Blanca*
M IND 33	Indonesia	*No. 734-5*
M MAL 2	Malaysia	*Black Twig*
M MEX 59	Mexico	—
M NGA 2	Nigeria	*TMS 30572*
M PAN 51	Panama	—
M PAR 110	Paraguay	*Tacuara Sayyu*
M PTR 19	Puerto Rico	*No. 9588*
M TAI 1	Thailand	*Rayong 1*
M VEN 25	Venezuela	*Querepa Amarga*
HMC 1	Colombia	—

SOURCE: All accessions are from the "core of the core" world core germplasm collection maintained by the Centro Internacional de Agricultura Tropical (CIAT; Cali, Colombia).

SAMPLING

Collection of wild *Manihot* populations involved two trips in Brazil during the months of November and December, 1996 and 1997. Populations of *M. e. flabellifolia* were collected along the southern and eastern borders of the Amazon basin, in the Brazilian states of Acre, Rondônia, Mato Grosso, Goiás, and Tocantins (Figure 9.3). No *M. e. flabellifolia* populations were found in states north and east of Tocantins (Pará, Maranhão, Piauí, Bahia), where the vegetation grades into either rainforest or xeric thornscrub.

For each sampled *M. e. flabellifolia* population, young leaf tissue was collected from up to ten individuals and dried in silica desiccant. Undisturbed *M. e. flabellifolia* populations are estimated to typically comprise fewer than 15 individuals. In total, 157 *M. e. flabellifolia* individuals representing 27 populations were included in our analyses.

Manihot pruinosa populations were collected where this species co-occurs with *M. e. flabellifolia*, in the states of Tocantins, Goiás, and eastern Mato Grosso (Figure 9.3). Thirty-five individuals from six *M. pruinosa* populations were included in our analyses.

To represent the genetic diversity within the crop, 20 cassava landraces were sampled from the cassava "world core collection" maintained by the Centro Internacional de Agricultura Tropical (CIAT; Cali, Colombia). This germplasm collection includes cassava lines from around the world; the 20 selected landraces come from the "core of the core" collection designated as most representative of the crop's diversity (Table 9.1). Fresh leaf tissue from each selected line was dried in silica, as with the wild *Manihot* populations.

Data Collection

DNA SEQUENCING

Genomic DNA was extracted from the silica-dried leaf tissue using a CTAB protocol (Hillis et al. 1996). For each accession, the three low-copy nuclear genes were amplified by polymerase chain reaction (PCR), using forward and reverse primers specific to each gene (Strand et al. 1997; Olsen 2000). Technical details of DNA sequencing, haplotype identification, and haplotype genealogy (gene tree) construction are described elsewhere (Olsen and Schaal 1999; Olsen 2000, 2002).

MICROSATELLITES

The five microsatellite loci used in analyses were selected from 14 loci that had been previously identified in cassava (Chavarriaga-Aguirre et al. 1998). Loci were chosen on the basis of high polymorphism and unambiguous banding patterns in a sample of 12 to 36 accessions from the study system. All five loci (GAGG5, GA134, GA16, GA12, and GA140) are composed of GA dinucleotide repeats. Methods of PCR and allele scoring are described in Olsen and Schaal (2001).

Analyses

DNA SEQUENCES

On an evolutionary time scale, the divergence between a crop and its wild relative(s) represents a rapid, recent split. Examination of haplotype sharing between a crop and wild species is therefore likely to provide better insight into a crop's origins than inferences based on genealogical relationships among haplotypes (see Olsen and Schaal 1999; Schaal and Olsen 2000). For the low-copy nuclear genes, we focused on the degree of haplotype sharing between cassava, *M. e. flabellifolia*, and *M. pruinosa*, and the geographical locations of any cassava haplotypes detected in the wild populations.

MICROSATELLITES

The high allelic diversity associated with microsatellite markers makes them amenable to clustering (phenetic) analyses based on estimates of genetic distances among populations. We were specifically interested in seeing which wild populations cassava would cluster with, and whether the resulting evolutionary inferences would corroborate those based on the DNA sequence data. We used a maximum likelihood algorithm to construct a distance tree based on allele

TABLE 9.2
Summary of Variation in the Nuclear Loci across Taxa[a]

		Number of Variants		
DNA Region	Polymorphism Type	G3pdh	BglA	Hnl
Exon	Replacement substitutions[b]	1	3	17
	Synonymous substitutions[c]	9	4	15
Intron	Substitutions[b]	47	29	9
	Insertions/deletions[b]	7	5	2
	Total polymorphisms:	64	41	43

[a] The table is modified from Table 1 of Olsen and Schaal (1999) and Tables 7-1 and 7-3 of Olsen (2000).

[b] These result in an amino acid change in the protein encoded by the gene.

[c] These do not result in amino acid changes.

frequencies within each of the wild populations and the group of cassava accessions (details of methods are described in Olsen and Schaal 2001).

Results

DNA Sequences

Levels of nucleotide variation in the three low-copy nuclear genes are shown in Table 9.2. For two of the three genes (*G3pdh* and *BglA*), levels of non-coding variation far exceed polymorphisms resulting in amino acid replacements; this pattern is expected for genes encoding functional proteins (see above). For *Hnl*, the excess of amino acid replacement polymorphisms suggests that this copy of the gene may not encode a functional enzyme. Whether it is a functional gene or not, however, the mutations that have accumulated at this locus are still informative for inferring genetic relationships. Because greater genetic variation was observed in *G3pdh* than in the other two nuclear genes (Table 9.2), our analyses focused primarily on this gene.

The *G3pdh* gene tree is shown in Figure 9.4. Twenty-eight *G3pdh* haplotypes were observed in the study system across all taxa. Twenty-four of these haplotypes were detected in *M. esculenta* overall (*M. e. flabellifolia* and cassava together). Six of these were observed in cassava, and only one rare cassava haplotype (ε, observed a single time in a heterozygote) was not observed in the wild subspecies (see Table 9.3). Thus, the genetic variation observed in the crop is almost entirely a subset of that found in *M. e. flabellifolia*. This pattern strongly suggests that the crop is derived from *M. e. flabellifolia* and that *M. e. flabellifolia* is not a feral crop derivative.

When the *M. e. flabellifolia* samples are considered in a geographical context, all populations possessing cassava *G3pdh* haplotypes occur along the southern border of the Amazon basin, in the states of Acre, Rondônia, and Mato Grosso (Figure 9.5). No cassava haplotypes were detected in *M. e. flabellifolia* populations in the eastern portion of the

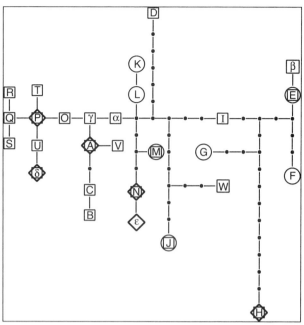

FIGURE 9.4 The *G3pdh* haplotype genealogy (gene tree). Letters correspond to haplotypes in published sequences (Genbank accessions AF136119 to AF136149). Each line between haplotypes is a mutational step corresponding to a specific nucleotide polymorphism in the gene sequence. Small black circles represent infer intermediate haplotypes. Shapes around letters indicate whether haplotypes were detected in *Manihot esculenta* ssp. *flabellifolia* (squares), cassava (diamonds), or *M. pruinosa* (circles). The figure is modified from Fig. 3 of Olsen and Schaal (1999).

species range. This finding points to the southern Amazonian transition zone as the crop's likely cradle of domestication.

None of the six cassava *G3pdh* haplotypes was detected in *M. pruinosa* populations, suggesting that the crop's origins do not extend to this species (Figure 9.5). However, three instances of haplotype sharing were observed between *M. e. flabellifolia* and *M. pruinosa* (Figure 9.4). This pattern could reflect occasional hybridization between the wild taxa, or it may be a consequence of the recent evolutionary divergence between *M. pruinosa* and *M. e. flabellifolia* (incomplete lineage sorting; see discussions in Olsen and Schaal 1999, 2001; Olsen 2002). In either case, the lack of haplotype sharing between cassava and *M. pruinosa* suggests that the latter has not played a significant role as a direct progenitor of cassava. In fact, the *M. e. flabellifolia* populations containing cassava haplotypes all occur west of the *M. pruinosa* species distribution (Figure 9.5), further minimizing any potential role of this species in the crop's origin.

Haplotype variation at the other two nuclear genes corroborates the patterns observed with the *G3pdh* data. For both *BglA* and *Hnl*, cassava's haplotype variation is entirely a subset of that found in *M. e. flabellifolia*, representing 16.7% and 12.1% of the total *M. e. flabellifolia* haplotype variation for the two genes, respectively (Table 9.3). As with *G3pdh*, cassava haplotypes are distributed primarily along the southern

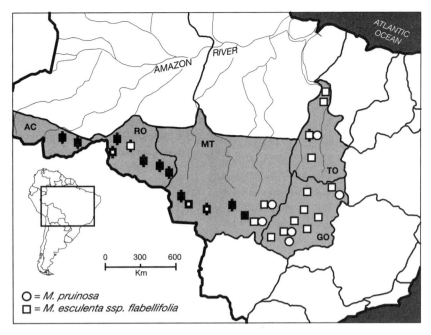

FIGURE 9.5 Wild populations most closely related to cassava. Shapes with partial or full shading indicate populations containing cassava *G3pdh* haplotypes. Fully shaded populations contain cassava haplotypes in at least two of the three nuclear gene data sets. Vertical bars indicate populations grouped with cassava in the microsatellite analysis. The figure is modified from Fig. 8–1 of Olsen (2000).

TABLE 9.3
Allele Sharing between Cassava and *Manihot esculenta* ssp. *flabellifolia*, by Locus and across All Loci (mean ± SE)[a]

Gene	Number of Alleles in Cassava	Number of Alleles in flabellifolia	Cassava Alleles as a Proportion of Those Found in flabellifolia
DNA sequences			
G3pdh	6[b]	24	0.250
BglA	5	30	0.167
Hnl	4	33	0.121
Microsatellites			
GAGG5	2	7	0.286
GA134	2	19	0.105
GA16	3	16	0.188
GA12	3	15	0.200
GA140	5	16	0.313
Total	30	160	0.188 ± 0.03

[a] The table is modified from Table 8–1 of Olsen (2000).

[b] One cassava *G3pdh* haplotype (ε) was not observed in any wild population.

border of the Amazon basin (Figure 9.5), which is west of the *M. pruinosa* species range. Thus, both the *BglA* and *Hnl* data sets confirm that cassava was likely domesticated solely from *M. e. flabellifolia*, and specifically from populations occurring along the southern border of the Amazon basin.

Microsatellites

As with the DNA sequence data, microsatellite allele variation within cassava was again a subset of that found in *M. e. flabellifolia* (Table 9.3). Of the 73 alleles observed across all 5 loci, 15 occurred in cassava, all of which were also observed in *M. e. flabellifolia* accessions. A maximum likelihood distance tree based on allele frequencies in each population is shown in Figure 9.6. The cassava accessions are clustered with a group of 12 *M. e. flabellifolia* populations, all but one of which are located in the westernmost portion of the sampled range (Figure 9.5). Interestingly, the single northeastern *M. e. flabellifolia* population included in this cluster may be an artifact of pre-Holocene *M. e. flabellifolia* population distributions (see Olsen 2002). Thus, the microsatellite data yield the same overall pattern as observed with the three nuclear gene sequences: the crop appears to be derived from southern Amazonian *M. e. flabellifolia* populations.

Manihot pruinosa populations are not clustered into a discrete (monophyletic) group on the microsatellite distance tree, but instead are intermingled with *M. e. flabellifolia* populations (Figure 9.6). As with the DNA sequences, this lack of genetic differentiation between the two species likely reflects recent common ancestry, and possibly occasional interspecific hybridization as well (discussed by Olsen and Schaal 2001). Nonetheless, as with the DNA sequences, the microsatellite distance tree clearly indicates that cassava is not closely related to these *M. pruinosa* populations. It should be noted that the distance tree in Figure 9.6 is not rooted by an outgroup. Thus, the *M. pruinosa* populations should not be

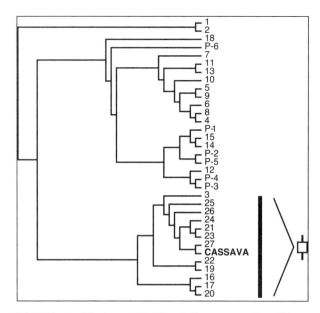

FIGURE 9.6 Maximum likelihood distance tree for wild populations and cassava accessions. Numbers correspond to population labels in Figure 9.3 herein and to Table 1 of Olsen (2002). A prefix of P indicates *M. pruinosa* populations. The vertical bar indicates the cluster of wild populations most closely grouped with cassava (see also Figure 9.5). The figure is modified from Fig. 3B of Olsen and Schaal (2001).

construed to be derived from *M. e. flabellifolia*, as would be the implication on a rooted tree.

Conclusions

Taken together, the DNA sequence and microsatellite analyses present a consistent view of genetic relationships in the study system. Two broad conclusions may be inferred from these data; they are summarized below.

1. **Cassava has a single wild progenitor, *M. esculenta* ssp. *flabellifolia*.** The genetic variation found in domesticated cassava is almost entirely a subset of that detected in wild *M. e. flabellifolia* populations. With the exception of a single rare *G3pdh* haplotype not observed in any wild population, all genetic variation observed in the crop was also found in *M. e.* ssp. *flabellifolia* populations (Table 9.3). Across all loci, cassava accessions contain between 10.5% (GA134) and 31.3% (GA140) of the total variation in *M. esculenta* (Mean: 18.8 + 3.0%; Table 9.3). These values fall within the range typically observed in crop/progenitor comparisons (e.g., Doebley 1989, Gepts 1993). This pattern strongly suggests that the crop is derived directly and solely from *M. e. flabellifolia*. Thus, the genetic variation in *M. e. flabellifolia* populations is sufficient to account for cassava's genetic diversity, and there is no need to invoke a hybrid origin of the crop.

   We included the potentially hybridizing species *M. pruinosa* in our sampling specifically to test whether cassava's origins extend beyond *M. e. flabellifolia*. Our data suggest that hybridization may, in fact, occur between wild populations of the two species. However, this hybridization is unlikely to have played a role in the crop's origin. Neither the DNA sequences nor the microsatellites indicate a close relationship between cassava and *M. pruinosa*. Moreover, the crop's likely geographical origin of domestication falls outside the species range of *M. pruinosa* (Figure 9.5).

2. **Cassava's origins appear to lie along the southern border of the Amazon basin.** The three low-copy nuclear genes and microsatellite loci all indicate that cassava accessions are genetically most similar to southern Amazonian *M. e. flabellifolia* populations (Figure 9.5). When considered collectively, the data suggest that cassava's closest wild relatives lie within a region of transitional vegetation that includes the Brazilian states of Mato Grosso, Rondônia, and Acre. The fairly substantial sampling for this study (33 wild populations, 20 crop landraces; N= 212) and the extensive genetic analysis (3 DNA sequence datasets, 5 microsatellite markers) provide the most conclusive and most specific evidence to date on cassava's origin of domestication.

If correct, our conclusions place cassava's origin within a previously identified zone of crop domestication that has been proposed for the peanut, chili pepper (e.g., *Capsicum baccatum*), and the jack bean *(Canavalia plagiosperma)* (Piperno and Pearsall 1998). Moreover, archaeological evidence suggests that this same region may have been an important site of early (preceramic) agricultural settlements (Miller 1992; Meggers, B., e-mail to Olsen, 11 August 2000), as well as later, technologically advanced farming communities (Erickson 2000). Taken together, these inferences suggest that the southern Amazonian border region may have been an important location for the development of lowland tropical agriculture in the New World. This zone of crop domestication extends south of the area sampled for this study, into lowland Bolivia (Piperno and Pearsall 1998). Although the distribution of *M. e. flabellifolia* outside of Brazil is poorly defined, additional populations are expected to occur in Bolivia, and examination of these more southern populations could provide important further insight into cassava's geographical origins.

Caveats

DNA sequences and microsatellite markers are most useful for addressing two basic evolutionary questions: (1) the degree of relatedness among populations or species, and (2) overall levels of genetic diversity within populations and species. To the extent that these questions have formed the basis of our analyses, DNA sequence data and microsatellites are extremely well suited to examining origins of domestication. On the other hand, these two types of genetic marker share some disadvantages which should be pointed out.

Perhaps most problematic for domestication studies is that these markers are not well suited for the absolute dating of times of divergence. Such calculations require the assumption that genetic mutations occur at a constant, clocklike rate; this assumption is often violated (e.g., Olsen 2002). Thus, neither DNA sequences nor microsatellites are particularly appropriate for determining the actual time frame when a crop was domesticated. In this respect, these data are best applied in conjunction with archaeological data, which can better provide a chronological context.

Another potential complication involves distinguishing between a true progenitor population and a population into which there has been gene flow from the crop. Both may show close genetic similarity to the crop, and in both cases the crop's genetic diversity may be a subset of that in the wild population. For the specific case of cassava, we tentatively dismiss this possibility that gene flow from the crop into wild populations has led us severely astray. Cassava plants flower less often than the wild species and do not appear to survive well outside of cultivation (Rogers 1965; Allem 1994; Olsen personal observation). In addition, cassava is cultivated primarily by stem cuttings, which must be planted to grow, so that incidental spread of the crop by seed would be expected to be lower than for seed-propagated crops (but see Elias and McKey 2000; and Elias et al. 2001 for evidence of reproduction by seed in traditional cassava fields). These factors would reduce opportunities for gene flow into *M. e. flabellifolia*, whether by cross-pollination or by seed dispersal. Moreover, for our specific conclusions on the cassava's geographical origin, we observed no obvious reason why there would be gene flow from the crop only in the area of the southern Amazon and not in the eastern portion of the sampling range (which is more populous, and where cassava has been observed growing in close proximity to *M. e. flabellifolia* (Olsen personal observation)). Thus, the lack of ambiguity in our data strengthens our confidence in these conclusions.

A final caveat involves sampling. The robustness of our conclusions depends on adequate sampling of wild populations. For the cassava study, our sampling included most of the known distribution of the candidate progenitor, *M. e. flabellifolia*. However, populations are also likely to occur further south in Bolivia, in the heart of the previously identified zone of domestication (see above). Collection and analysis of Bolivian populations of *M. e. flabellifolia* may well reveal that more southerly populations of this species were also important in the domestication of cassava.

In addition, it is possible that *M. esculenta* populations also occur along the northern border of the Amazon. Populations from this area have been variously classified as *M. e. flabellifolia* (Allem 1994) or as a different species, most notably *M. tristis* Muell.-Arg. (Rogers and Appan 1973; Allem 1987; Roa et al. 1997). At least one recent molecular study suggests that *M. tristis* is less closely related to cassava than the *M. e. flabellifolia* populations examined here (Roa et al. 1997). Nevertheless, additional sampling of populations from the northern Amazon would be useful for ruling out the possibility that these populations played a role in cassava's origin.

Implications and Future Research

Our data reject the "compilospecies" hypothesis for cassava in favor of a single origin of domestication. The crop appears to be derived from *M. e. flabellifolia*, with a cradle of domestication along the southern border of the Amazon basin. Thus, for cassava, the simpler domestication scenario is better supported. In this respect, cassava's domestication parallels that of maize, which was also thought to be a hybrid until molecular and other data proved otherwise (Iltis 1983; Doebley et al. 1984, Doebley et al. 1987).

There are three areas of research which would significantly help in further clarifying the origin and evolution of cassava. First, the entire genus *Manihot* is in dire need of systematic revision. Morphological characters are notoriously unreliable in this genus, which makes it a prime candidate for analyses using the molecular tools that have become available in the decades since the most recent revision. A well defined phylogeny would be particularly useful for putting to rest the notion that cassava is derived from Mexican and Central American species. Old theories die hard. Secondly, more studies are needed to document the role of interspecific hybridization in the evolution of *Manihot*. Reports of natural hybridization remain largely anecdotal and are based on morphological intermediacy, which is likely to be unreliable in *Manihot* (see the discussion in Olsen and Schaal 2001). Understanding the role of hybridization could be useful for both crop improvement efforts and for basic knowledge of the evolutionary dynamics in this recently evolved genus. Finally, analysis of the previously mentioned Bolivian and northern Amazonian populations that resemble the crop would resolve the question of whether these populations have played any role in cassava's domestication.

Acknowledgments

The authors express sincere thanks to Luiz Carvalho, Antonio Costa Allem, and the staff of the Centro Nacional de Pesquisa de Recursos Genéticos e Biotecnologia (CENARGEN; Brasília, Brazil) for invaluable assistance in collecting wild populations; to the Centro Internacional de Agricultura Tropical (CIAT; Cali, Colombia) for assistance in selection of cassava samples; and to members of the Washington University Program in Evolutionary and Population Biology for many helpful discussions. Funding for this work was provided by grants from The Explorers Club, the Jean Lowenhaupt Botany Fund, and the National Science Foundation (Doctoral Dissertation Improvement Grant DEB 9801213) to Olsen, and by grants from the Rockefeller and Guggenheim Foundations to Schaal.

References

Allem, A. C. 1987. *Manihot esculenta* is a native of the Neotropics. *Plant Genetic Resources Newsletter* 71: 22–24.

Allem, A. C. 1992. *Manihot* germplasm collecting priorities. In *Report of the first meeting of the international network for cassava genetic resources*, W. M. Roca and A. M. Thro (eds.), pp. 87–110. Cali, Colombia: Centro Internacional de Agricultura Tropical.

Allem, A. C. 1994. The origin of *Manihot esculenta* Crantz (Euphorbiaceae). *Genetic Resources and Crop Evolution* 41: 133–150.

Allem, A. C. 1999. The closest wild relatives of cassava (*Manihot esculenta* Crantz). *Euphytica* 107: 123–133.

Bertram, R. B. 1993. *Application of molecular techniques to genetic resources of cassava (Manihot esculenta Crantz—Euphorbiaceae): Interspecific evolutionary relationships and intraspecific characterization*. PhD diss., University of Maryland at College Park. Ann Arbor: University Microfilms.

Best, R. and G. Henry. 1992. Cassava: Towards the year 2000. In *Report of the first meeting of the international network for cassava genetic resources*, W. M. Roca and A. M. Thro (eds.), pp. 3–11. Cali, Colombia: Centro Internacional de Agricultura Tropical.

Chavarriaga-Aguirre, P., M. M. Maya, M. W. Bonierbale, S. Kresovich, M. A. Fregene, J. Tohme, and G. Kochert. 1998. Microsatellites in cassava (*Manihot esculenta* Crantz): Discovery, inheritance and variability. *Theoretical and Applied Genetics* 97: 493–501.

Cock, J. H. 1985. *Cassava: New potential for a neglected crop.* London: Westfield.

Doebley, J. 1989. Isozymic evidence and the evolution of crop plants. In *Isozymes in plant biology*, D. E. Soltis and P. S. Soltis (eds.), pp. 165–191. Portland, Ore.: Dioscorides Press.

Doebley, J., M. M. Goodman, and C. W. Stuber. 1984. Isoenzymatic variation in *Zea* (Gramineae). *Systematic Botany* 9: 203–218.

Doebley, J., W. Renfroe, and A. Blanton. 1987. Restriction site variation in the *Zea* chloroplast genome. *Genetics* 117: 139–147.

Elias, M. and D. McKey. 2000. The unmanaged reproductive ecology of domesticated plants in traditional agroecosystems: An example involving cassava and a call for data. *Acta Oecologia* 21: 223–230.

Elias, M., L. Penet, P. Vindry, D. McKey, O. Panaud, and T. Robert. 2001. Unmanaged sexual reproduction and the dynamics of genetic diversity of a vegetatively propagated crop plant, cassava (*Manihot esculenta* Crantz), in a traditional farming system. *Molecular Ecology* 10: 1895–1907.

Erickson, C. L. 2000. An artificial landscape-scale fishery in the Bolivian Amazon. *Nature* 408: 190–193.

FAO 2000. Championing the cause of cassava. Food and Agriculture Organization of the United Nations: (http://www.fao.org/NEWS/2000/000405-e.htm).

Fregene, M. A., J. Vargas, J. Ikea, F. Angel, J. Tohme, R. A. Asiedu, M. O. Akoroda, and W. M. Roca. 1994. Variability of chloroplast DNA and nuclear ribosomal DNA in cassava (*Manihot esculenta* Crantz) and its wild relatives. *Theoretical and Applied Genetics* 89: 719–727.

Gepts, P. 1993. The use of molecular and biochemical markers in crop evolution studies. In *Evolutionary biology*, Volume 27, M. K. Hecht, R. J. MacIntyre, and M. T. Clegg (eds.), pp. 51–94. New York: Plenum Press.

Hillis, D. M., B. K. Maple, A. Larson, S. K. Davis, and E. A. Zimmer. 1996. Nucleic acids IV: Sequencing and cloning. In *Molecular systematics*, D. M. Hillis, C. Moritz, and B. K. Maple (eds.), pp. 321–381. Sutherland, Mass.: Sinauer Press.

Hughes, J., Z. Keresztessy, K. Brown, S. Suhandono, and M. A. Hughes. 1998. Genomic organization and structure of α-hydroxynitrile lyase in cassava (*Manihot esculenta* Crantz). *Archives of Biochemistry and Biophysics* 356: 107–116.

Iltis, H. H. 1983. From teosinte to maize: The catastrophic sexual transmutation. *Science* 222: 886–894.

Liddle, S., Z. Keresztessy, J. Hughes, and M. A. Hughes. 1998. A genomic cyanogenic b-glucosidase gene from cassava (accession no. X94986). *Plant Physiology* 117: 1526–1526.

Mann, C. 1997. Reseeding the green revolution. *Science* 277: 1038–1043.

Miller, E. T. 1992. Archaeology in the hydroelectric projects of Eletronorte: Preliminary results. Brasília: Centrais Eletricas do Norte do Brasil S.A..

Olsen, K. M. 2000. *Evolution in a recently arisen species complex: Phylogeography of Manihot esculenta Crantz (Euphorbiaceae)*. Ph.D. diss., Washington University, St. Louis. Ann Arbor: University Microfilms.

Olsen, K. M. 2002. Population history of *Manihot esculenta* (Euphorbiaceae) inferred from nuclear DNA sequences. *Molecular Ecology* 11: 901–911.

Olsen, K. M. and B. A. Schaal. 1999. Evidence on the origin of cassava: Phylogeography of *Manihot esculenta*. *Proceedings of the National Academy of Sciences, USA* 96: 5586–5591.

Olsen, K. M. and B. A. Schaal. 2001. Microsatellite variation in cassava and its wild relatives: Further evidence for a southern Amazonian origin of domestication. *American Journal of Botany* 88: 131–142.

Pearsall, D. M. 1992. The origins of plant cultivation in South America. In *The origins of agriculture: An international perspective*, C. W. Cowan, P. J. Watson, and N. L. Benco (eds.), pp. 173–205. Washington, D.C.: Smithsonian Institution Press.

Piperno, D. R. and D. M. Pearsall. 1998. *The origins of agriculture in the lowland neotropics*. New York: Academic Press.

Piperno, D. R., A. J. Ranere, I. Holst, and P. Hansell. 2000. Starch grains reveal early root crop horticulture in the Panamanian tropical forest. *Nature* 407: 894–897.

Prance, G. T. 1987. Vegetation. In *Biogeography and quaternary history in tropical America*, T. C. Whitmore and G. T. Prance (eds.), pp. 28–54. Oxford: Clarendon Press.

Renvoise, B. S. 1972. The area of origin of *Manihot esculenta* as a crop plant—A review of the evidence. *Economic Botany* 26: 352–360.

Roa, A. C., M. M. Maya, M. C. Duque, J. Tohme, A. C. Allem, and M. W. Bonierbale. 1997. AFLP analysis of relationships among cassava and other *Manihot* species. *Theoretical and Applied Genetics* 95: 741–750.

Rogers, D. J. 1963. Studies of *Manihot esculenta* Crantz and related species. *Bulletin of the Torrey Botanical Club* 90: 43–54.

Rogers, D. J. 1965. Some botanical and ethnological considerations of *Manihot esculenta*. *Economic Botany* 19: 369–377.

Rogers, D. J. and S. G. Appan. 1973. *Manihot and Manihotoides (Euphorbiaceae): A computer-assisted study*. New York: Hafner Press.

Sauer, J. D. 1993. *Historical geography of crop plants: A select roster.* Boca Raton, Fla.: CRC.

Schaal, B. A. and K. M. Olsen. 2000. Gene genealogies and population variation in plants. *Proceedings of the National Academy of Sciences, USA* 97: 7024–7029.

Schaal, B. A., K. M. Olsen, and L. J. C. B. Carvalho. In press. Evolution, domestication, and agrobiodiversity in the tropical crop cassava. In *Darwin's harvest: New approaches to the origins, evolution, and conservation of crops*, T. J. Motley (ed.). New York: Columbia University Press.

Strand, A. E., J. Leebens-Mack, and B. G. Milligan. 1997. Nuclear DNA-based markers for plant evolutionary biology. *Molecular Ecology* 6: 113–118.

Ugent, D., S. Pozorski, and T. Pozorski. 1986. Archaeological manioc *(Manihot)* from coastal Peru. *Economic Botany* 40: 78–102.

CHAPTER 10

Relationship between Chinese Chive *(Allium tuberosum)* and Its Putative Progenitor *A. ramosum* as Assessed by Random Amplified Polymorphic DNA (RAPD)

FRANK R. BLATTNER AND NIKOLAI FRIESEN

Introduction

Chinese chive, also called Chinese leek, is the second-most economically important crop species of the onion genus *Allium* in eastern Asia, and it is widely cultivated throughout China, Korea, Vietnam, and Japan. It has been introduced to most other Asian countries and more recently to the Caribbean islands, the United States, and some parts of Europe (Hanelt 2001). Leaves and flower scapes are used as vegetables or in salad (often called "garlic sprouts" in Chinese restaurants), and young inflorescences make a tasty soup.

Hanelt (2001) proposed that domestication of Chinese chive took place in northern China more than 3,000 years ago, because this crop is mentioned in the classic Chinese *Book of Poetry*, which was compiled during the Chou dynasty at the beginning of the first millennium BC. Populations of an *Allium* closely resembling the modern crop species, and thus thought to be ancestral to the cultigen, occur in steppes and dry meadows in southern Siberia, Mongolia, northern China, and North Korea. These wild plants are treated by some taxonomists as a distinct species, *A. ramosum* L., with the crop being referred to as *A. tuberosum* Rottl. ex Spreng. (For additional taxonomic details see Stearn 1944.) However, the division of these taxa is controversial, and Hanelt (1988, 2001), in his latest accounts on these *Allium* species, subsumed all forms within *A. ramosum*.

Wild and cultivated forms are slightly distinct with respect to morphology, and they differ in life history traits. The wild populations flower in summer (June to July) and possess narrow tepals and short filaments, whereas the cultivated forms flower later in the year (August to October) and have broader tepals as well as long filaments. Although Hanelt (1988) reports substantial morphological variability and, particularly in Mongolia, the occurrence of morphologically transitional types in the wild forms, we will informally treat wild and cultivated forms here as two species, just to simplify our reference to the two forms. The taxonomic consequences of our research, in other words, whether it is desirable to formally divide the plants into two species or merge them into one, will be discussed at the end of this chapter.

The archaeological record of early *Allium* crops is scarce and does not provide a reliable estimation of domestication areas and wild progenitors of the crop species. Therefore, modern collections of the cultivated species and their putative wild relatives are the major sources of information to reveal location, time, and mode of domestication. Here, we describe an analysis of cultivated *A. tuberosum* germplasm from Eastern Asia and *A. ramosum* genotypes from the wild in our attempt to define the geographical vicinity of Chinese chive domestication and to elucidate the mode of this process (i.e., if a single domestication event occurred or if separate domestications took place in parallel, in multiple regions). To study relationships among the wild and cultivated accessions, we used an anonymous genetic marker approach called *random amplified polymorphic DNA analysis* (RAPD), which enables one to detect small differences among the genomes of the individuals surveyed.

Prior Research

Studies designed to reveal the origins of *Allium* crops are rare, possibly due to the relatively limited economic importance of these cultigens. In addition, the morphological similarities of closely related *Allium* taxa hamper identification at a glance. Thus, several incorrectly identified plants are maintained in living collections at botanical gardens throughout the world. Furthermore, there is the potential for intertaxon hybridization due to the allowance of open pollination in gardens. Consequently, scientists should be suspicious of seeds obtained from these institutions. The inclusion of garden material often leads to peculiar and conflicting results in investigations of closely related species in studies of *Allium* phylogeny (Friesen et al. 1999; Fritsch et al. 2001; Klaas and Friesen 2002). The origin of the studied material, therefore, is crucial for the interpretation of data in this genus. Consequently, we have created a living collection of *Allium* taxa, in which most plants are propagated vegetatively from material originally collected in the wild.

Molecular studies on major *Allium* crops have been performed for *A. ampeloprasum* L. (leek; Kik et al. 1997); *A. cepa* L. (common onion; Havey 1992; van Raamsdonk et al. 2000; Fritsch et al. 2001); *A. cornutum* Clementi et Visiani (triploid onion) and *A. oschaninii* B. Fedtsch. (French gray shallot) (Friesen and Klaas 1998); *A. fistulosum* L. (Japanese bunching onion; Friesen et al. 1999); *A. sativum* L. (garlic; Maass and Klaas 1995); and *A. schoenoprasum* L. (chives; Friesen and Blattner 2000). In these studies, different molecular marker techniques were used to identify progenitor species (or populations) and to reveal the geographic origins of the wild plants that were initially involved in the domestication process. Clear identification of crop progenitor was possible for only two species. For *A. fistulosum*, the wild bunching onion (*A. altaicum* Pall.) from southern Siberia and Mongolia was found to be the wild progenitor to the crop. For the French gray shallot, wild populations of *A. oschaninii* were revealed as ancestral to domesticated germplasm in the species. In most other species, no clear results could be obtained, partly due to identification problems with the plant material involved (e.g., Dubouzet et al. 1997), and partly because only one or a small number of accessions of the putative wild progenitor were included in the studies (e.g., van Raamsdonk et al. 2000). When few accessions are used, it is not always possible to distinguish between a wild species being the closest relative (i.e., sister group) to the crop or being the direct progenitor. To resolve this problem, it is necessary to include several accessions of the wild species, with these accessions representing the majority of genetic variation within the taxon. Only then can a phylogenetic tree reveal the relationships among a crop species and its closest relatives, allowing one to draw conclusions concerning the process of domestication (Heun et al. 1997; Friesen et al. 1999; Badr et al. 2000). One caveat, however, is that it is possible that the wild species (or ancestral populations) from which domestication started may be extinct, or might not yet be known to the scientific community (Fritsch et al. 2001).

Recognizing Evidence of Domestication in *Allium*

Morphological and Physiological Changes

In crops where seeds were the objects of human interest, strong selection took place on traits related to yield and harvesting. Thus, unintentional domestication and selection for favorable traits occurred automatically as soon as hunter-gatherer communities started to sow seeds. The major cereals are a good example of this mode of domestication. The main trait related to domestication in cereals is the shift from fragile to tough rachis, which allows harvesting of all seeds on a spike at once (Zohary and Hopf 2000). Plants with a fragile rachis contributed notably fewer seeds to the next plant generation due to the loss of seeds through broken ears before harvesting. Thus, this unfavorable trait was eliminated automatically from populations as soon as the crop was under cultivation. The process of cereal domestication consisted of automatic co-evolutionary changes without meticulous planning. These are the same types of changes that occur in other mutualistic relationships, such as the relationships between flowers and their pollinators.

Different mechanisms of plant domestication may be operating when plant parts other than seeds are the target of human interest. In the case of *Allium* taxa, it is essential to consider separately the production of plant material for human sustenance and the need to sustain the crop through propagation units (e.g., seeds or bulbs). In *A. tuberosum*, of which both leaves and young inflorescences are eaten, we find an uncoupling of leaf production from the production of inflorescences and, hence, seeds. Harvesting of plant parts during a vegetative period and then allowing the plants to set seeds for reproduction later on in the year is not an unusual scenario for various crops (e.g., asparagus, lettuce, and most herbs). Therefore, the different flowering times of *A. tuberosum* and its putative wild progenitor, *A. ramosum*, might be directly related to human impact on the earliest cultivated plants. Alternatively, perhaps the trait that allowed *A. tuberosum* to survive the gathering of vegetative parts by humans more easily resulted in domestication of an already preadapted plant species.

Differences in flowering time are mostly undetectable in the archaeological record (Zohary and Hopf 2000). Apart from these phenological differences, *A. tuberosum* and *A. ramosum* are morphologically very similar and difficult to distinguish. Consequently, we cannot expect to get relevant information about the domestication process from preserved plant remains. Phylogenetic and population genetic data are needed to better understand the domestication of *A. tuberosum*.

Genomic Changes

Differences in the processes of domestication leave specific marks on the genome of a crop. These differences can be predicted (Figure 10.1) and compared with the outcome of molecular analyses. Three major assertions can be made for domestication via a single domestication event from a population in a defined geographical area: (1) The crop should be the closest relative of its putative wild progenitor; (2) Accessions of the crop species should be nested as a single clade (i.e., branch) within the populations of its wild progenitor in a phylogenetic tree, and should show highest similarity to the wild populations from the area where domestication took place; and (3) Genetic diversity within the crop species should be lower than in the wild species, because domestication generally results in a severe genetic bottleneck due to the inclusion of only a small part of the naturally occurring genetic variation of the species. Though the above-mentioned scenario is the most common type for crop evolution (Diamond 1997; Zohary 1999), deviations from the predicted results can point to alternative mechanisms of domestication, incorrect assumptions, or post-domestication events. For example, deviant results may occur when: (1) Incorrect assumptions are made about the phylogeny of close relatives of the crop; (2) The crop has hybrid origins (van Raamsdonk 1995); (3) The crop has a polytypic origin, that is, parallel domestication of the same wild species in different areas (Salgado et al. 1995); or (4) Post-domestication geneflow occurs between wild and domesticated populations where these grow in close proximity (Blattner and Badani Méndez 2001).

To analyze the genetic structure of *A. tuberosum* and *A. ramosum*, we used an anonymous marker approach to screen large parts of the genome for taxon differences and similarities. Anonymous markers are a method within the class

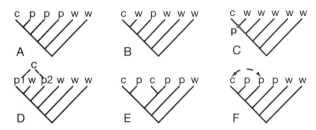

FIGURE 10.1 Possible evolutionary relationships of crop species and their wild relatives (c = crop, p = progenitor, w = other species). (A) The crop originated via a single domestication event from within its wild progenitor. (B) Wrong prior assumptions about the direct progenitor result in an unexpected sister group of the crop species. (C) The progenitor of the crop species is either unknown or extinct, so no direct progenitor can be found. Instead, the closest living relative of the crop species occurs as a sister group to it. (D) The crop originated via hybridization of two different progenitor species. (E) Parallel domestication in different areas results in two (or more) independent crop lines. (F) Geneflow among the crop and populations of its wild progenitor introduces additional genetic material into the crop's gene pool.

of molecular techniques that make use of the polymerase chain reaction (PCR; Mullis and Faloona 1987). The protocol of PCR is relatively simple. It basically comprises four steps: (1) melting of double-stranded genomic DNA at high temperatures, resulting in single strands that act as templates; (2) reduction of temperature and annealing of two short single-stranded oligonucleotide primers (about 20 nucleotides in length) to complementary regions of the single strands on both ends of the targeted sequence; (3) raising of temperature to about 70 °C, where the primers are elongated by a thermostable DNA polymerase; and (4) denaturation of the newly synthesized double-stranded target DNA. Steps (2) through (4) are repeated 30 to 40 times, which results in a nearly exponential increase of the target DNA, as every newly synthesized DNA can act as a template in the following cycle.

The major drawback of the method is the requirement of prior sequence information for the targeted DNA region in order to design the complementary primers. Depending on genome region and plant group, the primer sequences can vary enormously, even among related species, which often prevents successful amplification reactions. Without specific sequence information, it is possible to amplify only relatively conserved parts of the genome with universal primers, that is, primers that bind to evolutionary conserved genes and thus are useful in most plant families. PCR of variable regions without prior primer sequence information became possible with the invention of universal anonymous markers, which bind with multiple matching sequences arbitrarily, all over the genome.

The most familiar anonymous marker techniques are random amplified DNA (RAPD; Williams et al. 1990; and a similar technique described by Welsh and McClelland 1990 as arbitrary primed PCR (AP-PCR)); inter simple sequence repeats (ISSR; Gupta et al. 1994; Zietkiewicz et al. 1994); random amplified microsatellite polymorphisms (RAMP; Wu et al. 1994), which is a combination of the first two methods; and amplified fragment length polymorphisms (AFLP; Vos et al. 1995). The differences among these methods rely mostly on the nature of the primer-binding sites, which might be completely arbitrary (RAPD), or mediated by short nucleotide repeat motifs (ISSR) or by restriction enzyme recognition sites (AFLP). All methods have in common the fact that they produce a variable number of PCR fragments, which originate from loci randomly distributed in the genome. The presence or absence of a specific PCR product is usually scored as an independent binary character, based on the assumption that the difference represents a mutation in the primer-binding site or the restriction site (but see the subsequent discussion of the possibility that they vary by insertion or deletion mutations). Here we briefly describe the RAPD procedure.

The RAPD method uses a single PCR primer comprising only ten nucleotides, which is relatively short. The sequence of the primer is arbitrary, and due to its short length it binds often in the genome. At positions where two primer-binding sites with inverse orientation are at a convenient distance of up to 2,000 base pairs on opposite DNA strands, the PCR procedure will result in an amplification product. Depending on genome size, genome composition, and primer sequence, a single RAPD reaction will typically produce 5 to 25 DNA fragments of different lengths. After PCR, this fragment mixture is sorted according to fragment size via electrophoresis on an agarose gel, stained with a fluorescent dye that specifically attaches to DNA, and visualized under UV light. Band scoring is mostly done by eye, using an enlarged photograph of the gel. RAPDs have been used in a large number of studies (see reviews in Bachmann 1994, 1997; Karp et al. 1996; Rieseberg 1996; Wolfe and Liston 1998). The advantages of RAPD analyses are time and cost efficiency, and that only minimal laboratory equipment is required. In addition, only small amounts of DNA are needed, and detection of the DNA fragments does not involve radioactivity. Thus, a relatively high number of polymorphisms can be easily detected when several commercially available RAPD primers are used on a specific sample of accessions.

However, several limitations of the RAPD technique exist and should be considered before starting an analysis. Two major concerns are unproven band homology and possible non-Mendelian inheritance of fragments (Bachmann 1994; Rieseberg 1996). In the analysis of RAPD data, it is assumed that co-migrating bands (i.e., bands of identical length) represent homologous loci in the genome. Limitations in length measurement are due to the resolution of agarose gels. As in every electrophoretic technique, physical limitations prevent precise scoring of long fragments. Slight length differences might go unnoticed or might not be visible during gel examination. Additionally, the possibility of the occurrence of nonhomologous fragments with equal lengths rises with increasing genetic distances among the studied organisms (e.g., when comparing species instead of infraspecific lines). Both drawbacks can be reduced by using another gel matrix (e.g., polyacrylamide gels provide a higher resolution) and by testing band homology via Southern hybridization (i.e., excising a band from the gel and using it as a hybridization probe). However, these processes lead to a more time-consuming RAPD approach and abolish the major advantages of the technique, and therefore they are rarely performed.

Concerns about the heritability of RAPD markers are caused mainly by two sources of errors. Variation in reaction conditions can produce faint or even invisible bands. This might be due to imperfect primer pairing that allows amplification of the fragment only under otherwise optimal reaction conditions (influenced by DNA purity, PCR protocol, and/or brand of thermostable DNA polymerase and thermocycler). To avoid these problems, rigorous standardization of the RAPD procedure is necessary. Genetic reasons for non-Mendelian inheritance of RAPD markers are either (1) their origin from extranuclear DNA (chloroplast and mitochondrial genomes), which is mostly uniparentally inherited, or (2) due to the dominant nature of RAPD markers. With dominant markers, heterozygous alleles cannot easily be detected, because the absence of one allele is masked by the presence of the second one. Also, polymorphisms due to length differences (e.g., via insertions or deletions in a specific fragment instead of primer-binding site mutations) cannot easily be scored. This will result in an incorrect estimation of relatedness, because these characters are nonindependent (Bachmann 1997; Isabel et al. 1999).

The limitations of the RAPD method have caused a heated debate on the utility of RAPDs. However, most of the disadvantages listed above hold for other anonymous markers as well. In all cases, problems can be reduced via improvement and standardization of the reaction conditions (e.g., Benter et al. 1995); several repetitions of the analysis, with the inclusion only of reproducible fragments in the data matrix; the use of a high number of primers to get enough characters (usually character numbers should be three times higher than the number of accessions studied); and band scoring only within a single gel. The last protocol particularly limits the number of accessions within a study and might thus diminish the value of RAPD analysis (Isabel et al. 1999). Internal size standards included in every lane of a gel might help to overcome this handicap in the future. However, the basically noncumulative quality of all anonymous markers prevents their use as universal tools opposite to DNA sequences. Sequence data have the great advantage of being stored in open databases that can be extended with every newly submitted sequence. For anonymous marker data, no such database storage tools are available, nor are they workable.

In summary, we conclude that although not every RAPD fragment may be reproducible in different laboratories, the overall outcome of two studies will be comparable *if* RAPD analyses are carefully conducted. When ample resources are available, we always opt for a cumulative, sequence-based marker technique for comparative taxon studies. But when this technique is economically not reasonable, RAPDs provide a possible alternative.

Materials and Methods

A total of 29 accessions of *A. tuberosum* and *A. ramosum* from the living collection of the Department of Taxonomy of the IPK Gatersleben were investigated (Table 10.1). The analyzed accessions represent the entire geographical ranges of the species and mostly belong to the predominantly occurring tetraploid plant types ($2n = 32$), though we also included diploids ($2n = 16$) and triploids ($2n = 24$). Two accessions of *A. oreiprason* Schrenk were included as outgroups. Analysis of a noncoding part of the nuclear ribosomal DNA for the entire genus *Allium* (Friesen et al. in press) revealed that *A. ramosum* is the closest relative of *A. tuberosum*, and that *A. oreiprason* is the sister group of both taxa. Consequently, we are sure that we analyzed a natural group of species.

DNA was isolated from one plant per accession, using the NucleoSpin Plant Kit according to the instructions of the manufacturer (Macherey-Nagel, Düren, Germany). The concentration of the extracted DNA was checked on an agarose gel. Ten microliters (µL) of the isolated DNA were dissolved in 150 µL of water, and 4 µL (approximately 50 ng) of this DNA solution were used for PCR amplification.

Prior to the analysis of the entire set of accessions, about 50 RAPD primers were tested on a small set of plants. Final amplifications were carried out using 11 arbitrary decamer primers (A19, AB04, AB18, AC02, C07, C09, C13, D01, D03, G13, and G19, obtained from Operon Technologies, Alameda, Cal.) that provided clear and reproducible bands in the initial screening. The amplification conditions were optimized according to Friesen et al. (1997). Twelve microliters of each RAPD reaction mixture were separated on 1.5% agarose gels in 0.5x TBE buffer, followed by staining with ethidium bromide (Sambrook et al. 1989). Clearly visible RAPD bands (an example of a RAPD gel is given in Figure 10.2) were scored manually for presence (1) or absence (0), using enlarged photographs of the gels. Differing band intensities were not taken into account to avoid errors introduced by competition among priming sites during the initial PCR cycles (Bachmann 1997). Only reproducible bands of two independent amplification reactions were included in the data analyses.

From the resulting binary data matrix, distance and character-based analyses were calculated. Pairwise genetic distances were calculated using the Nei's coefficient, which considers only the shared presence of a fragment as a character (Nei and Li 1979). The absence of a band could be due to various reasons, and therefore was not taken into account. Phenograms were prepared applying UPGMA (unweighted pair group method using arithmetic averages) and neighbor-joining (NJ) cluster analyses (Saitou and Nei 1987) of the genetic distance matrix (with PAUP*; Swofford 2002).

TABLE 10.1
Accessions Used in the Comparison of *Allium ramosum* and *A. tuberosum*

#	Acc. No.[a]	Species	Origin	2n[b]
1	Tax 1419	A. ramosum	Russia, Tuva	32
2	Tax 1695	A. ramosum	Mongolia, Bayan-Chongor	32
3	Tax 1699	A. ramosum	Mongolia, Erdenesant	32
4	Tax 1836	A. ramosum	Russia, Yacutia	32
5	Tax 2014	A. ramosum	China	24
6	Tax 2115	A. ramosum	Northern Kazakhstan	32
7	Tax 2339	A. ramosum	Mongolia, Ulan-Bator	32
8	Tax 2347	A. ramosum	Mongolia, CherenBayan-Uul	32
9	Tax 2356	A. ramosum	Mongolia, Tumencogt-Uul	32
10	Tax 2363	A. ramosum	Mongolia, Erdenecagaan	32
11	Tax 2371	A. ramosum	Mongolia, Bayan-Changai	32
12	Tax 2378	A. ramosum	China	24
13	Tax 2735	A. ramosum	Kazakhstan	32
14	Tax 2755	A. ramosum	Russia, Buryatia	32
15	Tax 2759	A. ramosum	Russia, Tuva, Ersin	32
16	Tax 0582	A. tuberosum	Japan, Tsukuba	32
17	Tax 1482	A. tuberosum	Nepal	32
18	Tax 1969	A. tuberosum	Korea	32
19	Tax 1970	A. tuberosum	Korea	32
20	Tax 1971	A. tuberosum	Korea	32
21	Tax 2033	A. tuberosum	VIR, possibly from China	16
22	Tax 2426	A. tuberosum	Russia, Rusansky Chrebet (Pamir)	32
23	Tax 2453	A. tuberosum	India, Ladakh	32
24	Tax 2454	A. tuberosum	India, Agra	32
25	Tax 2499	A. tuberosum	China, NW Yunnan	32
26	Tax 3301	A. tuberosum	Pakistan, Gilgit	32
27	Tax 3866	A. tuberosum	India	32
28	Tax 4246	A. tuberosum	China	32
29	Tax 5557	A. tuberosum	Vietnam	24
30	Tax 3653	A. oreiprason	Kazakhstan, Transili-Alatau	16
31	Tax 5000	A. oreiprason	Kyrgyzstan, Tallas-Alatau	16

[a] The numbers following "Tax" are the accession numbers of the *Allium* collection of the Department of Taxonomy, Institute of Plant Genetics and Crop Plant Research (IPK) Gatersleben, Germany. All material was collected in the wild or, for the crop species, in the countries listed under "Origin." The only exception is accession Tax 2033, which is material from a germplasm collection of the Vavilov Institute (VIR, St. Petersburg, Russia) with unclear origin.

[b] Chromosome numbers: 16 = diploid samples; 24 = triploid samples; 32 = tetraploid samples.

Furthermore, the genetic distance matrix was subjected to a principal coordinate analysis (PCoA) in which, starting from the distances, new independent axial coordinates representing most of the variability of the original data were calculated (with NTSYSpc; Rohlf 1998). The accessions were then plotted as points in a three-dimensional continuous space defined by the first three coordinates. Maximum parsimony analyses (MP) of the binary data matrix were also calculated (in PAUP*; Swofford 2002), using either branch-and-bound or the heuristic (with 200 random addition sequences) search options, MULPARS, ACCTRAN, and TBR branch swapping. Bootstrap analysis (Felsenstein 1985) with 1,000 resamples was used to examine the statistical support of branches found in the MP and NJ trees.

Two different datasets were used to reveal the relationships among *A. tuberosum* and *A. ramosum* accessions. In the first dataset, we included nine representative accessions of the two species, together with the outgroups. We used this analysis to root the tree of the ingroups. A second dataset included all 29 ingroup accessions, but not *A. oreiprason*. This allowed us to score the differences among the ingroup populations in more detail, while avoiding the inclusion of false-homologous fragments in the more distantly related outgroup.

Results

In the dataset including the outgroup *A. oreiprason*, we found 136 polymorphic RAPD bands. Analyses of these bands resulted in a phylogenetic hypothesis that puts into question

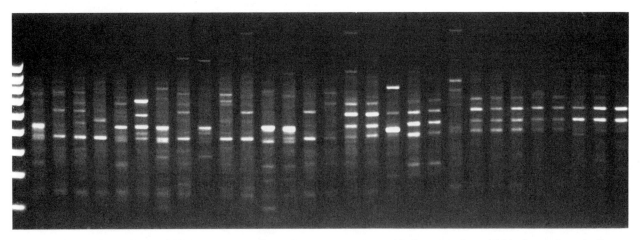

FIGURE 10.2 RAPD reaction of 29 *A. ramosum* and *A. tuberosum* accessions with Operon primer AB04, electrophoretically separated on a 1.5% agarose gel. In the first lane on the left side, a size standard (100-bp ladder) is included, which allows sizing of the RAPD fragments. The order of the samples on the gel (from left to right) is the same as that in Table 10.1.

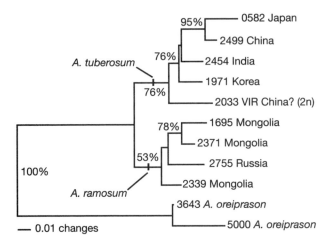

FIGURE 10.3 Phenogram of a neighbor-joining analysis of 137 RAPD characters of accessions of wild *A. ramosum* and the crop plant *A. tuberosum*, together with the closely related *A. oreiprason* as outgroup taxon. Numbers along the branches depict bootstrap values (%). *Allium ramosum* and *A. tuberosum* occur as sister groups, not as the crop nested within its proposed wild progenitor.

the putative progenitor–descendant relationship between the two ingroup species. All methods (NJ, UPGMA, and MP) resulted in very similar trees (only the NJ tree is shown in Figure 10.3), which placed *A. tuberosum* as a sister group of *A. ramosum*, corresponding to hypothesis C in Figure 10.1. Cladistic analysis (MP) resulted in one most parsimonious tree (not shown) with a length of 224 steps and a consistency index of 0.6071. The bootstrap support of this topology was relatively high, with values of 82% for the clade comprising *A. tuberosum* and 88% for *A. ramosum* in the cladistic analysis (slightly lower in the NJ analysis, see Figure 10.3). The nearly identical magnitudes of genetic variation within the crop and the wild species, as indicated by the branch lengths in Figure 10.3, are surprising. Within *A. tuberosum*, the diploid accession (Tax 2033) of unclear geographical origin occurs at a basal position.

In the second dataset, which included all 29 ingroup accessions, but not *A. oreiprason*, we found 127 polymorphic RAPD bands. Analyses of these data using phenetic (NJ and UPGMA), cladistic (MP), and statistical (PCoA) methods produced nearly identical results, similar to the analyses of the first dataset: Accessions of *A. ramosum* and *A. tuberosum* formed two genetically distinct and clearly separated groups (the NJ tree is shown in Figure 10.4a). Within the *A. ramosum* branch, a triploid accession from China (Tax 2014) occurs at a basal position (Figure 10.4a). Within *A. tuberosum* the deepest split separates the putative Chinese diploid (Tax 2033) from all other accessions. The next branches are represented by material from Korea and China. Samples from other countries occur at derived positions in the tree (Figure 10.4a). Bootstrap support of the internal branches is higher for *A. tuberosum* than for *A. ramosum*, even though genetic distances within both species were nearly equal.

The PCoA plot (Figure 10.4b) based on the second dataset shows the distribution of genetic variation along the first three coordinal axes. Genetic diversity in *A. ramosum* is mostly seen along the Z-axis, whereas the accessions of *A. tuberosum* vary mainly in their positions along the X- and Y-axes. In *A. tuberosum*, the PCoA plot reflects geographical differentiation among the accessions: mostly Eastern Asian accessions are in the foreground (i.e., on the left side of the Y-axes), and samples from the Indian subcontinent are further back, grouping on the right side of the Y-axes (Figure 10.4b). This result is similar to that of the NJ analysis (Figure 10.4a). Neither analysis points to a severe genetic bottleneck in the crop, which is often associated with single domestication events.

Interpretation of Results Relative to Domestication

The implications drawn from the domestication history of Chinese chive are mostly attributed to two assumptions concerning our RAPD results: (1) Tree topologies reflect phylogenetic relationships, and (2) branch lengths are reliable

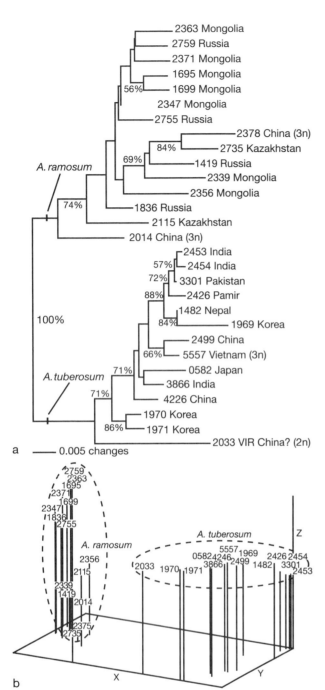

FIGURE 10.4 Unrooted neighbor-joining analysis of 127 RAPD characters of 29 accessions of wild *A. ramosum* and the crop *A. tuberosum*. (A) Phenogram: branch lengths represent genetic distances. Bootstrap values (%) are given along the branches. (B) Principal coordinate analysis of the same dataset. Both analyses show that the crop and its putative wild progenitor are clearly differentiated, and that genetic diversity within the domesticate is nearly of the same magnitude as within the wild species.

measures of genetic divergence. A comparison of our results to the hypothetical expectations of a phylogenetic analysis of crops and their ancestors (Figure 10.1) leads to the conclusion that the wild progenitor of *A. tuberosum* cannot be

A. ramosum. The crop is not nested within the wild species, as would be expected in Figure 10.1a. Instead, we found a sister group relationship, meaning that both species shared a common ancestor from which they developed as two independent lineages (Figure 10.1c). This progenitor clearly was not included in our study and might be unknown, or already extinct.

Another possible interpretation of the results is that we missed the population of *A. ramosum* from which domestication began. However, our representative sampling covered the geographical distribution of the species. Also, this interpretation conflicts with the second line of evidence—the branch lengths. Some of the longest branches in the phylogenetic trees (Figures 10.3 and 10.4a) served to separate *A. tuberosum* from *A. ramosum*. These species are also clearly separated in the PCoA plot (Figure 10.4b). In spite of their morphological similarities, the clear genetic differentiation makes a conspecific status for these species highly unlikely. Furthermore, branch-length differences are important in revealing genetic bottlenecks. If a species experienced a severe genetic reduction in its history (e.g., due to far-reaching extinction or founder events), we would expect to find relatively limited variation within the present-day gene pool of this species, when compared to a species that did not experience such a genetic bottleneck (Eyre-Walker et al. 1998; Mimura et al. 2000; Iqbal et al. 2001). This generally results in shorter branches in population-based trees of species with reduced genetic diversity. In the trees produced from our analyses, *A. tuberosum* accessions did not differ markedly in their branch lengths from those of *A. ramosum*, which points to a relatively uniform distribution of genetic diversity in both species.

Discussion

Our results of phenetic, cladistic, and multivariate analyses of RAPD data revealed an unexpected relationship between wild *A. ramosum* and domesticated *A. tuberosum*. In spite of their great morphological similarity, which led Hanelt (2001) to merge both taxa into *A. ramosum*, molecular data clearly separated wild and crop populations. In our opinion, this is an indication that *A. ramosum* and *A. tuberosum* consist of two long-separated gene pools with no (or only very restricted) geneflow between interspecific populations. Therefore, we propose to maintain both taxa as separate species, in other words, to use *A. ramosum* only for the wild, early-flowering plants and *A. tuberosum* for the domesticate.

Additionally, the clear division of these species makes a progenitor-derivative relationship between them highly unlikely. Instead, we hypothesize that there was a common progenitor somewhere in the past from which both species evolved. The basal branches in the *A. tuberosum* group of the phylogenetic tree (Figure 10.4a) were formed by material from China and Korea, with the only diploid of this study occurring at the deepest branch. This pattern points to northern China as the most likely place where the (diploid?) progenitor populations of the crop were subjected to domestication. If we take into account that the flora of this area (and of Mongolia) is far from being thoroughly studied, it even seems possible that some progenitor populations might still persist there. Yang et al. (1998) reported a (wild?) diploid *A. tuberosum* from Shangxi in China, which also indicates that this area may be a possible shelter of relatively basal types of the species. However, the large amount of genetic diversity within *A. tuberosum* calls into question a single domestication from only one progenitor population. Most species with a unique domestication event are characterized by short internal branches in phylograms (Heun et al. 1997; Badr et al. 2000), that is, by relatively small genetic differences compared to their wild relatives. In Chinese chive, no such genetic bottleneck could be detected.

The question remains how domestication took place in Chinese chive. Without a known crop progenitor, most of the propositions regarding the mode of domestication remain rather speculative. However, some statements seem to be possible, considering the current knowledge of *A. tuberosum*. High genetic variability within a crop can be the result of either multiple parallel domestications, which means that different populations contributed to the crop's gene pool, or post-domestication hybridization events between the crop and adjacent wild populations. The fact that possibly no wild *A. tuberosum* populations exist, at least not on a large scale, could be an indication that many wild populations served as the gene pool for the domesticate. This pattern occurs when wild plants are gathered for consumption throughout their entire range, which can lead to the extinction of the wild populations (Wawrosch et al. 2001). In this scenario, locally kept garden populations, mostly founded when demand on wild plants became severe, are the diffuse starting point for the crop's evolution. Geneflow among domesticated and wild plants can also result in absorption of the wild gene pool into the crop (Freyre et al. 1996). In this case, extinction of the wild populations is possible, though not inevitable, via the transmission of traits not favorable in wild plants. Against the background of current gathering behavior exhibited by many native human tribes in Eastern Asia, the first hypothesis—extinction of the wild progenitor by human collecting—seems more likely to us.

Future research in *A. tuberosum* will concentrate on inclusion of the newly found diploid material from Shangxi province. These plants might resemble the putative wild progenitors of the crop in morphology and karyotype. It would be interesting to test the position of these plants in a phylogeographic analysis (Schaal and Olsen 2000) relative to the domesticated germplasm. Furthermore, due to its ancestral ploidy level, it might be a starting point to study the evolution of the complex ploidy patterns (di-, tri-, and tetraploids) in *A. tuberosum*.

References

Bachmann, K. 1994. Molecular markers in plant ecology. *New Phytologist* 126: 403–418.

———. 1997. Nuclear DNA markers in biosystematic research. *Opera Botanica* 132: 137–148.

Badr, A., K. Müller, R. Schäfer-Pregl, H. El Rabey, S. Effgen, H. H. Ibrahim, C. Pozzi, W. Rohde, and F. Salamini. 2000. On the origin and domestication history of barley (*Hordeum vulgare*). *Molecular Biology and Evolution* 17: 499–510.

Benter, T., S. Papadopoulos, M. Pape, M. Manns, and H. Poliwoda. 1995. Optimization and reproducibility of random amplified polymorphic DNA in humans. *Analytical Biochemistry* 230: 92–100.

Blattner, F. R. and A. G. Badani Méndez. 2001. RAPD data do not support a second centre of barley domestication in Morocco. *Genetic Resources and Crop Evolution* 48: 13–19.

Diamond, J. 1997. *Guns, germs and steel: The fates of human societies*. London: Jonathan Cape.

Dubouzet, J. G., K. Shinoda, and N. Murata. 1997. Phylogeny of *Allium* L. subgenus *Rhizirideum* (G. Don ex Koch) Wendelbo according to dot-blot hybridization with randomly amplified DNA probes. *Theoretical and Applied Genetics* 95: 1223–1228.

Eyre-Walker, A., R. L. Gaut, H. Hilton, D. L. Feldman, and B. S. Gaut. 1998. Investigation of the bottleneck leading to the domestication of maize. *Proceedings of the National Academy of Sciences, USA* 95: 4441–4446.

Felsenstein, J. 1985. Confidence limits on phylogenies: An approach using the bootstrap. *Evolution* 39: 783–791.

Freyre, R., R. Rios, L. Guzman, D. G. Debouck, and P. Gepts. 1996. Ecogeographic distribution of *Phaseolus* spp. (Fabaceae) in Bolivia. *Economic Botany* 50: 195–215.

Friesen, N. and F. R. Blattner. 2000. RAPD analysis reveals geographic differentiations within *Allium schoenoprasum* L. (Alliaceae). *Plant Biology* 2: 297–305.

Friesen, N. and M. Klaas. 1998. Origin of some minor vegetatively propagated *Allium* crops studied with RAPD and GISH. *Genetic Resources and Crop Evolution* 45: 511–523.

Friesen, N., R. M. Fritsch, and K. Bachmann. 1997. Hybrid origin of some ornamental *Alliums* of subgenus *Melanocrommyum* verified with GISH and RAPD. *Theoretical and Applied Genetics* 95: 1229–1238.

Friesen, N., S. Pollner, K. Bachmann, and F. R. Blattner. 1999. RAPDs and noncoding chloroplast DNA reveal a single origin of the cultivated *Allium fistulosum* from *A. altaicum* (Alliaceae). *American Journal of Botany* 86: 554–562.

Friesen, N., R. M. Fritsch, and F. R. Blattner. In press. Phylogeny and new intrageneric classification of Allium (Alliaceae) based on nuclear rDNA ITS sequences. In *Monocots: Comparative biology and evolution*, J. T. Columbus, E. A. Friar, C. W. Hamilton, J. M. Porter, L. M. Prince, and M. G. Simpson (eds.). Claremont, Rancho Santa Ana Botanic Garden.

Fritsch, R. M., F. Matin, and M. Klaas. 2001. *Allium vavilovii* M. Popov et Vved. and a new Iranian species are the closest among the known relatives of the common onion *A. cepa* L. (Alliaceae). *Genetic Resources and Crop Evolution* 48: 401–408.

Gupta, M., Y.-S. Chyi, J. Romero-Severson, and J. L. Owen. 1994. Amplification of DNA markers from evolutionarily diverse genomes using single primers of simple-sequence repeats. *Theoretical and Applied Genetics* 89: 998–1006.

Hanelt, P. 1988. Taxonomy as a tool for studying plant genetic resources. *Die Kulturpflanze* 36: 169–187.

———. 2001. Alliaceae. In *Mansfeld's Encyclopedia of Agricultural and Horticultural Crops*, P. Hanelt and Institute of Plant Genetics and Crop Plant Research (eds.), pp. 2250–2269. Berlin: Springer-Verlag.

Havey, M. J. 1992. Restriction enzyme analysis of the chloroplast and nuclear 45S ribosomal DNA of *Allium* sections *Cepa* and *Phyllodon* (Alliaceae). *Plant Systematics and Evolution* 183: 17–31.

Heun, M., R. Schäfer-Pregl, D. Klawan, R. Castagna, M. Accerbi, B. Borghi, and F. Salamini. 1997. Site of einkorn wheat domestication identified by DNA fingerprinting. *Science* 278: 1312–1314.

Iqbal, M. J., O. U. K. Reddy, K. M. El-Zik, and A. E. Pepper. 2001. A genetic bottleneck in the "evolution under domestication" of upland cotton *Gossypium hirsutum* L. examined using DNA fingerprinting. *Theoretical and Applied Genetics* 103: 547–554.

Isabel, N., J. Beaulieu, P. Theriault, and J. Bousquet. 1999. Direct evidence of biased gene diversity estimates from dominant random amplified polymorphic DNA (RAPD) fingerprints. *Molecular Ecology* 8: 477–483.

Karp, A., O. Seberg, and M. Buiatti. 1996. Molecular techniques in the assessment of botanical diversity. *Annals of Botany* 78: 143–148.

Kik, C., A. M. Samoylov, W. H. J. Verbeek, and L. W. D. van Raamsdonk. 1997. Mitochondrial DNA variation and crossability of leek *(Allium porrum)* and its wild relatives from the *Allium ampeloprasum* complex. *Theoretical and Applied Genetics* 94: 465–471.

Klaas, M. and N. Friesen. 2002. Molecular markers in *Allium*. In *Allium Crop Science—Recent Advances*, H. Rabinovich and L. Curah (eds.), pp. 159–185. Wallingford, UK: CABI Publishing.

Maass, H. I. and M. Klaas. 1995. Infraspecific differentiation of garlic (*Allium sativum* L.) by isozyme and RAPD markers. *Theoretical and Applied Genetics* 91: 89–97.

Mimura, M., K. Yasuda, and H. Yamaguchi. 2000. RAPD variation in wild, weedy and cultivated azuki beans in Asia. *Genetic Resources and Crop Evolution* 47: 603–610.

Mullis, K. B. and F. A. Faloona. 1987. Specific synthesis of DNA in vitro via a polymerase-catalyzed chain reaction. *Methods in Enzymology* 155: 335–350.

Nei, M. and W.-H. Li. 1979. Mathematical model for studying genetic variation in terms of restriction endonucleases. *Proceedings of the National Academy of Sciences, USA* 76: 5269–5273.

Rieseberg, L. H. 1996. Homology among RAPD fragments in interspecific comparisons. *Molecular Ecology* 5: 99–105.

Rohlf, F. J. 1998. *NTSYSpc: Numerical taxonomy system. Version 1.8.* Setauket, N.Y.: Exeter Publishing Ltd.

Salgado, A. G., P. Gepts, and D. G. Debouck. 1995. Evidence for two gene pools of the lima bean, *Phaseolus lunatus* L., in the Americas. *Genetic Resources and Crop Evolution* 42: 15–28.

Sambrook, J., E. F. Fritsch, and T. Maniatis. 1989. *Molecular cloning: A laboratory manual*, 2nd edition. Cold Spring Harbor, N.Y.: Cold Spring Harbor Laboratory Press.

Saitou, N. and M. Nei. 1987. The neighbor-joining method: A new method for reconstructing phylogenetic trees. *Journal of Molecular Evolution* 4: 406–425.

Schaal, B. A. and K. M. Olsen. 2000. Gene genealogies and population variation in plants. *Proceedings of the National Academy of Sciences, USA* 97: 7024–7029.

Stearn, W. T. 1944. Nomenclature and synonymy of *Allium odorum* and *A. tuberosum*. *Herbertia* 11: 227–245.

Swofford, D. L. 2002. *PAUP*. Phylogenetic analysis using parsimony (*and other methods)*, Version 4. Sunderland, Mass.: Sinauer Associates.

van Raamsdonk, L. W. D. 1995. The cytological and genetical mechanisms of plant domestication exemplified by four crop models. *Botanical Reviews* 61: 367–399.

van Raamsdonk, L. W. D., M. Vrielink-van Ginkel, and C. Kik. 2000. Phylogeny reconstruction and hybrid analysis in *Allium* subgenus *Rhizirideum*. *Theoretical and Applied Genetics* 100: 1000–1009.

Vos, P., R. Hogers, M. Bleeker, M. Reijans, T. van de Lee, M. Hornes, A. Frijters, J. Pot, J. Peleman, M. Kuiper, and M. Zabeau. 1995. AFLP: A new technique for DNA fingerprinting. *Nucleic Acids Research* 23: 4407–4414.

Wawrosch, C., P. R. Malla, and B. Kopp. 2001. Micropropagation of *Allium wallichii* Kunth, a threatened medicinal plant of Nepal. *In Vitro Cellular and Developmental Biology—Plant* 37: 555–557.

Welsh, J. and J. McClelland. 1990. Fingerprinting genomes using PCR with arbitrary primers. *Nucleic Acids Research* 18: 7213–7218.

Williams, G. K., A. R. Kubelik, K. L. Livak, J. A. Rafalaski, and S. V. Tingey. 1990. DNA polymorphisms amplified by arbitrary primers are useful as genetic markers. *Nucleic Acids Research* 18: 6531–6535.

Wolfe, A. D. and A. Liston. 1998. Contribution of PCR-based methods to plant systematics and evolutionary biology. In *Molecular systematics of plants* II, D. E. Soltis, P. S. Soltis, and J. J. Doyle (eds.), pp. 43–86. Dordrecht, Netherlands: Kluwer Academic Publishers.

Wu, K.-S., R. Jones, L. Danneberger, and P. A. Scolnik. 1994. Detection of microsatellite polymorphisms without cloning. *Nucleic Acids Research* 22: 3257–3258.

Yang, L., J.-M. Xu, X.-L. Zhang, and H.-Q. Wan. 1998. Karyotypical studies of six species of the genus *Allium*. *Acta Phytotaxonomica Sinica* 36: 36–46.

Zietkiewicz, E., A. Rafalski, and D. Labuda. 1994. Genome fingerprinting by simple sequence (SSR)-anchored polymerase chain reaction amplification. *Genomics* 20: 176–183.

Zohary, D. 1999. Monophyletic vs. polyphyletic origin of the crops on which agriculture was founded in the Near East. *Genetic Resources and Crop Evolution* 46: 133–142.

Zohary, D. and M. Hopf. 2000. *Domestication of plants in the Old World*, 3rd Edition. Oxford, UK: Oxford University Press.

CHAPTER 11

Using Multiple Types of Molecular Markers to Understand Olive Phylogeography

CATHERINE BRETON, GUILLAUME BESNARD, AND
ANDRÉ A. BERVILLÉ

Olive and Its History

The olive *(Olea europaea* ssp. *europaea* var. *europaea)* has played an important role in the Mediterranean Basin for hundreds of years (Figure 11.1). It is not only a key economic crop; the olive's hardiness, hardness, and longevity have come to symbolize values that Mediterranean cultures hold central. The wild form known as oleaster *(Olea europaea* ssp. *europaea* var. *sylvestris)*, with its ability to survive in harsh environments, is also a key component of the countryside across the region.

Olive production is difficult and the harvest is challenging because of drought, frost, biology of flowering, and pruning. As an outcrossed wind-pollinated species, a majority of olive cultivars are self-incompatible, and some are male-sterile (Besnard et al. 2000). Despite these difficulties, olive cropping continues to increase in acreage, largely because of the high potential return to olive growers from the production of olive oil. Although there has been growing competition from seed oils in recent years, olive oil, with its well known health benefits, continues to be a highly valued product both within and outside the Mediterranean Basin. As a result, most olive cultivars are grown for oil (90% of production), a few are grown for table fruit, and some are grown for mixed uses. Oil olives are usually harvested when they overripen, or turn dark red to black. Commercial olive oil labeled "extra virgin" is obtained by crushing fresh olives at about 20° C and then separating the clear oil from the pulp and water. Table olives, which are usually bigger than oil olives, are either canned for extended storage or rapidly preserved with various spices. Green olives are harvested before turning color (before maturation), whereas black and red olives are fully ripe.

Thousands of cultivars have been described based on fruit, pit, and leaf shapes and colors; tree architecture; and phenology (i.e., flowering time). However, some synonymies among olive cultivar names have been revealed with molecular markers (Besnard et al. 2001d). The species *Olea europaea* belongs to a widespread genus that is mostly subtropical or subequatorial in distribution. The Mediterranean olive (subspecies *europaea*) is one of various subspecies, the others of which are distributed in Africa and Asia (Green 2002). Olive trees are known for their frost sensitivity, and in some Mediterranean areas they are periodically destroyed by exceptional frosts (below −15° C). Consequently, the traditional inference has been that the ssp. *europaea* diffused from warmer African regions north into the Mediterranean Basin.

Scenarios regarding the early history of the olive are abundant in the literature. In Greek mythology, the olive was brought to Greece by Athena (Estin and Laporte 1987), its fruit ripening when no other fruits were available during the winter. Botanists once thought that domesticated olive (var. *europaea*) was introduced to the Mediterranean Basin and that self-sustaining plants called oleasters (var. *sylvestris*) were feral forms escaped from the cultivars (Chevalier 1948; Green 2002). Furthermore, historians (e.g., Amoureux 1784) proposed that Phoenicians, Greeks, and Romans not only introduced domesticated olive from the eastern to the western Mediterranean, but that these travelers later brought back locally selected western cultivars to the East.

Feral forms of cultivated olive are able to sustain themselves in woodland areas. Morphologically, they appear similar to the truly wild oleasters. Only the relatively large size of the fruit reveals their origins as cultivar escapes. However, some cultivars have small fruits, which makes it difficult to distinguish their escapes (i.e., feral oleasters) from wild oleasters. Moreover, juvenile trees and those that have been severely pruned or grazed exhibit distinct leaf shapes and tree architectures, which often confound identification.

Olive *(O. e. europaea)* occurs at the northern limit of *Olea* L. taxa, which mostly have tropical or subtropical distributions. Within the genus *Olea*, the subgenus *Olea* is divided into two sections—section *Ligustroides* Benth. & Hook. and section *Olea*. Species belonging to section *Ligustroides* occur in central and southern Africa (Green and Wickens 1989). Section *Olea* contains the *O. europaea* complex, in which six subspecies have been recognized (Green and Wickens 1989; Vargas et al. 2001; Green 2002). Whereas *O. e. europaea* occurs in the Mediterranean region, the other five *O. europaea* subspecies are distributed in Africa and Asia. Some of these subspecies are closely related to olive and therefore have been considered candidates for progenitor to the crop. Those in greatest proximity to the Mediterranean are in northern Africa (Besnard et al. 2002a). They include *O. e. laperrinei* (Batt. & Trab.) Ciferri, which is present in the Saharan mountains (Wickens 1976; Quézel 1978); *O. e. maroccana* (Greut. & Burd.) Vargas et al., which occurs in southwestern Morocco (Médail et al. 2001); *O. e. cerasiformis* (Webb & Berth.) Kunk. & Sund., an endemic in Madeira; and *O. e. guanchica* Vargas et al., an endemic in the Canary Islands (Hess et al. 2000).

According to pollen grain remains (Quézel 1978), a wild *Olea* taxon existed in the Mediterranean Basin at the end of

FIGURE 11.1 Oleasters near Tamanar, Morocco.

the Tertiary Era. During the Quaternary Era, the situation was similar, but glaciating periods greatly disturbed the distribution of the taxon. As early as 19,000 BP, oleaster was present in the eastern Mediterranean (Kislev et al. 1992), where its fruits were gathered, presumably for food. Several botanists and historians think that oleaster occurred in this region much earlier (Zohary and Spiegel-Roy 1975). For example, palynological data suggest the presence of O. europaea in the Near East around 30,000 BP (Van Zeist and Woldring 1980; Neef 1990), although other less thermophilic tree species, such as oak, were absent (Brewer et al. 2002). Domestication of O. europaea in this region occurred by 6000 BP (Galili et al. 1997).

Scientists have traditionally believed that (1) wild oleasters are a homogenous group confined to the eastern Mediterranean; (2) cultivars were derived from a single oleaster source in the Near East and then carried westward to the rest of the Mediterranean Basin; and (3) oleasters in the West are all of feral origin. These hypotheses are rooted in the long-held belief that feral oleasters can be identified simply by morphology. However, we now know that identification of feral oleasters can be confounded by the age of the tree, as well as by the effects of grazing and pruning. Lumaret and Ouazzani (2001) claimed that they could recognize wild oleasters using nuclear markers correlated with domestication traits, and that their results suggested that oleaster was present in the western Mediterranean before the arrival of cultivated olive. Other recent studies (e.g., Terral and Arnold-Simard 1996; Terral 2000; Besnard and Bervillé 2000; Besnard et al. 2001a, 2001b) have also put some of the traditional hypotheses in question. For example, additional olive domestication sites have been documented based on archaeological remains for Spain (6000 BP) by Terral (2000) and for Corsica (at least 5000 BP) by Magdelaine and Ottaviani (1984).

New Tools for Studying Olive Origins

The utility of molecular tools for evolutionary studies arises from the insensitivity of the genetic markers to environmental factors. Several markers based on DNA amplification technology have been used to explore various independent portions of the olive genome (e.g., Besnard and Bervillé 2000, 2002; Besnard et al. 2002b, 2002c; Bronzini de Caraffa et al. 2002), including DNA from the nucleus, chloroplast (cpDNA), and mitochondria (mtDNA).

The technique of random amplified polymorphic DNA (RAPD) consists of the amplification of short nuclear DNA fragments (200 base pairs (bp) to 2.5 kb) adjacent to a 10-bp-long primer. Fragments that differ in length within and between samples are identified by the differential mobilities of the fragments (which appear as bands) in an electrified, porous gel. A single primer often produces multiple bands for a single sample. Consequently, these data are treated as band presence/absence characters.

Simple sequence repeats (SSRs) consist of one-, two- or three-bp motifs, repeated a variable number of times (e.g., about ten). Specific repeats are identified by long (ca. 20 bp) flanking primer pairs that delimit a very short DNA fragment (100–250 bp). Again, fragment length polymorphisms among samples are identified by their mobilities on a gel. However, given the specificity of the long primers, usually only a single region within a genome is amplified. Thus, in the case of nuclear SSRs in a diploid taxon, the single- or double-banded patterns are interpreted as one or two alleles at a single locus for homozygous and heterozygous individuals, respectively.

As with SSRs, restriction fragment length polymorphisms (RFLPs) are typically used to compare DNA fragment length polymorphisms in organelles (e.g., chloroplasts or mitochondria). Although the RFLP flanking sequences (i.e., hybridization probes) are much longer than SSR primer sequences, and are therefore quite specific, the procedure is more costly than RAPDs or SSRs. The RFLP data from multiple probe and restriction enzyme combinations can be combined to recognize distinct genetic patterns, called chlorotypes or mitotypes. Cytoplasmic markers such as mitochondrial RFLPs and chloroplastic SSRs trace maternal lineages only, since these organelles are usually passed to offspring from the female parent.

In this chapter we examine nuclear, chloroplastic, and mitochondrial DNA-based molecular diversity of populations and cultivars of O. e. europaea from around the Mediterranean Basin. With these data, we address questions concerning the early distribution of oleaster and the origins and spread of domesticated olive.

Materials and Methods

Plant Material

From the Mediterranean Basin, 235 trees representing 27 oleaster populations were sampled for molecular markers (Table 11.1). Population size varied from 5 to 22 individuals per population, except for Corsica, where it varied from 1 to 10. For geographic analyses, populations were scored as "East" or "West" (Table 11.1). The north-south dividing line between these Mediterranean Basin regions follows natural

TABLE 11.1
List of Oleaster Populations Analyzed, Their Locations, the Markers Tested, and the Mitotype Results

Pop No.	Population Location[a]	Country	RAPD[b]	SSR[b]	Mitotypes[c]
1	Urla, Izmir*	Turkey	+	+	7 ME1
2	Harem, Oronte Valley*	Syria	+	+	13 ME1
3	Al Ascharinah, El Ghab*	Syria	+		12 ME1
4	Mont Carmel, Haifa*	Israel	+	+	18 ME1
5	Cyrenaique	Libya	+	+	4 ME1; 11 MOM
6	*Zaghouan*	Tunisia	+	+	6 MCK; 1 MOM
7	Mont Belloua, Kabylia	Algeria	+	+	6 MCK; 1 MOM
8	Tamanar, Essaouira	Morocco	+	+	10 MOM
9	Immouzzer, High Atlas	Morocco	+		3 MCK; 2 MOM
10	Torviczon, Andalucia	Spain	+	+	4 ME1; 13 MOM
11	Asturias	Spain	+		3 MCK; 2ME1; 2 MOM
12	Cap des Mèdes, Porquerolles, Var	France	+		1 MCK; 3 ME1; 18 MOM
13	La Repentence, Porquerolles, Var	France	+		1 MCK; 10 MOM
14	Mont Boron, Nice, Alpes Maritimes	France	+	+	4 MCK; 2 ME1; 16 MOM
15	*Ostricone*, Corsica	France	+	+	2 MCK; 9 MOM
16	Ogliastro, Corsica	France	+	+	10 ME1
17	Filitosa, Corsica	France	+	+	2 MCK; 9 MOM
18	Bonifacio, Corsica	France	+	+	5 MOM
19	Messine, Sicily	Italy	+	+	17 MOM
20	Ali, Sicily	Italy	+	+	2 ME1; 10 MOM
21	Corte, Corsica	France		+	8 ME1
22	Sarrola Carcopino, Corsica	France		+	5 ME1; 5 MOM
23	Lama, Corsica	France		+	3 ME1; 7 MOM
24	Oletta, Corsica	France	+	+	1 MOM
25	Casta, Corsica	France	+	+	1 MOM
26	Pigna, Corsica	France	+	+	1 MOM
27	Ogliastrello 2, Corsica	France	+	+	1 MOM

[a] Locations with an asterisk are from the eastern Mediterranean Basin.
[b] All populations were tested for mitotypes; populations tested for RAPDs or SSRs are marked with a "+" in the appropriate column.
[c] Number of individuals with each mitotype (MCK, ME1, MOM).

geographic barriers to geneflow, namely the Adriatic Sea and the Libyan and Egyptian deserts.

Reference cultivars (#1–102 in Table 11.2) and 19 cultivated forms of undetermined cultivar affiliation (#103–121) that had been initially sampled for RAPD variation (Besnard et al. 2001b) were included in this study. Cultivars were chosen to represent maximum diversity in olive. We added to the dataset additional domesticated types from Corsica (#122–127) because of the potential importance of this island in early human migrations. One or more individuals were tested per cultivar. In all, 122 cultivars were tested for nuclear RAPDs, 40 were tested for chloroplastic SSRs, and 127 were tested for mitotypes (Table11.2). For geographic analyses, cultivars were assigned to country or region based on the origin of the collection or as inferred from the cultivar's name (Table 11.2).

One to two samples each of three other *Olea europaea* subspecies (ssp. *laperrinei*, *maroccana*, and *guanchica*) were also examined to help clarify the origins of ssp. *europaea*.

Molecular Data

Total DNA preparation was performed as previously described in Besnard and Bervillé (2000). The RAPD amplification and electrophoresis procedures are described by Quillet et al. (1995). Eight RAPD primers were used: A1, A2, A9, A10, C9, C15, E15, and O8. Chloroplast SSR procedures followed those described previously by the authors (Besnard et al. 2002c), with primers (OeUA-DCA 01, 03–09, 11, 13–15, and 18) used according to Sefc et al. (2000). Individuals were characterized for mitochondrial DNA polymorphism using the RFLP methods described in Besnard and Bervillé (2000).

Data Analyses

In this chapter we present the geographic distributions of mitotypes and a discussion of a previously published dendrogram based on the RAPD data (Besnard et al. 2001a, 2001d). More detailed analyses of the mitotype and RAPD datasets are given in Besnard et al. (2001b, 2001c, 2001d) and Besnard and Bervillé (2000).

TABLE 11.2
List of the Cultivars Studied[a]

#	Cultivar or Landrace Name	RAPD	Mt DNA	SSR	#	Cultivar or Landrace Name	RAPD	Mt DNA	SSR	#	Cultivar or Landrace Name	RAPD	Mt DNA	SSR
1	Kalamon	+	+		44	Cailletier	+	+	+	87	Picual	+	+	+
2	Vallanolia	+	+		45	Cayon	+	+		88	Villalonga	+	+	
3	Gaïdourolia	+	+		46	Salonenque	+	+		89	Manzanilla	+	+	
4	Koroneiki	+	+		47	Verdanel	+	+		90	Sevillenca	+	+	
5	Carolia	+	+		48	Poumal	+	+		91	Galega	+	+	
6	Amygdalolia	+	+		49	Redouneil	+	+		92	Chemlal (Cordoba)	+	+	+
7	Uslu	+	+		50	Négrette	+	+	+	93	Chemlal	+	+	
8	Domat	+	+		51	Noirette	+	+		94	Chemlal Mechtrass	+	+	
9	Ayvalik	+	+		52	Grapié	+	+		95	Azeradj	+	+	
10	Sofralik	+	+		53	Aglandau	+	+		96	Taksrit	+	+	
11	Souri	+	+	+	54	Celounen	+	+		97	Chetoui	+	+	
12	Souri Mansi	+	+		55	Courbeil	+	+		98	Zarazi	+	+	+
13	Nabali Mohassen	+	+		56	Coucourelle	+	+	+	99	Meski	+	+	
14	Barnea	+	+		57	Cayet Rouge	+	+	+	100	Barouni	+	+	
15	Kaissy	+	+		58	Rascasset	+	+	+	101	Chemlali	+	+	
16	Zaity	+	+	+	59	Malaussena	+	+		102	Picholine Marocaine	+	+	+
17	Merhavia	+	+		60	Aubenc	+	+		103	Palmyre 1 (Syria)	+	+	
18	Toffahi	+	+		61	Reymet	+	+		104	Palmyre 2 (Syria)	+	+	
19	Oblica	+	+		62	Curnet	+	+	+	105	Palmyre 3 (Syria)	+	+	
20	Ascolana Tenera	+	+	+	63	Colombale	+	+		106	Palmyre 4 (Syria)	+	+	
21	Pendolino	+	+		64	Poulo	+	+		107	Palmyre 5 (Syria)	+	+	
22	Frantoio	+	+	+	65	Verdale de l'Hérault	+	+		108	Palmyre 6 (Syria)	+	+	
23	Giarraffa	+	+	+	66	Amellau	+	+	+	109	Palmyre 7 (Syria)	+	+	
24	Nocellara del Belice	+	+		67	Corniale	+	+	+	110	Palmyre 8 (Syria)	+	+	
25	Dolce Agogia	+	+	+	68	Rougette de Pignan	+	+		111	Palmyre 9 (Syria)	+	+	
26	Leccino	+	+	+	69	Vermillau	+	+		112	Derkouch 1 (Syria)	+	+	
27	San Felice	+	+		70	Verdelé	+	+		113	Derkouch 2 (Syria)	+	+	
28	Moraiolo	+	+	+	71	Dorée	+	+		114	Derkouch 3 (Syria)	+	+	
29	Cassanese	+	+	+	72	Pigale	+	+		115	Beth Hakerem (Israel)	+	+	
30	Leucocarpa	+	+		73	Picholine Rochefort	+	+		116	Alaon 1 (Spain)	+	+	
31	Zaituna	+	+		74	Tanche	+	+		117	Alaon 2 (Spain)	+	+	
32	Santagatese	+	+		75	Sauzin	+	+	+	118	Montpellier 1 (France)	+	+	
33	Tonda Iblea	+	+	+	76	Blanquetier de Nice	+	+		119	Monpellier 2 (France)	+	+	
34	Biancolilla	+			77	Grossane	+	+	+	120	Monpellier 3 (France)	+	+	
35	Passalunara	+			78	Filayre Rouge	+	+	+	121	Montpellier 4 (France)	+	+	
36	Moresca	+	+	+	79	Sabina (St Giuliano)	+	+	+	122	Corse S (France)		+	+
37	Ogliarola Messinese	+	+		80	Sabina 300	+	+	+	123	Corse G (France)		+	+
38	Nocellara Etnea	+	+	+	81	Zinzala 302	+	+	+	124	Zinzala Sud (France)		+	+
39	Picholine	+	+	+	82	Capanacce	+	+	+	125	Filitosa 12 (France)		+	+
40	Lucques	+			83	Cornicabra	+	+	+	126	Bonifacio 9 (France)		+	+
41	Bouteillan	+			84	Lechin de Sevilla	+	+		127	Bonifacio (France)		+	+
42	Olivière	+	+	+	85	Arbequina	+	+						
43	Blanquetier d'Antibes	+	+		86	Empeltre	+	+						

[a] Country or region of origin: Greece (#1–6), Turkey (7–10), Near East (11–17, 103–115), Egypt (18), Yugoslavia (19), Italy (20–38), continental France (39–78, 118–121), Corsica (79–82, 122–127), Spain (83–90, 116–117), Portugal (91), Algeria (92–96), Tunisia (97–101), Morocco (102). Cultivated forms with unknown names (#103–121) are named with reference to the original locality from which they were collected. A "+" in a column means that the cultivar was tested with the corresponding marker.

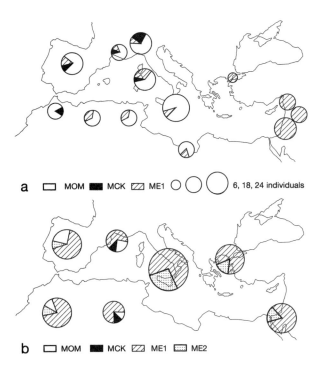

FIGURE 11.2 Map of oleaster and cultivar mitotypes in the Mediterranean Basin. (a) Distribution of mitotypes in oleasters. From small to large, the three circle sizes represent 6, 18, and 24 sampled individuals, respectively. (b) Regional distribution of mitotypes in cultivars.

From the SSR dataset, the centroid values of Nei's genetic distances (Nei 1978) were used to create a dendrogram of oleaster populations. Clustering was performed using Ward's algorithm (Ward 1963). Other analyses of the SSR data included multiple correspondence analyses (MCA), which were performed on two subsets of the SSR dataset—all individuals and the cultivars only.

Results

Oleasters

Distinct mitotypes were revealed in ssp. *laperrinei* (mitotype = ME1), ssp. *guanchica* (MCE), and ssp. *maroccana* (MMA). Three mitotypes were found in oleaster (ssp. *europaea* var. *sylvestris*)—MCK, ME1, and MOM. ME1 was found in all eastern Mediterranean Basin populations and in about half of the western populations (Figure 11.2a; Table 11.1). MOM and MCK were found exclusively in the West, specifically in northwestern Africa and western Europe. Eastern oleaster populations typically appeared homogeneous, possessing only ME1. In contrast, western populations were often heterogeneous. Most western populations possessed MOM; almost half of those populations also possessed MCK; and three populations (#11, 12, and 14) possessed all three mitotypes (Figure 11.2a; Table 11.1). Among the 13 populations from Corsica and Sicily, 11 possessed MOM, 2 possessed MCK, and 5 possessed ME1 (Table 11.1).

RAPDs and SSRs also revealed an "East vs. West" structure in oleaster diversity. We found a gradient between the east and the west in the frequencies of some RAPD and SSR markers (Besnard et al. 2001b; Table 11.3). Also, some markers were unique to the east, while others were unique to the west (Table 11.3).

In the SSR-based dendrogram of oleaster populations, we obtained three main groupings corresponding to the eastern Mediterranean, the western Mediterranean, and Sicily and Corsica (Figure 11.3). In the plot produced by the MCA of all individuals (cultivars and oleasters), we found that eastern (e.g., #1, from Turkey) and centrally located (#5, from Libya) populations clustered near the top and right side of the plot (Figures 11.4a, c, and e); Sicilian (#20), Corsican (#17), and western populations clustered on the left side of the plot (Figures 11.4b, d, and e); and the cultivars clustered primarily in the lower right corner of the plot (Figure 11.4e).

Olive Cultivars

In cultivars, four mitotypes were revealed, including ME2, which was not found in oleaster (Figure 11.2b). Most of the cultivars possessed ME1 (74%), whereas 10%, 8%, and 8% possessed ME2, MOM, and MCK, respectively. Correlation between geography and genetic structure appeared weaker for cultivars than it did for oleasters. For example, ME1 was found in cultivars from throughout the Mediterranean Basin. However, most cultivars containing MOM and MCK were from western regions (Figure 11.2b), whereas ME2 was most frequent in eastern countries (Besnard et al. 2001d).

A subset of the RAPD bands found in oleasters was found in olive cultivars (Table 11.3). A previously constructed dendrogram based on RAPD data placed cultivars into 24 small groups within two major groupings (Figure 2B in Besnard et al. 2001d). Although the major groupings did not

TABLE 11.3
Distribution of Molecular Markers in Oleasters and Olive Cultivars

Type of Marker	Oleasters				Cultivars
	Number Observed	Total	Specific to East[a]	Specific to West[a]	
Mitotypes[b]	4	3	1	2	4
RAPD Bands[c]	57	57	6	12	45
SSR Alleles[d]	173	167	12	33	99

[a] "West" and "East" refer to the Mediterranean Basin.
[b] Besnard et al. (2002b).
[c] Besnard et al. (2001a, 2001c).
[d] Breton & Bervillé unpublished data.

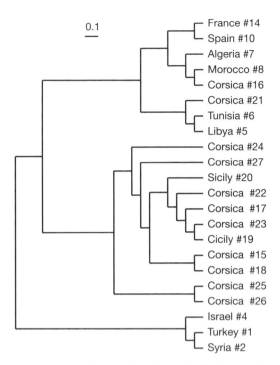

FIGURE 11.3 Dendrogram based on the SSR dataset for oleasters. Population codes are in Table 11.1. The bar indicates scale.

have a clear geographic structure, the smaller groups corresponded partially to country or region of origin. The larger of the two major groupings contained cultivars from all countries and included all four mitotypes. The minor group, which also included all four mitotypes, consisted primarily of cultivars from the West, including Corsican cultivars such as "Sabina" (#79 and 80) and "Zinzala" (#81). The only cultivar in this group presumed to be from the East was "Koroneiki" (#4) from the island of Crete, in Greece. However, a cultivar's name may not be an accurate indicator of the cultivar's region of origin.

The SSR-based MCA plot of cultivars placed all cultivars, except for most of those from Corsica, together on the left side of the plot (Figure 11.4f).

Comparison of Oleaster and Olive

Genetic diversity with respect to RAPDs and SSRs was found to be higher in oleasters than in the cultivars (Table 11.3). However, the cultivars possessed all three oleaster mitotypes, plus a unique mitotype: ME2.

Looking across both cultivars and oleasters, we see that ME1 and ME2 are characteristic of the eastern Mediterranean, and MOM and MCK are essentially western mitotypes (Figures 11.2a and b). Some RAPD and SSR markers are also region-specific (Table 11.3). For example, C15-1200 is an eastern-specific RAPD band, whereas A1-1080 is a western-specific RAPD band. Some cultivars possess western-specific oleaster RAPD bands as well as western-specific mitotypes, suggesting that these cultivars were selected directly from oleasters in the western Mediterranean.

In the SSR-based MCA plot of all individuals, most cultivars were separated from oleasters, the former being restricted to the lower right corner of the plot (Figure 11.4e). Thus, the SSR alleles are fairly good indicators of domestication status. However, many of those Corsican cultivars that clustered distinctly in Figure 11.4f clustered among the western oleasters in Figure 11.4e. These cultivars include "Zinzala" (#81 and 124) and "Sabina" (#80), which also share uncommon western-type RAPD and SSR markers, as well as MOM, with two oleaster populations (#25 and 26). Thus, these cultivars may have been directly selected from these populations. Other oleaster trees that clustered near to cultivars in Figure 11.4e (in the lower right corner) may represent escapes from cultivation. Such oleasters typically possessed eastern-specific RAPD and mitotype markers, as well as SSR alleles typical of the cultivars.

Discussion: Molecular Diversity

The Ancestor(s) of Oleaster

In addition to the SSR data presented here, ongoing research on chloroplastic genetic variation has recently included the RFLP technique, which produced chlorotype data (Besnard et al. 2002c) similar to the mitotype data described in this chapter. The chlorotype data indicate that *O. e. europaea* possesses chlorotypes from two distinct lineages, and therefore one of the lineages may be of hybrid origin. One maternal lineage is probably *O. e. laperrinei*, which not only shares a chlorotype with ssp. *europaea*, but also shares the mitotype ME1.

The oleaster mitotypes MOM and MCK were not found in the other subspecies. However, ssp. *maroccana* possessed MMA, which is similar to MCK; perhaps the latter mitotype evolved from the former (Médail et al. 2001). The origin of MOM is still unclear. It could have evolved within oleaster or it could have been transferred from another, possibly lost, infraspecific taxon of *Olea europaea*.

Oleaster Diversity

Genetic diversity in oleaster, as indicated by RAPDs, SSRs, and mitotypes, appears highly correlated with geography, with a clear differentiation between populations in the eastern Mediterranean and those in the West. This disjunction suggests that oleasters survived in separate refugia during Quaternary glaciations. Three of these refugia are characterized by specific mitotypes: (1) Northwestern Africa, where MCK was preserved; (2) The western Mediterranean, where MOM was preserved; and (3) the Near East, where ME1 was preserved. The relatively ubiquitous ME1 may have been preserved in one or more of the western refugia as well. RAPD and SSR data indicate that there could be a fourth refuge zone encompassing Corsica and Sicily, as discussed in the next section. RAPD variation suggests that

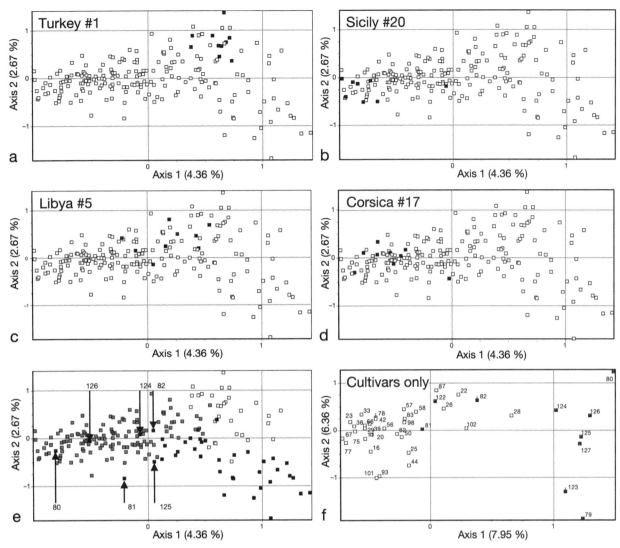

FIGURE 11.4 Results from multiple correspondence analyses based on SSR data. Analysis of 214 individuals representing oleaster populations and olive cultivars. (a) Oleaster population #1 (from Turkey), highlighted in black. (b) Oleaster population #20 (Sicily), highlighted in black. (c) Oleaster population #5 (Libya), highlighted in black. (d) Oleaster population #17 (Corsica), highlighted in black. (e) Western oleasters in gray, eastern oleasters in white, and cultivars in black. Arrows indicate the position of the Corsican cultivars, referenced by the cultivar codes in Table 11.2. (f) Cultivars only. Corsican cultivars are shown in black. The numbers correspond to the cultivar codes in Table 11.2.

only weak geneflow has occurred among these refugia since the glaciations.

Genetic diversity is higher in the West than in the East, suggesting that oleaster diversified in the West. We believe the data best support the hypothesis that oleaster arose and diversified in southern Morocco, where it could have been in contact with ssp. *maroccana* (and hence received a MMA/MCK mitotype relative), and later migrated eastward, with only the ME1 mitotype arriving in the Near East (Figure 11.5a).

Corsican Oleasters and Cultivars

Near the end of the Quaternary glaciations, Corsica was in contact with continental Italy, Sardinia, and Sicily (Vigne 1992). Sicily has been proposed as a refuge for other tree species during this period (Quézel 1995). Migration of endemic oleasters from Sicily to Corsica could have occurred early during Quaternary recolonization, when sea levels were still low (Taberlet at al. 1998). Our RAPD and SSR data indicate distinct genetic profiles for and close relationships among some Corsican oleasters and cultivars (e.g., "Sabina" and "Zinzala"). A recent RAPD study performed by Bronzini de Caraffa et al. (2002) on a larger sample of wild populations from Corsica and Sicily revealed a similar association among oleasters and cultivars of this region. Consequently, it appears that some olive cultivars may have been selected directly from ancient oleasters indigenous to these islands. In fact, this region may have been a separate center of olive domestication, as suggested by archaeological evidence of olive domestication by 5000 BP in Corsica (Magdelaine and Ottaviani 1984).

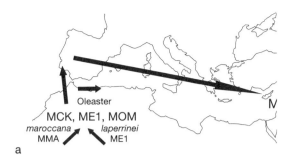

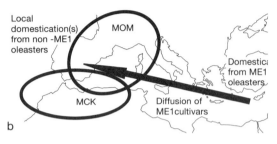

FIGURE 11.5 Maps of possible routes followed by oleasters and cultivars in the Mediterranean Basin. (a) Possible routes (arrows) followed by oleaster mitotypes from the West to the eastern Mediterranean Basin. (b) Westward diffusion of cultivars from the primary center of olive domestication in the East.

The genetic distinctiveness of oleasters in Corsica was probably enhanced by the isolation of the populations on an island (Bronzini de Caraffa et al. 2002). In time, this isolation would have been diminished by human migrations as cultivars were exchanged throughout the Mediterranean. Historically, Corsica has been invaded by Phoenician, Roman, Genoese, Italian, and French peoples. Thus, its importance as a migratory crossroads as well as its intermediate position in the Mediterranean Basin could account for the high levels of genetic diversity observed on this island (Bronzini de Caraffa et al. 2002).

Cultivar Diversity

Most cultivars are clearly of eastern oleaster origin, possessing eastern-type RAPDs and mitotypes. These cultivars occurred in the larger of two branches in the dendrogram based on the RAPD data. The smaller branch contained some Corsican cultivars that were also revealed as distinct by the SSR data (Figure 11.4f). The genetic profile of these cultivars—with oleaster-type SSRs and western-type RAPDs and mitotypes—suggests that they were selected directly from western oleasters.

Although cultivars possessing only western-type RAPDs and mitotypes were most likely selected from western oleasters, cultivars with western-type RAPDs, but also containing the ubiquitous ME1, were probably also selected in the West. ME1 could have been preserved in western as well as eastern oleaster refugia during the Quaternary. Or, the presence of ME1 in western cultivars could be the result of geneflow with eastern cultivars that were carried by humans from the eastern Mediterranean to the West (Figure 11.5b).

The Advantage of Using Multiple Molecular Markers for Studying Domestication

Not all genetic markers tell the same story. Nuclear DNA, which exists as two copies contributed to progeny by both parents, undergoes recombination. In contrast, the cytoplasmic genomes (i.e., of chloroplasts and mitochondria), which are passed on to progeny by a single parent (usually the maternal parent), do not experience this type of gene shuffling. Nuclear DNA markers, such as the RAPDs examined in this study, often exhibit more variation at a particular taxon level than do cytoplasmic markers (e.g., chloroplastic SSRs and mitotypes).

Also, there are differences in specificity for the various marker types. Of the marker types used in this study, RAPD primers were the least specific in targeting DNA sequences, whereas the primers producing the RFLP-based mitotypes were the most specific. This means that one would expect to find less variation in a taxon for the mitotypes than for the SSRs or RAPDs.

The different rates of evolution for these various markers in the three genomes allow us to reconstruct an evolutionary history for a taxon at multiple levels—for example, between the taxon and its closest relatives, and within the taxon at various levels. For olive *(O. e. europaea)*, we were able to distinguish: (1) most cultivated (var. europaea) from wild (var. *sylvestris*) germplasm with SSRs; (2) a geographic divergence of East vs. West within the subspecies with RAPDs, and to some extent mitotypes; and (3) the relationships with other subspecies and the geographic origins of ssp. *europaea* with mitotypes. Consequently, we could piece together a history for olive that would not have been possible with just one type of marker.

Also, with the combined data we can now more clearly distinguish wild from feral oleasters, the latter typically lacking oleaster-type SSR and RAPD markers (Table 11.4). The feral forms include escapes from cultivation, as well as hybrids between cultivars and wild oleasters.

Summary

Oleasters originated and spread in the western Mediterranean, becoming more diversified (e.g., obtaining MOM and a MMA/MCK relative) as a result of contact with other subspecies (e.g., *O. e. laperrinei* and *O. e. maroccana*). From a western center of diversity, one lineage of oleasters, which possessed ME1, spread to the Near East. Thus, by the time of the Quaternary glaciations, several mitotypes were scattered throughout the Mediterranean Basin. Some of these genetically differentiated populations survived in refugia until the end of the glaciations. Subsequent expansion of oleasters was probably limited by natural geographic barriers (e.g.,

TABLE 11.4
Key for Prediction of the Wild versus Feral Status of Any Oleaster Tree[a]

Information on the Tree			Predicted Status	Observed
Location	Mitotype	RAPD Profile and Specific Markers	Wild/Feral	Yes/No
East	ME1	East, with oleaster-specific markers	Wild	Yes
East	ME1	East, no oleaster-specific markers	Feral[b]	Yes
East	ME1	East-West, with or without oleaster-specific markers	Feral[c]	Yes
East	ME2	—	—	No
West	MOM,MCK	West, with oleaster-specific markers	Wild	Yes
West	MOM,MCK	West, no oleaster-specific markers	Feral[d]	Yes
West	MOM,MCK	East-West, with or without oleaster-specific markers	Feral[c]	Yes
West	ME1	East, no oleaster-specific markers	Feral[b]	Yes
West	ME1	West, with oleaster-specific markers	Wild	Yes
West	ME2	—	—	No

[a] Prediction based on the location of the tree, its mitotype, its RAPD profile (East vs. West), and its specific SSR and RAPD markers (i.e., whether or not it possesses markers specific to oleasters).
[b] Escaped from eastern cultivars.
[c] Hybrids between cultivars and oleasters.
[d] Escaped from western cultivars.

the Adriatic Sea, and the Libyan and Egyptian deserts), which served to maintain a strong, geographically based genetic structure in oleasters until human migrations.

Olive domestication occurred in the Near East, where it is well documented by 6000 BP (Zohary and Spiegel-Roy 1975). The first cultivars, which possessed ME1, were carried by humans to western portions of the Mediterranean Basin, where geneflow increased the diversity in both western oleasters and in western cultivar selections. Another source of genetic diversity in cultivars may stem from additional centers of olive domestication from genetically differentiated oleasters. Recently, archaeological evidence has been found of olive domestication by 6000 BP in Spain (Terral 2000) and by 5000 BP in Corsica (Magdelaine and Ottaviani 1984). Consequently, we think that at least two domestications occurred more or less simultaneously on both sides of the Mediterranean Basin.

The mitochondrial DNA polymorphisms, in combination with chloroplast SSRs and nuclear RAPDs, were an excellent set of molecular markers with which to unravel olive phylogeography.

Future Research Needs

To determine exactly when and where olive cultivars originated and spread in the Mediterranean Basin will require combined data from archaeology, genetics, history, and palynology. The present diversity and distribution of cultivars have undoubtedly been perturbed by the movement of seeds and cuttings by humans who practiced coastal trade around the Mediterranean Sea during both prehistoric and historic times. Knowing cultivar age is important for establishing genealogical relationships among these cultivars. We plan, therefore, to establish a chronology of mitotype movements, using seeds found in archaeological sites with known dates. Such seeds are usually found charred or in amphora, conditions leading to highly altered nuclear DNA that is difficult to amplify. However, we believe that mitochondrial DNA, which is much more abundant in remains than nuclear DNA, could probably be amplified and compared.

Various herbarium samples of olive have been collected during the two last centuries. Some of those from eastern Africa were collected by Italian, English, and French botanists, who each named their specimens independently. Consequently, a review of these specimens could reveal cultivar synonymies. Fruit color for many specimens may have faded, however, making these determinations more difficult (Green 2002).

Acknowledgments

Thanks are due to P. Baradat, B. Khadari, and F. Dosba for stimulating discussions. This work was supported by EU contract FAIR CT950689. We would like to thank Dr. Deena S. Decker-Walters for her help in finalizing and improving our manuscript.

References

Amoureux, P.-J. 1784. *Traité de l'olivier*. Ed Société Royale des Sciences, Chez la Veuve Gontier, Montpellier: Libraire à la Loge.

Besnard, G. and A. Bervillé. 2000. Multiple origins of Mediterranean olive *(Olea europaea* L. ssp. *europaea)* based upon mitochondrial DNA polymorphisms. *Comptes Rendus de l'Académie des Sciences, Sciences de la vie / Life Science* 323: 173–181.

———. 2002. On chloroplast DNA variations in the olive (*Olea europaea* L.) complex: Comparison of RFLP and PCR polymorphisms. *Theoretical and Applied Genetics* 104: 1157–1163.

Besnard, G., B. Khadari, P. Villemur, and A. Bervillé. 2000. Cytoplasmic male sterility in the olive (*Olea europaea* L.): Phenotypic, genetic and molecular approaches. *Theoretical and Applied Genetics* 100: 1018–1024.

Besnard, G., P. Baradat, and A. Bervillé. 2001a. Genetic relationships in the olive (*Olea europaea* L.) reflect multilocal selection of cultivars. *Theoretical and Applied Genetics* 102: 251–258.

Besnard, G., P. Baradat, C. Breton, B. Khadari, and A. Bervillé. 2001b. Olive domestication from structure of oleasters and cultivars using RAPDs and mitochondrial RFLP. *Génétics, Sélection, Evolution* 33 (Suppl. 1): S251–S268.

Besnard, G., P. Baradat, D. Chevalier, A. Tagmount, and A. Bervillé. 2001c. Genetic differentiation in the olive complex (*Olea europaea*) revealed by RAPDs and RFLPs in the rRNA genes. *Genetic Resources and Crop Evolution* 48: 165–182.

Besnard, G., C. Breton, P. Baradat, B. Khadari, and A. Bervillé. 2001d. Cultivar identification in olive (*Olea europaea* L.) based on RAPD markers. *Journal of the American Society for Horticultural Science* 126: 668–675.

Besnard, G., P. S. Green, and A. Bervillé. 2002a. The genus *Olea*: Molecular approaches of its structure and relationships to other *Oleaceae*. *Acta Botanica Gallica* 149: 49–66.

Besnard, G., B. Khadari, P. Baradat, and A. Bervillé. 2002b. Combination of chloroplast and mitochondrial DNA polymorphisms to study cytoplasm genetic differentiation in the olive complex (*Olea europaea* L.). *Theoretical and Applied Genetics* 105: 139–144.

———. 2002c. *Olea europaea* phylogeography based on chloroplast DNA polymorphism. *Theoretical and Applied Genetics* 104: 1353–1361.

Brewer, S., R. Cheddadi, J. L. Beaulieu, and Data Contributors. 2002. The migration of deciduous *Quercus* throughout Europe since the last glacial period. *Forest Ecology and Management* 156: 27–48.

Bronzini de Caraffa, V., J. Maury, C. Gambotti, C. Breton, A. Bervillé, and J. Giannettini. 2002. Mitochondrial DNA variation and RAPD mark oleasters, olive and feral olive from western and eastern Mediterranean. *Theoretical and Applied Genetics* 104: 1209–1216.

Chevalier, A. 1948. L'origine de l'olivier cultivé et ses variations. *Revue Internationale de Botanique Appliquée et d'Agriculture Tropicale* 28: 1–25.

Estin, C. and H. Laporte. 1987. *Le livre de la mythologie Grecque et Romaine*. Paris, France: Ed Gallimard, Collection Découverte Cadet.

Galili, E., D. J. Stanley, J. Sharvit, and M. Weinstein-Evron. 1997. Evidence for earliest olive-oil production in submerged settlements off the Carmel Coast, Israel. *Journal of Archaeological Science* 24: 1141–1150.

Green, P. S. 2002. A revision of *Olea* L. (Oleaceae). *Kew Bulletin* 57: 91–140.

Green, P. S. and G. E. Wickens. 1989. The *Olea europaea* complex. In *The Davis & Hedge Festschrift*, K. Tan (ed.), pp. 287–299. Edinburgh: Edinburgh University Press.

Hess, J., J. W. Kadereit, and P. Vargas. 2000. The colonization history of *Olea europaea* L. in Macaronesia based on internal transcribed spacer 1 (ITS-1) sequences, randomly amplified polymorphic DNAs (RAPD), and inter simple sequence repeats (ISSR). *Molecular Ecology* 9: 857–868.

Kislev, M. E., D. Nadel, and I. Carmi. 1992. Epipalaeolithic (19,000 BP) cereal and fruit diet at Ohalo II, Sea of Galilee, Israel. *Review of Palaeobotany and Palynology* 73: 161–166.

Lumaret, R. and N. Ouazzani. 2001. Ancient wild olives in Mediterranean forests. *Nature* 413: 700.

Magdelaine, J. and J. C. Ottaviani. 1984. L'occupation pre et proto historique de l'Abri de Scaffa Piana près de Saint Florent. *Bulletin de la Société des Sciences Historiques et Naturelles de la Corse* 647: 39–48.

Médail, F., P. Quézel, G. Besnard, and B. Khadari. 2001. Systematics, ecology and phylogeographic significance of *Olea europaea* L. ssp. *maroccana* (Greuter & Burdet) P. Vargas et al., a relictual olive tree from South West Morocco. *Botanical Journal of the Linnean Society* 137: 249–266.

Neef, R. 1990. Introduction, development and environmental implications of olive culture: The evidence from Jordan. In *Man's role in the shaping of the eastern Mediterranean landscape*, S. Bottema, G. Entjes-Nieborg, and W. van Zeist (eds.), pp. 295–306. Rotterdam: A. A. Balkema.

Nei, M. 1978. Estimation of average heterozygosity and genetic distance from a small number of individuals. *Genetics* 89: 583–590.

Quézel, P. 1978. Analysis of the flora of Mediterranean and Saharan Africa. *Annals of the Missouri Botanical Garden* 65: 479–534.

———. 1995. La flore du bassin Méditerranéen: Origine, mise en place, endémisme. *Ecologia Mediterranea* 21: 19–39.

Quillet, M. C., N. Madjidian, Y. Griveau, H. Serieys, M. Tersac, M. Lorieux, and A. Bervillé. 1995. Mapping genetic factors controlling pollen viability in an interspecific cross in *Helianthus* sect. *Helianthus*. *Theoretical and Applied Genetics* 91: 1195–1202.

Sefc, K. M., M. S. Lopes, D. Mendoça, M. Rodrigues Dos Santos, M. Laimer da Câmara Machado, and A. da Câmara Machado 2000. Identification of microsatellites loci in olive (*Olea europaea*) and their characterization in Italian and Iberian trees. *Molecular Ecology* 9: 1171–1173.

Taberlet, P., L. Fumagalli, A. G. Wust-Saucy, and J. F. Cosson. 1998. Comparative phylogeography and postglacial colonization routes in Europe. *Molecular Ecology* 7: 453–464.

Terral, J. F. 2000. Exploitation and management of the olive tree during prehistoric times in Mediterranean France and Spain. *Journal of Archaeological Sciences* 27: 127–133.

Terral, J. F. and G. Arnold-Simard. 1996. Beginnings of olive cultivation in eastern Spain in relation to Holocene bioclimatic changes. *Quaternary Research* 46: 176–185.

Van Zeist, W. and H. Woldring. 1980. Holocene vegetation and climate of northwestern Syria. *Palaeohistoria* 22: 111–125.

Vargas, P., F. Muñoz Garmendia, J. Hess, and J. Kadereit. 2001. *Olea europaea* ssp. *guanchica* and ssp. *maroccana* (Oleaceae), two new names for olive tree relatives. *Anales Jardìn Botànico de Madrid* 58: 360–361.

Vigne, J.-D. 1992. Zooarcheological and biogeographical history of the mammals of Corsica and Sardinia since the last ice age. *Mammal Review* 22: 87–96.

Ward, J. H., Jr. 1963. Hierarchical grouping to optimize an objective function. *Journal of the American Statistical Association* 58: 236–244.

Wickens, G. E. 1976. The Flora of Jebel Marra (Sudan) and its geographical affinities. *Kew Bulletin, Additional Series* 5: 1–368.

Zohary, D. and P. Spiegel-Roy. 1975. Beginnings of fruit growing in the Old World. *Science* 187: 319–327.

CHAPTER 12

Origins of Polyploid Crops
The Example of the Octoploid Tuber Crop *Oxalis tuberosa*

EVE EMSHWILLER

Overview

Although most animals and many plants are diploid, with two sets of chromosomes, a large proportion of plants are polyploid, with three or more sets of chromosomes. Polyploidy is a feature of many crop species (Zeven 1980), and it can add additional complexity to the challenges of identifying the wild ancestors of a domesticate (Doebley 1992). Most polyploids are of hybrid origin. Thus, the study of a polyploid crop's origins includes not only identifying the wild ancestors of the cultigen, but also determining (1) the number of distinct wild progenitors that contributed their genomes to the polyploid crop, (2) their identities, (3) whether multiple levels of ploidy exist in the crop and its wild relatives, (4) whether polyploidization occurred before or after domestication, and (5) whether the polyploid arose once or in multiple polyploidization events.

This chapter focuses on domestication in polyploid crops, using the tuber crop "oca," *Oxalis tuberosa* Molina (Figure 12.1), as a case study. Oca is cultivated primarily in the central Andean region of South America and plays an important role in the diet and traditional farming systems of rural highland communities. The study of the origins of polyploidy in *O. tuberosa* and the relationships of the crop to its wild allies illustrates techniques that can be used to study polyploid origins. I focus primarily on the use of DNA sequence data to determine the number and identities of wild taxa that contributed genomes to octoploid oca. In addition, I briefly review another source of data utilized in this research: the use of flow cytometry to survey ploidy levels among wild *Oxalis* species related to the cultigen.

Terminology of Polyploidy

Various terms are used to distinguish among kinds of polyploids. At the simplest level, polyploids are classified according to the number of genomes (sets of chromosomes) in each cell and are called triploids (with three sets, designated $3x$), tetraploids ($4x$), pentaploids ($5x$), hexaploids ($6x$), heptaploids ($7x$), octoploids ($8x$), etc. Most odd-number polyploids (e.g., $3x$, $5x$) are sterile and persist over time only if they are capable of some form of asexual reproduction, such as vegetative propagation (Allard 1960; Grant 1971). Thus, crops such as native Andean potatoes, which are propagated vegetatively, exist at odd ($3x$ and $5x$) as well as even ($2x$ and $4x$) ploidy levels, whereas most seed-propagated crops—such as tetraploid cotton and hexaploid wheat—have even ploidy numbers.

Polyploids are also distinguished as either autopolyploids or allopolyploids (Grant 1971). Typical autopolyploids are derived from a single species (even a single individual, in the case of artificial polyploids), whereas typical allopolyploids (e.g., cotton, tobacco, and artificially produced polyploids such as triticale) are derived from hybridization among distinct species. These taxonomic differences are usually reflected in the behavior of the chromosomes in the polyploid. The chromosomes in typical allopolyploids pair two-by-two as they do in diploids, joining exclusively with the homologous chromosome from the same ancestral genome. On the other hand, the multiple homologous chromosomes in typical autopolyploids may pair at random or may join together abnormally at meiosis, forming some groups of more than two chromosomes while leaving other chromosomes unpaired (Jackson and Casey 1982).

The genome complements of polyploids are sometimes designated with letters to indicate their numbers and levels of divergence, such as AAAA for an autotetraploid and AABB for an allotetraploid. However, allopolyploidy and autopolyploidy really denote parts of a continuum of levels of divergence of their ancestral genomes. Polyploids with ancestors from closely related species or divergent populations within the same species may be referred to as segmental allopolyploids (Stebbins 1947, 1950), indicating the intermediate level of divergence of the parental genomes (designated AAA'A'). Autoallopolyploids are higher-level allopolyploids (above $4x$) that also combine aspects of autopolyploidy, in that at least one of the parental genomes is represented more than twice. For instance, the sweet potato is a hexaploid that appears to be autoallopolyploid, with a genome complement of AAAABB (Buteler et al. 1999). The random pairing of chromosomes associated with autopolyploidy has genetic consequences that are beyond the scope of this chapter. However, the effects on segregation of genes and genetic markers in progeny can affect the choice of appropriate markers, analytical methods, and the interpretation of results when studying the origins of a polyploid.

Prior Research

When I began my research on oca as a dissertation project (Emshwiller 1999a), under the direction of Jeff Doyle, the amount of prior information on oca on which to base a

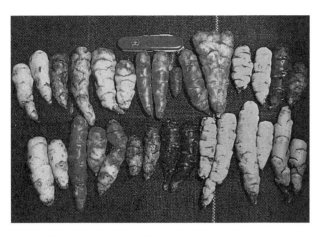

FIGURE 12.1 Diversity of tuber morphology and pigmentation in *Oxalis tuberosa* cultivated by a single household in the Campesino community of Viacha, Pisac District, Cusco Department, in southern Peru.

study of its origins was very limited. Like many crops that are currently important only in a limited region, oca has not been the subject of as much research as some of the crops that have worldwide importance. Because some areas of the Andean region have been little explored by botanists, the wild relatives of the crop are poorly known. Some wild allies are known from only a single population, while others are still being discovered. Archaeological evidence is also scanty, as it is for most tuber crops, which tend to be poorly preserved in comparison with other kinds of remains.

Speculations Concerning the Habitat of Oca's Domestication

Perhaps because of the paucity of empirical evidence, there was much speculation about the area of domestication of oca and the other "minor" Andean tubers. While agreeing that oca's origins would be found in the Andean region, various speculations differed with respect to the kind of environment in which the crop originated. Vavilov (1992) and Hawkes (1983) suggested that the Andean environment had the proper conditions for the evolution of tubers, especially in having seasonally dry environments that favor plants with storage organs (Hawkes 1983). The hypothesis of an origin in high Andean valleys, perhaps in southern Peru, was favored by Hodge (1949, 1951) and Brücher (1969), whereas the high levels of morphological variability in cultivated oca in Bolivia and southern Peru led Cook (1925) and León (1967) to specify that oca was domesticated on the Altiplano near Lake Titicaca. In contrast, Sauer (1952) and Harris (1969) suggested that the ecotone found on the steep eastern Andean slopes, between the cloud forest and the *puna* (high Andean steppes), would favor the domestication of root and tuber crops (vegeculture). In apparent agreement, Debouck and Libreros Ferla (1995) included oca among crops that have wild relatives in neotropical montane forests, although they pointed out the lack of information on its origins.

Archaeology

Reports of preserved remains of oca tubers in archaeological contexts are very few, as tubers tend to be poorly preserved and difficult to identify from small fragments. There are also some representations of oca in pre-Colombian art. However, most such artworks were presumably made long after the time of a crop's domestication, so they can provide only an indication of the crop's importance and minimum area of diffusion at a particular time period, not pinpoint the time or place of domestication. The archaeological evidence of oca ranges from Colombia to northwestern Argentina, and from a putative date of 10,000 years ago to after the European arrival (Table 12.1). However, the reported ages of oca are not direct dates from accelerated mass spectrometry (AMS) of the tuber fragments, but rather dates of the surrounding strata. The small number of reports and the lack of direct dating means that very little can be concluded about oca's domestication from these data, other than that there is nothing to suggest an origin outside the Andean region.

What might be the oldest remains of oca yet found are those from Guitarrero Cave, in the central highlands of Peru (Smith 1980). This site was excavated before the development of modern archaeobotanical techniques such as flotation and AMS dating, but reanalysis of some of the remains indicates that the principal use of the cave occurred between 9500 and 10,000 years ago (Lynch et al. 1985). Tubers identified as *Oxalis* were found in all strata, including the oldest (Smith 1980), but the oca remains themselves have not yet been reanalyzed by AMS. This is important, because results of direct AMS dating have shown that beans from Guitarrero Cave are not as old as previously thought (Kaplan and Lynch 1999). Although the morphology of starch grains of oca and other Andean tubers can help with identification of their remains (Martins 1976; Cortella and Pochettino 1995), Smith (1980) found few starch grains in tuber remains from Guitarrero Cave, and those that were present were not diagnostic. The plant materials found in the early strata, if their age and identity were confirmed, would show that oca was already established as an important element of the diet over 9000 years ago (see Lynch 1980), regardless of whether or not it was domesticated. The tantalizing suggestion that oca may be among the oldest of cultivated plants in the Andean region should be more rigorously tested by confirming the identities and ages of these remains, as well as their cultivated status.

Unfortunately, other reports of oca in the archaeological record are very few (Table 12.1), especially in older strata, and none of these other remains were directly dated. Reports of older remains of oca include a few tubers dated from about 4000 to 3800 years ago in the Central Coast of Peru (Martins 1976), and a single calcified oca tuber found in strata in Colombia was dated in different publications at about either 3860 or 2725 years old (Correal Urrego 1990a, 1990b). Remains from later prehistoric sites are also sparse. With the exception of some oca remains from about 1270 years ago

TABLE 12.1
Archaeological Reports of Oca Tuber Remains or Representations in Art

Site and Evidence[a]	Region	Reported Age	Reference
Guitarrero Cave in Callejón de Huayllas (T)	Central highlands of Peru	Up to 9000–10,000 BP	Smith 1980
Punta Grande in Ancón-Chillón region (T)	Central coast of Peru	3800–4000 BP	Martins 1976
Sabana de Bogotá (T, calcified)	Central Colombia	~3860 or 2725 BP	Correal Urrego 1990a, 1990b
Huachichocana in Valliserrana region (T)	Northwestern Argentina	~1270 BP	Pearsall 1992
Putaca in Tarma area (T)	Central highlands of Peru	Middle Horizon (i.e., ~1400–1000 BP)	Hastorf 1983; Hastings 1986 cited in Hastorf 1993
Pacheco (A, painted pottery)	Coastal Peru	Middle Horizon (i.e., ~1400–1000 BP)	Yácovleff and Herrera 1934
Jauja region in Mantaro valley (T)	Central highlands of Peru	Late Horizon (Inca period) ~530–470 BP	Hastorf 1983; D'Altroy and Hastorf 1984
Pachacamac (T)	Coastal Peru	Inca period (~530–470 BP)	Towle 1961
Cusco (A, painted wooden *qero* vessel)	Southern highlands of Peru	Post-conquest (17th century)	Vargas 1981; Flores Ochoa et al., 1998

[a] (T) indicates tuber remains; (A) indicates representational art and is followed by the kind of item.

in northwestern Argentina (Pearsall 1992), the remaining reports are all from Peru and dated to the Middle Horizon (i.e., about 1400–1000 BP) or younger (see Table 12.1).

Taxonomy

Another challenge for this project was the confused taxonomy of *Oxalis*. The large genus *Oxalis* is found worldwide, with the greatest diversity of forms occurring in South America. Morphological diversity in *Oxalis* can be bewildering, as it is in *Manihot* (Olsen and Schaal, Chapter 9 of this volume), with phenotypic plasticity and suspected cases of hybridization that impede efforts to delimit species. Until the publication in 2000 of the treatment of *Oxalis* by Lourteig, the traditional taxonomy of Knuth (e.g., 1914, 1930, 1931, 1935, 1936, 1937, 1940) was the only work that covered the entire genus, including the species related to oca. However, Knuth's taxonomy is widely regarded as being quite artificial, since the 38 sections into which he classified the more than 800 species of *Oxalis* are based on a few vegetative characters. Lourteig (2000) has greatly improved the infrageneric taxonomy, and her work has turned out to be much more congruent with the cytological and molecular data, albeit with some conflicts. However, her classification was not yet available when my work began.

Although she did not explicitly address the origin of domesticated oca, Lourteig (2000) reduced the species name *O. unduavensis* (Rusby) Knuth to a subspecies of *O. tuberosa*. This taxonomic change implies that she considered *O. tuberosa* ssp. *unduavensis* (Rusby) Lourt. to be oca's closest wild relative, perhaps also implying that she considered it to be oca's progenitor. However, my observations of living material of *O. unduavensis* in the field and cultivated in common garden-greenhouse conditions lead me to uphold it as a species distinct from *O. tuberosa*. In addition to having morphological differences in leaflet shape, number of flowers in an inflorescence, and longer flexible stems, as acknowledged by Lourteig (2000), *O. unduavensis* differs from *O. tuberosa* in other floral, pigmentation, and pubescence characters, and especially in its lack of tubers. It also differs in ploidy level and molecular data, as discussed below.

Wild Tuber-Bearing Oxalis Populations

Several authors have denied the existence of oca growing outside of cultivation, or of wild tuber-bearing taxa of *Oxalis* (e.g., León 1967). Unfortunately, many collectors do not describe the underground portions of plants, and some labels on herbarium specimens do not mention the cultivated versus wild status of the plants. Thus, herbarium collections are not as informative about this question as they should be. Populations of wild tuber-bearing *Oxalis* do indeed occur in several areas of the Andes (Figure 12.2), and some of these have been named as distinct species. Although these names have been reduced to synonymy with other taxa by Lourteig (2000), I believe at least some of these names should be retained.

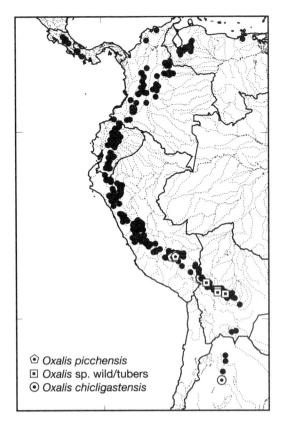

FIGURE 12.2 Distribution of wild species in the *O. tuberosa* alliance (after Emshwiller 2002a, Fig. 5). Known populations of three wild tuber-bearing taxa are indicated. *Oxalis chicligastensis* was only recently rediscovered in its type locality, and has not yet been analyzed molecularly.

One of the countries where wild tuber-bearing *Oxalis* occurs is Bolivia. Unlike *O. unduavensis*, these Bolivian plants resemble *O. tuberosa* in their leaflet shape, erect stems, pubescence, and in bearing tubers, although the tubers are smaller and the plants less robust than cultivated oca. Although a few of the specimens listed by Lourteig (2000) as *O. tuberosa* (either as subspecies *tuberosa* or *unduavensis*) may represent such populations, they are clearly distinct from *O. unduavensis* and may also be distinct from *O. tuberosa*. These wild tuber-bearing Bolivian populations should probably be considered a separate (as yet unnamed) species, or possibly a new subspecies of *O. tuberosa*, depending on whether they are confirmed to be distinct at the specific level.

Tuber-bearing *Oxalis* populations are also found in Peru and Argentina (Figure 12.2). The original botanical description (Knuth 1931) of *O. picchensis* R. Knuth mentioned that this species has tiny tubers. Although Lourteig (2000) reduced *O. picchensis* to synonymy with *O. petrophila* R. Knuth, the latter species has small corms—unlike the true tubers found in *O. picchensis*—and also differs in floral and pubescence characters. Also reduced to synonymy with *O. tuberosa* by Lourteig (2000) were two taxon names that were both based on types from northwestern Argentina: *O. chicligastensis* R. Knuth and *O. melilotoides* Zuccarini var. *argentina* Grisebach. These two names probably refer to free-living populations of the same nearly glabrous, tuber-bearing entity. Studies are in progress to determine whether this entity is, as I suspect, conspecific neither with *O. tuberosa* nor with the Bolivian wild taxon.

Cytology

Cytological work by de Azkue and Martínez (1990) provided a starting point in the search for the origins of oca. They reported on a morphologically similar group of a dozen *Oxalis* species, including *O. tuberosa*, all having 8 as their base chromosome number (i.e., the number of chromosomes in each set, symbolized by $x = 8$). They called this group of species the "*Oxalis tuberosa* alliance." Cultivated oca was reported to be octoploid, with 64 chromosomes (de Azkue and Martínez 1990). However, de Azkue and Martínez (1990) pointed out that this cytological grouping conflicted with the classification of Knuth (e.g., 1930), who had placed the members of the cytological grouping in several different sections, and most of those sections also included species with different base chromosome numbers.

The *O. tuberosa* alliance appears to comprise a larger group of species than those originally included by de Azkue and Martínez (1990). A few other *Oxalis* species have been reported to have $x = 8$ (see Table 1 in Emshwiller 2002b), and there are additional species that resemble the alliance species morphologically, but which have not yet had chromosome counts. These morphological similarities were also noted in annotations on herbarium sheets by George Eiten, as well as by Brücher (1969). Many of these additional species have been confirmed as members of the same clade (i.e., branch on an evolutionary tree) based on molecular data, so I include these additional species in a broader "*O. tuberosa* alliance." The alliance probably includes about three to five dozen species, depending on how those species are delimited. The members of the alliance are found in the northern and central Andean regions, as well as in a few high-elevation volcanic areas in Central America (Figure 12.2; see also Emshwiller 2002a).

There have been some conflicting chromosome counts reported for *O. tuberosa*, some of which differ markedly (see citations in Table 2 in Emshwiller 2002b). Among these are reports of chromosome numbers as low as 14 (Heitz 1927; Talledo and Escobar 1995) and of different ploidy levels in cultivated oca (e.g., Guamán 1997). Nevertheless, the count of 64 chromosomes for domesticated oca has been confirmed by other workers in more than 100 accessions (de Azkue and Martínez 1990; Medina Hinostroza 1994; Valladolid et al. 1994; Valladolid 1996; Vinueza Vela 1997).

Brücher (1969) mentioned tubers on an Argentinean plant identified as *O. melilotoides* Zucc., which he reported to be tetraploid with 32 chromosomes. He speculated that oca might have originated from a species similar to *O. melilotoides* through autopolyploidy. As mentioned above, *O. melilotoides* var. *argentina* may be the same entity as the tuber-bearing *O. chicligastensis*.

Recognizing Evidence of Domestication in Polyploids

Domestication and Polyploidy

Domestication involves genetic changes in populations of plants (or animals) as a result of human activities. Yet in spite of the fact that so many crops are polyploid, there is no clear universal link between polyploidy and domestication. There is some thought that heterozygosity, gene dosage effects, or physiological aspects of polyploids may predispose them for domestication. However, an analysis of the proportion of polyploidy in crops compared to the proportion among wild species in their respective higher taxonomic groups found no significant differences (Hilu 1993).

One question in the study of the origins of polyploid crops is the temporal relationship of polyploidization and domestication, and whether human activities had anything to do with the origins of the polyploid. In some cases, the origin of a polyploid appears to be associated with domestication, while in other cases the polyploid arose long before it was domesticated. Allotetraploid cotton originated in the Americas long before the arrival of humans in the Western Hemisphere (Wendel 1995), and thus polyploidization long preceded domestication in this case. In contrast, hexaploid bread wheat is unknown in the wild, and presumably arose after the domestication of tetraploid wheat from its wild forms (Feldman et al. 1995; Feldman and Millet 2001). Even if humans were not directly responsible for polyploidization of a crop, they may have had some role to play. It is possible that in some cases people might have brought the ancestors of the polyploid into contact if their original ranges did not overlap, giving those ancestors an opportunity to hybridize. Alternatively, humans may have allowed the survival of polyploids that formed naturally but would not have survived without human interference (Zeven 1980).

In addition to the complexities of the contributions of multiple genome donors in case of allopolyploidy, polyploid crops may have originated in a complex multistage process. Several crops (e.g., wheat, potato, sugarcane) have multiple levels of polyploidy within domesticated forms, and it is now understood that most natural polyploids formed not once but in multiple events (reviewed in Soltis and Soltis 1993, 1999).

Regardless of whether or not there is a direct link between polyploidy and domestication, the question of the origins of any polyploid domesticate must necessarily address the origins of the polyploid itself. Not only do we need to know the identity of the wild species that was domesticated, but also the identity of the diploid species that contributed genomes to the polyploid. In the case of oca, cytological and morphological data had suggested that the genomes of octoploid oca came from ancestors within the *O. tuberosa* alliance. In order to better understand the origins of the cultigen from within that group, we need to determine how many and which species contributed genomes to octoploid *O. tuberosa*.

Molecular Data and the Study of Polyploid Crops

Morphological data can be misleading in the case of hybrid origins, since morphological traits in hybrids may be intermediate between those of the parents, resemble one parent more than the other, or transform into unique forms that are dissimilar to either parent (McDade 1990; Rieseberg and Ellstrand 1993). Both polyploidy and domestication can result in even more transformations in morphology, resulting in forms that scarcely resemble those of the progenitors. Molecular data have an advantage in studies of the origins of polyploids and hybrids, in that these data can potentially provide evidence of each of the parental genomes.

In general, DNA sequence-based studies involve determining the nucleotide sequence for a segment of DNA at a particular locus (pl. *loci*), that is, a genetically defined site, that may be located either on one of the chromosomes (nuclear DNA) or on the chloroplast or mitochondrial genomes. For studies of polyploid origins, nuclear gene regions are necessary to provide evidence of all of the ancestral genomes. Although the chloroplast genome is used in many plant evolutionary studies, it does not have the power to reveal all the parents of a polyploid. The chloroplast genome is usually inherited maternally in flowering plants. So, in the case of an allopolyploid, it will provide evidence of only one of the progenitor species, rather than all of them. Chloroplast DNA can nonetheless be useful when used in concert with nuclear DNA sequences (e.g., Doyle et al. 1990; Soltis and Soltis 1990; Potter et al. 2000), because it can help to distinguish which was the "seed parent" of the hybrid (e.g., Erickson et al. 1983; Wendel 1989; Ge et al. 1999; Liu and Musial 2001) and may provide evidence of multiple origins of polyploidy (e.g., Soltis and Soltis 1989; Segraves et al. 1999; see also reviews in Soltis et al. 1992; Soltis and Soltis 1999). Evidence from chloroplast DNA has not yet been used in the study of oca's origins because there was a lack of any variation among the members of the alliance in the regions of the chloroplast genome that were initially tested.

There are unique expectations for polyploids in molecular systematic studies. As long as a single-copy nuclear gene is used, it is expected that a polyploid will have one copy of the gene in each set of chromosomes (i.e., a single copy in each genome). Slight differences among those copies allow researchers to distinguish them from each other and to compare them with the corresponding loci in the wild relatives. When the DNA sequences are analyzed phylogenetically to determine their evolutionary relationships (i.e., by construction of a phylogenetic tree of the sequences, also known as a gene tree), the results will depend on whether or not the polyploid is of hybrid origin. Because an autopolyploid has genomes that are all derived from a single progenitor species, all the homologous chromosomes in the polyploid, and the genes on them, are derived from the same homologous pair (Figure 12.3a). The DNA sequence of a particular gene in an autopolyploid will usually group

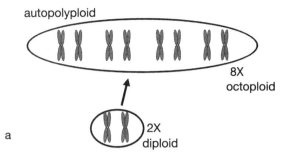
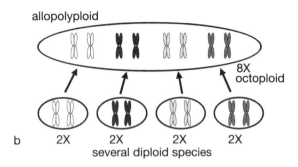
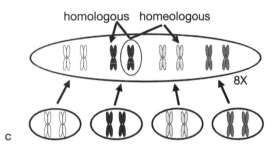

FIGURE 12.3 Simplified diagrams of the derivation of homologous and homeologous (partially homologous) chromosomes in an octoploid. Only one chromosome of each set is shown, and any intermediate levels of polyploidy are eliminated. The expectations when DNA sequences from these chromosomes are included in an evolutionary analysis are explained in the text. (a) Autopolyploid. Because an autopolyploid is derived from a single diploid species, all of its homologous chromosomes are derived from the same homologous pair in the ancestral species. (b) Allopolyploid. The chromosomes in an allopolyploid are derived from different ancestral species. (c) Allopolyploid. The chromosomes (and genes), in an allopolyploid may be either homologous (derived from the same ancestral species) or homeologous (derived from a different species) to each other.

with that of the progenitor species on the gene-based phylogenetic tree. In an allopolyploid, with genomes derived from multiple species or well-differentiated populations (Figure 12.3b), the multiple sequences within each individual plant are expected to group with those of each of the progenitor species, and so they may appear in different parts of the gene tree. Some of the genes and chromosomes of an allopolyploid, then, are homologous (i.e., derived from the same diploid progenitor), whereas others are homeologous (i.e., derived from different diploid progenitors; Figure 12.3c).

One expectation of allopolyploids is that all individuals will have each of the homeologous versions of the gene in their cells—a condition termed "fixed heterozygosity." However, recent studies are bringing to light ways that polyploid evolution can differ from this straightforward scenario. These complicating factors have been well reviewed elsewhere (Soltis and Soltis 1999; Wendel 2000; Pikaard 2001). I mention here only that they include the possibilities of multiple origins of polyploids; loss of homeologous loci; and loss of sequences or chromosomal segments from the polyploid within a few generations of its formation, through gene conversion, concerted evolution, or various other mechanisms (Parokonny et al. 1994; Song et al. 1995; Escalante et al. 1998; Liu et al. 1998a; Liu et al. 1998b; Ozkan et al. 2001; Shaked et al. 2001; Kashkush et al. 2002). The loss of these genetic regions may be an important mechanism for restoring diploid-like chromosome-pairing behavior in polyploids (Feldman et al. 1997).

Materials and Methods

Choosing Sequences for Study

Despite the advantages that nuclear loci provide for this kind of evolutionary study, at the time this project began there were few such loci available. The internal transcribed spacer (ITS) of nuclear ribosomal DNA (Figure 12.4) was one of the first nuclear loci for which universal primers were developed (Baldwin et al. 1995), so it was the first locus applied to the question of oca's origins. Nuclear ribosomal DNA, including the ITS regions, occurs in hundreds or thousands of copies per cell. These multiple copies are usually, but not always, identical to each other as the result of a process called "concerted evolution," so they are often treated as if they were single-copy loci. Unfortunately, concerted evolution either may be incomplete, resulting in multiple different copies within each cell, or may act to "erase" homeologous copies in a polyploid (Wendel et al. 1995). Thus, the ITS regions do not truly have the advantages of a single-copy locus.

Later, a second nuclear locus, a part of the gene for chloroplast-expressed glutamine synthetase (ncpGS) that contains four introns (Figure 12.5), was utilized in the study of *Oxalis* (Emshwiller and Doyle 1999, 2002). Although this form of the glutamine synthetase protein is active in the chloroplast, the gene that encodes it is located in chromosomal DNA of the nucleus rather than in the chloroplast genome. The chloroplast-expressed form diverged so long ago from the other members of the glutamine synthetase multigene family that it was possible to design primers to amplify only the gene for the chloroplast-expressed version. With these specific primers, the ncpGS locus had the necessary advantages of a single-copy nuclear locus.

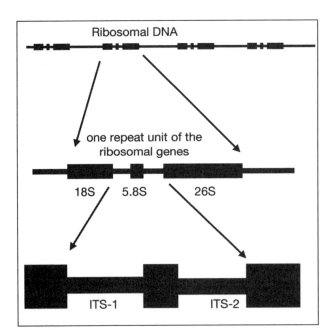

FIGURE 12.4 The internal transcribed spacer (ITS) region of nuclear ribosomal DNA. Ribosomal DNA is made up of hundreds to thousands of copies of repeats of the ribosomal genes, separated by noncoding spacers, which are usually all identical. The ITS region comprises the two spacers on either side of the 5.8S segment, which codes for part of the RNA in ribosomes.

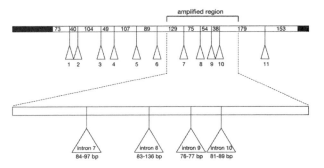

FIGURE 12.5 The amplified portion of chloroplast-expressed glutamine synthetase (ncpGS; adapted from Emshwiller and Doyle 1999, 2002). The upper portion indicates the location of the amplified region compared to the location of introns and the size of exons in *Medicago* cytosolic GS (Tischer et al. 1986). Lower portion shows the sizes of the intron regions found in the *Oxalis* species sampled to date.

Both of the DNA regions used in this study have the important characteristic of possessing a large proportion of noncoding DNA—the spacers themselves in the case of the ITS region, and the four introns in the case of the part of ncpGS utilized. Non-coding DNA regions are usually more variable than the coding regions around them because they are less constrained in their evolution. Thus, they are expected to have more of the differences among species that can be used to infer evolutionary relationships.

Accessions Sampled

Species in the broader *Oxalis tuberosa* alliance are found throughout the Andean region from Venezuela to northwestern Argentina, as well as in Central America (Figure 12.2 and Emshwiller 2002a). Sampling for DNA sequencing included all species that were originally included in the *O. tuberosa* alliance by de Azkue and Martínez (1990), as well as many of the taxa that appeared to be morphologically similar to alliance members. Other less-related Andean *Oxalis* species, which could be distinguished from the alliance members on the basis of morphology (Emshwiller 1999a, 1999b), were included as outgroups. A larger selection of diverse outgroups from the genus *Oxalis* was included in the ITS study in order to test the monophyly of the alliance. Most of the samples for both studies were selected from among those that I collected with colleagues in Bolivia and Peru. Most of the alliance members and the species that appeared morphologically to be members of the alliance have populations in those two countries, although there are exceptions. Collection information is available in Emshwiller and Doyle (1998, 2002, in supplementary data) and in Emshwiller (1999a, 2002b).

Accessions of cultivated *O. tuberosa* sampled were selected to be morphologically distinct (i.e., not duplicates of the same clone). The sample for the ncpGS study included three accessions from Bolivia and six accessions from southern Peru. Two wild tuber-bearing taxa were included: two accessions of *Oxalis picchensis* R. Knuth from inter-Andean valleys in southern Peru, and four accessions of the as-yet-unnamed taxon from the humid eastern Andean slopes of Bolivia.

Sequencing and Analyses

Sequencing protocols for ITS and ncpGS are described in Emshwiller and Doyle (1998, 1999, 2002). Initial attempts to determine the DNA sequence of ncpGS in octoploid *O. tuberosa* found clear signs of multiple sequences within individual plants. The technique of molecular cloning was used to separate the different ncpGS sequence types within individuals (Emshwiller and Doyle 2002). Three individual plants of oca were selected for molecular cloning, and because the wild tuber-bearing taxon of Bolivia also had multiple sequences within individuals, one plant of this taxon was also cloned. Subsequently, additional plants (six oca and three of the wild tuber-bearing taxon of Bolivia) were sampled by direct sequencing to confirm the presence or absence of similar sequences in a somewhat larger sample (Emshwiller and Doyle 2002). Very few of the cloned ncpGS sequences were identical, but they could be grouped into classes of similar sequences (see Results, below).

Sequences were aligned with other sequences of the corresponding locus, either manually (ncpGS data) or using the program Clustal V (Higgins and Sharp 1989; ITS data). The more divergent *Oxalis* taxa included as outgroups in the ITS study were not unambiguously alignable across the entire ITS region. Thus, two separate analyses were performed: one with the entire ITS region of the alliance and a limited selection of closer outgroups; another including all outgroup taxa, but only the more easily alignable parts of the ITS region (Emshwiller and Doyle, 1998).

Data for phylogenetic analyses consisted primarily of single base-pair (bp) differences (i.e., substitutions) among sequences. Gaps in the ncpGS alignment (as a result of insertions or deletions of one or more contiguous base pairs) were treated as additional binary characters. Characters were analyzed using a parsimony criterion with either PAUP* (Swofford 1991; ITS data) or NONA (Goloboff 1998; ncpGS data) software. Additional details of tree construction are described in Emshwiller and Doyle (1998, 2002).

Flow Cytometry Data

Additional data for this study were produced by flow cytometry. This was used as an independent data source to confirm the octoploidy of *Oxalis tuberosa* and to rapidly survey the ploidy levels among wild members of the alliance, only a small proportion of which had published chromosome counts. Flow cytometry permits an estimate of the nuclear DNA content of a plant sample (reviewed in Doležel 1997) by comparing the fluorescence of propidium iodide-stained nuclei with that of a standard of known DNA content (e.g., chicken red blood cells). The methods of Doležel and Göhde (1995) and Galbraith, et al. (1997) were modified as described in Emshwiller (2002b).

Results

ITS Data

Data from ITS sequencing corroborated the morphological and cytological data indicating that oca originated from ancestors within the *Oxalis tuberosa* alliance (Emshwiller and Doyle 1998). The inclusion of a wide range of morphologically diverse species from across the genus *Oxalis* in the analyses of ITS sequence data helped to confirm that the alliance is a monophyletic group (i.e., all species are derived from a single common ancestor). The primary ITS sequence type of cultivated oca was found within the alliance clade on the ITS phylogenetic tree, supporting the idea that oca's origins would be found within the alliance. In addition, one of the three oca plants sequenced showed faint signs of a second sequence type, also from within the alliance clade. There was no indication of any contribution from species outside of the alliance.

A group of Bolivian cloud forest species, including *O. andina* Britton, was supported by ITS data as the sister group (i.e., closest group of relatives) of the *O. tuberosa* alliance. However, there was no evidence of involvement of these species in the origins of oca, so I do not consider them part of the *O. tuberosa* alliance. The close relationship of these species to those of the alliance is reflected in the classification of Lourteig (2000), who included them in the same section as most members of the alliance.

The ITS data were unable to provide as much information about the origins of oca as had been hoped, because in spite of high levels of divergence across the genus overall, there were very few differences among the ITS sequences of the different species of the alliance. Notably, several wild species all shared the same ITS sequence as the primary one in cultivated oca, including not only the wild tuber-bearing taxon from Bolivia and *O. unduavensis*, but also *O. oulophora* Lourteig. The faint secondary sequence in oca was also shared with a large number of other alliance members. Additional wild *Oxalis* taxa from Peru that had not been included in the original ITS study (Emshwiller and Doyle 1998) were sampled later (Emshwiller 2002a), but no additional informative variation was found in this larger sample. Thus, the ITS data were not variable enough to identify the ancestors of the cultigen with precision, and other sources of data would clearly be needed.

ncpGS Data

A pilot project with the ncpGS gene (Emshwiller and Doyle 1999) indicated that there was potentially more variation among ncpGS sequences in the alliance than among their ITS sequences, and this variation did indeed provide more information about oca's origins (Emshwiller and Doyle 2002). Variation among *Oxalis* ncpGS sequences provided more characters that supported the monophyly of the alliance, notably including a deletion of 31 base pairs, and also helped to resolve evolutionary relationships among some of the species in the alliance (Emshwiller and Doyle 2002). Data from both ncpGS and ITS sequences supported the naturalness of the alliance, including not only the members as originally described by de Azkue and Martínez (1990), but several additional morphologically similar species as well.

More important for this study was the ncpGS variation found within individual plants of oca. However, the interpretation of the results was complicated by the fact that the cloning method used can produce mistakes in the sequences. The kind of mistake known as "*Taq* error," in which the enzyme inserts an incorrect nucleotide, is less problematic than the second kind of mistake, known as "PCR recombination" (Cronn et al. 2002). The latter can produce a sequence that does not exist in the plant by mixing up different parts of the real sequences. Among the cloned sequences of oca there seemed to be both kinds of errors (Emshwiller and Doyle 2002). Although it was not possible to determine with absolute certainty which of the sequences were the real ones, it was possible to use several criteria to eliminate those that appeared to be recombined.

Those cloned ncpGS sequences that did not appear to be recombinants were grouped into several classes of similar, but not identical, sequences (Emshwiller and Doyle 2002) designated B, C, and D (Table 12.2). Because some of the differences among the sequences were shared with wild *Oxalis* species, these differences were used to help group the cloned sequences into classes, to distinguish which of them were

TABLE 12.2
Diagnostic Differences among the Three ncpGS Sequence Classes[a]

Aligned Site(s)	Ancestral State[b]	Class B	Class C	Class D
249	C[c]	=	T	T
359–365	absent	=	=	7-bp insertion*
425	C	T	=	=
505	present	=	1-bp deletion	=
515	C	=	=	T
601	T	=	=	A
606	A	=	=	T
620–621	present	2-bp deletion	=	=
648	A	=	=	G*
653	T	=	A*	=
657	G	=	A	A
720	G	=	C	C

[a] Differences (e.g., base pair [bp] substitutions and insertion/deletion sequences) are found within individual plants of oca and shared with O. picchensis (class C) and the putative wild progenitor of Bolivia (classes B and D).

States that remained unchanged from the ancestral state are designated with an equal sign so that the derived states distinguishing the different sequence classes are easier to observe.

Asterisks (*) indicate states that were unique to the corresponding sequence class (i.e., not observed in any other taxon sampled herein). All other character states were shared with other *Oxalis* taxa, and thus they helped determine the placement of these sequence classes on the ncpGS phylogenetic tree. The class B sequences were the only ones sampled that had the combination of both a thymine (T) at site 425 and a two-base deletion at sites 620–621. Several other members of the alliance (including *O. unduavensis*) had one or the other of these states, but not both.

[b] Ancestral states in the ancestor of these classes (at the base of the *O. tuberosa* alliance) were inferred by most parsimonious optimization of states on the tree.

[c] Bases: A (adenine), C (cytosine), G (guanine), and T (thymine).

likely to have been recombined, and to place them on the phylogenetic tree. Classes B and D were present in all nine oca plants sampled (three plants in the cloned sample and six additional oca accessions that were sequenced directly). The same two sequence classes were shared by all four plants of the Bolivian wild tuber-bearing taxon. Notably, the class C sequences were found in all three oca plants whose sequences were cloned, but were absent in one of the six plants sequenced directly. The class C sequence was shared with the other wild tuber-bearing species, *Oxalis picchensis* of southern Peru. These two wild tuber-bearing taxa, *O. picchensis* of southern Peru and the unnamed taxon of Bolivia, were the only wild *Oxalis* species sampled that shared the same ncpGS sequence classes as oca. Neither species had all three of the sequence types found in oca. The currently available ncpGS data support these two wild tuber-bearing taxa as the best candidates for the ancestral species of octoploid *O. tuberosa*.

Oxalis unduavensis did not share the same sequence as *O. tuberosa*. Instead, it shared an ncpGS sequence that was found in identical or nearly identical form in several wild species of the alliance that occur in moist cloud forests of the eastern Andes, from northern Peru to Bolivia (Emshwiller and Doyle 2002).

The fact that sequence classes B and D are found in all of the oca plants sampled means that these classes exhibit fixed heterozygosity, which supports the idea that these classes represent the homeologous sequences and that oca is allopolyploid (see Figure 12.3). Fixed heterozygosity by itself, especially at this level of sampling, cannot be taken as complete confirmation of allopolyploidy (discussed in Emshwiller and Doyle 2002). However, the phylogenetic context provides stronger support for this idea, because these three different sequence classes are clearly separated in their positions on the ncpGS phylogenetic trees (Figure 12.6), and no diploid species was found to be heterozygous for sequences from these different parts of the tree. As is often the case in phylogenetic analyses, the algorithms found multiple solutions, each having a somewhat different set of hypothesized evolutionary relationships among the sequences. Dotted lines in Figure 12.6 indicate branches that do not appear in all of the 208 equally parsimonious trees. There is variation in the position of the class B sequences of oca and the wild Bolivian taxon with respect to those of the other species, but the three sequence classes were nonetheless always separated in their positions in each of the trees found. Regardless of these rearrangements, the fact remains that the class B sequences of oca share a set of character states with those of the wild tuber-bearing plants from Bolivia, and this set is not shared by the other ncpGS sequences on the tree. Thus, the match between these sequences holds, as does the separation of the class B sequences from the class C and D sequences of oca in each of the different trees. The separation of the sequence classes in different parts of the ncpGS-based phylogenetic tree supports the idea that oca is an allopolyploid.

Because there seem to be three, rather than four, different sequence classes, oca may be an autoallopolyploid, a kind of allopolyploid that also shares aspects of autopolyploidy (see Terminology of Polyploidy at the beginning of this chapter). The available ncpGS data support the two wild tuber-bearing taxa, *O. picchensis* and the unnamed taxon of Bolivia, as the best candidates for ancestors (genome donors) of octoploid oca. The hypothesis suggested by these data is that octoploid oca may have originated through hybridization of these two taxa (Figure 12.7), although there are several scenarios by which this may have occurred.

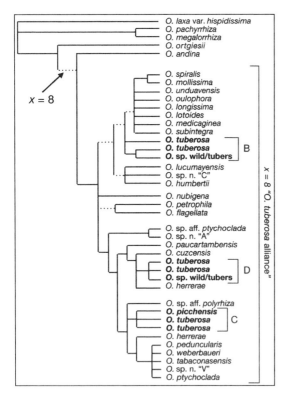

FIGURE 12.6 One of 208 trees that resulted from analysis of ncpGS sequences from members of the *O. tuberosa* alliance, including cloned sequences from three plants of cultivated oca (adapted from Emshwiller and Doyle 2002). For simplicity, some of the multiple clones and some of the duplicate sequences from different populations of the same species have been eliminated (in cases where the different sequences, albeit not identical, would join the same place on the tree). Dashed lines indicate branches that do not appear in all of the 208 trees. Brackets with the letters B, C, and D indicate the sequence classes (see Table 12.2) found within individual oca plants and shared with the two wild tuber-bearing taxa—*O. picchensis* and the unnamed Bolivian taxon (indicated by *O.* sp. wild/tubers).

However, the absence of the *O. picchensis*-like class C ncpGS sequence in one of the oca plants sampled by direct sequencing suggests the possibility that oca's origins may be more complex than this simple hypothesis. Alternative scenarios that are congruent with the ncpGS data include the possibilities of different levels of ploidy in cultivated oca, multiple origins of polyploidy, introgression of the *O. picchensis*-like ncpGS sequence through wild-crop geneflow, or loss of that sequence after formation of the polyploid (discussed in Emshwiller 1999a; Emshwiller and Doyle 2002).

Flow Cytometry Data

Information about the ploidy levels of oca's putative progenitors can help distinguish among some of the alternative hypotheses mentioned earlier. Results of the flow cytometry data indicated that the vast majority of the wild *Oxalis* species of the alliance that were sampled were diploids with a

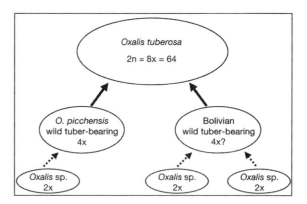

FIGURE 12.7 One of the scenarios that was congruent with the ncpGS sequencing results. Octoploid oca might be an autoallopolyploid derived through hybridization between tetraploid *O. picchensis* and the Bolivian wild tuber-bearing taxon (of unknown ploidy level, but probably allopolyploid). The bottom row indicates the unknown diploid progenitors of the wild tuber-bearing taxa.

relatively small genome size, ranging from about 0.79 to 1.34 picograms (pg) per diploid nucleus. The plants determined to be diploid included two plants of *O. unduavensis* from different localities in Bolivia. All of the 10 living accessions of cultivated oca measured (from different areas in Bolivia) were confirmed to be octoploid (mean ~3.6 pg per nucleus). Two wild *Oxalis* species, one of them *O. picchensis*, fell in the tetraploid range (both ~1.67 pg per nucleus). Because *O. picchensis* had a single sequence type for ncpGS, it is probably autopolyploid. This information permitted the elimination of certain hypotheses about oca's origins that were congruent with the ncpGS data (discussed in Emshwiller 2002b; Emshwiller and Doyle 2002). Unfortunately, living material of the Bolivian wild tuber-bearing taxon was not available, so it was not possible to confirm its ploidy level. The Bolivian taxon is probably an allopolyploid, because it also showed fixed heterozygosity of class B and D ncpGS sequences.

Inferences from the Data Concerning Domestication

The flow cytometry results confirmed that domesticated oca is octoploid, although these results do not eliminate the possibility that there may be other ploidy levels within cultivated material. DNA sequence results for both ITS and ncpGS loci confirmed that oca originated from ancestors within the *Oxalis tuberosa* alliance, and that this group of species is monophyletic. The presence of the class B and D ncpGS sequences in all oca plants sampled, in combination with their separation in different parts of the *O. tuberosa* alliance subclade, provides strong evidence that oca is an allopolyploid. The presence of these same two sequence classes in the wild tuber-bearing taxon of Bolivia suggests that it too is an allopolyploid (of unknown ploidy level), and that it may be a progenitor of *O. tuberosa*. The contribution of *O. picchensis* to the evolution of oca is less certain. The sharing of the class C ncpGS sequence type between this wild tuber-bearing tetraploid and eight out of nine oca plants sampled could have several causes. Oca could be an autoallopolyploid, with genomes contributed in a multistage process by both *O. picchensis* and the Bolivian taxon (Figure 12.7). However, the fact that this sequence type was not present in all oca sampled suggests that alternatives to this simple scenario may be more likely. Additional independent sources of data, such as other nuclear loci, should help to distinguish among these possibilities. These results of molecular analysis of the origins of *O. tuberosa* highlight some of the complications that are possible in studying the origins of polyploid crops.

It may be tempting to think that these sequence data could be used to determine the timing of oca's polyploidization or of its domestication, but such inferences are not possible from these data. Olsen and Schaal (Chapter 9, this volume) discuss the pitfalls involved in the dating of times since divergence based on "molecular clock" assumptions. Several factors are thought to affect the rate at which mutations accumulate in lineages, including generation time, GC content, chromosomal location, and so forth (Bulmer et al. 1991; Gaut and Weir 1994; Gaut 1998; Matassi et al. 1999; Bromham 2002). In the case of an allopolyploid, the situation is complicated by the fact that genomes have first diverged (at the time of splitting of the ancestral species) and then have come back together in a single nucleus at the time of polyploidization. From the time of origin of the polyploid, the sequences in the polyploid and its diploid progenitor will begin to diverge, but evolution of genomes within polyploids may also change in ways that are only beginning to be understood (Wendel 2000). Some calculations have been made to estimate the time of divergence of ancestral genomes and subsequent polyploidization in long-established polyploids (Wendel and Albert 1992; Wendel 1995; Gaut and Doebley 1997). In the case of oca, however, there is limited divergence among the various species in the *O. tuberosa* alliance, suggesting that this group may represent a relatively recent radiation. With the accumulation of few differences among the ancestral genomes, there would be even fewer since the time of polyploidization and domestication.

Discussion and Future Research

Comparisons with Prior Hypotheses about Oca's Origins

In his discussion of the oca tuber remains found in Guitarrero Cave, Smith (1980) acknowledged that they may have been wild rather than cultivated, but he suggested that knowledge that the cave was outside the range of the wild *Oxalis* that bear such tubers "would indicate a strong possibility that the plants were introduced by man through cultivation." Now that the molecular data have indicated wild species as progenitor candidates, the distributions of these taxa can be compared with the location of the archaeological remains of oca. The Callejón de Huaylas in the department of Ancash, in which Guitarrero Cave is located, is indeed well north of known locations of any populations of these putative progenitors. This supports the notion that the oca tubers found in the cave are of cultivated origin. Caution is necessary in drawing conclusions from these distributions, however. Information on species distributions in the Andean region is rudimentary (Emshwiller 2002a), and patterns of vegetation in the Andes have been changed by both climatic and anthropogenic factors during the past 10,000 years.

Information about the distributions of the putative progenitors can also help to distinguish among the speculations that have been put forward about the area or kind of habitat in which oca was domesticated. In spite of the high degree of diversity that has been observed in cultivated oca in that region, the Altiplano near Titicaca is not supported as the area of origin of the crop, because neither the putative progenitors nor any other members of the alliance are found in that area. Reports of tubers on wild *Oxalis* plants in this region appear to refer to nonhomologous structures.

The other two habitat types that were hypothesized to be the areas of oca's domestication were the seasonally dry inter-Andean valleys and the ecotone between puna and cloud forest on eastern Andean slopes. Interestingly, the putative ancestors of oca seem to include species from each of these habitats. Both *O. tuberosa* and the putative progenitor taxon of Bolivia are supported as allopolyploids with sequence classes that are found in two different subclades within the alliance. These two subclades inhabit different ecological habitats, although their ranges are overlapping (Emshwiller 2002a). The class B sequences join a clade that includes mostly species in moist forests of the eastern Andean slopes, whereas the class C and D sequences join a clade that includes mostly species found in drier inter-Andean valleys. If, as I suspect, the wild Bolivian taxon has an intermediate ploidy level, probably tetraploid or perhaps hexaploid, then this initial polyploidization step occurred before domestication. One might speculate that it was the joining of genomes from different ecological zones that led to the initial survival of the polyploid, its subsequent habitat expansion, and its successful domestication.

With current data it is not possible to distinguish whether the octoploid level was reached before or after domestication. The existence of either wild octoploids or domesticated oca of different ploidy levels would clarify this question, but these have not been confirmed. As mentioned earlier, the existence of both $4x$ and $6x$ domesticated wheat and the absence of any wild $6x$ populations (Feldman et al. 1995), in concert with the archaeological evidence, clearly indicates that $6x$ bread wheat arose after domestication. I am not aware of any reports of wild octoploid populations of *O. tuberosa*. On the other hand, there are reports of cultivated oca of different ploidy levels in unpublished theses by Bolivian students (e.g., Guamán 1997). However, as there are no photographs of oca chromosomes presented, it is difficult to evaluate these reports. Guamán (1997) shows a photograph of chromosomes of a wild tuber-bearing *Oxalis* (the putative progenitor of the Bolivian taxon?) that she states is triploid, but it appears to me to have more than 24 chromosomes, and it may in fact be tetraploid (the latter interpretation is also more likely because these wild populations include seedlings, and triploids are usually sterile). Because octoploid oca has been confirmed by many independent counts and now by flow cytometry, it is important that any claims of variation in ploidy level in *O. tuberosa* be confirmed independently. If these different ploidy levels were confirmed, this variation would favor the idea that octoploid oca arose after domestication at a lower ploidy level. On the other hand, until there is a thorough survey of ploidy levels in the wild progenitor populations, the possibility that octoploidy arose before domestication cannot be eliminated. Even so, there always remains the possibility that cytotypes have become extinct.

In the absence of evidence from ploidy variation in extant domesticated oca or its progenitors, it would be tempting to think that the question of whether polyploidization preceded domestication could be inferred from molecular data, as was possible for tetraploid cotton (Wendel and Albert 1992; Wendel 1995). However, in addition to the aforementioned caveats mentioned about molecular clocks, the variation levels among DNA sequences of oca and its allies are too low to permit that kind of inference.

The molecular and flow cytometry results allow a comparison with Lourteig's (2000) sinking of *O. unduavensis* as a subspecies of *O. tuberosa*. Examination of the specimens that she lists reveals that she includes under the *O. tuberosa* name not only the type of *O. unduavensis* and other specimens from the same area, but also at least one specimen that is probably from one of the wild tuber-bearing populations of Bolivia. Although *O. unduavensis* bears a resemblance to *O. tuberosa*, it does not bear tubers, and it differs from both *O. tuberosa* and the Bolivian tuber-bearing plants in several other floral and vegetative morphological features. *Oxalis unduavensis* is also diploid, rather than octoploid (Emshwiller 2002b). Although it shares the same ITS sequence as that in oca, the identical sequence was also shared by two other species: the wild tuber-bearing taxon of Bolivia and *O. oulophora*. The ncpGS data exclude *O. unduavensis* as an ancestor of oca because it does not share any of the sequence types found in octoploid oca. In my opinion, the name *O. unduavensis* should be maintained at the species level, and it should exclude the wild tuber-bearing *Oxalis* in Bolivia. If these tuber-bearing populations are confirmed to be distinct from oca, then they are an unnamed species; if not so confirmed, then these populations, rather than *O. unduavensis*, should be considered a subspecies of *O. tuberosa*.

Future Studies

There is a need for additional sampling of populations and species to address unresolved questions about the origins of oca and the evolution of the *O. tuberosa* alliance. Additional sampling is needed to find the diploid progenitors of tetraploid *O. picchensis* and of the Bolivian taxon, and also to confirm whether yet-unsampled species or populations may have the same sequence types as oca. Of course, one can never rule out the possibility that other progenitor candidates are now extinct. Although the majority of the putative alliance members have populations in Bolivia and Peru that were available to be sampled, there remain some species that resemble the members of the alliance morphologically—most of which are included in the same section by Lourteig (2000)—that have not yet been sampled for molecular data. Among these remaining unsampled taxa, I do not expect that samples from the Northern Andes will provide more information on oca's origins. Increased sampling from northwestern Argentina, on the other hand, may indeed help clarify oca's domestication. A recent expedition to the type locality of *Oxalis chicligastensis*, in collaboration with A. Grau of the University of Tucumán, found that this taxon does bear tubers, although these were not mentioned by the original collector or in the original description of the species (Knuth 1936). As discussed earlier, this taxon was sunk into synonymy with *O. tuberosa* by Lourteig (2000), and it may represent the same taxon that Brücher (1969) reported to be tetraploid. Studies are currently in progress to determine its ploidy level and relationship to oca, and to the wild tuber-bearing taxon of Bolivia.

Increased sampling of additional independent sources of data is also needed. Additional nuclear loci will help to distinguish among the various hypotheses that were consistent with the ncpGS data. Whether or not the phylogenetic trees based on different loci are congruent should test the working hypothesis that the Bolivian wild tuber-bearing taxon is a progenitor of oca, and should clarify the role of *O. picchensis* in the evolution of oca. Initial tests to assess the utility of DNA sequences of malate synthase (Lewis and Doyle 2001) in the *O. tuberosa* alliance found levels of variation that may surpass those of ncpGS data (E. Emshwiller, unpublished data). If variation in the chloroplast genome can be detected within the *O. tuberosa* alliance, this would indicate which is the maternal genome of the polyploids.

The use of accessions from national and international ex-situ germplasm banks will allow the sampling of variation in a larger number of oca accessions from other areas of the Andes. Studies are currently in progress with fluorescent AFLP (amplified fragment length polymorphism) markers to study several aspects of the human influence on the evolution of *O. tuberosa*. Like microsatellites (Olsen and Schaal, Chapter 9 of this volume), these markers provide high levels of variation and wide genomic coverage, but they lack genealogical information. AFLP data are being used to help distinguish among the alternative scenarios that are congruent with the results based on DNA sequence data. Other questions that AFLP data will be used to address include (1) whether the distinct use categories in which Andean farmers classify oca cultivars have different evolutionary histories, (2) how humans have moved the clones of this vegetatively propagated crop among regions, and (3) how the clones of the crop as distinguished by AFLP correspond to the folk taxonomy.

Finally, the possibility that polyploidy in oca may have arisen more than once is hinted at by differences among the ncpGS sequences within a class, as well as by the preliminary AFLP data. Because multiple origins are now understood to be common in polyploids (Soltis and Soltis 1999), this possibility, if confirmed, would help to explain the great diversity of forms of oca cultivars.

Acknowledgments

I am especially grateful to my former graduate advisor, Jeff Doyle, for his invaluable advice and constructive criticism during the dissertation phase of this research program. Financial support for my research has been provided by the generosity of a grant from Abbott Laboratories to The Field Museum, and by the A.W. Mellon Foundation, the Fulbright Commission of Peru, the U.S. National Science Foundation, the U.S. Department of Education, the L. H. Bailey Hortorium, and the Cornell Graduate School. I thank INRENA and IBTA for plant collection permits for Peru and Bolivia, respectively. Thanks also to the following people (listed in chronological order) for help during field work in Bolivia, Peru, and Argentina: M.G. Asbun Claros; J. Miller; M. L. Ugarte, F. Terrazas, S. Guamán, T. Villarroel, and J. Almanza; R. Vargas; B. Eriksen and U. Molau; A. Valladolid; G. Meza and P. Cruz; A. Castelo; A. Tupayachi, P. Núñez, and R. Urrunaga; H. Flores, R. Estrada, R. Ortega, C. Arbizu, M. Ramírez, and A. Andia; F. Vivanco; N. Arce and L. Torrez; M. Vallenas; I. Sánchez Vega; M. Hermann; A. Grau, J. G. López, and D. M. Ávila.

References

Allard, R. W. 1960. *Principles of plant breeding*. New York: John Wiley & Sons.

Baldwin, B. G., M. J. Sanderson, J. M. Porter, M. F. Wojciechowski, C. S. Campbell, and M. J. Donoghue. 1995. The ITS region of nuclear ribosomal DNA: A valuable source of evidence on angiosperm phylogeny. *Annals of the Missouri Botanical Garden* 82: 247–277.

Bromham, L. 2002. Molecular clocks in reptiles: Life history influences rate of molecular evolution. *Molecular Biology and Evolution* 19: 302–309.

Brücher, H. 1969. Poliploidia en especies sudamericana de *Oxalis*. *Boletín de la Sociedad Venezolana de Ciencias Naturales* 28: 145–178.

Bulmer, M., K. H. Wolfe, and P. M. Sharp. 1991. Synonymous nucleotide substitution rates in mammalian genes—Implications for the molecular clock and the relationship of mammalian orders. *Proceedings of the National Academy of Sciences, USA* 88: 5974–5978.

Buteler, M. I., R. L. Jarret, and D. R. LaBonte. 1999. Sequence characterization of microsatellites in diploid and polyploid *Ipomoea*. *Theoretical and Applied Genetics* 99: 123–132.

Cook, O. F. 1925. Peru as a center of domestication: Tracing the origin of civilization through the domesticated plants. *Journal of Heredity* 16: 33–46, 95–110.

Correal Urrego, G. 1990a. *Aguazuque—evidencias de cazadores, recolectores y plantadores en la altiplanicie de la cordillera oriental*. Fundación de Investigaciones Archeológicas Nacionales. Bogotá, Colombia: Banco de la República.

———. 1990b. Evidencias culturales durante el pleistoceno y holoceno de Colombia. *Revista de Arqueología Americana* 1: 69–89.

Cortella, A. R. and M. L. Pochettino. 1995. Comparative morphology of starch of three Andean tubers. *Starch* 47: 455–461.

Cronn, R., M. Cedroni, T. Haselkorn, C. Grover, and J. F. Wendel. 2002. PCR-mediated recombination in amplification products derived from polyploid cotton. *Theoretical and Applied Genetics* 104: 482–489.

D'Altroy, T. N. and C. A. Hastorf. 1984. The distribution and contents of Inca state storehouses in the Xauxa Region of Peru. *American Antiquity* 49: 334–349.

de Azkue, D. and A. Martínez. 1990. Chromosome numbers of the *Oxalis tuberosa* alliance (Oxalidaceae). *Plant Systematics and Evolution* 169: 25–29.

Debouck, D. G. and D. Libreros Ferla. 1995. Neotropical montane forests: A fragile home of genetic resources of wild relatives of New World crops. In *Biodiversity and conservation of neotropical montane forests*, S. P. Churchill, H. Balslev, E. Forero, and J. L. Luteyn (eds.), pp. 561–577. New York: New York Botanical Garden.

Doebley, J. 1992. Molecular systematics and crop evolution. In *Molecular systematics of plants*, P. S. Soltis, D. E. Soltis, and J. J. Doyle (eds.), pp. 202–222. New York: Chapman and Hall.

Doležel, J. 1997. Application of flow cytometry for the study of plant genomes. *Journal of Applied Genetics* 38: 285–302.

Doležel, J. and W. Göhde. 1995. Sex determination in dioecious plants *Melandrium album* and *M. rubrum* using high-resolution flow cytometry. *Cytometry* 19: 103–106.

Doyle, J. J., J. L. Doyle, and A. H. D. Brown. 1990. Analysis of a polyploid complex in *Glycine* with chloroplast and nuclear DNA. *Australian Systematic Botany* 3: 125–136.

Emshwiller, E. 1999a. *Origins of domestication and polyploidy in the Andean tuber crop Oxalis tuberosa Molina (Oxalidaceae)*. Ph.D. diss., Cornell University. Ann Arbor: University Microfilms.

———. 1999b The relationships of Peruvian *Oxalis* species to cultivated oca. *Arnaldoa* 6: 117–139.

———. 2002a. Biogeography of the *Oxalis tuberosa* alliance. *Botanical Review* 68: 128–152.

———. 2002b. Ploidy levels among species in the "*Oxalis tuberosa* alliance" as inferred by flow cytometry. *Annals of Botany* 89: 741–753.

Emshwiller, E. and J. J. Doyle. 1998. Origins of domestication and polyploidy in oca (*Oxalis tuberosa*: Oxalidaceae): nrDNA ITS data. *American Journal of Botany* 85: 975–985.

———. 1999. Chloroplast-expressed glutamine synthetase (ncpGS): Potential utility for phylogenetic studies with an example from *Oxalis* (Oxalidaceae). *Molecular Phylogenetics and Evolution* 12: 310–319.

———. 2002. Origins of domestication and polyploidy in oca (*Oxalis tuberosa*: Oxalidaceae). 2. Chloroplast-expressed glutamine synthetase data. *American Journal of Botany* 89: 1042–1056.

Erickson, L. R., N. A. Straus, and W. D. Beversdorf. 1983. Restriction patterns reveal origins of chloroplast genomes in

Brassica amphidiploids. *Theoretical and Applied Genetics* 65: 201–206.

Escalante, A., S. Imanishi, M. Hossain, N. Ohmido, and K. Fukui. 1998. RFLP analysis and genomic *in situ* hybridization (GISH) in somatic hybrids and their progeny between *Lycopersicon esculentum* and *Solanum lycopersicoides*. *Theoretical and Applied Genetics* 96: 719–726.

Feldman, M. and E. Millet. 2001. The contribution of the discovery of wild emmer to an understanding of wheat evolution and domestication and to wheat improvement. *Israel Journal of Plant Sciences* 49, supplement: S25–S35.

Feldman, M., B. Liu, G. Segal, S. Abbo, A. A. Levy, and J. M. Vega. 1997. Rapid elimination of low-copy DNA sequences in polyploid wheat: A possible mechanism for differentiation of homeologous chromosomes. *Genetics* 147: 1381–1387.

Feldman, M., F. G. H. Lupton, and T. E. Miller. 1995. Wheats—*Triticum* spp. (Gramineae—Triticinae). In *Evolution of crop plants*, J. Smartt and N. W. Simmonds (eds.), pp. 184–192. Essex, UK: Longman Scientific and Technical.

Flores Ochoa, J. A., E. Kuon Arce, and R. Samanez Argumedo. 1998. *Qeros—Arte Inka en vasos ceremoniales*. Lima, Peru: Banco de Crédito del Perú.

Galbraith, D. W., G. M. Lambert, J. Macas, and J. Doležel. 1997. Analysis of nuclear DNA content and ploidy in higher plants. In *Current protocols in cytometry*, J. P. Robinson, Z. Darzynkiewicz, P. Dean, A. Orfao, P. Rabinovitch, C. C. Stewart, H. Tanke, and L. Wheelesset. (eds.), pp. 7.6.1–7.6.22. New York: John Wiley & Sons, Inc.

Gaut, B. S. 1998. Molecular clocks and nucleotide substitution rates in higher plants. In *Evolutionary biology*, Volume 30, M. K. Hecht, R. J. Macintyre, and M. T. Clegg (eds.), pp. 93–120. New York and London: Plenum Press.

Gaut, B. S. and J. F. Doebley. 1997. DNA sequence evidence for the segmental allotetraploid origin of maize. *Proceedings of the National Academy of Sciences, USA* 94: 6809–6814.

Gaut, B. S. and B. S. Weir. 1994. Detecting substitution-rate heterogeneity among regions of a nucleotide sequence. *Molecular Biology and Evolution* 11: 620–629.

Ge, S., T. Sang, B. R. Lu, and D. Y. Hong. 1999. Phylogeny of rice genomes with emphasis on origins of allotetraploid species. *Proceedings of the National Academy of Sciences, USA* 96: 14400–14405.

Goloboff, P. 1998. Nona computer program. Tucumán, Argentina. Distributed by the author.

Grant, V. 1971. *Plant speciation*. New York: Colombia University Press.

Guamán, S. 1997. *Conservación, in situ caracterización y evaluación de la biodiversidad de oca* (Oxalis tuberosa) *y papalisa* (Ullucus tuberosus) *en Candelaria (Chapare) y Pocanche (Ayopaya)*. Ingeniero Agrónomo thesis, Universidad Mayor de San Simón, Cochabamba, Bolivia.

Harris, D. R. 1969. Agricultural systems, ecosystems and the origins of agriculture. In *The domestication and exploitation of plants and animals*, P. J. Ucko and G. W. Dimbleby (eds.), pp. 3–15. London: Duckworth.

Hastings, C. M. 1986. *The eastern frontier: Settlement and subsistence in the Andean margins of Central Peru*. Ph. D. diss., University of Michigan, Ann Arbor: University Microfilms.

Hastorf, C. A. 1983. *Prehistoric agricultural intensification and political development in the Jauja region of Central Peru*. Ph. D. diss., University of California—Los Angeles. Ann Arbor: University Microfilms.

———. 1993. *Agriculture and the onset of political inequality before the Inka*. Cambridge: Cambridge University Press.

Hawkes, J. G. 1983. *The diversity of crop plants*. Cambridge, Mass.: Harvard University Press.

Heitz, E. 1927. Über multiple und aberrante Chromosomenzahlen. *Abhandlungen aus dem Gebiete der Naturwissenschaften* 21: 47–57.

Higgins, D. G. and P. M. Sharp. 1989. Fast and sensitive multiple sequence alignments on a microcomputer. *CABIOS* 5: 151–153.

Hilu, K. W. 1993. Polyploidy and the evolution of domesticated plants. *American Journal of Botany* 80: 1494–1499.

Hodge, W. H. 1949. Tuber foods of the old Incas. *Natural History* 68: 464–470.

———. 1951. Three native tuber foods of the Andes. *Economic Botany* 5: 185–201.

Jackson, R. C. and J. Casey. 1982. Cytogenetic analyses of autopolyploids: Models and methods for triploids to octoploids. *American Journal of Botany* 69: 487–501.

Kaplan, L. and T. F. Lynch. 1999. *Phaseolus* (Fabaceae) in archaeology: AMS radiocarbon dates and their significance for pre-Colombian agriculture. *Economic Botany* 53: 261–272.

Kashkush, K., M. Feldman, and A. A. Levy. 2002. Gene loss, silencing and activation in a newly synthesized wheat allotetraploid. *Genetics* 160: 1651–1659.

Knuth, R. 1914. Ein beitrag zur systematik und geographischen verbreitung de Oxalidaceen. *Engler's Botanische Jahrbücher* 50, supplement: 215–237.

———. 1930. Oxalidaceae. In *Das pflanzenreich IV; Regni vegetabilis conspectus*, A. Engler (ed.), pp. 1–481. Leipzig, Germany: W. Engelmann.

———. 1931. Oxalidaceae novae, post editionem monographiae meae (a. 1930) detectae. *Repertorium Specierum Novarum Regni Vegetabilis* 29: 213–219.

———. 1935. Oxalidaceae 2. *Repertorium Specierum Novarum Regni Vegetabilis* 38: 194–199.

———. 1936. Oxalidaceae 3. *Repertorium Specierum Novarum Regni Vegetabilis* 40: 289–293.

———. 1937. Oxalidaceae. In: F. L. E. Diels, Beitrage zur Kenntniss der Vegetation und Flora von Ecuador. *Bibliotheca Botanica* 116: 99–100.

———. 1940. Oxalidaceae 4. *Repertorium Specierum Novarum Regni Vegetabilis* 48: 1–4.

León, J. 1967. Andean tuber and root crops: Origin and variability. In *Proceedings of the International Symposium on Tropical Root Crops held at University of the West Indies*, E. A. Tai, W. B. Charles, E. F. Iton, P. H. Haynes, and K. A. Leslie (eds.), pp. 118–123. St. Augustine, Trinidad: Department of Crop Science, University of the West Indies.

Lewis, C. E. and J. J. Doyle. 2001. Phylogenetic utility of the nuclear gene malate synthase in the palm family (Arecaceae). *Molecular Phylogenetics and Evolution* 19: 409–420.

Liu, B., J. M. Vega, and M. Feldman. 1998a. Rapid genomic changes in newly synthesized amphiploids of *Triticum* and *Aegilops*. II. Changes in low-copy coding DNA sequences. *Genome* 41: 535–542.

Liu, B., J. M. Vega, G. Segal, S. Abbo, M. Rodova, and M. Feldman. 1998b. Rapid genomic changes in newly synthesized amphiploids of *Triticum* and *Aegilops*. I. Changes in low-copy noncoding DNA sequences. *Genome* 41: 272–277.

Liu, C. J. and J. M. Musial. 2001. The application of chloroplast DNA clones in identifying maternal donors for polyploid species of *Stylosanthes*. *Theoretical and Applied Genetics* 102: 73–77.

Lourteig, A. 2000. *Oxalis* L. subgénero *Monoxalis* (Small) Lourteig, *Oxalis* y *Trifidus* Lourteig. *Bradea* 7(2): 1–629.

Lynch, T. F. 1980. Guitarrero cave in its Andean context. In *Guitarrero Cave—Early man in the Andes*, T. F. Lynch (ed.), pp. 293–320. New York: Academic Press.

Lynch, T. F., R. Gillespie, J. A. J. Gowlett, and R. E. M. Hedges. 1985. Chronology of Guitarrero Cave, Peru. *Science* 229: 864–867.

Martins, R. 1976. *New archaeological techniques for the study of ancient root crops in Peru*. Ph.D. diss., University of Birmingham, Birmingham, UK.

Matassi, G., P. M. Sharp, and C. Gautier. 1999. Chromosomal location effects on gene sequence evolution in mammals. *Current Biology* 9: 786–791.

McDade, L. A. 1990. Hybrids and phylogenetic systematics. I. Patterns of character expression in hybrids and their implications for cladistic analysis. *Evolution* 44: 1685–1700.

Medina Hinostroza, T. C. 1994. *Contaje Cromosómico de la Oca* (*Oxalis tuberosa* Molina) *conservada in vitro*. Ingeniero Agrónomo thesis, Universidad Nacional del Centro del Perú, Huancayo.

Ozkan, H., A. A. Levy, and M. Feldman. 2001. Allopolyploidy-induced rapid genome evolution in the wheat (*Aegilops–Triticum*) group. *Plant Cell* 13: 1735–1747.

Parokonny, A. S., A. Kenton, Y. Y. Gleba, and M. D. Bennett. 1994. The fate of recombinant chromosomes and genome interaction in *Nicotiana* asymmetric somatic hybrids and their sexual progeny. *Theoretical and Applied Genetics* 89: 488–497.

Pearsall, D. M. 1992. The origins of plant cultivation in South America. In *The Origins of Agriculture: An international perspective*, C. W. Cowan and P. J. Watson (eds.), pp. 173–205. Washington, D. C.: Smithsonian Institution Press.

Pikaard, C. S. 2001. Genomic change and gene silencing in polyploids. *Trends in Genetics* 17: 675–677.

Potter, D., J. J. Luby, and R. E. Harrison. 2000. Phylogenetic relationships among species of *Fragaria* (Rosaceae) inferred from non-coding nuclear and chloroplast DNA sequences. *Systematic Botany* 25: 337–348.

Rieseberg, L. H. and N. C. Ellstrand. 1993. What can molecular and morphological markers tell us about plant hybridization? *Critical Reviews in Plant Sciences* 12: 213–241.

Sauer, C. O. 1952. *Agricultural origins and dispersals*. New York: American Geographical Society.

Segraves, K. A., J. N. Thompson, P. S. Soltis, and D. E. Soltis. 1999. Multiple origins of polyploidy and the geographic structure of *Heuchera grossulariifolia*. *Molecular Ecology* 8: 253–262.

Shaked, H., K. Kashkush, H. Ozkan, M. Feldman, and A. A. Levy. 2001. Sequence elimination and cytosine methylation are rapid and reproducible responses of the genome to wide hybridization and allopolyploidy in wheat. *Plant Cell* 13: 1749–1759.

Smith, C. E. 1980. Plant remains from Guitarrero Cave. In *Guitarrero Cave—Early man in the Andes*, T. F. Lynch (ed.), pp. 87–119. New York: Academic Press.

Soltis, D. E. and P. S. Soltis. 1989. Allopolyploid speciation in *Tragopogon*: Insights from chloroplast DNA. *American Journal of Botany* 76: 1119–1124.

———. 1990. Chloroplast DNA and nuclear rDNA variation: Insights into autopolyploid and allopolyploid evolution. In *Biological approaches and evolutionary trends in plants*, S. Kawano (ed.), pp. 97–117. San Diego, Cal.: Academic Press.

———. 1993. Molecular data and the dynamic nature of polyploidy. *Critical Reviews in Plant Sciences* 12: 243–273.

———. 1999. Polyploidy: Recurrent formation and genome evolution. *Trends in Ecology and Evolution* 14: 348–352.

Soltis, P. S., J. J. Doyle, and D. E. Soltis. 1992. Molecular data and polyploid evolution in plants. In *Molecular systematics of plants*, P. S. Soltis, D. E. Soltis, and J. J. Doyle (eds.), pp. 177–201. New York: Chapman and Hall.

Song, K., P. Lu, K. Tang, and T. C. Osborn. 1995. Rapid genome change in synthetic polyploids of *Brassica* and its implications for polyploid evolution. *Proceedings of the National Academy of Sciences, USA* 92: 7719–7723.

Stebbins, G. L. 1947. Types of polyploids: Their classification and significance. *Advances in Genetics* 1: 403–429.

———. 1950. *Variation and evolution in plants*. New York: Colombia University Press.

Swofford, D. L. 1991. *PAUP*: Phylogenetic Analysis Using Parsimony*, Version 3.1.1. Champaign, Ill.: Illinois Natural History Survey.

Talledo, D. and C. Escobar. 1995. Citogenética de *Oxalis tuberosa*: Ciclo celular y número cromosómico. *Biotempo* (Universidad Ricardo Palma, Lima, Peru) 2: 33–46.

Tischer, E., S. DasSarma, and H. M. Goodman. 1986. Nucleotide sequence of an alfalfa *Medicago sativa* glutamine synthetase gene. *Molecular and General Genetics* 203: 221–229.

Towle, M. A. 1961. *The ethnobotany of pre-Colombian Peru*. Viking Fund Publications in Anthropology. New York: Wenner-Gren Foundation for Anthropological Research.

Valladolid, A. 1996. *Niveles de ploidía de la oca (Oxalis tuberosa Mol.) y sus parientes silvestres*. M.Sc. thesis, Universidad Nacional Agraria La Molina, Lima, Peru.

Valladolid, A., C. Arbizu, and D. Talledo. 1994. Niveles de ploidía de la Oca (*Oxalis tuberosa* Mol.) y sus parientes silvestres. *Agro Sur* (Universidad Austral de Chile, Facultad de Ciencias Agrarias) 22 (número especial): 11–12.

Vargas, C. C.1981. Plant motifs on Inca ceremonial vases from Peru. *Botanical Journal of the Linnean Society* 82: 313–325.

Vavilov, N. I. 1992. *Origin and geography of cultivated plants*. (Translation by D. Löve of Russian text originally published during 1920–1940). Cambridge: Cambridge University Press.

Vinueza Vela, J. F. 1997. *Evaluación y caracterización citogenética de 20 entradas de oca (Oxalis tuberosa Molina) recolectadas en Ecuador*. Ingeniero Agrónomo thesis, Universidad Central del Ecuador, Cutuglahua, Pichincha, Ecuador.

Wendel, J. F. 1989. New World tetraploid cottons contain Old World cytoplasm. *Proceedings of the National Academy of Sciences, USA* 86: 4132–4136.

———. 1995. Cotton-*Gossypium* (Malvaceae). In *Evolution of crop plants*, J. Smartt and N. W. Simmonds (eds.), pp. 358–366. Essex, UK: Longman Scientific and Technical.

———. 2000. Genome evolution in polyploids. *Plant Molecular Biology* 42: 225–249.

Wendel, J. F. and V. A. Albert. 1992. Phylogenetics of the cotton genus (*Gossypium*)-Character-state weighted parsimony analysis of chloroplast-DNA restriction site data and its systematic and biogeographic implications. *Systematic Botany* 17: 115–143.

Wendel, J. F., A. Schnabel, and T. Seelman. 1995. Bidirectional interlocus concerted evolution following allopolyploid speciation in cotton *(Gossypium)*. *Proceedings of the National Academy of Sciences, USA* 92: 280–284.

Yácovleff, E. and F. L. Herrera. 1934. El mundo vegetal de los antiguos Peruanos. *Revista del Museo Nacional-Lima* 3: 235–323.

Zeven, A. C. 1980. Polyploidy and plant domestication: The origin and survival of polyploids in cytotype mixtures. In *Polyploidy-Biological relevance*, W. H. Lewis (ed.), pp. 385–407. New York: Plenum Press.

SECTION THREE

ARCHAEOLOGICAL DOCUMENTATION OF ANIMAL DOMESTICATION

Melinda A. Zeder, Section Editor

CHAPTER 13

Archaeological Approaches to Documenting Animal Domestication

MELINDA A. ZEDER

Introduction

Documenting animal domestication in the archaeological record is somewhat more complicated than it is for plants. Detecting initial domestication in plants generally involves identifying a range of direct morphological responses to domestication (see Chapter 2). Whether in the form of the largely automatic responses to domestication, such as changes in seed germination and dispersal mechanisms or changes that result from more conscious human selection for increased productivity (e.g., the development of larger or more numerous fruits), the domestic partnership between plants and humans leaves its mark in a number of distinct morphological changes, detectable on both the macro- and micromorphological level. New selective pressures on animals undergoing domestication, in contrast, are more likely to focus on behavioral attributes that may be only indirectly linked to any morphological change. Of all the behavioral traits that make certain animal species more attractive candidates for domestication than others (e.g., a social structure based on a dominance hierarchy that can be co-opted by humans, tolerance for penning or living in crowded conditions, and breeding in captivity; see Clutton-Brock 1999; Diamond 2002), it is the possession of a more placid, tractable, less wary nature that is the single most important factor making certain individuals within these target species potentially more suitable domesticates. Thus while the process of domestication in plants directly selects for morphological changes that can be readily identified in the archaeological record, new selective pressures associated with animal domestication are likely to focus first on a reduction of aggression and wariness, which will be much more difficult to detect archaeologically.

As a result, the archaeological documentation of animal domestication has tended to rely not only on various morphological markers, but also on a wide range of nonmorphological markers that can be identified as reflecting human attempts to manage animal populations. The four chapters in this section feature the work of researchers who have been actively developing innovative ways of documenting domestication in different animal species: goats (Chapter 14), pigs (Chapter 15), South American camelids (Chapter 16), and horses (Chapter 17). In each chapter, a variety of different traditional and emerging approaches to documenting animal domestication are considered as they pertain to these species and recommendations are made for future work. In this overview chapter, I draw from these case study examples, as well as from other relevant recent research, to consider some of the challenges and future opportunities for documenting animal domestication in the archaeological record.

Morphological Markers

Morphological markers of animal domestication fall into two general categories: those that reflect genetically driven, selective responses to domestication passed from one generation to the next, and those that represent plastic responses in individual animals that are more episodic in nature.

Genetically Driven Markers

Both Kruska (1988) and Hemmer (1990) have argued that domestic animals display a constellation of behavioral, physiological, and morphological characteristics that can be directly tied to selection for less aggressive behavior, as well as to a decline of "environmental appreciation" in animals that have become increasingly dependent on humans. Most notably, domestic animals display an array of more "juvenile," or "paedomorphic," behavioral traits that include greater gregariousness, less wariness, greater playfulness, and intensified sexuality. In turn, these more paedomorphic behaviors seem to be linked to a series of paedomorphic physiological and morphological characteristics, such as earlier onset of sexual maturity and more frequent receptivity, smaller brain size and other changes in neurological organization, shortening of the snout, tooth-size reduction and changes in tooth number, smaller bodies, and lessening of sexual dimorphism.

Of these later morphological traits, changes in the facial structure and in the size and placement of teeth have been particularly useful in the archaeological documentation of animal domestication in certain species. In his doctoral dissertation and a series of path-breaking articles, Darcy Moray (1986, 1990, 1992, 1994) demonstrated that the principal differentiation between dogs and wolves lies in the juvenilization of the skull (expressed in a shorter face, steeped forehead, and wider cranial dimensions), resulting in the characteristic tooth crowding and reduction in tooth size often used as a marker of initial domestication in dogs. Moray related these morphological changes to changes in developmental rate and timing, which he maintained were essentially automatic evolutionary responses to new selective pressures growing out of an evolving mutualistic relationship with humans. Similar changes in the facial region and in teeth have been used to document domestication in pigs, where they are usually linked to sexual precocity and associated paedomorphic behaviors and physiological changes that come about with domestication (Boettger 1958 in Flannery 1983). In this volume, Albarella et al. demonstrate that changes in tooth size in pigs, especially a progressive reduction in the length of molars (most strongly expressed in the third molar), is one of the earliest morphological changes seen in pigs undergoing domestication. Both dogs and pigs have omnivorous diets that overlap with those of humans, and it is possible that the opening chapter of the domestication of these species was set in motion when less wary individuals approached human settlements to feed off human refuse. This kind of automatic, self-selection for reduced aggression may, in turn, be linked to the manifestation of other paedomorphic morphological features (like snout shortening, tooth crowding, and reduction in the length of molars) seen early on in domestication of dogs and pigs.

Other morphological markers used to document animal domestication may have come about when humans assumed responsibility for selecting breeding partners in managed herds, thereby neutralizing the strong selective advantage of a variety of morphological features that wild animals use to attract and compete for mates. Most notable are the changes in the size and shape of horns seen in essentially all domestic bovid species. Once humans controlled mate selection in these animals, horn forms that may have been disadvantageous in the wild could manifest themselves without having a negative impact on breeding success. In fact, the large horns of male animals that are energetically costly to grow and carry may have been actively selected against once humans took over the role of selecting mating partners. There may even be some linkage between less aggressive behavior and smaller horns, especially in males. Thus, there are clear changes in the size and shape of horns in animals like sheep and goats (Chapter 14) and in cattle (Grigson 1978) that can be linked directly to the process of domestication. However, as discussed in the chapter by Zeder, these changes may take some time to manifest themselves and are often difficult to see in highly fragmented archaeological material.

Change in body size is the most widely used marker of animal domestication today. Generally, this involves a decrease in body size, which is argued to be a definitive, quick response to domestication experienced by wide range of primary domestic species including cattle (Grigson 1969), pigs (Bökönyi 1974; Hongo and Meadow 1998; Chapter 15), sheep and goats (Uerpmann 1978, 1979; Meadow 1989), and dogs (Morey 1994). This reduction in body size in domestic animals has been attributed to a variety of causal mechanisms. Some see body-size reduction as one of the number of morphological traits linked to the selection for tamer, less aggressive animals (Hemmer 1990; Kruska 1988; Moray 1994). Others have attributed the reduction of body size in domesticates to the removal of the selective advantage of large body size for breeding success (Zohary et al. 1998), a factor that would have a particularly profound impact on males. Body-size reduction has also been argued to have been first a plastic response to impoverished diets that later became genetically encoded as a selective advantage for smaller body size (Meadow 1989).

Recent studies, however, have called into question the utility of body-size reduction as an initial marker of animal domestication in goats, and likely in other animals as well (Zeder 2001, 2003, in press, Chapter 14; Vigne et al. 2005). In a study of modern skeletal collections, the impact of domestic status in modern goats was found to be limited to a reduction in the length of long bones in males (an attribute of little use in archaeological assemblages where intact long bones are seldom found) and to a slight decrease in the robusticity in the breadth of male postcranial bones. In modern goats, there is no apparent difference in body size between wild and domestic females. Instead, sex is the single most important factor affecting size in goats, with males being absolutely larger than females in all dimensions once they are older than one year. Geography seems to be the second most important factor affecting size in goats, with a strong north-south decrease in body size associated with increasing temperatures. In a parallel study of archaeological collections that bracket the transition from hunting to herding of goats in the eastern Fertile Crescent region of today's Iran and Iraq, evidence that other researchers had mistakenly interpreted as clearly supporting body-size reduction in one of the earliest archaeological assemblages of domestic goats was instead caused by a demographic shift in the adult portion of a managed herd (made up mostly of females), taphonomic bias against the recovery of the bones of males killed at young ages, and distortions introduced by commonly used osteometric methods. Clear signs of a downward shift in body size do not appear in domestic goats until at least 1,000 years after other solid evidence of herd management. However, it is difficult to say whether the smaller size of these goats is directly attributable to new selective factors introduced by domestication, to the movement of new varieties of domestic goats into the region, or to other factors shown to have caused a similar reduction in the size of both wild goats and gazelles (an animal never domesticated) over the course of the Holocene.

Other chapters in this section also highlight some of the challenges to size as a marker of initial animal domestication. South American camelids, which paradoxically seem to get larger with domestication, present a special problem in this regard since there are four different-sized camelid species in the Andes. The smallest of these species is the vicuña, followed by its domestic descendent, the alpaca, followed by the guanaco, with the guanaco's domestic descendent, the llama, as the largest of the four. And while there is apparently little sexual dimorphism in camelid species, Mengoni Goñalons and Yacobaccio present compelling evidence for a high degree of regional variation in the size of all four forms (with size reduction running south to north, along with increasing temperatures). There is also evidence for temporal variation in the size of wild camelid species following the last Ice Age, similar to that seen by Zeder in wild goats and gazelles, and by other researchers in a variety of different wild species (Davis 1981; Guthrie 2003). This high degree of taxonomic, spatial, and temporal variation in body size makes it difficult to sort out size changes in South American camelids that might be related to domestication. Mengoni Goñalons and Yacobaccio do, however, point to the sudden appearance of a particularly large variety of camelid in northwestern Argentina at about 4400 BP as part of their case for the appearance of domesticated llama in this region.

For Olsen, uncertainty about the size of the wild progenitor of the domestic horse, the huge geographic range of horses, and the likely presence of both domestic and wild horses in Eurasian archaeological assemblages render size essentially useless as a marker of initial horse domestication. In contrast, Albarella et al. still use body-size reduction as a marker of initial domestication in pigs, interpreting a classic "peak and tail" size distribution of pigs in assemblages from several European sites as evidence for the presence of a large number of small-bodied domesticates and a few larger-bodied wild pigs. They also, however, raise cautions about the impact of region and sex on size, and they highlight several important allometric variations in size changes in teeth and postcrania in pigs that suggest different causal mechanisms are responsible for the nature and timing of size change in different body parts of pigs.

This last point underscores a key issue in the use of morphological change as a marker of animal domestication. Clearly, there are a number of factors that may cause morphological change in different features at different points in the domestication process. The impact of these factors will also vary depending on the biology of different species and the nature of their interaction with humans. Changes in the crania and dentition seen in pigs and dogs, for example, seem to reflect the leading edge of the development of a mutualism between these animals and humans that precedes any deliberate attempt by humans to manipulate breeding partners. It is possible that such changes will be more strongly expressed in species like dogs and pigs that embark on the path toward domestication as commensals drawn to human habitations to scavenge on refuse and stored foodstuff. Although selection for reduced aggression clearly played a role in the domestication of other species, the process of their domestication may have skipped the initial "getting-to-know-you" phase seen in pigs and dogs, which may be why paedomorphic cranial changes so clearly expressed in these species are not seen in species like sheep and goats. Evidence for relaxation of selection for morphological characteristics associated with breeding success in the wild, like horn form and possibly size in bovid species, will not happen until humans begin to deliberately control mate selection. Genetic isolation from wild progenitor populations will also result in a wide range of possible morphological changes. These changes will result from random genetic drift, as well as from more directed adaptations to the new and varied environments encountered by domestic populations once they are moved away from the site of initial domestication, which for an increasing number of species seems to be within the natural habitat of their wild progenitors.

These different causal factors will kick in at different points along the domestication process. The mutualism between humans and pigs and dogs may have developed thousands of years before humans began to deliberately control breeding and shield them from predators, which likely occurred at a much earlier point during the domestic relationship with pigs than with dogs. Genetic isolation from wild progenitor populations and the movement of managed herds into new and challenging environments seems to have taken many hundreds of years in a number of domestic species (see Chapter 14).

Changes in body size, thus, may be related to a range of factors, including reduced aggression, controlled breeding, diet quality, and reproductive isolation. Moreover, depending on the nature of the causal factors, body-size changes may be expressed in some dimensions but not in others (e.g., changes in length of limbs versus robustness); they may have more impact on one sex than another (e.g., greater body-size reduction in males than in females); or they may vary depending on the conditions encountered in the new environments into which domesticates are moved (e.g., selective pressures for smaller body sizes in areas with impoverished resources versus areas with more optimal pasturage). It is also possible that different causal factors will have similar morphological consequences (e.g., both selection for less aggressive individuals and control of mate selection could result in body-size reduction in males), but will differ in the degree of impact at different points in the domestication process. At the same time, there are other factors that affect body size in both wild and domestic species that have nothing to do with domestication (e.g., climate changes, habitat destruction). The work presented in the four chapters in this section highlights the complexity of the use of genetically controlled morphological markers in documenting animal domestication. These case-study chapters underscore the need to think critically about how the process of domestication may have affected the species at hand, and how and when this process might be expected to result in detectable morphological change.

Plastic Responses to Domestication

Morphological change can also come about as a plastic response to conditions experienced by animals under human control. Morphological changes caused by genetic selection operate on a population level and are passed on from one generation to the next. In contrast, plastic responses to domestication operate on individual animals and will vary in degree and nature depending on highly localized factors that may change over time. A beveling on the lower second premolars of horses in the Eurasian steppe has been interpreted as definitive evidence of bit wear and therefore domestication (Brown and Anthony 1998). Olsen's chapter, however, raises real concerns about the utility of such a marker in documenting early horse domestication, noting that "bit wear" morphology can be found on the teeth of Pleistocene-age horses from the New World, which cannot have been caused by bits. Albarella et al. point to dental irregularities caused by stress and malnutrition as a sign of human (mis)management in pigs. But their study also indicates that the frequency of hypoplasias in the teeth of pigs increases relatively late in the process of pig domestication, well after other signs of domestication are clearly expressed. Their work highlights, once again, the episodic nature of such markers and the need to discriminate between pathologies resulting from domestication and those that occur in wild populations.

Others have argued that tethering or penning of domesticates resulted in pathologies that can be used as markers of domestication, like the deformed lower foot bones of caprines seen at sites in the southern Levant that Köhler-Rollefson (1989) attributes to restricting early domesticates' movement. Penning and overcrowding of domesticates can also result in high mortality, especially in young animals, because of the rapid spread of infectious disease. Wheeler (1984a, 1984b, 1995, 1998), for example, has proposed that the dramatic increase in the number of neonatal camelids at the site of Telamachay in the Peruvian Andes was a sign of such infection and thus a signal of initial domestication of small camelids (probably alpacas) at about 6000 BP. Mengoni Goñalons and Yacobaccio, however, find shifting evidence of high neonatal mortality over the long course of occupation at the site of Quebrada Seca 3 in northwest Argentina, occupied from about 9000 to 8000 BP, which they relate to variations in the seasonal occupation of the site and the opportunistic hunting of newborns during certain seasons of the year.

Recently, a number of researchers have begun to explore the potential of dietary shifts as detected by isotopic analysis as a marker of initial domestication in animals (e.g., Balasse et al. 2000, for sheep and goats, and Chapter 15, for pigs). Again, it is important to remember that dietary conditions of early domestic animals will vary depending on the environment in which they are raised and the management practices used by their human masters. Factors like climate change or human encroachment can radically affect the diet of wild animals as well, and care needs to be taken in distinguishing human-controlled dietary shifts from those resulting from factors not related to domestication. Therefore, it would be a mistake to attempt to apply any single isotopic or other chemically detected shift in diet as a definitive marker of domestication in any species in any single region, and certainly not as a widely applicable marker across species and regions.

Nonmorphological Markers

Given the difficulties in the use of morphological markers to document animal domestication, archaeozoologists have developed a number of nonmorphological markers aimed at detecting animal domestication. Like the more plastic morphological markers discussed above, these markers seek to detect evidence of human management or control of animals, especially those that may precede any detectable, genetically driven morphological change.

Demographic Profiling

Demographic profiling was one of the first of these nonmorphological markers to be applied to documenting animal domestication (Coon 1951; Dyson 1953; Perkins 1964; Hole et al. 1969). The technique is based on the assumption that the age and sex of animals taken by hunters interested in maximizing the return from their hunt will differ from the age and sex of those harvested by herders interested in promoting the long-term growth of their herds. This marker has fallen out of favor in recent years, especially by those attempting to document the domestication of primary livestock species in the Near East. As outlined by Zeder in this volume, the decline in the popularity of this technique can, in part, be attributed to doubts about how well the theoretically derived harvest strategies model the actual practices of ancient hunters and herders. Of particular concern has been whether certain selective hunting practices that target young animals can be distinguished from a herder's emphasis on the slaughter of young males, and whether herders uniformly follow an expected pattern of young male slaughter and prolonged female survivorship (e.g., Collier and White 1976; Meadow 1989). The principal impediment to the effective use of demographic profiling as a leading-edge marker of animal domestication, however, has been limitations in the techniques used to reconstruct these profiles. There has been little standardization in how long-bone fusion and dental eruption and wear patterns are used to construct slaughter profiles in archaeological animal populations. Closely related species, which nonetheless may have been exploited quite differently (e.g., sheep and goats, alpacas and llamas), have been lumped together in constructing harvest profiles because of difficulties in distinguishing between the two species using fragmented material. Most importantly, until recently, it has not been possible to construct separate harvest profiles for male and female animals, which is essential in order to distinguish certain hunting practices from herding.

Recently Zeder has taken up Hesse's (1978) long-ignored early efforts to bring greater resolution to the construction of harvest profiles of ancient goat populations (Zeder and Hesse 2000; Zeder 1999, 2003, 2005, and Chapter 14). Drawing from baseline information provided by a study of modern skeletal collections, she has devised a method for computing sex-specific harvest profiles that is capable of tracking the slaughter of male and female animals with some precision. Using this technique has made it possible to demonstrate clear signs of human management in the form of selective slaughter of young male goats and prolonged female survivorship in the highland Zagros region of the Near East at about 10,000 BP—a signal that precedes any evidence of morphological change in goats by 500 to 1,000 years. This technique has yet to be applied to other species, although it should be useful in documenting domestication in any other species that exhibit clear size separation between males and females (e.g., sheep, cattle, and pigs).

Other chapters in this section highlight some of the complexities to be addressed before broader application of this approach will be possible. Albarella et al. cite evidence for an increase in the proportion of young in Neolithic pig populations in southeastern Anatolia as one of the lines of evidence pointing to initial domestication. Indeed, the postcranial "peak and tail" size distributions found in a number of European collections, which they interpret as indicative of a mix of smaller domestic and larger wild pigs, are consistent with the female-biased size distributions seen in early domestic goats in the Zagros region and suggest the promise of applying age and sex profiling to pigs. Albarella et al., however, raise several important caveats about the application of demographic profiling to documenting pig domestication. In particular, they highlight the need for a better understanding of the population dynamics of wild pigs in modeling likely hunting strategies, and the difficulty this absence of understanding poses in trying to distinguishing between selective hunting of pigs and herding using demographic profiles.

Mengoni Goñalons and Yacobaccio also point to an increase in younger camelids in archaeological assemblages from the Andes as one leg of their case for camelid domestication. Here, the more effective use of demographic profiling to document South American camelid domestication has been hampered by inconsistencies and lack of resolution in the current methods for determining age at death using camelid long bones and teeth, the difficulty in distinguishing among the four different camelid species in the region, and the apparent lack of clear-cut sexual dimorphism within these species. Moreover, the use of camelids for fiber production also has an impact on culling strategies that, depending on how early these more specialized pastoral systems arose, might be difficult to differentiate from various selective hunting strategies.

The use of demographic profiling for documenting horse domestication also has its own special complexities. Mongol herds today provide useful model harvest strategies of the selective slaughter of stallions at 2.5 years of age and the prolonged survivorship of mares, much like that seen in modern sheep and goat herds. However, Olsen argues that castration and horseback riding were early developments in the history of horse domestication, and that both would result in more prolonged survivorship of males than expected under harvest strategies directed solely at meat and milk offtake. The ritual practice of male horse sacrifice and internment both in human burials and in ritual pits outside of houses might further distort the classic harvest-for-meat profile. To further complicate matters, she suggests that domestic horses were used to aid in the hunting of wild horses and other large game, both for capturing wild horses and for transporting their meat back to home bases. Midden deposits from these sites would, then, contain large numbers of both herded and hunted horses, making it especially difficult to distinguish a clear signal of the harvest of domestic animals. Again, the apparent lack of diagnostic differences between male and female horses (both in size and other features) and problems with the methods for computing age profiles represent special challenges to the use of demographic data for documenting initial domestication in horses.

In all these instances, more effective use of demographic profiling as a means of documenting animal domestication will require the following three things. First, researchers need to build more detailed, more nuanced models of harvest profiles under a variety of hunting and herding strategies, drawing upon a better understanding of the population dynamics of both wild and managed herds. These models must be constructed on a species-by-species basis, taking into account the species-specific population dynamics for wild populations and for domestic herds under a number of different possible resource extraction strategies. Second, modern baseline studies, similar to those recently carried out for sheep and goats (Chapter 14), are needed to provide sold empirical data on the impact of sex and age on the size of skeletal elements, on the sequence and rate of bone fusion and tooth eruption and wear, and on the morphological distinctions of closely related, sympatric species. Empirical verification of the degree of dimorphism in the size of male and female animals is especially important given the impressionistic nature of many previous assessments of size variation. Finally, application of these techniques requires more thorough study of larger, more carefully collected assemblages of animal bones than is commonly done using today's methods.

Zoogeography and Abundance

The appearance of a potential domestic species outside its presumed natural range is often taken as a signal of human involvement in the movement of animals, either as already domesticated herds or as captive wild animals undergoing domestication. As highlighted in the chapters in this section, however, zoogeographic data can sometimes be problematic in making the case for animal domestication. The timing of the reappearance of horses in the Near East and in Europe in the Early and Mid-Holocene, for example, has been used to bracket the date of their initial domestication in the Eurasian steppe. And yet, since horses were found in both the Near East and Europe during the Pleistocene, Olsen maintains that it is hard to know if a Holocene-age horse in either of these regions represents a member of a relic wild population or an introduced domesticate. Often the attribution of these horses as either wild or domestic in these regions is based solely, and somewhat circularly, on a researcher's preconception of the likely time of horse domestication and diffusion out of the Eurasian steppe. Moreover, without direct dating, it is difficult to know if a specimen is contemporary with the strata in which it was found or with a later intrusion into earlier strata.

Zoogeographic data are particularly problematic in the case of a species like the pig, with a broad natural distribution. Albarella et al. report that the introduction of presumed non-native pigs to islands situated at the margins of this wide geographic range (to Gotland, 80 km off the coast of Sweden, to Cyprus, 60–70 km from Turkey, and to New Guinea), can, and has, been alternatively argued as evidence for the import of domestic pigs, as a game-stocking strategy with no follow-on domestication, or as the radiation of wild pigs without human assistance.

Moreover, it is also difficult to know whether present-day distributions of progenitor populations, subjected to millennia of hunting and human encroachment, are adequate proxies for the distribution of these species during the Early Holocene. Present-day relic populations of guanaco and vicuña, for example, clearly cannot be used to delimit their likely range 10,000 years ago (Wheeler 1995). It is similarly difficult to project the distribution of sheep and goats in the Near East based on modern-day distributions. Instead, approximations of the range of these animals during initial phases of their domestication are best provided by studies of Late Pleistocene and very Early Holocene assemblages that predate this transition (i.e., Uerpmann 1979; Horwitz and Ducos 1998).

In the case of cattle, whose wild progenitor, the auroch, is now extinct, zoogeographic arguments are based solely on hypothetical reconstructions of the geographic range of wild cattle based on paleontological and archaeological remains (which show that auroch ranged from Europe through Anatolia and the Near East and across Northern Africa; see Churcher 1972, 1983; Gautier 1984, 1987, 1988), coupled with assumptions about the probable ecological requirements of these extinct animals over their wide geographic range. The presence of fragmentary remains of cattle in Early Holocene contexts in the Sahara age sites at about 9000 BP in regions presumed too arid for wild cattle, for example, has been interpreted as evidence for the seasonal use of this arid region by mobile cattle pastoralists and, because of the early date, for an independent center of cattle domestication in North Africa (Wendorf et al. 1984; Gautier 1984, 1987; Close and Wendorf 1992). Subsequent researchers, however, have challenged these conclusions (Smith 1984, 1986; and see Marshall 1994), questioning both the identification and dating of these fragmentary remains, as well as assumptions made by Wendorf et al. about the ecological limitations of extinct North African wild cattle.

The sudden increase in the abundance of a potential domestic species is often seen as a potent marker of domestication. When this increase takes place within the presumed natural habitat of the species, however, it is difficult to say whether it signals the domestication of the species or simply an intensification of hunting. While there is good evidence for an increase in the use of both small and large camelids throughout the Andes beginning at about 8500 BP, Mengoni Goñalons and Yacobaccio see this increase as signaling the beginning of a long-term process in which a progressively more intensive focus on camelid hunting evolved into their domestication. They then rely on a variety of independent morphological and other nonmorphological markers to determine the point during this process when the threshold of domestication was crossed. Similarly, Olsen plots the increasing proportion of horses in archaeological assemblages from Eastern Europe to Kazakhstan to identify the likely epicenter of horse domestication, but relies on other lines of inference to make the case for domestication itself.

Efforts to document the domestication of sheep and goats in the Near East have drawn heavily on both zoogeographic and abundance markers. Early work in the Zagros region of the eastern Fertile Crescent effectively used zoogeographic arguments to buttress the case for the domestic status of goats on the arid, lowland Deh Luran Plain (Hole et al. 1969), and, less convincingly, to make a case for the domestic status of sheep in highland Iraq (Perkins 1964). These markers have played a much larger role in more recent work, which focuses on the piedmont and steppe regions of the Taurus Mountains in the central Fertile Crescent and on its far western arm in the Levant. Beginning about 9000–8500 BP, a many-millennia-long concentration on gazelles was replaced in this region by a dramatic increase in the importance of caprovines, especially goats. Yet depending on a somewhat murky understanding of the natural distribution of wild sheep and goats across the region at this time, various researchers interpret the increased use of caprines at certain sites as evidence for the introduction of domesticates into the region (Helmer 1992; Horwitz and Ducos 1998; Legge and Rowley-Conwy 2000; Bar-Yosef 2000), or as part of a case for local domestication developing out of intensive exploitation of long-established populations of wild sheep and goats (Horwitz 2003, Peters et al. 1999).

Other Evidence of Human Control

Different types of more circumstantial evidence are also often marshaled to make a case for the presence of domestic animals. Remnants of pens or corrals, sometimes with associated dung pellets, have been used as evidence for domestication of camelids both in South America (Mengoni Goñalons and Yacobaccio, Chapter 16, this volume) and in the Old World (Compagnoni and Tosi 1978). Miller (2002) argues that an increase in the proportion of weedy species in archaeobotanical assemblages in the Near East at about 8500 BP signals the use of dung as a major source or fuel and, by extension, the presence of domestic sheep and goats. Perhaps one of the most extensive uses of such indirect evidence is found in Olsen's contribution, in which she cites a wide range of data sets (skeletal part distributions, bone tool types, changes in lithic sources and technology, site architecture, soil micromorphology, and the possibility of detecting residues of mares' milk) to build a strong circumstantial case for the use of domestic horses by the Botai culture people of the Eurasian steppe in the fourth millennium BC.

Geographic Focus of Research

One theme that cuts across the contributions in this section of the volume is the important role geographic focus plays in shaping our understanding of animal domestication. Nowhere is the impact of geographic context clearer than in the study of South American camelid domestication. Until recently, the case for camelid domestication in South America was based almost exclusively on data derived from a handful of Peruvian sites in the Central Andes that have been carefully analyzed and widely published in English language venues (Wing 1972, 1975; Wheeler Pires-Ferreira 1975; Wheeler Pires-Ferreira et al. 1976; Wheeler et al. 1977; Kent 1982; Moore 1989). The focus of this large body of work has resulted in a Peru-centric view of the process of camelid domestication that sees this region as the epicenter of camelid domestication, with other parts of the Andes playing a more marginal role. Mengoni Goñalons and Yacobaccio's comprehensive survey of the largely Latin American literature on sites from northwest Argentina, Chile, and Bolivia, however, suggests that the process was more widespread. They produce a convincing body of evidence in the form of abundance data, age profiles, and morphological change, showing that people across the central and south-central Andes were all embarking on a similar process that eventually resulted in the domestication of alpacas and llama, perhaps multiple times in multiple places (see also Wheeler 1995 and Chapter 23).

The shifting geographic focus of research in the Near East has also played a profound role in shaping our perception of the domestication of all four primary livestock species (sheep, goats, cattle, and pigs) and the relationship of animal domestication to the domestication of crop plants. Early research in the Fertile Crescent focused on the Zagros region of Iran and Iraq, where the earliest evidence of the domestication of sheep and goats seemed to coincide with, or perhaps even slightly precede, the domestication of cereal grains (Hole 1989). With the shift in research away from politically troubled regions like Iran and Iraq in the 1970s and the explosive growth of research in the Levant, particularly in Israel, Jordan, and the piedmont and steppe regions of northern Syria, a very different picture emerged. Here, there was very early evidence for cereal domestication but no corresponding evidence for animal domestication. Although people in this region may have been auditioning the gazelle for a possible role as a future domesticate, gazelles are not behaviorally suited to domestication. Domestic animals do not appear in the Levant until as much as a millennium after the initial domestication of certain types of barley and wheat, when goats begin to play a major role in local subsistence economies—alternatively attributed to either a local process of domestication (Horwitz 2003) or to the introduction of already domesticated animals from outside the region (Bar Yosef 2000). Emerging evidence from recent work in the foothills and mountain valleys of the Taurus Mountains is pushing back dates for initial domestication of sheep, goats, and pigs, promising, once again, to rewrite the story of the domestication of plants and animals in the region (Peters et al. 1999; Vigne and Buitenhuis 1999; Hongo and Meadow 1998; Ervynck et al. 2002). Indeed, as Zeder points out in this volume, as research emphasis shifts more to the natural habitat of progenitor species of eventual domesticates, and as the realization grows that morphological markers of domestication may be quite delayed in expression and that other, nonmorphological markers may better capture the leading edge of domestication in animals, the story of animal domestication in the Near East will likely stretch even further back in time. It no longer seems particularly profitable to look for any one center of domestication of either plants or animals in the Near East, or even to worry much about whether animals or plants were domesticated first. After the last Ice Age and the ensuing changes in plant communities, people across the entire arc of the Fertile Crescent seem to have adopted an increasingly sedentary, more territorially circumscribed focus on cereals, pulses, and nut trees, as well as on associated herbivore species. The unfolding of this process, however, varied greatly from region to region, depending on the mix of available species, local climatic and environmental conditions, and a variety of demographic, social, and cultural factors that were themselves highly localized. Telling this story, then, requires a much more encompassing, less myopic regional focus than that shaped solely by where research is currently being conducted. It also requires the acceptance of the real possibility that the process of domestication of various animal (and possibly plant) species encompassed more that individual, small areas in the Near East.

The need to broaden the geographic scope of the documentation of animal domestication is driven home by the chapter on pig domestication. Unlike the authors of the other three chapters in this section, Albarella et al. do not focus on a single case-study example, nor do they explicitly set out to identify the earliest evidence for initial pig domestication. Instead, they broaden the lens of inquiry, considering evidence for the origin and diffusion of domestic pigs and the development of different swine herding strategies across the entire range of this widespread species, from England to Japan. Their chapter also underscores the utility of using multiple complementary lines of inquiry in exploring the wide range of relationships between humans and pigs across this broad geographic range. And although Olsen's chapter focuses on the Botai culture of the eastern Urals and northern Kazakhstan, she does not argue that this region was the center of horse domestication on the Eurasian steppe. Instead, she uses this case example to demonstrate how multiple lines of evidence can be brought together to study this complex process, building a paradigm for documenting horse domestication that can be broadly applied to the likely multiple instances of horse domestication that took place across a broad geographic range.

Multiple Markers

The use of multiple markers in the archaeological documentation of animal domestication is another theme that cuts across the chapters in this section, with the mix of markers used varying from species to species. Albarella et al.'s chapter on pig domestication makes the greatest use of morphological markers—both those caused by genetic selection and those that reflect more plastic responses to domestication.

Mengoni Goñalons and Yacobaccio use a more diverse mix of morphological and nonmorphological markers. Olsen's case for horse domestication is based almost exclusively on nonmorphological markers, especially on a wide range of more circumstantial indicators for the use of domestic horses by the Botai. Zeder's chapter reviews the use of various morphological and nonmorphological markers as they have been applied to the documentation of sheep and goats, focusing on an evaluation of the efficacy of two of the most commonly used markers in detecting the initial stages of goat domestication.

Together, these chapters make the point that there is no single prescription for documenting animal domestication that can be universally applied to all species in all regions. There is no one-size-fits-all approach to documenting animal domestication. Instead, researchers need to tailor the methods they use both to the species and to the cultural context at hand. Above all, this must be done with an explicit understanding of how they propose that the domestic partnership between humans and the species progressed and how that relationship would have manifested itself in the various markers they use to document it.

What the studies here also underscore is that different markers will be manifested at different points of the domestication process. In some species, high-resolution demographic profiling techniques may be capable of recognizing the transition as intensive hunting strategies are replaced by herd management. Certain morphological changes may capture the very first steps toward an evolving mutualism between humans and an animal species that precedes any deliberate attempt to control breeding, which itself sets another set of morphological responses into motion. Genetic isolation and adaptation to new environments may result in additional morphological change at a different point in the domestication process. Evidence of disease or dietary stress may not come until an even greater degree of control over animals is asserted. Exploitation of a wider range of resources provided by the animal (e.g., milk, fiber, labor) may leave its own range of morphological and nonmorphological markers.

Documenting Domestication

This raises the question of when along this developmental process a wild animal can be considered to have become a domestic animal. Albarella et al. maintain that the end point of the domestication process is the complete control over an animal population's survival, reproduction, and nutrition. But they stress that there is a long, unfolding process through which a wild population theoretically free of any direct or indirect human influence is transformed into a domestic one, totally dependant on humans. Moreover, they also maintain that both ends of this theoretical continuum—complete freedom from human influence or complete control by humans—are rarely attained in the real world. Any "wild" species that comes into contact with humans, especially those hunted by humans, is influenced in some way by human presence. Moreover, in all but the most recent examples of highly mechanized livestock production, there is a large amount of variability in the degree to which humans are able to completely isolate a "domestic" species from the natural world. Rather than worry about arriving at a neatly defined dividing point between wild and domestic, they prefer to focus on the process itself, using a range of markers to track the dynamic, evolving, mutualistic relationship between humans and animals that lies at the heart of domestication.

Zeder also emphasizes the fact that domestication is a process not a moment, and that documenting domestication requires developing markers able to trace that process. However, she acknowledges that drawing the line between wild and domestic will vary with one's personal conceptual approach to the study of domestication. Those that approach the topic from the point of view of the impact of domestication on the animal will require evidence of genetic isolation from wild populations, usually as marked by morphological change. On the other hand, those that approach the topic from the human end of the domestic partnership will emphasize human goals of controlling animal populations to provide more secure and predictable access to desired animal resources. In this case, demographic profiles that demonstrate human management of breeding would be sufficient to claim that domestication had been achieved, even in the absence of morphological, or genetic, change. More important than quibbling over definitional issues is the need to approach the subject with a clear understanding of how the relationship between humans and the species of interest likely evolved, and how it might be best detected using a wide range of possible markers. Collectively, the chapters in this section point the way as to how this might be done, not only for the species represented here but for animal domesticates in general.

References

Balasse, M., A. Tresset, H. Bocherens, A. Mariotti, and J.-D. Vigne. 2000. Un abattage "post-lactation" sur des bovins domestiques Néolithiques. Étude isotopique des restes osseux du site de Bercy (Paris, France). *Ibex, Journal of Mountain Ecology* 5: 39–48/*Anthropozoologica* 31: 39–48.

Bar Yosef, O. 2000. The context of animal domestication in southwestern Asia. In Vol. 2 of *Archaeozoology of the Near East*, M. Mashkour, A. M. Choyke, and H. Buitenhuis (eds.), pp. 185–195. Groningen: ARC Publications.

Boettger, C. 1958. *Die haustiere Afrikas*. Jena: G. Fischer.

Bökönyi, S. 1974. *History of domestic animals in central and eastern Europe*. Budapest: Akadèmiai Kiadó.

Brown, D. and D. Anthony. 1998. Bit wear, horseback riding and the Botai site in Kazakstan. *Journal of Archaeological Science* 25: 331–347.

Churcher, C. S. 1972. *Late Pleistocene vertebrates from archaeological sites in the Plain of Kom Ombo, Upper Egypt*. Life Sciences Contribution, Royal Ontario Museum 82.

———. 1983. Dakhleh Oasis Project paleontology: Interim report on the 1982 field season. *Journal of the Society for the Study of Egyptian Antiquities* 13: 178–187.

Close, A. and F. Wendorf. 1992. The beginnings of food production in the eastern Sahara. In *Transitions to agriculture in prehistory*, A. B. Gebauer and T. D. Price (eds.), pp. 63–72. Madison, WI: Prehistory Press.

Clutton-Brock, J. 1999. *Domesticated animals*, 2nd edition. London: British Museum of Natural History.

Collier, S. and J. P. White. 1976. Get them young? Age and sex inferences on animal domestication. *American Antiquity* 41: 96–102.

Compagnoni, B. and M. Tosi. 1978. The camel. In *Approaches to faunal analysis in the Middle East*, R. H. Meadow and M. A. Zeder (eds.), pp. 92–106. Cambridge, MA: Peabody Museum.

Coon, C. S. 1951. *Cave explorations in Iran, 1949*. Museum Monographs. Philadelphia: The University Museum.

Davis, S. J. M. 1981. The effects of temperature change and domestication on the body size of Late Pleistocene to Holocene mammals of Israel. *Paleobiology* 7: 101–114.

Diamond, J. 2002. Evolution, consequences, and future of plant and animal domestication. *Nature* 418: 34–41.

Dyson, R. H. 1953. Archaeology and the domestication of animals in the Old World. *American Anthropologist* 55: 661–673.

Ervynck, A., K. Dobney, H. Hongo, and R. Meadow. 2002. Born free! New evidence for the status of pigs from Çayönü Tepesi, Eastern Anatolia. *Paléorient* 27: 47–73.

Flannery, K. V. 1983. Early pig domestication in the Fertile Crescent: A retrospective look. In *The hilly flanks. Essays on the prehistory of Southwest Asia*, T. C. Young, P. E. L. Smith, and P. Mortensen (eds.), pp. 163–188. Studies in Ancient Oriental Civilization 36. Chicago: Oriental Institute, University of Chicago.

Gautier, A. 1984. Archaeozoology of the Bir Kiseiba region, Eastern Sahara. In *Cattle-keepers of the Eastern Sahara*, F. Wendorf, R. Schild, and A. E. Close (eds.), pp. 49–72. Dallas: Southern Methodist University.

———. 1987. Fishing, fowling, and hunting in Late Palaeolithic times in the Nile Valley in Upper Egypt. *Palaeoecology of Africa* 18: 429–440.

———. 1988. The final demise of *Bos ibericus*? *Sahara* 1: 37–48.

Grigson, C. 1969. The uses and limitations of differences in absolute size in the distinction between the bones of aurochs *(Bos primigenius)* and domestic cattle *(Bos Taurus)*. In *The domestication and exploitation of plants and animals*, P. J. Ucko and G. W. Dimbleby (eds.), pp. 277–294. Chicago: Aldine-Atherton, Inc.

———. 1978. The craniology and relationships of four species of *Bos. Journal of Archaeological Science* 5: 123–152.

Guthrie, R. D. 2003. Rapid body size decline in Alaskan Pleistocene horses before extinction. *Nature* 426: 169–171.

Helmer, D. 1992. *La domestication des animaux par les hommes préhistoriques*. Paris: Masson.

Hemmer, H. 1990. *Domestication: The decline of environmental appreciation*. Cambridge: Cambridge University Press.

Hesse, B. 1978. *Evidence for husbandry from the early Neolithic site of Ganj Dareh in western Iran*. PhD dissertation, Columbia University. Ann Arbor, MI: University Microfilms.

Hole, F. 1989. A two-part, two-stage model of domestication. In *The walking larder: patterns of domestication, pastoralism, and predation*, J. Clutton-Brock (ed.), pp. 97–104. London: Unwin Hyman.

Hole, F., K. V. Flannery, and J. A. Neely. 1969. *Prehistory and human ecology on the Deh Luran Plain*. Memoirs of the Museum of Anthropology, No. 1. Ann Arbor: University of Michigan Press.

Hongo, H. and R. Meadow. 1998. Pig exploitation at Neolithic Çayönü Tepesi (southeastern Anatolia). In *Ancestor for the pigs: Pigs in prehistory*, S. Nelson (ed.), pp. 77–98. Research Papers in Science and Archaeology 15. Philadelphia: MASCA.

Horwitz, L. 2003. Temporal and spatial variation in Neolithic caprine exploitation strategies: A case study of fauna from the site of Yiftah'el (Israel). *Paléorient* 29: 19–58.

Horwitz, L. K. and Ducos, P., 1998. An investigation into the origins of domestic sheep in the southern Levant. In *Archaeozoology of the Near East, III*, H. Buitenhuis, L. Bartosiewicz, and A. M. Choyke (eds.), pp. 80–95. ARC Publications No. 18. Groningen: ARC.

Kent, J. D. 1982. The domestication and exploitation of the South American camelids: Methods of analysis and their application to circum-lacustrine archaeological sites in Bolivia and Peru. PhD dissertation, Department of Anthropology, Washington University, St. Louis. Ann Arbor, MI: University Microfilms.

Köhler-Rollefson, E. 1989. Changes in goat exploitation at 'Ain Ghazal between the Early and Late Neolithic: A metrical analysis. *Paléorient* 15: 141–146.

Kruska, D. 1988. Mammalian domestication and its effect on brain structure and behavior. In *Intelligence and evolutionary biology*, H. J. Jerison and I. Jerison (eds.), pp. 211–250. New York: Springer-Verlag.

Legge, A. J. and P. A. Rowley-Conwy 2000. The exploitation of animals. In *Village on the Euphrates: From foraging to farming at Abu Hureyra*, A. M. T. Moore, G. C. Hillman, and A. J. Legge (eds.), pp. 423–474. Oxford: Oxford University Press.

Marshall, F. 1994. Archaeological perspectives on East African pastoralism. In *African pastoralist systems*, E. Fratkin, K. Galvin, and E. Roth (eds.). Boulder, CO: Lynne Rienner.

Meadow, R. 1989. Osteological evidence for the process of animal domestication. In *The walking larder: Patterns of domestication, pastoralism, and predation*, J. Clutton-Brock (ed.), pp. 80–90. London: Unwin Hyman.

Miller, N. F. 2002. Tracing the development of the agropastoral economy in southeastern Anatolia and northern Syria. In *The dawn of farming in the Near East*, R. T. J. Cappers and S. Bottema (eds.), pp. 85–94. Studies in Early Near Eastern Production, Subsistence, and Environment 6. Berlin: Ex Oriente.

Moore, K. M. 1989. Hunting and the origins of herding in Peru. PhD dissertation, University of Michigan, Ann Arbor. Ann Arbor: University Microfilms.

Moray, D. 1986. Studies of Amerindian dogs: Taxonomic analysis of canid crania from the northern plains. *Journal of Archaeological Science* 13: 119–145.

———. 1990. Cranial allometry and the evolution of the domestic dog. PhD dissertation, University of Tennessee. Ann Arbor, MI: University Microfilms.

———. 1992. Size, shape, and development in the evolution of the domestic dog. *Journal of Archaeological Science* 19: 181–204.

———. 1994. The early evolution of the domestic dog. *American Scientist* 82: 336–347.

Perkins, D. 1964. Prehistoric fauna from Shanidar, Iraq. *Science* 144: 1565–1566.

Peters, J., D. Helmer, A. von den Driesch, and S. Segui. 1999. Animal husbandry in the northern Levant. *Paléorient* 25: 27–48.

Smith, A. B. 1984. Origins of the Neolithic in the Sahara. In *From hunters to farmers: The causes and consequences of food production in Africa*, J. D. Clark and S. A. Brandt (eds.), pp. 84–92. Berkeley: University of California Press.

———. 1986. Cattle domestication in North Africa. *African Archaeological Review* 4: 197–203.

Uerpmann, H.-P. 1978. Metrical analysis of faunal remains from the Middle East, In *Approaches to faunal analysis in the Middle East*, R. H. Meadow and M. A. Zeder (eds.), pp. 41–45. Peabody Museum Bulletin, No. 2. Cambridge, MA: Peabody Museum.

———. 1979. *Probleme der Neolithisierung des Mittelmeeraumes*. Biehefte zum Tübinger Atlas des Vordern Orients, Reihe B, Nr. 28. Wiesbaden: Dr. Ludwig Reichert.

Vigne, J.-D. and Buitenhuis, H. 1999. Les premiers pas de la domestication animale à l'Ouest de l'Euphrate: Chypre et l'Anatolie Centrale. *Paléorient* 25: 49–62.

Vigne, J.-D., J. Peters, and D. Helmer. 2005. New archaeozoological approaches to trace the first steps of animal domestication: General presentation, reflections and proposals. In *The first steps of animal domestication: New archaeozoological approaches*, J.-D. Vigne, D. Helmer, and J. Peters (eds.), pp. 1–16. London: Oxbow Books.

Wendorf, F., and R. Schild (assemblers); A. E. Clode (ed.). 1984. *Cattle keepers of the eastern Sahara: The Neolithic of Bir Kiseiba*. Dallas: Southern Methodist University.

Wheeler, J. 1984a. La domesticación de la alpaca (*Lama pacos* L.) y la llama (*Lama glama* L.) y el desarrollo temprano de la ganadería autóctona en los Andes Centrales. *Boletín de Lima* 36: 74–84.

———. 1984b. On the origin and early development of camelid pastoralism in the Andes. In *Animals and Archaeology 3: Early herders and their flocks*, J. Clutton-Brock and C. Grigson (eds.), pp. 395–410. BAR International Series, 202. Oxford: BAR International Series.

———. 1995. Evolution and present situation of the South American camelidae. *Biological Journal of the Linnean Sociey* 54: 271–295.

———. 1998. Evolution and origin of the domesticated camelids. *The Alpaca Registry Journal* 3: 1–16.

Wheeler, J. C., C. R. Cardoza, and D. Pozzi-Escot. 1977. Estudio provisional de la fauna de las capas II y III de Telarmachay. *Revista Nacional de Lima* 43: 97–102.

Wheeler Pires-Ferreira, J. C. 1975. La fauna de Cuchimachay, Acomachay A, Acomachay B, Telarmachay y Utco 1. *Revista del Museo Nacional* 41: 120–127.

Wheeler Pires-Ferreira, J. C., E. Pires-Ferreira, and P. Kaulicke. 1976. Preceramic animal utilization in the central Peruvian Andes. *Science* 194: 483–490.

Wing, E. S. 1972. Utilization of animal resources in the Peruvian Andes. In *Andes 4: Excavations at Kotosh, Peru, 1963 and 1964*, I. Seiichi and K. Terada (eds.), pp. 327–351. Tokyo: University of Tokyo Press.

———. 1975. Hunting and herding in the Peruvian Andes. In *Archaeozoological Studies*, A. T. Clason (ed.). Amsterdam: North Holland Press.

Zeder, M. A. 2001. A metrical analysis of a collection of modern goats (*Capra hircus aegargus* and *Capra hircus hircus*) from Iran and Iraq: Implications for the study of caprine domestication. *Journal of Archaeological Science* 28: 61–79.

———. 2003. Hiding in plain sight: The value of museum collections in the study of the origins of animal domestication. In: *Documenta archaeobiologiae 1: Deciphering ancient bones. The research potential of bioarchaeological collections. Yearbook of the State Collection of Palaeoanatomy, München, Germany*, G. Grupe and J, Peters (eds.), pp. 125–138. Rahden/Westf: Verlag M. Leidorf GmbH.

———. 2005. A View from the Zagros: New perspectives on livestock domestication in the Fertile Crescent. In *New methods and the first steps of animal domestications*, J.-D. Vigne, D. Helmer, and J. Peters (eds.), pp. 125–146. London: Oxbow Press.

Zeder, M. A. and Hesse B. 2000. The initial domestication of goats (Capra hircus) in the Zagros Mountains 10,000 years ago. *Science* 287: 2254–2257.

Zohary, D., Tchernov, E., and Horwitz, L. K. 1998. The role of unconscious selection in the domestication of sheep and goats. *Journal of Zoology* 245: 129–135.

CHAPTER 14

A Critical Assessment of Markers of Initial Domestication in Goats *(Capra hircus)*

MELINDA A. ZEDER

Introduction

Goats, along with their close relatives sheep, have long been thought to be the earliest domesticated livestock species. The wild progenitor of the domestic goat *(Capra hircus)* is the bezoar goat *(C. aegargrus)* (Zeuner 1963: 130; Epstein 1971; Schaller 1973: 27; Clutton-Brock 1999: 77), a species that lives in high, rocky mountain regions extending from the Taurus Mountains of Turkey into Pakistan (Figure 14.1). The geographical range of the bezoar parallels that of the Asiatic mouflon *(Ovis orientalis)*, the likely progenitor of domestic sheep *(O. aries)* (Zeuner 1963: 169; Schaller 1973: 41; Nadler et al. 1973; Clutton-Brock 1999: 70), which occupy somewhat lower, more rolling elevations along the same mountain arc. Attempts to document the domestication of these species date back at least 50 years to the earliest systematic investigations aimed at understanding when, where, how, and why herding and farming first began in the Fertile Crescent (Braidwood and Howe 1960). In fact, nearly all of the methods currently employed today to document animal domestication were developed first for the study of sheep and, especially, goat domestication in this region. Despite the decades-long application of these methods, debate continues over the relative utility of the various markers used to document the earliest stages of the domestication process (Zeder 2005). As a result, sheep and goats offer an excellent opportunity to assess the value of different archaeological approaches to documenting animal domestication in general.

This chapter begins with a critical review of the different approaches developed over the years to mark animal domestication in the archaeological record as they have been applied to the study of sheep and goat domestication. Of central importance in this review is the question of causality: specifically, how the relationship between humans and a target animal species undergoing domestication caused the proposed resultant "signature" or "marker" of domestication observed in the archaeological record. Building on this review, the chapter then moves on to a reconsideration of the evidence for goat domestication in the Zagros Mountains of present-day Iran and Iraq. Particular emphasis is placed on evaluating the utility of two of main markers of animal domestication—body-size reduction and demographic profiling—that represent distinctly different conceptual approaches to the study of domestication. This discussion in turn leads to a reexamination of the process of domestication in animals, and to a call for major changes in methods and markers used to document animal domestication in the archaeological record.

Proposed Markers of Domestication in Goats

Archaeological signatures or markers of initial domestication in goats (and other species) can be divided into two major categories: (1) those that reflect the evolutionary impact of domestication on the animal, and (2) those that reflect human goals in managing animal populations.

Animal-Oriented Markers

Animal-oriented markers of initial domestication are those that signal the evolutionary divergence of domestic animals from wild progenitors and the response of managed animals to new, selective pressures introduced when humans assume control over the breeding, movement, feeding, and protection from predators. These include a range of morphological changes in the form, size, proportions, and even the internal structure of bone, all thought to represent a genetically driven response to domestication. This class of markers also includes changes in the genotype of managed animals that are the focus of the next section of this book. A number of the proposed archaeological markers of domestication in sheep and goats fall into this category.

CHANGES IN HORNS

One of the most distinctive morphological differences observed between wild and domestic caprovines involves changes in horn morphology. Wild bezoar goats have scimitar-shaped horns that are quadrilateral in cross section, with a strong dorsal keel, especially in males, and a flattened nucal edge (Bökönyi 1977: 18). Wild goats also show a marked degree of sexual dimorphism in horn size. An older male bezoar may have horns up to 900 mm or more in length, while horns in an older female may only grow to about 250 mm in length (based on dorsal length of horn sheath from base to tip; Zeder unpublished data). In contrast, horns of domestic goats show a considerable amount of medial flattening, causing a characteristic medially flattened shape when the horn core is viewed in cross section. They may also display varying degrees of helical twisting. Domestic goat horns are also considerably smaller than those of wild goats, especially in males. And while horns of domestic males are larger than those of both wild and domestic female goats

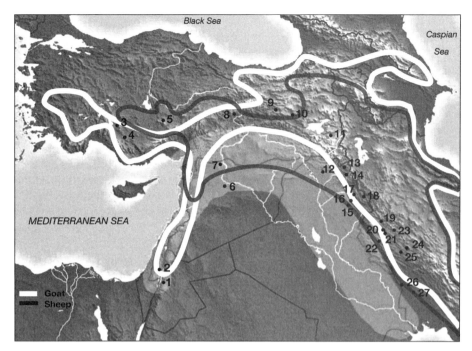

FIGURE 14.1 Map of the Fertile Crescent showing the distribution of wild goats *(Capra aegargrus)* and wild sheep *(Ovis orientalis)*, and the archaeological sites mentioned in the text. 1= Jericho, 2 = 'Ain Ghazal, 3 = Erbaba, 4 = Suberde, 5 = Aşikli Höyük, 6 = Abu Hereyra, 7 = Halula, 8 = Navali Çori, 9 = Çayönü Höyük, 10 = Hallan Chemi, 11 = Tamtama, 12 = M'lefaat, 13 = Zawi Chemi Sanidar, 14 = Shanidar, 15 = Matarrah, 15 = Karim Shahir, 18 = Palegawra, 19 = Warwasi, 20 = Sarab, 21 = Asiab, 22 = Guran, 23 = Ganj Dareh, 24 = Pa Sangar, 25 = Yafteh, 26 = Ali Kosh, and 27 = Chogha Bonut.

(Bökönyi 1977: 18), the difference between male and female horn length in domestic goats is much less marked than in wild goats. The reduction in the length of male horns results in a dramatic decrease in sexual dimorphism in the horns of domestic goats. The walls of the horns in domestic males are also reported to be thinner than in wild males (Bökönyi 1977: 18). Domestication also has had a distinct but somewhat less dramatic impact on the horns of sheep, consisting primarily of a reduction in the size of male horns and a tendency for hornlessness in domestic females (Hesse 1978: 184).

This change in horn form can be directly and convincingly linked to human intervention in the selection of breeding partners for managed caprovines. In their study of cranial morphology and social behavior in sheep and goats, Shaffer and Reed (1972) argue that large, heavy horns give wild males a distinct selective advantage in competition for females. Competition for females involves a display of horns, aggressive head butting, and horn-to-horn ramming. The victor in these contests gains essentially exclusive access to breeding females (Reed 1983: 152). Males with large, heavy horns have a distinct advantage in these contests and thus a clear selective advantage for passing on their genes to the next generation.

While head-butting behavior is still found in domestic goats (Shaffer and Reed 1972: 12), human selection of mating partners greatly relaxes the strong selective pressure for large horns in males, allowing forms that might have been maladaptive in the wild to manifest themselves (Hesse 1978: 192). Once their selective advantage in mate competition was neutralized through human control of mating, the large, heavy horns of wild males, which are energetically costly to grow and to carry, would likely become a distinct disadvantage that would be selected against in domestic caprovines.

Given the clear changes in horn morphology in domestic goats, it is not surprising that horn morphology was the first marker used to document domestication in this species. In his study of Neolithic-age levels at Jericho in southern Levant, Zeuner (1955) found changes in the cross sections of goat horn cores that he maintained were indicative of the presence of at least some domestic goats at the site. Similarly, Reed used horn morphology to argue for the domestic status of the goats in Aceramic Neolithic levels at Jarmo in highland Iraq (Reed 1959). Bökönyi (1977: 19) found three horn cores at the highland Iranian site of Asiab that he felt were transitional between wild and domestic morphotypes. Flannery produced even stronger evidence of changes in goat horn morphology at Ali Kosh in lowland Iran. In his studies of the course of Aceramic Neolithic occupation at this site, he was able to trace changes in horn form from the quadrilateral, wild morphotype, to the more "lozenge" or almond-shaped "transitional" forms, to cores with the distinct medial flattening characteristic of domestic goats (Hole et al. 1969). Horns with some helical twisting were found in the latest Ceramic Neolithic levels at Ali Kosh and at later sites in the Deh Luran Plain.

In contrast, evidence for sheep domestication based on horns has been more equivocal, consisting primarily of the presence of cranial fragments of hornless females at early sites. The recovery of the skull of a hornless female sheep in basal levels of Ali Kosh, for example, was used to argue for the presence of domestic sheep during the earliest occupation of this site (Hole et al. 1969: 278–281).

Despite other strong evidence for human management (see below), Hesse, on the other hand, found little evidence of medial flattening or size reduction in horns in his study of the large goat population from Ganj Dareh in highland Iran (Hesse 1978: 212). While some horn cores from Ganj Dareh displayed a slight degree of flattening similar to that observed by Bökönyi at nearby Asiab, none had the clear degree of flattening or evidence of twisting seen in later levels at Ali Kosh. None of these forms, moreover, would seem out of place in a wild population where some degree of medial flattening in horns is sometimes found (Hesse 1978: 214). Hesse was similarly reluctant to declare that hornless cranial bones from a young female sheep found at Ganj Dareh belonged to a domesticate, citing the known occurrence of this trait in a small proportion of wild females (Hesse 1978: 184). Hesse's reluctance to claim domestic status for sheep and goats from Ganj Dareh on the basis of horn morphology sounded an important cautionary note about the use of horn morphology as a marker of sheep and goat domestication.

Because traits in horn morphology that become dominant in domestic goat populations can be found in low numbers in wild ones, the simple presence of a few goat horn cores that show some tendency toward flattening or a single hornless female sheep cannot be taken as proof positive of the presence of domestic caprovines at the site. The standard of evidence has to be set higher. If not a predominance of clear "domestic" morphotypes, then at least a high or an increasing percentage of domestic forms should be expected in order to make a reasonable case for domestic status.

Moreover, goat horn form is essentially a qualitative characteristic, and categorization of horn cores as falling into either wild, transitional, or domestic morphotypes is an uncomfortably subjective process. Making such assessments is further hampered by the fact that cross-sectional shape varies over the length of the core. Archaeologically recovered horn cores are often in quite fragmentary condition, making controlled comparisons between cores quite difficult. Horn size is also closely linked to the age of an animal, especially in males. Without knowing the age at death of the animal, it is difficult to say whether an apparent reduction in the length or breadth of horn cores in an archaeological assemblage is a reflection of horn-size reduction resulting from domestication or simply a shift toward use of younger animals. As seen by the lack of changes in horn morphology at Ganj Dareh, the gradual pace of change in horn form at Ali Kosh, and the late appearance of helical twisting, the use of horn core morphology as a marker of initial domestication in caprovines is further limited by the apparently slow rate of change in horn form in managed sheep and goats.

In summary, horn morphology is a well supported marker of domestication in sheep and goats in that it can be clearly and directly linked both to the relaxation of selective pressures for large horns in wild males and to the introduction of new selective pressures on animals under human management. The utility of this marker for detecting the initial stages of domestication, however, is limited by the difficulty of detecting subtle changes in size and shape in fragmentary archaeological material, the problem of controlling for age at death when evaluating the size of horns, and the apparently slow rate of change in horn morphology in early domestic caprovines. As a result, horn morphology is rarely used as a marker of initial domestication in caprovines in more recent research.

BONE STRUCTURE

In the 1970s, Drew, Daly, and Perkins proposed that domestication causes changes in internal bone structure that could be used to mark domestication in sheep and goats, as well as in other species of domestic animals (Drew et al. 1971; MASCA 1970, 1973). They claimed that examination of histological thin-sections of weight-bearing bones revealed systematic differences in the orientation of hydroxyapatite crystals in the bones of wild and domesticated animals. They maintained that, when viewed through crossed polarizers, the bones of domesticated animals showed strong blue or yellow interference colors on all articular surfaces (depending on the orientation of the slow ray of the gypsum plate), indicating that the c-axes of the apatite in these bones were uniformly aligned perpendicular to the surface of the bone. The bones of wild animals did not show this strong degree of orientation; the c-axes of the apatite were instead essentially randomly oriented throughout most of the bone (exhibiting dramatic magenta interference colors when viewed under polarized light). Although they claimed to have consistent results for a variety of wild and domestic species, both archaeological and modern (MASCA 1973), their most widely reported results (Drew et al. 1971) came from two archaeological assemblages in southeastern Anatolia: Suberde, where caprovines were argued to be wild on the basis of harvest profiles, and Erbaba, where demographic data were interpreted as indicating the presence of domestic animals (Perkins and Daly 1968).

This "blue-rim" phenomenon was heralded as a new marker of initial domestication that could be broadly applied to any bone and would yield indisputable evidence of domestication in a wide range of domestic species. Drew et al., however, did not offer much in the way of an explanation for why the bones of domesticated animals might have experienced a change in microstructure, claiming only that it was "a response to stress in the weight-bearing bones of the bodies of domesticated animals which, through lack of exercise, poor nutrition, or genetic deterioration, lacked sufficient material in their bones to form the sturdy bones characteristic of ... wild animals" (Drew et al. 1971: 282). The mention of "genetic deterioration" implies that they felt this change was a genetically driven response to domestication—that the conditions of domestication somehow selected for animals with the strong, preferred orientation of hydroxyapatite crystals in their weight bearing bones. However, they also cast this change in bone microstructure as a plastic response to the impoverished conditions under which domesticates were kept. Unfortunately, nothing more was offered about the possible cause and effect relationship between domestication and crystalline orientation in bones. Just how lack of exercise or poor nutrition might result in more uniform orientation of hydroxyapatite crystals in the bones of domesticated animals, whether as an evolutionary or ecophenotypic response to domestication, was never explained. Nor was there any effort to demonstrate that crystalline orientation played a role in the tensile strength of bone, or that the bones of wild and domestic animals actually varied in this way.

Subsequent attempts to replicate Drew et al.'s results using a range of modern animal species of known wild and domestic status were not successful (Watson 1975; Zeder 1978; Østergård 1980). None of these studies found any evidence of systematic microstructural differences between a number of species of wild and domestic animals, including sheep, goats, cattle, wolves, and dogs. The most plausible explanation for the differences in the microstructure of the archaeological caprine bones studied by Drew et al. was offered by Watson (1975), who argued that that the pattern they observed in archaeological material was caused by postdepositional taphonomic factors, not by domestication. Specifically, Watson argued that the birefringence in Drew et al.'s thin-sections was actually caused by quasi-crystalline collagen, not apatite, and that the difference between archaeological samples of caprovines was a reflection of differential collagen preservation at the two archaeological sites featured in their work.

There is some reason to believe that differences in the types and amounts of activity of wild and domestic animals might have some impact on bone microstructure (Zeder 1978). Finding any such differences, however, will require a better understanding of the different activity levels of wild and domestic animals and of how such differences might effect bone microstructure, as well as how to best detect and measure these differences. As yet this work has not been undertaken, and the use of microstructure to document wild and domestic status of bones has been largely abandoned. While the "blue-rim" phenomenon may no longer be considered a viable marker of domestication in caprovines or other species, it does serve as an excellent cautionary tale about the dangers of constructing post-hoc explanations for how proposed markers reflect the process of domestication without considering other alternative explanations that are either unrelated to domestication or only tangentially so.

SIZE REDUCTION

Size reduction was first proposed as a method for documenting domestication in caprovines by Hans-Peter Uerpmann in the 1970s. Noting the apparently clear evidence of body-size reduction in other early domesticates (i.e., cattle and pigs, Reed 1961; Bökönyi 1974), Uerpmann undertook an ambitious project to see if a similar pattern could be found in sheep and goats (Uerpmann 1978, 1979). Uerpmann located curated assemblages of archaeological caprovine bones from across the Fertile Crescent that spanned the transition from hunting to herding, recording multiple measurements on a wide range of skeletal elements from the sheep and goats in these assemblages. The number of measurable bones in these different assemblages was often quite small and fragmentary, making it difficult to amass a large enough sample of any single element or measurement to build a statistically reliable sample. Working under the assumption that the bones in an individual animal co-vary in size (Uerpmann 1978: 41), Uerpmann devised a method for normalizing measurements of different elements. He did this by computing the difference between the size of an archaeologically recovered bone and the same dimension in a standard skeleton (usually a modern animal). He then combined the results of these normalized comparisons into a single size profile. To further enhance sample size, Uerpmann also combined the normalized measurements of sheep and goats from various sites all across the Fertile Crescent into broad, temporal groups that spanned 1,000 to 5,000 years or more.

Comparing these temporal size profile composites, he found that both sheep and goats showed a marked downward shift in body size after about 9500 BP uncalibrated, with sheep displaying this change somewhat later than goats (Uerpmann 1979). Uerpmann acknowledged that there are multiple factors that affect body size in ungulates—including climatic conditions, which can cause both temporal and spatial variation in body size, and sexual dimorphism, which is strongly expressed in many ungulate species. He argued, however, that the apparently quite sudden downward shift in the size of sheep and goats after about 9500 BP uncalibrated, and the subsequent seeming lack of change in body size of domestic sheep and goats in later prehistoric assemblages, indicated that body-size reduction in these early sheep and goats was directly caused by their domestication, and not by either climate change or a shift in herd demography.

He did not, however, outline in any detail why he thought domestication caused this change, saying only that body-size reduction was likely a genetic response to a shift in environmental conditions of animals kept under human control. Richard Meadow further explored the causality of body-size reduction in domesticates in his influential review of osteological evidence for domestication (Meadow 1989). He argued that body-size reduction in domesticates initially had ecophenotypic causes; for example, lower overall levels of nutrition and early weaning in animals under human care may have caused a reduction in body size in early domesticates. He also posited, however, that there was subsequent genetic selection for reduced body size in domesticates. Smaller body size in females may have made them better able to cope with the lower nutritional plane experienced by domesticates kept in more marginal environments. In males, smaller body size may have resulted from the relaxation of selective factors that favored large individuals in the wild, a similar mechanism to that proposed to account for changes in horn size and morphology once humans began to select breeding partners.

More recently, Zohary et al. (1998) have attempted to predict the likely changes in morphology, behavior, and physiology that would result from the domestic partnership with humans. They argue that human protection from predation, culling of young males, and changes in mobility and terrain might all have impacts on body size. In particular they project that a number of these factors would result in a shortening of limbs relative to body size and a loss of skeletal robustness that would be especially evident in males.

Today, body-size reduction, detected through some variation of the size index method for normalizing and amalgamating metric data (Ducos 1969; Meadow 1984, 1999; Helmer 1992), is the most widely used marker of caprovine domestication, considered by many as proof positive of domestic status. One of the first applications of the technique was on the goat remains from the site of Ganj Dareh in the Zagros region. Although skeptical of Hesse's use of demographic data to argue for the domestic status of the goats from Ganj Dareh (see below), Uerpmann and a number of subsequent researchers still accepted this population as domestic on the grounds of an apparent reduction in body size seen when normalized size profiles of the Ganj Dareh goats were compared to normalized data from benchmark Epi-Paleolithic and earlier sites (Uerpmann 1979; Helmer 1992; Bar-Yosef and Meadow 1995). More recent work in the central and western Fertile Crescent clearly reflects the dominance of this marker and the methods developed to detect it. Comparison of normalized metric data from different sites in the upper Euphrates drainage, for example, is the primary evidence used to buttress Peters's argument for the presence of domestic sheep and goats at about 9400 BP uncalibrated at the site of Nevali Çori in southeastern Anatolia (Peters et al. 1999). Size reduction is also the core of Helmer's case for the somewhat latter appearance of sheep and goats at Tell Halula in northeastern Syria. Legge and Rowley-Conwy (2000) cite the similarity in normalized size profiles between the goats of Abu Hurerya in the steppe lands of the northern Levant and those from highland Ganj Dareh in the Zagros to argue for the domestic status of the goats at Abu Hurerya site at about 8600 BP uncalibrated. Horwitz (Horwitz 1989, 1993; Horwitz et al. 1999) points to an apparent reduction in the body size of goats, as detected by the size index method, to assign full domestic status to goats in the southern Levant at around 8300 BP uncalibrated.

Despite the widespread adoption of body-size reduction as the leading-edge marker of initial domestication in sheep and goats, there remain a number of open questions about its utility in documenting domestication in these species. Most importantly, the cause and effect mechanisms that link domestication to body-size reduction are still less than adequately mapped. Zohary et al. (1998) have presented the most thoughtful attempt to explain why domestication might cause size reduction in early domestic ungulates, but the exact selective pressures that might have caused these changes in body size still remain largely unspecified. Moreover, until the research featured below, there has been no effort to empirically test whether modifications in body size attributed to domestication can be found in animals with known life histories. Nor has there been any systematic attempt to distinguish domestication-induced body-size reduction from changes in size caused by other factors—particularly size variation caused by varying climatic and environmental conditions or sexual dimorphism.

The rate of size reduction expected to result from domestication has not been established. Researchers using this marker generally cite Bökönyi's (1976) estimate, based on modern animals undergoing domestication experiments, of about 30 generations as the length of time it would take for definitive morphological changes to manifest themselves in early domesticates (Meadow 1989: 85). In smaller-bodied ungulates like sheep and goats, this would be about 60 years—essentially instantaneous as far as the archaeological record is concerned. However, Bökönyi never specified just what these modern experiments were, or the species that were involved in them, so it is hard to know how well this experimental situation applies to the initial stages of animal domestication, when there was likely little intentional human selection for specific morphological traits and repeated back-crossing with wild animals. Changes in horn size and morphology seem to have happened quite gradually in goats. If at least some of the factors that selected for changes in the size and shape of horns are the same as those that selected for smaller body size (i.e., human control of mate selection), then changes in body size, if they occur at all, might be similarly slow in manifesting themselves.

Human-Oriented Markers

While animal-oriented markers reflect the evolutionary impact of domestication on the animal, human-oriented markers, in contrast, reflect human control of the breeding, care, and movement of animals in order to promote greater security or predictability in access to animal resources. There is wide range of markers developed to document animal domestication that focuses more on the human side of the domestic partnership. Perhaps the best known of these markers are demographic profiles that reflect harvest strategies aimed at promoting the long-term propagation of a managed herd. But this category also includes zoogeographic and abundance measures, architectural features (i.e., corals and other animal shelters), artifacts associated with the use of animal resources (i.e., churns or milk strainers), and artistic renderings that feature domesticates or herding activities. Plastic, ecophenotypic responses to management (i.e., pathologies resulting from tethering, nutritional stress, or the use of animals as beasts of burden) might also be considered as markers of human control, as might changes in chemical composition of skeletal elements resulting from changes in the diet. As with animal-oriented markers, many of the markers of animal domestication that fall in this second category were first developed to study domestication in sheep and goats.

ZOOGEOGRAPHY AND ABUNDANCE

Zoogeographic markers of domestication rely on the assumption that a potential domesticate found outside the natural geographic range of its wild progenitor was brought there by humans, either as a fully domesticated animal or as part of the process of domestication. Zoogeographic arguments for domestication are often associated with evidence of an increase in the abundance of a presumed non-native species, signaling a possible shift away from an emphasis on hunting native wild species to herding introduced non-native domesticates. An increase in the abundance of a potential domestic species within or near the presumed natural habitat of its wild progenitor is also sometimes interpreted as a sign of local domestication.

Perkins used these markers to support his controversial claim of sheep domestication at 11,000 BP uncalibrated in northwestern Iraq (Perkins 1964). Analysis of faunal remains from the deep deposits at Shanidar Cave pointed to a strong, almost exclusive emphasis on hunting wild goats that reached back into the Middle Paleolithic. Epi-Paleolithic levels in the cave, however, seemed to show a dramatic increase in sheep, which were also dominant at the nearby and roughly contemporary open-air site of Zawi Chemi Shanidar. Perkins maintained that this increase in sheep exploitation in an area that he felt was unsuitable for wild sheep supported his claim that the Epi-Paleolithic sheep from Shanidar and Zawi Chemi Shanidar were domestic. A similar argument was used by Flannery to buttress his case for the domestic status of goats at lowland Ali Kosh (Hole et al. 1969: 266). The high proportion of goats in the assemblage from basal levels at this site, located far from the highland natural habitat of goats, indicated that wild goats had been intentionally brought to this marginal area.

The last 20 to 30 years has seen a shift in the focus of research away from the eastern Fertile Crescent and toward the western and central Fertile Crescent regions of the southern and northern Levant. This change in regional focus has prompted a dramatic increase in the use of zoogeographic and abundance markers in the documentation of sheep and goat domestication. Wild goats and sheep were endemic to the mountains and foothills of the Zagros and figure prominently in the diet of Late Pleistocene and Early Holocene hunters there, limiting the utility of both zoogeographic evidence and abundance measures in documenting caprovine domestication. In contrast, gazelles play a much more prominent role in the subsistence economy of early hunters in the steppe regions and coastal plains of the northern and southern Levant, with wild cattle, equids, and deer also important game animals in different ecozones within this broad region. Caprovines are either not part of the endemic fauna, or were only minor contributors to Epi-Paleolithic and earlier economies. Beginning about 9000 to 8500 BP, there was a sudden shift away from the use of gazelles and other indigenous wild-game animals toward an increasing focus on goats (Peters et al. 1999; Horwitz et al. 1999; Legge and Rowley-Conwy 2000; Horwitz 2003). In some cases, this shift has been interpreted as evidence for the introduction of managed goats domesticated elsewhere, as has been argued for the Syrian steppe sites of Abu Hureyra by Legge (1996: 256) and Tell Halula by Helmer (1992). Bar Yosef (2000) also attributes the increased use of goats in the southern Levant to the introduction of already domesticated animals, while Horwitz (1989, 1993, 2003) interprets this same evidence as supporting her case for the local domestication of indigenous wild goats in the southern Levant.

Similarly, zoogeographic evidence has been used to argue that domestic sheep were introduced into the southern Levant at about 8500–8000 BP uncalibrated and not domesticated locally (Horwitz and Ducos 1998). An increase in the taxonomic abundance of sheep at the site of Abu Hureyra (Legge 1996; Legge and Rowley-Conwy 2000) and at Tell Halula (Helmer 1992) at about the same time has also been interpreted as evidence of the introduction of domesticated sheep into the steppe regions of the northern Levant. Peters uses such increases to buttress his argument for the local domestication of both sheep and goats at Nevali Çori in the foothills of the Taurus Mountains in southeastern Anatolia at about 9400–9200 BP uncalibrated (Peters et al. 1999: 35). At the site of Shillourokambos in Cyprus, thought to be well outside the natural range of all of the primary Near Eastern livestock species, the appearance of morphologically wild sheep and goats (as well as cattle, pigs, and deer) at about 9500 BP uncalibrated is seen as evidence of a kind of game-stocking strategy. The younger age profile of the caprovines at this site is interpreted as a form of proto-domestication that later resulted in their obtaining full domestic status as marked by morphological change (see below, Vigne and Buitenhuis 1999; Vigne et al. 2000).

The problems inherent in the use of zoogeographic and abundance data in documenting domestication are well known. Present-day distributions of wild progenitor species have been strongly affected by overhunting and the loss of habitat to agricultural and urban settlement, and are likely quite different today from what they were during the Early Holocene. Thus, while the area around Zawi Chemi Shanidar may not be optimal territory for sheep today, this may not have been the case 11,000 years ago. Zoogeographic evidence for the introduction of domesticates into a region can only be made when that region is clearly far removed from both the current range of wild progenitor species and outside the known ecological parameters for those species, as is the case for goats at lowland Ali Kosh and in the steppe areas of the northern Levant. But there is a real possibility (perhaps likelihood) that the appearance of non-native animals in an assemblage marks the movement of already domesticated animals out of the area of initial domestication and into new regions. And while the appearance of a likely domesticate outside its presumed natural habitat may help track the diffusion of domesticates and herding technology, it is not useful in identifying the location or the process of initial domestication.

The use of relative abundance data to mark initial stages of the domestication process faces similar problems, especially in assemblages from sites within or close to the likely natural habitat of wild progenitors. Without additional evidence of domestic status (i.e., morphological change or domestic demographic patterns), it is difficult to tell if an increase in the use of a potential domestic species signals the adoption of herd management or the intensification of hunting strategies. Moreover, it is entirely possible that domesticates initially made only a small contribution to subsistence economies otherwise dominated by the exploitation of wild plant and animal resources, and that an increased emphasis on these domestic resources came about well after their initial domestication.

DEMOGRAPHIC MARKERS

The use of demographic markers to document initial animal domestication is based on the assumption that hunters and herders are guided by different strategies in their selection of the types of animals they harvest, and that these different strategies shape the age and sex profile of harvested animals in distinctive ways. For the hunter, prey selection is more likely oriented around maximizing the immediate off-take of the hunt, focusing on individual animals that provide an optimal return for the effort expended in stalking, killing, transporting, and processing them (Kelly 1995). In some cases, this goal is met by using opportunistic prey strategies that focus on animals closest at hand and easiest to kill. The age and sex profile of prey animals under such a strategy will reflect the demographic composition of the herd population most commonly preyed upon (Hole et al. 1969: 286). This profile will vary depending on seasonal movements of the herd and subgroups within the herd (i.e., nursery herds of lactating females and young or bachelor herds of sexually mature males; see Hesse 1978: 338). Hunters may also more selectively target animals, especially those that will provide the largest yield of meat per individual (i.e., large adult males; see Stiner 1990).

The herder's harvest strategy, on the other hand, is shaped by the underlying goal of promoting the long-term propagation of the herd. Off-take is naturally of interest, but so is making sure there are animals on hand for future consumption. This goal is best met in caprovines by selectively culling males at young ages (which both ensures a steady supply of meat and helps preserve pasture resources needed to sustain herds), and delaying slaughter of females and a few older males for breeding purposes (Redding 1981).

Early efforts to employ demographic profiling to document sheep and goat domestication can be traced back to the 1950s, when Coon (1951) used a young age profile to argue for the presence of domestic caprovines at Belt Cave, and Dyson (1953) proposed demographic data as a means of documenting animal domestication in the Old World. Perkins (1964) employed demographic data as one leg of his argument for a very early date for sheep domestication at Shanidar Cave and Zawi Chemi Shanidar in northwestern Iraq. As already discussed, he also used demographic data to determine the status of the sheep from Suberde and Erbaba in southeastern Anatolia (Perkins and Daly 1968). In both cases, Perkins focused on the fusion status of two bones (the metacarpal and the metatarsal) known to fuse at about 24 months of age in caprovines (Silver 1969) to discriminate between juvenile and adult animals. Bökönyi (1972: 22) also emphasized demographic data in his study of animal remains from sites in the Kermanshah Valley of the Zagros Mountains of Iran, tracking changes in the proportions of goats that fall into categories of juveniles, subadults, adults, and senile animals (without specifying the bones used to distinguish age categories) to argue for the domestic status of goats from the sites of Sarab, Shiahbad, and Dehsaver, and the wild status of goats from earlier Asiab. However, Bökönyi also interpreted the strong focus on adult male goats at Asiab as an indication of a transitional herding strategy in which superfluous and aggressive older males were killed so that young individuals and females could be captured and tamed (Bökönyi 1973: 71, 1977: 20). Hesse (1978: 338–339), on the other hand, suggested that this harvest profile reflects more opportunistic seasonal hunting of bachelor herds of adult males likely to have been near the site during the probable season of occupation (August through April). More recently, Redding has suggested that a similar concentration on adult male sheep at the site of Hallan Chemi suggests a strategy that drew males into a region where local herds were being intensively hunted (Redding 2005).

In the late 1960s, Flannery employed a more sophisticated approach to the use of demographic data than had been used in earlier studies as the centerpiece of his argument for the domestic status of caprovines at Ali Kosh (Hole et al. 1969). Drawing from studies of the behavior and population structure of modern wild caprovines, Flannery built a model of harvest profiles expected with hunting as opposed to herding. This was the first time such an approach had been employed in the application of demographic information to document animal domestication. He then demonstrated that the harvest profile for sheep and goats from the Upper Paleolithic site of Yafteh Cave in highland Iran was consistent with what would be expected from hunting, while profiles from two different components at lowland Ali Kosh and the roughly contemporary mid-elevation site of Tepe Guran were consistent with a herder's strategy of harvesting young animals. Flannery was also the first to use the full range of Silver's (1969) long-bone fusion sequence to compute a much more refined, calibrated harvest profile. Unlike earlier schemes that presented demographic data in somewhat amorphous categories of juvenile, subadult, or adult computed using a very limited number of skeletal elements, the adoption of Silver's fusion sequence allowed Flannery to use a much more inclusive suite of skeletal elements to more precisely estimate the shape of the harvest curve over the first four years of the caprine life span (the age at which long-bone fusion is complete).

Ten years later, Hesse (1978) took a similar approach in his study of the animal remains from the highland Iranian site of Ganj Dareh. Hesse drew on a wide range of literature on ungulate behavioral ecology to model different human and nonhuman prey strategies. He then used these models as a point of comparison in his interpretation of harvest profiles of Ganj Dareh caprovines. Hesse also refined Flannery's methods for computing harvest profiles, applying more recently accumulated data on the timing and sequence of long-bone fusion in goats (e.g., Noddle 1974), as well as data on tooth eruption and wear that extended the range of harvest profiles to about 10 years of age, into the upper limit of caprine life expectancies (Payne 1973).

Prior to Hesse's study, it was only possible to estimate the gross proportions of male and female animals used by ancient hunters and herders by comparing the numbers of male and female horn cores in the assemblage. No one had found a way to determine if there were sex-specific differences in the ages of harvested animals. Hesse discovered a marked bimodality in the size of Ganj Dareh goats (as well as in the smaller sample of sheep from the site) that he hypothesized was attributable to the strong degree of sexual dimorphism thought to occur in caprovines. In addition, he noted that the bones of the larger, presumably male, goats were more likely to be unfused than were the bones of smaller animals that he presumed to be females. Hesse interpreted this pattern as a sign that male goats at Ganj Dareh had been killed at younger ages than the females. The sample of sheep from the site showed no such difference in long-bone fusion patterns. Thus, by using the distinctive sex-linked size difference in caprovines, Hesse was able, for the first time, to combine both age and sex data to get a sense of the differential age profiles of male and female animals.

At the time of Hesse's analysis of the Ganj Dareh assemblage, little of the faunal assemblage from the site could be assigned to a stratigraphic level. Nonetheless, he was still able offer some insights into possible diachronic changes in harvest profiles over the course of the occupation at Ganj Dareh. Specifically, he found what appeared to be differences between the age and sex profiles of goats harvested by occupants of the lowest level (Level E) at the site, where only a series of hearths and depressions were found, and those generated by residents in the upper (D–A) levels, which contained a series of levels of mud-brick architecture. When Hesse conducted his study, available radiocarbon dates suggested that basal Level E predated the upper levels of Ganj Dareh by as much as 1,000 years. The harvest profile of goats that could be assigned to the limited exposure of the basal level at the site seemed to contain a combination of fully adult females and young males, with an emphasis on animals fewer than six months old. Hesse felt that this harvest profile was consistent with the expected off-take from a nursery herd composed of lactating adult females and nursing young. The location of the site, he argued, would have put hunters in close proximity to the habitat zone likely to be frequented by such nursery herds during most of the year.

On the other hand, the larger sample of goat bones recovered from Levels D–A yielded a harvest profile that indicated a relatively stronger focus on young males between the ages of about 6 to 24 months, and a somewhat increased higher percentage of slaughter of females older than 24 months. This pattern did not fit the demographics of any configuration of a wild goat population—bachelor herds, nursery herds, or the combined herd. But it did fit well with expected harvest profiles produced by herdsmen seeking to maximize off-take and promote herd propagation. This pattern was interpreted as evidence for the domestic status of the goats from later Ganj Dareh.

Employing demographic profiling as a marker of domestication has not been without its critics. In part, this criticism was aimed at the assumption made in some of the early demographic approaches that hunters would kill a random and representative sample of animals in a wild herd (resulting in a harvest profile that approximated the overall sex and age composition of a living herd of wild ungulates), and that this profile would be readily distinguishable from a herder's emphasis on culling young animals (Collier and White 1976). Critics pointed out that there are a number of different hunting strategies that would result in harvest profiles that were not representative of the overall composition of the wild population, including those that might mimic the emphasis on young animals expected in domestic herds (i.e., hunting a nursery herd; Higgs and Jarman 1972; Jarman and Wilkinson 1972). Moreover, increased hunting pressure on wild herds can lower the age profile of the overall herd from which hunters select their prey (Uerpmann 1979). Other researchers have worried that harvest profiles in a managed herd might deviate from the expected pattern of young male kill-off and prolonged survivorship of female breeding stock resulting from catastrophic herd losses, specialized production of targeted animal resources, or unspecified social factors that might result in counterintuitive approaches to culling (Meadow 1989: 83).

While it is unreasonable to assume that wild ungulates display a "universal" demographic pattern that will be directly reflected in the range of prey taken by hunters, as shown above, it should be possible to model a range of different alternative prey strategies for different species based on a more focused, species-specific understanding of possible seasonal and long-term demographic variations in herd composition. It should also be possible to distinguish these different alternative hunting harvest profiles from a herding harvest profile, given the fundamental differences in the goals that guide prey selection of hunters vs. herders. It is true, for example, that both a selective focus on nursery herds and intense hunting pressure might result in harvest profiles that show an emphasis on young animals. But in both cases, however, the young animals taken would consist of male and female animals, not just young males as in a herder's harvest profile. Moreover, the range of ages of these young animals likely would differ from that expected with herding (i.e., very young animals in the nursery herd example and a wider range of younger animals under conditions of hunting pressure vs. the emphasis on males between about one to two years expected with herding). It is also true that there is a range of social, economic, or purely natural factors that might cause the harvest profile from a managed herd to deviate from the expected pattern of young male kill-off and prolonged female survival. However, the conservation of a core breeding stock is of necessity a central feature in all successful herding strategies that cannot be ignored without risking the long-term prospects for herd propagation. This strategy should be detectable with higher-resolution harvest profiles. Clearly then, the successful use of demographic profiling as a marker of animal domestication requires careful modeling of the harvest profiles that might be expected under a range of different hunting and herding strategies, based on a clear understanding of the demographic dynamics of wild and domestic herds.

Equally important is the development of improved methods capable of producing higher-resolution harvest profiles—specifically, finely calibrated species and sex-specific harvest profiles. Curiously, despite Hesse's promising methodological advances toward realizing this goal, his techniques were not adopted by other researchers. Nor has anyone, including Hesse, attempted to further develop his method for combining osteometric and long-bone fusion data to produce sex-specific harvest profiles until the study presented below. In fact, demographic profiling has seldom been featured as a primary marker for documenting domestication in caprovines in the more recent literature. Instead, demographic profiling today usually takes a backseat to size reduction as a marker of domestication. Age and sex data, when presented at all, are usually treated as ancillary evidence used to support claims of domestication based on apparent body-size reduction (cf. Peters et al. 1999: 39), and there is a general lack of uniformity or comparability between age classes used by different researchers, with little attention paid to specifying the methods or elements used to compute them. When an emphasis on younger animals is found prior to clear evidence of size reduction, it is more likely to be interpreted as evidence of incipient proto-domestication, or as indicating a selective hunting strategy leading to domestication (Vigne and Buitenhuis 1999; Vigne et al. 2000; Horwitz 2003: 53), and not clear-cut evidence of the achievement of "full" domestic status. For the past 30 years, efforts to document animal domestication have clearly been dominated by an approach that equates "domestication" with evidence of genetic isolation as manifested by morphological change in the form of size reduction.

OTHER MARKERS OF HUMAN CONTROL

Other markers of human control used to document animal domestication include architectural or artifactual evidence for the restriction of movement of animals or the exploitation of animal resources, as well as physical evidence of restraints or shifts in nutritional intake caused by human management. For sheep and goats, such markers have included impressions of goat hooves made in fresh mud-brick at the site of Ganj Dareh in highland Iran, which Hesse (1978: 314) took as a sign that goats were thoroughly habituated to humans and were left to roam freely about human settlements without fear of their escaping back into the wild. Similarly, the presence of a sheep figurine at the site of Sarab that appeared to have woolly fleece was interpreted as an early sign of the shift in pelage from the wild to the domestic form (Bökönyi 1977: 25). Pathologies in the bone and jaws noted in sheep and goats at Sarab (Bökönyi 1977: 38) and at the site of Ain Ghazal in Jordan (Köhler-Rollefson 1989) have also been interpreted as signs of human control of domesticates. More recently, a number of researchers are using isotopic signatures in bones and teeth to look for shifts in the dietary intake that might be associated with animal management (cf., Mashkour et al. 2005). In all cases, however, such markers are best viewed as sporadic and somewhat tangential supporting evidence of domestication.

Evaluating the Utility of Body-Size Reduction and Demographic Profiling as Markers of Initial Goat Domestication

The above review of past efforts to document sheep and goat domestication in the Near East underscores the lack of any clear consensus on the most effective method for recognizing this process in the archaeological record. There has been little or no consistency in the ways in which different markers of domestication in sheep and goats have been employed. More importantly, there has been virtually no attempt to systematically evaluate the efficacy of these markers in detecting initial stages of the domestication process. Instead, the investigation of this key transition in human history has, in recent years, been marked by a generally ad hoc, nonstandardized, and rote application of methods developed almost 50 years ago. As a result, there is still no clear understanding of how the domestication of these key livestock species in the Near East actually unfolded.

Over the past several years I have carried out a comprehensive analysis of archaeological and modern collections of sheep and goat skeletal elements from the eastern Fertile Crescent designed to evaluate the relative value of the two primary markers employed to recognize the initial domestication of caprovines in the Near East: body-size reduction and demographic profiling (Zeder 1999, 2001, 2003, 2005, in press; Zeder and Hesse 2000). As noted above, body-size reduction is the predominant method used today to document domestication in sheep and goats, as well as in other livestock species. Indeed, it seems that in the minds of many using this marker, domestication, genetic change, and body-size reduction are inextricably linked in a direct and immediate cause and effect relationship, making body-size reduction the optimal way to document the evolutionary impact of domestication on animals. On the other hand, although it has generally been underemployed in recent years, demographic profiling is the most prominent of the human-oriented markers, in that it directly monitors changing harvest strategies from the hunter's interest in short-term off-take maximization to the herder's goal of promoting long-term optimization of herd propagation and security. A comparison of these two methods, then, promises not just methodological improvements, but also the possibility of a far more important reorientation of thinking about the process of domestication itself.

Analysis of Modern Caprovines

An essential aspect of evaluating the utility of these two markers of animal domestication has been the analysis of a large, modern collection of mostly wild sheep and goats, and a more limited sample of gazelles, with the goal of providing a solid empirical basis for the archaeological study of body-size variation and herd demography in ancient sheep and goats. Modern skeletal collections used in this part of the study are curated by the Field Museum of Natural History in Chicago and the Smithsonian Institution's National Museum of Natural History in Washington, D.C.

Analysis of these modern skeletal collections has included (1) a comprehensive osteometric analysis aimed at evaluating the impact of various factors on body size (Zeder 2001, 2005); (2) a study of long-bone fusion and tooth eruption and wear sequences to resolve discrepancies between various standards developed for determining the age at death of archaeological specimens (Zeder in press); (3) a critical and comprehensive evaluation of the morphological criteria used to discriminate between the bones of sheep and goat (Zeder and Lapham, in prep); and (4) tissue sampling for DNA analysis (conducted by Gordon Luikart, Grenoble). Forty-one goat specimens were included in this study, consisting of 37 wild goats *(Capra aegargus)* and 4 domestic goats *(Capra hircus)*. There were 21 males and 20 females in the sample, ranging in age from just a few months to over 10 years of age (Zeder in press). Regional distribution of the sample included over six different collecting areas that extended from the colder and wetter regions of north-central Iran in the Taurus Mountains south of the Caspian, to the far northwestern border of Iran in the northern Zagros Mountains, and then down the spine of the Zagros to its hot, arid far southern reaches near the Persian Gulf (Figure 14.2).

BODY SIZE

A major goal of the analysis of the modern skeletal collections was to assess the relative importance of four factors in determining body size in sheep and goats: sex, age, geography, and domestic status. Surprisingly, while much has been written over the years about the possible impact of these various factors on the size of ungulates, and sheep and goats in particular, a solid, empirically based evaluation of their relative impact on body size using modern animals with known life histories had never been undertaken.

Of the four factors considered, sex was clearly the most important variable affecting body size in the collections of modern wild and domestic goats from Iran and Iraq studied

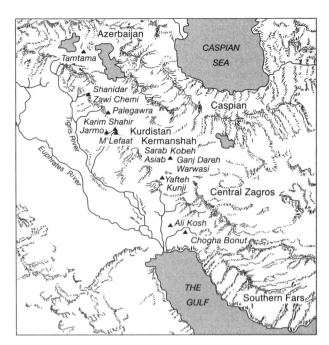

FIGURE 14.2 Map of Iran and Iraq showing locations of modern and archaeological samples (after Zeder 2003, Fig. 1).

here. Male goats were distinctly larger than females in all skeletal elements and for all dimensions measured (nine different long bones were included in this osteometric analysis, with a total of 71 different possible measurements per skeleton; Figures 14.3 and 14.4). Moreover, for each bone and each dimension there was remarkable consistency in the relationship between males and females. Females showed less variability, with female dimension measurements more tightly grouped around the mean value. Males, on the other hand, had a much greater range of variation and a wider distribution around the male mean. The split between female and male values was almost always at about 40 percent of the total range of variation for the combined sample of male and female elements, especially in breadth measurements (Figure 14.3).

Age played only a limited role in determining skeletal dimensions, especially breadth measurements. At one year of age, males were markedly larger than females, with little apparent change in breadth measurements in animals older than one year of age (Zeder 2001, Fig. 3). In fact, even the unfused bones of younger males were absolutely larger than the fully fused bones of older females (Figure 14.3). Geography was the second most powerful variable influencing body size, with a strong tendency toward smaller body size with increasing temperature and aridity (Figure 14.5a–14.5c).

Domestic status, on the other hand, had little impact on size, especially in females. This is particularly clear in breadth measurements (Figure 14.3), where wild and domestic females were indistinguishable from one another, and domestic males, although at the small end of the range of variation of wild males, were larger than all but the largest wild females from

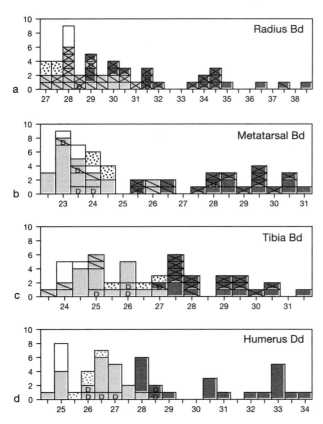

FIGURE 14.3 Modern goat breadth and depth measurements of selected long bones. The *x*-axis shows dimension in mm; the *y*-axis shows number of specimens. Dark gray shaded bars are males older than one year of age. Stippled bars are males one year of age or younger. Light gray shaded bars are females older than one year of age. White bars are females one year of age. Diagonal hatches mark fusing bones; crosshatches mark unfused bones; D marks bones of domestic specimens. (a) Radius Bd (n = 48); (b) metatarsal Bd (n = 52); (c) tibia Bd (n = 48); (d) humerus Dd (n = 50) (after Zeder 2005, Fig. 2).

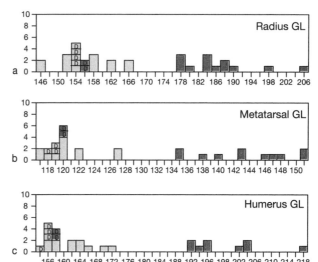

FIGURE 14.4 Modern goat length measurements of selected long bones. The *x*-axis shows dimension in mm; the *y*-axis shows number of specimens. Dark gray shaded bars are males older than one year of age. Light gray shaded bars are females older than one year of age. D marks bones of domestic specimens. (a) Radius GL (n = 32); (b) metatarsal GL (n = 28); (c) humerus GL (n = 26) (after Zeder 2005, Fig. 3).

the northernmost collecting localities. Length measurements also failed to discriminate between wild and domestic females (Figure 14.4). However, the length of the bones of male domesticates was substantially shorter than that of wild males, and some wild females, and only slightly longer than those of female domestic goats.

Thus, contrary to claims that domestic status is a major factor influencing body size in goats, this study indicates that domestication has essentially no impact on body size in modern female goats, and only a limited one in males. One might argue that the collection of modern wild and domestic goats used in this study is not a good proxy for ancient goats in evaluating the impact of domestication on body size. Modern wild goats in the sample may have experienced some reduction in size because of habitat destruction, while the domestic goats measured may have increased in body size as a result of improved breeding and nutrition (Uerpmann, H.-P., conversation with the author, August 1998). There is little to support this critique, however, since the domestic goats in this sample represent unimproved breeds from nomadic and village herds raised in the same region as many of the modern wild goats. Another possible reason offered for the lack of size difference between the modern wild and domestic animals in this sample is that there has been a considerable amount of introgression between the two populations. And while introgression is certainly possible between these modern populations of wild and domestic goats, it would be difficult to argue that more opportunity for introgression exists today than in the past, especially during the initial stages of domestication. The analysis of modern skeletal collections indicate, then, that it is sex, not domestic status, that plays the primary role in determining body size in goats, with environment (probably resulting from variation in temperature and aridity) playing a strong, but secondary, role in shaping body size. The only clearly discernable impact of domestication on body size in modern goats that was observed was a reduction in the length of male long bones. The potential utility of this difference in distinguishing between wild and domestic goats in archaeological samples, however, is quite limited since intact, unbroken long bones are rarely recovered in archaeological assemblages. There is also a slight but difficult to detect reduction in breadth measurements in domestic males, which are nonetheless still larger than in both wild and domestic females.

DEMOGRAPHY

Another goal of the study of the modern baseline skeletal collections was to refine the methods for determining age at death of archaeological caprovines (Zeder in press). Of particular interest was determining the sequence and timing

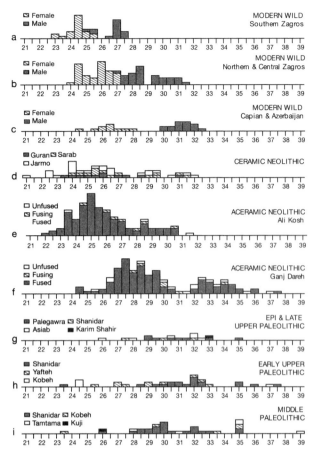

FIGURE 14.5 Diachronic view of changes in the length of the second phalanx (GL) of goats from the Middle Paleolithic to the present day. The *x*-axis shows dimension in mm; the *y*-axis shows number of specimens. (a) Modern wild goats from the southern Zagros: dark gray bars are males, light gray bars are females, (n = 36); (b) modern wild goats from the northern and central Zagros; dark gray bars are males, light gray bars are females, (n = 70); (c) modern wild goats from the Caspian region and Azerbaijan: dark gray bars are males, light gray bars females (n = 96); (d) goats from Ceramic Neolithic sites (ca. 8000–70,000 BP uncalibrated): dark gray bars are goats from Tepe Guran, light gray bars goats from Tepe Sarab, white bars goats from Jarmo (n = 50); (e) goats from Aceramic Neolithic levels at Ali Kosh (8500–8000 BP calibrated): dark gray bars are fused bones, light gray bars fusing bones, white bars unfused bones (n = 123); (f) goats from Aceramic Neolithic levels at Ganj Dareh (8900 BP calibrated): dark gray bars are fused bones, light gray bars fusing bones, white bars unfused bones (n = 132); (g) goats from Epi- and Late Upper Paleolithic sites (15,000–9000 BP uncalibrated): dark gray bars are goats from Palegawra, light gray bars goats from Shanidar level B, white bars goats from Asiab, stippled bars goats from Karim Shahir (n = 18); (h) goats from Early Upper Paleolithic sites (30,000–15,000 BP uncalibrated): dark gray bars are goats from Shanidar level C, light gray bars goats from Yafteh Cave, white bars goats from Kobeh Cave levels C–O, (n = 34); (i) goats from Middle Paleolithic sites (ca. 60,000–45,0000 BP uncalibrated): dark gray bars are goats from Shanidar level D, light gray bars goats from Kobeh Cave levels P–CC, white bars goats from Tamtama Cave, stippled bars goats from Kunji Cave (n = 34) (after Zeder 2005, Fig. 4).

of long-bone fusion and tooth eruption and wear in sheep and goats, and the degree of correlation between methods of aging based on long-bone fusion and dental patterns in these two taxa. While a number of researchers have analyzed modern collections to refine aging methods (Payne 1973; Noddle 1974; Deniz and Payne 1982; Grant 1982; Bullock and Rackham 1982; Moran and O'Connor 1994; Davis 2000), these studies have uniformly used either improved domestic breeds or feral animals descended from modern breeds—samples that may not be entirely suitable proxies for caprovines from archaeological assemblages in the Near East. The large sample of mostly wild animals from the presumed heartland of caprovine domestication analyzed here, in contrast, offers an unparalleled opportunity to refine and recalibrate these basic aging methods. The results of this baseline study are reported in detail in Zeder in press. The new sequence and estimated ages of long-bone fusion in goats is presented in Table 14.1.

Archaeological Case Study

The archaeological assemblages analyzed in this study are curated by the National Museum of Natural History, the Field Museum, the Royal Copenhagen Museum, and the University Museum of Pennsylvania. They represent virtually all the curated caprovine collections from Iran and Iraq that could be located and studied spanning the period from about 50,000 years ago to about 7500 BP uncalibrated (Figure 14.2). The oldest assemblages were recovered from Paleolithic cave sites located in the highlands of the Zagros Mountains. They include assemblages from Middle Paleolithic deposits dating to about 50,000–45,000 BP (Shanidar Level D, Kobeh, Tamtama, and Kunji caves), as well as two phases of the Upper Paleolithic, the Early Upper Paleolithic or Baradostian period at about 35,000 years ago (Shanidar Level C, Yafteh, and Kobeh Cave) and the Late Upper Paleolithic or Zarzian period at about 15,000 years ago (Palegawra and Pa Sangar caves). Essentially all analysts accept the goats from these assemblages as wild hunted animals, based on both the biometric and demographic analyses (Perkins 1964; Uerpmann 1979; Hole et al. 1969; Evins 1982).

Epi-Paleolithic assemblages (ca. 12,000–10,000 years ago) included in the study come from the site of Asiab in highland Iran and from Karim Shahir in highland Iraq, both of which are believed to represent hunting of wild herds, again based on both size and demographic markers (Uerpmann 1979; Stampfli 1983; Bar Yosef and Meadow 1995; Legge 1996; and Bökönyi 1977, with qualifications). Regrettably, the Epi-Paleolithic remains from Shanidar Level B and from Zawi Chemi Shanidar, originally studied by Perkins, have only just recently been located and have not yet been analyzed for this project.

Likely domestic goats come from two large Aceramic Neolithic assemblages in Iran: Ganj Dareh in highland Iran, originally analyzed by Hesse (1978), and Ali Kosh in lowland Iran, originally analyzed by Flannery (Hole et al 1969). As

TABLE 14.1
Revised Fusion Sequence and Ages for Goats

Age Group	Bone	Order of Fusion	Estimated Age[a]
A	Astragalus[b]	1	0–6
A	P. Radius	2	0–6
B	D. Humerus	3	6–12
B	Pelvis	3	6–12
B	Scapula	3	6–12
C	2nd Phalanx	4	12–18
C	1st Phalanx	5	12–18
D	D. Tibia	6	18–30
D	D. Metacarpal	7	18–30
D	D. Metatarsal	7	18–30
E	Calcaneus	8	30–48
E	P. Femur	9	30–48
E	D. Femur	9	30–48
E	P. Ulna	9	30–48
E	D. Radius	9	30–48
E	P. Tibia	9	30–48
F	P. Humerus[c]	10	48+
G	P. Humerus[d]	11	48++

SOURCE: From Zeder in press.
[a] Age in months.
[b] Age at which astragalus assumes adult texture.
[c] Bones in early fusion.
[d] Bones in late fusion or fully fused.

discussed above, Hesse used a novel demographic profiling technique capable of providing sex-specific information on harvest strategies to argue for the domestic status of the goats from the upper levels at Ganj Dareh. Other researchers, dubious of the utility of demographic markers in documenting domestication, have argued that goats from all levels at Ganj Dareh were domestic, based on an apparent body-size reduction when compared to benchmark size profile composites of goats from Epi- and Paleolithic assemblages (Uerpmann 1979; Helmer 1992; Bar Yosef and Meadow 1995; and Legge 1996). As noted above, Flannery used zoogeographic, demographic, and morphological criteria to argue for the domestic status of the goats at Ali Kosh. Osteometric data from the Ali Kosh assemblage were not available prior to this study. Another small assemblage from lowland Iran included in this analysis comes from the site of Chogha Bonut, recently excavated by Abbas Alizadeh, which was originally studied by Redding. Also included are assemblages from Tepe Guran and Sarab in highland Iran, and Jarmo in highland Iraq, all of which date to the Ceramic Neolithic and are uniformly accepted as domestic, based on both biometric and demographic data (Bökönyi 1977; Reed 1961, 1983; Stampfli 1983; Flannery n.d.).

All of these sites were excavated during the early days of radiocarbon dating, long before the development of small-volume accelerator mass spectrometer (AMS) radiocarbon dating, which allows direct dating of domestic candidate specimens. The poor resolution of the earlier, indirect, large-sample traditional radiocarbon dates raised a number of unresolved questions about the chronology of sites in the Zagros region, especially those bracketing the transition to early agriculture (Hole 1987). Of particular concern has been the relative temporal placement of Ganj Dareh and Ali Kosh, the two leading contenders for the earliest evidence of animal domestication in the region. Some of the early large sample radiocarbon dates suggested that basal levels at Ganj Dareh were more than 1,000 years earlier in time than the overlying upper village levels, placing Level E at about 10,000 BP uncalibrated and upper levels D–A at about 9000 BP uncalibrated (Hesse 1978: 84; Hole 1987). At Ali Kosh, a confusing array of radiocarbon dates suggested basal levels were as old as 9500 BP uncalibrated, while upper Ceramic Neolithic-age levels were thought to date to about 7600 BP uncalibrated (Hole et al 1969: 331–334; Hole 1987).

Clarifying the temporal sequence of these sites was essential for tracing the developmental trajectory of animal domestication in the Zagros. Did goat domestication begin within the highland natural habitat of wild goats where Ganj Dareh was located and then move out of the highlands as a developed technology (cf., Braidwood 1960)? Or did it originate in marginal areas outside the natural habitat of wild goats by displaced former occupants of this region trying to replicate the wild resource base found in the highlands (cf., Hole and Flannery 1967; Hole et al 1969; Binford 1971)?

A series of more than 60 AMS radiocarbon dates obtained as part of this study has brought new clarity to the temporal sequence of sites in the Zagros region (Hesse and Zeder 2000; Zeder 2003, 2005; Table 14.2). The 12 dates obtained for Ganj Dareh are all tightly clustered around 8900–8700 BP uncalibrated (9900–9700 BP calibrated), indicating that all levels at the site (A–E) were occupied over a fairly brief span of time lasting no more than about 200 years. Basal levels at Ali Kosh, on the other hand, can now be reasonably firmly dated to about 8500–8400 BP uncalibrated (ca. 9500 to 9400 calibrated BP), about 500 years later than Ganj Dareh. The Epi-Paleolithic site of Asiab, once thought contemporary with basal levels at Ganj Dareh, is now dated to 9500 BP uncalibrated (ca. 10,700 BP calibrated), roughly 700 years earlier than Ganj Dareh. Ceramic Neolithic-age sites in both lowland and highland Iran seem to have been founded about 8000 BP uncalibrated (9000 BP calibrated).

BODY SIZE

A diachronic view of size variation in Zagros goats from the Middle Paleolithic to the Ceramic Neolithic is presented in Figure 14.5. In contrast to the standard practice of presenting metric data as amalgamations of normalized data, raw metric data are presented here (in this case the greatest length of the second phalanx). Also in a departure from standard practice that restricts metric analysis to fused bones (following von den Driesch 1976: 4), Figure 14.5 also includes measurements of fusing and unfused bones.

TABLE 14.2
AMS Dates on Bones from Zagros Sites

	Beta Analytic Number	Level	Depth (cm)	14C BP[b]	14C Cal BP[c]	2-Sigma Cal BP[d]	1-Sigma Cal BP[e]
Yafteh	B-177135[a]		250–260	30,300 ± 320			
	B-177136[a]		270–280	32,400 ± 380			
	B-177120		270–280	18,580 ± 80	22,060	22,600–21,550	22,500–21,640
	B-177121		270–280	18,980 ± 80	22,520	23,070–22,000	22,980–22,090
Palegawra	B-159546		10–20	5130 ± 50	5910	5980–5970 5950–5740	5920–5890 5800–5770
	B-159543		20–40	12,510 ± 90	15,100	15,530–14,150	15,480–14,240
	B-159545		60–80	8790 ± 70	9860	10,150–9560	10,100–10,090 9920–9690
	B-159542		80–100	11,210 ± 110	13,160	13,760–13,700 13,470–12,900	13,420–13,000
	B-159544		80–100	10,170 ± 70	11,910	12,340–11,550 11,490–11,430	12,290–12,220 12,140–11,650
Asiab	B-159555		30–45	9480 ± 80	10,710	11,110–10530	11,050–10,960 10,770–10,590
	B-159554		45–60	9370 ± 60	10,570	10,720–10,420	10,670–10,520
	B-159552		75–90	7790 ± 60	8580	8660–8420	8610–8460
Ganj Dareh	B-108239	B	165–180	8930 ± 60	9940	10,005–9870	9975–9905
	B-108238	A	180–200	8780 ± 50	9850	9910–9585	9880–9805 9780–9660
	B-108240	B	220–240	8780 ± 50	9850	9910–9585	9880–9805 9780–9660
	B-108241	B	240–260	8720 ± 50	9650	9875–9525	9845–9725 9695–9565
	B-108242	B	280–300	8940 ± 50	9945	10,000–9890	9975–9915
	B-108244	D	430–460	8840 ± 50	9890	9945–9820 9765–9665	9915–9860
	B-108243	C	460–480	8920 ± 50	9935	9990–9875	9960–9905
	B-108246	E	580–585	8870 ± 50	9905	9960–9845 9725–9695	9935–9875
	B-108245	D	580–600	8940 ± 50	9945	10,000–9890	9975–9915
	B-108247	E	665–675	8830 ± 50	9880	9940–9805 9780–9660	9910–9850
	B-108248	E	700–710	8900 ± 50	9920	9980–9865	9950–9895
	B-108249	E	765–768/70	8840 ± 50	9890	9945–9820 9765–9665	9915–9860
Ali Kosh	B-137020[a]	Mohammed Jaffar	50–60	7100 ± 70	7940	8020–7775	7970–7845
	B-177122[a]	Mohammed Jaffar	90–100	7550 ± 40	8370	8400–8300	8390–8350
	B-118719[a]	Mohammed Jaffar	70–80	8130 ± 70	8995	9245–8940	9185–9110 9090–8970
	B-118720[a]	Mohammed Jaffar	130–140	8130 ± 70	9000	9360–8715	9215–8965
	B-118721[a]	Ali Kosh	180–200	8720 ± 100	9650	9935–9480	9875–9525
	B-118722[a]	Ali Kosh	210–230	8110 ± 80	8985	9245–8750	9040–8960

TABLE 14.2 (continued)

	Beta Analytic Number	Level	Depth (cm)	^{14}C BP[b]	^{14}C Cal BP[c]	2-Sigma Cal BP[d]	1-Sigma Cal BP[e]
	B-177124[a]	Ali Kosh	230	8050 ± 40	9000	9030–8970 8910–8870 8830–8790	9020–8990
	B-137021[a]	Ali Kosh	250–270	8450 ± 70	9485	9555–9270	9530–9425
	B-118723[a]	Ali Kosh	280–300	8490 ± 90	9465	9565–9350	9505–9415
	B-118724[a]	Ali Kosh	380–400	8340 ± 100	9375	9485–9000	9440–9220
	B-108256	Bus Mordeh	540–560	8000 ± 50	8945	8985–8620	8965–8705
	B-122721[a]	Bus Mordeh	630–650	8540 ± 90	9482	9650–9385	9530–9445
	B-137024[a]	Bus Mordeh	680–710	8410 ± 50	9465	9520–9380 9370–9300	9490–9425
	B-177126[a]	Bus Mordeh	680	8530 ± 40	9520	9550–9490	9540–9510
Chogha Bonut	B-177134			8040 ± 40	9020	9100–8990	9100–8900
	B-177132			8070 ± 40	9010	9040–8980	9020–9000
	B-177133			8120 ± 40	9020	9130–9000	9050–9010
Guran	B-147111	D		7630 ± 60	8400	8530–8350	8430–8380
	B-147112	F		7260 ± 40	8030	8160–7970	8140–8010
	B-177131	H		7810 ± 40	8590	8640–8460	8610–8550
	B-147113	H		7950 ± 40	8770	9000–8630	8980–8660
	B-147114	K		7080 ± 60	7930	7990–7780	7960–7840
	B-147115	L		7940 ± 40	8760	9000–8620	8980–8820 8800–8650
	B-177116	L		8130 ± 40	9030	9140–9000	9100–9020
	B-147116	N		3690 ± 40	4060	4150–3900	4090–3970
	B-147117	P		7890 ± 40	8640	8970–8910 8870–8830 8790–8590	8740–8610
	B-147118	Q		8070 ± 40	9010	9040–8980	9020–9000
	B-147119	R		8000 ± 50	8990	9020–8650	9000–8770
	B-147122	T		8170 ± 40	9100	9260–9020	9140–9030
	B-147120	U		8060 ± 40	9010	9030–8980 8820–8800	9020–9000
	B-147121	V		7820 ± 50	8600	8710–8450	8630–8550
	B-177117	V		8280 ± 40	9280	9420–9130	9400–9360 9310–9250
Sarab	B-159547	1A		7470 ± 70	8330	8400–8160	8360–8190
	B-159548	3		7950 ± 60	8770	9010–8600	8990–8640
	B-159550	4		8070 ± 60	9010	9120–8770	9030–8990
	B-159549	5		7800 ± 60	8580	8710–8430	8620–8510

[a] Date based on carbonized bone; all other dates based on collagen.
[b] Uncalibrated conventional ^{14}C age of specimens in ^{14}C BP (± 1δ).
[c] Intercept between the conventional ^{14}C age and the dendrocalibrated calendar time scale, in calendar years BP (Pretoria calibration procedure program, Beta Analytic).
[d] Two-sigma dendrocalibrated age range for specimens, in calendar years BP.
[e] One-sigma dendrocalibrated age range for specimens, in calendar years BP.

Contrary to previous studies that maintained the Ganj Dareh goats had undergone body-size reduction, Figure 14.5 clearly shows that there is no difference in the size of goats from Ganj Dareh (Figure 14.5f) when compared to wild goats recovered from Epi-Paleolithic and older sites (Figure14.5g–i). In fact, when the actual, unnormalized, measurements of bones (both fused and unfused) are directly compared, no case can be made for change in the body size of goats in the Zagros highlands over the roughly 30,000-year period that separated the occupation of Ganj Dareh from the occupation of Middle Paleolithic cave sites.

There is a clear reduction in the size of the goats from lowland Ali Kosh (Figure 14.5e) when compared with goats from highland sites (Figure 14.5f–i). Interestingly, however, a similar difference in body size can be seen when modern wild goats from the same geographic region as the upland archaeological sites (Figure 14.5b) are compared with the modern wild goats from the arid southern collecting localities (Figure 14.5a), with hot, arid environments comparable to that of Ali Kosh in lowland Iran. As a result, it is difficult to judge whether the smaller size of the goats from Ali Kosh is an artifact of domestication or simply reflective of clinal variation in size resulting from increasing aridity in a manner similar to that seen in modern wild goats.

Goats from Ceramic Neolithic–age sites in the highland Zagros (Figure 14.5d) are also clearly smaller than earlier goats from this same region (Figure 14.5f–i). These later highland goats are roughly comparable in size to those from lowland Ali Kosh (Figure 14.5e). Once again, however, it is difficult to confidently attribute this temporal shift in body size directly and exclusively to the process of domestication. This is especially the case when one considers that modern wild goats in the region (Figure 14.5a–c) show considerable and continuing reduction in body size when compared to ancient goats, both wild and presumed domestic. Modern wild goats from the northern and central Zagros (Figure 14.5b) are substantially smaller than goats from the same region dating to about 9000 BP uncalibrated and older (Figure 14.5f–i), and are about the same size as the presumed domestic goats from Ceramic Neolithic sites in the same region (Figure 14.5d). Modern wild goats from the southern Zagros (Figure 14.5a) exhibit a similar reduction in body size when compared both with Ceramic Neolithic upland goats and with 8500-year-old goats from the arid, lowland site of Ali Kosh (Figure 14.5e).

This deep-time diachronic view of size variation in wild and domestic goats in the Zagros underscores how difficult it is to isolate the possible impact of domestication from other geographic and temporal factors that clearly play a role in shaping body size in goats. In this regard, it is interesting to look at patterns of size variation in another indigenous ungulate that was never domesticated—the gazelle. A biometric analysis of modern and archaeological gazelles from the Zagros shown in Figure 14.6 provides a similar diachronic view of the size variation of gazelles as compared with that presented for goats in Figure 14.5. Modern gazelles from the region, like the modern goats, show a strong pattern of clinal

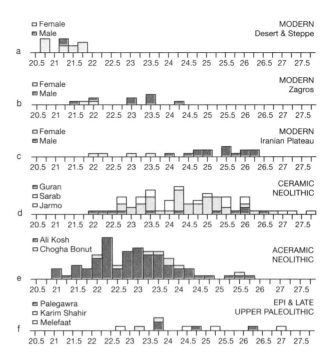

FIGURE 14.6 Diachronic view of changes in the depth of the distal humerus (Dd) of gazelles from the Epi- and Late Paleolithic to the present day. The x-axis shows dimension in mm; the y-axis shows number of specimens. (a) Modern gazelles from desert and steppe regions of Iran and Iraq: dark gray bars are males, light gray bars females (n = 13); (b) modern gazelles from the Zagros Mountains: dark gray bars are males, light gray bars females, (n = 10); (c) modern gazelles from the Iranian Plateau: dark gray bars are males, light gray bars females, (n = 19); (d) gazelles from Ceramic Neolithic sites (ca. 8000–7000 BP uncalibrated): dark gray bars are gazelles from Tepe Guran, light gray bars gazelles from Tepe Sarab, white bars gazelles from Jarmo, (n = 70); (e) gazelles from Aceramic Neolithic lowland sites (8500–8000 BP uncalibrated): dark gray bars are gazelles from Ali Kosh, light gray bars gazelles from Chogha Bonut, (n = 86); (f) gazelles from Epi- and Late Upper Paleolithic sites: dark gray bars are gazelles from Palegawra, light gray bars goats from Karim Shahir, white bars gazelles from Melefaat, (n = 9) (after Zeder 2005, Fig. 5).

variation of decreasing body size correlated with increasing temperature and aridity (Figure 14.6a–c). Similarly, modern gazelles show a strong dimorphism in body size with a range of variation in male and female size strikingly like that seen in modern goats (Figure14.6c). As is the case with goats, there is also obvious clinal variation in body size of ancient gazelles from Epi- and Late Upper Paleolithic upland sites (Figure 14.6f) when compared to gazelles from Aceramic-age sites in lowland Iran (Figure 14.6e). This pattern of decreasing body size correlated with increasing temperature and aridity is directly analogous to the trend noted for modern gazelles, as well as that for modern and ancient goats. Moreover, as with goats, gazelles show a general temporal trend toward body-size reduction. The reduction in body size in gazelles over the past 8,000 to 9,000 years is especially evident when gazelles from Aceramic Neolithic–age sites in the lowland of Iran (Figure 14.6e) are compared with modern gazelles from steppe and desert regions in Iran and Iraq (Figure 14.6a), and when ancient upland gazelles (Figure 14.6f) are compared with modern gazelles from the same region (Figure 14.6b).

Interestingly, the observed reduction in body size in upland goats and gazelles did not occur at the same time. Upland goat populations experienced a major decline in size in the Ceramic Neolithic after about 8000 BP uncalibrated (Figure 14.5d). Gazelles from these same sites (Figure 14.5d), in contrast, show no reduction in body size when compared with gazelles from earlier upland sites (Figure 14.6f). The downward trend in gazelle body size must have begun sometime after the occupation of these Ceramic Neolithic–age sites (after about 7000 BP uncalibrated), more than 1,000 years later than a comparable downward shift is seen in the size of goats in the Zagros. The difference in timing in the reduction of body size in goats and gazelles argues against a single cause for the size reduction in these species (e.g., a uniform, region-wide response in ungulates to climatic change or human-induced habitat disruption). It suggests instead that these species responded independently to factors that affect body size. Just what these factors were is difficult to say. Possible factors that might have affected body size in gazelles and goats include species-specific responses to climate change, habitat degradation, and over-hunting. There are other factors, however, that may pertain only to size variation in goats, such as genetic isolation of managed goats, introgression between wild and domestic animals, selective breeding, and the introduction of smaller-bodied varieties of domestic goats into the region. Whatever the cause or causes, the end result has been a dramatic overall decrease in body size both in domestic goats and in their wild progenitors—and in gazelles, a species that never underwent domestication—between 8000–7000 BP uncalibrated and the present day.

DEMOGRAPHY

One of the more interesting biometric patterns evident in goat-bone assemblages like those from Ali Kosh and Ganj Dareh is a striking bimodality found in all bones and all dimensions measured (Figures 14.5e–f, 14.7, and 14.8). This bimodal distribution in the metric data of ancient wild goats is comparable to the sex-related bimodality documented in modern baseline comparison populations (Figure 14.3). Even though modern goats are smaller overall than ancient goats, in both ancient and modern samples the mode comprising smaller animals makes up about 40 percent of the total range of variation in all bones, while the mode comprising larger animals represents about 60 percent of the range of variation. The coefficients of variance for these two modes are remarkably similar in both ancient and modern populations (Zeder 2001, Table 2 and Table 4). The statistical similarity between the ancient and modern biometric data supports Hesse's hypothesis that the bimodality observed in the Ganj Dareh goat biometrics represents a marked size difference between male and female goats.

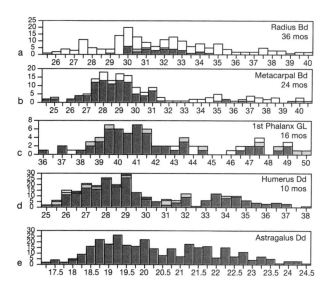

FIGURE 14.7 Breadth and depth measurements of goat long bones from Ganj Dareh arranged by age at fusion. The x-axis shows dimension in mm; the y-axis shows number of specimens. Dark gray bars are fused specimens. Light gray bars are fusing specimens. White bars are unfused specimens. (a) Radius Bd: fusion at 36 months (n = 187); (b) metacarpal Bd: fusion at 24 months (n = 158); (c) first phalanx GL: fusion at 16 months (n = 73); (d) humerus Dd: fusion at 10 months (n = 263); (e) astragalus Dd: fused at birth (n = 313) (after Zeder 2005, Fig. 6).

Another significant feature of the goat biometric data from Ganj Dareh and Ali Kosh is the higher frequency of unfused elements among the bones of larger animals, especially in later-fusing bones (Figure 14.7a–b and Figure 14.8a–b). The comparative baseline study of modern goat skeletal material demonstrated that the unfused bones of young males older than one year of age are invariably larger than the bones of females, fused or unfused. Given the total size separation in the bones of males and females in modern goats regardless of fusion, it is safe to conclude that the unfused elements of larger animals in the archaeological samples represent males killed before the age of fusion of that bone. The low representation of bones of larger animals is largely attributable to the friability of unfused elements, which makes them less likely than fused elements to be preserved and recovered, and, if recovered, makes them more difficult to identify as to species or to measure. The better representation of larger animals in bones like the astragalus that do not fuse and the subsequent progressive decrease in the representation of larger animals in later-fusing bones supports the conclusion that males are underrepresented in the biometric data from Ganj Dareh and Ali Kosh, primarily as a result of poor preservation rather than as an actual imbalance of male and female animals in the ancient herd.

Following protocols discussed in Zeder 2001, this distinctive bimodality can be used to separate ancient assemblages into male and female subpopulations. The revised fusion sequence and calibration indices derived from the analysis of the modern goats' skeletal materials (Table 14.1; Zeder

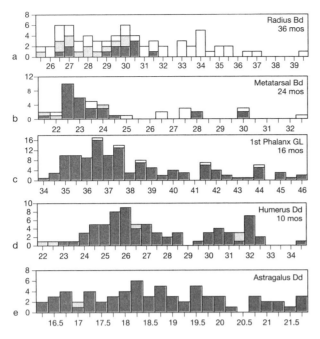

FIGURE 14.8 Breadth and depth measurements of goat long bones from Ali Kosh arranged by age at fusion. The *x*-axis shows dimension in mm; the *y*-axis shows number of specimens. Dark gray bars are fused specimens. Light gray bars are fusing specimens. White bars are unfused specimens. (a) Radius Bd: fusion at 36 months (n = 183); (b) metacarpal Bd: fusion at 24 months (n = 158); (c) first phalanx GL: fusion at 16 months (n = 73); (d) humerus Dd: fusion at 10 months (n = 264); (e) astragalus Dd: fused at birth (n = 314) (after Zeder 2005, Fig. 7).

in press) can then be used to compute sex-specific demographic profiles for each of these subsamples (Table 14.3 and Figure 14.9). In the likely wild, hunted populations from Middle and Upper Paleolithic levels at Shanidar Cave (Figure 14.9a) and from Epi-Paleolithic Asiab (Figure 14.9b), the sex-specific harvest profiles indicate that most males killed were older than the four-year age limit of the long-bone fusion sequence. Females also tended to be killed at somewhat younger ages, especially at Shanidar Cave. An emphasis on the hunted males is also suggested by the higher proportion of bones that fall into the larger-bodied, male subpopulations at these sites (Table 14.4). The low percentage of unfused or fusing bones among the larger male elements also confirms that an emphasis on older males dominated prey strategies at these two sites (Table 14.5). This pattern is consistent with a hunting strategy that seeks to optimize short-term off-take by preying on animals with the greatest amount of meat per individual.

Assemblages from Ganj Dareh and Ali Kosh (Figure 14.9c–d), on the other hand, both show distinctively different harvest profiles for males and females, with an emphasis on the slaughter of young males and a delay in the slaughter of females. Also, and in clear contrast to the assemblages from Asiab and Shanidar, female animals are much better represented in the Ganj Dareh and Ali Kosh assemblages (Table 14.4). As noted above, this pattern is quite likely the result of taphonomic factors that select against the preservation of the bones of young male animals, rather than an emphasis on females in ancient harvest strategies. The percentage of unfused and fusing bones in the subpopulation of larger-bodied males at both Ganj Dareh and Ali Kosh is also substantially higher in the female subpopulation (Table 14.5).

Closer inspection of the Ganj Dareh and Ali Kosh demographic profiles suggests some possible differences in herd management strategies at the two sites. Ganj Dareh residents seem to have harvested both young males and older females with greater intensity at somewhat earlier ages than did residents of Ali Kosh. The slaughter of male animals at Ganj Dareh increases after about 12 months of age, and is extremely high up to about 24 months, with only about 20 percent of the males surviving beyond two years. In contrast, slaughter of males at Ali Kosh does not begin in earnest until around 18 months of age, with nearly 50 percent of males surviving beyond two years of age. Female slaughter differs more in intensity than in timing, with less than 30 percent of the females surviving beyond four years at Ganj Dareh and nearly 50 percent of the females surviving beyond this point at Ali Kosh. The higher proportion of unfused and fusing bones in both the male and female subpopulations of the goats from Ganj Dareh, when compared to Ali Kosh (Table 14.5), also suggests a younger harvest profile for both males and females at this site.

The harvest profile for the total sample of goats from Ganj Dareh, however, does not conform to that expected with the selective hunting of nursery herds of lactating females and young, which Hesse thought he detected in the earliest basal Level E occupation of the site. Moreover, there no longer appears to be any change in goat harvest profiles over time at Ganj Dareh. When Hesse performed his study, only a limited sample of bones could be firmly attributed to stratigraphic levels at the site. Thanks in large measure to work he did after he completed his dissertation and generously shared with the author, it is now possible to assign a much larger proportion of the Ganj Dareh assemblage to individual levels at the site. Sex-specific harvest profiles computed for the goats from basal Level E are quite similar in most respects to those computed for the goats from upper levels D–A (Figure 14.10, Tables 14.3–14.5), with the single exception that kill-off of males in Level E appears to intensify after one year of age, while in upper levels it does not intensify until between the ages of about 18 to 24 months. The profile of female harvest remains essentially the same over time. It is difficult to say whether the observed difference in the timing of male harvest is attributable to a change in herding strategy over the occupation at this site, or (perhaps more likely) is an artifact of the relatively small sample size in Level E of bones in the age class that tracks fusion at about 13 to 18 months (only 14 bones). But it is clear that the harvest profile generated from this expanded sample of elements now ascribable to Level E differs dramatically from the profile Hesse originally obtained with a much smaller sample, in which he detected an emphasis on males and females younger than six months of age and prolonged female survivorship beyond the limits

TABLE 14.3
Long-Bone Fusion Scores for Goats and Gazelles from Archaeological Sites

Age Class	Shanidar C&D Goats			Asiab Goats			Ganj Dareh Goats Total			Ganj Dareh Goats Level E			Ganj Dareh Goats Level A–D			Ali Kosh Goats			Jarmo Goats			Ali Kosh Gazelle		
	F	M	All[a]	F	M	All[b]	F	M	All[a]	F	M	All[a]	F	M	All[a]	F	M	All[a]	F	M	All[b]	F	M	All
A	100	100	100	-	97	98	96	98	81	88	100	92	94	97	91	98	100	96	99	99	99	100	99	99
B	100	-	100	100	100	100	91	94	78	91	83	76	83	97	77	95	96	88	98	100	98	96	100	98
C	79	99	72	100	100	63	93	74	56	91	57	52	94	86	53	98	93	82	99	88	93	97	98	96
D	50	83	36	100	100	64	74	21	33	72	23	34	74	20	27	82	54	57	96	88	81	89	75	75
E	0	80	46	50	100	67	27	4	19	25	6	20	28	7	18	47	25	34	59	46	53	53	66	67
# Bones	13	63	414	11	26	93	872	514	4838	142	71	815	565	345	4222	380	191	1675	564	251	1232	203	372	804

NOTE: Follows Zeder 2001 and in press. Male ("M") and female ("F") samples are composed primarily of goats, and only bones identified as goat or sheep/goat. "All" samples include both measured and unmeasured bones identified as either goat or sheep/goat in assemblages composed primarily of goats, and only bones identified as goats in assemblages with larger proportions of sheep.

[a] Combined sample includes bones identified as goat or sheep/goat.
[b] Combined sample includes only bones identified as goat.

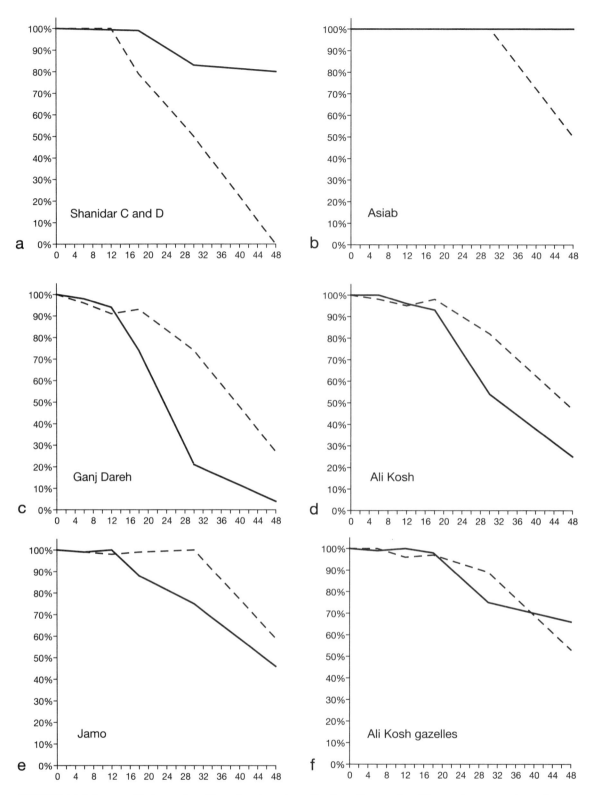

FIGURE 14.9 Sex-specific harvest profiles of goats and gazelles from Zagros sites. The *x*-axis represents age in months; the *y*-axis represents percentage surviving. Solid lines are males. Dashed lines are females. (a) Goats from Shanidar levels C and D: females = 13 specimens, males = 63 specimens; (b) goats from Asiab: females = 11 specimens, males = 26 specimens; (c) goats from Ganj Dareh: females = 872 specimens, males = 514 specimens; (d) goats from Ali Kosh: females = 380 specimens, males = 191; (e) goats from Jarmo: females = 564 specimens, males = 251 specimens; (f) gazelles from Ali Kosh: females = 203 specimens, males = 372 specimens (after Zeder 2005, Fig. 8).

TABLE 14.4
Proportions of Male and Female Goats in Sites from the Zagros

Site	Percentage Males	Percentage Females	Total NISP Measured Goats
Shanidar C and D	83	17	76
Asiab	70	30	37
Ganj Dareh Total	37	63	1386
Ganj Dareh—Level E	33	67	213
Ganj Dareh—Levels A–D	38	62	910
Ali Kosh	30	70	571
Jarmo	32	68	815
Ali Kosh—Gazelles	65	35	575

TABLE 14.5
Proportions of Unfused or Fusing Bones among Male and Female Goats in Sites from the Zagros

Site	Percentage Unfused + Fusing Males	NSIP Male	Percentage Unfused + Fusing Females	NISP Female
Shanidar C and D	6%	13	23%	54
Asiab	0%	17	9%	11
Ganj Dareh Total	54%	400	35%	725
Ganj Dareh—Level E	67%	60	35%	132
Ganj Dareh—Levels A–D	54%	300	38%	512
Ali Kosh	36%	165	17%	359
Jarmo	20%	126	12%	370
Ali Kosh—Gazelles	8%	334	12%	198

of detection using long bone fusion. The sex-specific harvest profiles for all levels at the site, then, are consistent with the long-range goals of a herder interested in promoting security and growth of his herd. Given the new dating of the site, which compresses the total occupation of Ganj Dareh to about 200 years, this result is not surprising.

Thus, at both Ganj Dareh and Ali Kosh there is strong evidence for management of goat herds with the goal of promoting the overall security and growth of the herd. The somewhat delayed slaughter both of young males and of older females at Ali Kosh is more consistent with modern domestic slaughter strategies and may, then, represent a later evolution of herd management practices.

Later sites in the Zagros uplands show a similar pattern. The measurable goat remains from Jarmo, for example, which are drawn primarily from ceramic-bearing upper levels at the site, also show a difference in the harvest profiles of male and female animals that is consistent with herd management (Figure 14.9e, Table 14.3). The data from Jarmo appear to show an even greater delay in the slaughter of both male and female animals at this site than that seen at either Ganj Dareh or Ali Kosh. An older harvest pattern also appears evident in the lower proportions of unfused and fusing elements among both male and female goats at Jarmo (Table 14.5). However, this pattern may not be indicative of the practice of different herd management strategies at Jarmo, but is perhaps more likely a reflection of the lack of systematic collection of smaller and more fragmentary unfused elements at this site excavated more than 50 years ago. Despite this likely recovery bias, male elements are nearly twice as likely to be either unfused or fusing than are female elements (Table 14.5), indicating a clear, general pattern of young male slaughter and prolonged female survivorship at Jarmo consistent with herd management.

Finally, it is useful to examine sex-specific profiles computed for gazelles from Ali Kosh (Figure 14.9f)—again, a species never domesticated. These harvest profiles show little difference between the harvest of males and females, with more than 50 percent of both sexes surviving beyond the limits of detection of long-bone fusion data. The higher percentage of larger, male gazelle elements compared with female elements at Ali Kosh (Table 14.4) is indicative of an emphasis on male animals, although it is not as strongly exhibited as is the emphasis on wild male goats seen at the earlier sites of Shanidar and Asiab. The low percentage of unfused or fusing bones in both the male and female gazelle subpopulations (Table 14.5) further suggests an emphasis on adult animals, which is more strongly expressed among the males. Again, this pattern is consistent with a selective hunting strategy that maximizes off-take by focusing on prime-age animals.

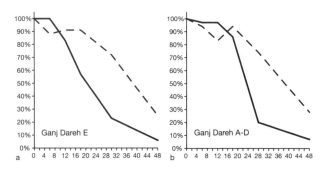

FIGURE 14.10 Sex-specific harvest profiles of goats from different levels at Ganj Dareh. The *x*-axis represents age in months; the *y*-axis represents percentage surviving. Solid lines are males. Dashed lines are females. (a) Goats from Ganj Dareh Level E: females = 142 specimens, males = 71 specimens; (b) goats from Ganj Dareh Levels D–A: females = 565 specimens, males = 345 specimens.

Goat Domestication in the Zagros

A focus on prime-age animals, especially males, seems to have dominated goat hunting strategies in the Zagros from the Middle Paleolithic into the Epi-Paleolithic. Sometime between the occupation of Asiab and Ganj Dareh (between about 9500 to 8900 BP uncalibrated, 10,500 to 9900 BP calibrated), there was a change in the underlying goal of highland goat exploitation. Long-practiced harvest strategies that emphasized short-term maximization of immediate return were replaced by sustainable harvest strategies that promoted the long-term availability of animals for future consumption. As elsewhere in the Fertile Crescent, this change in animal exploitation strategies is likely linked to the increasingly sedentary and territorially circumscribed exploitation of plant resources, like cereals, pulses, and nut trees, that became more widely and more densely distributed across the region after the end of the last ice age (Hillman 1996; Zeder 2003, 2005). Reduction of mobility and the growth of increasingly more sedentary communities focused on these newly abundant plant resources probably resulted in localized pressure on wild herds. This pressure may have in turn been met with an adjustment in harvest strategies toward those that enhanced the predictability of supply of animal resources by culling expendable young males and preserving older females for breeding stock. Little if any morphological change would be expected in these early managed animals since there was likely little change in general environmental conditions, ample opportunity for back-crossing with wild animals, and the possibility of restocking of managed herds from wild ones.

As much as 500 years later, managed herds and their herders expanded out of this highland homeland of initial domestication and into adjacent, less-optimal areas like the Deh Luran Plain, where Ali Kosh is situated. Moving into these lowland areas undoubtedly represented a substantial departure from the environmental conditions that prevailed in their native highland habitats. It also ended any potential for interbreeding between managed and wild goats. The more conservative harvest profile at Ali Kosh, in which slaughter of both young males and older females is delayed, may be a response to the loss of easy access to wild animals for restocking.

The genetic isolation from wild populations, plus the impact of tighter human control of breeding, undoubtedly resulted in the changes in horn morphology documented at Ali Kosh. These factors, coupled with more arid environmental conditions and poorer pasture opportunities, also may have contributed to a reduction in the size of these animals. It is perhaps equally plausible, however, that the founder population of the Ali Kosh goats was composed of smaller-bodied wild goats derived from the more southern reaches of the highland natural habitat of goats, and that domestication played no role in the size of these animals.

Nearly 1,000 years after the occupation of Ganj Dareh during the Ceramic Neolithic, highland managed goat herds were composed of animals similar in size to those at lowland Ali Kosh. Once again, however, the smaller size of these animals needs not to have been a direct consequence of domestication. The Ceramic Neolithic in the Zagros was clearly a time of considerable change. Not only were ceramics introduced into the region, but there is also evidence of increased contact across the Fertile Crescent, as marked by the widespread exchange of exotic materials like obsidian, cowrie shells, and turquoise (Mellart 1975; Blackman 1984; Bar Yosef and Meadow 1995). There is also a dramatic shift in animal exploitation in the Zagros. Ceramic Neolithic-age faunal assemblages in the Zagros show an increased importance of sheep, possibly domesticated sheep introduced into the Zagros from outside the region. There was also a marked increase in the importance of gazelles in sites at all elevations in the Zagros (Zeder 2003, 2005). Clearly, this was a time of significant change involving considerable movement of goods, technology, people, and animals across the entire Fertile Crescent. It is quite possible, then, that the smaller size of the highland managed goats after 8000 BP reflects the intermingling of upland populations with smaller-bodied goats introduced from other regions, or their replacement by a smaller variety of domestic goat altogether.

Conclusions

Body-Size Reduction as a Marker of Domestication

This reconsideration of the evidence for goat domestication in the Zagros calls into question the utility of size reduction as a leading-edge marker of initial animal domestication. There is no evidence for decrease in body size in the goats from highland Ganj Dareh despite incontrovertible evidence that they were being managed in a way consistent with the modern strategies for managing domestic herds. The apparent decrease in body size of the Ganj Dareh goats that earlier researchers interpreted as a rapid response to the selection for smaller body size in animals undergoing domestication can now be seen to be an artifact of a shift in the sex ratio of the adult portion of managed herds. Large adult males will likely dominate in the sample of animals harvested by hunters. The remains of animals harvested by herders, in contrast, will be dominated by adult females, given the under-representation of the bones of male animals slaughtered at a young age in these assemblages. Thus, when metric data from a managed herd, like that from Ganj Dareh, is compared with the assemblage representing a hunted population, like Asiab, the differential representation of males and females in the adult animals harvested by hunters and herders will make it seem as if the herded population is composed of smaller animals, leading to the erroneous conclusion that body-size reduction had occurred.

The misleading impression of size reduction in a managed herd is exacerbated when, as is standard practice following von den Driesch (1978: 4), unfused and fusing bones are excluded from osteometric analysis. The practice of excluding these bones from osteometric analysis virtually ensures that even the reduced number of males present in the assemblage from a managed herd is eliminated from the sample of measured bones. Moreover, normalizing and then homogenizing the metric data from this unrepresentative sample into a single size profile, as is also standard practice, further exaggerates the bias for the bones of smaller-bodied adult females. Both these standard methodological practices distort true size variation, giving the false impression of body-size reduction when a managed population is compared with a hunted population. The types of skeletal elements contributing to the normalized size profile can also make it difficult to identify the real causes of size variation in ancient populations. A normalized size profile from a managed population made up primarily of early-fusing bones (where both males and females are more evenly represented), for example, will look quite different from a profile for a managed population made up of late-fusing bones (which will be dominated by females). Combining specimens from broadly separated regions into temporal benchmark size profiles (following Uerpmann 1979), or comparing normalized size profiles across broad, environmentally varied regions (e.g., Peters et al. 1999), adds regional/environmental bias as an additional complicating factor that further obscures any accurate characterization of size variation in faunal assemblages.

While the claim for size reduction in early managed goats from Ganj Dareh can be discounted as an artifact of shifting demographics, taphonomic bias, and methodological distortion, there is solid biometric evidence for smaller body size in later managed goat populations from lowland Ali Kosh and Ceramic Neolithic–age goats from highland sites in the Zagros. These goats were clearly smaller than both the managed and wild goats from earlier sites in the highlands. However, the smaller size of these animals may not have been caused by domestication, but may instead reflect geographic variation in the size of goats raised under different climatic regimes or the immigration of smaller varieties of domestic goats. The substantial reduction in the body size both of wild goats and of gazelles (a species never subjected to domestication) over the past 8,000 to 10,000 years makes it especially hard to cling to the notion that domestication is the primary factor contributing to the decrease in body size in managed goats.

In fact, the only indications of a difference in body size between domestic and wild goats documented in this study are a marked reduction of the length of bones in domestic males (a difference of little utility in archaeological assemblages) and a slight reduction in breadth measurements in domestic males, which are still larger than both wild and domestic females. Modern domestic females are entirely undistinguishable from wild females on the basis of both length and breadth measurements. These differences are consistent with Zohary et al.'s prediction that selective factors introduced by domestication will operate primarily on body size in males, as manifested by a decrease in the degree of sexual dimorphism in males, especially in the length of limbs relative to body size (Zohary et al. 1998). A similar pattern is seen in horn length in wild and domestic male goats (Bökönyi 1977: 18; Zeder unpublished data). Thus, it is possible to make the case that the selective factors introduced by domestication will select for shorter-legged males and an overall decrease in the degree of sexual dimorphism in both horn length and long-bone breadth. These changes in body size of males will be hard to detect in archaeological assemblages, however, given the rarity of complete long bones in archaeological assemblages and the difficulty of discriminating a subtle decrease in breadth dimensions in domestic males from a range of other, much more powerful factors demonstrated to affect size in both wild and domestic ungulates. In light of this evidence, the claim that body-size reduction is a direct and definitive marker of domestication in goats, or in other species, is hard to support. Although it has long been employed as the primary marker of animal domestication, body-size reduction has little or no value as a leading-edge indicator of animal domestication.

Demography as a Marker of Initial Domestication in Goats

It would seem an inescapable conclusion that demographic profiling is the optimal method for marking the initial stages of domestication in goats, and likely in other domestic livestock. In particular, sex-specific harvest profiles, constructed by combining osteometric data with refined data on the sequence and calibration of long-bone fusion, are capable of distinguishing between herd management and a range of selective and nonselective hunting practices. Employing this method requires changes in current osteometric techniques. Of primary importance is the need to include all measurable elements—regardless of their state of fusion—in the osteometric analysis. Metric data also should be evaluated in their unaltered "raw" form on a bone-by-bone basis.

This technique will be most effective with large assemblages, such as those from Ganj Dareh and Ali Kosh. Regrettably, curated assemblages from early excavations in the Near East tend to be quite small and somewhat haphazardly collected.

As this study has shown, however, even small, poorly collected assemblages can be fruitfully examined. If the sample is too small to generate a reliable harvest profile, ratios of male and female elements and differentials in the percentages of unfused bones in male and female subpopulations (as shown in Tables 14.4 and 14.5) can still provide useful information on harvest strategies. This approach holds even more promise for more recently collected assemblages, and especially for those generated by ongoing and future excavations where attention is paid to recovering a representative sample of faunal remains from larger archaeological exposures.

Unfortunately, dental data, capable of tracking survivorship over a longer portion of an animal's likely lifespan, as yet have limited potential for the construction of sex-specific harvest profiles. Despite recent efforts (Payne 1985; Halstead and Collins 2002), techniques for discriminating between sheep and goats using teeth and mandible morphology still are not very well established. Nor are there clear size differences in the teeth and jaws that can be used to discriminate between male and female animals.

While the current study focused on goats, this technique also holds promise for the study of other livestock species, especially those where there is a significant degree of sexual dimorphism. Applying this technique to other species will require similar baseline studies to determine empirically the degree and nature of dimorphism in the species, preferably using modern skeletal collections with known life histories. More refined data on long-bone fusion sequences and timing also must be collected. Finally, species and sex-specific modeling of the herd structure and harvest strategies employed under different likely hunting and herding regimes for these species is also needed.

Implications for Understanding Domestication

For almost a quarter of a century, body-size reduction has been the primary marker of animal domestication used to document domestication in goats, as well as in other livestock species. The widespread adoption of this marker signifies at least a tacit (perhaps at times unconscious) acceptance of an underlying conceptual approach to animal domestication that emphasizes the evolutionary impact of domestication on the animal. To many adopting this view, genetic change, morphological change, and body-size reduction have become practically synonymous, making evidence of body-size reduction an incontrovertible marker of domestication, capable of cleanly separating wild and domestic populations. It now seems that body-size reduction is neither a direct nor an immediate signal of domestication. Indeed, evidence of size reduction that has mistakenly been interpreted as a response to domestication more likely may be attributable to a variety of other causes: shifting demographic patterns, geographic variation and environmental change, immigration of animals from outside regions, sample bias, and flawed methodologies. And while changes in horn morphology can be linked more directly to selective pressures resulting from human management, the slow rate of change in horn form in managed goats and the difficulty in detecting this change using fragmentary horn core fragments from archaeological assemblages, limits the utility of this marker in detecting the initial stages of the domestication process.

To a certain extent, the lack of definitive, quick-response morphological markers of animal domestication should not come as a surprise. Unlike in plants, where the evolving domestic partnership with humans selects for changes in morphological traits (i.e., in seed germination and dispersal mechanisms), in animals, behavioral characteristics play a more primarily role. It has long been recognized that there are behavioral characteristics that make certain taxa more attractive candidates for domestication than others (e.g., lack of aggression, herd dominance structure, and flight reflex; see Clutton-Brock 1999). It is also likely that there are variations in behavior within taxa that make certain individuals more suitable domestic partners than others, and that the behavioral traits of these individuals will be passed on to subsequent generations of managed animals. Selective pressures on animals undergoing domestication, then, will, likely first affect behavior and only secondarily, and perhaps quite tangentially, morphology.

Selection that operates on behavior rather than morphology clearly will be more difficult to detect in the archaeological record. As has been shown, it is hard to predict what the morphological changes linked to the selection for behavioral traits would be, and harder still to clearly identify these archaeologically. Indicators of increased tractability of animals, like the hoofprints in the bricks at Ganj Dareh, are rarely found and should be considered highly circumstantial evidence of behavioral changes at best. A case might also be made that a demographic profile indicating human manipulation of mate selection and control of a breeding stock implies an increased degree of tractability in managed animals over their wild counterparts, but this again would be only circumstantial evidence of the selection for animals behaviorally more suited to human control.

Obtaining definitive proof of these behavioral changes may be possible only through genetic analysis of ancient DNA. To do this, first the genes that control for these selected behaviors must be identified, and then techniques must be developed that will allow these genes to be isolated in highly degraded archaeological samples (see Chapter 1). As discussed in Bradley's overview for the section on genetic documentation of animal domestication (Chapter 18), we are still a long way from being able to do either of these things. More dramatic changes in the genetic structure of early domesticates may not have occurred until these animals were moved away from the initial site of domestication, cut off from the wild gene pool, and subjected to a new range of selective pressures. And while these changes may be more detectable with ancient DNA, as with morphological changes related to domestication, this kind of genetic break with wild forms likely occurred well after the initial phases of the domestication process.

This raises the question of whether morphologically, and perhaps even genetically, unaltered populations that were clearly managed can be considered domestic. Certainly a view that requires genetic alteration to achieve the domestic status would not accept these as domesticated animals. Instead, these populations would be classified as "proto-domesticated" or in a stage of "incipient domestication" (following Vigne et al. 1999; Horwitz 2003).

A conceptual approach that emphasizes human motivations in animal domestication, in contrast, would have no difficulty in classifying such a population as domesticated. Under this approach, documenting domestication requires demonstrating that a herder's goals shaped the demographic profile of a target species in an archaeological assemblage. Human manipulation of herd demographics to meet these goals usually results in genetic isolation of managed animals, which, in turn, may result in genetically driven morphological change. But for those who emphasize the human role in domestication, tracking the changes in harvest strategies from those that meet the hunter's short-term, off-take maximization goals to those that promote the herder's long-term interest in herd propagation is central to documenting domestication, even in the absence of genetic isolation and subsequent morphological change.

Genetically driven approaches tend to draw a more definitive boundary between wild and domestic, treating domestication more as a moment than a process. This human-oriented approach, on the other hand, widens the lens of inquiry to a much broader middle ground between hunting and herding (cf., Smith 2001). It may not be possible to identify definitively the moment when the threshold between wild and domestic was crossed as people traversed this middle-ground territory. However, the technique presented here for reconstructing sex-specific harvest strategies surely provides a better tool for charting the process of domestication than the continued reliance on morphological markers that are only delayed, indirect manifestations of domestication at best.

Exploring this middle ground will also require a shift in the geographic focus of much of the work on caprovine domestication in the Fertile Crescent. Because morphological change was needed to demonstrate domestication, recent work on sheep and goat domestication in the Near East has tended to focus on areas where the first apparent morphological manifestations of domestication can be found. As a result, this research captures a later stage of the domestication process. Whether or not one accepts morphologically unaltered managed animals as domestic, it is critical to recognize that the process of domestication begins with a long period of human manipulation of herd structure without any morphologically, or at this point genetically, detectable manifestations. This means that the regional focus of research needs to move farther into the natural habitats of wild progenitor species, and its temporal focus needs to expand to encompass a deeper period of time.

Ultimately, it is a matter of personal preference whether one requires clear-cut indications of genetic isolation from wild progenitors before declaring that domestication has occurred, or accepts evidence of herd management as a marker of domestication. Those that require evidence of genetic isolation, however, should recognize that it will be difficult, if not impossible, to demonstrate definitively that this separation has been achieved using current archaeological and genetic markers. Whatever the conceptual approach to domestication, successful documentation of the domestication in sheep and goats in the Near East, and likely in other species, requires a shift both in the regional and in the temporal focus of research, and the careful application of new techniques capable of tracking the changing patterns of human animal exploitation that drive the process. It also requires a more critical evaluation of the markers used to document domestication and the methods developed to detect these markers, making sure that both markers and methods are clearly and explicitly linked to the process of domestication however it is defined.

Acknowledgments

This research was supported by funds from the National Museum of Natural History's Research Initiative Fund and by a Smithsonian Scholarly Studies Grant. Many thanks are owed to the Field Museum of Natural History staff in the departments of zoology and geology for facilitating access to skeletal collections of modern and archaeological caprines. This research could not have been completed without the assistance of a number of people, including Amy Aisen, Susan Arter, Naomi Cleghorn, Lesley Gregoricka, Margaret Heirs, Heather Lapham, Justin Lev-Tov, Colleen McLinn, Sarah McClure, Anastasia Poulos, Douglas Park, Scott Rufolo, and Bruce Smith.

References

Bar-Yosef, O. 2000. The context of animal domestication in southwestern Asia. In *Archaeozoology of the Near East*, Volume IV, M. Mashkour, A. M. Choyke, and H. Buitenhuis (eds.), pp. 185–195. Groningen: ARC Publications.

Bar-Yosef, O. and R. Meadow. 1995. The origins of agriculture in the Near East. In *Last hunters, first farmers: New perspectives on the transition to agriculture*, D. Price and A.-B. Gebauer (eds.), pp. 39–94. School of American Research Advanced Seminar Series. Santa Fe: SAR Press.

Binford, L. 1971. Post-Pleistocene adaptations, In *Prehistoric agriculture*, S. Struever (ed.), pp. 22–49. American Museum Sourcebooks in Anthropology. New York: Natural History Press.

Blackman, M. J. 1984. Provenience studies of Middle Eastern obsidian from sites in highland Iran. In *Archaeological chemistry III*, J. Lambert (ed.), pp. 19–50. Washington, D.C.: American Chemical Society.

Bökönyi, S. 1972. Zoological evidence for seasonal or permanent occupation of prehistoric settlements. In *Man, settlement, and urbanism*, P. J. Ucko, R. Tringham, and G. W. Dimbleby (eds.). London: Duckworth.

———. 1973. Some problems of animal domestication in the Middle East. In *Domestikationsforschung und Geschichte der Haustiere*, J. Matolisci (ed.), pp. 121–126. Budapest: Akadèmiai Kiadó.

———. 1974. *History of domestic animals in central and eastern Europe*. Budapest: Akadèmiai Kiadó.

———. 1976. Development of early stock rearing in the Near East. *Nature* 264: 19–23.

———. 1977. *Animal remains from the Kermanshah Valley, Iran*. BAR Supplementary Series, No. 34. Oxford: BAR.

Braidwood, R. J. 1960. The agricultural revolution. *Scientific American* 203: 130–148.

Braidwood, R. J. and B. Howe. 1960. *Prehistoric investigations in Iraqi Kurdistan*. The Oriental Institute of the University of Chicago Studies in Ancient Oriental Civilization, No. 31. Chicago: University of Chicago Press.

Bullock, D, and J. Rackham 1982. Epiphyseal fusion and tooth eruption of feral goats from Moffatdale, Dumfries and Galloway, Scotland. In *Ageing and sexing animal bones from archaeological sites*, B. Wilson, C. Grigson, and S. Payne (eds.) pp. 55–72. BAR British Series 109. Oxford: British Archaeological Reports.

Clutton-Brock, J. 1999. *Domesticated animals*, 2nd edition. London: British Museum of Natural History.

Collier, S. and J. P. White. 1976. Get them young? Age and sex inferences on animal domestication. *American Antiquity* 41: 96–102.

Coon, C. S. 1951. *Cave explorations in Iran, 1949*. Museum Monographs. Philadelphia: University Museum.

Davis, S. 2000. The effect of castration and age on the development of the Shetland sheep skeleton and a metric comparison between bones of males, females and castrates. *Journal of Archaeological Science* 27: 373–390.

Deniz, E. and Payne, S. 1982. Eruption and wear in the mandibular dentition as a guide to ageing Turkish angora goats. In *Ageing and sexing animal bones from archaeological sites*, B. Wilson, C. Grigson, and S. Payne (eds.), pp. 155–206. BAR British Series 109. Oxford: British Archaeological Reports.

Drew, I., P. Daly, and D. Perkins. 1971. Prehistoric domestication of animals: Effects on bone structure. *Science* 171: 280–282.

Ducos, P. 1969. Methodology and results of the study of the earliest domesticated animals in the Near East (Palestine). In *The domestication and exploitation of plants and animals*, P. J. Ucko and G. W. Dimbleby (eds.), pp. 265–275. London: Duckworth.

Dyson, R. H. 1953. Archaeology and the domestication of animals in the Old World. *American Anthropologist* 55: 661–673.

Epstein, H. 1971. *The origin of domesticated animals is Africa*, Volume 2. New York: Africana Publishing Corporation.

Evins, M. 1982. The fauna from Shanidar Cave: Mousterian wild goat exploitation in northeastern Iraq. *Paléorient* 8: 47–73.

Flannery, K. V. n.d. *Hunting and early animal domestication at Tepe Guran*. Unpublished manuscript.

Grant, A. 1982. The use of tooth wear as a guide to the age of domestic ungulates. In *Ageing and sexing animal bones from archaeological sites*, B. Wilson, C. Grigson, and S. Payne (eds.), pp. 91–108. BAR British Series 109. Oxford: British Archaeological Reports.

Halstead, P. and P. Collins. 2002. Sorting the sheep from the goats: Morphological distinctions between the mandibles and mandibular teeth of adult *Ovis* and *Capra*. *Journal of Archaeological Science* 29: 545–553.

Helmer, D. 1992. *La domestication des animaux par les hommes préhistoriques*. Paris: Masson.

Hesse, B. 1978. *Evidence for husbandry from the Early Neolithic site of Ganj Dareh in western Iran*. PhD dissertation, Columbia University. Ann Arbor, MI: University Microfilms.

Higgs, E. S. and M. R. Jarman. 1972. The origins of animal and plant husbandry. In *Papers in economic prehistory*, E. S. Higgs (ed.), pp. 3–13. Cambridge: Cambridge University Press.

Hillman, G. C. 1996. Late Pleistocene changes in wild plant-food available to hunter-gatherers of the northern Fertile Crescent: Possible preludes to cereal cultivation. In *The origins and spread of agriculture and pastoralism in Eurasia*, D. R. Harris (ed.), pp. 159–203. Washington: Smithsonian Institution Press.

Hole, F. 1987. Chronologies in the Iranian Neolithic. In *Chronologies in the Near East*, O. Aurenche, J. Evin, and F. Hours (eds.), pp. 353–379. BAR International Series 379. Oxford: BAR.

Hole, F. and K. V. Flannery. 1967. The prehistory of southwestern Iran: A preliminary report. *Proceedings of the Prehistoric Society* 33: 147–206.

Hole, F., K. V. Flannery, and J. A. Neely. 1969. *Prehistory and human ecology on the Deh Luran Plain*. Memoirs of the Museum of Anthropology, No. 1. Ann Arbor: University of Michigan Press.

Horwitz, L. K. 1989. A reassessment of caprovine domestication in the Levantine Neolithic: Old questions, new answers. In *People and culture in change*, I. Hershkovitz (ed.), pp. 153–181. BAR International Series, No. 508. Oxford: BAR.

———. 1993. The development of ovicaprine domestication during the PPNB of the southern Levant. In *Archaeozoology of the Near East I*, H. Buitenhuis and A. T. Clason (eds.), pp. 27–36. Leiden: Universal Book Service.

———. 2003. Temporal and spatial variation in Neolithic caprine exploitation strategies: A case study of fauna from the site of Yiftah'el (Israel). *Paléorient* 29: 19–58.

Horwitz, L. K. and Ducos, P., 1998. An investigation into the origins of domestic sheep in the southern Levant. In *Archaeozoology of the Near East, III*, H. Buitenhuis, L. Bartosiewicz, and A. M. Choyke (eds.), pp. 80–95. ARC Publications No. 18. Groningen: ARC.

Horwitz, L. K., E. Tchernov, P. Ducos, C. Becker, A. von den Driesch, L. Martin, and A. Garrard. 1999. Animal domestication in the southern Levant. *Paléorient* 25: 63–80.

Jarman, M. R. and P. F. Wilkinson. 1972. Criteria for animal domestication. In *Papers in economic prehistory*, E. S. Higgs (ed.), pp. 83–93. Cambridge: Cambridge University Press.

Kelly, R. 1995. *The foraging spectrum: Diversity in hunter-gatherer lifeways*. Washington, D.C.: Smithsonian Institution Press.

Köhler-Rollefson, E. 1989. Changes in goat exploitation at 'Ain Ghazal between the Early and Late Neolithic: A metrical analysis. *Paléorient* 15: 141–146.

Legge, A. 1996. The beginning of caprine domestication in southwest Asia. In *The origins and spread of agriculture and pastoralism in Eurasia*, D. R. Harris (ed.), pp. 238–263. Washington, D.C.: Smithsonian Institution Press

Legge, A. J. and P. A. Rowley-Conwy. 2000. The exploitation of animals. In *Village on the Euphrates: From foraging to farming at Abu Hureyra*, A. M. T. Moore, G. C. Hillman, and A. J. Legge (eds.), pp. 423–474. Oxford: Oxford University Press.

MASCA. 1970. Bone from cosmetic and wild animals: Crystallographic differences. *MASCA Newsletter* 6: 2.

———. 1973. Technique for determining animal domestication based on a study of thin sections of bone under polarized lights. *MASCA Newsletter* 9: 1–2.

Mashkour, M., H. Bocherens, and I. Moussa. 2005. Long distance movement of sheep and goats of Bakhtiari nomads tracked with intra-tooth variations of stable isotopes (^{13}C and ^{18}O). In *Health and diet in past animal populations: Current research and future directions*, J. Davies, M. Fabi, I. Mainland, M. Richards, and R. Thomas (eds.), pp. 113–124. London: Oxbow Books.

Meadow, R. H. 1984. Animal domestication in the Middle East: A view from the eastern margin. In *Animals and archaeology, Volume 3: Early herders and their flocks*, J. Clutton-Brock and C. Grigson (eds.), pp. 309–337. BAR International Series 202. Oxford: BAR.

———. 1989. Osteological evidence for the process of animal domestication. In *The walking larder: Patterns of domestication, pastoralism, and predation*, J. Clutton-Brock (ed.), pp. 80–90. London: Unwin Hyman.

———. 1999. The use of size index scaling techniques for research on archaeozoological collections from the Middle East. In *Historia Animalium ex Ossibus, Festschrift für Angela von den Driesch. Beiträge zur Paläoanatomie, Archäologie, Ägyptologie, Ethnologie und Geschichte der Tiermedizin*, C. Becker, H. Manhart, J. Peters, and J. Schibler (eds.), pp. 285–300. Rahden/Westf: Verlag Maire leidorf GmbH.

Mellart, J. 1975. *The Neolithic of the Near East*. New York: Charles Scribner's and Sons.

Moran, N. and T. O'Connor. 1994. Age attribution in domestic sheep by skeletal and dental maturation: A pilot study of available sources. *International Journal of Osteoarchaeology* 4: 267–285.

Nadler, C. F., K. V. Lorobitsina, R. S. Hoffmann, and N. V. Voronstov. 1973. Cytogenetic differentiation, geographic distribution, and domestication in Palearctic sheep *(Ovis)*. Zeitschrift für Säugetierkunde 38: 109–125.

Noddle, B. 1974. Age of epiphyseal closure in feral and domestic goats and ages of dental eruption. *Journal of Archaeological Science* 1: 195–204.

Østergård, M. 1980. X-ray diffractometer investigations of bone samples from domestic and wild animals. *American Antiquity* 45: 59–63.

Payne, S. 1973. Kill-off patterns in sheep and goats: The mandibles from Aşvan Kale. *Anatolian Studies* 23: 281–303.

———. 1985. Morphological distinctions between the mandibular teeth of young sheep, *Ovis*, and goats, *Capra*. *Journal of Archaeological Science* 12: 139–147.

Perkins, D. 1964. Prehistoric fauna from Shanidar, Iraq. *Science* 144: 1565–1566.

Perkins, D. and P. Daly. 1968. A hunter's village in Neolithic Turkey. *Scientific American* 219: 96–106.

Peters, J., D. Helmer, A. von den Driesch, and S. Segui. 1999. Animal husbandry in the northern Levant. *Paléorient* 25: 27–48.

Redding, R. 1981. *Decision making in subsistence herding of sheep and goats in the Middle East*. PhD dissertation, University of Michigan. Ann Arbor, MI: University Microfilms.

Redding, R. 2005. Breaking the mold: A consideration of variation and evolution in animal domestication. In *New methods and the first steps of animal domestications*, J. D. Vigne, D. Helmer, and J. Peters (eds.), pp. 41–49. London: Oxbow Press.

Reed, C. A. 1959. Animal domestication in the prehistoric Near East. *Science* 130: 1629–1639.

———. 1961. Osteological evidences for prehisotric domestication in southwestern Asia. Zeitschrift für Tierzüchtung und Züchtungsbiologie 76: 31–38.

———. 1983. Archaeozoological studies in the Near East: A short history (1960–1980). In *Prehistoric archaeology among the Zagros flanks*, L. Braidwood, R. J. Braidwood, B. Howe, C. A. Reed, and P. J. Watson (eds.), pp. 511–536. Oriental Institute Publications, No. 105. Chicago: The Oriental Institute.

Schaller, G. 1973. *Mountain monarchs: Wild sheep and goats of the Himalaya*. Chicago: University of Chicago Press.

Shaffer, V. M. and C. A. Reed. 1972. The co-evolution of social behavior and cranial morphology in sheep and goats (Bovidae, Caprini). *Fieldiana Zoology, 61*. Chicago: Field Museum of Natural History.

Silver, I. 1969. The ageing of domestic animals. In *Science in Archaeology*, 2nd edition, D. Brothwell and E. S. Higgs (eds.), pp 283–302. London: Thames.

Smith, B. D. 2001. Low level food production. *Journal of Archaeological Research* 9: 1–43.

Stampfli, H. R. 1983. The fauna of Jarmo, with notes on animal bones from Matarrah, the Amuq, and Karim Shahir. In *Prehistoric archaeology among the Zagros flanks*, L. Braidwood, R. J. Braidwood, B. Howe, C. A. Reed, and P. J. Watson (eds.), pp. 431–484. Oriental Institute Publications, No. 105. Chicago: The Oriental Institute.

Stiner, M. 1990. The use of mortality patterns in archaeological studies of hominid predatory adaptations. *Journal of Anthropological Archaeology* 9: 305–351.

Uerpmann, H.-P. 1978. Metrical analysis of faunal remains from the Middle East. In *Approaches to faunal analysis in the Middle East*, R. H. Meadow and M. A. Zeder (eds.), pp. 41–45. Peabody Museum Bulletin, No. 2. Cambridge, Mass.: Peabody Museum.

———. 1979. *Probleme der Neolithisierung des Mittelmeeraumes*. Biehefte zum Tübinger Atlas des Vordern Orients, Reihe B, Nr. 28. Wiesbaden: Dr Ludwig Reichert.

Vigne, J.-D. and H. Buitenhuis. 1999. Les premiers pas de la domestication animale à l'Ouest de l'Euphrate: Chypre et l'Anatolie Centrale. *Paléorient* 25: 49–62.

Vigne, J.-D., I. Carrére, J.-F. Saliége, A. Person, H. Bocherens, J. Guilaine, and F. Briois. 2000. Predomestic cattle, sheep, goat, and pig during the late 9th and the 8th millennium cal. B.C. on Cyprus: Preliminary results of Shillourokambos (Parekklisha, Limassol). In *Archaeozoology of the Near East, IVA*, M. Mashkour, A. M. Choyke, H. Buitenhuis, and F. Poplin (eds.), pp. 83–106. ARC Publication No. 32. Groningen: ARC.

von den Driesch, A. 1978. *A guide to the measurement of animal bones from archaeological sites*. Peabody Museum Bulletin, No. 1. Cambridge, Mass.: Peabody Museum.

Watson, J. P. N. 1975. Domestication and bone structure in sheep and goats. *Journal of Archaeological Science* 2: 375–383.

Zeder, M. A. 1978. Differentiation between the bones of caprines from different ecosystems in Iran by the analysis of osteological microstructure and chemical composition. In *Approaches to faunal analysis in the Middle East*. R. H. Meadow and M. A. Zeder (eds.), pp. 69–86. Peabody Museum Bulletin, No. 2. Cambridge, Mass.: Peabody Museum.

———. 1999. Animal domestication in the Zagros: A review of past and current research. *Paléorient* 25: 11–25.

———. 2001. A metrical analysis of a collection of modern goats *(Capra hircus aegargus* and *Capra hircus hircus)* from Iran and Iraq: Implications for the study of caprine domestication. *Journal of Archaeological Science* 28: 61–79.

———. 2003. Hiding in plain sight: The value of museum collections in the study of the origins of animal domestication. In: *Documenta archaeobiologiae 1: Deciphering ancient bones. The research potential of bioarchaeological collections. Yearbook of the State Collection of Palaeoanatomy. München, Germany*, G. Grupe and J, Peters (eds.), pp. 125–138. Rahden/Westf: Verlag M. Leidorf GmbH.

———. 2005. A view from the Zagros: New perspectives on livestock domestication in the Fertile Crescent. In *New methods and the first steps of animal domestications*, J. D. Vigne, D. Helmer, and J. Peters (eds.), pp. 125–146. London: Oxbow Press.

———. in press. Reconciling rates of long-bone fusion and tooth eruption and wear in sheep *(Ovis)* and goat *(Capra)*. In *ageing and sexing animals from archaeological sites*, D. Ruscillo (ed.). Oxford: Oxbow Press.

Zeder, M. A. and B. Hesse. 2000. The initial domestication of goats (Capra hircus) in the Zagros Mountains 10,000 years ago. *Science* 287: 2254–2257.

Zeder, M. A. and H. A. Lapham, in prep. Morphological criteria for distinguishing between the bones of sheep *(Ovis aries)* and goats *(Capra hircus)*.

Zeuner, F. 1955. The goats of early Jericho. *Palestine Exploration Quarterly*: 70–86.

———. 1963. *A history of domesticated animals*. London: Hutchinson.

Zohary, D, Tchernov, E., and Horwitz, L. K. 1998. The role of unconscious selection in the domestication of sheep and goats. *Journal of Zoology* 245: 129–135.

CHAPTER 15

The Domestication of the Pig *(Sus scrofa)*
New Challenges and Approaches

UMBERTO ALBARELLA, KEITH DOBNEY, AND
PETER ROWLEY-CONWY

Introduction

Pigs are the victims of their own success in two ways. First, wild forms are distributed over most of the Old World except for the very dry and the very cold regions. This contrasts with other animals like sheep, whose much more limited distribution constrains the search for domestication to a restricted area. It also means that archaeological finds outside that area must come from domestic animals. The wide distribution of pigs and their close relatives (Groves 1981; Oliver 1993) means that a simple geographical diagnosis of domestication is usually impossible.

Second, pigs are adaptable and generalized omnivores. They may, therefore, have a wider range of possible relationships with humans than do other species. At one end of the spectrum is unambiguous hunting, and at the other is close domestic control. But in between is a "gray area," which is perhaps wider and more complex than for most other species. Both urban and rural domestic pigs may at times wander freely and forage for themselves, returning to their owners each evening. In the medieval period, pigs were driven into woodlands to forage for acorns, a practice known as *pannage*. Such free-ranging behavior is not what we associate with conventional domestic animals. There can be greater complexities. Among the Etoro of New Guinea, domestic female pigs roam freely round the villages and stray into the surrounding forests, where they meet and interbreed with feral males. All breeding takes place this way since males in domestic litters are all castrated (Rosman and Rubel 1989). In fact, the unique biology and behavior of pigs present special challenges to the study of their domestication that have caused some to question whether the threshold we term "domestication" is really relevant to pigs. This chapter explores some of these challenges for the study of pig domestication and the growing range of new approaches that can be used to address them.

Prior Research

The earliest studied assemblages of animal bones included those from pigs. These came from various European sites such as Danish shell middens, Swiss lake dwellings, Italian *terramare* settlements, and others. By the end of the nineteenth century, it was clear that two main forms were represented: the wild boar and the domestic porker. The Danish zoologist Herluf Winge used both metrical and morphological criteria to separate them, establishing ground rules that are often still used today; for example, lower M3s greater than 40 mm in length are likely to come from wild boars, and those under 40 mm in length are likely to come from domestic individuals, although there is an overlap (Winge 1900).

Winge (1900) believed that prehistoric European domestic pigs (which, following earlier writers, he termed *Sus scrofa domesticus* or *S. s. palustris*) were descendants of the wild boar of Europe, northern Asia, and North Africa (*Sus scrofa ferus*). Modern domestic pigs in Southeast Asia were more similar to the local wild boars, which he believed might be a different species (*Sus vittatus*). Beyond this, however, he did not seek the geographical origins of domestication. Soon various ideas were put forward. Pira (1909: 373), working on Swedish material, argued for local domestication because the earliest domestic pigs were the closest to wild boars in both size and morphology. However, archaeology was finding ever-earlier agriculture in the Near East, with pigs as part of the package (Flannery 1983). As a result, the dominant view of the mid-twentieth century was that pigs were domesticated in the Near East and brought to Europe by immigrant farmers (e.g., Childe 1958: 34). A few dissenters, like the geographer C. O. Sauer (1952), preferred a Southeast Asian origin for domestic pigs. But nearly all researchers argued for a limited geographical origin for pig domestication.

The hypothesis that domestic pigs spread from a limited geographic area of initial domestication was challenged initially by Eric Higgs and colleagues in the later 1960s. They argued that domestication as a threshold event was an illusion, and that there was a multiplicity of potential, intermediate states for all animals and plants (Higgs and Jarman 1969). With regard to pigs, Winge's metrical division was challenged, and the case was made for intermediate or semidomestic pigs under extensive control; a trend toward closer relations could occur anywhere, not just in the previously recognized centers (e.g., Jarman 1976a; Zvelebil 1995). The situation can now best be described as being in a state of flux. It is becoming increasingly unclear whether we should be looking for domestication as a threshold event, how we would recognize it in the archaeological record, or even whether we can effectively define domestication in a broadly applicable way.

Defining Domestication

Over the years, a number of books and articles have attempted to devise a satisfactory, general definition of a domestic animal. Many authors have focused exclusively on cultural factors, while others have focused only on biological ones, with insufficient attention paid to integrating the two. In his seminal review of the role of animal behavior in domestication, Price (1984: 3) rightly states that "it is difficult to formulate a definition of domestication that is general enough to account for the wide variation observed in different species, in different captive environments, yet specific enough to be meaningful in terms of the biological processes involved." Nevertheless, he attempts to define domestication as "that process by which a population of animals becomes adapted to man and to the captive environment by some combination of genetic changes occurring over generations, and environmentally induced developmental events reoccurring during each generation."

This is a broad definition, which essentially emphasizes captivity (i.e., direct human control) as a basic requirement. Some have argued, however, that the very terms "domestic" and "wild" are, in fact, the extreme ends of a continuum, along which a whole host of environmental, biological, and cultural factors vary, and various combinations of these factors may have either gradually or rapidly altered the behavior and genetics of the animals (Ervynck et al. 2002: 50). As a result of this somewhat more complex view of human-animal relationships, a number of intermediate stages of domestication have been proposed, such as cultural control (Hecker 1982: 219; Hongo and Meadow 1998, 2000), predomestic (Vigne and Buitenhuis 1999), and intermediary stage (Ervynck et al. 2002). These imply that the terms "domestic" and "wild" merely describe the extremes of a spectrum defined as follows (after Ervynck et al. 2002: 50): "wild" populations not experiencing (in the most simplified case) any direct or indirect influence of human behavior; "domestic" populations being characterized by survival, reproduction, and nutrition under complete human control.

In this definition, an animal can only be considered to be domestic where there is a conscious and prolonged intervention by humans to control many or all aspects of its life cycle. Although this definition is one with which we would agree, it means that true domestication is still the end point of an ongoing process, where, in its early/intermediate stages, morphological, genetic, and demographic shifts can still occur.

Although a continuum of relationships may occur in all species, in pigs the actual domestic or wild status of individuals or populations can be particularly difficult (if not impossible) to identify. In most of the world, domestic pigs live in areas also populated by wild boars, and inevitably interbreeding occurs regularly. In addition, domestic pigs are often kept in free-range conditions and can escape, creating entirely feral populations. This means that pig populations cannot be classified so easily as wild or domestic, and other possible conditions must be considered. Mayer and Brisbin (1991) and Mayer et al. (1998) consider four different types of pig populations: wild, domestic, feral, and genetic hybrids, although intermediates between even these may occur.

If this is the situation in the modern world, it is likely that a similarly wide range of relations between humans and pigs must have occurred in the past. This especially could be so for the early stages of the domestication process, when control of animal populations might have been relatively loose, and genotypic and phenotypic changes in domestic animals were still minimal. The likely contiguity of wild and domestic pig populations has lead Hongo and Meadow (1998: 89) to propose that criteria for the identification of the origin of domestication in bovids may not be entirely applicable to suids. Thus, rigid definitions of what represents a wild or domestic pig are fraught with problems, which, in the end, largely boil down to personal preference or arguments over semantics.

Detecting Domestication in Pigs

It is clear from the discussion in the previous section that domestication is difficult to define for any species, and probably particularly so for the pig. It would, therefore, be naive to expect straight or easy answers from the archaeological record. At various points along the cline from wild to domestic, animals may change their biology, behavior, and attitude toward humans in many different ways and to varying degrees. It follows that changes in exploitation strategy resulting in a shift along the continuum cannot be analyzed in a univocal way, and that a diversity of approaches is required. In addition to the complexity of the domestication process per se, we also have to consider climatic, environmental, geographic, chronological, and cultural variables that provide the context for domestication.

The three authors of this chapter make up the Durham Pig Project, which is examining pig domestication and early husbandry around the world. Because of the complexities we have outlined above, we decided that we could not confine our analysis to a specific area (e.g., the Near East or China, likely to contain the earliest stages of domestication), a specific period (e.g., the Mesolithic/Neolithic transition), or a single methodological approach. Work carried out in the last few years by other authors and by ourselves suggests that medieval pigs can be as informative about the domestication process as their Neolithic counterparts, and that the onset of domestication in peripheral areas may also be illuminating about the earliest origin of the phenomenon.

In the following sections, we therefore present some results of multiple approaches and techniques used by others and by ourselves to document pig domestication. We emphasize that although each approach and technique has its own unique potential, we feel strongly that a combination of some or all of these is essential for a fuller understanding of the domestication process.

Zoogeographic Markers

One of the primary criteria used by zooarchaeologists to infer the presence of domestic animals is the appearance of a species outside its natural range, or in locations where it is unlikely to have reached without the intervention of humans (e.g., on remote islands). Davis (1987: 133) goes so far as to state, "The sudden arrival of a new species is often a sure sign that it was introduced as domestic stock by humans." The appearance of sheep in archaeological sites in southern France, Corsica, and South Africa, sheep and goats in Greece and Britain, horses in the Levant, cats in Cyprus, dogs in South America, and turkeys in Mesoamerica are all given as examples of and evidence for the introduction of domesticates.

For pigs, however, the high degree of morphological similarity between the various species of *Sus* means that in certain parts of the world it is difficult for us to ascertain which species of *Sus* we are dealing with in the archaeological record. In island Southeast Asia, for example, several different *Sus* species (*Sus scrofa*, *Sus celebensis*,[1] *Sus cebifrons*, and *Sus barbatus*) are found, yet apparently maintain significant differences in morphology, ecology, and behavior (Rothchild and Ruvinsky 1998). All these pig species were certainly exploited by Holocene peoples, and some like *Sus scrofa* (and possibly even *Sus celebensis*) were even transported by them outside their natural habitat. Unfortunately, continuing confusion over the present-day taxonomy of these island suids, and problems with the specific identification of their fossil remains, leave us very little idea which species are present in the archaeological record, let alone their wild, feral, or domestic status.

To make matters worse, interspecies hybridization between introduced *S. scrofa* and the other indigenous species cited above has been claimed (Lotsy 1922; Blouch and Groves 1990), and this could have occurred in the past. For example, Groves (1981) claims that the feral and domestic pigs in New Guinea at the time of European contact were in fact hybrids of *Sus scrofa* and *Sus celebensis*.[2] If correct, there could be little doubt that these animals were initially fully domesticated when introduced. However, this important hypothesis has yet to be conclusively proven, and is one where genetic analysis could be employed on both modern and ancient material to prove or disprove this theory.

Even where introduction by people is the most likely explanation for the presence of pigs, can we be sure that the animals were actually domestic? In the case of New Guinea, although there is still much debate as to precisely when pigs arrived,[3] their appearance is commonly thought to be a direct consequence of introduction as a domesticate by humans. However, Bulmer (conversation with Dobney, 12 September 2002) contends that the transport of pigs to New Guinea does not necessarily imply domestication, and that wild (or feral) pigs might just as easily have arrived on New Guinea without human assistance.

The spread of Neolithic farmers into certain parts of northwestern Europe appears to have also heralded the arrival of fully domesticated relatives of the indigenous wild boar (see below). With their genetic origins outside the area of introduction, these discrete populations should be recognizable both in terms of their morphology and their genetics. On the Baltic island of Gotland, pig bones are one of the most common mammal species' remains excavated from Middle Neolithic settlements (e.g., Österholm 1989; Lindqvist and Possnert 1997). Mandibles were also important as grave offerings. There was no land bridge between Gotland and the mainland when wild boars were recolonizing Scandinavia after the last glaciation. Although wild boars have been known to swim significant distances between islands, Gotland is some 80 km from mainland Sweden, so it is highly unlikely that wild boars could have colonized the island by themselves during the Early Holocene. All the ^{14}C-dated examples from "Mesolithic" contexts in Stora Förvar Cave have turned out to be Neolithic intrusions (Lindqvist and Possnert 1997), so the high frequency of pig bones found in Middle Neolithic sites on Gotland clearly indicates that humans introduced pigs sometime during the Neolithic. But does this imply that these pigs were domesticated? There has been much debate about this. Ekman (1974) argued that they were morphometrically wild. Jonsson (1986) believed they were domestic because they must have been under close control when actually shipped across to Gotland. Österholm (1989) pointed out that some funerary rituals required the mandibles of 20 or 30 pigs, and such numbers could not be hunted to order when needed, so the animals must have been domestic. However, Rowley-Conwy and Storå (1997) aged the mandibles and showed that they came from pigs killed at various times during the year, suggesting that the jaws were trophied before deposition in the graves, and thus need not have been domestic. Biometry has been used to argue that the pigs from Gotland were domestic (Benecke 1994), and it has also been used to make the case for their having been wild (Rowley-Conwy and Storå 1997). We are currently conducting our own biometrical analysis of this material, which suggests that these animals are intermediate in size between wild and domestic mainland pigs, and so the debate continues.

The limited range of wild mammals from Ireland suggests that most (including pigs) have been introduced by humans at varying times in the past (McCormick 1999). The presence of Mesolithic pigs suggests that this introduction occurred prior to the beginnings of agriculture, and that pigs were deliberately transported to Ireland by humans and released to found a hunted population. Their pre-Neolithic date means that we can be reasonably confident that they were indeed wild boars and not domesticated animals. A similar scenario has been suggested by Vigne and Buitenhuis (1999) for the introduction of pigs to Cyprus in the Early Pre-Pottery Neolithic period, commonly thought to predate pig domestication in the Near East. In this case, Vigne et al. suggest that the animals were already on the way toward domestication. They suggest that a loose relationship must have existed between transporter (human) and transported (pig), which signifies some kind of predomestication phase. This term involves "a form of husbandry without any apparent morphological change in the animals" (Vigne and Buitenhuis 1999: 55). In the case of pigs, therefore, the mere presence of remains does not provide definite proof of their domestic status. This requires the application of other, newer techniques such as those outlined below. In the meantime, we must remember there are a number of instances when humans have transported clearly wild animals to islands to found populations for hunting—the earliest case currently being the introduction of a marsupial, the cuscus (*Phalanger orientalis*), to New Britain at 19,000 BP (Flannery and White 1991).

Biometry

Despite the difficulties of defining what exactly a domestic population is, and the inadequacy of using size alone to address this question, biometry, however, remains the most widespread tool used by zooarchaeologists in considering domestication. There are reasons to believe that such a reliance on biometrical analysis has solid foundations, although some simplistic interpretations of metric data carried out in the past may need revision.

The main challenge to the metrical method came from the growing awareness of the potential range of animal-human relationships revealed by history and anthropology. As mentioned above, Higgs and his Cambridge-based research group in particular asked whether there ever was "a 'beginning' of either agriculture or domestication" (Higgs and Jarman 1969: 31), and this led them to seek out problem cases. Working on the Neolithic pigs from Molino Casarotto in northern Italy, Jarman (1976a: 528) concluded that the animals "bridge the accepted size ranges of wild pigs and Neolithic domestic pigs from sites such as Seeberg Burgäschisee-Süd. Furthermore, there is no indication that we are dealing with two separate populations of pigs as regards size, as no strongly bi-modal tendency is apparent in the size distribution of the bones." From this, Jarman concluded that no simple classification into wild or domestic was adequate to describe these pigs: Molino Casarotto contained a single pig population, intermediate between wild and fully domestic.

Jarman's work, however, was just as problematic as the simplistic view that it sought to replace (Rowley-Conwy 2003). First, he assumed that because he could not demonstrate whether the Molino Casarotto pigs were metrically wild or domestic, it followed that they were, therefore, behaviorally in between the two. This is not a valid extrapolation. Second, his metrical argument for the intermediate status of the pigs from Molino Casarotto derived from a comparison with Seeberg Burgäschisee-Süd. The two sites, however, are on different sides of the Alps and in totally different environmental zones. More recent work has shown that there is considerable environmentally linked size variation both in prehistoric pigs (Davis 1981; Rowley-Conwy 1995) and in recent wild boars in various regions, so Jarman's initial metrical conclusion may be flawed (Rowley-Conwy 2003).

One of the most significant recent advances in *Sus* biometrical methods is the work of Payne and Bull (1988). These authors revised the analysis of biometrical data from a number of important prehistoric sites from Europe and western Asia by adopting a size index scaling technique (see Meadow 1999). By taking into account to what extent measurements were smaller or larger than a particular standard—in this case, the mean taken from a sample of modern wild boar skeletons from Turkey—the authors could plot measurements of different bones on the same graph. Although this system was not new in zooarchaeology (e.g., see Ducos 1968; Uerpmann 1979; Meadow 1981), it was extensively applied to pig assemblages for the first time. This system has the advantage of increasing sample size by lumping measurements of different bones together. In addition, when different measurements are plotted separately but on the same scale, it is possible to detect variations in the relative dimensions of different body parts between pig populations.

The importance of the introduction of this method can hardly be overestimated, as biological change (resulting from domestication or subsequent selection) cannot be reduced to mere size diminution. Size variation often goes hand in hand with morphological change. For instance, typical characters that are used to separate domestic from wild pigs include the shortening of the snout—particularly notable on the relative dimensions of the lachrymal bone (Jonsson 1986: 125; Clutton-Brock 1987: 72). In addition, a reduction in relative brain size following domestication has been well documented in a range of species (Herre and Röhrs 1973; Kruska 1988). In fact, some of the biggest changes in the overall size of the brain, as well as in a variety of its selected functional systems, have been noted in pigs (Kruska 1988: Tables 13.1 and 13.2). Although rarely applicable to archaeological material (because complete skulls are very uncommon), this approach can be useful when using modern skeletons to make extrapolations about the past. For instance, Groves (1989) has hypothesized that the "wild" pigs from Sardinia and Corsica, because of their relatively large brain capacity, derive from wild, rather than domestic, animals introduced to the islands in prehistoric times. Consequently, these pigs should be regarded as genuinely wild rather than feral, in contrast to the feral status suggested for the mouflon (*Ovis musimon*) on the same islands.

To show the way these biometrical approaches have developed over time and are improving our understanding of pig domestication, we now present a number of examples taken from our ongoing pig domestication research. Figure 15.1 presents a first set of case studies, in which modern data are used as an aid for our interpretation of the archaeological material; it directly compares the greatest length of the astragalus of pigs from a modern wild population with results from several archaeological populations across Europe. The modern wild Turkish population (from Payne and Bull 1988) includes both females and males, juveniles and adults, and groups quite tightly, with a coefficient of variation of 5.7 (Figure 15.1a). Sværdborg I is an Early Mesolithic site where only wild boars would be expected; the large sample has a normal distribution and a coefficient of variation (cov) of 4.7,

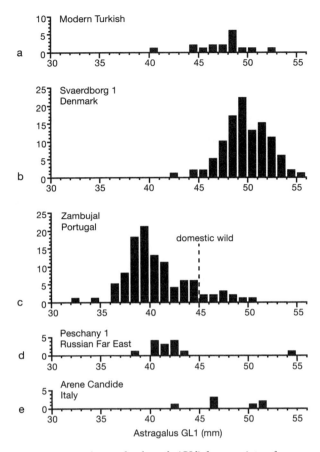

FIGURE 15.1 Astragalus length (GLl) for a variety of samples: (a) Modern Turkish (n = 18, cov = 5.7) (Payne and Bull 1988); (b) Sværdborg, Denmark, early Mesolithic (n = 115, cov = 4.7) (Rowley-Conwy unpublished data); (c) Zambujal, Portugal, Copper Age (n = 105, cov = 8.1) (von den Driesch and Boessneck 1976); (d) Peschany 1, Russian Far East, Iron Age (n = 14, cov = 8.8) (Rowley-Conwy 1999); (e) Arene Candide, Italy, Early Neolithic (n = 7, cov = 5.3) (Rowley-Conwy unpublished data).

supporting the interpretation that a single population is present (Figure 15.1b). The Portuguese Copper Age assemblage from Zambujal, however, is very different (Figure 15.1c). Here the distribution of normalized postcranial elements is strongly negatively skewed, which is generally interpreted as a mixture comprised primarily of many domestic animals that form a "peak" in the distribution, and a "tail" consisting of larger-bodied wild specimens extending to the right of the graph (von den Driesch and Boessneck 1976). The skewness cannot be the result of a concentration on one sex (e.g., smaller females) because, if this were the case, the coefficient of variation would resemble that of the Turkish control population or be even smaller. The Zambujal coefficient of variation, however, is 8.1, larger than that of the previous samples considered. This supports the interpretation that more than one population is indeed present. The larger animals, interpreted as wild, match the size of local Mesolithic wild boars, while the smaller ones, interpreted as domestic, are similar in size to late prehistoric domestic pigs from elsewhere in Iberia (Rowley-Conwy 1995), further supporting the interpretation of von den Driesch and Boessneck (1976).

"Peak and tail" distributions are quite common in Neolithic and Bronze Age Europe (Rowley-Conwy 2003). A similar pattern is seen in the small sample of postcranial bones from Peschany 1, an Iron Age site near Vladivostok in the Russian Far East (Rowley-Conwy 1999; Figure 15.1d), although the wide separation between the specimens that comprise the peak and tail in this case indicates that in these areas at least, the wild and domestic populations did not interbreed very much. This, in turn, implies that these domestic populations were under close control and were not kept under an extensive regime of the kind proposed by Jarman (1976a).

Not all areas of Europe show such clear-cut patterns, however. The difficulties in interpreting the intermediate size of the pigs from the island of Gotland were mentioned above. Another puzzling dataset comes from the Early Neolithic of Arene Candide in Italy (Figure 15.1e). The sample is very small but does not suggest the presence of more than one population. In such cases, the problem is more difficult. Metrically, the population could be wild, domestic, or in some extensive "in-between" state. It has been argued that these animals were wild (Rowley-Conwy 1997, 2000), but the possibility remains that they were domestic (Sorrentino 1999). Prehistoric North Italian pigs have attracted a fair amount of debate in the last 30 years and have become one of the key topics for our understanding of the phenomenon of pig domestication. Therefore, it is worth discussing the problem in greater depth, on the basis of further work we have carried out in the last few years.

To understand the intriguing evidence from Arene Candide, it is necessary to compare it with larger datasets from other sites. An opportunity is provided by the reanalysis of the Middle Neolithic animal bones from Rivoli, which, like the above-mentioned site of Molino Casarotto, was originally studied by Jarman (1976b). This restudy was carried out by Piper, supervised by one of us (UA) (Piper 2001; Albarella et al. in preparation b). Rivoli is near Lake Garda, east of Arene Candide; both are south of the main Alpine watershed (see Figure 15.2 for the location of sites in the Alpine region).

Figure 15.3a compares lower third molar measurements of pigs from Arene Candide and Rivoli. In addition, we have included the data from the Swiss Late Neolithic site of Seeberg Burgäschisee-Süd, as these were the data used by Jarman in his assessment of the case for swine domestication at Molino Casarotto. The material from Seeberg Burgäschisee-Süd has also been reanalyzed as part of our current project on pig domestication and early husbandry (Albarella et al. in preparation b). The Seeberg Burgäschisee-Süd teeth are much larger than those from Rivoli and Arene Candide, while those from the latter two sites seem to be of comparable size. However, the only measurable lower third molar from Early Neolithic Arene Candide plots in the lower part of the Seeberg Burgäschisee-Süd distribution. A simplistic analysis of this distribution could lead to the suggestion that the pigs from Seeberg Burgäschisee-Süd (and perhaps the Early Neolithic M3 from Arene Candide) are wild, and those from the Middle Neolithic of Arene Candide and Rivoli are domestic.

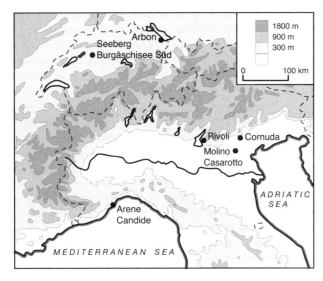

FIGURE 15.2 Map of the Alpine region and northern Italy showing the locations of sites mentioned in the text.

In fact, in the original publication, the pigs from Seeberg Burgäschisee-Süd were interpreted as being predominantly wild (Boessneck et al. 1963). This conclusion was questioned by Payne and Bull (1988: 35), who suggested that there were perhaps more domestic pigs represented at the site than originally suggested. Our own analysis (see below) further supports this latter view.

Figure 15.3b compares the measurements from Seeberg Burgäschisee-Süd with those from Arbon, another Late Neolithic Swiss site to the east that is slightly later than Seeberg Burgäschisee-Süd (Figure 15.2). Although the Seeberg Burgäschisee-Süd teeth are on average quite a lot larger than the Arbon ones, overlap occurs. If we interpret all the Seeberg Burgäschisee-Süd M3s as deriving from wild specimens, it would follow that all specimens in the top part of the Arbon distribution are also wild, with no clear separation from the domestic population. This is, however, extremely unlikely, as extensive work on the Arbon data (Sabine Deschler, Elisabeth Marti Graedel, and Jörg Schibler personal communication May 2001; Albarella et al. in preparation a) has proven that measurements of pig postcranial bones from this site have the peak and tail distribution typical of the European Neolithic, with a large predominance of domestic animals. An alternative explanation for Figure 15.3b is, then, that the domestic-wild divide, in fact, is placed higher up in the distribution, and that only the six largest specimens from Seeberg Burgäschisee-Süd and the two largest from Arbon are wild, with most of the rest being domestic. This means that the domestic pigs from Seeberg Burgäschisee-Süd are not only mostly larger than the Arbon ones, but also larger than those from Arene Candide and Rivoli. Whether this is because of different environmental conditions or the interbreeding of the Seeberg Burgäschisee-Süd domestic pigs with wild ones, it is clear that the populations from north and south of the Alps are different and that any comparison between the two areas must be interpreted cautiously. It is likely that

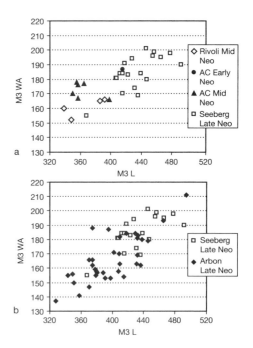

FIGURE 15.3 M3 measurements from Alpine Neolithic sites: (a) Rivoli Mid-Neolithic, Arene Candide Early and Mid-Neolithic (AC), and Seeberg Burgäschisee-Süd Late Neolithic (Seeberg); (b) Seeberg Burgäschisee-Süd (Seeberg), Arbon.

conditions north of the Alps were colder, and this may have affected animal size. In particular during the period of occupation of Seeberg Burgäschisee-Süd, there seems to have been a worsening of the climatic conditions, which seems to be associated with an intensification of hunting (Schibler et al. 1997).

One further element that must be taken into account is the degree to which these differences between sites can be the result of a different sexual composition of the pig assemblages under investigation. At Arbon, on the basis of the morphology of the canines and their alveoli, there seems to be a predominance of males—a ratio of 22:12. At Seeberg Burgäschisee-Süd, the sample is unfortunately too small to draw any firm conclusion. Only three mandibles could be sexed, and these were all from females. It must be clarified that here we are counting only mandibles and not isolated teeth, as these latter are much prone to a recovery bias between sexes. This evidence suggests that the average larger size of the Seeberg Burgäschisee-Süd pigs is unlikely to be the result of a sex bias since the Arbon population includes individuals of both and sexes and, if anything, the larger males are more numerous. At Rivoli, as at Seeberg Burgäschisee-Süd, the sample of sexed mandibles is too small to be significant, although isolated canines indicate a predominance of females. Although this is interesting because a recovery bias should favor the larger male tusks, these latter are sometimes used as tools, which might explain their rarity in the assemblage (i.e., they may have been disposed of outside the excavated area).

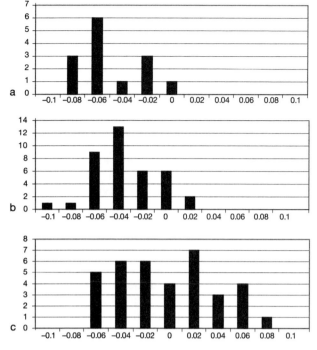

FIGURE 15.4 Variation in *Sus* tooth and bone measurements at the Mid-Neolithic site of Rivoli (northern Italy). Log ratio compared to a standard ("0") modern wild population from Turkey (Payne and Bull 1988). (a) Tooth lengths (n = 14); (b) tooth widths (n = 38); (c) postcranial bones (n = 36).

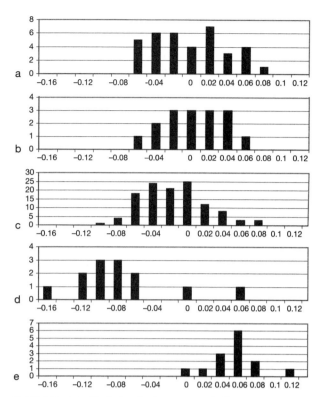

FIGURE 15.5 Variation in *Sus* postcranial bone measurements at three Neolithic sites in northern Italy. Log ratio compared to a standard ("0") modern wild population from Turkey (Payne and Bull 1988). (a) Rivoli, Mid-Neolithic (n = 36); (b) Arene Candide, Early Neolithic (n = 16); (c) Arene Candide, Mid-Neolithic (n =119); (d) Arene Candide, Late Neolithic (n = 13); (e) Cornuda, Late Neolithic (n = 14).

It is important to remember, however, that in pigs, teeth are much less sexually dimorphic than are postcranial bones (see Payne and Bull 1988). This means that variation in tooth size is unlikely to be linked to the sex of the animal, while sex may play a larger role in variation in the size of postcranial elements. More work is clearly needed to determine the differential degree to which sex affects the size of postcranial bones and teeth in pigs and how these differences influence the metric patterning we see in the archaeological record.

Returning to the Italian material and a closer look at the Rivoli data, since the sample size of M3s from the site is so small we have used a size index scaling technique to compare measurements of tooth lengths, tooth widths, and postcranial bones (Figure 15.4). This diagram clearly demonstrates that the size of the Rivoli pigs cannot be characterized easily. Tooth measurements from the Rivoli pigs tend to be somewhat smaller than the Turkish modern wild boar standard, while the postcranial elements of these ancient pigs are relatively larger. Consequently, relying exclusively on the small tooth size of the Rivoli pigs as an index of body size would be a mistake. Since it is highly unlikely that the Rivoli assemblage is represented by teeth of domesticates and postcranial bones of wild animals, we have to consider the possibility that the size of the teeth and the postcrania in pigs do not co-vary. Figure 15.5 compares normalized postcranial data from Rivoli with those from various levels at Arene Candide. As with the teeth (Figure 15.3), the postcranial measurements from Middle Neolithic Arene Candide are comparable to those from Middle Neolithic Rivoli (Figure 15.5a, c). This suggests that Arene Candide pigs also had relatively larger bones than teeth. However, there is a dramatic reduction in body size in the pigs from Late Neolithic levels at Arene Candide (Figure 15.5d) (see also Rowley-Conwy 1997). The Late Neolithic pigs can confidently be interpreted as domestic with two probably wild, large outliers. The separation into two distinct wild and domestic populations—a classic peak and tail pattern—that seems to have occurred by the Late Neolithic is not reflected in the earlier levels at Arene Candide or at Rivoli. On the basis of these metric data and other evidence, it was previously suggested (Rowley-Conwy 1997) that pigs may not have been domestic until the Late Neolithic at Arene Candide. This subsequent and more extensive analysis, however, makes it clear that that the situation is more complex.

All of these pigs are smaller than the extremely large pigs from the Italian Late Neolithic site of Cornuda. Since residents of Cornuda seem to have relied primarily on hunted game, especially deer (Riedel 1988), there is some reason to believe that these large pigs are mostly, if not exclusively, wild. If so, how do we classify the pigs from Early and Mid-Neolithic levels at Arene Candide and Rivoli that are intermediate in size between the presumed wild pigs from Cornuda and the presumed domestic pigs in Late Neolithic levels at Arene Candide? There are three possible hypotheses that can be offered to explain these patterns.

The first is that the Early and Middle Neolithic pigs from Arene Candide and Rivoli are wild. However, this hypothesis is unlikely because it would be difficult to explain their smaller absolute body size compared to that of the presumed wild pigs from Cornuda, as well as the relatively small size of their teeth. Also, this would imply that northern Italian wild pigs had smaller teeth than did domestic pigs from Switzerland, which is unlikely, even if we take into account possible climatic and environmental differences.

Alternatively, one might argue that the Early and Middle Neolithic pigs from Arene Candide and Rivoli are fully domestic. While this would seem a more likely explanation for these osteometric data, it would still cause major interpretative problems. If this were the case, the dramatic size diminution occurring in the Late Neolithic would be hard to explain simply by an intensification of the domestication process. In addition, can such large-boned animals really be consistent with a domestic type?

A third possible explanation, which on the basis of the data so far illustrated seems to be the most likely, is that the human inhabitants of Rivoli and Arene Candide kept their pigs in free-range conditions, which would encourage interbreeding with wild boars living in the woods around the site. By the Late Neolithic, pigs would be kept under closer control, which would explain size reduction, the genetic isolation of wild and domestic populations, and the increased occurrence of shed teeth found on site (Rowley-Conwy 1997)—the latter an indication that the animals were spending more time in or around the settlement. Isotopic and microwear samples being collected as part of our ongoing study of pig domestication will be closely examined for indications of dietary change between the Early/Mid- and the Late Neolithic pigs that might support this hypothesis.

These examples from the Alpine Neolithic provide some idea of the complexity of detecting early domestic pigs and understanding subsequent developments. As we have seen, simplistic interpretations based on limited data can easily lead to mistakes. In addition, both size and shape have to be considered, as various forms of husbandry can lead to the creation of animals of differing conformation. It has been observed that improved pig breeds found in the archaeological record from the seventeenth century onward have large bones and small teeth (Albarella and Davis 1996; Albarella 2002: Fig.7.2). We now have seen that such relative differences in parts of the body can also be found in prehistoric populations. It is, therefore, worth exploring in more detail whether the early domestication process could bring about, already in its early stages, a reduction in tooth size in comparison to bones.

There are many other possible approaches to shape variation in pigs, and these can all illuminate the domestication process. One method that has not widely been used, but has some history, is the analysis of pig molar morphotypes. Suids have complex molars that can show much intraspecific variation. Cusps can be short or long, and supplementary pillars can be absent or present. These might be considered nonmetric traits, presumably with a genetic origin, and might therefore discriminate between genetically distinct populations (something that could be corroborated by DNA studies). This methodology has been successfully applied by Kratochvíl (1981) to the medieval site of Mikulcice (Czech Republic), and more recently by Warman (2000) on mandibles from various modern breeds and by Fujita (2001) on Japanese wild boars. Current ongoing work by several colleagues and ourselves on Japanese (Jomon/Yayoi) and Polynesian (pre- and postcontact) pigs, has also shown that there is extreme variation in the detailed crown morphology of the permanent dentition, which must reflect genetic variation within and between these populations.

Tooth size in itself represents an insufficient criterion to analyze morphological change, as has also been proven by Mayer et al. (1998). They have demonstrated that, although lower M2 is probably the least useful cheek tooth to discriminate between pig populations on the basis of single measurements, it tends to vary allometrically between domestic and wild animals. In particular, large M2s tend to be relatively longer in wild animals than in domestic ones, while small M2s tend to be relatively shorter in wild than in domestic pigs. If archaeological M2 measurements are sufficiently numerous, it should be possible to construct regression lines for the width/length distribution. It should follow that M2s from wild populations should show a steeper regression line. This is an interesting approach that has so far not been much investigated by zooarchaeologists, but which deserves a greater degree of attention and analysis.

Even when we analyze only tooth metrics, relative differences between different teeth can be noted when comparing populations at different stages in the domestication process, a probable consequence of the gradual shortening of the snout. At the British medieval site of West Cotton, M1s were only marginally smaller than those from the British Neolithic, while the difference was greater in the M2s, and greater still in the M3s (Albarella and Davis 1994; Albarella 2002: Fig.7.6). In other words, teeth placed farther back in the jaw seem to be more affected by the size reduction caused by the intensification of the domestication process.

A similar pattern has been found by Davis (e-mail to Albarella 18 September 2003) in early pig populations from the eastern Mediterranean, suggesting that a progressively greater degree of size reduction as one moves back in the molar row can also separate wild and domestic pigs and therefore detect the origin of domestication. Davis compared the widths of the three lower molars of a relatively large sample of modern wild specimens from Israel and Syria with those from the Aceramic Neolithic of Khirokitia (Cyprus, ca. 6000 BC) and the Ceramic Neolithic of Nahal Zehora (Israel, fifth millennium BC). Both archaeological populations are interpreted as domestic. Molar teeth from modern wild boars are on average wider that those from both sites. However, while the degree of reduction in the width of the first molar was minimal (6% at Khirokitia and 2% at Nahal Zehora), third molars were subject to a reduction in width greater than 10% at both sites, with M2s providing intermediate values.

Echoing these results, there is also some indication that at the Early Neolithic site of Çayönü Tepesi in southeastern Turkey (10,200–7500 BP uncalibrated), third molars decrease in size (particularly in length) more than do other teeth (Ervynck et al. 2002: 66). This suggests that even in the early stages of domestication, the phenomenon of snout shortening can be detected through differences in tooth-size reduction (see also Flannery 1983). The data from the British medieval site of West Cotton show that the phenomenon intensifies as control increases, and perhaps with the selection of varieties or breeds.

The shortening of the snout (and in fact of the whole skull, with the consequent reduction in brain capacity) almost certainly represents an adaptive phenomenon, connected with changes in diet and lifestyle following domestication. The snout, rather than the tusks, is the most important tool used to dig for food (Nowak 1999: 1054). When such activity becomes less frequent, perhaps as a result of humans supplying food to pigs, this character is no longer selected for, and, in fact, becomes redundant. Since rooting for food can also occur in domestic pigs, these can also have skulls with relatively straight profiles and long snouts. The character is generally used to assess primitiveness in pigs, with highly selected modern breeds having extremely shortened skulls. However, snout shortening could also be linked with a suite of behavioral and physiological changes associated with neotonization, the possible result of human selection for nonaggressive behavior (Trut 1999).

Aging

Modern pigs are generally slaughtered at younger ages than in the past because, as predominantly meat-producers, they will be killed once they have reached full size (except for a few kept for reproduction). Growth is quicker in modern animals than in traditional, unimproved breeds. This is, however, a recent phenomenon, as improved, fast-growing breeds did not become common until late medieval or even early modern times. The speed of growth of a typical medieval pig may not have been much different from that of its Neolithic counterpart, and substantial differences in the age at slaughter of domestic pigs for most of their history are therefore not expected, unless they are based on reasons other than economic.

Variation in kill-off patterns between wild and domestic populations, however, are likely because of the different method of exploitation—in pigs as well as in other species. In particular, a greater number of older individuals would be expected in assemblages deriving from hunting, as the survival of a wild animal beyond its attainment of full size would not have occurred at human expense. Changes in mortality curves are regarded as one of the main criteria in the identification of the beginning of domestication (see Davis 1987).

There are, however, a number of problems with this approach. For example, when wild populations are hunted more intensively, more young animals are killed, creating a quasi-domestic mortality curve (Elder 1965; Rowley-Conwy 2001: 59–60). Such a hunting pattern is especially likely on the threshold of the domestication process when pressure on wild resources was likely quite strong (Davis et al. 1994). Benecke (1993), for example, interprets the young age profiles of pigs from Early Mesolithic to Late Neolithic sites from the Crimean Peninsula as indications of over-hunting and pressure of wild populations and not as a sign of swine domestication.

Seasonality of hunting may also affect the age pattern. Wild boars have quite large litters, and this fecundity means that many very juvenile animals are encountered shortly after the breeding season. In northern Europe, wild boars breed mostly in spring, so assemblages from sites occupied in summer therefore may contain more very juvenile specimens in their first summer. However, such hunting might not be sexually selective but be directed toward both male and female juveniles. Intensified hunting as a prelude to herding, however, might concentrate on juvenile males. A differential sex ratio could provide a possible means for distinguishing seasonality from intensification as the cause of hunting more juveniles, but there are difficulties. Very juvenile animals are the most difficult to sex. In fully domestic populations of sexually dimorphic animals, sex proportions among very juvenile specimens may be extrapolated from the proportions among the adults; for example, if most determinable adults are female, most indeterminable juveniles are probably male. But in hunting, not all members of the population end up on the archaeological site, so it would be difficult or impossible to extrapolate the sexual proportion of the juveniles from that of the adults.

A further problem is that sample sizes must be quite large if changes through time are to be detected, and the chronological sequence must span the period of the suggested onset of food production. This is quite rare for pig assemblages. One site that has produced an important sequence is Çayönü Tepesi in southeastern Turkey. Here there is indeed evidence that pigs were killed at progressively younger ages through the entire Neolithic sequence (10,200–7500 BP uncalibrated), which goes (as previously discussed) hand in hand with a decrease in certain dental measurements (most notably the length of third molar) showing progressive shortening of the snout (Hongo and Meadow 1998; Ervynck et al. 2002).

Although it is likely that these changes represent steps toward greater human control and a closer relationship with the animals, it is difficult to establish when in the Çayönü sequence fully domesticated pigs actually appear. Using multiple lines of evidence, it has been argued that they may appear only in the final phase of occupation (i.e., in the Pottery Neolithic), when there is the first obvious decrease in the width of the teeth and in the size of the postcranial elements. At the same time, there appears to be a marked increase in physiological stress on the pig population—as reflected in an increase in the frequency of dental enamel defects (Ervynck et al. 2002: 66 and later discussion).

Another site with both a useful chronological sequence and a large bone assemblage is Grotta dell'Uzzo, in northwestern Sicily. At this site, pigs older than four years were observed in Late Mesolithic levels but are no longer found in the Early Neolithic (Tagliacozzo 1993). As at Çayönü, a small size decrease occurs at the same time, although the number of measurable specimens is small and any conclusion must therefore be regarded as tentative. The analysis of a larger sample and of relative size change of different parts of the skeleton—presently in progress by the authors—may provide further information in due course.

Population Genetics

According to most traditional views, the domestic pig originated in the Near East and spread west to Europe and east to China. Some scholars, however, favor the idea that pigs may have been independently domesticated elsewhere (see above for discussion). Limited early studies of the karyotypes of modern wild boar and domestic pig indicate that significant variation exists between pigs in western Europe, Israel, Asia, and the Far East (Popescu et al. 1980; Bosma et al. 1984). These early studies highlighted the potential for cytogenetics to establish whether pig domestication occurred in many areas or just once. Since the 1980s, the rapid development of biomolecular techniques (using both modern and ancient DNA) has resulted in significant advances in our understanding of the molecular evolution of a range of organisms. However, it is only relatively recently that attention has turned toward domestic animals, and even more recently toward pigs.

Recent preliminary research into modern pig genetics has provided rather compelling evidence that appears to support a "multiple domestication" hypothesis. Sequences extracted from hair and blood samples of European and Asian wild boars (7 Japanese, 24 Italian, 15 Polish, and 3 Israeli) and various breeds of domestic pig (n = 74) showed three distinct mtDNA clades—one Asian (A) and two European (EI and EII) (Giuffra et al. 2000; Kijas and Andersson 2001). "A" included Japanese wild boars, Chinese Meishan domestic pigs, and some European domestic pigs. "EI" included the majority of European and all Israeli wild boars, as well as most European domestic pigs and one from the Cook Islands in the Pacific, while "EII" included only 3 Italian wild boars. The considerable genetic diversity noted in this study of modern *Sus scrofa* has been interpreted as providing conclusive evidence for (1) at least two pig domestication events occurring independently somewhere in western and eastern Eurasia, and (2) the later introgression of Asian genetic material into European domestic pigs (Giuffra et al. 2000: 1788).

Although the modern genetic data are extremely important for our broader understanding of pig domestication in Eurasia, they offer few clues with regard to specific geographic context. Nor do they provide any firm temporal framework beyond the crude (and untested) calculations of rates of gene mutation (the so-called molecular clock) that in this case merely indicate a possible divergence of the Asian and European subspecies of wild boar sometime around 900,000 BP (Kijas and Andersson 2001: 307). However, the new and rapidly developing field of ancient DNA (aDNA) research is beginning to provide us with more powerful interpretative tools in these respects.

Some of the only ancient biomolecular research carried out on pigs to date has been undertaken on material from the Japanese archipelago. Using DNA extracted from recent wild boars, as well as from zooarchaeological pig assemblages of primarily Jomon (12,000–2500 BP uncalibrated) and Yayoi (2400–2000 BP uncalibrated) date, researchers have begun to explore the genetic relationships that exist within and between populations of modern/recent and ancient pigs in Japan. Although this research has mainly focused on phylogeographic questions (Watanobe et al. 2001 and 2002; Morii et al. 2002), indirect evidence of pig domestication has also been elucidated. For example, ancient sequence data extracted from pig remains from a number of sites in Southern and central Honshu, suggest that domestic pigs were probably introduced to the Japanese mainland during Yayoi times (Morii et al. 2002: 326). Moreover, those showing sequences most closely related to East Asian domestic pigs were from sites in southern and western Japan (rather than from eastern ones), and all those from the Asahi site in central Japan were more closely related to Japanese wild boar (Morii et al. 2002). Although this current genetic evidence could be used to tentatively suggest that the indigenous Japanese wild boar (*Sus scrofa leucomystax*) appears not to be involved in local domestication events, this limited genetic dataset cannot yet be used to exclude this possibility.

Further research by the same team has also identified an East Asian domestic pig lineage in archaeological material from the Okinawan islands, further supporting their previous conclusion that some pigs were transported from the Asian continent during the early Yayoi-Heinan period (2000–1700 BP), and perhaps even earlier during Jomon times (Watanobe et al. 2002). This evidence appears to cast further doubt on (but again does not rule out) the possible local domestication of the dwarf subspecies of wild boar (*S. scrofa riukiuanus*) indigenous to the Ryuku islands—the remains of which have also been identified through aDNA analysis from archaeological sites in the region (Watanobe et al. 2002).

Ongoing research by our own research team (working with colleagues from England, Sweden, and New Zealand in collaboration with numerous other colleagues and institutions around the world) is currently attempting to understand more fully the phylogenetics of *Sus scrofa* throughout Eurasia. Our aim is to undertake a wide geographical and temporal survey of the extent of possible genetic variation in modern wild boars and archaeological pigs in order to help us understand in more detail specific questions regarding (1) the timing and possible location(s) of domestication (i.e., whether there were single or separate foci for domestication events, and even where these might have been); (2) the genetic diversity and phylogenetic relationships between wild and domestic pigs in the past; (3) the effects that the processes of isolation, domestication, and feralization had on that genetic diversity; (4) whether species other than *Sus scrofa* have been involved in domestication (as has been claimed for *Sus celebensis* in Indonesia) as well as claims for the existence of hybrids between *S. scrofa* and *S. celebensis* in Papua New Guinea; and (5) the colonization history and possible cultural affinities of animals and peoples in the past through palaeophylogeography.

New data resulting from the analysis of mitochondrial DNA extracted from the teeth of 223 recent wild boars/feral pigs across Eurasia, compared with an additional 471 previously published wild, domestic, and feral *Sus* sequences available on Genbank, have provided some exciting and somewhat controversial initial results (Larson et al. 2005) which can be briefly summarized as follows:

1. Perhaps the most fundamental insight is the fact that some *Sus* species designations (i.e., *Sus celebensis*, *S. verrucosus*, *S. barbatus*) based upon morphological criteria are not supported by our mitochondrial DNA data. Each of these groups clusters not with each other, but within the variation of *Sus scrofa* as a whole.
2. *Sus scrofa* as a species originated somewhere in Island Southeast Asia (Philippines, Indonesia).
3. There are a number of extant wild *Sus scrofa* lineages from which recent domestic animals do derive, clearly indicating that pig domestication occurred independently in several diverse geographic locations across Eurasia. Thus, in the Far East, there appears to be a likely minimum of three (but possibly more) wild lineages that were domesticated (two in China and additional ones in Burma/Thailand and northern India). From Wallacea (samples from Halmahera and Papua New Guinea), there is possible evidence for the independent domestication of an introduced ancient lineage of pigs from elsewhere in Island or South East Asia not so far sampled by us. But perhaps most exciting is clear evidence from Europe that points to the independent domestication of probably two wild lineages that form the basis for all modern European breeds (including those that were later improved by mixing with Asian types).
4. Intriguingly, all modern European breeds in our sample have either a European or an Asian "signature." None is even remotely similar to sampled recent wild boar lineages from Armenia, Iran, or Turkey, suggesting little or no importation of Near Eastern domestic pigs into Europe by early farmers.

The conclusions are all based on data deriving from recent or modern, wild or domestic boars and pigs. However, over 500 archaeological samples have also been collected and recently processed and sequenced from over 50% of these samples. These preliminary (and as yet unpublished) data largely support our original published conclusion based on moedern mtDNA, although they add further complexity to the existing picture.

Ancient Diet and Health

The changes in behavioral ecology that occur with domestication provide a promising, yet often neglected, area of research. In particular, changes both in the diet and in the health of animals are potentially very important research tools for identifying the processes of domestication. Adequate fodder must have been one of the defining variables for the success or failure of domestication experiments. Human control over the diet of domestic stock must inevitably have led to a dichotomy between the diets of wild and domestic populations, and this would perhaps have been most exaggerated in omnivores such as pigs. Pigs' omnivory might have led to a more straightforward transition to human control than would have occurred for herbivores such as sheep and goats. Hongo and Meadow (1998: 77) have suggested that the processes of pig domestication were more similar to those of the dog than to those of other artiodactyls. One major reason cited was the fact that both pigs and dogs are more generalized omnivores, more likely to have been drawn to human settlements to feed on refuse. In fact, Ervynck et al. (2002: 68) have suggested that possible early morphological and biometrical changes to the skulls of pigs at Çayönü were related to changes in the rooting behavior of wild individuals, perhaps attracted to human settlements by new food sources such as crops or human refuse.

Regarding health, if natural selection favors hardier individuals, one would expect that the frequency of conditions detrimental to health would normally be low in a "natural" or truly "wild" population (Ervynck et al. 2002: 69). At the other extreme (e.g., in a tightly managed population), higher densities of animals, changes in demographic structure, or poor husbandry strategies ought to lead to the disturbance of the animal's natural behavior and feeding regimes (Price 1984: 14). As a result, it could theoretically be expected that a significant rise in pathological conditions should also be a consequence of early domestication attempts. This has been perhaps most aptly demonstrated in horses, where specific skeletal (joint arthropathies) and dental ("bit wear") abnormalities, thought to be linked primarily to riding, have been used as criteria to signal early domestication (e.g., Anthony 1996; Levine et al. 2000; but see Chapter 17). Work on the health status of ancient pigs, however, has been much more limited.

These broad themes can be explored using a variety of approaches and techniques, and the following are examples of those we and others have recently applied to archaeological pigs.

FIGURE 15.6 *Sus* mandible from second millennium BC deposits at Chagar Bazar, northern Syria, showing abnormal wear and breakage of the tooth crowns (photos by Augusta McMahon).

FIGURE 15.7 *Sus* mandible from second millennium BC deposits at Chagar Bazar, northern Syria, showing heavy dental calculus deposits (photos by Augusta McMahon).

EVIDENCE FROM TOOTH WEAR AND DENTAL CALCULUS

The mouth is the place where the initial physical and chemical breakdown of food occurs. Teeth are the means of shearing, chopping, and masticating and, as such, will be affected to varying degrees by any major changes in the physical and chemical makeup of ingested food. Normal progressive wear on the teeth at a macroscopic level has provided a methodology by which to estimate the relative age at death of ancient pigs (Grant 1982). However, macroscopic and microscopic studies of tooth wear may also provide clues as to changes in the physical properties of the diet.

The study of microscopic tooth wear (dental microwear) has a relatively recent and varied pedigree in the dietary reconstruction of a wide range of extant and fossil species (e.g., Teaford and Runestad 1992; van Valkenburgh et al. 1990; Strait 1993; Solounias and Hayek 1993). Ward and Mainland (1999) showed that modern free-range/rooting pigs had a greater density of microwear features on the buccal and occlusal surfaces of their teeth compared to modern stall-fed pigs. This difference in wear patterns was attributed to the more abrasive diet of free-range/rooting pigs that ingested more soil during feeding. Although microwear studies on modern and archaeological pigs are few and far between, this study, and research on other domestic species such as sheep (e.g., Mainland 1998), has shown the potential of the method for highlighting major differences in diet between archaeological populations.

Major changes in dietary components can also result in chemical changes in the oral environment, which can also indirectly manifest themselves on the dental tissues by stimulating or suppressing different kinds of oral bacteria. A lowering of the pH of saliva, for example, will lead to an increase in cariogenic oral flora (those that cause caries lesions), while an increase in pH is more likely to lead to the calcification of elements of the oral flora, so forming dental calculus. Thus, changes in the frequency of caries or calculus within populations can potentially indicate shifts in diet most likely linked to human husbandry practices.

While major changes in diet would not be expected in the early phase of domestication, such occurrences cannot be ruled out, particularly if poor or initially inappropriate husbandry practices were employed. Also, when fully domesticated pigs were introduced to an area where indigenous wild boars were also present, major differences in the diet and health of these populations (manifested by the conditions outlined above) could be used to separate them in the zooarchaeological record. During our extensive study, the most marked examples of severe calculus formation, caries development, and abnormal wear have been recognized only from sites where fully domestic pigs were certainly present.

For example, pigs are an important component of vertebrate assemblages recovered from fifth to second millennium BC sites (e.g., Tel Brak, Leilan, Chagar Bazar) from the steppe region of the Khabur, northern Mesopotamia, ranging in frequency from between 20 and 60% of the domestic animal remains identified. Changes in the exploitation of pigs and caprines have been explained in a number of ways that primarily focus on environmental, socioeconomic, and political factors (e.g., Zeder 1998a and b; 2003a; Dobney et al. 2003). Recent dental analysis has shown that the pigs from the site of Chagar Bazar have a high frequency of abnormal wear (29% of pig mandibles from second-millennium deposits) and dental calculus (11%) (Figures 15.6 and 15.7). In numerous cases, large portions of the cheek teeth have been broken antemortem and then subsequently worn further by mastication. In many of the same mandibles, "severe" deposits of dental calculus have also been noted, usually on the buccal surfaces of the premolars. These pathologies can probably be explained by an unusual behavior or diet of the pigs at these sites.

In semi-arid regions such as northern Syria, where shade is limited, pig keeping would have been severely constrained by both ecological and maintenance factors. Their high water requirements, and an inability to utilize cellulose-rich pasture plants, means that pigs are best kept close to or within the settlement. So, in densely settled areas (like these early city-states), small-scale, enclosed pig keeping was likely the norm. As such, pigs would have been unable to carry out extensive rooting and would have been fed a range of domestic human refuse. The presence of these oral pathologies, in higher frequencies than normally found, must reflect a shift in diet and behavior related to human husbandry practices and may indicate examples of urban pig keeping.

DEVELOPMENTAL STRESS: LINEAR ENAMEL HYPOPLASIA

Mammal teeth can provide many clues to an individual's living conditions. Linear enamel hypoplasia (LEH) is a deficiency in enamel thickness occurring during tooth crown formation, typically visible on a tooth's surface as one or more grooves or lines (see Goodman and Rose 1990 for a detailed discussion of the hypoplastic lines in humans). The condition is generally caused by developmental stress (Sarnat and Moss 1985). The causes may vary, but nutritional deficiencies are certainly important factors. For humans, the analysis of LEH has been used successfully to assess the general health status of archaeological and recent populations (e.g., Goodman et al. 1988).

Recent studies on pigs have developed a recording protocol for this condition (Dobney and Ervynck 1998), recorded its frequency and chronology in numerous archaeological pig assemblages of varying date and location (Ervynck and Dobney 1999; Dobney and Ervynck 2000; Dobney et al. 2002), and even explored its use in identifying second farrowing (Ervynck and Dobney 2002). These analyses have shown that LEH is common in pigs, but is certainly not a random event. Its occurrence follows clear patterns that reflect causal relationships between events in life and seasonal conditions affecting the individual's food intake and energy balance.

Recent work on Neolithic (Pre-Pottery and Pottery) pigs from the site of Çayönü Tepesi has shown very low frequencies of LEH as compared to much later northern European Neolithic and medieval site assemblages (Ervynck et al. 2002: Figure 23)—something that should perhaps be expected from a healthy, wild population in a relatively undisturbed natural environment. However, there did appear to be a slight increase the frequency of LEH through the three Pre-Pottery Neolithic phases (indicated by slight changes in LEH frequency; see Ervynck et al. 2002: Figure 22), as well as a more obviously significant increase between the Pre-Pottery and Pottery Neolithic phases (Ervynck et al. 2002: Figure 23).

Could these changes in LEH frequency also reflect human intervention and early domestication attempts at Çayönü? It has been argued that the change during the Pre-Pottery phases may simply reflect individual differences in behavior within the wild population (Ervynck et al. 2002). Changes in feeding behavior could have led to increased physiological stress in those particular animals but do not necessarily reflect direct human influence over them, at least in the earliest Pre-Pottery phases. It is as yet unclear from the Çayönü sequence when humans began to play a more active role in pig management. However, it could be argued that the significant increase in LEH frequencies noted in the Pottery Neolithic may well mark this transition.

A more recent large-scale study of both recent and archaeological *Sus scrofa* remains from northwest Europe showed the frequency of LEH to be consistently low within all ancient and recent wild boar populations studied, in contrast to early domestic populations of Neolithic date, which show generally higher LEH frequencies (Dobney et al. 2004). As was the case at Çayönü, higher frequencies of LEH in these ancient pig populations have also been broadly interpreted as being the result of direct or indirect human interference related to domestication and husbandry.

ISOTOPES

Of all the approaches to the study of ancient diet, perhaps the greatest potential lies in the field of stable isotope analysis, where the most rapid developments are also taking place. This approach can tell us about the changing diets and adaptations of mammals, and more about the environment in which they lived. Techniques of analysis now allow the fairly routine collection of measurements of both $\delta^{13}C$ and $\delta^{15}N$ in a wide range of animal material, which in turn have provided a range of important dietary information for a range of species including humans (e.g., Fizet et al. 1995; Richards et al. 2001; Bocherens et al. 2001), and in some cases have distinguished wild from domestic animals (e.g., Balasse et al. 2000).

The relative importance of terrestrial vs. marine products ingested can also provide information about the trophic level of the animal in the food chain. Thus, within a single ecosystem, the diets of herbivores, carnivores, and omnivores can theoretically be identified. Pigs (like dogs and humans) are essentially omnivores, so they can show a range of $\delta^{13}C$ and $\delta^{15}N$ values, which range from those most commonly found in herbivores to those of carnivores. Wild boars are probably largely herbivorous, but domestic pigs are more likely to have become increasingly carnivorous through consuming human waste and excrement. Thus, there should be interpretable patterns through time that are related to human influence on the diet of pigs (Richards et al. 2002).

Research into the diet of pigs using stable isotopes is as yet relatively limited. However, preliminary and ongoing work by several colleagues has begun to test a number of general hypotheses concerning isotopic signatures for pig domestication and husbandry. For example, isotopic analysis has been undertaken on Jomon-age pig bones from the Ryuku and other islands in the Japanese archipelago in order to test the hypothesis that wild boar diet changed during the initial stages of domestication (Minagawa and Matsui 2002). Comparison of $\delta^{13}C$ and $\delta^{15}N$ values led the authors to conclude that two feeding strategies were developed for domestic pigs: one based mainly on human leftovers and the other on feeding pigs cultivated C_4 plants or marine foods (Minagawa and Matsui 2002: 106).

Preliminary results of $\delta^{13}C$ and $\delta^{15}N$ values from nine European sites are also available (Pearson 2001; Richards et al. 2002). At most of these sites, pigs indeed appear to be mainly herbivorous (i.e., more negative $\delta^{13}C$ and associated low $\delta^{15}N$ values). This suggests that if humans controlled pig diets, this control involved feeding pigs (or allowing pigs to feed on) mainly plant foods. This herbivorous diet is typically found in European wild boars (Briedermann 1990) and would presumably have been very similar for domestic pigs herded in woodland areas. Preliminary results suggest that only at later sites (i.e., Iron Age and medieval) are there pigs that show $\delta^{13}C$ and $\delta^{15}N$ values consistent with the consumption of more animal products and/or human waste.

The results from the Middle and Late Neolithic site of Arbon (Switzerland) (Richards et al. 2002) are also important. The $\delta^{13}C$ and $\delta^{15}N$ values from large specimens (presumed to be wild) and smaller specimens (presumed to be domestic) show very little difference, implying a very similar—mostly herbivorous—diet. These data would suggest that the diets of wild and domestic pigs exploited at the site were similar.

Isotopic analysis as a tool to assess the diet of ancient domestic animals is increasing in its importance to zooarchaeologists. The field is, however, a young and rapidly developing one, and there is still much we do not understand about the processes involved. More samples must be analyzed before results such as those outlined above can be more meaningfully interpreted. The results also cannot be interpreted in isolation: values from other species, and other related zooarchaeological information, needs to be considered. We need to establish just what we mean by a typical nondomesticated pig diet as reflected in stable isotopes, by looking at modern wild boars and (probably more usefully) clearly pre-domesticated individuals (i.e., Late Pleistocene and Early Holocene specimens). Ultimately, we need to carry out controlled feeding experiments to construct more reliable interpretative frameworks. It may be, however, that any shifts that took place in the diet of early domesticated pigs were too subtle to detect with the crude ("averaging") tool of stable isotopes in bone collagen. However, the microsampling of dental tissues offers the promise of exploring more subtle changes within the lifetime of a single individual. Alternatively, major changes in diet are perhaps not linked with domestication at all, but perhaps occurred much later during urbanization and the related intensification of husbandry techniques, aspreviously suggested for second millennium sites in northern Syria.

Establishing Temporal Context for Initial Pig Domestication

Establishing the temporal context of pig domestication is not as easy as it is for some other species for two reasons. First, as we have discussed, there are many potential types of pig-human relationships. Correctly identifying the relationship(s) at any archaeological site is only the first step; we must then decide which ones we admit within our definition of domestication. Secondly, we have also seen that the widespread distribution of wild boars has allowed a variety of origins to be suggested.

The site of Çayönü Tepesi has been mentioned several times in the course of this chapter, and it is perhaps one of the most important and earliest sites in Eurasia in which pig domestication has been studied. As previously discussed, its unique, long chronological sequence (covering the entire Pre-Pottery and Pottery Neolithic periods) provides one of the very few opportunities to create a temporal framework within which to document the process of domestication in some detail. Thus, visible changes to the skeleton and teeth (e.g., snout shortening and change in conformation), evidence for an increase in physiological stress, and a shift in demographic profile indicate a process that appears to have been very slow—occurring gradually over approximately 2,000 years and apparently complete by the start of the Pottery Neolithic (around 8000–7500 BP uncalibrated). Perhaps what is most interesting about the evidence from Çayönü is the fact that these changes were not all coeval, suggesting a more complex physiological and behavioral response, perhaps not originally driven by direct human intervention (Ervynck et al. 2002).

Finds from Hallan çemi Tepesi, a Turkish site in the eastern Taurus Mountains, has suggested an even earlier date for the first shift within a *Sus scrofa* population from completely "wild" behavior to a way of living closer to humans (Rosenberg et al. 1995, 1998). However, while the measurements of the five recovered lower third molars from Hallan çemi Tepesi are surprisingly small, the dataset is limited. In a review of *Sus* data from sites more recent than Çayönü or contemporary with its later Pre-Pottery phases, Peters et al. (1999) observed a decrease in the length of the third molar between PPNB material from Çayönü and LPPNB specimens from Gürcütepe (Peters et al. 1999: Figure 11). The authors claim this to be "unequivocal morphometrical evidence for the occurrence of domestic pigs" in the LPPNB as a somewhat rapid, punctuated event. However, the Çayönü data presented in Peters et al. (1999) represents an amalgam of teeth from the Channeled to the Cell Plan Building phases, which covers the whole PPNB. When data from the different subphases of the PPNB at Cayönü are examined separately, one can see clear evidence for decrease in the size of third molar occurring over the course of this long period (Ervynck et al. 2002).

Peters et al. (1999) also highlight data from other LPPNB sites such as Hayaz Tepe and Tell Hallula to substantiate their claim for the appearance of domestic pigs in that chronological period. However, why the *Sus scrofa* specimens from Hayaz Tepe and Tell Hallula should be labeled "domestic" is not clear from the review. In fact, a comparison of the postcranial data from Gürcütepe and older sites (not Çayönü) rather indicates a continuous (slight) size decrease through time instead of a sudden change between the Middle and Late PPNB (see Peters et al. 1999: Figure 10).

In addition to southwest Asia, there is tentative evidence for another domestication event occurring in China, and the site of Cishan points to a similar time period as that from southeastern Turkey (i.e., around 8000 BP). Although limited data exist to support this claim (Jing and Flad 2002;

Giuffra et al. 2000; Kijas and Andersson 2001), a more detailed analysis similar to that outlined above needs to be undertaken in order to confirm this and to see to what degree the specific processes involved are similar or different.

Once these issues are resolved, it is vital that selected specimens are directly dated by accelerator mass spectrometry (AMS) radiocarbon dating techniques. These should preferably be specimens actually showing some significant feature or trait, not just some bone from the same layer or context. This is because individual objects can move between archaeological contexts, before or during excavation. Naturally, we are eager to find early examples of important developments like domestication and are, therefore, sometimes uncritical in our acceptance of contextual data from complex sites (see Rowley-Conwy 1995).

Future Research

Our aim in this chapter has been to stress the multiple methods that should be applied to the questions of pig domestication. Future research will use these methods to test some of the ideas put forward above. Were there really only two separate hearths of domestication, in the Near and the Far East respectively? Our recent genetic research suggests not. The intervening areas are much less well known, and substantial swathes of the map remain a blank—our equivalent of the medieval cartographers' "here be dragons." Establishing what was going on in between the Near and Far East hearths is now a priority and will be a significant test of the validity of the "twin hearth" model.

What of other potential areas of domestication indicated by recent genetic research? The Baltic region may have seen the intensification of wild boar exploitation, and even local domestication (Zvelebil 1995), and a major goal is to establish whether or not this was actually the case. The Indian subcontinent is another area, currently unknown, from which interesting results may be confidently predicted. Finally, the status of the various modern *Sus* lineages in peninsular and island Southeast Asia needs to be sorted out as a matter of priority. This should go hand in hand with the study of archaeological materials in this region. The results will be of great importance to our understanding of human dispersal and the settlement of the Pacific, when pigs accompanied humans on the greatest diaspora ever undertaken by either species.

Notes

1. *Sus celebensis* (the Sulawesi warty pig) has been claimed to have been involved in a separate domestication event in Indonesia during the early Holocene (Groves 1981).
2. This is based on subtle variations in cranial morphology of a small sample of modern specimens he measured.
3. There are those that claim it to be by about 6000 BP or even earlier (Golson and Hughes 1976; Golson 1982), while others favor a more recent introduction (Bayliss-Smith 1996; Harris 1995) based on radiocarbon dates (Hedges et al. 1995). Bulmer (conversation with Dobney, 12 September 2002) argues that these dates are likely to be in error in light of the firm stratigraphic position of pig bones in Late Pleistocene or Early Holocene contexts in at least four sites.

Acknowledgments

We would like to thank the editors for inviting us to contribute to the volume, and Jörg Schibler and Simon Davis for giving us permission to refer to and use unpublished data. Our pig domestication research project is supported by several Wellcome Trust Bioarchaeology Fellowships (Keith Dobney) and a research grant from the Arts and Humanities Research Board (Peter Rowley-Conwy and Umberto Albarella). We are also very grateful to the many colleagues around the globe who are collaborating with us on the pig project, and who are providing invaluable data, insight, and expertise into this complex but intriguing subject.

References

Albarella, U. 2002. "Size matters": How and why biometry is still important in zooarchaeology. In *Bones and the man. Studies in honour of Don Brothwell*, K. Dobney and T. O'Connor (eds.), pp. 51–62. Oxford: Oxbow Books.

Albarella, U. and S. J. M. Davis. 1994. *The Saxon and medieval animal bones excavated 1985–1989 from West Cotton, Northamptonshire*. Ancient Monuments Laboratory Report 17/94. London: English Heritage.

———.1996. *Mammals and birds from Launceston Castle, Cornwall: Decline in status and the rise of agriculture*. York: Circaea, 12 (1): 1–156.

Albarella, U., K. Dobney, P. Rowley-Conwy, and J. Schibler. In preparation a. *The pigs of the Horgen culture: A review*.

Albarella, U., L. Piper, and L. Barfield. In preparation b. *Rivoli revisited: A restudy of the animal bones in the context of the northern Italian Neolithic*.

Anthony, D. 1996. Bridling horse power: The domestication of the horse. In *Horses through time*, S. Olsen (ed.), pp 57–82. Colorado and Dublin: Roberts Rinehart for the Carnegie Museum of Natural History.

Balasse, M., A. Tresset, H. Bocherens, A. Mariotti, and J.-D. Vigne. 2000. Un abattage "post-lactation" sur des bovins domestiques Néolithiques. Étude isotopique des restes osseux du site de Bercy (Paris, France). *Ibex, Journal of Mountain Ecology* 5:39–48/*Anthropozoologica* 31: 39–48.

Bayliss-Smith, T. 1996. People-plant interactions in the New Guinea highlands: Agricultural hearthland or horticultural backwater? In *The origin and spread of agriculture and pastoralism in Eurasia*, D. R. Harris (ed.), pp. 499–523. London: University College London Press.

Benecke, N. 1993. The exploitation of *Sus scrofa* (Linné, 1758) on the Crimean Peninsula and in southern Scandinavia in the Early and Middle Holocene. Two regions, two strategies. In *Exploitation des animaux sauvages à travers le temps*, J. Desse and F. Audoin-Rouzeau (eds.), pp. 233–245. Juan-les-Pins: éditions APDCA.

———. 1994. *Der Mensch und seine Haustiere. Die Geschichte einer jahrtausendealten Beziehung*. Stuttgart: Thesis.

Blouch, R. A. and C. P. Groves. 1990. Naturally occurring suid hybride in Java. *Zeitschrift für Saugetierkunde* 55: 270–275.

Bocherens, H., D. Billiou, A. Mariotti, M. Toussaint, M. Patou-Mathis, D. Bonjean, and M. Otte. 2001. New isotopic evidence for dietary habits of Neanderthals from Belgium. *Journal of Human Evolution* 40: 497–505.

Boessneck, J., J.-P. Jéquier, and H.-R. Stampfli. 1963. *Seeberg Burgäschisee-Süd. Teil 3. Der tierreste*. Bern: Acta Bernensia.

Bosma, A. A., N. A. Haan, and A. A. MacDonald. 1984. Karyotype variability in the wild boar (*Sus scrofa*). In *Symposium international sur le sanglier*, F. Spitz and D. Pepin (eds.), pp. 53–56. Toulouse: Institut National de la Recherche Agronomique.

Briedermann, L. 1990. *Schwarzwild*, 2nd edition. Berlin: Deutscher Landwirtschaftsverlag.

Childe, V. G. 1958. *The prehistory of European society*. Harmondsworth: Penguin.

Clutton-Brock, J. 1987. *A natural history of domesticated mammals*. Cambridge: Cambridge University Press.

Davis, S. J. M. 1981. The effects of temperature change and domestication on the body size of Late Pleistocene to Holocene mammals of Israel. *Paleobiology* 7:101–114.

———. 1987. *The Archaeology of animals*. London: Batsford.

Davis, S. J. M., O. Lernau, and J. Pichon. 1994. The animal remains: New light on the origin of animal husbandry. In *Le gisement de Hatoula en Judée occidental, Israël*, M. Lechevallier and A. Ronen (eds.), pp. 83–100. Paris: Memoires et travaux du centre de recherche Français de Jerusalem n.8.

Dobney, K. and A. Ervynck. 1998. A protocol for recording enamel hypoplasia on archaeological pig teeth. *International Journal of Osteoarchaeology* 8: 263–273.

———. 2000. Interpreting developmental stress in archaeological pigs: The chronology of linear enamel hypoplasia. *Journal of Archaeological Science* 27: 597–607.

Dobney, K., A. Ervynck, and B. La Ferla. 2002. Assessment and further development of the recording and interpretation of linear enamel hypoplasia in archaeological pig populations. *Environmental Archaeology* 7: 35–46.

Dobney, K., D. Jaques, and W. Van Neer. 2003. Diet, economy and status: Evidence from the animal bones. In *Excavations at Tell Brak, Volume 4. Exploring a regional centre in Upper Mesopotamia, 1994–1996*, R. Matthews (ed.), pp. 417–430. Cambridge: McDonald Institute and British School of Archaeology in Iraq.

Dobney, K., A. Ervynck, U. Albarella, and P. Rowley-Conwy. 2004. The chronology and frequency of a stress marker (linear enamel hypoplasia) in recent and archaeological populations of *Sus scrofa*, and the effects of early domestication. *Journal of Zoology* 264: 197–208.

Ducos, P. 1968. *L'origine des animaux domestiques en Palestine*. Bordeaux: Delmas.

Ekman, J. 1974. Djurbensmaterialet från stenålderslokalen Ire, Hangvar sn, Gotland. In *Gotlands mellanneolitiska gravar*, G. Janzon (ed.), pp. 212–246. Acta Universitatis Stockholmiensis, Studies in North European Archaeology 6. Stockholm: Almqvist and Wiksell.

Elder, W. H. 1965. Primeval deer hunting pressures revealed by remains from American Indian middens. *Journal of Wildlife Management* 29: 366–370.

Ervynck, A. and K. Dobney. 1999. Lining up on the M1: A tooth defect as a bio-indicator for environment and husbandry in ancient pigs. *Environmental Archaeology: Journal of Human Palaeoecology* 4: 1–8.

———. 2002. A pig for all seasons? Approaches to the assessment of second farrowing in archaeological pig populations. *Archaeofauna* 11: 7–22.

Ervynck, A., K. Dobney, H. Hongo, and R. Meadow. 2002. Born free!: New evidence for the status of pigs from Çayönü Tepesi, Eastern Anatolia. *Paléorient* 27: 47–73.

Fizet, M., A. Mariotti, H. Bocherens, B. Lange-Badré, B. Vandersmeersch, J. Borel, and G. Bellon. 1995. Effect of diet, physiology and climate on carbon and nitrogen stable isotopes of collagen in Late Pleistocene anthropic palaeoecosystem: Marillac, Charente, France. *Journal of Archaeological Science* 22: 67–79.

Flannery, K. V. 1983. Early pig domestication in the Fertile Crescent: A retrospective look. In *The hilly flanks. Essays on the prehistory of Southwest Asia*, T. C. Young, P.E.L. Smith, and P. Mortensen (eds.), pp. 163–188. Studies in Ancient Oriental Civilization 36. Chicago: Oriental Institute, University of Chicago.

Flannery, T. F. and J. P. White. 1991. Animal translocation. *National Geographic Research and Exploration* 7: 96–113.

Fujita, M. 2001. *Quaternary wild boars of Japan*. Unpublished PhD thesis. Graduate School of Science, Osaka City University.

Giuffra, E., J. M. H. Kijas, V. Amager, Ö. Carlborg, J.-T. Jeon, and L. Andersson. 2000. The origin of the domestic pig: Independent domestication and subsequent introgression. *Genetics* 154: 1785–1791.

Golson, J. 1982. The Ipomoean revolution revisited: Society and the sweet potato in the upper Wahgi valley. In *Inequality in New Guinea highland societies*, A. Strathern (ed.), pp. 109–136. Cambridge: Cambridge University Press.

Golson, J. and P. J. Hughes. 1976. The appearance of plant and animal domestication in New Guinea. *Journal de la Société des Océanistes* 36: 294–303.

Goodman, A. H. and J. C. Rose. 1990. Assessment of systemic physiological perturbations from dental enamel hypoplasias and associated histological structures. *Yearbook of Physical Anthropology* 33: 59–110.

Goodman, A. H., R. Brooke-Thomas, A. C. Swedland, and G. J. Armilagus. 1988. Biocultural perspectives on stress in prehistoric, historical and contemporary population research. *Yearbook of Physical Anthropology* 31: 169–202.

Grant, A. 1982. The use of tooth wear as a guide to the age of domestic ungulates. In *Ageing and sexing animal bones from archaeological sites*, B. Wilson, C. Grigson, and S. Payne (eds.), pp. 91–108. British Archaeological Reports, British Series 109. Oxford: BAR.

Groves, C. 1981. *Ancestors for the pigs: Taxonomy and phylogeny of the genus Sus*. Technical Bulletin No. 3, Department of Prehistory, Research School of Pacific Studies. Canberra: Australian National University.

———. 1989. Feral mammals of the Mediterranean islands: Documents of early domestication. In *The walking larder*.

Patterns of domestication, pastoralism, and predation, J. Clutton-Brock (ed.), pp. 46–58. London: Unwin Hyman.

Harris, D. R. 1995. Early agriculture in New Guinea and the Torres Strait divide. *Antiquity* 69: 848–854.

Hecker H. M. 1982. Domestication revisited: Its implications for faunal analysis. *Journal of Field Archaeology* 9: 217–236.

Hedges, R. E. M., R. A. Housley, C. R. Bronk, and G. J. Van Klinken. 1995. Radiocarbon dates from the Oxford AMS system: Archaeometry datelist 20. *Archaeometry* 37: 428.

Herre, W. and M. Röhrs. 1973. *Haustiere—zoologisch gesehen*. Stuttgart: Gustav Fisher.

Higgs, E. S. and M. R. Jarman. 1969. The origins of agriculture: A reconsideration. *Antiquity* 43: 31–41.

Hongo, H. and R. Meadow. 1998. Pig exploitation at Neolithic Çayönü Tepesi (southeastern Anatolia). In *Ancestor for the pigs: Pigs in prehistory*, S. Nelson (ed.), pp. 77–98. Research Papers in Science and Archaeology 15. Philadelphia: MASCA.

———. 2000. Faunal remains from Pre-Pottery Neolithic levels at çayönü, southeastern Turkey: A preliminary report focusing on pigs (*Sus* sp.). In *Archaeozoology of the Near East IV A*, M. Mashkour, A. M. Choyke, H. Buitenhuis, and F. Poplin (eds.), pp. 121–140. Groningen: ARC.

Jarman, M. R. 1976a. Prehistoric economic development in sub-Alpine Italy. In *Problems in economic and social archaeology*, G. D. G. Sieveking, I. H. Longworth, and K. E. Wilson (eds.), pp. 375–399. London: Duckworth.

———. 1976b. Rivoli: The Fauna. In *The excavations on the Rocca di Rivoli 1963–68*, L. H. Barfield and B. Bagolini (eds.), pp. 159–173. Verona: Museo Civico di Storia Naturale di Verona (II serie) Sezione Scienze dell'Uomo.

Jing, Y. and R. K. Flad. 2002. Pig domestication in ancient China. *Antiquity* 76: 723–732.

Jonsson, L. 1986. From wild boar to domestic pig. A reassessment of Neolithic swine of northwestern Europe. In *Nordic Late Quaternary biology and ecology*, L.-K. Königsson (ed.), *Striae* 24: 125–129.

Kijas, J. M. H. and L. Anderson. 2001. A phylogenetic study of the origin of the domestic pig estimated from the near-complete mtDNA genome. *Journal of Molecular Evolution* 52: 302–308.

Kratochvíl, Z. 1981. *Tierknochenfunden aus der grossmährischen Siedlung Mikulcice I. Das Hausschwein*. Praha: Academia Praha.

Kruska, D. 1988. *Mammalian domestication and its effect on brain structure and behavior*. NATO ASI series, Intelligence and Evolutionary Biology G17: 211–249.

Larson, G., K. Dobney, U. Albarella, M. Fang, E. Matisoo-Smith, J. Robins, S. Lowden, H. Finlayson, T. Brand, P. Rowley-Conwy, L. Andersson, and A. Cooper. 2005. Worldwide phylogeography of wild boar reveals multiple centers of pig domestication. *Science* 307: 1618–1621.

Levine, M. A., G. N. Bailey, K. Whitwell, and L. Jeffcott. 2000. Palaeopathology and horse domestication: The case of some Iron Age horses from the Altai Mountains, Siberia. In *Human ecodynamics and environmental archaeology*, G. N. Bailey, R. Charles, and N. Winder (eds.), pp 123–133. Oxford: Oxbow.

Lindqvist, C. and G. Possnert. 1997. The subsistence economy and diet at Jakobs/Ajvide, Eksta parish and other prehistoric dwelling and burial sites on Gotland in long-term perspective. In *Remote sensing*, Volume I, G. Burenhult (ed.), pp. 29–90. Theses and Papers in North European Archaeology 13a. Stockholm: Institute of Archaeology, University of Stockholm.

Lotsy, J. P. 1922. Die Aufarbeitung des Kühn'schen Kreuzungs materials im Institut für Tierzucht der Universität Halle. *Genetica* 4: 32–61.

Mainland, I. 1998. The lamb's last supper: The role of dental microwear analysis in reconstructing livestock diet in the past. In *Fodder: Archaeological, historical and ethnographic studies*, M. Charles, P. Halstead, and G. Jones (eds.). Special issue of *Environmental Archaeology* 1: 55–62.

Mayer, J. J. and I. L. Brisbin. 1991. *Wild pigs of the United States. Their history, morphology and current status*. Athens and London: University of Georgia Press.

Mayer, J. J., J. M. Novak, and I. L. Brisbin. 1998. Evaluation of molar size as a basis for distinguishing wild boar from domestic swine: Employing the present to decipher the past. In *Ancestor for the pigs: Pigs in prehistory*, S. Nelson (ed.), pp. 39–53. Research Papers in Science and Archaeology 15. Philadelphia: MASCA.

McCormick. F. 1999. Early evidence for wild animals in Ireland. In *The Holocene history of the European vertebrate fauna: Modern aspects of research*, N. Benecke (ed.), pp. 355–372. Archäologie in Eurasien 6. Berlin: Deutsches Archäologisches Institut.

Meadow, R. 1981. Early animal domestication in South Asia: A first report of the faunal remains from Mehrgarh, Pakistan. In *South Asian Archaeology 1979*, H. Härtel (ed.), pp. 143–179. Berlin: Dietrich Reimer Verlag.

———. 1999. The use of size index scaling techniques for research on archaeozoological collections from the Middle East. In *Historia animalium ex ossibus. Festschrift für Angela von den Driesch*, C. Becker, H. Manhart, J. Peters, and J. Schibler (eds.), pp. 285–300. Rahden/Westf.: Verlag Marie Leidorf GmbH.

Minagawa, M. and A. Matsui. 2002. Chemical evidence of domestication of *Sus scrofa* in Ryuku Islands and its implication with prehistoric inter-islands trade. Abstract (IS6-6) of paper presented at the Inter-Congress of the International Union of Anthropological and Ethnological Sciences 2002. Tokyo, Japan.

Morii, Y., N. Ishiguro, T. Watanabe, M. Nakano, H. Hongo, A. Matsui, and T. Nishimoto. 2002. Ancient DNA reveals genetic lineage of *Sus Scrofa* among archaeological sites in Japan. *Anthropological Science* 110: 313–328.

Nowak, R. M. 1999. *Walker's mammals of the world*, Volume II. Baltimore and London: John Hopkins University Press.

Oliver, W. L. R., ed. 1993. *Pigs, peccaries and hippos*. Gland (Switzerland): International Union for the Conservation of Nature and Natural Resources.

Österholm, I. 1989. *Bosättningsmönstret på Gotland under Stenåldern*. Theses and Papers in Archaeology 4. Stockholm: University of Stockholm, Institute of Archaeology.

Payne, S. and G. Bull. 1988. Components of variation in measurements of pig bones and teeth, and the use of measurements to distinguish wild from domestic pig remains. *ArchæoZoologia* 2: 27–65.

Pearson, J. A. 2001. *Stable isotope analysis at Çatalhöyük, a Neolithic site in Anatolia*. Poster presented at the Wellcome Trust Bioarchaeology Conference, London.

Peters J., D. Helmer, A. von den Driesch, and M. Saña Seguí. 1999. Early animal husbandry in the Northern Levant. *Paléorient* 25: 27–47.

Piper, L. 2001. *Born to be wild? The problem with pigs in the North Italian Neolithic: A re-analysis of the animal bone assemblage*

from Rocca di Rivoli. BA degree dissertation, University of Birmingham, UK.
Pira, A. 1909. Studien zur Geschichte der Schweinerassen, inbesondere derjenigen Schwedens. *Zoologischen Jahrbüchern* (supplement) 10: 233–426.
Popescu, C. P., J.-P. Quere, and P. Francheschi. 1980. Observations chromosomiques chez le sanglier français (*Sus scrofa scrofa*). *Annales de Génétique et de Séléction Animale* 12: 395–400.
Price E. O. 1984. Behavioral aspects of animal domestication. *Quarterly Review of Biology* 59: 1–32.
Richards, M. P., P. B. Pettitt, M. C. Stiner, and E. Trinkaus. 2001. Stable isotope evidence for increasing dietary breadth in the European Mid-Upper Paleolithic. *Proceedings of the National Academy of Sciences, USA* 98: 6528–6532.
Richards, M., K. Dobney, U. Albarella, B. Fuller, J. Pearson, G. Muldner, M. Jay, T. Molleson, and J. Schibler. 2002. Stable isotope evidence of *Sus* diets from European and Near Eastern sites. Abstract of paper presented at the 9th Congress of the International Council of Archaeozoology, Department of Archaeology, Durham, UK.
Riedel, A. 1988. *The Neolithic animal bone deposit of Cornuda (Treviso)*. Ferrara Universita' degli Studi.
Rosenberg M., R. Nesbitt, R. W. Redding, and B. L. Peasnall. 1998. Hallan çemi, pig husbandry, and post-Pleistocene adaptations along the Taurus-Zagros arc (Turkey). *Paléorient* 24: 25–41.
Rosenberg M., M. R. Nesbitt, R. W. Redding, and T. F. Strasser. 1995. Hallan çemi Tepesi: Some preliminary observations concerning Early Neolithic subsistence behaviors in Eastern Anatolia. *Anatolica* 21: 1–12.
Rosman, A. and P. G. Rubel. 1989. Stalking the wild pig: Hunting and horticulture in Papua New Guinea. In *Farmers as hunters*, S. Kent (ed.), pp. 27–36. Cambridge: University Press.
Rothchild, M. F. and A. Ruvinsky, eds. 1998. *The genetics of the pig*. Oxford and New York: CAB International.
Rowley-Conwy, P. 1995. Making first farmers younger: The West European evidence. *Current Anthropology* 36: 346–353.
———. 1997. The animal bones from Arene Candide. Final report. In *Arene Candide: Functional and environmental assessment of the Holocene sequence*, R. Maggi (ed.), pp. 153–277. Memorie dell'Istituto Italiano di Paleontologia Umana, new series 5. Rome: Ministero per i Beni Culturali e Ambientali.
———. 1999. East is east and west is west but pigs go on forever: Domestication from the Baltic to the Sea of Japan. In *Current and recent research in osteoarchaeology 2*, S. Anderson (ed.), pp. 35–40. Oxford: Oxbow Books.
———. 2000. Milking caprines, hunting pigs: The Neolithic economy of Arene Candide in its West Mediterranean context. In *Animal bones, human societies*, P. Rowley-Conwy (ed.), pp. 124–132. Oxford: Oxbow Books.
———. 2001. Time, change and the archaeology of hunter-gatherers: How original is the "Original Affluent Society"? In *Hunter-gatherers. An interdisciplinary perspective*, C. Panter-Brick, R. H. Layton and P. Rowley-Conwy (eds.), pp. 39–72. Biosocial Society Symposium Series 13. Cambridge: University Press.
———. 2003. Early domestic animals in Europe: Imported or locally domesticated? In: *The widening harvest*, A. Ammerman and P. Biagi (eds.). Boston: Archaeological Institute of America.
Rowley-Conwy, P. and J. Storå. 1997. Pitted ware seals and pigs from Ajvide, Gotland: Methods of study and first results. In *Remote sensing*, Volume I, G. Burenhult (ed.), pp. 113–127. Theses and Papers in North European Archaeology 13a. Stockholm: Institute of Archaeology, University of Stockholm.
Sarnat, H. and S. J. Moss. 1985. Diagnosis of enamel defects. *New York State Dental Journal* 51: 103–106.
Sauer, C. O. 1952. *Agricultural Origins and Dispersals*. Washington, D.C.: American Geographical Society.
Schibler, J., S. Jacomet, H. Hüster-Plogmann, and C. Brombacher. 1997. Synthesis. In *Ökonomie und ökologie neolitischer und bronzezeitlicher ufer-siedlungen am zürichsee*, J. Schibler, H. Hüster-Plogmann, S. Jacomet, C. Brombacher, E. Gross-Klee, and A. Rast-Eicher (eds.), pp. 329–361. Zürich und Egg: Monographien der Kantonsarchäologie Zürich 20.
Solounias, N. and L. A. C. Hayek. 1993. New methods of tooth microwear analysis and application to dietary determination of two extinct ungulates. *Journal of Zoology* 229: 421–445
Sorrentino, C. 1999. Faune terrestri. In *Il Neolitico nella Caverna delle Arene Candide (Scavi 1972–1977)*, S. Tiné (ed.), pp. 66–108. Collezione di Monografie Preistoriche ed Archeologiche X. Bordighera: Istituto Internazionale Studi Liguri.
Strait, S. 1993. Molar microwear in extant small-bodied faunivorous mammals: An analysis of feature density and pit frequency. *American Journal of Physical Anthropology* 92: 63–79.
Tagliacozzo, A. 1993. *Archeozoologia della Grotta dell'Uzzo, Sicilia*. Roma: Istituto Poligrafico e Zecca dello Stato.
Teaford, M. F. and J. A. Runestad. 1992. Dental microwear and diet in Venezuelan primates. *American Journal of Physical Anthropology* 88: 347–364.
Trut, L. N. 1999. Early *Canid* domestication: The farm-fox experiment. *American Scientist* 87: 160–169.
Uerpmann, H.-P. 1979. *Probleme der Neolithisierung des Mittelmeerraums*. Wiesbaden: Dr. Ludwig Reichert Verlag.
van Valkenburgh, B., M. F. Teaford, and A. Walker. 1990. Molar microwear and diet in large carnivores: Inferences concerning diet in the sabre-toothed cat, *Smilodon fatalis*. *Journal of Zoology* 222: 319–340.
von den Driesch, A. E. and J. Boessneck. 1976. Die Fauna vom Castro do Zambujal. In *Studien über frühe Tierknochenfunde von der Iberischen Halbinsel 5*, A. von den Driesch and J. Boessneck (eds.), pp. 4–129. Munich: Institut für Palaeoanatomie, Domestikationsforschung und Geschichte der Tiermedizin der Universität München.
Vigne, J.-D. and H. Buitenhuis. 1999. Les premiers pas de la domestication animale à l'Ouest de l'Euphrate: Chypre et l'Anatolie centrale. *Paléorient* 25: 49–62
Ward, J. and I. Mainland. 1999. Microwear in modern rooting and stall-fed pigs. *Environmental Archaeology* 4: 25–32.
Warman, S. 2000. *Morphometric investigation of dental variation to examine genetic relationships between pig populations*. PhD thesis, University of London.
Watanobe, T., N. Ishiguro, M. Nakano, H. Takamiya, A. Matsui, and H. Hongo. 2002. Prehistoric introduction of domestic pigs onto the Okinawa islands: Ancient mitochondrial DNA evidence. *Journal of Molecular Evolution* 52: 281–289.
Watanobe, T., N. Ishiguro, N. Okumura, M. Nakano, A. Matsui, H. Hongo, and H. Ushiro. 2001. Ancient mitochondrial DNA reveals the origin of *Sus scrofa* from Rebun Island, Japan. *Journal of Molecular Evolution* 55: 222–231.
Winge, H. 1900. Bløddyr-skaller. Knogler af Dyr. Plantelevninger. In *Affaldsdynger fra Stenalderen i Danmark*, A. P. Madsen,

S. Müller, C. Neergaard, C. G. J. Petersen, E. Rostrup, K. J. V. Steenstrup, and H. Winge (eds.), pp. 158–163. Copenhagen: C.A. Reitzel.

Zeder, M. A. 1998a. Pigs and emergent complexity in the Near East. In *Ancestors for the pigs*, S. Nelson (ed.), pp. 109–122. MASCA Research Papers in Science and Archaeology 15. Philadelphia: MASCA.

———. 1998b. Environment, economy and subsistence on the threshold of urban emergence in northern Mesopotamia. In Natural space, inhabited space in northern Syria (10th–2nd millennium BC), M. Fortin and O. Aurenche (eds.), pp. 55–67. Canadian Society for Mesopotamian Studies, Bulletin 33. Quebec: Canadian Society for Mesopotamian Studies; Lyon: La Maison de l'Orient Méditerranéen.

———. 2003a. Provisioning in urban societies: A view from northern Mesopotamia. In *The social construction of cities*, M. Smith (ed.), pp. 156–183. Washington, D.C.: Smithsonian Press.

Zvelebil, M. 1995. *Hunting, gathering, or husbandry? Management of food resources by the Late Mesolithic communities of temperate Europe*. MASCA Research Papers in Science and Archaeology 12, Supplement: 79–104.

CHAPTER 16

The Domestication of South American Camelids
A View from the South-Central Andes

GUILLERMO L. MENGONI GOÑALONS AND
HUGO D. YACOBACCIO

Introduction

South American camelids are the only large herd mammals that were domesticated in all the Americas. The origins of domestication and the development of native camelid herding are restricted to the Andes, particularly the Central and South-Central portion. In pre-European times, domesticated camelids were widely distributed from the highlands to the valleys, lowlands, and coast. They constituted a primary element in Andean economies and social life, and were pivotal for the expansion of early states starting with Tiwanaku and then with the Incas. There is no general agreement on the timing of this process or whether only one or several centers of domestication existed. In this chapter, we will consider both traditional and new archaeological tools for documenting domestication in South American camelids, and how the application of these tools to assemblages from the South-Central Andes is yielding a new perspective on the chronology and extent of this process.

The South American camelids are classified in two genera, *Lama* and *Vicugna*, based on their physical appearance and DNA data (Franklin 1982; Stanley et al. 1994; Wheeler 1995). At present, four existing species are recognized: two wild, the vicuña *(V. vicugna)* and the guanaco *(L. guanicoe)*, and two domesticated, the llama *(L. glama)* and the alpaca *(L. pacos)* (see Chapter 23). Vicuñas are the smallest (35–50 kg), followed by the alpaca (55–65 kg), then the guanaco (80–130 kg), and finally the llama (80–150 kg), which is the largest (Raedeke 1978, 1979; Larrieu et al. 1979; Franklin 1982, 1983; Rabinovich et al. 1984; Cajal 1985; Cunazza et al. 1995). Based on genetic studies, some researchers currently believe that the alpaca is derived from the vicuña and the llama from the guanaco, and changes in nomenclature have been proposed (Kadwell et al. 2001). Recent studies have yielded evidence of remarkable variability, not only in the size of domestic camelids across the region (e.g., Stahl 1988; Miller and Gill 1990), but also in the number of breeds with fiber characteristics that have no present counterparts (Wheeler et al. 1992; Wheeler et al., 1995; Wheeler 1996).

Camelids are producers of both primary and secondary products: meat, hide, fiber, and dung are among the most significant products they offer, including their use as beasts of burden in the case of the llama. Both in the present and in the past, they have been important in rituals and ceremonies, and were frequently represented in prehistoric pottery, rock art, and figurines. Guanaco and vicuña were hunted for their meat, grease, and hide, while in Inca times, at least, vicuñas were captured, sheared, and later released. Although both wild species are sympatric in some regions (e.g., the highlands of western Argentina), social groups stay naturally segregated and do not interbreed. In postconquest times, the alpaca has been bred mainly as a fine-fiber producer, with two varieties, *huacaya* and *suri*, both distinguished by pointed ears and droopy tails (Cardozo 1954). The *huacaya* has short and crimped fleece, while *suri* fiber is longer and wavy. The llama has a wider geographical distribution than the alpaca and is the most versatile form, as it has been used as a source of food, hide, and fiber, and also as pack animal. It also has two varieties, the *chaku* and the *ccara*, both with banana-shaped ears and raised tails (Cardozo 1954). The *chaku* has finer fiber than does the *ccara*, although both varieties can be used as beasts of burden (Calle Escobar 1984; Bonavia 1996). Recent studies in Argentina have shown a larger variety of coats and fleece types associated with other physical attributes (e.g., Lamas 1994). Most of the four camelid forms interbreed, giving birth to hybrids: e.g., *huarizos*, *misti*, and *paco-vicuña*.

Several overviews have been written during the last decades, each emphasizing different aspects of the process of domestication, its indicators, and the archaeological evidence available (Wing 1975a, 1975b, 1977a, 1977b, 1978, 1986; Novoa and Wheeler 1984; Kent 1987; Browman 1989; Lavallée 1990; Wheeler 1991, 1995, 1998; Bonavia 1996, 1999). These general overviews have primarily centered on the Central Andes (Peru), with few references to the South-Central Andes. As we will see in the following sections, the current picture of the origins of camelid domestication is largely shaped by the Central Andean focus of archaeological investigations over the past three decades, and may not give the full story of the domestication of South American camelids.

This chapter reviews the current information available for the Central and South-Central Andes, and includes a discussion of the principal indicators traditionally used for identifying domesticated forms. New criteria, including contextual information, new standards for osteometric analysis, and fiber analysis allow us to trace the process of camelid domestication. Based on this evidence, we propose a new chronology for the initial appearance of the llama (4500–4000 BP) and the existence of multiple centers of origin.

Prior Research: A View from the Central Andes

Since the 1960s, the Peruvian Central Andes have been the primary focus of archaeological and zooarchaeological research on the domestication of South American camelids. As a result, this region has been widely accepted as the heartland of camelid domestication, while the other regions of the Andes have been portrayed as secondary recipients of this new technology.

The origins of camelid domestication in the Andes were first addressed from a zooarchaeological perspective by Wing (1972) in her detailed study of the fauna at Kotosh, a site located in the upper valley of the Huallaga River (Peru), at an elevation of 2,000 m (see Figure 16.1). This information was complemented with that from Tarma, a site occupied during Inca times and located at a higher elevation (4,000 m) (Wing 1972). Wing used an overall increase in camelid utilization, a shift in the proportions of different camelid species utilized, and age profiles to argue for the appearance of llama and alpaca herding by 3400–2700 BP.

At that time, it was believed that all domesticated camelids present in valley sites were introduced from the puna, following a model of high mobility and pastoral transhumance (Lynch 1967) that was supported by evidence from throughout the Andes (e.g., Lynch 1967; Browman 1974; Núñez and Dillehay 1979). More recently, Lynch (1980: 310–311) introduced the idea of a more restricted transhumance (puna-upper valleys) that did not include the coast. The faunal information retrieved at cave sites located above 4,000 m from the Puna of Junín in Peru (see Figure 16.1) was particularly important in reinforcing this view. These sites include Uchcumachay, Pachamachay, Acomachay A and B, Telarmachay, and other related puna sites (Wing 1975c; Wheeler Pires-Ferreira 1975; Wheeler Pires-Ferreira et al. 1976; Wheeler et al. 1977; Kent 1982; Moore 1989).

Using indicators similar to those developed by Wing, plus newly developed tools for discriminating between wild and domestic camelid species, researchers in the later 1970s and 1980s were able to detect evidence that signaled the ongoing process of domestication at a much earlier date than previously thought. Initial work with assemblages from the cave sites of Uchcumachay (4,050 m), talus of Panaulauca (4,100 m), and Telarmachay (4,420 m) detected a progressive intensification of camelid exploitation between 7450 and 4450 BP. This pattern was interpreted as a long-term shift from more generalized hunting strategies that evolved first into more selective hunting of camelids and then to camelid domestication (Wheeler Pires-Ferreira et al. 1976; see also Wing 1989). Wheeler (1984a, 1984b, 1995, 1998) used the strikingly high mortality of neonatal animals in the assemblage from Telarmachay, plus the appearance of lower incisors with distinctive alpaca morphology, to argue for a management of domestic camelids at this site dating back to at least 6000 BP.

Kent's analysis of animal remains from later excavations at Pachamachay (4,030 m) and from the site of Chiripa (3,860 m) provides a remarkably long sequence of animal exploitation in the Central Andes stretching back to ca. 12,000 years ago

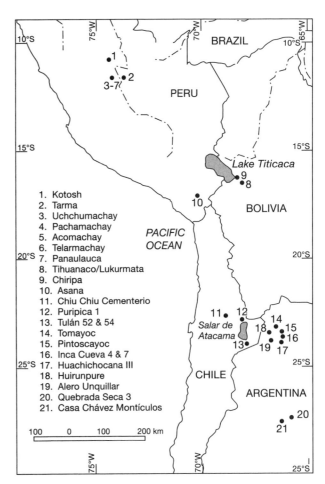

FIGURE 16.1 Map of the Andean area showing localities discussed in the text.

(Rick 1980). Contrary to the interpretation of material from earlier excavations at Pachamachay (Wheeler Pires-Ferreira et al. 1976), Kent found no evidence of intensification in camelid use over time. Nor did he find shifts in mortality patterns that might mark the onset of domestication. Camelids consistently comprised over 80 percent of the assemblage from the site, and mortality profiles were dominated by adults in all levels. Osteometric evidence, however, suggests the introduction of domesticated forms (alpaca and llama) possibly by 5,000 years ago, and certainly by 4150 BP (Kent 1982).

Moore's (1989) analysis of the assemblage retrieved at the main excavation area from Panaulauca (4,010 m) further underscores the complexity of camelid use in the Andes. Once again, this new analysis found that the intensification of camelid use seemed less marked than earlier studies had indicated. Camelids always dominate at over 85 percent of the assemblage of animal bones from all levels at the site. However, Moore did find a significant shift in the types of camelids used through time, with vicuña steadily decreasing and being replaced by a slightly larger small camelid (alpaca?) that became important in later phases (Moore 1989: 373, and see also Figures 8:3 and 8:12) at the Formative period. During the early Formative, the proportion of large camelids (considered to be guanaco, llama, or both) increased to 25 percent. Moreover, during this period there was an increase in the use of newborn animals, signaling a possible growing dependence on domesticated camelids at about 3600 BP.

Thus, multiple lines of evidence have been used to mark the transition from hunting to herding camelids in the Central Andes. At one site, Telarmachay, this transition has been dated to about 6,000 years ago, while analyses from other puna sites would put this transition at about 2,000 years later at about 4600–3600 BP.

Domestication and Its Indicators

Definitions of domestication vary depending upon whether it is defined from a human (e.g., Ducos 1978) or animal (e.g., Price 1984) point of view. In this chapter, we view domestication more from the human perspective, as a process through which animals are integrated into the domestic realm as property or prestige goods by controlling their reproduction and by providing them with the means for feeding and protection. We distinguish domestication from pastoralism, which we define as an economic system based on the use of domesticated animals as its core element. This is a particularly important distinction when speaking about South American camelids, not only because the initial domestication of camelids and the development of pastoral economies based on camelids may be separated by many hundreds of years, but also because detecting the process of animal domestication and the development of pastoral economy requires different types of archaeological indicators.

Human control over reproduction in domesticated animals may result in certain genetic or phenotypical changes that may be detected in the archaeological record, which we call direct measures of domestication. In South American camelids, direct measures of domestication include changes in dental morphology, in bone size and shape, and in fiber characteristics, as well as in DNA (see Chapter 23). Indirect measures of domestication are reflections of the economic strategies humans employ either in the production of domestic animal resources or in their use. Indirect measures focus not on individual specimens but on assemblage properties, such as species diversity, mortality profiles, part distributions, and contextual information, all of which are useful in detecting both camelid domestication and the advent of pastoral economies focusing on camelids. Examining these different direct and indirect measures over time and space provides mutually reinforcing pictures of the process both of domestication and of the development of pastoral economies.

Direct Measures

DENTAL MORPHOLOGY

Perhaps the greatest challenge in documenting domestication of South American camelids in the archaeological record is distinguishing between the two closely related wild progenitor species (guanaco and vicuña) and their domestic descendents (llama and alpaca). Fortunately, there are distinctive morphological characteristics on the incisors of these animals that can help (Wheeler 1982, 1991). This is especially the case in distinguishing guanacos and llamas from vicuñas and alpacas (Table 16.1). The incisors of guanacos and llamas (both deciduous and permanent) are spatulate in shape, with enamel covering all sides of the crowns. Both deciduous and permanent incisors of guanaco and llama also have well developed roots. In contrast, deciduous and permanent incisors in the vicuña and alpaca are parallel-sided, and enamel is restricted to the labial surfaces of the crowns. In the vicuña, the permanent incisors do not form a root.

Distinguishing wild from domestic forms on the basis of dental morphology is not as clear cut. In fact, guanaco and llama incisors are indistinguishable from another on the basis of morphology. It is also impossible to draw morphological distinctions between the deciduous incisors of vicuña and alpaca, which in both species are root forming and have enamel restricted to the upper labial surface of the crown. However, the morphology of the permanent incisors of the vicuña and alpaca can be readily distinguished. Permanent vicuña incisors lack roots, and enamel covers the entire labial surface, while alpaca permanent incisors retain juvenile traits of forming roots and having enamel only on the upper labial surface. There are exceptions to these patterns in contemporary camelids as noted by Kent (1982: 142, i.e., "alpacas with either open-rooted or parallel-sided incisors"), but it is not yet clear if these exceptions are the result of hybridization (Wheeler 1998).

Recent histological analyses on contemporary domestic dental specimens have pioneered attempts to refine these distinctions (Riviere et al. 1997) but have achieved only partial results since a study of wild specimens is still pending. Once again, the long history of hybridization in domestic camelids may make it difficult to use modern animals in developing clear-cut methods for distinguishing between various camelid species in the archaeological record.

OSTEOMETRY

Many of the efforts to develop archaeological indicators of camelid domestication have been based on observable differences in the sizes of the four South American camelid species. These efforts are founded on the assumption that body size should correspond to the size of bones (Moore 1989; Mengoni Goñalons and Elkin 1990), an assumption supported by an allometric study of a large sample of alpaca of different age groups that showed a strong correlation between individual body size and bone measurements (Wheeler and Reitz 1987).

Some researchers have focused on craniometric differences, like Otte and Venero (1979) for Peruvian vicuña and alpaca, or Puig and Cajal (1985) for vicuña and guanaco from Argentina (see Puig 1988 for a summary of craniometric characteristics that can be used to distinguish between the crania of the four South American camelid species). Because of the usually poor preservation of crania, however, most

TABLE 16.1
Matrix of Dental Morphology on South American Camelid Incisors

	Guanaco	Llama	Vicuña	Alpaca
Deciduous	Spatulate	Spatulate	Parallel-sided	Parallel-sided
	Entire crown	Entire crown	Upper labial	Upper labial
	Roots present	Roots present	Roots present	Roots present
Permanent	Spatulate	Spatulate	Parallel-sided	Parallel-sided
	Entire crown	Entire crown	Entire labial	Upper labial
	Roots present	Roots present	Roots absent	Roots present

zooarchaeological work aimed at drawing osteometric distinctions between South American camelids focuses on postcranial bones. Bones that tend to be well-preserved and, therefore, well-represented in the archaeological record are naturally favored in these analyses. Univariate analyses of the breadth and width measurement of the proximal first phalanx, for example, seem particularly effective in discriminating between various camelids (Miller 1979; Miller and Gill 1990; Miller and Burger 1995). Length measurements of these ubiquitous bones have not proven as useful, however, primarily because of difficulties in discriminating between the first phalanges of front and hind limbs, which are markedly different in length (Kent 1982). Bivariate analyses of astragali, calcaneum, and distal metapodials tend to corroborate the univariate analyses of first phalanx proximal breadth and depth measures (Miller 1979). Kent (1982) developed an innovative approach that used discriminant function analysis of a series of dimensions from many postcranial elements, but this technique has not been widely adopted by other researchers. Moore (1989) discovered proportional differences in the long bones useful in distinguishing between guanacos and llamas, as well as between vicuñas and alpacas. However, these techniques can be performed only on whole, articulated bones, which are rarely found in archaeological contexts.

There are several factors that make drawing osteometric distinctions between South American camelids particularly difficult. The two primary camelid genera of *Lama* and *Vicugna* do seem to sort out clearly into two distinct size groups of larger *(Lama)* and smaller *(Vicugna)* animals. However, each genus contains wild and domestic forms that differ in size, and the degree of overlap between the various domestic and wild forms is difficult to measure. This difficulty is exacerbated by the above-noted degree of interbreeding and resultant hybridization between these various forms. Luckily, there does not seem to be marked sexual dimorphism in South American camelids (Vilá 2000), as seen in other domesticated species (see Zeder 2001, and also Chapter 14), that would further complicate this already complicated puzzle.

However, other factors do present significant challenges for the use of size in documenting initial domestication in camelids. The first is that the impact of climatic changes between the Late Pleistocene and Early Holocene, known to result in significant diminution in the size of a number of other species around the globe (Davis 1981; Ducos and Horwitz 1997). While there is some indication that camelids of the Late Pleistocene-Early Holocene boundary were considerably larger than camelids later in the Holocene (Yacobaccio 1991; Rosenfeld 2002), the precise nature of the impact of post-Pleistocene climatic amelioration on the size of South American camelids is unclear.

Perhaps even more significant is the dramatic geographic variation in the size of camelids as one moves southward toward the tip of South America. This clinal variation in size is most clearly seen in the guanaco, the most widely distributed of the camelid species, which can be found today from Peru to Tierra del Fuego (Franklin 1982). Those populations living at low latitudes (Peru, northern Chile, and northwestern Argentina) are the smallest, while those at the higher latitudes to the south are by far the largest (Raedeke 1978; Larrieu et al. 1979; Rabinovich et al. 1984; Franklin 1982, 1983; Cajal 1985). A similar pattern is also suspected for vicuña, although further studies are still needed (Wheeler 1995). The strong clinal variation in the size of South American camelids is reminiscent of a pattern documented by Zeder (2001, and Chapter 14) for modern wild goats from Iran. A similar pattern is inferred for pigs in the Alps by Albarella et al. (Chapter 15). In all cases, the increase in body size in colder regions may be a function of Bergman's rule that predicts increasing body size with decreasing temperatures.

Failure to recognize the impact of regional variation on the size of camelids has proven to be a significant impediment to the use of osteometric analysis in detecting initial camelid domestication in the Andes. Most of the early work along these lines used modern standards composed of vicuñas, alpacas, and llamas from Peru and guanacos primarily from Tierra del Fuego or Patagonia (Wing 1972; Miller 1979; Kent 1982; Miller and Burger 1995). As a result, the widely accepted size gradient between camelid species in the Andes has been that vicuña are always the smallest, alpacas are larger, llamas even larger, and guanacos the largest of them all.

Very different results are obtained when one compares species from the same geographical region, thus eliminating an important bias in size variation and providing more reliable size classes as a reference. Although the available osteometric data for wild camelids is still scarce, some important points can be stressed. Figure 16.2 illustrates a bivariate plot of the proximal latero-medial width (*x*-axis) and

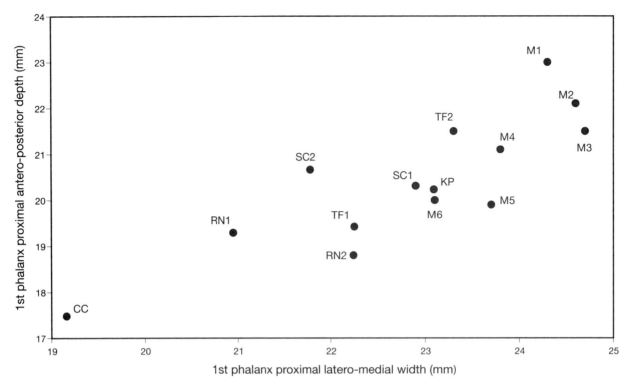

FIGURE 16.2 Size variation in contemporary guanaco based on the proximal width and proximal depth of the first phalanx (fore and hind toe averaged). *CC*, Cumbres Calchaquíes, northwestern Argentina; *RN1-2*, Río Negro, northern Patagonia, Argentina; *SC1–2*, Santa Cruz, southern Patagonia, Argentina; *TF1–2*, Tierra del Fuego; *KP*, Patagonia, taken from Kent (1982, Appendix IV); *M1–6*, Tierra del Fuego, taken from Miller and Burger (1995, Fig. 6). All measurements plotted, except those produced by Miller, were measured following Kent's protocol (Kent 1982).

the proximal antero-posterior width (*y*-axis) taken on first phalanges from several contemporary guanacos along a latitudinal range (26°–55° S) that runs from northwest Argentina and northern Patagonia to southern Patagonia and Tierra del Fuego. A geographical size variation is clear, showing that the guanacos from Patagonia and Tierra del Fuego are the largest and those from northwestern Argentina are the smallest. This pattern has several consequences: (1) the guanacos from Tierra del Fuego should not be used as a standard for comparison with archaeological material coming from the Andean region; (2) contemporary camelids from the same or a neighboring region from which the archaeological material is derived must be used as size standards; (3) upholding a size gradient that considers guanaco as the largest camelid is inaccurate when analyzing bones from Andean sites (e.g., Wing 1972; Kent 1982; Miller and Burger 1995); and (4) the correct size gradient for analyzing materials from the Central and South-Central Andean regions should run from vicuña, the smallest, on to alpaca and then guanaco, ending with llama, the largest. This pattern is clearly seen when metric data from Andean vicuñas, alpacas, and llamas (from Kent 1982) are compared to an Andean guanaco from northwest Argentina (Figure 16.3).

BONE MORPHOLOGY

Skeletal differences among South American camelids are hard to find. Working with a total of 10 skeletons of adult guanaco, vicuña, alpaca, and llama Adaro and Benavente (1990a, 1990b; Benavente and Adaro 1991; Benavente et al. 1993) defined 51 qualitative features that they considered showed "clear and precise" identification. However, the subjective nature of deciding whether a feature is "very developed," "less developed," or "little developed" makes it sometimes difficult to apply these distinctions with much confidence. Moreover, some of these features could be the result of individual differences resulting from mechanical factors, including robusticity of muscles (see Benavente 1997–1998; Cartajena et al. 2001), and may not be reliable for drawing clean taxonomic distinctions. The fragmentary nature of most archaeological assemblages adds another difficulty to employing this technique. Nevertheless, this line of research deserves to be further explored.

FIBER CHARACTERISTICS

Fleece from the four varieties of camelids varies in color, diameter, and length (Dransart 1991a, 1991b; Benavente et al. 1993; Reigadas 1994a, 1994b). Given the arid conditions in many parts of the Andes and the remarkable preservation of many otherwise perishable materials, fiber holds considerable promise for determining the variety of camelids used. Color seems a particularly useful attribute for distinguishing between wild and domestic forms. Guanacos are reddish-brown to brown and white, while vicuña are light fawn and white. By contrast, domesticated llamas and alpacas show a

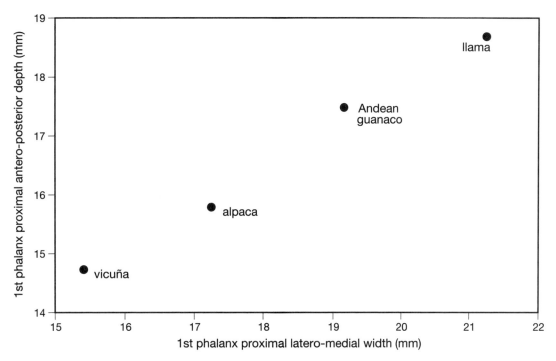

FIGURE 16.3 Size gradient in contemporary camelids using the Andean guanaco as standard. The measurements for the guanaco were taken from an individual from the Cumbres Calchaquíes, northwestern Argentina. Those for vicuña, alpaca, and llama are the averaged values for the fore and hind toe as presented by Kent (1982, Appendix IV).

variety of colors, such as black, white, brown, and gray. Also, the patterning in coat colors shows a great variation in llamas and alpacas, an attribute reflected in the rich classification system based on color developed by Andean herders (Flores Ochoa 1981).

Diameter also seems a good indicator for distinguishing wild and domesticated camelids. Coats of wild species are comprised of a mix of very fine fibers (around 12 μm in vicuñas and 16 μm in guanacos) and very coarse ones (greater than 60 μm). In contrast, intermediate fibers (i.e., between 20–40 μm) dominate in modern domesticated camelids, which tend to have more homogenous coats as a result of artificial selection (Calle Escobar 1984; Lamas 1994). In certain areas of the South-Central Andes (e.g., Puna of Jujuy, Argentina), some present-day herds of llamas exhibit very fine fiber diameters (i.e., between 20–23 μm), with values below the averages known from Peru (Lamas 1994). Recent studies carried out on 1,000-year-old prehispanic camelid mummies from El Yaral (Wheeler 1995, 1996) have shown the existence of breeds with very homogeneous coats (e.g., extra-fine in alpaca (17.9 μm) and fine in llama (22 μm)) that have no present counterpart in Peru. While these remarkable mummies clearly demonstrate the emphasis placed on breeding animals with fine coats suited for high-quality textile manufacture, it is not clear whether changes in fiber quality is a later development linked to the intensification of a camelid-based pastoral economy, rather than a marker of initial camelid domestication.

Indirect Measures

SPECIES DIVERSITY AND TEMPORAL TRENDS

An increase in the representation of camelids over time and a corresponding decrease in the overall diversity of species in archaeological assemblages frequently have been taken as leading indicators of the process of camelid domestication in the Central Andes (Wing 1972, 1980, 1986; Wheeler Pires-Ferreira et al. 1976; Wheeler 1984a, 1984b). In particular, an increase in camelids relative to cervids has been cited as a useful index for monitoring the intensification in camelid use that ultimately resulted in their domestication. The magnitude of the increase in relative abundance of camelids varies depending on elevation. In the lower-elevation valley sites, outside the natural range of wild camelids, camelid representation may increase from 0% to as much as 50% of an assemblage. In the puna, where these animals occur naturally, in clear hunter-gatherer contexts camelids may begin at 50% and increase to as much as 96% at sites engaged in a highly developed pastoral economy.

The problem with using intensification as a marker of camelid domestication is that intensification is often seen both as creating the conditions in which domestication might occur and as an indicator that the process has taken place. The sudden appearance of camelids into lower elevation areas outside their natural habitat, like highland valleys or coastal areas, most likely represents the introduction of already domesticated camelids. However, in the higher-elevation natural habitat of these animals, where initial domestication most likely occurred, species diversity and representation of camelids in archaeological assemblages by themselves cannot distinguish a selective hunting strategy that focuses on camelids from a reliance on domesticated camelids.

MORTALITY PATTERNS

Mortality patterns are a commonly used tool for determining whether a camelid assemblage represents a hunted prey population or the slaughter of domesticated herd animals. Mortality profiles have also sometimes been used to determine season of death, and therefore slaughtering practices and seasonality of occupations that also shed light on the transition from hunting to herding. Given the different species involved and the diverse array of resources they offer, camelids present a special challenge to those using mortality patterns to reconstruct culling strategies. An emphasis on the exploitation of camelids for fiber or for use as beasts of burden may result in very different mortality patterns than strategies aimed at promoting meat production. Being able to model expected mortality patterns with expected economic strategies that emphasize the exploitation of regenerative resources like fiber and labor is particularly important in monitoring the development of complex, specialized pastoral economies of later periods in Andean history. For the initial phases of domestication, however, it is more likely that a generalized strategy that emphasized the propagation of the herd, with meat being the primary resource of interest, was employed. Such a strategy would most likely emphasize the slaughter of young males with prolonged survivorship of females, and a few males, through their prime reproductive years. Thus, an emphasis on young camelids has often been taken as an indicator of management of breeding behavior to promote herd propagation, which is a leading-edge marker of domestication (i.e., Wing 1972; Moore 1989).

But not all mortality patterns reflect the conscious strategies of human hunters or herders. They can also be an indicator of the overall health of an animal population and the conditions under which animals lived. Wheeler, for example, linked the increasingly high representation of young, neonatal camelids at Telarmachay with human management of camelid populations. The proportion of neonates in layers from this site dating from between 9,000 to 6,000 years ago is about 36% (a figure similar to the proportion of neonates in contemporary wild camelid populations). By around 6,000 years ago, this figure rose to 57%, reaching a peak of 73% by 3,800 years ago. Wheeler interprets the unusually high neonatal mortality in these later levels as the result of a bacterial infection caused by *Clostridium perfringens* Type A, an infection that today is a major killer in camelid herds kept under unsanitary corralling conditions (Wheeler 1985, 1998). Coupled with a steady increase in the intensity of camelid use and the presence of incisors with distinctive alpaca morphology in layers dated to about 6,000 years ago, the very high neonatal mortality at Telarmachay is interpreted by Wheeler as a clear marker of initial camelid management and domestication. As yet there is no evidence for corrals of that age in the Puna of Junín or other Andean areas that would lend further support to this hypothesis.

Camelid mortality profiles have been constructed using both dental eruption and wear patterns and long-bone fusion. Early attempts at reconstructing these patterns from long-bone fusion used fairly gross categories of "juvenile" for unfused bones and "adult" for fused bones (Wing 1972, 1975a, 1978). Since postcranial bones fuse at different ages, such an approach risks including early fusing elements from young animals in the "adult" category and later-fusing elements of older animals in the "juvenile" category. Moreover, these categories are too broad to detect differential mortality of neonatal and yearling animals or the difference between culling strategies that focus on prime-age animals as opposed to elderly animals. Over the years, several researchers have presented more refined sequences for both dental eruption and wear and long-bone fusion that allow for the reconstruction of much more accurate, detailed, and informative mortality patterns (Hesse 1982a; Kent 1982; Moore 1989; Wheeler 1999).

CONTEXTUAL INFORMATION

Different kinds of evidence can provide contextual information indicating the presence of domesticated animals, including corrals, dung layers, textiles, and art representation. Corrals and dung layers may be indicating practices of enclosing animals for particular management purposes (e.g., slaughtering, shearing, or marking). And in many cases, rock art or geoglyphs found in many localities throughout the Andean region show realistic depictions of several aligned animals led by a person or animals carrying goods, suggesting the representation of caravans. Although these indirect indicators can be ambiguous in some cases, they are still very important and should be considered when available in conjunction with direct indicators.

Recent Research in the South-Central Andes

As we have discussed, the picture of camelid domestication drawn to date has been based largely on research conducted in the Central Andes, in particular from the analyses and reanalyses of assemblages from several rock shelter sites in the Puna of Junín in central Peru. Together, this work has provided evidence of an in situ developmental trajectory in which specialized hunting of camelids developed into camelid management and domestication. It is important to ask, however, whether the identification of the Central Andes as the heartland of camelid domestication is an accurate characterization of this process, or an artifact of the intensive archaeological investigations and pioneering zooarchaeological analyses undertaken here.

Recent research outside this region in the South-Central Andes of southern Peru, northern Chile, and northwestern Argentina widens the lens of the investigation of South American camelid domestication, adding an important new perspective on the process and timing of camelid domestication in South America. Although early research in the South-Central Andes tended to see animal domestication as a secondary, and derivative, result of the onset of agriculture (e.g., Núñez 1974), research of the 1980s and 1990s focused on a growing understanding of the social and economic complexity among hunter-gatherer populations in the puna and the changing nature of camelid exploitation that accompanied these changes (e.g., Aschero 1984, 1994; Yacobaccio 1985, 1991, 2001; Aschero and Podestá 1986; Mengoni Goñalons 1986; Núñez 1992).

In particular, archaeological investigations in Chile, northwestern Argentina, Bolivia, and southern Peru have detected a process of increasing social and economic complexity among hunter-gatherer groups marked by decreasing residential mobility or even sedentism, complex burial patterns, prestige technology, and elaborate ceremonial structures. From 5300 BP onward, substantial sites with stone-made habitation structures appeared in the region (Núñez 1981). Some of them, like Tulán 52 and Puripica 1 in northern Chile, have between 20 to 40 circular structures interspersed with courtyards, covering a surface of about 400 to 540 m². Evidence of domestic activities was found in the structures and, in one case, storage pits; great quantities of mortars and pestles were found in the courtyards. The evidence from northwestern Argentina shows the inhumation of isolated human heads at Morro del Ciénego Chico or selected body parts at Inca Cueva 4, layer 1a, that marks the beginning of a practice associated with rising socioeconomic complexity and bounded territories (Yacobaccio 2000). Also, burials with rich offerings appear at high-altitude locations during this period, for example at Huachichocana III, layer E2. These offerings are generally long-distance trade items like Pacific Ocean shells, feathers from lowland birds such as guacamayo (*Ara militaris*), and psychotropic drugs (cebil, *Anadenanthera colubrina*) (Fernández Distel 1986). At Inca Cueva 7, an assemblage, dated to 4080 BP, included prestige technology such as pyro-engraved flutes, bone flutes, decorated bone spatulae, hardwood sticks decorated with geometric designs, pipes made of puma (*Felis concolor*) long bones, baskets, a host of textiles, and pyro-engraved domestic gourds (*Lagenaria sicearia*) (Aguerre et al. 1973). Ceremonial structures appear from levels IX to VIII (5000–4400 BP) at the Asana site in the highlands of southern Peru. Following Aldenderfer (1998), these structures are defined by prepared clay floors, altars, stone circles and ovals, trenches, clay-surfaced basins, surface hearths, miniature ovals, and circles of posts, although showing changes through time, suggesting that "the ceremony and the ritual that took place within them moving across a continuum from open and public in the earliest levels to close and private in level VIII times" (Aldenderfer 1998: 256). Together, these developments suggest the emergence of a hierarchical society, with increasingly more developed notions of territory, expanded trade contacts, more elaborate social structure, and ceremonial practice.

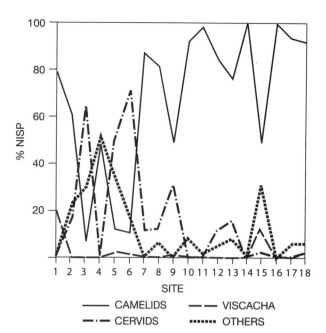

FIGURE 16.4 Temporal trends in the use of camelids for the South-Central Andes. 11,000–8500 BP: 1. Asana; 2. Tuina; 3. San Lorenzo; 4. Tambillo; 5. Pintoscayoc; 6. Inca Cueva-Cueva 4, Layer 2; 7. Huachichocana III, level E3; 8. Quebrada Seca 3, lower layers; 8500–5300 BP: 9. Hornillos 2; 10. Quebrada Seca 3, middle layers; 11. Chiu Chiu Cementerio; 5300–3000 BP: 12. Tulán 52; 13. Puripica 1; 14. Tomayoc; 15. Inca Cueva-Cueva 7; 16. Huachichocana III, level E2; 17. Quebrada Seca 3, upper layers; 18. Alero Unquillar. Viscacha (*Lagidium* sp.) is a medium-size rodent.

Against this backdrop of emergent social and economic complexity, the question of the trajectory of camelid domestication in the South-Central Andes becomes especially significant. Much of the more recent work on camelids in the South-Central Andes has been conducted by Latin American researchers publishing in venues not widely available outside the region. But this work provides multiple lines of evidence for tracing the process of camelid domestication and the later development of a pastoral economy based on camelids.

Intensification

As in the Central Andes, the zooarchaeological record in the South-Central Andes shows a long-term trend of intensification of camelid use that parallels the Central Andean pattern in degree and timing. The representation of camelids in a sample of 18 sites from southern Peru, northern Chile, and northwestern Argentina, ranging in age from 11,000 to 3000 BP, shows this pattern well (Figure 16.4, Table 16.2). Camelids average 48.9 percent of the identifiable remains from sites dating to the 11,000–8500 BP range (1–8), with a great deal of variability at each locality, perhaps showing a generalized, opportunistic strategy for obtaining animal resources. By 8500–5300 BP (9–11) camelids increase to 70.3 percent, with little variability in the profile of exploited species from site to site across this broad region. Camelids are almost always more than 85 percent of the assemblages from sites

TABLE 16.2
Archaeological Sites in the South-Central Andes

Site	Level	Country	Location	Elevation	Dates	Type	Reference	% Camelids[a]	% Small Camelids[b]
Asana	PXXXIII-PX	S Peru	Moquegua	3,400 m	9500–8000 BP	Logistical camp	Aldenderfer 1998	80%	na
Tuina 1	II–IV	N Chile	Loa	2,800 m	10,800–9000 BP	Temporary camp	Núñez 1983	61%	na
San Lorenzo 1	IV–IX	N Chile	Atacama	2,500 m	10,000 BP	Temporary camp	Núñez 1983	7%	na
Tambillo	—	N Chile	Atacama	2,300 m	9590–8590 BP	Base camp (?)	Hesse 1982a, 1982b	48%	na
Pintoscayoc	6	NW Argentina	Jujuy	3,650 m	10,700 BP	Temporary camp	Hernández Llosas 2000	10%	na
Inca-Cueva 4	2	NW Argentina	Jujuy	3,650 m	10,600–9200 BP	Base camp	Yacobaccio 1994	10%	presence
Huachichocana III	E3	NW Argentina	Jujuy	3,400 m	10,200–8600 BP	Temporary camp	Fernández Distel 1986	86%	0%
Quebrada Seca 3	Lower	NW Argentina	Catamarca	4,050 m	9050–8300 BP	Temporary camp	Elkin 1995	81%	44%
Hornillos 2	2	NW Argentina	Jujuy	4,020 m	6300 BP	Temporary camp	Yacobaccio et al. 2000	49%	na
Quebrada Seca 3	Middle	NW Argentina	Catamarca	4,050 m	8300–6160 BP	Temporary camp	Elkin 1995	92%	90%
Asana	IX–VIII	S Peru	Moquegua	3,400 m	4600 BP	Base camp	Aldenderfer 1998	na	na
Chiu Chiu Cementerio	—	N Chile	Atacama	2,300 m	4100 BP	Base camp	Cartajena 1994	98%	2.5%
Túlan 52	II–IV	N Chile	Atacama	3,200 m	4300 BP	Base camp	Hesse 1982a, 1982b	86%	32%
Puripica 1	II–IV	N Chile	Atacama	3,250 m	4500 BP	Base camp	Hesse 1982a, 1982b	76%	58%
Inca Cueva 7	EII	NW Argentina	Jujuy	3,600 m	4080 BP	Ceremonial	Aschero and Yacobaccio 1998–1999	50%	0%
Inca Cueva 7	EIII	NW Argentina	Jujuy	3,600 m	4030 BP	Corral	Aschero and Yacobaccio 1998–1999	—	—
Asana	III–I	S Peru	Moquegua	3,400 m	3640 BP	Base camp	Aldenderfer 1998	na	na
Huachichocana III	E2	NW Argentina	Jujuy	3,400 m	3400 BP	Burial	Fernández Distel 1986	100%	0%
Tomayoc	III	NW Argentina	Jujuy	4,170 m	3480–3250 BP	Temporary camp	Lavallée et al. 1997	100%	na
Quebrada Seca 3	Upper levels	NW Argentina	Catamarca	4,050 m	6160–4510 BP	Temporary camp	Elkin 1995	94%	99%
Alero Unquillar	1–2	NW Argentina	Jujuy	3,700 m	3500 BP	Transient camp	Yacobaccio et al. 1997	93%	0%
Casa Chavez Montículos	VIII–Vc	NW Argentina	Catamarca	3,600 m	2120 BP	Base camp	Olivera and Elkin 1994	89%	20%
Túlan 85	—	N Chile	Atacama	2,300 m	2600 BP	?	Dransart 1991a, 1991b	—	—
Huirunpure	E2	NW Argentina	Jujuy	4,020 m	2040 BP	Temporary camp	Yacobaccio et al. 1997	92%	50%

na = not available.
[a] Percentage of camelids in total faunal assemblage.
[b] Percentage of small camelids in camelid assemblage.

dating to 5300–3000 BP (12–18), reaching 100 percent of the archaeofaunas from some sites, while exploitation of other animal resources declines dramatically. Thus, as in the Central Andes, over several millennia of intensive interactions camelids become the overwhelmingly dominant animal resource in the South-Central Andes.

Osteometric Data

Excavations in two regions in northwestern Argentina have yielded important osteometric data that contribute to the emerging picture of camelid domestication in the South-Central Andes. These regions are the Puna of Jujuy and the Puna of Catamarca, where a number of excavated sites provide a record of camelid exploitation ranging from 10,000 years ago to 2000 BP.

Several caves and rock shelters were located in dry puna environments to the east and west of the Quebrada de Humahuaca in Jujuy at altitudes ranging from 3,400 to 4,020 m (Figure 16.1, Table 16.2). Some camelid bones larger than those of the present North Andean guanaco were found in the oldest layers dated between 10,000 and 7400 BP at Pintoscayoc, Inca Cueva 4, Huachichocana III, and at Quebrada Seca 3 (Yacobaccio 1991; Yacobaccio and Madero 1992; Elkin 1996; Rosenfeld 2002). These measurements were taken from fragmented first and second phalanges and metapodials. While regrettably too small a sample to be statistically significant, no indicator suggests we are dealing with an extinct species. Most probably, these specimens mark an upper size range for the guanaco during the Late Pleistocene–Early Holocene, a similar pattern observed for other species (Davis 1981).

As discussed above (see also Chapter 23), recent genetic studies have shown the vicuña and the guanaco as the wild ancestors of the alpaca and llama, respectively. This means that the two domesticated camelids in the Andes are currently larger than their progenitors. It is possible, then, that at some point during the process of domestication, camelids larger than present guanacos (i.e., llamas) appeared. There is mounting evidence for such a development in the South-Central Andes, as well as in the Central Andes (see below). In order to evaluate a possible trend in size change through time, we have summarized the metric data available.

In Figure 16.5, we have compiled all the metric information available for guanaco from northwestern Argentina and northern Chile. In constructing this figure, we have followed Meadow's (1999) log-ratio technique, in which individual measurements of archaeological specimens are compared to the same measurement from a known standard animal, in this case a North Andean guanaco. Those specimens that fall to the left of the axis are smaller than the standard, and those to the right are larger. The bars represent the absolute frequencies of each size category, where one score is one individual bone. For the period 11,000–8500 BP, sites included are Inca Cueva 4, Pintoscayoc, Huachichocana III, and Quebrada Seca 3. For the 8500–5300 BP period, we used data from Pintoscayoc and Quebrada Seca 3. For the period 5300–3000

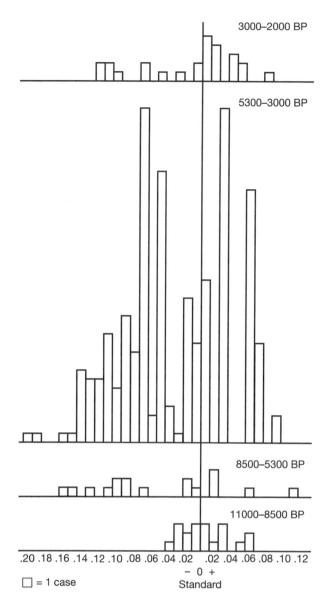

FIGURE 16.5 Histogram showing the log difference between measurements of modern North Andean guanaco and archaeological specimens from several sites located in the South-Central Andes.

BP, the data come from Tulán 52, Puripica 1, Inca Cueva 7, Alero Unquillar, and Quebrada Seca 3. In the last period, 3000–2000 BP, sites included are Huirunpure and Casa Chávez Montículos (see Table 16.2 for references).

During the Mid-Holocene, in northwestern Argentina and northern Chile, (8500–5300 BP), small camelids were dominant, while large camelids (likely guanacos) also were present. For this period, the existence of very few sites is associated with scanty metric information derived form relatively few bones.

In the next period (5300–3000 BP), information is derived from several sites and the samples are much larger. These samples show a wide range of variability and can be grouped into different size categories. On the left of the figure, there is group of small camelids that fall well apart from the guanaco standard. This group is here interpreted as vicuñas. This interpretation is supported by the identification of vicuña incisors at many of the sites. No alpaca teeth were identified in these samples. There can be no doubt that vicuñas occupied an important economic role as prey animal, with small camelids ranging between 32 to 99 percent of the camelid samples from sites dating to this time (see Table 16.2). A second size group is observed around the standard of the modern guanaco, suggesting that at this time guanacos had an average size similar to the size of the present ones. At most sites, both small and large camelids appear together in the same site layers. A third group, composed of samples found at sites both in Chile and in Argentina, is composed of individuals larger than the present guanaco. The appearance of a relatively large number of these large camelids at a number of sites both in northwestern Argentina and in northern Chile at this time has not been noted previously. The biggest animals identified in these samples belong to layers dated around 4400 BP. We believe that these large camelids probably represent the initial steps of llama domestication. As discussed below, this interpretation is supported by other indicators such as mortality patterns and contextual information (corrals and dung layers).

There are also changes in the relative dimension of some of the limbs of these larger camelids that suggest a change in the shape of these bones accompanies the increase in size. This feature is especially apparent in specimens from northwestern Argentinean sites of Inca Cueva 7 (IC7), Alero Unquillar (UNQ), and Huirunpure (HUI) dated 4100–2000 BP. In Figure 16.6, we present the data for three measurements of the distal metacarpal (maximum width of the distal end (mc6), maximum depth of the lateral condyle (mc9), and the maximum depth of the medial condyle (mc10)) from these three sites and compare them with a modern North Andean guanaco standard. In all but one of these archaeological large camelids (UNQ), the average depth of the metacarpal is comparatively greater than the wild standard (North Andean guanaco), and in all cases the width is proportionally smaller.

In sum, these data signal the appearance of a bigger form of camelid, larger than present guanaco and matching the size of current-day large llamas such as pack-llamas or *kcara*, which are the upper range for this species. These larger camelids were widely distributed across the South-Central Andes, from the highlands of northwestern Argentina to the Salar of Atacama in northern Chile from about 4400 BP onward. In another sector of the South-Central Andes, osteometric analysis on camelid distal humeri and proximal metatarsal widths detected the presence of large camelids, presumably llamas, at two rural archaeological sites located south of Lake Titicaca in Bolivia dating to about 3500 BP (Webster 1993).

Dental Morphology

At Tomayoc, in the Puna of Jujuy (4,170 m), two incisors identified as alpaca were found in layers dated to 3300–3200 BP (Lavallée et al. 1997). However, the criteria used to

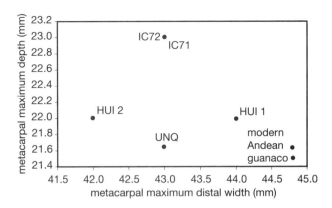

FIGURE 16.6 Bivariate plot of measurements of the distal metacarpal of selected large camelid specimens from northwestern Argentinean sites dated from 4100 to 2000 BP taken following Kent's protocols (Kent 1982). For the Andean guanaco, we present the measurements of two individuals from northwestern Argentina (Cumbres Calchaquíes and Nevados del Aconquija). Measurements selected are maximum width of the distal end, and the averaged maximum depth of the lateral and medial condyles: IC7 1 and 2, Inca Cueva 7; *UNQ*, Alero Unquillar; *HUI* 1 and 2, Huirunpure.

identify these teeth as alpaca were not reported and, as noted above, deciduous vicuña incisors and permanent alpaca incisors share several traits (see Table 16.1). Moreover, the presence of alpaca at this site seems rather unlikely, given environmental restrictions of the southern dry puna that would seem to preclude the keeping of alpaca. Until now, alpaca have not been recorded in assemblages from later periods and only mentioned in historical times.

Bone Morphology

Morphological characteristics for distinguishing between different camelids have been applied to several sites in the Loa River area, as well as in the Salar de Atacama of northern Chile (Benavente and Adaro 1991; Cartajena 1994; Cartajena and Concha 1997; Núñez et al. 1999). For example, at Chiu Chiu Cementerio, a residential site with stone constructions dated 4100 BP, guanaco, llama, and vicuña were identified on the basis of morphological indicators (Cartajena 1994). Large camelids outnumber small ones at this site (see Table 16.2).

Mortality Patterns

There is a great deal of variability in camelid mortality profiles assemblages in the South-Central Andes. Applying techniques developed to study livestock domestication in the Near East, Hesse combined osteometric analysis and mortality profiling to address the question of camelid domestication in the southern Andes (Hesse 1982a, 1982b, 1984, 1986). Osteometric analysis of camelid remains from the sites of Tulán 52 (3,200 m) and Puripica 1 (3,250 m) in the Salar de Atacama of Chile revealed two distinct populations of large and small animals. Large camelids made up

238 ARCHAEOLOGY AND ANIMAL DOMESTICATION

about 68% of the camelid sample from Túlan 52, and about 42% at Puripica 1. Mortality profiles of the larger camelids at Puripica 1 showed a heavy emphasis on young animals that Hesse interpreted as indicating the management of domestic llama by 4800–4300 BP. In contrast, mortality profiles of the large camelids at Tulán 52 indicated an emphasis on adult animals and thus seemed to reflect the activities of ancient hunters.

At Chiu Chiu Cementerio, where large camelids dominate, mortality patterns also point to an emphasis on adult animals (87.5% of the total). The great majority of all the camelids (small and large) are adult individuals (Cartajena 1994: 37), showing that at this critical period (4400–3500 BP) there is great variability in mortality profiles.

Mortality data from the long sequence at the site of Quebrada Seca 3, in northwest Argentina, do not provide any evidence of the development of management of the small camelids (presumably vicuña) that dominate the assemblage after 8300 BP (Elkin 1996). Both dental and long-bone fusion data were used to divide the camelid sample into two age classes: newborn (<1 yr) and juvenile/adults. Although the percentage of newborns changes from one layer to another (between 20 percent and 50 percent), there is no clear temporal trend over the 5,000-year occupation at the site. Thus, rather than a decline in the health of camelid herds resulting from a change in management strategy as suggested by Wheeler (1984a, 1984b, 1995, 1998) at Telarmachay, the shifting proportions of newborns in different layers at Quebrada Seca 3 probably represent variations in the seasonal occupation of the site and the opportunistic hunting of newborns during certain seasons of the year.

Fiber

Analysis of fiber remains found throughout the long sequence at Quebrada Seca 3 is also difficult to interpret. Reigadas (1992, 1994a, 1994b) identified both vicuña and guanaco fleece in almost all the levels. However, there were also samples of camelid fiber with characteristics analogous in color, diameter, and medullation to those of some contemporary llamas in levels dating to as early as 9100 BP. These samples showed similarities to an "intermediate llama type," a breed presently used by local herders for production of both meat and fiber (Lamas 1994). Fiber with similar characteristics to that recovered at Quebrada Seca 3 was also found in levels at Inca Cueva 4 in the Puna of Jujuy dating to 10,600–9200 BP. One possible explanation for the presence of these fibers at this early time is that they represent fleece types found among wild camelids (probably guanaco) that were later selected for in early domestic llama.

Analysis of yarns and fleeces from several sites in the Quebrada Tulán by Dransart (1991a, 1991b, 1999) points to the presence of stock at Tulán 54 with fleece characteristic of domestic camelids by 3100 BP, and increased use of domestic camelids by 2600 BP. At the base camp of Chiu Chiu Cementerio, fiber of vicuña, guanaco, and llama were identified (Cartajena 1994).

Contextual Indicators

Evidence of corrals and the penning of camelids can also be found at sites in the South-Central Andes. In the first occupation of Inca Cueva 7, a small cave located in the Argentine puna (dated to 4080–4030 BP), dung pellets cover the surface of the cave floor and a stone wall encloses the mouth of the cave (Aschero and Yacobaccio 1998–1999). At Asana, an open-air site located in southern Peru with layers dated to 3640 BP, dung-derived soil deposits are outlined by a series of post-molds that have been interpreted as forming the oldest open-air corral found in the Andes (Aldenderfer 1998). These two cases are the oldest evidence of enclosures for the entire high Andes.

Comparison of Camelid Exploitation in the Central and South-Central Andes

Taken together, these different lines of evidence point to a trajectory of intensification and domestication of camelids in the South-Central Andes taking place parallel to similar developments in the Central Andes. Beginning about 8400 BP, there was a region-wide intensification in the exploitation of camelids and a corresponding decrease in the exploitation of other species that peaked during the period 5300–3000 BP, when camelids are routinely 85–100 percent of faunal assemblages from the region. From 4400–2000 BP a large variety of camelids, larger than present guanacos, are found at sites across a broad region, including Late Archaic sites in the Salar de Atacama and the Puna of Argentina, as well as Early Formative sites at Lake Titicaca. We suggest that these large camelids represent a transitional form between hunted guanacos and herded llamas. Later on, these large forms seem to have undergone some reduction in the average size of individuals in the population, and an increase in overall metric variability.

Mortality data for large camelids from northern Chile and evidence for corralling camelids in both cave and open-air sites in northwest Argentina and southern Peru further indicate that these animals were managed. A picture then emerges of the development of a system of protective herding in the South-Central Andes, growing out of a gradual period of increasing intensification and specialization in hunting camelids that crystallizes with the domestication of the llama sometime between 4400 and 3000 BP. This process is set in the context of decreasing mobility of hunter-gatherer groups and corresponding increases in social, ideological, and economic complexity. The later part of this period (from 3000 to 2000 BP) was characterized by continued intensification in domestic camelid use (although wild camelids were still hunted), including the development of more specialized uses of camelids in textile production, associated with the appearance of highland agriculture and the incorporation of ceramic technology.

This pattern is strikingly similar to that seen in the Central Andes where the majority of indicators for camelid domestication converge somewhere between 4600 and 3000 BP.

For example, alpaca and llama are documented at the site of Pachamachay during the phase dated to 4150–3450 BP (Kent 1982). The long sequence at Panalauca shows the persistent importance of hunted vicuña until the onset of the Early Formative, ca. 3600 BP, when domesticated camelids were introduced (Moore 1989). Moore notes the presence of particularly large camelids in the assemblage in levels dating to between 4590 and 3570 BP, as well as a trend toward size increase that begins in early phases and increases in intensity between 5750 and 4590 BP. These large camelids are quite similar in size to those from sites in the South-Central Andes, which are interpreted here as llamas. Moore (1989) emphasizes the existence of a statistically significant size increase in bones of the lower hind limb, especially in the distal depth of the metacarpal, which is also a feature noted above in the large camelids from northwestern Argentina shown in Figure 16.6.

The only way in which the Central Andes sequence deviates from that emerging for the South-Central Andes is the apparent early appearance of domestic camelids at Telarmachay, where both a pattern of high neonatal mortality and the presence of alpaca incisors occurred in a phase dated to between 6000 and 5500 BP (Wheeler 1984a, 1984b, 1985, 1994, 1995). The disparity between the evidence for early camelid domestication at Telarmachay and the more delayed appearance of domestic forms at other sites in the Puna de Junín has been attributed recently to the persistence of camelid hunting and the presence of both hunter-gatherer and pastoral groups in this puna region (Lavallée 1995).

Although it is entirely likely that hunting of wild camelids continued well after initial domestication, it is important to note that the temporal framework for the development of camelid domestication in the Central Andes rests on a very different foundation from that in the South-Central Andes. The chronology of some of the sites in the South-Central Andes, which were excavated recently, is anchored to radiocarbon dates derived from materials found in closed contexts with camelid bones. Although direct dating of camelids, especially the large specimens, has not been performed, and although some of the criteria commonly used for accepting or rejecting these dates may not have been routinely applied, the overall chronological framework for these developments in the South-Central Andes is quite refined and secure. In contrast, the age and timing of the development of camelid domestication in the Central Andes is based on a much looser chronological framework of archaeologically defined cultural phases that, although taking into account radiocarbon dates, may span several centuries or even millennia, giving this temporal framework a low resolution. Thus, it is impossible to say precisely when events occurred within broad periods that may cover more than 1,000 years. Clearly, more refined radiocarbon dating techniques need to be applied to these older collections before arguments of temporal primacy can be advanced.

Thus, when data from the South-Central Andes are considered alongside those from the Central Andes, we see a much broader spatial context for the development of camelid domestication in South America, occurring within a possibly a much tighter temporal framework. In both regions, there are parallel developments including intensification of camelid exploitation, changes in culling practices, and efforts to restrict the movement of managed animals, with most of the data pointing to the period between 4600 and 4000 BP for the appearance of domesticated llamas. The wide geographic spread of this evidence, which comes from localities ranging over a vast geographical area 2,300 km long (ca. between lat 10° S and lat 26° S), raises the possibility that there were multiple centers of llama domestication across a vast region that includes the Central Peruvian Andes, as well as the South-Central Andes of southern Peru, northwest Argentina, western Bolivia, and northern Chile. We could argue further that the process of alpaca and llama domestication may have occurred independently at different times and places within the Andes.

Directions for Future Research

Only continued analysis of assemblages across this large geographic region will sort out the story of South American camelid domestication. Larger samples from sites that span the key period from 8500 to 4600 BP are needed. More systematic application of techniques of osteometric analysis is essential. In particular, it is critical that analysts working with this material recognize the need for regional comparability in developing modern standards and in drawing comparisons between archaeological assemblages. Application of more refined techniques of mortality profiling, especially those that combine osteometric data with age data, are also key to tracing the shifts in exploitation strategies that accompany the transition from hunting to herding of different camelid species. Finally, chronological placement of these developments requires direct radiocarbon dating of camelid remains from these sites.

The process of South American camelid domestication, involving multiple species spread over a large and environmentally varied area, is clearly complex and difficult to monitor archaeologically. Recent work in the South-Central Andes has succeeded in broadening the focus of the inquiry from its initial, narrower concentration on Central Peru. Continued refinement of the pioneering methods developed by researchers working in both the Central and South-Central Andes promises a more detailed and refined picture of this complex process in the future.

Acknowledgments

We are extremely grateful to María José Figuerero Torres for her careful reading and thoughtful editing of our first manuscript as well as for her assistance in the construction of the bibliographical data base. We also appreciate the valuable comments made by Bibiana Vilá. Both contributed to improve and clarify our ideas, although we alone are responsible for the opinions here presented. Andrés Izeta generously supplied us with the measurements of one of the individuals of Andean guanaco we used as a standard. Cristian Kaufmann also provided us with selected measurements he took from guanacos of northern Patagonia.

References

Adaro, L. and M. A. Benavente. 1990a. Identificación de patrones óseos de camélidos sudamericanos. *Avances en Ciencias Veterinarias* 5: 9–86.

———. 1990b. Determinación de indicadores óseos de camélidos sudamericanos. *Actas del VIII Congreso de Medicina Veterinaria (Valdivia)*: 198–200.

Aguerre, A. M., A. A. Fernández Distel, and C. A. Aschero. 1973. Hallazgo de un sitio acerámico en la quebrada de Inca Cueva (Provincia de Jujuy). *Relaciones de la Sociedad Argentina de Antropología* 7: 197–235.

Aldenderfer, M. S. 1998. *Montane foragers: Asana and the South-Central Andean archaic*. Iowa City: University of Iowa Press.

Aschero, C. A. 1984. El sitio ICC4: Un asentamiento precerámico en la Quebrada de Inca Cueva (Jujuy, Argentina). *Estudios Atacameños* 7: 62–72.

———. 1994. Reflexiones desde el Arcaico Tardío (6000–3000 AP). *Rumitacana. Revista de Antropología* 1: 13–17.

Aschero, C. A. and M. Podestá. 1986. El arte rupestre en asentamientos precerámicos de la puna Argentina. *RUNA* 16: 29–57.

Aschero, C. A. and H. D. Yacobaccio. 1998–1999. 20 años después: Inca Cueva 7 reinterpretado. *Cuadernos del Instituto Nacional de Antropología y Pensamiento Latinoamericano* 18: 7–18.

Benavente, A. 1997–1998. Determinación de especies animales en la arqueología: Un enfoque zooarqueológico. *Revista Chilena de Antropología* 14: 105–112.

Benavente, M. A. and L. Adaro. 1991. Selección de algunos indicadores óseos actuales de la familia camelidae sudamericana y su contraste con muestras arqueológicas. *Revista Chilena de Anatomía* 9: 134.

Benavente, M. A., L. Adaro, P. Gecele, and C. Cunazza. 1993. *Contribución a la determinación de especies animales en arqueología: Familia camelidae y taruca del norte*. Santiago: Universidad de Chile-Departamento Técnico de Investigación.

Bonavía, D. 1996. *Los camélidos sudamericanos: Una introducción a su estudio*. Lima: IFEA-UPCH-Conservation International.

———. 1999. The domestication of Andean camelids. In *Archaeology in Latin America*, G. G. Politis and B. Alberti (eds.), pp. 130–147. London: Routledge.

Browman, D. L. 1974. Pastoral nomadism in the Andes. *Current Anthropology* 15: 188–196.

———. 1989. Origins and development of Andean pastoralism: An overview of the past 6,000 years. In *The walking larder: Patterns of domestication, pastoralism, and predation*, J. Clutton-Brock (ed.), pp. 256–268. London: Unwin Hyman.

Cajal, J. L. 1985. Origen, evolución y nomenclatura. In *Estado actual de las investigaciones sobre camélidos en la República Argentina*, J. L. Cajal and J. N. Amaya (eds.), pp. 5–19. Buenos Aires: Secretaría de Ciencia y Técnica.

Calle Escobar, R. 1984. *Animal breeding and production of American camelids*. Lima: Talleres Gráficos de Abril.

Cardozo, G. A. 1954. *Los auquénidos*. La Paz: Editorial Centenario.

Cartajena, I. 1994. Determinación de restos óseos de camélidos en dos yacimientos del Loa Medio (II Región). *Estudios Atacameños* 11: 25–52.

Cartajena, I. and I. Concha. 1997. Una contribución a la determinación taxonómica de la familia Camelidae en sitios Formativos del Loa Medio. *Estudios Atacameños* 14: 71–84.

Cartajena, I., M. A. Benavente, P. Gecele, I. Concha, and J. M. Benavente. 2001. The transport function in camelids: An archaeozoological approach. In *Progress in South American camelid research*, M. Gerken and C. Renieri (eds.), pp. 159–165. Wageningen: Wageningen Press.

Cunazza, P., C. S. Puig, and L. Villalba. 1995. Situación actual del guanaco y su ambiente. In *Técnicas para el manejo del guanaco*, S. Puig (ed.), pp. 27–50. Gland: UICN.

Davis, S. J. M. 1981. The effects of temperature change and domestication on the body size of Late Pleistocene to Holocene mammals of Israel. *Paleobiology* 7: 101–114.

Dransart, P. Z. 1991a. Fibre to fabric: The role of fibre in camelid economies in prehispanic and contemporary Chile. PhD dissertation, Linacre College, Oxford University, Oxford.

———. 1991b. Llamas, herders and the exploitation of raw materials in the Atacama Desert. *World Archaeology* 22: 304–319.

———. 1999. La domesticación de los camélidos en los Andes centro-sur: Una consideración. *Relaciones de la Sociedad Argentina de Antropología* 24: 125–138.

Ducos, P. 1978. "Domestication" defined and methodological approaches to its recognition in faunal assemblages. In *Approaches to faunal analysis in the Middle East*, R. H. Meadow and M. A. Zeder (eds.), pp. 53–68. Peabody Museum Bulletin 2. Cambridge, Mass.: Peabody Museum.

Ducos, P. and L. R. Horwitz. 1997. The influence of climate on artiodactyl size during the Late Pleistocene-Early Holocene of the Southern Levant. *Paléorient* 23: 229–247.

Elkin, D. C. 1995. El uso del recurso fauna por los primeros habitantes de Antofagasta de la Sierra (Puna de Catamarca). *Actas del I Congreso de Investigación Social*: 202–209.

———. 1996. Arqueozoología de Quebrada Seca 3: Indicadores de subsistencia humana temprana en la Puna Meridional argentina. PhD dissertation, Facultad de Filosofía y Letras, UBA, Buenos Aires.

Fernández Distel, A. A. 1986. Las Cuevas de Huachichocana, su posición dentro del precerámico con agricultura incipiente del Noroeste Argentino. *Beiträge zur Allgemeinen und Vergleichenden Archäologie* 8: 353–430.

Flores Ochoa, J. A. 1981. Clasificación y nominación de camélidos sudamericanos. In *La tecnología en el mundo andino*, H. Lechtman and A. M. Soldi (eds.), pp. 195–215. Mexico D. F.: Universidad Nacional Autónoma de México.

Franklin, W. L. 1982. Biology, ecology, and relationship to man of the South American Camelids. In *Mammalian biology in South America*, M. A. Mares and H. H. Genoways (eds.), pp. 457–489. Pittsburg: University of Pittsburg.

———. 1983. Contrasting socioecologies of South America's wild camelids: The vicuña and the guanaco. In *Advances in the study of mammalian behavior*, J. F. Eisemberg and D. G. Kleiman (eds.), pp. 573–629. The American Society of Mammalogists, Special Publication No. 1.

Hernández Llosas, M. I. 2000. Quebradas Altas de Humahuaca a travès del tiempo: El caso Pintoscayoc. *Estudios Sociales del NOA*, Año 4, No. 2: 167–224. Tilcara.

Hesse, B. 1982a. Animal domestication and oscillating climates. *Journal of Ethnobiology* 2: 1–15.

———. 1982b. Archaeological evidence for camelid exploitation in the Chilean Andes. *Säugetierkundliche Mitteilungen* 30: 201–11.

———. 1984. Archaic exploitation of small mammals and birds in northern Chile. *Estudios Atacameños* 7: 42–61.

———. 1986. Buffer resources and animal domestication in prehistoric northern Chile. *ArchaeoZoologica* Mélanges: 73–85.

Kadwell, M., M. Fernandez, H. F. Stanley, J. C. Wheeler, R. Rosadio, and M. W. Bruford. 2001. Genetic analysis reveals the wild ancestors of the llama and alpaca. *Proceedings of the Royal Society of London B*. 268: 2575–2584.

Kent, J. D. 1982. The domestication and exploitation of the South American camelids: Methods of analysis and their application to circum-lacustrine archaeological sites in Bolivia and Peru. PhD dissertation, Department of Anthropology, Washington University, St. Louis. Ann Arbor, MI: University Microfilms.

———. 1987. The most ancient south: A review of the domestication of the Andean camelids. In *Studies in the Neolithic and urban revolutions*, L. Manzanilla (ed.), pp. 169–184. Oxford: BAR International Series 349.

Lamas, H. E. 1994. Avances en la caracterización y diferenciación en la morfología y morfometría de los camélidos domésticos en un sector del altiplano argentino. In *Zooarqueología de camélidos*, D. C. Elkin, C. Madero, G. L. Mengoni Goñalons, D. E. Olivera, M. C. Reigadas, and H. D. Yacobaccio (eds.), 1: 57–72. Buenos Aires: Grupo Zooarqueología de Camélidos.

Larrieu, E., N. R. Oporto, and R. Bigatti. 1979. Avances en estudios reproductivos en guanacos de Río Negro (Argentina). *Revista Argentina de Producción Animal 3*: 134–149.

Lavallée, D. 1990. La domestication animale en Amérique du Sud. *Bulletin Institute Française d´Études Andines* 19: 25–44.

———. 1995. *Promesse d'Amérique. La préhistoire de l'Amérique du Sud*. Paris: Hachette.

Lavallée, D., M. Julien, C. Karlin, L. C. García, D. Pozzi-Escot, and M. Fontugne. 1997. Entre desiertos y quebrada: Primeros resultados de las excavaciones realizadas en el abrigo de Tomayoc (Puna de Jujuy, Argentina). Bulletin Institute Française d´Études Andines 26: 141–175.

Lynch, T. F. 1967. *The nature of the Central Andean Preceramic*. Occasional Papers of the Idaho State Museum, Volume 21. Pocatello: Idaho State Museum.

———. 1980. *Guitarrero Cave*. New York: Academic Press.

Meadow, R. H. 1999. The use of size index scaling techniques for research on archaeozoological collections from the Middle East. In *Historia Animalium ex Ossibus, Festschrift für Angela von den Driesch*, C. Becker, H. Manhart, J. Peters, and J. Schibler (eds.), pp. 285–300. Beiträge zur Paläoanatomie, Archäologie, Ägyptologie, Ethnologie und Geschichte der Tiermedizin. Rahden/Westf: Verlag Maire leidorf GmbH.

Mengoni Goñalons, G. L. 1986. Vizcacha *(Lagidium viscacia)* and taruca *(Hippocamelus* sp.) in early south andean economies. *ArchaeoZoologica* Mélanges: 63–71.

Mengoni Goñalons, G. L. and D. C. Elkin. 1990. Camelid zooarchaeological research in Argentina: Present status and perspectives. Paper read at Sixth International Conference of the International Council for Archaeozoology (ICAZ), Washington, D.C.

Miller, G. R. 1979. An introduction to the ethnoarchaeology of the Andean camelids. PhD dissertation, University of California–Berkeley. Ann Arbor, Mich.: University Microfilms.

Miller, G. R. and R. L. Burger. 1995. Our father the cayman, our dinner the llama: Animal utilization at Chavín de Huantar, Peru. *American Antiquity* 60: 421–458.

Miller, G. R., and A. R. Gill. 1990. Zooarchaeology at Pirincay: A Formative period site in highland Ecuador. *Journal of Field Archaeology* 17: 49–68.

Moore, K. M. 1989. Hunting and the origins of herding in Peru. PhD dissertation, University of Michigan, Ann Arbor. Ann Arbor, Mich.: University Microfilms

Novoa, C. and J. C. Wheeler. 1984. Lama and alpaca. In *Evolution of domesticated animals*, I. L. Mason (ed.). London: Longman.

Núñez, L. 1974. *La agricultura prehistórica en los Andes Meridionales*. Santiago: Editorial Orbe.

———. 1981. Asentamientos de cazadores tardíos de la Puna de Atacama: Hacia el sedentarismo. *Chungara* 8: 137–168.

———. 1983. *Paleoindio y Arcaico en Chile: Diversidad, secuencia y procesos*. México: Ediciones Cuicuilco.

———. 1992. Ocupación arcaica en la Puna de Atacama: Secuencia, movilidad y cambio. In *Prehistoria sudamericana—Nuevas perspectivas*, B. J. Meggers (ed.). Washington, D.C.: Taraxacum.

Núñez, L. and T. D. Dillehay. 1979. *Movilidad giratoria, armonía y desarrollo en los Andes Meridionales, patrones de tráfico e interacción económica*. Antofagasta: Dirección General de Investigaciones Tecnológicas. Universidad del Norte.

Núñez, L., M. Grosjean, and I. Cartajena. 1999. Un ecorefugio oportunístico en la puna de Atacama durante eventos áridos del Holoceno Medio. *Estudios Atacameños* 17: 125–174.

Olivera, D. E. and D. C. Elkin. 1994. De cazadores y pastores: El proceso de domesticación de camélidos en la Puna Meridional Argentina. In *Zooarqueología de Camélidos*, D. C. Elkin, C. Madero, G. L. Mengoni Goñalons, D. E. Olivera, M. C. Reigadas, and H. D. Yacobaccio (eds.), 1: 95–124. Buenos Aires: Grupo Zooarqueología de Camélidos.

Otte, K. C. and J. L. Venero. 1979. Análisis de la craneometría diferencial entre la vicuña *(Vicugna vicugna)* y la alpaca *(Lama guanicoe pacos)*. *Studies on Neotropical Fauna and Environment* 14: 125–152.

Price, E. O. 1984. Behavioural aspects of animal domestication. *Quarterly Review of Biology* 59: 1–32.

Puig, S. 1988. Craneología y craneometría de camélidos: Diferenciación interespecífica y determinación de la edad. *Xama* 1: 43–56.

Puig, S. and J. L. Cajal. 1985. Descripción general, craneometría y dentición de los camélidos. In *Estado actual de las investigaciones sobre camélidos en la República Argentina*, J. L. Cajal and J. N. Amaya (eds.), pp. 53–63. Buenos Aires: Secretaría de Ciencia y Técnica.

Rabinovich, J. E., J. L. Cajal, M. J. Hernández, S. Puig, R. Ojeda, and J. N. Amaya. 1984. *Un modelo de simulación en computadoras digitales para el manejo de vicuñas y guanacos en Sudamérica*. Buenos Aires: Secretaría de Ciencia y Técnica.

Raedeke, J. K. 1978. *El guanaco de Magallanes, Chile*. Distribución y biología. Santiago: CONAF.

———. 1979. Population dynamics and socioecology of the guanaco *(Lama guanicoe)* of Magallanes, Chile. PhD dissertation, University of Washington. Ann Arbor, Mich.: University Microfilms.

Reigadas, M. C. 1992. La punta del ovillo: Determinación de domesticación y pastoreo a partir del análisis microscopio de fibras y folículos pilosos de camélidos. *Arqueología* 2: 9–52.

———. 1994a. Caracterización de tipos de camélidos domésticos actuales para el estudio de fibras arqueológicas en tiempos de transición y consolidación de la domesticación animal.

In *Zooarqueología de Camélidos*, D. C. Elkin, C. Madero, G. L. Mengoni Goñalons, D. E. Olivera, M. C. Reigadas, and H. D. Yacobaccio (eds.), 1: 125–154. Buenos Aires: Grupo Zooarqueología de Camélidos.

———. 1994b. Incidencia de los factores de variación en las especies de camélidos y tipos domésticos especializados en el NOA. Un paso más allá de la taxonomía en la explicación del proceso de domesticación. *Estudios Atacameños* 11: 53–72.

Rick, J. W. 1980. *Prehistoric hunters of the high Andes*. New York: Academic Press.

Rivera, M. 1980. La agriculturación del maíz en el norte de Chile: Actualización de problemas y metodología de investigación. *Actas del V Congreso Nacional de Arqueología Argentina* 1: 157–180.

Riviere, H. L., E. J. Gentz, and K. I. Timm. 1997. Presence of enamel on the incisors of the llama *(Lama glama)* and alpaca *(Lama pacos)*. *Anatomical Record* 249: 441–449.

Rosenfeld, S. A. 2002. Análisis zooarqueológico de Pintoscayoc 1 (Jujuy, Argentina): Explotación faunística y procesos de formación de sitio en las tierras altas andinas durante el Holoceno Temprano. Licenciate thesis, University of Buenos Aires.

Stahl, P. W. 1988. Prehistoric camelids in the lowlands of western Ecuador. *Journal of Archaeological Science* 15: 355–65.

Stanley, H., F. M. Kadwell, and J. C. Wheeler. 1994. Molecular evolution of the family Camelidae-A mitochondrial DNA study. *Proceedings of the Royal Society, London B*. 256: 1–6.

Vilá, B. 2000. Comportamiento y organización social de la vicuña. In *Manejo sustentable de la vicuña y el guanaco*, B. González, F. Bas, C. Tala, and A. Iriarte (eds.), pp. 175–192. Santiago: Servicio Agrícola y Ganadero-Pontificia Universidad Católica de Chile-Fundación para la Innovación Agraria.

Webster, A. D. 1993. Camelids and the rise of the Tiwanaku state. PhD dissertation, University of Chicago. Ann Arbor, Mich.: University Microfilms.

Wheeler, J. C. 1982. Aging llamas and alpacas by their teeth. *Llama World* 1: 12–17.

———. 1984a. La domesticación de la alpaca (*Lama pacos* L.) y la llama (*Lama glama* L.) y el desarrollo temprano de la ganadería autóctona en los Andes Centrales. *Boletín de Lima* 36: 74–84.

———. 1984b. On the origin and early development of camelid pastoralism in the Andes. In *Animals and archaeology 3: Early herders and their flocks*, J. Clutton-Brock and C. Grigson (eds.), pp. 395–410. BAR International Series, 202. Oxford: BAR International Series.

———. 1985. De la chasse a l'elevage. In *Telarmachay: Chasseurs et pasteurs préhistoriques des Andes*, D. Lavallée, M. Julien, J. C. Wheeler, and C. Karlin (eds.), 1: 61–79. Paris: Éditions Recherches sur les Civilisations, ADPF.

———. 1991. Origen, evolución y status actual. In *Avances y perspectivas del conocimiento de los camélidos sudamericanos*, S. Fernández Baca (ed.), pp. 11–48. Santiago: Oficina regional de la FAO para América Latina y el Caribe.

———. 1994. The domestic South American Camelidae: Past, present and future. In *European symposium on South American camelids*, M. Gerken and C. Renieri (eds.), pp. 13–28. Camerino: Universita degli Studi di Camerino.

———. 1995. Evolution and present situation of the South American Camelidae. *Biological Journal of the Linnean Society* 54: 271–295.

———. 1996. El estudio de restos momificados de alpacas y llamas precolombinas. In *Zooarqueología de camélidos*, D. C. Elkin, C. Madero, G. L. Mengoni Goñalons, D. E. Olivera, M. C. Reigadas, and H. D. Yacobaccio (eds.), 2: 75–84. Buenos Aires: Grupo Zooarqueología de Camélidos.

———. 1998. Evolution and origin of the domesticated camelids. *Alpaca Registry Journal* 3: 1–16.

———. 1999. Patrones prehistóricos de utilización de los camélidos sudamericanos. *Boletín de Arqueología PUCP* 3: 297–305.

Wheeler, J. C. and E. J. Reitz. 1987. Allometric prediction of live weight in the alpaca (*Lama pacos* L.). *Archaeozoologia* 1: 31–46.

Wheeler, J. C., C. R. Cardoza, and D. Pozzi-Escot. 1977. Estudio provisional de la fauna de las capas II y III de Telarmachay. *Revista Nacional de Lima* 43: 97–102.

Wheeler, J. C., A. J. F. Russel, and H. Redden. 1995. Llamas and alpacas: Pre-conquest breeds and post-conquest hybrids. *Journal of Archaeological Science* 22: 833–840.

Wheeler, J. C., A. J. F. Russel, and H. F. Stanley. 1992. A measure of loss: Prehispanic llama and alpaca breeds. *Archivos de Zootecnia (Córdoba)* 41: 467–475.

Wheeler Pires-Ferreira, J. C. 1975. La fauna de Cuchimachay, Acomachay A, Acomachay B, Telarmachay y Utco 1. *Revista del Museo Nacional* 41: 120–127.

Wheeler Pires-Ferreira, J. C., E. Pires-Ferreira, and P. Kaulicke. 1976. Preceramic animal utilization in the Central Peruvian Andes. *Science* 194: 483–490.

Wing, E. S. 1972. Utilization of animal resources in the Peruvian Andes. In *Andes 4: Excavations at Kotosh, Peru, 1963 and 1964*, I. Seiichi and K. Terada (eds.), pp. 327–351. Tokyo: University of Tokyo Press.

———. 1975a. Hunting and herding in the Peruvian Andes. In *Archaeozoological studies*, A. T. Clason (ed.). Amsterdam: North Holland Press.

———. 1975b. La domesticación de animales en los Andes. *Allpanchis Phuturinqa* 8: 25–44.

———. 1975c. Informe preliminar acerca de los restos de fauna de la cueva de Pachamachay en Junín, Perú. *Revista del Museo Nacional* 41: 79–80.

———. 1977a. Animal domestication in the Andes. In *Origins of agriculture*, C. A. Reed (ed.), pp. 837–859. The Hague: Mouton.

———. 1977b. Caza y pastoreo tradicionales en los Andes Peruanos. In *Pastores de puna. Uywasmichiq punarunakuna*, J. F. Ochoa (ed.), pp. 121–130. Lima: Instituto de Estudios Peruanos.

———. 1978. Animal domestication in the Andes. In *Advances in Andean archaeology*, D. L. Browman (ed.), pp. 167–188. The Hague: Mouton.

———. 1980. Faunal remains. In *Guitarrero Cave. Early man in the Andes*, T. F. Lynch (ed.). New York: Academic Press.

———. 1986. Domestication of Andean mammals. In *High altitude tropical biogeography*, F. Viulleumier and M. Monasterio (eds.), pp. 246–264. New York: Oxford University Press.

———. 1989. Human use of canids in the Central Andes. In *Advances in Neotropical mammalogy*, J. Eisenberg and K. Redford (eds.), pp. 265–278. Gainesville: Sandhill Crane Press.

Yacobaccio, H. D. 1985. Almacenamiento y adaptación en el Precerámico andino. *Runa* 25: 117–131.

———. 1991. Sistemas de asentamiento de cazadores-recolectores tempranos en los Andes Centro-Sur. PhD dissertation, University of Buenos Aires.

———. 1994. Hilos conductores y nudos gordianos: Problemas y perspectivas en la arqueología de cazadores-recolectores puneños. Rumitacana. *Revista de Antropología* 1: 19–21.

———. 2000. Inhumación de una cabeza aislada en la Puna argentina. *Estudios Sociales del NOA* 4: 59–72.

———. 2001. La domesticación de camélidos en el noroeste argentino. In *Historia Argentina Prehispánica*, E. E. Berberían and A. E. Nielsen (eds.), pp. 7–40. Córdoba: Editorial Brujas.

Yacobaccio, H. D. and C. M. Madero. 1992. Zooarqueología de Huachichocana III (Prov. de Jujuy, Argentina). *Arqueología* 2: 149–188.

Yacobaccio, H. D., C. M. Madero, M. P. Malmierca, and M. C. Reigadas. 1997. Isótopos estables, dieta y orígenes del pastoreo. *Arqueología* 7: 105–109.

Yacobaccio, H. D., M. Lazzari, G. Guráieb, and G. Ibañez. 2000. Los cazadores en el borde oriental de la puna. *Arqueología* 10: 11–38.

Zeder, M. A. 2001. A metrical analysis of a collection of modern goats *(Capra hircus aegagrus* and *Capra hircus hircus)* from Iran and Iraq: Implications for the study of caprine domestication. *Journal of Archaeological Science* 28: 61–79.

CHAPTER 17

Early Horse Domestication on the Eurasian Steppe

SANDRA L. OLSEN

Introduction

The questions of when, where, and why horses were first domesticated are still hotly debated. Textbooks have handily marked the location of initial horse domestication as Dereivka (Telegin 1986), a Copper Age settlement in Ukraine, dating to between 4470 and 3530 BC (Rassamakin 1999: 162–163). However, the so-called cult stallion of Dereivka, with teeth heavily worn from clutching a metal bit, was recently radiocarbon dated directly and found to be an Iron Age intrusion from the first millennium BC (Anthony and Brown 2000). When the Dereivka stallion was removed from the record as the first individual classified as a domestic horse, both the timing and location of this event were suddenly reopened for debate.

Molecular studies (Vilà et al. 2001, and Chapter 23) now suggest that it is fruitless to seek a single point of origin, since their data indicate multiple successful efforts at horse domestication in different regions. During the Pleistocene and Holocene, both body size and cranial morphology varied in wild horse populations, depending in part on the region, climate, and ecozone. As a result, variation also should be expected in early domesticates across the broad zone where the process probably occurred. If the mitochondrial DNA studies are correct, the lack of distinctive diagnostic traits in the early stages of domestication may be related, as well, to frequent crossbreeding between domestic and wild stallions and mares.

Beyond the possibility of multiple independent occurrences, there are a number of other reasons why it has proven so difficult to document the process of horse domestication. Unlike dogs, cattle, sheep, goats, and pigs, it is unclear whether the bones of early domestic horses ever developed reliable and consistent morphological or size differences that would allow them to be distinguished from their wild progenitors. Horses are not strongly sexually dimorphic and lack horns, like cattle, and hypertrophied canines, like pigs, which were reduced through the process of domestication.

Documenting early horse domestication is made even more difficult by the poor understanding of the geographic distribution of possible wild progenitors and the complexities of mortality patterns in domestic horse populations. An examination of previous research highlights a number of the problems that make the archaeological documentation of horse domestication particularly difficult. This review is followed by a consideration of the evidence for horse domestication in northern Kazakhstan that serves as a model for future research.

Possible Wild Progenitors

The first problem that confronts researchers seeking to document domestication in horses is determining the wild progenitor of the domestic horse. The abundant fossil record of European Pleistocene horses would seem to provide an excellent data set for identifying the ancestor of the domestic horse. However, there were many types of caballine equids (horses and their ancestors, as opposed to zebrines, asinines, and hemionines) in the Pleistocene, including both small and large forms (Forsten 1988). Some are clearly separate species, but there is considerable intraspecific and individual variation as well. For example, great diversity has been noted among the Late Pleistocene Magdalenian populations (Skorkowski 1956), and it is unclear whether this simply expresses individual characteristics or reflects sympatric subspecies (Meunier 1962).

In the past, many scholars believed that different wild equids gave rise to different breeds of domestic horses. The enormous draft horses, for example, were thought to have had their origins in northern Europe, where species such as *Equus germanicus* lived in the Pleistocene. However, it is now thought that all the Pleistocene heavy horses became extinct by the end of the Ice Age (Epstein 1971: 417). Littauer (1963) has also shown that the titans that used to pull heavy farm wagons were a late development, and that most of the so-called great horses used to carry the cumbersome medieval armor in the fourteenth and fifteenth centuries were, although sturdy, only about 13–15 hands (130–150 cm) tall.

One of the best summaries of Pleistocene and Early Holocene horses is provided by Forsten (1988), who argued that, of all the different varieties of horses in the Pleistocene, only one relatively small species of wild horse survived into the Holocene. In her opinion, the domestic horse emerged from that single, small progenitor, which many refer to as *Equus ferus* (Nobis 1971). Current researchers generally accept this thesis and look toward a single species with a broad geographic distribution as the wild progenitor of the domestic horse, instead of toward multiple lineages. As noted above, however, it is likely that domestication occurred independently more than once within that species' range. Two relatively modern species of caballine equids are normally used for comparison with possible early domesticates: the tarpan, or wild European horse, and the Przewalski, or wild Asiatic horse. The tarpan (*Equus ferus*) was recorded in historic times in northern Germany, Lithuania, Poland, Ukraine, and western Russia, but may have been much more widespread in the past. It was described in numerous references from the eighteenth and nineteenth centuries as a small animal, having a mouse-dun coat with a light underbelly, sooty to black limbs from the knees and hocks down, a short, frizzled mane, and a tail with short dock hair. In fact, in most features the tarpan was very similar in appearance to the Przewalski horse (*E. przewalskii*), except that the coat was grayer and apparently turned very light in the winter. Russian zoologist Heptner reported that the last tarpan died in 1918 or 1919 in captivity (Bökönyi 1974a: 71). In the wild, the last tarpans were killed in Ukraine in 1851 (Zeuner 1963: 305).

The only available skeletal material attributed to the tarpan consists of one complete skeleton and a cranium lacking a mandible from another individual, housed in the Zoological Institute in St. Petersburg, Russia. There are so many accounts of tarpans stealing domestic mares and forming harems (Bökönyi 1974a: 72), however, that recent specimens should not be considered genetically pure. Many populations attributed to the tarpan in historic accounts may simply have been feral horses or hybrids. The erect mane tends to be the most reliable "wild" characteristic, but hybrids can also have fairly short manes.

The wild Asiatic or Przewalski horse is much better known, with extant populations now under human control (Mohr 1971; Bökönyi 1974a; Boyd and Houpt 1994). Existing Przewalski horses were originally derived from 12 individuals from the Djungar Basin, a barren desert of western Mongolia and China, but interbreeding with domestic horses in captivity has tainted their gene pool to some degree. Like the tarpan, Przewalski stallions were notorious for stealing mares from Mongol horse herds. The modern herds of Mongolian horses also suggest that hybridization occurred in the other direction as well, perhaps through the capture and adoption of Przewalski horses in the past. Horses with dun coats, zebra striping on the legs, and prominent eel stripes down their backs are very common today in domestic herds in Mongolia. Their artificially clipped manes give them an even stronger resemblance to the wild form.

Today, few would claim that the Przewalski horse is the ancestor of the domestic horse. Although Przewalski horses can interbreed with domestic individuals and produce viable offspring, they have a different number of chromosomes (in *Equus przewalskii*, $2n = 66$ and in *E. caballus*, $2n = 64$, and in hybrid *E. przewalskii* x *E. caballus*, $2n = 65$), reflecting what is known in genetics as a Robertsonian translocation (Ryder 1994). Under these circumstances, two chromosomes break and their two long arms fuse to form a new, large chromosome containing essential genes, while the two short arms, which apparently do not contain significant genes, either fuse with each other or are lost without joining (Mange and Mange 1980). Although there are examples of this type of chromosomal variation at the conspecific level in mammalian taxa, it usually occurs between two closely related species. In light of this evidence, most experts would agree that the Przewalski is a sister species to the ancestor of the domestic horse. Recent mtDNA research supports this conclusion (Vilà et al. 2001). The tarpan, as problematic as it is, is still the best candidate for the ancestor of the domestic horse.

Prior Research

There is a long history of efforts to document the domestication of the horse in the archaeological record, using a wide range of conventional and unconventional criteria. Some of the most significant lines of research are discussed below.

Cranial Morphology

The classification of the crania of Late Pleistocene and Holocene wild horses is quite problematic. Different authors classify the fossil and subfossil remains in different ways, and each has his or her own interpretation of which group of fossil horses gave rise to domestic horses (see Epstein 1971 for a comprehensive summary of the early attempts to classify horse crania).

Some of the first work on cranial morphology of recent wild horses was done by Duerst (1908), Ewart (1907), and Antonius (1913). Whereas these researchers all classified horses into groups that were assigned environmental terms (e.g., steppe, desert or plateau, and forest), the basis for grouping was actually cranial morphology rather than habitat. There was no general consensus regarding groupings of these horse remains, and often more than one type could be found in the same locality (Epstein 1971). Some groups were merely hypothetical (see Groves 1986) and lacked any actual specimens. The implication was still quite strong, however, that local populations diverged because of their unique environment and climate.

Undoubtedly, difficulties in the systematic classification of horses arise when comparisons are drawn among the handful of remains representing an enormous geographic area, from western Europe to Siberia, that include a wide range of habitats and a long temporal span. However, even within an assemblage from a single locality, multiple forms have been detected, and these are often interpreted as representing sympatric subspecies (Meunier 1962). When variation resulting from age, sex, and individual characteristics is added into the mix, classification becomes rather daunting. Despite the variety of forms detected by researchers studying cranial morphology, it is important to remember that, despite the enormous size range of modern horse breeds (withers heights ranging from 60 to 200 cm), their cranial morphology is not as diverse as that seen in dogs or pigs (Epstein 1971; Wayne 1986).

The tarpan, according to Lundholm (1949), could be generally distinguished from the wild Asiatic horse by its flatter forehead, wavy cranial profile, short and low nose, supraorbitals vaulted above the forehead, a depression between the orbits, and another depression between the middle and anterior third of the nasal. Bökönyi (1974a: 72) remarked that the tarpan's skull was "broad and spacious, with orbits rising above the plain of the forehead and with a markedly short facial part." It is important to note that the reconstituted "tarpans" found in the forest of Bialowieza, in present-day Poland, may be derived from domestic Polish Konik horses and should never be considered reliable for metric analysis. Those in the St. Petersburg collections that are used by most scholars, including Lundholm and Bökönyi, are to date our best representations of true tarpans.

Eisenmann (1996: 28, 30) demonstrated that the Przewalski horse has a relatively shorter Vomer-Basion distance (a measurement reflecting the length of the basicranium) than "tarpans," Arabian horses, wild fossil horses, and subfossil domestic horses. Instead it resembles the kiang (a wild Asiatic hemione) in this character. Interestingly, Lundholm (1949) has shown that there are significant craniometric differences between the captured wild Przewalski horses and those raised in zoos. The zoo specimens exhibited a decrease in basal skull length and tooth row length, as well as in cranial and snout width, but postorbital width did not change (Forsten 1988). These alterations may provide clues to possible changes that would have occurred in *Equus ferus* under captivity, but also complicate comparisons between modern Przewalski skulls, which are mostly from zoo specimens, and archaeological specimens.

The ideal comparative cranial material would be the skulls of *Equus ferus* from the Mesolithic and Neolithic of the forest steppe of Ukraine, Russia, and northern Kazakhstan. Unfortunately, however, there are few reasonably complete specimens available from these periods (Kosintsev in press). The ephemeral nature of shallow campsites of the earlier periods generally does not lend itself to the preservation of crania. Most complete crania are derived from horse sacrifices in pits and burials dating to the Eneolithic or later, presumably after domestication had begun.

In order to understand better the craniomorphological changes that have occurred through the process of domestication, a thorough morphometric analysis should be conducted first on the rather large collections of Bronze and Iron Age horse skulls from kurgan burials in the former Soviet Union. Some work has been done in this direction already by Vitt (1952), Bibikova (1986a), Kuzmina (1997), Eisenmann (1996), and others. Eisenmann (1996) perhaps best points out the problems with any osteometric analysis of caballine equids when she says that no distinct phylogenetic or taxonomic patterns can be discerned in Quaternary horses, whether based on adaptive or nonadaptive characters. Instead, traits seem to combine in a "reticular pattern, a mosaic of characters that are combined at different places and times in different ways" (Eisenmann 1996: 30).

Bökönyi (1974b: 236) maintained that domestication brought on few changes in the horse's cranium, but he listed some features that may be important. He cited a decrease in cranial capacity, a broadening of the forehead, facial foreshortening, a narrowing of the muzzle, and a decrease in tooth dimensions. It is unclear when these cranial changes first became evident, although many claim to have observed differences in Iron Age horses (Vitt 1952; Bibikova 1986a).

Bibikova (1986a) performed a detailed study on the Dereivka cult stallion, which is now known to date to the Iron Age. Because it is clearly a domesticate based on its date and well-developed bit wear, it is interesting to see how it compares to other horses of the region, both wild and domestic. The Dereivka skull showed many of the traits that Bökönyi maintained were the products of domestication: a short, broad brain case, a broad forehead, a broad snout, and a maxillary cheek tooth row shorter than that of Przewalski horses and the "tarpan." But while Bökönyi predicted facial foreshortening, the Dereivka skull had a long narrow face. It should be pointed out that the face of a gelding tends to be longer and narrower, and often has a more convex profile than that of a stallion (Littauer 1971: 294). Geldings normally do develop the standard, large male canine teeth like those in the Dereivka skull, so it is entirely possible that the Iron Age Dereivka skull could have been derived from a castrated individual. Thus, except for the long, narrow face, the Dereivka skull does support Bökönyi's characters for domestic horses and is comparable to the other late archaeological horses and the modern domestic Kyrgyz breed.

The short, but broad, brain case and the broad forehead may be useful domestic trends, although there is no proof yet that these characteristics are seen consistently in other domestic horses. Facial length and muzzle width may vary in domestic horses as breeds develop, perhaps as early as the Iron Age. It is obvious, then, that there is a range of variation in both wild and domestic animals, and, in most characters, there is considerable overlap.

Although individual tooth size and cheek tooth row length may be useful for distinguishing the domestic horse from the Przewalski horse, it is unclear whether dental metrics can differentiate early domesticates from their immediate progenitors. Dental size variation is related to age and occlusal wear patterns, as well as to genetics. The cheek tooth row can be noticeably short in older individuals, as it is in one of the two existing tarpan skulls, so it is wise to avoid using this measurement in senile samples. The enamel patterns of cheek teeth are generally similar in all populations of caballine horses (Forsten 1988), and what variation there is often may be attributed to individual age and wear patterns.

Body Size

Despite the lack of clear morphological criteria for distinguishing wild from domestic horses, many archaeologists working in Ukraine, Russia, and Kazakhstan claim to be able to separate domestic horses from their untamed ancestors. Sometimes these claims are based on metric analysis, but sample sizes for any given site are usually very small, and there is little consistency in how body-size variations are interpreted. Benecke (1994), for example, maintains that an increase in variation in body size should be considered a marker of domestication in horses, similar to that seen in the probable domestic populations from Kazakhstan and Russia dating to the Bronze and Iron ages (Akhinzhanov et al. 1992; Kosintsev in press). Uerpmann (1990), on the other hand, believes that a similar degree of variation in body size among Neolithic equine populations in fourth millennium Germany falls within the natural limits of a wild population.

As with skulls, however, postcranial remains of *E. ferus* from Mesolithic and Early Neolithic contexts in the Ukraine, Russia, and Kazakhstan are extremely rare (Kosintsev in press). Thus, there is only a limited sample of remains from definitely wild Early to Middle Holocene specimens from the forest steppe available for comparison with later, possibly domestic horses.

Another complicating factor is that during the initial phases of domestication, the middens of sites within the natural geographic range of *E. ferus* very likely contained remains of both domestic livestock and wild hunted horses. It would be difficult to attribute any observed variation in these small assemblages to a difference between wild and domestic horses, however, given the wide range of ages likely to be represented and size differences among stallions, geldings, and mares.

To muddy the waters further, a temporal shift in size can occur under natural conditions as a result of changes in mean annual temperature and precipitation, alterations in season length and intensity, and other factors related to the change in climatic conditions known to have occurred throughout the Holocene (see Guthrie 2003). Eisenmann (1984) has suggested that there is a close link between bone proportions, especially in the feet, and the local environment (e.g., open vs. closed country, hard vs. soft ground, and dry vs. humid climate). Bökönyi (1974b: 242) saw a correlation between climate change in Bronze Age Hungary and a shift from horses with narrow hoofs to stockier horses with broad hoofs as the climate became cooler and wetter. Clearly, determining whether local size changes reflect adaptation of wild populations to changing environments or the development or introduction of a new breed of domestic horses has proven to be complicated.

Because the early domestic horses must have relied either solely or primarily on natural vegetation, body size in early domesticates also may have responded to natural differences in available pasturage across the wide geographic range in which horses were kept. If the herds were fed domestic grain as fodder through the winter, however, then it is possible that they could have grown to greater size. Today, Kazakh and Mongolian pastoralists continue to graze their herds on natural vegetation through the winter and to gather wild grasses for fodder only in the most critical times. Significant increase in the body size of these animals would make horses less efficient grazers in the forest-steppe and would decrease the sustainable herd size. On the other hand, significant reduction in body size related to nutrition would hamper endurance and speed in riding and lessen their capacity to carry heavy loads. Maintaining body volume is also advantageous given the severe winters of the region.

Mortality Patterns

Mortality patterns, so helpful in discerning early herd management in other domestic herbivores (Redding 1981; Chapter 14), are more complicated in horses. For most livestock, there is a strong incentive for culling young males for consumption as soon as they have obtained most of their adult volume in order to reduce the investment in grazing land and fodder for these animals. In addition, it is much easier to control herds and to carry out genetic selection if only a few of the best mature males are kept for breeding purposes. The adult females (needed to breed, rear the young, and provide milk) significantly outnumber the adult males in a typical livestock herd. Under such a regime, livestock management is signaled in the archaeological record by high numbers of juvenile males in the death assemblage and prolonged female survivorship. There are several serious methodological impediments to using this marker to document horse domestication.

Although the bodies of males tend to be slightly bigger than those of females, the degree of size-linked sexual dimorphism in horses is less strongly expressed than it is in other livestock species. Clear bimodality between the sizes of male and female horses can only be expected in metric analyses conducted on unusually large collections. Determining sex in horse remains, then, is based on two characters: the presence of large canines in males and distinctive differences in the shape of the pelvis in male and female animals. In the stallion, the internal surface of the pubis is convex, whereas it is concave in the mare and in horses gelded very young (Riegel and Halola 1996: 152). The presence of robust canines can be used to distinguish adult stallions and geldings from females that either have no canines or only very small vestigial canines. This feature, however, is of no use in sexing juvenile specimens since there are no deciduous canines. Generally, sexing by dentition is restricted to individuals over the age of 4 or 5 years, after the canines have erupted (except that this author has observed that some females may obtain their diminutive canines as early as 2.5 years). This means that it is not possible to determine how many juvenile males vs. females were culled based on dentition. We can, however, look at the overall percentage of juveniles in the assemblage and the ratio of adult males to females.

Ageing horse remains can also be problematic. Most ageing schemes are based on patterns of dental eruption and wear. Because most eruption tables have been constructed by biologists looking at living horses (Silver 1963; Getty 1975; Habermehl 1975), the ages of tooth eruption given are generally when the tooth first erupts through the gums, and not when it erupts through the bone, and thus will tend to overestimate age at death in archaeological specimens. There are also problems with some of the common methods for computing age at death based on the crown heights of cheek teeth, a technique often used to age horses from archaeological sites (see Levine 1982). First, there are significant questions about the accuracy and resolution of estimated ages based on crown-height methods (Brown and Anthony 1998: 338–339). Moreover, the practice of treating both whole jaws and each isolated tooth as separate specimens risks counting one individual several times. Another inherent problem in dealing with isolated teeth is that it is extremely difficult to distinguish between third and fourth premolars or first and second molars. At times, it even can be hard to separate any of these four teeth from the others. If the problematic cheek teeth are lumped together, then the age determination for each tooth is quite a large span of years, but if they are excluded from the sample, then considerable data will be missing. Calculating crown heights on isolated P3-4 and M1-2 is subject to errors amounting to three or more years.

It is also difficult to reconstruct reliable mortality profiles for horses in Eurasian steppe sites because sample sizes are often quite small. Moreover, even if the assemblages merit ageing, the results from one site are often not comparable to the next, either because one zooarchaeologist groups ages differently from another or the dates and locations of the sites are quite different. Since domesticated horses seem to have arrived at different places at different times and may have been used in different ways with varying economies, it is not easy to combine results from all sites across the Eurasian steppes together.

The use of mortality data in identifying horse domestication is further complicated by the complex and shifting demographic patterns in wild horse herds and the diverse exploitation strategies practiced by both horse hunters and herders. Wild and feral horses form two types of social groups (Berger 1986). The primary group is a family band composed of a stallion, his harem of mares, and their offspring. Both sons and daughters normally leave their natal band between the ages of one and three years (Berger 1986: 129). The second social group among wild horses consists of two or more males that are too old to stay with their families but have not succeeded in acquiring mares. Bachelor groups are not stable and vary through the year. The typical size is around 4 individuals, although bachelor groups have been reported to have as many as 17 individuals at a time (Berger 1986: 131).

It is expected that a communal hunting strategy normally would focus on family bands because they contain more individuals and are more cohesive than the bachelor groups. Stalking by one or two hunters would be more appropriate for dispatching single bachelors, or those in small, loosely affiliated groups. Most communal hunts would be likely to yield a high percentage of both male and female juveniles, a relatively large number of mares, and one stallion. Unfortunately, the typical harvest pattern for most livestock produces a similar pattern: high numbers of juveniles (mostly males, which could not be distinguished using archaeological remains), relatively high numbers of adult females, and low numbers of adult males.

On the other hand, there are many reasons why a high frequency of juvenile males, as expected in most livestock culling strategies, might not be found in a mortality profile in an assemblage of domestic horses. First, it is clear from the frozen horse mummies at Pazyryk (Rudenko 1970) that, at least by the Iron Age, castration was a common practice. This would eliminate some of the need to cull juvenile males, especially if geldings were as highly valued as Rudenko indicates. The question that remains is when this practice began to be employed for horses. If horse domestication started in the Eurasian steppe after people in this region had received domestic cattle and sheep from the Near East, then they would probably have been familiar with the practice of castrating already.

Second, for horses, adult males take on important roles beyond that of breeding. In addition to providing meat, milk, bones, and hides, at some point in their history horses were used for riding, carrying loads, pulling wheeled vehicles, and plowing (although the use of horses as draft animals was probably considerably delayed compared to other uses). Males would almost certainly be preferred in all of the nonfood uses of horses. Despite Rudenko's (1970) comments on the preference for geldings over stallions, there is evidence that stallions were highly prized. Even a cursory examination of classical depictions of cavalry horses and chariot horses on pottery, metal vessels, and the like from Scythian, Greek, Roman, and Egyptian art collections will reveal a strong preference for intact stallions. Littauer and Crouwel (1984) have also remarked on the predominance of stallions in the depictions of ancient Near Eastern chariot teams. Mongolians today prefer to ride stallions when herding or hunting.

Adult males may also be better represented than expected in faunal assemblages in the Eurasian steppes after domestication began because of the practice of ritual horse sacrifice, which seems to have focused primarily on males in some regions. This practice is most likely correlated with the value of males as mounts from an early date onward and later for pulling chariots and other vehicles. Sacrificial remains are common in human burials from the Copper Age through the Iron Age and in deep domestic pits within Copper Age settlements. In these contexts, preservation is reasonably good compared to open-air middens containing animals slaughtered for food. For the Bronze and Iron ages, excavators have concentrated heavily on human burials containing animal sacrifices in preference over middens, artificially boosting the ratio of adult males in the faunal collections. The discussion of the Copper Age Botai culture below will demonstrate that even near the inception of horse domestication, there was a preference for the use of stallions in ritual deposits.

Kosintsev (in press) has recently gathered fairly comprehensive data on mortality patterns from Bronze and Iron Age sites

TABLE 17.1
Mortality Profiles for Bronze and Iron Age Horses

Age	Early Bronze Age Sergyeevka[a] % Killed	Middle-Late Bronze/Iron Age Sites[b] % Killed Mean	% Killed Range
0–1 yr	6	29	13–35
1–5 yr	29	15	8–20
5–12 yr	58	47	35–60
12+ yr	7	9	2–17

SOURCE: Kosintsev (in press).
[a] 2500–2920 BC.
[b] Sintashta, Srubnaya, Alakul Fedorovo, and western Siberia.

TABLE 17.2
Mortality Profile for Copper Age Botai Horses

Age	% Killed at Dereivka
0–5 yr	25
5–8 yr	50
8–16 yr	20
16+ yr	5

SOURCE: Levine (1993, 1999a).

on the Eurasian steppe (Table 17.1). Since horse domestication was certainly accomplished by this time, these mortality patterns provide a useful point of comparison with earlier sites where the status of the horses as wild or domestic is still open to question. Early Bronze Age (2500–2920 BC calibrated) mortality patterns from the site of Sergeevka, Kazakhstan (Levine and Kislenko 1997), indicate a relatively low mortality rate for animals in their first year of life (around 6%). Slaughter focuses instead on animals in the 1–5-year range and the 5–12-year range. The mortality rate for horses below 1 year of age increases dramatically in later sites to a mean figure of about 29%. At the same time, there is a relative decrease (from 29% down to 15%) in mortality for animals in the 1–5-year range in later sites. The frequencies of animals in the older 5–12 and 12+ age groups remain about the same through time.

The high mortality rate of very young animals in the later Eurasian sites is not consistent with natural mortality rates in wild and feral herds. The annual mortality rates for feral horses under one year of age in the United States range from about 8% to 14%, with an unusually high mortality of 25% reported for only one locality (Berger 1986: 82, Keiper and Houpt 1984, Feist and McCullough 1975, Welsh 1975).

The high mortality rate of horses from the Middle Bronze Age through the Iron Age is also not consistent with modern pastoral slaughter strategies where the ideal is to slaughter 2.5-year-old animals in the late autumn or early winter. While adult weight is reached at about 5 years of age in horses, there is little difference between the weight of a 3-year-old animal and a 5 year old. At 2.5 years, a horse on the Eurasian steppe has reached most of its adult weight (about 90%), and in the autumn its fat content is relatively high, a crucial factor for people dwelling in regions with harsh, prolonged winters. The meat of animals slaughtered at this time can also be readily frozen and kept through early spring. Horses killed before their first birthday, in contrast, have only reached 12–50% of their adult meat weight.

It is possible that the high mortality of neonates and yearlings in the Bronze and Iron ages is a signal of a distressed herd management system. It may reflect the harvesting of half-grown horses out of necessity or the loss of young horses in the winter because of severe ice storms, heavy snows, droughts, and epidemics, all known to cause catastrophic deaths of young animals in modern Kazakh herds. Rice (1963) has reported that ice storms recur every 10–12 years in northern and central Kazakhstan, killing nearly half of all livestock. In the particularly harsh winter of 1891–1892, for example, 47% of the horses, 32% of the cattle, sheep, and goats, and 22% of the camels were lost. Young animals are especially vulnerable to harsh weather, although they are frequently taken indoors for protection. But it seems unlikely that catastrophic loss of young animals can account for the persistent and widespread pattern of young animals in sites dating to the Middle-Late Bronze Age and Iron Age in the Eurasian steppe.

Alternatively, this pattern may be attributable to an increase in the importance of mare's milk during this later period. Fermented mare's milk, also known as *koumiss* (Kazakh) or *airag* (Mongolian), has very high vitamin content and may well have been greatly prized by Bronze and Iron age peoples. Production of mare's milk is most plentiful in the summer months from May to August but diminishes in the autumn. If mares were bred to maximize milk production, foals would have become expendable when milk production decreased in the autumn. Foals may not have reached full adult body weight, but their slaughter at this time would reduce the problems of keeping more horses than necessary through the winter. Early slaughter of young animals might also reduce birth spacing so that a second foal could be conceived soon afterward. This would maximize koumiss production.

It is interesting to compare these patterns to that reported by Levine (1999a) for the Copper Age site of Dereivka. Levine has argued that the mortality pattern for these horses shows an emphasis on older animals that is consistent with hunting wild horses, particularly a stalking strategy that focuses on animals of prime age. However, judging from the graphical representation of horse mortality patterns at Dereivka provided by Levine (1993: Figure 12), reproduced here in Table 17.2, it would seem that about 24–25% of the horses were killed before five years of age, and another 50% were killed between five and eight years of age. This pattern is consistent with that observed in an earlier study of a larger sample of the Dereivka material by Bibikova (1986b), who reported that young horses constitute about 23% of the Dereivka horses. Noting the similarity between the Dereivka assemblage and modern Mongol herds, where individuals in the one-to-two-year age range constitute about 26% of the total domestic horse stock, Bibikova felt the mortality pattern at this Copper Age settlement was consistent with a domestic herd.

It is difficult to directly compare the Dereivka mortality patterns with those reported by Kosintsev for later sites since different age classes were used to construct these profiles. However, there do appear to be strong similarities between the mortality profiles at all these sites. At all sites there is an emphasis on animals below 5 years of age, with only 25% of the horses surviving beyond 8 years of age at Dereivka and fewer than 10% surviving beyond 12 years at the Bronze and Iron Age sites.

Levine also bases her argument for the wild status of the Dereivka horses on the predominance of males in the sexable jaws at the site. Nine out of 10 sexable jaws studied by Levine were males, a figure similar to that found by Bibikova in her study of a larger sample from the site (15 of 17 sexable jaws). This emphasis on males is interpreted by Levine as evidence for the hunting of bachelor herds or the hunting of "inexperienced" stallions defending harems. It is quite possible, however, that the majority of these jaws belonged to well-preserved crania from sacrificial stallions or geldings purposefully buried in ritual pits, similar to ritual practices seen in Copper Age contexts in Kazakhstan (Olsen 2003). Crania of horses slaughtered for subsistence probably would not be as well represented as these carefully buried specimens, since they might have been broken open to extract brains or discarded in open-air middens. Thus, the predominance of males in the Dereivka assemblage observed by Levine may actually be a reflection of ritual deposits and not of general harvest practices.

Geographic Range and the Spread of Domestic Horses

One promising avenue that will require additional work in the future is the reconstruction of the geographic distribution of the probable wild progenitor, *Equus ferus*, during the Holocene (Bökönyi 1974b: 236; Grigson 1993). If it can be shown convincingly that wild horses became extinct in certain regions of Europe and the Near East by the Middle Holocene, then the reappearance of horses in these regions could signal the diffusion of domestic horses outward from their place (or places) of origin in the Eurasian steppe. Tracking the distribution of horse remains in these marginal areas, then, could help bracket the period of initial horse domestication.

There are, however, some serious impediments to using zoogeographic evidence in this way. First, there is considerable disagreement about the boundaries of the range of wild horses in the Early to Middle Holocene. The data needed to reconstruct these boundaries have yet to be assembled, and there are many gaps in our knowledge. Second, many of the horse remains attributed to Mesolithic and Neolithic contexts come from very early excavations in which there is a strong possibility of modern intrusions. Recent efforts to directly date some of these remains using small sample radiocarbon dating techniques certainly will clarify the temporal placement of these specimens. However, even in well-dated assemblages, it is hard to assess whether the sudden reappearance of horses in some regions after a significant hiatus reflects the introduction of domestic stock or the expansion of small, relict wild populations resulting from climate change.

Nevertheless, it is useful to summarize the information available on some of the key finds in the Near East and Europe (Table 17.3). In the Near East, a single possible wild horse specimen was dated to between 7400 and 5000 BC at the site of Abu Hureyra in northern Syria (Legge and Rowley-Conwy 2000: 429; Moore 2000: 259). Horse remains were also found in the somewhat later site of Tappeh Zageh in north central Iran at about 5000 BC (Mashkour et al. 1999). The likely domestic horse does not appear in Mesopotamia until about the mid-third millennium BC (Grigson 1993) and about 500 years later in southern Iran (Zeder 1986).

Horse remains have been recovered from a number of sites in the Caucasus dated between 5500 and 4000 BC (Anthony 1996: 72), but do not appear in Anatolia until the late Chalcolithic period (4000–3000 BC) (Anthony 1996; Boessneck and von den Driesch 1976). Although Boessneck and von den Driesch (1976) maintained that these animals belonged to relict wild populations, Bökönyi (1991) argued that they represented domestic horses. Equally problematic is the status of the isolated but indisputable horse remains found in a number of Chalcolithic and Early Bronze Age contexts in the southern Levant (Grigson 1993; Davis 1976; Lernau 1978; Zeder, conversation with Olsen, September 2003).

There are similar problems with the attribution of either wild or domestic status to the horse remains from western Europe and western Central Europe. Horse remains are rare in this region from the end of the Pleistocene until around 3500 BC. However, there seem to be sparse relict populations of wild horses in the Early Holocene from Hungary to England, as shown in Table 17.3. Lundholm (1949: 171) had no record of them in the Mesolithic and Neolithic in Italy or in the Balkan Peninsula, and this is still true today. Greenfield's (in press) work also supports a lack of horses before the Bronze Age in the Balkans. It is particularly difficult to know how to interpret the status of horses found in small numbers in Mesolithic and Neolithic sites in Central and western Europe. In many, if not most, cases, attribution of either wild or domestic status to these specimens is based solely on their temporal placement (rarely fixed through direct dating) and the preconceived ideas of the analyst about the likely date of horse domestication. The domestic

TABLE 17.3
Selection of Early Horse Remains from Sites in the Near East and Europe

Region	Sites	Dates Uncalibrated, BC	Dates Calibrated, BC	References
Syria	Abu Hureyra	7400–5000		Legge and Rowley-Conwy 2000
	Tell Brak	2300		Clutton-Brock and Davis 1993
Negev	Shiqmim	3240 ± 75		Grigson 1993
Central Iran	Tappeh Zagheh	5000		Mashkour et al. 1999
Southern Iran	Malyan	1800		Zeder 1991, 1986
Anatolia	Malatya, Elazig, Demircihöyük	3300–3000		Anthony 1996
	Norsun Tepe	3000–2500		Boessneck and von den Driesch 1976; Bökönyi 1991
Egypt	Buhen	1675		Clutton-Brock 1974
Caucasus	Arukhlo, Tsopi, Alimeklek	5500–4000		Anthony 1996
Hungary	Many Neolithic Sites	6000–5000		Vörös 1981
Denmark	Ertebölle/Ellerbek	5500–4000		Davidsen 1978
	Braband So	3550 ± 75		Davidsen 1978
	Lindskov	2570 ± 65		Davidsen 1978
Netherlands	Swifterbant	3525–3280		Clason 1991
No. Germany/ Poland	Potsdam-Schlaatz		9390–9250	Benecke in press
	Gaukönigshofen		7960–7760	Benecke in press
	Hohen Viecheln	5500		Gehl 1961
	Rüde, Bondebrück	4000–3500		Lüttschwager 1967
	Dąbki 9		4940–4800	Benecke in press
	Eilsleben		4830–4720	Benecke in press
	Rosenhof		4720–4540	Benecke in press
	Siniarzewo 1		4430–4250	Benecke in press
Southern Germany	Bruchsal-Aue	4236 ± 64– 3529 ± 74		Steppan in press
France	Charcognier	7000–5500		Bignon, O. e-mail to Olsen 5/16/2003
	Bercy		3900	Bignon, O. e-mail to Olsen 5/16/2003
Italy	Le Cerquiete-Fianello		2510 ± 110	Curci and Tagliacozzo 1995
	Querciola-Sesto Fiorentino	2130 ± 150, 1690 ± 200		Corridi and Sarti 1989–1990
England	Seamer Carr	7750 ± 180		Clutton-Brock and Burleigh 1991b
	Grimes Graves	1690 ± 210		Clutton-Brock and Burleigh 1991a
Ireland	Newgrange	1900		Wijngaarden-Bakker 1974

horse is thought to have reached western Europe surprisingly slowly, as attested by directly dated horse remains appearing in England at about 1700 BC (Clutton-Brock and Burleigh 1991a) and the first arrival of the horse in Ireland at about 1900 BC (Wijngaarden-Bakker 1974).

The gradual succession of forests in Europe with the warming trend in the Early to Middle Holocene (Flint 1957: 286; Butzer 1971: 532) may have been a key factor in the reduction of horse populations there, as areas previously consisting of open steppe began to close up. Similarly, some have suggested that the later return of horses into these areas was the result of climate change rather than the diffusion of domestic horses (Steppan 1998). Recent pollen evidence from western Siberia to eastern Europe supports the idea that the period of initial domestication and diffusion of horses was marked by considerable climatic change (Kremenetski et al. 1997). These data indicate climatic conditions from about 4000–2500 BC, a time period that likely

FIGURE 17.1 Map showing major sites in the Eurasian steppe.

witnessed horse domestication in Eurasia, were both wetter and less continental than either preceding or subsequent periods. The climate became colder and drier from about 2500 to 1500 BC, coinciding with the likely development of nomadic pastoralism on the Eurasian steppe. It is difficult, then, to distinguish the diffusion of domestic horses out of the Eurasian steppe and into Central and western Europe from climate-induced shifts in the zoogeography of wild populations. It is also possible that climatic change played a role in the spread of equine husbandry.

Increase in Frequencies of Horse Remains

Bibikova (1986b) has proposed that domestication of horses will be signaled by a marked increase in the ratio of horse elements to the overall number of ungulate remains, whether domestic or wild. The use of abundance data to indicate domestication in any species must be undertaken with care, however. Natural changes in climate or local environment can lead to shifts in wild species' frequencies in a region, independent of any advances in animal control. It is important to remember that even Paleolithic sites can have a high concentration of one species as a result of selective hunting practices or seasonal herd migration patterns, as at the Upper Paleolithic kill site of Solutré (Olsen 1989, 1995), where most of the faunal remains are from horses. On its own, then, a sizable ratio of a particular animal in the assemblage does not constitute proof of domestication. Instead, this information can provide support for other more direct lines of evidence or contribute to the amassed secondary evidence. This is especially the case when abundance patterns are examined across larger geographic and temporal ranges.

Bibikova used Neolithic or Copper Age (Eneolithic) sites in Romania, Moldova, Ukraine, and western Russia to examine trends in the abundance of horse remains as a possible indicator of domestication. Although sample sizes for most of the sites are small, the overall pattern is very informative. In her Western Zone (from Bucharest to the Dnepr River), prior to the Late Tripolye period (ca. 2500 BC calibrated) the ratio of horses was generally well below 15%. In her Eastern Zone (from the Dnepr to the southern Ural Mountains), the percentages were consistently higher. Most significantly, at Dereivka, horses made up 63.27%, and at the Early Yamnaya (Early Bronze Age, ca. 3500 BC) settlement of Repin, on the Don River, the percentage soared to 79.8% (Figure 17.1). The sample sizes are relatively good for these two sites. At Dereivka, the total number of ungulate bones is 3,564, 2,255 of which are horse bones (Bibikova 1986b: 169). The minimum number of horses is given as 44 individuals (Bibikova 1986b: 174), however, the methods for calculating MNI may differ from the standard procedure used in the West. At Repin, 652 of the 817 ungulate bone fragments are from horses. As we shall see, even farther to the east, in northern Kazakhstan, the numbers reach 99% at the large settlement of Botai, 3700–3100 BC (calibrated). There are no domestic sheep or cattle in Copper Age deposits at Botai sites, which represent the greatest dependence on horsemeat of any culture through time. These temporal (fourth millennium) and geographic trends (eastward to Kazakhstan) toward an increase in proportions of horses almost certainly reflect the presence of equine livestock. Once nomadic pastoralism involving cattle and sheep takes hold in the Middle Bronze Age, however, the percentages of horses diminish.

Artifactual and Architectural Evidence

Evidence for early horse domestication might also include horse tackle, corrals, and other cultural indicators of horse control. The most cited early evidence of possible horse tackle consists of six perforated antler tines from Dereivka, in Ukraine (Telegin 1986: 82–83). Five of these pieces have one hole and one has two holes, resembling later bridle cheek pieces from European Bronze Age sites (Clark 1941; Britnell 1976; Longley 1980). Late TRB (Funnel Beaker) sites in Germany produced single-holed antler tine artifacts similar to those from Dereivka (Telegin 1986: 85). The TRB culture extended from Ukraine to the Netherlands and dates from 4350 to 2670 BC (Milisauskas 1978). Antler cheek pieces are found with Iron Age horse mummies at Pazyryk (ca. fourth century BC) that are similar in dimension to the Dereivka perforated antler tines, although they are somewhat thinner. Dereivka also produced a number of perforated antler artifacts made from the beam (or base) of the antler that were clearly hafted to wooden shafts and used as tools, so it is difficult to say that the perforated antler tine artifacts from the site were bridle cheek pieces. Unlike these other antler artifacts, however, the perforations in the antler tine pieces were only about .4 to 1 cm in diameter and appear to have been worn or polished as though continually rubbed by some soft material such as leather or hemp cord. Dereivka did not produce any artifacts that might be identified as mouthpieces. However, here it is important to remember that the cult stallion with clear signs of bit wear is now known to be an Iron Age intrusion and that early mouthpieces may well have been made of soft material that would leave no trace in the archaeological record (see below).

More secure evidence of cultural control comes in the form of disk-shaped cheek pieces fashioned from antler that are found directly associated with horse skulls in Russian "chariot burials" dating to the Middle Bronze Age (ca. 2000–1700 BC) (Gening et al. 1992). The lack of any preserved bits in this burial context confirms the suggestion that early mouthpieces were made of rope or leather (Littauer and Crouwel 2001). Comparable metal-studded disk cheek pieces and bits date to the second half of the second millennium BC in the Near East, Egypt, and Greece (Littauer and Crouwel 1986, 1988, 2001). A donkey with a metal bridle dated to about 1700 BC was recovered at Tel Haror in the northern Negev (Littauer and Crouwel 2001).

The absence of parts of tackle in the archaeological record, however, is not proof of the absence of domesticated horses, since horses can be ridden without a bridle or with a simple leather or hemp bridle that would unlikely be preserved. The Greeks are nearly always shown riding bareback, often with no bridle (Markman 1969), and the Plains Indians of North America often used a rawhide loop around the diastema, or bar, between the incisors and cheek teeth on the lower jaw (Wilson 1978). The earliest evidence for the use of saddles comes from Pazyryk (fourth century BC); the earliest horseshoes come from the Roman era (Clutton-Brock 1996); and the earliest stirrup is depicted on a *bas relief* at the Indian site of Mathura dating to about 50 BC (Littauer 1981).

The search for horse corrals in the archaeological record is only beginning but should be marked by postholes that outline a relatively large enclosure. There should also be chemical, botanical, and micromorphological evidence of horse manure in the soil in these enclosures (see below).

Finally, Gheorghiu (1994) claims harnesses can be seen on some of the stone horse-head scepters dating to the Eneolithic (ca. 4500 BC) that are found in sites in western Russia, the Ukraine, and southeastern Europe. However, this interpretation is highly speculative and based on a generous reading of highly stylized carvings. It is interesting to note, however, that these scepters seem to mark the spread of Pontic-Caspian steppe peoples into southeastern Europe (Mallory 1996) and may in fact signal the diffusion of domestic horses out of the region of initial domestication.

Pathologies Associated with Domestication

Levine (1999a) has investigated the relationship between riding and pathologies in the horse skeleton. Although it has been documented that weight-bearing and saddles cause pathologies of the back, many modern riders scoff at the notion that riding bareback or with a soft pad or blanket would create significant problems for the horse's skeleton. Depending on the conformation of the horse's back, as well as the ability and weight of the rider, riding need not result in any pathologies in either the horse or its rider. With bareback riding, the rider shifts his or her weight from one side of the horse's back to the other, so circulation is not inhibited in any part of the horse. The padding in human buttocks acts as a hydraulic cushion that protects his or her ischial tuberosities (the part of the pelvis that humans sit on) as well. A flexible leather pad also helps reduce impact for both rider and horse. Skeletal pathologies resulting from riding should be rather rare in horses prior to the invention of the true saddle with a wooden frame. It might be more profitable to look for anomalies in the human skeleton that signal muscle hypertrophy caused by habitual riding or an increase in mended fractures occurring as a result of falls from horseback (Molleson and Blondiaux 1994).

On the other hand, it is well established that a saddle with a wooden tree can cause pathologies. Preventing injuries through precise saddle fitting and padding is an enormous industry today. The distribution of the rider's weight on a rigid frame creates pressure points, which can eventually cause muscle atrophy in the horse's back. A poor saddle can lead to problems in the regions of the horse's shoulders, back, pelvis, hock, stifle, and leg. Baker and Brothwell (1980: 131) report that riding and workhorses are sometimes afflicted with *spondylosis deformans* of the vertebrae, but they do not specify whether this damage is caused by a saddle or not. These lesions typically occur in the lumbar region in domestic horses, whereas they are more common in the thoracic area in wild horses.

Saddles with wooden frameworks are not known from contexts dating before the Iron Age and Roman period, however. Saddles found in the frozen tombs of Pazyryk, in Siberia, had wooden spacers consisting of flat boards with tongues at either end that were inserted into leather saddle cushions (Rudenko 1970: 132–133). These were found elsewhere among the Scythians and seem to have been widespread. It would not be surprising if these boards applied pressure that could lead to muscle atrophy and perhaps eventually have impact on the skeleton itself. The Romans were the first known to have used a treed saddle with horns or pommels (Hyland 1990; Clutton-Brock 1996: 89).

Traction is likely to lead to increased rugosity of muscle attachments or arthritis of the horse's limbs. Equids were used to pull wagons, carts, and chariots by the middle of the third millennium BC. Yokes initially developed for use on oxen are quite unsuited to equid anatomy and almost certainly would have caused the development of pathologies in the neck and shoulder region. The Greeks, Romans, and Chinese (second century AD) initiated experimentation with various means of harnessing draft horses, but this was long after domestication began.

Perhaps the most widely known pathology used as a marker of domestication in horses is the wear on the lower second premolar caused by the use of a bit. This form of attrition is caused by a horse habitually clenching a bit between its teeth for brief intervals. After metal bits were developed, this pattern became fairly visible. Evidence of bit wear is usually seen on the mesial half of the lower second premolar, but occasionally occurs over most of the occlusal surface of this tooth and often on the upper second premolar. The most distinctive aspect of the wear caused by a metal bit is the absence or reduction of the "Greaves' Effect" (Greaves 1973). The Greaves' Effect is a natural process caused by food abrasion on the occlusal surfaces of artiodactyl and

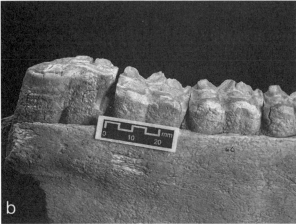

FIGURE 17.2 Bit wear on lower second premolar from Malyan: (a) right mandible lingual aspect; (b) left mandible buccal aspect.

perissodactyl teeth that results in differential erosion of the softer dentine between the more resistant enamel ridges. Frequent contact with a metal bit causes much faster erosion of the rigid enamel plates so that they wear at a rate comparable to dentine. The result is a facet with a very smooth, even surface, at least initially. Older horses that have been bitted for long periods sometimes develop a cupping of the second premolar or a steep step down in either the middle part or near the mesial end of the tooth. This is the kind of wear seen on a horse specimen from the highland Iranian urban site of Tal-e Malyan dating to about 2400–1800 BC (Figure 17.2; Zeder 1986, 1991) and on a premolar of a stallion from Buhen Egypt dated to 1675 BC (Clutton-Brock 1974, 1992: 83). The blue-green staining and bit wear on the premolars of donkeys from Tell Brak in Syria (ca. 2300 BC) indicate the use of copper or bronze bits (Clutton-Brock and Davis 1993).

Bit wear can also be seen on the so-called cult stallion that until recently was held to be the most convincing piece of evidence for claiming early horse domestication at Dereivka in the Ukraine (Anthony and Brown 1991). This specimen was originally dated by stratigraphic association to around 4000 BC, well before there was any artifactual evidence for metal bits. Recent direct dating of the specimen reveals it to be an Iron Age intrusion from between 790 to 200 BC (calibrated) (Anthony and Brown 2000). The first solid evidence for the use of metal bits in Russia is seen in the ninth century BC, so it is likely that the wear pattern on the Dereivka cult stallion was caused by a metal bit.

The question remains whether soft bits made from leather or hemp, likely used before the invention of the metal bit, leave similar wear patterns that could be used to detect the riding of horses. In the late 1980s, Anthony and Brown (1989) began a series of longitudinal actualistic experiments comparing the effects of a range of materials for bits on living horses, including leather, hemp, horsehair, and bone. According to their experiments, nonmetallic bits leave a reduced bevel on the mesial portion of the lower second premolar that lacks the characteristic abrasion pattern left by metal bits. A few of the Copper Age horse teeth from Botai (3700–3100 BC calibrated) studied by Anthony and Brown show this kind of wear (Brown and Anthony 1998), which was interpreted as a likely indictor of early horseback riding. As discussed below, subsequent work calls these results into question.

Ritual Horse Sacrifice

There is a long tradition of burying horses and other domestic livestock with humans as funerary offerings all across Ukraine, Russia, and Kazakhstan stretching back to the Neolithic. The earliest evidence of horse sacrifice performed to accompany human interments comes from Late Neolithic graves at Varfolomievka (5570–4840 BC calibrated), in the Volga-Ural steppes north of the Caspian (Telegin and Potekhina 1987), where horse bones were placed in ochre-stained pits above the human remains. The same was true at the Khvalynsk I cemetery (5242–4580 BC calibrated), in the Middle Volga region of Russia (Agapov et al. 1990; Anthony and Brown 2000); the early Eneolithic site of S'yezzhe, in the Samara River valley (Vasiliev and Matveeva 1979); and Nikol'skoe (4950–4000 BC calibrated) on the Dnieper River (Anthony and Brown 2000) (Figure 17.1). Wild animals are unusual in human graves in this region, and it is clear that the sheep and cattle remains interred with humans were domesticated. The horses ritually buried with humans were treated in exactly the same manner as were cattle and sheep, often with skulls and foot bones being placed near the human body in what Piggott (1962) referred to as "head and hoof" burials. The treatment of horses and obviously domesticated bovid livestock in a similar manner is generally taken as a signal of the horse's domestic status, although this association should be considered only circumstantial evidence. This would push the date for the earliest domesticated horses back to the middle of the sixth millennium BC, which would be very early indeed.

The Process of Horse Domestication

While evidence of riding might be assumed to be a good indicator of horse domestication, Levine (1999b) has argued that people might have captured wild individuals, broken, tamed, and ridden them without maintaining a breeding population. In this way, highly nomadic hunters would have been able to have the benefit of horse transport without the trouble of breeding horses and caring for newborns and suckling young. Thus, the argument goes, while horseback riding may indicate the taming of horses, it is not necessarily a sign of their domestication. There are two serious problems with this scenario.

First, the resolution of the archaeological record is unlikely to allow us to locate a boundary between taming and early domestication. It is, after all, a process that, as zooarchaeologists recognize, flows through our arbitrary classes of wild, captive, tame, and domesticated. The archaeological record exacerbates the problem further through deterioration of perishables related to horse control (rawhide or hemp tackle) and the more delicate neonatal and juvenile bones expected in domestic herds.

The second more serious problem with this thesis is that it seems unlikely that wild mares and stallions would be captured, tamed, and even perhaps trained for riding but prevented from breeding with each other. This scenario requires thorough isolation of the two sexes in separate corrals to avoid any contact during estrus. One mistake and the tame population would suddenly become a breeding one. Overcoming the powerful natural urges of horses to breed is especially difficult given that, even today, domestic horses in Kazakhstan, Mongolia, and Russia are released from the corrals in the evening and driven overland to graze through the night on natural pasture. Only a couple of horse herders may drive as many as 250 horses at a time, and even with the aid of persistent herding dogs, control is never absolute.

This debate over whether horses were tamed or truly domesticated raises the deeper questions of what constitutes a domestic population and how it can be detected in the archaeological record. Different scientists use different criteria for the recognition of a domestic population. Some require evidence of total genetic isolation from wild herds, but in reality, this degree of isolation is rarely achieved until the wild population is totally extinct. Dogs still breed with wolves, for example, but none would deny that dogs are domesticated. For the early domesticates, this standard is much too rigid. Prior to extirpation from their natural habitat, wild Asiatic stallions, *Equus przewalskii*, frequently stole the Mongols' domestic *Equus caballus* mares and bred with them. The same was true for the tarpan, *Equus ferus*, the likely wild progenitor of the domestic horse, in its natural range until its extinction. Realistically, without extinction, isolation can occur only if the domestic herds are moved outside the natural geographic range of the wild progenitors. Although domestic herds did eventually spread out to areas not inhabited by wild equids in the Middle to Late Holocene, they also thrived in the heart of the wild horse's habitat, Ukraine, Russia, and northern Kazakhstan.

Some zooarchaeologists demand either morphological or metric distinctions on the skeleton as proof that domestication has taken place, but such changes may not always be present. There are many cases of separate species within a family of wild animals that cannot be distinguished on either of these grounds, and yet they do not interbreed and are recognized as valid taxa by biologists. As demonstrated above, despite over a century of morphometrics on equid skeletons, there is still no consensus on which criteria are reliable for distinguishing between wild and domestic horses, and there is considerable variation even in the Pleistocene among caballine equids. Archaeologists place unnecessary restrictions on themselves if they limit themselves to metric or morphological markers to detect domestication in animals. Reliance on morphological and size change as the only criteria used to document domestication may delay recognition of horse domestication by as much as two or even three millennia after the initial event took place.

Culling is a useful trait of livestock domestication that holds more promise for recognizing incipient horse breeding, but the keeping of highly valued stallions and geldings for riding may have happened fairly soon after domestication began. Controlling herds of domestic horses without being able to ride would be difficult, although theoretically possible. Herds had to be released to graze far and wide over the natural landscape. To round the animals up in open steppe with nothing more than pedestrians and dogs would have led to chaos and the loss of many individuals from the herd. Feral animals would have found it easy to strike out on their own, especially in the case of young males, or to join a band of wild horses in the case of females. Ridden domestic stallions would have been excellent at rounding up the herd, since it is their natural instinct to keep control of their family band.

Although there are problems using bit wear from a soft bridle as a reliable line of evidence for early horse domestication (see discussion below), there is still a wide range of secondary data that can be useful in documenting horse domestication. These data are described in the discussion of research in northern Kazakhstan that follows.

As seen in the review of previous efforts of researchers working on this problem across the Eurasian continent, clearly documenting where and when horses were first domesticated is not an easy task. At present, there is no smoking gun, no single definitive criterion for demarcating the domestication of the horse in the archaeological record. The solution to documenting horse domestication lies in attacking the problem with a holistic strategy that looks at a wide range of secondary, more circumstantial evidence for the keeping of horses. When the secondary, indirect evidence reaches critical mass, then it becomes more and more likely that horse domestication had, indeed, taken place. The multidisciplinary approach to the study of horse exploitation of the Botai-age peoples in the central Eurasian steppe of modern day Kazakhstan discussed below demonstrates the utility of applying multiple lines of investigation in order to document early horse domestication.

Integrated Multidisciplinary Research at Botai Culture Settlements

The Geometric Impressed Pottery Province (GIPP) of the eastern Urals and northern Kazakhstan (Olsen et al. 2006) shows real promise for providing valuable insight into the process of horse domestication. The Botai culture of northern Kazakhstan (Zaibert 1993) is now seen as one of the most crucial sources of information for documenting this landmark in human history. This is not, however, because the Botai region necessarily represents the first or only site of horse domestication. Instead, the Botai culture provides the optimal case study for this difficult task because Botai sites are located in the heart of the native geographic range of the wild horse and date to the fourth millennium, sometime soon after it is thought horse domestication began. Moreover, the Botai based their whole economy on the horse, and their large, permanent settlements have yielded enormous collections of horse remains. As a result, Botai sites provide an ideal opportunity for developing a multidisciplinary, holistic approach to the problem.

The Botai culture is represented by four known settlements: Botai, Krasnyi Yar, Vasilkovka, and Roshchinskoe (Figure 17.1). Of these, most of the work has been done on the largest site, Botai, which has been radiocarbon dated to 3700–3100 BC (calibrated) (Levine and Kislenko 1997). Excavations have been conducted regularly at Botai since the 1980s by the University of North Kazakhstan, under the direction of Victor Zaibert (1993). The nine-hectare settlement, located on a small tributary of the Ishim River, has at least 158 houses, but many more have been washed away by the encroaching Iman-Burluk River. Over half of the site has been excavated.

Two of the other sites, both located near the modern city of Kokshetau, have been investigated to a lesser extent. Two pithouses at Krasnyi Yar and one at Vasilkovka were excavated in the 1980s (Kislenko and Tatarintseva 1999; Zaibert 1993). Their assemblages compare very closely with those of Botai. Recently, a joint team of archaeologists from the Carnegie Museum of Natural History and the Presidential Cultural Center of Kazakhstan has excavated one house, plus adjacent areas at each of these settlements. Based on detailed remote sensing, it is possible to determine that Krasnyi Yar had at least 54 houses and Vasilkovka had 44 (Olsen et al. 2006). The fourth site, Roshchinskoe, is the least known and has had very little testing or mapping conducted.

The faunal assemblage of the eponymous site of Botai has never been studied in its entirety, but it is estimated to total over 300,000 bone fragments. Given that most faunal assemblages from the steppe consist of a few hundred to a few thousand elements, this collection is unique. Only a tiny portion of the Botai assemblage is from animals other than the horse, so it is safe to say that 99% of the identifiable remains are derived from horses. This is a significantly higher percentage than at Dereivka, where 63% of the faunal assemblage is horse (Telegin 1986). Dereivka itself has a notably greater focus on horses than most other contemporaneous sites in the Ukrainian and Russian steppe.

Previous work on the Botai horse remains includes the metric analysis of 10,000 postcranial bones by Akhinzhanov et al. (1992) and numerous partial and complete crania by Kuzmina (1997), inspection of 36 lower second premolars for possible bit wear by Anthony and Brown (1998), mortality patterns and pathological investigation by Levine (1999a, 1999b), and studies of the taphonomy, butchery, ritual treatment, mortality patterns, dog burials, and bone artifacts by Olsen (1996, 2000a, 2000b, 2001, 2003). Since 1993, my research has focused on understanding the horse-based lifestyle and economy of the Botai people using a variety of methods, including remote sensing of whole villages; excavation of houses, middens, and pits; faunal analysis; and studies of related artifacts in an effort to pursue multiple lines of evidence in the documentation of early horse domestication (Olsen et al. 2006).

Mortality Patterns at Botai

Levine (1999a) was the first to perform a detailed examination of the dentition from Botai using the crown-height method discussed above. She concluded that there were few fatalities before 2.5 years in the sample of 529 cheek teeth (MNI 29) that she studied. According to her findings, most individuals were slaughtered between the ages of 3 and 8 years, with about a quarter of the harvested animals dying between the ages of 8 and 16. Based on mandibles, the ratio of males to females was 7:6; based on pelves, it was 17:20 (Table 17.4). Because this profile resembled that in a living wild herd, she interpreted these mortality data as the result of a hunting strategy using communal drives. Thus Levine (1999a) concluded that the horses at Botai, in Kazakhstan, were wild animals hunted by means of stalking, just as she had proposed to be the case at the slightly earlier site of Dereivka, in Ukraine.

In 1996, I studied all the Botai maxillae and mandibles excavated to date, as well as the isolated teeth. Estimated dental ages in my study were based on dental eruption and occlusal wear (Hillson 1986: 215; Kainer and McCracken 1994: Plate 48, 49). Minimum numbers of individuals for the whole dental assemblage were obtained by recording all intact, identifiable teeth (both those in jaws and isolated teeth), a total sample of 748 teeth. For isolated teeth, as mentioned above, it is not always possible to distinguish among third and fourth premolars and first and second molars. In such cases, they were grouped and the sums were divided appropriately to yield estimated MNI figures, but these never exceeded those for specifically identifiable teeth in the overall MNI calculations.

Sex ratios obtained for mandibles in my study resemble those reported by Levine based on mandibles and pelves (Table 17.4) and show a fairly even proportion of adult males and females in the sample. The maxillary data differ in exhibiting a very high ratio of males to females. The high

TABLE 17.4
Comparison of Proportions of Adult Female and Male Horses at Botai

Sex Indicator	Analyst	MNI Female	MNI Male	% Male
Mandibles	Levine	6	7	54%
Pelves	Levine	20	17	46%
Mandibles	Olsen	12	13	52%
Maxillae	Olsen	1	8	88%

TABLE 17.5
Distribution of Botai Horses by Sex/Age Categories, Based on Mandible and Maxilla MNIs

Age	Juvenile, Sex Indeterminate, MNI	Female MNI	Male MNI
5–14 mo	1		
9 mo–2.5 yr	1		
2–3 yr	3		
3.5 yr	1		
4 yr		1	2
5 yr		1	1
6 yr		4	5
7 yr		5	3
8 yr		1	1
>8 yr			1
Totals	6	12	13

proportion of adult male maxillae, however, is likely a reflection of the differential deposition and preservation of stallion crania in ritual pits. Mandibles are known to have been included on occasion in these ceremonial contexts, but with less frequency than crania. Thus, the lower jaw data are more likely to represent overall slaughter patterns both for subsistence and ritual, while the maxillary data show the influence of ritual practices on the assemblage.

The age profile that I have compiled for the Botai horses based on dental eruption and occlusal wear (Table 17.5) shows that most of the horses died as mature adults, rather than through culling of juveniles. In a domestic herd, this pattern would imply an emphasis on the use of secondary products or services supplied by adults. In the case of horses, the females were probably maintained for breeding and milking, while the males would have been kept for riding and as packhorses. Samples are small, but there is some slight (probably not significant) tendency in these data for males to be killed at younger ages than females. This might represent a minor amount of culling of young males. In modern times, only a small number of stallions are ridden, primarily by the horse herders, in comparison to the much greater numbers of mares that are maintained for breeding and milking. In prehistoric times, however, riding would have been an important form of transportation for hunting, herding, and traveling in general.

My mortality pattern for Botai differs from Levine's based on crown heights partly because different methods were employed to determine age. Levine compensates for perceived differential destruction of deciduous teeth by artificially inflating the numbers of immature individuals, but deciduous teeth are not significantly different from permanent teeth in their mineral content and are therefore not more susceptible to taphonomic destruction. Levine's method of dealing with the ambiguities of isolated second and third premolars and first and second molars also differs from my method of calculating MNIs on teeth whose positions can be more confidently identified. Finally, the use of crown heights (in contrast to the use of eruption and occlusal wear) does make it possible to estimate the ages of older adults beyond the age of eight, which is why Levine has a broader range of years in her data. In large measure, however, my age profile generally agrees with Levine's. Both indicate that there is an emphasis on mature males and females between 3 and 8 years of age.

It is also useful to examine the ratio of juvenile teeth to adult teeth for both incisors and the more durable and recoverable second premolar (P2) (Table 17.6). Based on the P2 alone, the proportion of juveniles under the age of 2.5 years in the sample would seem relatively high—at about 35%. However, based on the I1, which erupts about the same time, this figure is significantly lower—at about 22%. Because incisors are easily lost during excavations, they are less likely to be truly representative of actual patterns and are more susceptible to random variation than are the larger, more durable premolars. This is illustrated clearly by the complete absence of deciduous second and third incisors. The MNIs for jaws are lower than for P2s because there are so many more isolated teeth, so in this case, the readily identifiable second premolars provide extremely useful and reliable assessments of proportions of very young vs. animals over the age of 2.5 years. Mandibles were regularly broken and the teeth were removed in order to make thong-smoothers. To a large extent, this practice certainly contributes to the paucity of ageable mandibles in this collection. Thus the P2 estimate of 35% for juveniles under 2.5 years of age at Botai should be considered a more accurate number.

However it is computed, the percentage of young in the Botai sample is not high enough to indicate a very strong emphasis on the culling of juveniles. The adult sex profile also does not reflect a prevalence of adult females, as is expected in an assemblage derived from a focus on breeding and milking. However, as discussed above, the Botai horse assemblage is likely to contain both domestic and hunted wild horses. Moreover, if domesticates were important for riding and as pack animals, in addition to meat and milk, then more adult males would be expected in the assemblage. Finally, the intentional burial of skulls of ritually sacrificed adult males leads to better preservation of these animals than juveniles and adult females, so it is not surprising that the sex ratio for maxillae is biased toward males at Botai. Although an emphasis on adults, as detected by Levine and myself, might

TABLE 17.6
Comparison of Proportions of Juvenile and Adult Horses at Botai

Age Indicator	Age Separating Juvenile and Adult	MNI Juvenile	MNI Adult	% Juvenile
Mandible	4 yr	6	25	19%
Maxilla	4 yr	0	9	0%
P2	2.5 yr	8	15	35%
I1	2.5 yr	8	28	22%
I2	3.5 yr	0	29	0%
I3	4.5 yr	0	23	0%

FIGURE 17.4 Modern horse cranium in Mongolia with pole-axing wound.

FIGURE 17.5 Horse cranium with possible pole-axing wound from Botai.

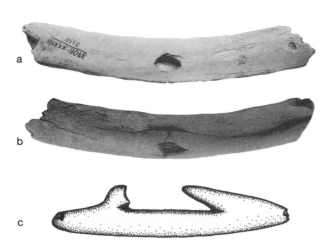

FIGURE 17.3 Botai hunting tools and traces: (a) wound in horse rib, medial surface; (b) same wound, lateral surface; (c) harpoon.

be consistent with hunting, it could equally reflect a mixed population of wild and domestic animals. Also, the retention of adult females for breeding and milking and adult males for riding could lead to higher numbers of adults in domestic horse populations than would be typical for other livestock that are not used for milk and riding. The differential preservation of stallion skulls in ritual pits is also clearly a factor in increasing the proportion of adults in the record.

Slaughter Methods

Studies of the faunal material from Botai included a detailed examination of bones in order to understand the distribution and nature of wounds incurred during slaughter (Olsen 2003). Based on the shape and size of their apertures, three wounds in horse bones (Figure 17.3) and one in an aurochs bone appear to have been made by harpoons, rather than by stone projectile points. This evidence suggests that wild horses were still in the region and were occasionally hunted, as they were in the preceding Neolithic.

Richard Marlar, a blood pathologist at the University of Oklahoma, performed blood residue analysis on 12 stone projectile points from Krasnyi Yar and found hemoglobin residues on three specimens. All three tested positive for human blood, which is rather unusual. One also tested positive for suid (most likely wild boar) and wild (most likely) or domestic sheep. The Argali sheep, *Ovis ammon*, exists in the mountains of Kazakhstan, although no remains have been found in Botai sites to date. The third point tested positive for human, suid, canine (fox, wolf, or dog), ibex or goat, and bovine (aurochs, bison, or domestic cattle). Ibexes, like Argali sheep, still can be found in the mountains of Kazakhstan but have not been identified in the faunal remains. Surprisingly, no horse blood was identified on the stone points, but perhaps harpoons were preferred weapons for hunting wild equines.

One common method of slaughtering animals in the steppe, pole-axing, was most likely reserved for domestic livestock. Whereas a single hunter can use a bow and arrow to bring down a wild beast, pole-axing usually involves three persons or at least one person and two posts firmly set in the ground. Normally, two people hold the ends of lassos that are wrapped around the neck of the horse and pull them taut on either side of the standing beast. In this position, it is not easy for the horse to step backward or move out of harm's way. A third person then approaches and strikes the animal between the eyes with a pole-ax, ideally killing it in one stroke. Pole-axed cattle and horse skulls are common in kurgans (burial tumuli) from the Bronze Age across Russia and Kazakhstan. The weapon typically leaves a large, depressed fracture in the frontal bones, like the one shown on a modern skull of a horse from Mongolia that was killed in this manner (Figure 17.4). One cranium in the Botai collection (Figure 17.5) shows a round, depressed fracture with the broken tissue still *in situ*; however, the injury is more lateral than normal and is located on the maxilla, instead of the frontal. It is possible that the blow was struck just at the moment the animal averted its head. Thus, although this cranium seems to bear the kind of

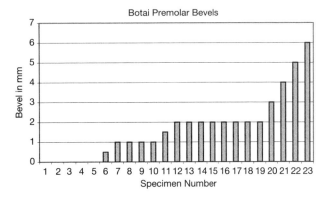

FIGURE 17.6 Graph of Botai P2 bevel measurements taken by Olsen.

fracture caused by pole-axing, the placement of the fracture is not entirely consistent with this practice. Therefore, this skull cannot be considered solid evidence for a method of slaughter typically associated with domestic or at least tame animals.

Pathologies

Neither Levine's (1999b: 53) nor my own study of the Botai material discovered any pathologies associated with riding or the use of horses as beasts of burden or for traction. As discussed above, however, riding horses without a saddle or with only a soft pad is unlikely to have a detectable impact on a horse's back. The wheel was unknown in this region in the fourth millennium, so horses probably would not have been used for draft; however, it is possible that they were used to haul heavy loads of stone or meat on their backs as pack animals. The dentition indicates that most horses were killed before their ninth year, so there was little time for arthritis and similar work-related maladies to develop.

As noted above, an earlier study of Botai horse dentition conducted by Brown and Anthony in 1992 detected some wear that they associated with the use of a soft bit (1998: 344). This has been presented as evidence of horseback riding at Botai. They reported examining 36 lower second premolars from a minimum of 20 individuals. Of these, at least 19 were older than three years of age, the minimum age for reliable detection of bit wear. Five of the premolars (26%) showed a bevel on the mesial part of the occlusal surface measuring 3 mm or more. None of these teeth exhibited any sign of either macro- or microscopic abrasion or other alteration indicative of bit wear. Based on Anthony and Brown's actualistic experiments with different types of soft bridles, they attributed these wear patterns to the use of a soft bit, probably made of hemp.

My own examination of 23 lower second premolars from Botai (including the 5 Brown and Anthony felt showed bit wear) found only 4 with beveling sufficient enough to qualify as soft bit wear using Brown and Anthony's criteria (Figure 17.6; Table 17.7). Seven of these teeth had bevels that measured 2 mm at their deepest point, which is less than

TABLE 17.7
Botai P2 Bevel Measurements Taken by Olsen

Tooth Number	Side	Bevel (mm)
3579	L	0
B82	L	0
491	L	0
33-P-37	R	0
32-2C 75b	R	0
29564	L	0.5
2025	L	1
800	L	1
549	L	1
B84-88-HM	R	1
32-P-37	R	1.5
518	L	1.75
589	L	2
2361	L	2
3583	L	2
32 no spec. #	R	2
532	R	2
32-2ZH-75b	R	2
A & B 14	L	2
A & B 21	L	3
A & B 2	R	4
A & B 7	R	5
A & B 37	L	6

the 3 mm minimum required by Brown and Anthony to qualify as bit wear. Five had no bevel or actually curled upward instead of dipping down. One tooth with an extreme 6-mm-deep bevel (Brown and Anthony's specimen no. 37) also had slightly exaggerated lingual-buccal wear along the entire tooth, suggesting some kind of malocclusion with the upper jaw. My observations of the "beveled" teeth show that the normal Greaves' Effect, of uneven wear in the dentine and enamel of the occlusal surface, was still found on all of them.

Recent investigations have identified a potential problem with restricting the identification of soft bit wear to the single characteristic of a slope on the mesial part of the occlusal surface. Upon examining lower second premolars of horses from four North American Pleistocene localities dating to many millennia before the arrival of humans, I found several examples of Brown and Anthony's "soft bit wear." *Equus lambei* (Specimen USNM 8426, curated in the Department of Paleobiology, National Museum of Natural History), from Gold Run Creek, Yukon Territory, in the Klondike region, dating to approximately 24,000 BC, has what appears to be a sharp decline (8.5 mm) that is obvious in the lingual view (Figure 17.7). The mandible has full dentition with normal wear along the whole tooth row. Another mandible of this species from near Ruby, Alaska (USNM 11705) has a bevel on

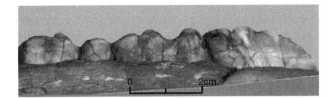

FIGURE 17.7 Lingual view of lower second premolar of *Equus lambei*.

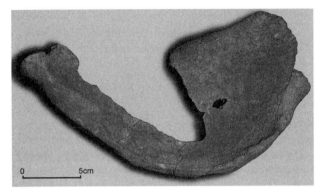

FIGURE 17.8 Thong-smoother made on a horse mandible from the Botai culture settlement of Krasnyi Yar.

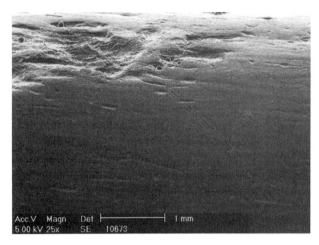

FIGURE 17.9 Scanning electron micrograph of the notch of a thong-smoother from Botai.

the P2 of 4.5 mm that includes most of the mesial third of the occlusal surface. Both of these specimens exhibited the normal Greaves' Effect. Examination of a collection of horse remains from the Big Horn Basin of Wyoming dating from 18,170 to 15,620 BC (curated by the University of Kansas and studied by Martin and Gilbert 1978) showed a high degree of variation in the wear of the P_2 under totally natural conditions. This means that the presence of a beveled second premolar alone can no longer be taken as a reliable marker of bit wear, whether from a soft bit or from any other kind. For now, the soft bit wear evidence for early horseback riding at Botai should be put aside.

Horse Tackle

At present, the "bit wear" on teeth from Botai identified by Brown and Anthony (1998) is at the very least ambiguous and is likely to be simply natural dietary wear. However, there is clear evidence that people there were manufacturing rawhide thongs that might well have been used in a wide range of activities relating to both riding and capturing horses. The most common bone implement at the site of Botai is a *thong-smoother* (Figure 17.8), an implement used to straighten and stretch a strip of rawhide (Olsen 2001). Made from horse mandibles, thong-smoothers constituted 32 percent of the assemblage of bone artifacts from Botai, with at least 135 horses required to produce them. The high polish and fine striations seen in notches on these tools (Figure 17.9) indicate that narrow strips of soft, pliable material were pulled back and forth across the edge of the notch. The widths of the notches range from 1.1–6.5 cm, with an average of 4.3 cm. Thongs made with these smoothers could have served many functions, but thongs are essential in horse control for bridles, whips, lassos, hobbles, and pole snares. Although thongs also would have been useful as harpoon lines, the small number of harpoons found at Botai (N = 13) in comparison to the large number of thong-smoothers (N = 270) suggests that there was a much more significant purpose for rawhide thongs.

Body-Part Distribution

Skeletal-part distribution data may also shed light on the status of horses from Botai culture sites. It has long been known that hunters of large game often leave parts of their prey back at the kill site (Reitz and Wing 1999; Outram and Rowley-Conwy 1998). Perkins and Daly (1968) coined the term *Schlepp Effect* for this condition, because it was the result of hunters on foot reducing the load they had to schlepp, or drag, back home by field-dressing the carcass. To lessen the transport load, hunters will abandon bones of their prey that are of low utility for either meat or marrow, or those that can be easily stripped of meat. However, hunters may choose to transport certain bones back to their base camps for later manufacture into bone tools or for ceremonial purposes. Given that horses fall into the large game category, it seems likely that, without some form of transport, horse hunters would seek to reduce transport weight by field butchering wild horses and abandoning skeletal elements not needed for food, manufacture, or ritual at the kill site. In fact, as long ago as 1947, Arcikhovski predicted that an assemblage of domesticated horses should include relatively complete skeletons, while that of wild horses would be lacking certain elements, particularly vertebrae and scapulae (cited in Bökönyi 1974b: 237).

A butchering experiment conducted by Bruce Bradley and myself on a horse carcass, using replicas of Botai stone tools, indicated that, with the exception of the intercostal muscles of the rib cage, most of the meat could be removed easily from

TABLE 17.8
Artifact Raw Material from Horse Elements at Botai

Element	Number of Artifacts	MNI by Element
Mandible	270	135
Hyoid	7	4
Scapula	10	5
Rib	158	5
Innominate	6	3
Radius	1	1
Metapodial (3/4)	108	27
Metapodial splint	64	8
First phalanx	44	11
Second phalanx	2	1
Astragalus	1	1
Total	671	135

the carcass in the field. The most efficient treatment of the ribs was to cook the meat on the bone and eat it directly off the ribs, as we do in restaurants today. So, unless the ribs were cooked and eaten by hunters in the field, the only high meat yield elements that would necessitate being carried back to the home base would be the ribs. The use of bone marrow, however, also played a role in determining which bones were transported back from a kill site. Outram and Rowley-Conwy's (1998) study of meat and marrow content of horse bones indicated that the most important marrow-yielding bones in horses are the mandible, humerus, radius, femur, and tibia. The metapodials and proximal phalanges also produce small amounts of marrow.

Artifacts made from horse bones at Botai culture sites include mandible thong-smoothers, notched rib stamps for decorating pottery, scapula paddles for smoothing ceramic vessels, decorated first and second phalanges, and gouges, leather punches, cylinder stamps, and harpoons made from metapodials (Table 17.8). Other elements with nonfood-related value would have been crania used in ritual contexts.

Taking all these factors into consideration, the elements most likely to be left behind at the kill site of wild horses would be the vertebrae. It is relatively easy to remove the large muscles of the back like the *longissimus dorsi*, which runs from the sacrum and the ilium all the way to the neck on both sides of the vertebral column. The vertebral column also has low marrow content, has little utility in tool manufacture, and is weighty and difficult to separate into portions that would be less awkward to carry. The large, cumbersome pelvis, which is tightly connected to the vertebral column, would also seem a likely element to leave behind at the kill site after the meat was removed.

Analysis of cutmarks and breakage patterns on the Botai horse remains (Olsen 2003) indicates that Botai people were intensively using most parts of the horse, including brains, tongues, meat, marrow, sinew, and skins. Ribs were snapped at the necks in a way that might facilitate transport. However, fully articulated vertebral columns are found frequently at both the sites of Botai and Krasnyi Yar, as are pelves. The high proportion of these parts in the Botai settlements suggests either that some horses were killed and butchered at or near the village (perhaps as domesticates) or that alternative means of transportation, such as packhorses, were available to hunters. Related to this latter idea, it is interesting to note that the crania of wild aurochs, with horn cores up to a meter in length, were found at Krasnyi Yar, along with vertebrae and other large bones that might have been expected to have been left at the kill site. Carrying these heavy, bulky elements clearly would have been difficult without the aid of some form of transport, suggesting that perhaps domestic horses aided humans in bringing back carcasses of hunted large game, including wild horses and aurochs.

Evidence for Lithic Transport

There is also an interesting change in lithic sources and use in Botai culture sites compared to earlier Neolithic sites in the region that supports the notion that domestic horses may have been used in transport. The Neolithic site of Zhusan, just 600 m from Krasnyi Yar, produced finished blades and blade tools made from local quartzite available at a quarry 9 km from the two sites. Bradley found 30 shallow pits surrounded by flaking debris at the Zhartas quarry site (Olsen et al. 2006). Blade precores and waste products also found there indicate that blades, probably of Neolithic age, were prepared at the Zhartas quarry site. Small bifaces, typical of the later Botai lithic assemblages, were also present at Zhartas, and may indicate continued use of the quarry.

In addition to quartzite tools, the Botai culture sites of Krasnyi Yar and Vasilkovka produced a much wider array of raw materials from a number of unknown sources. Jasper, flint, and especially a fine-grained quartzite are plentiful in Botai lithic assemblages, despite the fact that several surveys have failed to locate the sources of these materials. Detailed surface geology maps do not indicate potential local outcrops either. Jasper is common, however, in the Ural Mountains far to the west (Matyushin 2003), raising the possibility that some or all of these materials were derived from more distant sources.

Bradley's analysis of the Krasnyi Yar assemblage (Olsen et al. 2006) suggests that some of the stone was transported to the village as large, heavy pieces that were then knapped into bifaces, scrapers, and other flake tools. A cache full of large flint flakes and knapping stations with high concentrations of debitage, for example, were found just outside a house excavated at Krasnyi Yar. Moreover, the biface technology characteristic of the Botai culture results in heavier tools and is clearly a less efficient use of raw material than the earlier Neolithic pressure-blade technology. The diversification of stone sources used by the Botai (some of which may be quite distant from their homeland), the importation of large pieces of lithic raw material for on-site manufacture,

FIGURE 17.10 Horse cranium and cervical vertebrae in a pit outside a house at Krasnyi Yar.

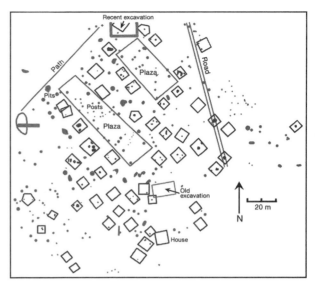

FIGURE 17.11 Plan of the Botai culture settlement of Krasnyi Yar.

and the shift to a technology that used more stone all suggest a significant change in transport costs for raw materials. The use of domestic horses in the transport of stone would certainly have had an impact on this industry. Thus, the changes in lithic technology and raw material use provide additional circumstantial evidence that might be marshaled to support the case for the use of domestic horses by the Botai.

Horse Rituals

A close connection between humans and horses is also shown in the prominent use of horses in burial and ritual contexts. The only human burial from Botai (and, for that matter, the only burial reported from any of the Botai sites) contained four humans (two men, one woman, and a 10–11-year-old child) surrounded by at least 14 horses laid out in an arc (Rikushina and Zaibert 1984). The horse remains consisted of skulls (N = 14), pelves (at least nine), articulated vertebral columns (at least five), some ribs, and a few limb elements. While it is conceivable that this burial represents the ceremonial sacrifice of 14 wild horses, as noted above, the general pattern of animal sacrifice on the Eurasian steppes has focused on domesticated animals (Jones-Bley 1997; Mallory 1981, 1996).

More commonly found in Botai sites are ritual deposits of horse skulls and articulated cervical vertebra in pits around the outsides of houses (Figure 17.10). At the site of Krasnyi Yar, for example, we found five pits containing horse skulls modified into partial plates made up of the nasals, frontals, and parietal bones. The function of these modified skulls is not clear. Another common Botai ritual involved the placement of whole dogs or dog skulls next to horse skulls, necks, pelves, or feet in pits on the west sides of houses (Olsen 2000a, 2000b). The ritual association of dogs with horses evokes the close connection between these species among steppe peoples today, where dogs are used alongside horses to hunt game and also to help herd domestic horses. This horse/dog association in Botai ritual contexts is also seen in other contemporaneous and later cultures of the steppe (Telegin 1986: 31–33), as well as in Bronze and Iron Age sites across Europe (Grant 1984: 222; Merrifield 1987; Green 1992) and East Asia (Mair 1998), where there is no question of the domestic status of the horse.

Botai Settlement Patterns

The size and layout of Botai settlements lend support to the thesis that these people were early horse pastoralists. Settlements of the preceding Neolithic hunters consist primarily of short-term camps marked by lithic scatters, or small settlements consisting of a couple of above-ground houses (Kislenko and Tatarintseva 1999; Shorin 1999; Zaibert 1992). In contrast, Copper Age Botai settlements are large villages, consisting of 44 to 158 or more houses, usually arrayed in rows running along a northwest-southeast axis (Olsen et al. 2006). Maps made using remote sensing at Krasnyi Yar and Vasilkovka clearly show carefully planned layouts of rows of square houses, with their corners pointing to the cardinal directions (Figures 17.11 and 17.12). Possible rectangular plazas evidenced in the northern portions of both sites may reflect the importance of communal activities, and the numerous series of postholes demarcating enclosures are suggestive of corrals. Conservative population estimates

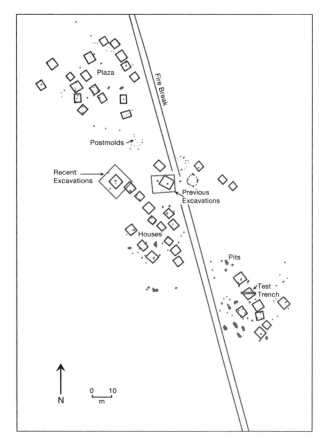

FIGURE 17.12 Plan of the Botai culture settlement of Vasilkovka.

project that from at least 110 to over 400 people lived at each of the villages of Botai, Krasnyi Yar, and Vasilkovka. Radiocarbon dates and artifact types suggest that these sites were occupied contemporaneously. If the fourth village of Roshchinskoe is included, then perhaps as many as 1,000 people lived in the relatively small Botai region.

The stark contrast between the previous Neolithic camps and the permanent, well-planned Botai villages points strongly to a change in subsistence strategies. In fact, it is difficult to imagine how such large concentrations of sedentary people could have survived the protracted and severe winters of the Siberian Plain if they were reliant solely on hunting wild horses, without any other domestic livestock or agricultural resources. The region immediately around the village would have become quickly depleted of wild horses and other game, and villagers would have been forced to venture far from their homes to obtain critical meat resources. Domestic horse herds could have provided a reliable supply of meat and milk, as well as aided in hunting and transporting people, meat, and stone. All of these assets would contribute to a significant degree in making these large, permanent villages both feasible and stable.

Presence of Horse Manure

Micromorphological analysis of soils in the fill of a pithouse excavated at Botai in 1995 revealed large quantities of horse manure lying on the collapsed roof (French and Kousoulakou 2003). The researchers who analyzed the soil column from our excavations interpreted these remains as the results of stable or corral cleaning. However, another possibility is that the Botai people used manure for insulating their houses. The Kazakhs still use horse and cow manure insulation on the roofs of their outbuildings to keep animals warm in the winter and cool in the summer. Although it is conceivable that this manure represents the droppings of wild horses collected on the steppe, the thick concentration of manure seen in the house from Botai would seem more likely to have been generated by managed herds.

Manure may also be detected through fecal biomarkers like sterols and bile acid, which can persist long after all morphological remnants have disappeared (Bull et al. 1999; Evershed et al. 1997). Future micromorphological and chemical analysis of soil samples will explore the use of manure as a building material at these sites and as evidence for the possible use of enclosures as corrals.

Evidence of Milking

One of the more promising lines of future investigation is the analysis of lipid residues on the insides of potsherds from Krasnyi Yar and Vasilkovka to determine whether horse milk was stored in vessels. Unfermented mare's milk is generally not drunk, because it is a strong laxative and purgative (Toomre 1994). Fermented mare's milk, on the other hand, is an important source of vitamins and other nutrients for the modern Kazakh, Kyrgyz, Bashkir, Mongol, Yakut, and other Eurasian steppe peoples. Koumiss, as it is known in Kazakh, or airag, in Mongolian, is reported to have been consumed on the Eurasian steppe at least since Herodotus's time (fifth century BC).

Koumiss contains five times more vitamin C than does cow's milk (Kosikowski 1982: 43), but also yields vitamins A, B1, B2, B12, D, and E and contains between 1% and 2.5% ethyl alcohol (Kosikowski 1982: 43). Mare's milk is extremely lean (1–2% fat) and is therefore normally not used to make yogurt, kefir, cheese, dried curds, or other dairy products. Mares provide milk over a roughly six-month period from April through September. The modern, high-yielding Yakut breed from central Siberia produces 1,200–1,700 kg of marketable milk per mare over one lactation period (Dmitriez and Ernst 1989), and today Kazakhstan is the leading nation in koumiss production. Fermented mare's milk is generally consumed shortly after it is made, unless it is allowed to convert into a more potent alcoholic beverage.

Dudd and Evershed (1998) have shown that lipids from both adipose tissue and milk can be preserved for thousands of years in pottery and that the products of different species can be identified using high temperature gas chromatography (HTGC) and HTGC/mass spectrometry (HTGC/MS). As part of the continuing work on Botai horse pastoralism, Alan Outram of Exeter University has collected samples of modern horse adipose tissue and milk from the vicinity of the site of Krasnyi Yar. Using these modern samples, Richard Evershed, at the University of Bristol, found that it is possible to distinguish between the two products. Dudd et al. (2003) have already identified horse adipose lipids in potsherds from Botai. Continued analysis will attempt to find out if lipids from mare's milk are present in Botai vessels as well. The finding of milk fat residues would be one of the strongest pieces of evidence for horse domestication, since it is highly unlikely that wild mares would tolerate being milked.

Conclusions

The multidisciplinary, holistic investigation performed on the Botai culture settlements in northern Kazakhstan provides substantial support for early horse domestication in this region during the Copper Age (3700–3100 BC calibrated). The equo-centric, or horse-centered, economy, establishment of large, permanent settlements with possible corrals and livestock enclosures, presence of horse manure in house fills, horse/dog ritual associations, horse slaughter patterns, horse sacrifice, abundance of thong-making tools, presence of low-utility skeletal elements with high transport costs, and changes in lithic sources and manufacturing practices all coalesce to produce a compelling argument for the existence of domestic horses in the Botai sites. Future efforts at finding residues of milk lipids on pottery and manure in enclosures may provide even more compelling proof for horse domestication. When all these lines of evidence are brought together, a strong argument can be made that the Botai were sedentary horse pastoralists who used domestic horses as sources of meat, and perhaps of milk. A good case can also be made that the Botai rode horses to aid in managing herds of domestic horses and to enhance their success in hunting wild horses and some aurochs, elk, red deer, and saiga antelope.

The probable mixed use of horses for both food and riding makes sorting out the economic focus of domestic horse exploitation a complex matter. The likely mixing of the remains of domestic and hunted wild horses in Botai middens further complicates efforts to document early horse domestication using traditional methods like size, morphology, and mortality patterns. Clearly, a combination of traditional direct, as well as indirect, circumstantial evidence must be employed to build the case for horse domestication.

This is not to say that the Botai were the first to develop horse domestication. In fact, early indications are that either people from the Urals moved into this region in the Copper Age, bringing domestic horses with them, or that the indigenous Neolithic people adopted horse domestication from neighbors to the west. Moreover, as mentioned above and discussed in detail in Chapter 23 of this volume, recent DNA studies (Vilà et al. 2001) suggest that horse domestication occurred more than once and perhaps many times along the Eurasian steppe. Therefore, it would not be surprising if other steppe cultures produce similar evidence in the near future. Ongoing work with the Botai culture of northern Kazakhstan provides important direction for those seeking to document this complex process across this broad region.

Acknowledgments

I would like to offer my gratitude to the National Science Foundation (Grants #BS 9816476 and #BCS 0415441), National Geographic Society, and Carnegie Museum of Natural History for their tremendous support for this research. My special thanks go out to the staff of the Presidential Cultural Center of Kazakhstan and the Kokshetau History Museum, Bruce Bradley, Alan Outram, our field team, all the students from Kazakhstan and America who have helped gather field data, and to my late friend Mary Littauer for the many years of advice she gave me. Finally, and most especially, I would like to thank Maral Ghabdulina and honor the memory of her husband, Kimal Akishev, for sharing their exhaustive knowledge of the prehistory of Kazakhstan and for making it possible for our projects there to happen.

References

Agapov, S. A., I. B. Vasiliev, and V. I. Pestrikova. 1990. *Khvalynskii eneoliticheskii mogil'nik*. Kuibyshev: Saratovskogo Universiteta, Kuibyshev.

Akhinzhanov, S. M., L. A. Makarova, and T. N. Nurumev. 1992. *K istorii Skotovodstva i okhoty v Kazakhstane*. Almaty: Akademia Nauk Kazakhskoi CCP.

Anthony, D. 1996. Bridling horse power: The domestication of the horse. In *Horses through time*, S. L. Olsen (ed.), pp. 57–82. Boulder, CO: Roberts Rinehart Publishers.

Anthony, D. and D. Brown 1989. Looking a gift horse in the mouth: Identification of the earliest bitted equids and microscopic analysis of bit wear. In *Early animal domestication and its cultural context*, P. J. Crabtree, D. Campana, and K. Ryan (eds.), pp. 98–116. MASCA Research Papers in Science and Archaeology, Volume 6, special supplement. Philadelphia: MASCA, The University Museum.

———. 1991. The origins of horseback riding. *Antiquity* 65: 22–38.

———. 1998. Bit wear, horseback riding and the Botai site in Kazakstan. *Journal of Archaeological Science* 25: 331–347.

———. 2000. Eneolithic horse exploitation in the Eurasian steppes: Diet, ritual, and riding. *Antiquity* 74: 75–86.

Antonius, O. 1913. *Equus abeli* nov. spec.: Ein Beitrag zur Genaueren Kenntnis unserer Quartärpferde. *Beiträge zur Paläontologie und Geologie Österreich-Ungarns und des Orients* 26: 241–301.

Baker, J. and D. Brothwell. 1980. *Animal diseases in archaeology*. London: Academic Press.

Benecke, N. 1994. Zur Domestikation des Pferdes in Mittel- und Osteuropa. Einige neue archäozoologische Befunde. In *Die Indogermanen und das pferd*, B. Hänsel and S. Zimmer (eds.), pp. 123–144. Budapest: Archaeolingua Alapítvány.

———. in press. Late prehistoric exploitation of horses in Central Germany and neighboring areas—the archaeozoological record. In *Horses and humans: The evolution of the human-equine relationship*. S. L. Olsen, S. Grant, A. Choyke, and L. Bartosiewicz (eds.). BAR International Series (number unassigned), Oxford.

Berger, J. 1986. *Wild horses of the Great Basin: Social competition and population size.* Chicago: University of Chicago Press.

Bibikova, V. I. 1986a. A study of the earliest domestic horses of eastern Europe. Appendix 2, Part 1. In *Dereivka: A settlement and cemetery of Copper Age horse keepers on the Middle Dnieper,* D. Telegin (ed.), pp. 135–149. BAR International Series, vol. 287. Oxford: BAR.

———. 1986b. On the history of horse domestication in southeast Europe. Appendix 3. In *Dereivka: A settlement and cemetery of Copper Age horse keepers on the Middle Dnieper,* D. Telegin (ed.), pp. 163–182. BAR International Series, vol. 287. Oxford: BAR.

Boessneck, J. and A. von den Driesch. 1976. Pferde in 4./3 Jahrtausend v. Chr. In Ostanatolien. *Saügetierkundliche Mitteilungen* 31: 89–104.

Bökönyi, S. 1974a. *The Przevalsky horse.* Plymouth: Souvenir Press.

———. 1974b. *History of domestic mammals in Central and eastern Europe.* Budapest: Akadémiai Kiadô.

———. 1991. Late Chalcolithic horses in Anatolia. In *Equids in the ancient world,* Volume 2, R. Meadow and H.-P. Uerpmann (eds.), pp. 123–131. Wiesbaden: Dr. Ludwig Reichert Verlag.

Boyd, L. and K. A. Houpt (eds.). 1994. *Przewalski's horse: The history and biology of an endangered species.* Albany: State University of New York Press.

Britnell, W. J. 1976. Antler cheekpieces of the British Late Bronze Age. *Antiquaries Journal* 56: 24–34.

Brown, D. and D. Anthony. 1998. Bit wear, horseback riding and the Botai site in Kazakstan. *Journal of Archaeological Science* 25: 331–347.

Bull, I. D., I. A. Simpson, P. F. van Bergen, and R. P. Evershed. 1999. Muck 'n' molecules: Organic geochemical methods for detecting ancient manuring. *Antiquity* 73: 86–96.

Butzer, K. W. 1971. *Environment and archeology: An ecological approach to prehistory,* 2nd edition. Chicago: Aldine. .

Clark, G. 1941. Horses and battle-axes. *Antiquity* 15: 50–70.

Clason, A. T. 1991. Horse remains from Swifterbant, the Netherlands. In *Equids in the ancient world,* Volume II, R. Meadow and H.-P. Uerpmann (eds.), pp. 226–232. Wiesbaden: Dr. Ludwig Reichert Verlag.

Clutton-Brock, J. 1974. The Buhen horse. *Journal of Archaeological Science* 1: 89–100.

———. 1992. *Horse power.* Cambridge, MA: Harvard University Press.

———. 1996. Horses in history. In *Horses through time,* S. L. Olsen (ed.), pp. 83–102. Boulder, CO: Roberts Rinehart, Boulder.

Clutton-Brock, J. and R. Burleigh. 1991a. The skull of a Neolithic horse from Grime's Graves, Norfolk, England. In *Equids in the ancient world,* Volume 2, R. Meadow and H.-P. Uerpmann (eds.), pp. 242–249. Wiesbaden: Dr. Ludwig Reichert Verlag.

———. 1991b. The mandible of a Mesolithic horse from Seamer Carr, Yorkshire, England. In *Equids in the ancient world,* Volume 2, R. Meadow and H.-P. Uerpmann (eds.), pp. 238–241. Wiesbaden: Dr. Ludwig Reichert Verlag.

Clutton-Brock, J. and S. Davis. 1993. More donkeys from Tell Brak. *Iraq* 55: 209–221.

Corridi, C. and L. Sarti. 1989–1990. Sulla presenza di *Equus* nell'Eneolitico italiano. *Rivista di Scienze Preistoriche* 42: 339–348.

Curci, A. and A. Tagliacozzo. 1995. Il pozzetto rituale con scheletro di cavallo dell'abitato eneolitico di Le Cerquete-Fianello (Maccarese-RM). *Origini: Preistoria e Protostoria delle Civiltà Antiche* 18: 297–350.

Davidsen, K. 1978. *The final TRB culture in Denmark.* Arkæologiske Studier V. Copenhagen: Akademisk Forlag.

Davis, S. 1976. Mammal bones from the Early Bronze Age city of Arad, northern Negev, Israel: Some implications concerning human exploitation. *Journal of Archaeological Science* 3: 153–164.

Dmitriez, N. G. and L. K. Ernst. 1989. *Animal genetic resources of the USSR.* Animal Production and Health Paper. Rome: FAO.

Dudd, S. N. and R. P. Evershed. 1998. Direct demonstration of milk as an element of archaeological economies. *Science* 282: 1478–1481.

Duerst, U. J. 1908. The horse of Anau in its relation to history and the races of domestic horses. In *Explorations in Turkestan, expedition of 1904, publication 73,* R. Pumpelly (ed.), pp. 339–442. Washington, D.C.: Carnegie Institution of Washington.

Eisenmann, V. 1984. *Sur quelques caractères adaptatifs du squelette d'Equus (Mammalia, Perissodactyla) et leurs implications paléoécologiques.* Bulletin du Muséum National d'Histoire Naturelle, 4ᵉ série, 6, section C (2): 185–195.

———. 1996. Quaternary horses: Possible candidates to domestication. In *Proceedings of the XII Congress, International Union of Prehistoric and Protohistoric Sciences.* 6: 27–36.

Epstein, H. 1971. *The origin of domestic animals of Africa,* Volume II. New York: Africana Publishing Corporation.

Evershed, R. P., P. H. Bethell, P. J. Reynolds, and N. J. Walsh. 1997. 5ß-Stigmastanol and related 5ß-stanols as biomarkers of manuring: Analysis of modern experimental material and assessment of the archaeological potential. *Journal of Archaeological Science* 24: 485–495.

Ewart, J. C. 1907. The multiple origins of horses and ponies. *Transactions of the Highland and Agricultural Society of Scotland,* Edinburgh.

Feist, J. and D. McCullough. 1975. Reproduction in feral horses. *Journal of Reproduction and Fertility,* Supplement 23: 13–18.

Flint, R. F. 1957. *Glacial and Pleistocene geology.* New York: Wiley and Sons.

Forsten, A. 1988. The small caballoid horse of the Upper Pleistocene and Holocene. *Journal of Animal Breeding and Genetics* 105: 161–176.

French, C. and M. Kousoulakou. 2003. Geomorphological and micromorphological investigations of palaeosols, valley sediments and a sunken floored dwelling at Botai, Kazakhstan. In *Prehistoric steppe adaptation and the horse,* M. Levine, A. C. Renfrew, and K. Boyle (eds.), pp. 105–114. Cambridge, UK: McDonald Institute for Archaeological Research.

Gehl, O. 1961. Die Säugetiere. In *Hohen viecheln: Ein mittelsteinzeitlicher wohnplatz in Mecklenburg,* E. Schuldt (ed.), pp. 40–63. Berlin: Akademie-Verlag.

Gening, V. F. 1979. The cemetery at Sintashta and the early Indo-Iranian peoples. *Journal of Indo-European Studies* 7:1–30.

Gening, V. F., G. B. Zdanovich, and V. V. Gening. 1992. *Sintashta.* Chelyabinsk: Iuzhno-Ural'skoe Knizhnoe Izdatel'stvo.

Getty, R. (ed.). 1975. *Sisson and Grossman's the anatomy of the domestic animals.* Philadelphia: W. B. Saunders.

Gheorghiu, D. 1994. Horse head scepters: First images of yoked horses. *Journal of Indo-European Studies* 22: 221–249.

Grant, A. 1984. Survival or sacrifice? A critical appraisal of animal burial in Britain in the Iron Age. In *Animals and archaeology:*

4. Husbandry in Europe, C. Grigson and J. Clutton-Brock (eds.), pp. 221–227. BAR International Series 227. Oxford: BAR.

Greaves, W. S. 1973. The inference of jaw motion from tooth wear facets. *Journal of Paleontology* 47: 1000–1001.

Green, M. 1992. *Animals in Celtic life*. London: Routledge.

Greenfield, H. in press. The social and economic context for domestic horse origins in southeastern Europe: A view from Ljuljaci in the central Balkans. In *Horses and humans: The evolution of human-equine relationships*, S. L. Olsen, S. Grant, A. Choyke, and L. Bartosiewicz (eds.). Oxford: BAR.

Grigson, C. 1993. The earliest domestic horses in the Levant? New finds from the fourth millennium of the Negev. *Journal of Archaeological Science* 20: 645–655.

Groves, C. 1986. The taxonomy, distribution, and adaptations of recent equids. In *Equids in the ancient world*, R. H. Meadow and H.-P. Uerpmann (eds.), pp. 11–66. Wiesbaden: Dr. Ludwig Reichert Verlag.

Guthrie, R. D. 2003. Rapid body size decline in Alaskan Pleistocene horses before extinction. *Nature* 426: 169–171

Habermehl, K. H. 1975. *Die alterbestimmung bei haus und labotieren*. Berlin: Verlag Paul Parey.

Hillson, S. 1986. *Teeth*. Cambridge Manuals in Archaeology. Cambridge: Cambridge University Press.

Hyland, A. 1990. *Equus: The horse in the Roman world*. London: Batsford.

Jones-Bley, K. 1997. Defining Indo-European burial. In *Varia on the Indo-European past: Papers in memory of Marija Gimbutas*. Journal of Indo-European Studies Monograph 19: 194–221.

Kainer, R. A. and T. O. McCracken. 1994. *The coloring atlas of horse anatomy*. Loveland, CO: Alpine Publications.

Keiper, R. and K. Houpt. 1984. Reproduction in feral horses: An eight-year study. *American Journal of Veterinary Research* 45: 991–95.

Kislenko, A. and N. Tatarentseva. 1999. The eastern Ural steppe at the end of the Stone Age. In *Late prehistoric exploitation of the Eurasian steppe*, M. Levine, Y. A. Kislenko, and N. Tatarintseva (eds.), pp. 183–216. Cambridge: McDonald Institute Monographs.

Kosikowski, F. 1982. *Cheese and fermented milk foods*. Brooktondale, NY: F. V. Kosikowski & Associates.

Kosintsev, P. in press. The horse and man in Neolithic to Early Iron Age in the territory from the Volga to the Irtysh Rivers. In *Horses and humans: The evolution of human-equine relationships*, S. L. Olsen, S. Grant, A. Choyke, and L. Bartosiewicz (eds.). Oxford: BAR.

Kremenetski, C., P. Tarasov, and A. Cherkinsky. 1997. Postglacial development of Kazakhstan pine forests. *Geographie physique et Quaternaire*, 51: 391–404.

Kuzmina, I. E. 1997. *Loshadi Severnoi Evrazii ot Pliotsena do Sovremennosti*. St. Petersburgh, Russia: Rossiiskaya Akademiya Nauk, Trudy Zoologicheskovo Institututa, Tom 273.

Legge, A. and P. Rowley-Conwy. 2000. The exploitation of animals. In *Village on the Euphrates*, A. Moore, G. Hillman, and A. Legge (eds.), pp. 423–471. Oxford: Oxford University Press.

Lernau, H. 1978. Faunal remains, strata III-I. In *Early Arad*, R. Amiran (ed.), pp. 83–113. Jerusalem: Israel Exploration Society.

Levine, M. 1982. The use of crown height measurements and eruption-wear sequences to age horse teeth. In *Ageing and sexing of animal bones from archaeological sites*, B. Wilson, C. Grigson, and S. Payne (eds.), pp. 223–250. BAR British Series 109. Oxford: BAR.

———. 1993. Social evolution and horse domestication. In *Trade and exchange in prehistoric Europe*, C. Scarre and F. Healy (eds.), pp. 135–141. New York: Oxford University Press.

———. 1999a. Botai and the origins of horse domestication. *Journal of Anthropological Archaeology* 18: 29–78.

———. 1999b. The origins of horse husbandry on the Eurasian steppe. In *Late prehistoric exploitation of the Eurasian steppe*, M. Levine, Y. Rassamakin, A. Kislenko, and N. Tatarintseva (eds.), pp. 5–58. Cambridge: McDonald Institute Monographs.

Levine, M. and A. M. Kislenko. 1997. New Eneolithic and early Bronze Age radiocarbon dates for north Kazakhstan and south Siberia. *Cambridge Archaeological Journal* 7: 297–300.

Littauer, M. A. 1963. How great was the "Great Horse"? A reassessment of the evidence. *Light Horse* 13: 350–352.

———. 1971. V. O. Vitt and the horses of Pazyryk, *Antiquity* 45: 293–294.

———. 1981. Early stirrups. *Antiquity* 55: 99–105.

Littauer, M. and J. Crouwel. 1984. Ancient Iranian horse helmets. *Iranica Antiqua* 19: 41–51.

———. 1986. A Near Eastern bridle bit of the second millennium BC in New York. *Levant* 18: 163–167.

———. 1988. A pair of horse bits of the second millennium BC from Iraq. *Iraq* 50: 169–171.

———. 2001. The earliest evidence for metal bridle bits. *Oxford Journal of Archaeology* 20: 329–338.

Longley, D. 1980. *Runneymede Bridge 1976: Excavations on the site of a Late Bronze Age settlement*. Research Volume of the Surrey Archaeological Society No. 6. Castle Arch. Guildford, England: Surrey Archaeological Society.

Lundholm, B. 1949. Abstammung und Domestikation des Hauspferdes. *Zoologiska Bidrag från Uppsala* 27: 3–292.

Mair, V. 1998. *Canine conundrums: Eurasian dog ancestor myths in historical and ethnic perspective*. University of Pennsylvania Sino-Platonic Papers 87. Philadelphia: University of Pennsylvania.

Mallory, J. P. 1981. The ritual treatment of the horse in the early Kurgan tradition. *Journal of Indo-European Studies* 9: 205–226.

———. 1996. *In search of the Indo-Europeans: Language, archaeology, and myth*. London: Thames and Hudson.

Mange, A. P. and E. J. Mange, 1980. *Genetics: Human aspects*. Philadelphia: W. B. Saunders.

Markman, S. 1969. *The horse in Greek art*. New York: Biblo and Tannen.

Martin, L. and B. M. Gilbert. 1978. Excavations at Natural Trap Cave. *Transactions of the Nebraska Academy of Sciences* 6: 107–116.

Mashcour, M., M. Fontugne, and C. Hatte. 1999. Investigations on the evolution of subsistence economy in the Qazvin Plain (Iran) from the Neolithic to the Iron Age. *Antiquity* 73: 65–76.

Matyushin, G. 1986. The Mesolithic and Neolithic in the southern Urals and Central Asia. In *Hunters in transition*, M. Zvelebil (ed.), pp. 133–150. Cambridge: Cambridge University Press.

———. 2003. Exploitation of the steppes of Central Eurasia in the Mesolithic-Eneolithic. In *Prehistoric steppe adaptation and the horse*, M. Levine, C. Renfrew, and K. Boyle (eds.), pp. 367–393. Cambridge, UK: McDonald Institute for Archaeological Research.

Merrifield, R. 1987. *The archaeology of ritual and magic*. New York: New Amsterdam Books.

Meunier, K. 1962. Zur diskussion über die typologie des hauspferdes und deren zoologisch-systematische bedeutung. *Zeitschrift für Tierzüchtung und Züchtungsbiologie* 76: 225–237.

Milisauskas, S. 1978. *European prehistory*. New York: Academic Press.

Mohr, E. 1971. *The Asiatic wild horse*. London: J. A. Allen and Co., Ltd.

Molleson, T. and J. Blondiaux. 1994. Riders' bones from Kish, Iraq. *Cambridge Archaeological Journal* 4: 312–316.

Moore, A. 2000. The excavation of Abu Hureyra 2. In *Village on the Euphrates*, A. Moore, G. Hillman, and A. Legge (eds.), pp. 189–259. Oxford: Oxford University Press.

Nobis, G. 1971. *Vom wildpferd zum hauspferd*. Fundamenta, Reihe B, Band 6. Köln: Bohlau Verlag.

Olsen, S. L. 1989. Solutré: A theoretical approach to the reconstruction of Upper Palaeolithic hunting strategies. *Journal of Human Evolution* 18: 295–327.

———. 1995. Pleistocene horse-hunting at Solutré: Why bison jump analogies fail. In *Ancient peoples and landscapes*, E. Johnson (ed.), pp. 65–75. Lubbock: Texas Tech University Press.

———. 1996. Prehistoric adaptation to the Kazak steppes. In *The colloquia of the XIII International Congress of Prehistoric and Protohistoric Sciences, Volume 16: The prehistory of Asia and Oceania*, G. Afanas'ev, S. Cleuziou, J. Lukacs, and M. Tosi (eds.), pp. 49–60. Forlì, Italy: A.B.A.C.O. Edizioni.

———. 2000a. Expressions of ritual behavior at Botai, Kazakhstan. In *Proceedings of the eleventh annual UCLA Indo-European conference*, K. Jones-Bley, M. Huld, and D. Volpe (eds.), pp.183–207. Journal of Indo-European Studies Monograph Series No. 35. Washington, D.C.: Institute for the Study of Man.

———. 2000b. The sacred and secular roles of dogs at Botai, north Kazakhstan. In *Dogs through time: An archaeological perspective*, S. Crockford (ed.), pp. 71–92. BAR International Series 889. Oxford: BAR.

———. 2001. The importance of thong-smoothers at Botai, Kazakhstan. In *Crafting bone: Skeletal technologies through time and space*, A. Choyke and L. Bartosiewicz (eds.), pp. 197–206. BAR International Series 937. Oxford: BAR.

———. 2003. The exploitation of horses at Botai, Kazakhstan. In *Prehistoric steppe adaptation and the horse*, M. Levine, C. Renfrew, and K. Boyle (eds.), pp. 83–104. Cambridge: McDonald Institute Monographs.

Olsen, S. L., B. Bradley, D. Maki, and A. Outram. 2006. Copper Age community organization in northern Kazakhstan. In *Proceedings of the University of Chicago conference on Eurasian archaeology*, D. Peterson (ed.). Monograph Series of Colloquia Pontica. Leiden: Brill Academic Publishers.

Outram, A. and P. Rowley-Conwy. 1998. Meat and marrow utility indices for horse (*Equus*). *Journal of Archaeological Science* 25: 839–849.

Perkins, D. and P. Daly. 1968. The potential of faunal analysis: An investigation of the faunal remains from Suberde, Turkey. *Scientific American* 219: 96–106.

Piggott, S. 1962. Head and hoofs. *Antiquity* 36: 110–118.

Rassamakin, Y. 1999. The Eneolithic of the Black Sea steppe: Dynamics of cultural and economic development 4500–2300 BC. In *Late prehistoric exploitation of the Eurasian steppe*, M. Levine, Y. Rassamakin, A. Kislenko, and N. Tatarintseva (eds.), pp. 59–182. Cambridge: McDonald Institute Monographs.

Redding, R. 1981. The faunal remains. In *An early town on the Deh Luran Plain, excavations at Tepe Farukhabad*, H. T. Wright (ed.), pp. 233–261. Memoirs of the Museum of Anthropology, University of Michigan, 13. Ann Arbor: University of Michigan Press.

Reitz, E. J. and E. S. Wing. 1999. *Zooarchaeology*. Cambridge: Cambridge University Press.

Rice, P. 1963. *Changes in the Kazak pastoral economy: History and analysis*. MA thesis, Ohio State University, Columbus.

Riegel, R. J. and S. E. Hakola. 1996. *Illustrated atlas of clinical equine anatomy and common disorders of the horse*, Volume 1. Marysville, OH: Equistar Publications.

Rikushina, G. B. and V. F. Zaibert. 1984. Predvaritel'noe soobshchenie o skeletnykh ostatkakh lyudei s eneoliticheskovo poceleniya Botai. In *Bronzovyi Vek Uralo-Irtyshskovo mezhdurech'ya*, pp. 121–134. Chelyabinsk, Russia: Ministerstvo Bysshevo I Srednevo Spetsial'novo Obrazovaniya RSFSR.

Rudenko, S. 1970. *Frozen tombs of Siberia: The Pazyryk burials of Iron-Age horsemen*. M. W. Thompson (trans.). Berkeley: University of California Press.

Ryder, O. 1994. Genetic studies of Przewalski's horses and their impact on conservation. In *Przewalski's horse: The history and biology of an endangered species*, L. Boyd and K. A. Houpt (eds.). Albany: State University of New York Press.

Shorin, A. F. 1999. *Eneolit Urala i Copredel'nykh Territoriii: Problemy kul'turogeneza*. Yekaterinburg, Russia: Rossiiskaya Akademiya Nauk.

Silver, I. A. 1963. The ageing of domestic animals. In *Science and archaeology*, D. Brothwell and E. Higgs (eds.), pp. 250–268. New York: Basic Books.

Skorkowski, E. 1956. Systematik und abstammung des pferdes. *Zeitschrift für Tierzüchtung und Züchtungs-biologie* 68: 42–74.

Steppan, K. 1998. Climatic fluctuations and Neolithic adaptations in the 4th millennium BC: A case study from south-west Germany. In *Papers from the EAA third annual meeting at Ravenna 1997, Volume I: Pre-and protohistory*, M. Pearce and M. Tosi (eds.), pp. 38–45. BAR International Series 717. Oxford: BAR.

———. in press. Neolithic human impact and wild horses in Central Europe. In *Horses and humans: The evolution of human-equine relationships*, S. Olsen, S. Grant, A. Choyke, and L. Bartosiewicz (eds.). Oxford: BAR.

Telegin, D. 1986. *Dereivka: A settlement and cemetery of Copper Age horse keepers on the Middle Dnieper*. BAR International Series, Vol. 287. Oxford: BAR.

Telegin, D. Y. and I. D. Potekhina. 1987. *Neolithic cemeteries and populations in the Dnieper basin*. BAR International Series, Vol. 383. Oxford: BAR.

Toomre, J. S. 1994. Koumiss in Mongol culture: Past and present. In *Milk and milk products from medieval to modern times*, P. Lysaght (ed.), pp. 130–139. Edinburgh: Canongate Academic Press.

Uerpmann, H.-P. 1990. Die domestikation des pferdes im Chalkolithikum West- und Mitteleuropas. *Madrider Mitteilungen* 31: 109–153.

Vasilev, I. and G. Matveeva. 1979. Mogil'nik u s. S'ezzhee na R. Samara. *Sovietskaya Arkheologiia* 4: 147–166.

Vilà, C., J. Leonard, A. Götherström, S. Marklund, J. Sandberg, K. Lidén, R. Wayne, and H. Ellegren. 2001. Widespread origins of domestic horse lineages. *Science* 291: 474–477.

Vitt, V. O. 1952. Loshad Pazyrykskikh kurganov. *Sovietskaya Arkheologiia* 16: 163–205.

Vörös, I. 1981. Wild equids from the Early Holocene in the Carpathian Basin. *Folia Archaeologica* 32: 37–68.

Wayne, R. K. 1986. Cranial morphology of domestic and wild canids: The influence of development on morphological change. *Evolution* 40: 243–261.

Welsh, D. 1975. *Population, behavioural, and grazing ecology of the horses of Sable Island, Nova Scotia*. PhD dissertation. Dalhousie University.

Wijngaarden-Bakker, L. 1974. The animal remains from the Beaker settlement at Newgrange, Co. Meath: First report. *Proceedings of the Royal Irish Academy*. 74: 313–383.

Wilson, G. L. 1978. *The horse and dog in Hidatsa culture*. Lincoln, Nebraska: J and L Reprint Co, Reprints in Anthropology 10. Reprinted from Anthropological Papers of the American Museum of Natural History 15, 2, 1924.

Zaibert, V. F. 1992. *Atbasarskaya Cultura*. Ekaterinburg: Rossiiskaya Academiya Nauk.

———. 1993. *Eneolit Uralo-Irtyshskovo*. Almaty, Kazakhstan: Academiya Nauk Respubliki Kazakhstan.

Zeder, M. A. 1986. The equid remains from Tal-e Malyan. In *Equids of the Old World*, Volume 1, R. Meadow and H.-P. Uerpmann (eds.), pp. 367–412. Beihefte zum Tubinger Atlas des Vorderern Orients. Tubingen, Germany: University of Tubingen.

———. 1991. *Feeding cities: Specialized animal economy in the ancient Near East*. Washington, D.C.: Smithsonian Institution Press.

Zeuner, F. E. 1963. *A history of domesticated animals*. New York: Harper and Row, New York.

SECTION FOUR

GENETIC DOCUMENTATION OF ANIMAL DOMESTICATION

Daniel G. Bradley, Section Editor

CHAPTER 18

Documenting Domestication
Reading Animal Genetic Texts

DANIEL G. BRADLEY

Archaeogenetics

Archaeogenetics is the meeting point of several disciplines. At its base, of course, it is the attempt to marry genetic data to archaeological themes and make meaningful, perhaps even innovative inference about the past. However, even within the discipline of genetics it occupies a new position—one where the subdisciplines of evolutionary phylogeny, theoretical population genetics, molecular ecology, and medical/veterinary genetics have met. From these eclectic origins a coherence is developing in the approaches needed for the difficult task of making fine distinctions between alternative hypotheses within shallow time depths.

The last 15 years have seen the accessibility of genetic data expand exponentially. The typing of single base-pair differences within handfuls of samples was a task that took a (moderately dedicated) PhD student such as myself a couple of weeks in the 1980s. Now commercial companies, which are capable of generating hundreds of thousands of genotypes in a day, are heralding a new era of data density.

One of the most productive developments has been the wide utilization of richly informative haplotypes. A haplotype is a package of genetic material that incorporates multiple variable sites or markers, and which can be considered as a unitary, heritable package that is uncomplicated by recombination through the generations. The whole mitochondrial DNA molecule and the vast bulk of the mammalian Y chromosome are passed through the generations as haplotypes. Markers very close together or individual gene sequences on autosomes may sometimes also be considered as such and analyzed accordingly.

The examination of these genetic packages as evolutionary units and the reconstruction of their ancestral relationships is the province of phylogeny, where it is expected that the evolutionary trajectories of haplotypes will match those of their species hosts. However, the simplicity of haplotypic-taxonomic identity is not necessarily informative or true within species. Perhaps the most extreme example lies within African cattle. In that continent, mtDNA seem almost exclusively to be of *Bos taurus* origin, whereas the majority of most genomes is *Bos indicus* (MacHugh et al. 1997). Inference from a single genetic system may be misleading. Also, with intraspecific study of haplotypic variation, linkages with other types of data are needed. These other attributes include geography, morphology, and ecology. The partnering of these varied sources of information with genetic variation falls into the burgeoning field of phylogeography, where aspects of the past of an organism are read from modern distributions of phylogenetically related sequence variants.

Phylogeographic Discontinuities

One of the clearest illustrations of the power of a phylogeographic approach within species lies in the interpretations of genetic diversity in multiple wild European species (Hewitt 2000). Intriguingly, a recurring feature of these distributions of sequences, or haplotypes, has been geographic discontinuities, or suture zones. Thus, hedgehogs, water voles, or crested newts sampled in central Europe can possess mtDNA haplotypes that differ by several percent (i.e., that are more than hundreds of thousands of years divergent) from those sampled just tens of kilometers further east. Conversely, the differences between these populations and those hundreds of kilometers further south or north can be minimal. These recurring patterns showing east-west discontinuity in phylogeography and north-south continuity within species are interpreted as legacies of the periodic ice ages, which forced flora and fauna to repeatedly retreat to southern refugia in Iberia, the Italian peninsula, and the Balkans. Within these separate ecological islands the populations accumulated divergence from each other, sometimes over several glaciation cycles. With the retreat of ice sheets during the Holocene, species recolonized northward again in parallel streams from these refugia. Geographic and phylogenetic discontinuities are legacies of expansions outward from a discrete number of southern glaciation havens.

Phylogeographic discontinuity is also a recurring theme in domestic animal diversity. However, in this case the restricted origins from which divergent genomes have emerged are interpreted as resulting from human rather than geological activity, that is, from domestication events. It seems that within many species, the initial capture and breeding of wild populations took place in restricted episodes. These were often two episodes, or several, that were separated geographically, and perhaps also temporally. The easiest visualization of these is through simple reconstructed phylogenies of mtDNA control region sequences sampled in domestic ungulates (Figure 18.1). In each species, there are clusters of closely related sequences that are separated by longer internal branches. Each of these clusters probably represents the domestic descendents of a localized capture of the wild progenitor species. Also, each tends to show somewhat different geographic distributions from sister clades—reflecting the

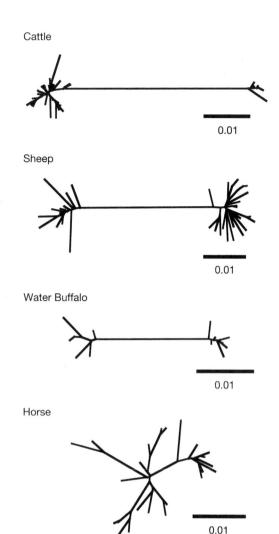

FIGURE 18.1 Neighbor-joining networks linking mtDNA sequences from four domestic animal samples. Phylogenies were constructed using sequence divergences calculated without correction and discarding positions displaying insertions or deletions (after MacHugh and Bradley 2001, Fig. 2). Each branch tip represents a complete control region sequence (except water buffalo, which are cytochrome b sequences), and the phylogeny scales in proportion to divergence are given by horizontal bars. These illustrate the phylogenetic discontinuities that are a motif of domestic animal diversity patterns. The discrete clusters within each phylogeny are separated by lengthy internal branches corresponding to hundreds of thousands of years of divergence. Each cluster probably corresponds to a series of captures from the wild in one or a few domestication centers localized in time and space. In cattle and water buffalo, the phylogenetic distinction between clusters is matched by geographic distributions separated on an east-west axis. Sheep have recently had a third discrete lineage cluster described (Chapter 20), a pattern matched by goat (Figure 19.5b). These both show geographically distinct distributions of clusters that are less clear. Horse, a relatively late domesticate and a highly mobile species, shows little phylogenetic or phylogeographic discontinuity (see Chapter 24). This is undoubtedly the legacy of a domestication process that was less limited in space and time—possibly extending over much of the Eurasian steppe.

differing imprint of the separate domestication events in different regions.

It is the discovery of this phylogenetic motif recurring with striking persistence that is the strongest result from domestic animal archaeogenetics. The implication is that, for the majority of domestic animal species, the process of domestication was both restricted and multiple. Thus, in many species two extreme hypotheses about domestication may be discounted. The first is that domestication was an unfocused activity in which variants from the wild were captured continuously through history and throughout the range of a progenitor species. The discrete clades of domestic diversity suggest too restricted a sampling of a wild pool for this scenario to hold. Second, the duality (often multiplicity) of phylogeographic distributions point eastward, away from an exclusive focus on the Near East as the sole primary center of domestication. It should be remembered that the discovery of separate divergent mtDNA haplotypes in a domestic gene pool also points toward a duality in whole genomes—with important consequences for exploitation and conservation (Hall and Bradley 1995).

There are some interesting differences in the genetic patterns observed in the several species studied. Roughly, mtDNA diversity can be divided into three categories: (1) those species that show strong phylogenetic discontinuity and clearly alternative phylogeographic distributions (cattle, water buffalo); (2) those with strong phylogenetic discontinuity but weak phylogeography (sheep, goat, llama/alpaca, dog, ass, yak); and (3) those that give neither clearly separated phylogenetic clusters nor strong geographic patterns of haplotype distribution (horse).

In cattle (Chapter 22) the phylogeography of the two major clusters is well defined. One centers on the Indian subcontinent and the other dominates absolutely in Europe and Africa. Pig also shows an east-west division in cluster geography, although one that is complicated by some importation of East Asian genetics in recent centuries during the creation of modern European breeds (Giuffra et al. 2000; Watanobe et al. 1999; see also Chapter 15). Water buffalo discontinuity reflects the well-known subspecies and east-west geographic distinction between river and swamp buffalo.

Within South American camelids, Wheeler et al. (Chapter 23) show that two separate families of mtDNA haplotypes may be discerned, from multiple analyses, as corresponding to two domestic species (llama and alpaca), despite substantial blurring of the distribution of lineages by interbreeding. When the findings of Luikart et al. (Chapter 20) and Bruford et al. (Chapter 21) are compared, a striking similarity between sheep and goat is apparent. Both species are dominated by a single, geographically dispersed cluster. However, within each species there are two additional clusters, less numerous and more geographically restricted. Both chapters suggest that the main cluster is of Fertile Crescent origin and that the additional clusters are legacies of separate domestications elsewhere. The relatively unclear geographic distinction between clusters is attributed to the tradeability of both species. In other words, the migration and trade of these smaller currency units over the millennia may have obscured earlier clear distribution patterns. Recent work in goat illustrates an important caveat for many of the results presented in this volume—the possibility of as yet unsampled lineages. In many species the key regions of South and Central Asia are relatively poorly sampled. Both Sultana et al. (2003) and Joshi et al. (2004) have uncovered new, rare mtDNA lineages in Pakistan and India, respectively, that are sufficiently divergent from the major types to merit their assignment to biologically separate wild captures.

Patterns of dog mtDNA diversity are also dominated by a single predominant and widely distributed phylogenetic cluster with divergent minor additional lineages. An advantage of canine study is that the wild progenitor species (gray wolf) is accessible for sampling, and three smaller dog clusters have discernibly separate origins because of their phylogenetic dispersal among wild lineages. Dog lineages are not clearly distributed geographically, with the exception of clade II, which is found solely within two Scandinavian breeds, suggesting a localized wolf input (reviewed in Chapter 19). The predominance of a single numerous and more diverse clade within dog, sheep, and goat may point toward a primary domestication event within each, with the additional lineage clusters resulting either from wild introgressions or from later or more restricted separate domestications.

The domestic ass shows two divergent families of sequences that have overlapping distributions, both in the region of origin in Northeast Africa and also throughout much of its range (Chapter 24; Beja-Periera et al. 2004). However, the mtDNA phylogenetics of the other domestic equid, horse, are an exception to the general rule. Despite levels of diversity similar to those in other species and a reasonably wide sampling, no genetically distinct clusters of lineages have emerged. This is indicative of domestication with less restriction over time and space, in which many separate lineages have been incorporated. However, it should be added that some weak phylogeographic signal has been suggested, with some lineages being found predominantly within Northern European ponies—perhaps indicating some localized western domestication (Jansen et al. 2002).

Anchoring Genetic Data to Archaeological Narrative

Phylogeography has been described as a subject with two orthogonal dimensions—the horizontal one of the geography of genetic diversity and the vertical one representing the time through which this geography emerges (Avise 2000). The error associated with reading the past from modern geographic distributions is obvious: These distributions have in many cases been complicated by multiple layers of migration and trade superimposed on the original structure created by domestication events. Disruption of ancient gene geography is greater within animals associated with transit (e.g., horse) and also within those which form smaller, more tradeable units (e.g., small ruminants). However, it remains possible to discern structure in many instances. The imprecision associated with estimates from modern data concerning the second dimension, time, generally exceeds that of geographic inference.

The interpretation of the gross genetic and geographic patterns described above is simplified by the scale of divergence between the major lineage clusters within species (Figure 18.1). These divergences may be calibrated using the concept of the molecular clock, which usually involves comparisons with other species of known divergence time, but gives estimates on the scale of hundreds of thousands of years—clearly outside the timescale of domestic history. However, when one tries to calibrate divergence for finer inference (i.e., that within clusters), one reaches the limits of accuracy of this tool. Put simply, unlike other dating tools in archaeology, time estimates using genetic data such as mtDNA control region sequence or allele frequency differences often have standard errors of similar magnitude to the archaeological distinctions they are employed to make. For example, when Troy et al. (2001) dated expansion times within lineage clusters in cattle, we found estimates consistent with the domestication process 10,000 years ago. However, the range (about 5000 to 20,000 BP) does not preclude an expansion signature that is earlier, and perhaps postglacial in origin. Also, using estimates of divergence time alone it is difficult to assert absolutely that the differences between African and European cattle are pre-domestic (Chapter 22).

Perhaps the best illustration of the complexity of applying the molecular clock in archaeological time depths is in the interpretation of canine mtDNA data (discussed in Chapter 19). Estimating the time of earliest dog domestication, using the diversity among mtDNA control region sequences within the major domestic clade, initially gave a radical inference of 100,000 years ago or more (Vilà et al. 1997). However, subsequent results from New World sequences indicate that the calibration may have been too high. Also, the method (assuming a single domestic progenitor haplotype for the major clade) has an alternative—assuming polyphyly—which yields estimates closer to 15,000 years ago and is concordant with more common archaeozoological opinion (Savlionenen et al. 2002; Chapter 19). Genetic evidence for the timing of dog domestication remains inconclusive.

The alternative approach to absolute dating using a molecular clock calibration from external reference is to compare patterns within the genetic data which may be linked with some probability to known archaeology. One such type of pattern is that which emerges from population expansion. Smooth single-mode curves in sequence mismatch histograms within a species or lineage (see Chapter 20) are indicative of past population expansion and are seen in many domestic data collections. The extent of diversity within a lineage, as interpreted by these curves, is related to the age of the expansion, which may be interpreted as contemporaneous with the domestication process. Interestingly, this implies that the minor lineages in goat, sheep, and dog are the products of later domestication or incorporation than that experienced by the major lineage in each. In cattle, one eastern sublineage of *Bos taurus* seems only half as diverse as those in the west and is most likely a late local adoption, probably into existing herds (Mannen et al. 1998). It should be added that another illustration of the limitations of absolute genetic dating is that it is difficult to assert with certainty that expansion patterns are solely the result of domestication rather than earlier phenomena as well, such as the favorable ecological conditions that met the wild progenitors during the Holocene.

Geographic trends within genetic diversity may sometimes be profitably linked to ecological or archaeological patterns within the same region. A classic example of this comes from human genetics, where Piazza et al. (1995) correlated the first principal component in European gene frequency (a genetic gradient stretching from the southwest to the northeast) with the dates for the first emergence of cereal agriculture within each region of the continent. A significant result allowed the conclusion that the major genetic trend was at least partly linked to the spread of the first European farmers. Hanotte et al. (2002) have similarly correlated the second principal component of microsatellite allele frequency in African cattle with the geographic variation of time depths of the advent of cattle pastoralism through the continent. Although attractive, this method of anchoring genetic patterns to archaeology is difficult and has the particular complexity of spatial autocorrelation. That is, two variables distributed in space, such as attributes of archaeological sites and genetic distances, may correlate simply because of their common geography (or another spatially distributed influence such as ecology), rather than as a result of a causal relationship.

The Future Is Ancient

Anchoring modern genetic patterns to archaeological frameworks is largely not secure enough to answer many interesting questions. For example, to what extent are the phylogeographic discontinuities and signatures of population expansion within modern data collections the legacy of the domestication process, rather than of allopatric separations and populations recoveries within the wild progenitors brought about by glaciation advance and retreat? We remain unaware, in many cases, of the extent to which modern variation is representative of prior genetic diversity. Can we be sure that the regional continuity through time, on which much inference is based, is secure? Are there lost pools of genetic diversity from earlier domestic populations? Important controversies, such as the age of dog domestication, remain unresolved.

The most promising future data source for many animal domesticates is from amplification of dated archaeological remains. Ancient DNA analysis of faunal material is an order of magnitude easier than with human remains. This is because recoverability of DNA from animal bones seems to be greater, and especially because the probability of contamination is greatly reduced (one is less likely to inadvertently amplify one's own mtDNA using PCR primers specific to a ruminant). Several contributions in this volume (e.g., chapters on genetic documentation of domestication in dogs, horses, goats, and cattle) illustrate the utility of data from ancient domesticates or wild ancestors for providing additional anchors to the archaeological scaffold. For example, aurochsen mtDNA from Britain is divergent from that of modern European cattle, illustrating a break between wild and domestic animals in the West that could only have occurred through importation of domesticates. Luikart et al. (Chapter 20) show some continuity in goat mtDNA lineages within the Near East. Ancient dog sequences from the Americas give the only opportunity to study unambiguously precolonial variation that is local in origin (Chapter 19; but see also Koop et al. 2000).

There are limitations to aDNA, however. Unfortunately, DNA is more poorly preserved in warmer climates, giving a geography of aDNA feasibility that is somewhat inverse to that of archaeological interest in the origins of agriculture (e.g., Edwards et al. 2004). There will be exceptions, however, and the feasibility of analyzing well-preserved Near Eastern samples is illustrated by Luikart et al. (Chapter 20). Another limitation is the one restricting most aDNA research to mtDNA. Whereas the bulk of phylogeography has been conducted with this chromosome, it is clear that the inference from its diversity patterns may be incomplete, even misleading. Luikart et al. show a welcome coherence between their mtDNA and Y-chromosome results in goat. In contrast, however, nuclear markers (microsatellites) give clearly divergent results in examination of the mixed ancestry of llama from mtDNA (Chapter 23). Another striking example is the introgression of maternal lineages of *Bos indicus* into Africa—lineages that are markedly out of step with the rest of the genome (Chapter 22). Any effort to time this major Asian influx by analysis of dated bones will require the use of nuclear sequences that distinguish the two cattle types. Population genetic phenomena may even confuse the concordance between different threads of haplotype ancestry in interspecific comparisons. Verkaar and colleagues (2004) recently showed that mtDNA and Y-chromosome phylogenetic signals are uncoupled in the *Bovini*, pointing toward a hybridization event in the lineage leading to European bison.

Nuclear markers are in a minority in ancient DNA analyses. Conventionally, mtDNA sequence is more accessible, probably because of higher copy number. However, some autosomal data have begun to trickle through. For example, Edwards et al. (2003) have used three microsatellite markers in twelfth-century Dublin cattle samples to illustrate that they are local rather than Viking in origin—a conclusion not possible from mtDNA analyses of the same samples. Advances in DNA extraction techniques may also make the retrieval of short nuclear DNA sequences more feasible in a wider sample range. New, quick, and inexpensive SNP typing technologies will also have an impact.

Archaeogenomics

Nuclear DNA analysis is likely to increase in importance for both ancient and modern domestic animal analysis for another reason. MtDNA, Y-chromosome haplotypes, and microsatellites can give rich genealogical information, but the variations examined are likely to be selectively neutral. This is useful, in that they may then be presumed to give an unbiased (if limited) sample of lineage history. However, in terms of the profound genetic changes expressed in animal behavior, morphology, or disease response that resulted from domestication, they tell us nothing. The generation of genome sequences for cattle, chicken, pig, and dog is now on the horizon and is coupled with ever-increasing understanding of gene function. Perhaps the most exciting current research aim is the uncovering of the central genetic changes that were involved in post-domestic alterations in animal populations. Some of these changes may have been coupled with morphological change, but many would not have been (i.e., changes in behavior), and hence would be invisible in the archaeological record. Genetic analysis of both extant and ancient material will the only way of recording the nature and presence of these changes.

One recent and interesting example is the survey of milk genes carried out by Albano Beja-Pereira and colleagues (2003). Here, genetic diversity in six genes that code for major milk proteins was found to be significantly higher in cattle populations from north-central Europe than in other regions of the continent and, surprisingly, higher than in Near Eastern cattle (Figure 18.2). Higher than expected allelic diversity in such loci may be interpreted as a signature of past selection acting on these genes, and indeed, specific tests based on allele frequency spectra confirmed this as a probable cause. That such selection may have occurred on these loci in particular and that this may have had a link with dairy culture is further suggested by two concordances. First, this north-central European region is also the global maximum and probable origin for a genetically determined human trait—the ability to digest lactose in adulthood. This trait has a frequency of over 90% in Denmark, but falls to less than 40% in parts of southern Europe (Sahi 1994). This is a feature of human populations with a history of dairy economy that included the drinking of raw milk. It would have created a powerful selective pressure against those individuals who suffered adverse reaction to what would have been a major dietary component (Holden and Mace 2002). Interestingly, the archaeology of the early Neolithic TRB (Funnel Beaker) culture on the north European plain, which is suggested as an early milk-drinking culture, also shows a settlement distribution centered on this area (Zvelebil 2000). These data are suggestive of gene-culture evolution in two species that have had intimately intertwined histories in north-central Europe. Interestingly, a huge proportion of the cattle alive today have an ancestry that stretches back to the black and white dairy cattle of the European Low Countries—perhaps one feature of a dairying legacy with localized continuity through thousands of years.

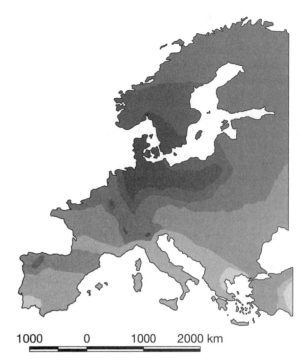

FIGURE 18.2 Synthetic map showing the first principal component resulting from the allele frequencies at six milk protein genes in 70 breeds across Europe and Turkey (after Beja-Pereira et al. 2003, Fig. 1b). Shading illustrates a gradation of values that have a geographic peak in north-central Europe and troughs in southern regions. This gradation coincides with that of gene diversity, with values in north-central Europe significantly exceeding those in southern Europe and in Turkey—in contrast to patterns seen in other genetic systems. These data also show evidence for selective effects with the allele frequency spectra in north-central Europe, where a majority of breeds sampled showed significant deviation from a model of neutral evolution (Ewens-Watterson test). Interestingly, this geographical pattern mirrors closely (and correlates significantly with) the distribution of human lactase persistence, and is concordant with the region of the early Neolithic cattle pastoralist (Funnel Beaker) culture.

Population genetics is undergoing a paradigm change from genotyping to genome typing (Luikart et al. 2003), as witnessed by a recent study of human SNP diversity in which patterns of variation were compared among populations using over 33,000 markers (Shriver et al. 2004). The advent of genome sequences and high-throughput typing technologies are sure to extend this shift into domestic animal studies; instead of using a small fraction of the genome as a proxy to infer its history, surveys of variation at an appreciable fraction of the total genes present will be feasible. This exciting future of archaeogenomics will enable samples of gene geography with sufficient power to draw fine-scaled inferences about past-population demography that elude us at present. Importantly, genome-wide surveys will help to detect genes that show exceptional population genetic behavior, yielding evidence for past selection effects. Such data will help to identify those special loci that were actors, rather than mere spectators, in the interaction between herders and their flocks.

References

Avise, J. C. 2000. *Phylogeography. The history and formation of a species*. Cambridge, Mass.: Harvard University Press.

Beja-Pereira, A., G. Luikart, P. R. England, D. G. Bradley, O. C. Jann, G. Bertorelle, A. T. Chamberlain, T. P. Nunes, S. Metodiev, N. Ferrand, G. Erhardt. 2003. Gene-culture coevolution between cattle milk protein genes and human lactase genes. *Nature Genetics* 35: 311–313.

Beja-Pereira, A., P. R. England, N. Ferrand, S. Jordan, A. O. Bakhiet, M. A. Abdalla, M. Mashkour, J. Jordana, P. Taberlet, and G. Luikart. 2004. African origins of the domestic donkey. *Science* 304: 1781.

Edwards, C. J., J. Connellan, P. F. Wallace, S. D. Park, F. M. McCormick, I. Olsaker, E. Eythorsdottir, D. E. MacHugh, J. F. Bailey, and D. G. Bradley. 2003. Feasibility and utility of microsatellite markers in archaeological cattle remains from a Viking Age settlement in Dublin. *Animal Genetics* 34: 410–416.

Edwards, C. J., D. E. MacHugh, K. M. Dobney, L. Martin, N. Russell, L. K. Horwitz, S. K. McIntosh, K. C. MacDonald, D. Helmer, A. Tresset, J.-D. Vigne, and D. G. Bradley. 2004. Ancient DNA analysis of 101 cattle remains: Limits and prospects. *Journal of Archaeological Science* 31: 695–710.

Giuffra, E., J. H. M. Kijas, V. Amarger, Ö. Carlborg, J. T. Jeon, and L. Andersson. 2000. The origins of the domestic pig: Independent domestication and subsequent introgression. *Genetics* 154: 1785–1791.

Hall, S. J. G. and D. G. Bradley. 1995. Conserving livestock breed diversity. *Trends in Ecology and Evolution* 10: 267–270.

Hanotte, O., D. G. Bradley, J. W. Ochieng, Y. Verjee, E. W. Hill, and J. E. Rege. 2002. African pastoralism: Genetic imprints of origins and migrations. *Science* 296: 336–339.

Hewitt, G. 2000. The genetic legacy of the Quaternary ice ages. *Nature* 405: 907–913.

Holden, C. and R. Mace. 2002. Pastoralism and the evolution of lactase persistence. In *Human biology of pastoral populations*, W. R. Leonard and M. H. Crawford (eds.), pp. 280–307. Cambridge: Cambridge University Press.

Jansen, T., P. Forster, M. A. Levine, H. Oelke, M. Hurles, C. Renfrew, J. Weber, and K. Olek. 2002. Mitochondrial DNA and the origins of the domestic horse. *Proceedings of the National Academy of Science, USA* 99: 10905–10910.

Joshi, M. B., P. K. Rout, A. K. Mandal, C. Tyler-Smith, L. Singh, and K. Thangaraj. 2004. Phylogeography and origin of Indian domestic goats. *Molecular Biology and Evolution* 21: 454–462.

Koop, B. F., M. Burbidge, A. Byun, U. Rink, and S. J. Crockford. 2000. Ancient DNA evidence of a separate origin for North American indigenous dogs. In *Dogs through time: An archaeological perspective*, S. J. Crockford (ed.), pp. 271–285. Oxford: British Archaeological Reports.

Luikart, G., P. R. England, D. Tallmon, S. Jordan, and P. Taberlet. 2003. The power and promise of population genomics: From genotyping to genome typing. *Nature Reviews Genetics* 4: 981–994.

MacHugh, D. E. and D. G. Bradley. 2001. Livestock genetic origins: Goats buck the trend. *Proceedings of the National Academy of Sciences, USA* 98: 5382–5384.

MacHugh, D. E., M. D. Shriver, R. T. Loftus, P. Cunningham, and D. G. Bradley. 1997. Microsatellite DNA variation and the evolution, domestication and phylogeography of taurine and zebu cattle *(Bos taurus* and *Bos indicus)*. *Genetics* 146: 1071–1086.

Mannen, H., S. Tsuji, R. T. Loftus, and D. G. Bradley. 1998. Mitochondrial DNA variation and evolution of Japanese black cattle *(Bos taurus)*. *Genetics* 150: 1169–1175.

Piazza, A., S. Rendine, E. Minch, P. Menozzi, J. Mountain, and L. L. Cavalli-Sforza. 1995. Genetics and the origin of European languages. *Proceedings of the National Academy of Sciences, USA* 92: 5836–5840.

Sahi, T. 1994. Genetics and epidemiology of adult-type hypolactasia. *Scandinavian Journal of Gastroenterology* 29 Suppl.: 7–20.

Savolainen, P., Y. P. Zhang, J. Luo, J. Lundeberg, and T. Leitner. 2002. Genetic evidence for an East Asian origin of domestic dogs. *Science* 298: 1610–1613.

Shriver, M. D., G. C. Kennedy, E. J. Parra, H. A. Lawson, V. Sonpar, J. Huang, J. M. Akey, and K. W. Jones. 2004. The genomic distribution of population substructure in four populations using 8525 autosomal SNPs. *Human Genomics* 1: 274–286.

Sultana, S., H. Mannen, and S. Tsuji. 2003. Mitochondrial DNA diversity of Pakistani goats. *Animal Genetics* 34: 417–421.

Troy, C. S., D. E. MacHugh, J. F. Bailey, D. A. Magee, R. T. Loftus, P. Cunningham, A. T. Chamberlain, B. C. Sykes, and D. G. Bradley. 2001. Genetic evidence for Near-Eastern origins of European cattle. *Nature* 410: 1088–1091.

Verkaar, E. L., I. J. Nijman, M. Beeke, E. Hanekamp, and J. A. Lenstra. 2004. Maternal and paternal lineages in cross-breeding bovine species. Has wisent a hybrid origin? *Molecular Biology and Evolution* 21: 1165–1170.

Vilà, C., P. Savolainen, J. E. Maldonado, I. R. Amorim, J. E. Rice, R. L. Honeycutt, K. A. Crandall, J. Lundeberg, and R. K. Wayne. 1997. Multiple and ancient origins of the domestic dog. *Science* 276: 1687–1689.

Watanobe, T., N. Okumura, N. Ishiguro, M. Nakano, A. Matsui, M. Sahara, and M. Komatsu. 1999. Genetic relationship and distribution of the Japanese wild boar *(Sus scrofa leucomystax)* and Ryukyu wild boar *(Sus scrofa riukiuanus)* analysed by mitochondrial DNA. *Molecular Ecology* 8: 1509–1512.

Zvelebil, M. 2000. The social context of the archaeological transition in Europe. In *Archaeogenetics: DNA and the population history of Europe*, C. Renfrew and K. Boyle (eds.), pp. 57–80. Cambridge, UK: MacDonald Institute for Archaeological Research.

CHAPTER 19

Genetic Analysis of Dog Domestication

ROBERT K. WAYNE, JENNIFER A. LEONARD, AND CARLES VILÀ

Introduction

Dogs are the most phenotypically diverse mammal. In fact, the difference in cranial and skeletal proportions among dog breeds exceeds that among wild canids (Wayne 1986a, 1986b). Considerable differences in behavior and physiology also are evident (Hart 1995). The origin of this diversity is uncertain. Darwin suggested that, considering the great diversity of dogs, they were likely founded from more than one species (Darwin 1859), such as gray wolves and the three extant species of jackal. This sentiment has been periodically voiced by researchers (e.g., Lorenz 1954; Coppinger and Schneider 1995). The more common view is that dogs originated only from the gray wolf *(Canis lupus)* (Morey 1994; Vilà et al. 1997a). However, many questions remain. Does phenotypic diversity reflect a genetically diverse founding pool? Has recurrent hybridization with wild canid species augmented genetic diversity throughout the history of dog domestication? Have there been multiple domestication events? Alternatively, is the genetic diversity of dogs consistent with a single founding event that involved a small number of founders, without further contributions from wild species? If the founding event was limited to a few individuals, then genetic diversity and subsequent morphological variation have been principally augmented through mutation. Clearly, mutation has been an important process in creating unusual breed phenotypes (Chase et al. 2002; Fondon and Garner 2004). For example, the dachshund has an achondroplasia-like syndrome likely due to a simple point mutation in a functional gene, as in humans (Passos-Bueno et al. 1999). Darwin called such anomalous breeds that arise through spontaneous mutation "sports" (Darwin 1859). However, the amount of segregating variation in the founder population is also important, as witnessed by the frequent formation of breeds by crossing phenotypically distinct forms (e.g., Ash 1927; Stockard 1941). Therefore, an essential question is whether phenotypic diversity reflects a diverse ancestral gene pool. In general, knowledge of the evolutionary history of domestic dogs and of their relationships to wild canids provides insight into the mechanisms that have generated the extraordinary diversity of form and function in the dog.

The origin of domesticated species or breeds is seldom well documented. The number, timing, and geographic origin of founding events may be difficult to determine from the patchy archaeological record (cf. Vilà et al. 1997a, 1997b), and historical records of breed origins, even those that have formed in the last few hundred years, seldom are well documented (Ash 1927; American Kennel Club 1997). In this chapter we discuss how molecular genetic data can be used to make inferences about the domestication history of the dog. First, we review prior genetic research on dogs, using older but well-established molecular techniques. We then discuss the application of new, modern sequencing methods and their limitations. We contrast molecular techniques with inferences based on morphologic approaches, following with a description and discussion of studies utilizing mitochondrial DNA sequencing techniques. Finally, we provide a perspective on important issues that need resolution and outline research that is required to address new questions raised by the genetic findings.

Previous Research

Previous genetic data consistently supported an origin of dogs from gray wolves, although the results were not definitive. Dogs and gray wolves are most similar with regard to single-copy DNA and albumins (Seal et al. 1970; Sarich 1977; Wayne et al. 1989). Moreover, they share protein variants as deduced by allozyme electrophoresis (Figure 19.1; Ferrell et al. 1978; Wayne and O'Brien 1987). Gray wolves and dogs have identical chromosome complements (Wurster-Hill and Centerwall 1982; Wayne et al. 1987a, 1987b). However, few allozyme alleles are shared uniquely between dogs and gray wolves, because these alleles are also found in closely related canids such as golden jackals *(Canis aureus)* and coyotes *(C. latrans;* Wayne and O'Brien 1987). Similarly, all members of the genus *Canis* (coyotes, wolves, jackals) are karyologically indistinguishable from each other, and from the dhole *(Cuon alpinus)* and the African hunting dog *(Lycaon pictus;* Wurster-Hill and Centerwall 1982).

More promising modern approaches utilized highly variable markers, such as short tandem repeat loci, often called microsatellites (Figure 19.1). Microsatellite loci have high levels of polymorphism and heterozygosity, and therefore would be more likely to reveal alleles that are uniquely shared between some populations and species (Bruford and Wayne 1993). However, although many more alleles are shared between dogs and wolves than between dogs and other canids, and genetic distances are smaller, the close relationships of dogs, wolves, coyotes, and golden jackals do not allow for definitive conclusions to be made about the origin of the dog from these data (see Roy et al. 1994; García-Moreno et al. 1996). Because the rate of sequence

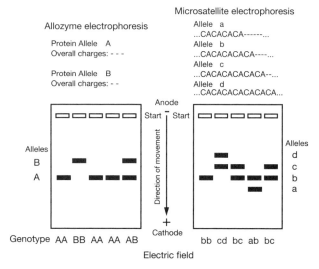
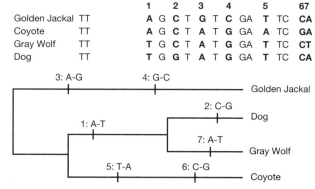

FIGURE 19.1 Allozyme and microsatellite electrophoresis. These techniques consist of differentiating alleles at a locus based on their mobility in an electric field. Allozymes are differentiated by the overall charge of all the amino acids that constitute each one of the sequences of a given protein in an individual. Microsatellites are repetitive DNA sequences (in the example, repetition of the pair of nucleotides CA) that are differentiated by their size (number of repetitive elements). In general, microsatellite loci have greater variability (e.g., higher heterozygosity and more alleles) than allozyme loci because of their higher mutation rate.

FIGURE 19.2 Example of the use of DNA sequence for the construction of phylogenetic trees. The mutations that differentiate the sequences are indicated in bold and identified by numbers. These mutations are used to build the phylogenetic tree, and the position of each mutation is indicated on the corresponding branch in the tree, together with the nucleotide change involved. The outgroup (golden jackal, in our case) corresponds to a taxon closely related to the group under study. Its sequence is used to give polarity to the tree and provide information on the branching order. In this example, dog and gray wolf differ by two substitutions, and each of them differs from coyote by four substitutions.

evolution is generally high for mitochondrial genes, similar arguments can be made for predicting that mitochondrial DNA sequencing would be a more definitive approach for reconstructing the history of domestication than past methods that utilized nuclear genes. Initial studies clearly suggested a close ancestry of gray wolves and dogs. However, these first studies involved indirect sequencing techniques and a limited sample of both species (Wayne et al. 1992; Gottelli et al. 1994).

Mitochondrial Control Region Sequencing

The mitochondria are cytoplasmic organelles containing multiple copies of a small circular DNA genome, about 16,000 to 18,000 base pairs in length in mammals. The mitochondrial genome codes for proteins that function in the electron transport system, as well as specifying messenger and transfer RNA. Hundreds or thousands of mitochondria may occur within the cell, and thus mitochondrial genes are several times more abundant than their nuclear counterparts (Rand 2001). This property facilitates genetic analysis in samples where DNA is degraded, such as in ancient remains, or where cell concentrations are low (Hofreiter et al. 2001). Additionally, mitochondrial DNA sequences have a very high mutation rate. In mammals the rate is often five to ten times faster than that of nuclear genes (Li 1997; Pesole et al. 1999). Consequently, closely related species and populations may have accumulated diagnostic mitochondrial DNA mutations in the absence of changes in the nuclear genome. Finally, with only a few exceptions, mtDNA is maternally rather than biparentally inherited, and there is no recombination (Birky 2001). Therefore, phylogenetic analysis of mitochondrial DNA sequences within species provides a history of maternal lineages that can be represented as simple branching phylogenetic trees (Avise 1994, 2000).

Beginning in the late 1980s the advent of the polymerase chain reaction (PCR), in combination with new DNA sequencing techniques, made population-level sequencing studies feasible. These techniques progressed rapidly, and today large-scale DNA sequencing projects are possible, involving thousands of DNA bases and large population samples. DNA sequence data allow a more precise reconstruction of historical demographic events, such as colonization and geneflow (Avise 1994, 2000). Both restriction fragment analysis (e.g., Lehman et al. 1991; Wayne et al. 1992; Randi et al. 1995; Pilgrim et al. 1998) and, more recently, mtDNA sequencing by PCR (Vilà et al. 1999a; Randi et al. 2000; Wilson et al. 2000) have been used to quantify mitochondrial DNA variability in wolflike canids and dogs.

Many of the DNA base changes, called nucleotide substitutions, do not cause a change in the functional product of the gene. For example, as much as 90% of mitochondrial DNA sequence differences in protein coding genes between dog and gray wolf cause no changes in the protein product, and hence are not influenced by natural selection (Wayne et al. 1997). Rather, such changes accumulate over time and may be uniquely shared among taxa (Figure 19.2). Hence, nucleotide changes are markers of common ancestry, and if

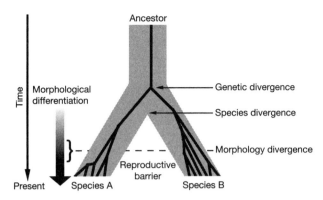

FIGURE 19.3 Difference between the species divergence time and the estimates obtained using genetic and morphological data. Species are indicated as gray bands, and the black lines represent the genetic relationships between sequences from each species. Morphological differentiation between species increases from the time of reproductive isolation.

they accumulate in a regular fashion with time, they may be used to estimate divergence time and time of origin.

As an example, consider hypothetical sequence changes in hemoglobin, the protein responsible for binding oxygen in red blood cells. Suppose all coyotes, wolves, and dogs share 90 nucleotides in a fragment of 100 in the hemoglobin gene, and the gray wolf and dog share, in addition, 8 more nucleotides. Consequently, the dog and wolf differ by two unique substitutions. Such nested changes allow construction of a phylogenetic tree in which each node and branch is defined by a unique set of nucleotide changes. Moreover, if the fossil record suggested that wolves and coyotes diverged about 1 million years ago, and they differ by 10 nucleotide changes, then the two nucleotide differences between dogs and wolves imply a divergence time of 200,000 years ago. Of course, this assumes that nucleotide substitutions accumulate in a clocklike fashion with little variation. Although the use of molecular data to estimate divergence time is a common practice, considerable debate surrounds the accuracy of divergence dates (Avise 1994, 2000).

The process of domestication needs to be studied with markers that have the highest mutation rates in the mitochondrial genome, because divergence times are extremely short on an evolutionary time scale. The left domain of the mitochondrial control region is often used because it is highly polymorphic and not transcribed; hence it presumably is under fewer selective constraints (e.g., Loftus et al. 1994; Luikart et al. 2001; Vilà et al. 1997a, 1999a).

Molecules vs. Morphology

The archaeological record suggests that the first domestic dogs are found in the Middle East or Central Europe about 14,000 to 15,000 years ago (Nobis 1979; Olsen 1985; Dayan 1994; Clutton-Brock 1995; Sablin and Khlopachev 2002). Most early dogs are morphologically distinct from gray wolves. Often they are smaller in body size and have wider crania, a more prominent stop on the face, and a shortened, crowded jaw (Olsen 1985; Morey 1992). The small body size of Asian wolves and the shared presence with early dogs of some traits led to the suggestion that Asian wolves are the direct ancestor of the dog (Olsen and Olsen 1977). However, the oldest dog remains unmistakably identified so far have recently been discovered in western Russia (Sablin and Khlopachev 2002) and date to 13,000 to 17,000 years BP. These early dogs do not show a reduction in body size when compared to the local populations of gray wolves, and they have cranial proportions similar to the Great Dane. These new remains suggest an early dog similar in size to the large European wolf, but having distinct differences in skull morphology that are unlike those previously reported. This finding implies that size and morphology are not a dependable guide to the relationships between early dogs and wolves. Rather, skeletal differences between wolves and dogs may be a response to new selective pressures on protodogs related to coexistence with humans.

Behavioral changes likely took place as well, such as increased docility and dependency on human handlers (see Chapter 13). In contrast, selection for specific phenotypic traits are unlikely to have directly caused changes in the mitochondrial DNA sequences used to understand the history of domestication (see the previous discussion). Mutations that do occur in mitochondrial sequences may better mark the passage of time during which dogs have been isolated reproductively from their wild brethren (Figure 19.3). Consequently, morphologic and molecular estimates of divergence time may correspond if the change in environment and selection that is marked by the appearance of new morphologic traits is coincident with the first reproductive isolation of dogs and wild canids. If protodogs and wolves had been reproductively isolated tens of thousands of years ago, but had not changed in morphology through selection or drift until about 15,000 years ago, then genetic and morphologic measures of origination time would not agree (Figure 19.3). Thus, morphology and DNA sequences provide different information with regard to domestication, and discordance of dates does not mean that one or the other is flawed.

Molecular Genetic Analysis of Dog Domestication

Origin and Timing of Dog Domestication

We now focus on three studies that we and our collaborators have undertaken to better understand the process of dog domestication (Vilà et al. 1997a, 1999b; Leonard et al. 2002). In addition, we incorporate findings from other studies that use control region sequencing to reconstruct the recent history of dogs.

To begin with, Okumura et al. (1996) were the first to sequence the control region of a large sample of dogs, but they

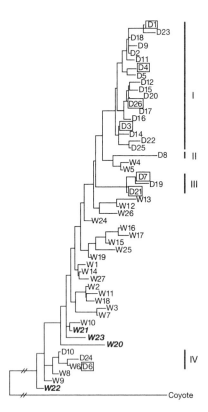

FIGURE 19.4 Neighbor-joining relationship tree of wolf (W) and dog (D) mitochondrial DNA control region sequences (261 base pairs in length; Vilà et al. 1997a). Dog haplotypes are grouped in four clades, I to IV. Boxes indicate haplotypes found in the 19 Xoloitzcuintle discussed later in this chapter (Vilà et al. 1999b). Bold characters indicate haplotypes found in New World wolves (W20 to W25). The figure is modified from Vilà et al. (1999b).

focused on dogs from Asia. These authors observed that dog mitochondrial DNA sequences grouped into a discrete number of clades. Tsuda et al. (1997) similarly studied a large sample of dogs from Asia, but also included some samples from wolves. Consequently, they were able to convincingly demonstrate a close relationship between wolves and dogs.

Along with our collaborators, we surveyed control region sequence variation in a geographically broad sampling of domestic dogs and gray wolves, and several other wild canid species (Vilà et al. 1997a). Our study included 162 wolves from 27 populations throughout Europe, Asia, and North America and 140 dogs representing 67 breeds. We found that the dog and wolf sequences differed by only 0 to 12 substitutions in 261 base pairs, whereas dogs always differed from other canids by at least 20 substitutions. Therefore, the gray wolf was strongly supported as the closest living ancestor of the dog, and ancestry from other canids such as the golden jackal was refuted. Moreover, one of the sequences found in dogs was identical to one in gray wolves (D6 and W6 in Figure 19.4), suggesting a recent hybridization event. The mean sequence diversity within dogs was 5.30% ± 0.17. This value was surprisingly large and very similar to the value of 5.31% ± 0.11 found in gray wolves. As discussed in the introduction to this chapter, the great extent of genetic variation in the dog suggests they were not founded by a small number of gray wolves—or if they were, interbreeding between dogs and their wild brethren maintained high levels of variation (Tsuda et al. 1997).

To determine the evolutionary relationship of sequences in dogs and gray wolves, we constructed phylogenetic trees (Figure 19.4). Four sequence groupings or clades are suggested by the topology of the sequence tree (Figure 19.4; I to IV as in Vilà et al. 1997a). These clades have been independently discovered by other researchers (Okumura et al. 1996) and suggest either that wolves were domesticated in several places and/or at different times, or that there was one domestication event followed by several episodes of admixture between dogs and wolves. Regardless, the results imply that dogs have a diverse origin and have had genetic contributions from more than one gray wolf population. Further, from an evolutionary perspective, no sequence in dogs showed closer kinship to jackals, coyotes, or Ethiopian wolves *(Canis simensis)* than to gray wolf sequences. This extensive survey of genetic variation in the dog and wolf clearly established gray wolves as the closest living kin of dogs, and most likely as their exclusive ancestor.

Most of the sequences found in dogs, including ancient breeds such as the Australian dingo and New Guinea singing dog, are found in dog Clade I. This clade includes 18 of the 26 haplotypes found in dogs by Vilà et al. (1997a; Figure 19.4). The genetic divergence within this clade is larger than in the other three clades, and sequences within Clade I differ by as much as 1% in DNA sequence. This level of divergence was confirmed with a fuller analysis of 1,030 base pairs (bp) of control region sequence in a subset of dog and gray wolf samples. In this analysis the same four sequence clades were supported, as well as the high divergence among dog sequences in Clade I. We tested for differences in the DNA substitution rates of sequences from dogs, wolves, and coyotes to determine if the rate of sequence change was the same in these taxa. We found no evidence for differences in rate (Tajima 1993). Consequently, if wolves and coyotes diverged about 1 million years ago and have control region sequences that are 7.5% different (Vilà et al. 1997a), dogs and gray wolves may have diverged 1/7.5 as long ago, or about 135,000 years BP. Similarly, Okumura et al. (1996) found that dog sequence clades shared a common ancestry approximately 76,000 to 121,000 years ago. In that study, however, no control region sequences from wild canids were used as a reference point.

Although these estimates are imprecise and have large confidence intervals, these molecular results imply an ancient origin of domestic dogs from wolves. This origin is much older than suggested by the archaeological record, a possibility at least not contradicted by the fossil record, since wolves and humans coexisted for as much as 500,000 years (Clutton-Brock 1995). As discussed earlier, this early date implies a morphologic shift in dogs about 14,000 to 15,000 years ago, perhaps due to changing conditions associated with the shift from hunter-gatherer cultures to more sedentary societies (Clutton-Brock 1995; Vilà et al. 1997b).

Dog sequences are imbedded in four distinct sequence clades. Therefore, at least four origination/interbreeding events are implied by the genetic results, each with a separate ancestry from gray wolves (Figure 19.4). Recently one additional dog haplotype that does not fall into any of these four lineages has been identified, indicating a fifth event (Savolainen et al. 2002). Once dogs were domesticated and spread over a wide geographic area, interbreeding between wolves and dogs could capture wolf sequences through backcrossing of hybrids and domestic dogs. This sequence capture would be genetically indistinguishable from an origination event. For example, in Clade IV a wolf sequence is identical to a dog sequence, suggesting a more recent interbreeding or origination event. The number of origination/interbreeding events is likely to be more than that suggested by the sequence phylogeny, for at least two reasons. First, because mitochondrial DNA is maternally inherited, interbreeding events between male wolves and female dogs would not be preserved in the dog mtDNA gene pool. Second, the wolflike mitochondrial DNA sequence from dog and wolf interbreeding events may have been lost by chance during the history of domestication. Because mitochondrial DNA is maternally inherited without recombination, female offspring of a pair may not reproduce, although nuclear genes are transmitted through male progeny (Avise 1994).

In a recent study, Savolainen et al. (2002) find that the mitochondrial DNA diversity present in East Asian dogs is much larger than in other areas, implying that domestication occurred there. Additionally, by dividing Clade I into multiple groups centered on the most abundant haplotypes and assuming that these are ancestral, the authors conclude that an origination date of approximately 15,000 years ago is conceivable, contradicting our previous ancient estimate. Although the underlying assumptions of these analyses may be appropriate for natural populations, we feel that this is a dubious exercise for domestic dog populations in which the frequencies of haplotypes are greatly influenced by the rise and fall of civilizations and human migrations. Further, Savolainen et al. assumed that Clade I is polyphyletic (with multiple origins), whereas we (Vilà et al. 1997a) assumed that it is monophyletic. Although this recent date is more consistent with dates deduced from archaeological findings, the study fails to explain the early presence of morphologically differentiated dogs in western Russia, close to the border with Ukraine (Sablin and Khlopachev 2002), aged at 13,000 to 17,000 years BP.

Other problems are intrinsic to the study by Savolainen et al. For example, the diversity of East Asian dogs appears inflated by the presence of a very large proportion of dogs of mixed origin, whereas European and West Asian populations are represented almost exclusively by purebred dogs. Due to the inbreeding practices involved in the origin of breeds, part of the genetic diversity of the ancestral population is expected to be lost in purebred lines, and hence they should be less variable in aggregate than a similar population of mixed-bred individuals. High variability may be characteristic of most aboriginal mixed-bred populations, as suggested by the high diversity observed by Leonard et al. (2002) in ancient American dogs (19 haplotypes in 24 samples, and most of them unique). Therefore, we maintain that the evidence for the location and time of origin of domestic dogs is not conclusive. Combined archaeological and genetic studies involving a larger, less-biased sampling of dogs worldwide—including ancient and extant intermixed populations such as those in India, Africa, and South America—will be necessary to support specific reconstructions of dog origins.

Origin of Dog Breeds

Within breeds, sequence diversity is relatively high. Even with small sample sizes, most breeds have multiple mtDNA sequences (Table 19.1), implying that several females were involved in the development of the breed (Vilà et al. 1997a). For example, five different haplotypes were found in eight German shepherds, and three haplotypes were found in three Siberian huskies. These high levels of mtDNA variability were unexpected, because many breeds are thought to be highly inbred and to have originated from only a few founders. In fact, due to the random loss of haplotypes indicated above, the number of founding females for each breed may be much greater than suggested by the number of different sequences alone. Ample genetic diversity within breeds is also supported by analysis of protein alleles (Simonsen 1976; Ferrell et al. 1978) and hypervariable microsatellite loci (Gottelli et al. 1994; García-Moreno et al. 1996; Lingaas et al. 1996; Zajc et al. 1997; Zajc and Sampson 1999; Parker et al. 2004). Consequently, the moderate-to-high genetic diversity of dog breeds indicates that they were derived from a diverse gene pool and have not often been intensively inbred.

In general, control region sequences are not unique to specific breeds, and if they are, they are not fixed within breeds (Table 19.1). Only the genotype D8 is unique, but not fixed, in two Scandinavian breeds—the Norwegian elkhound and the Jämthund—and suggests a potential independent origin from wolves (Table 19.1; Vilà et al. 1997a; see also Savolainen et al. 2002). Similarly, microsatellite and allozyme surveys have shown that breeds rarely contain unique alleles, although they may be differentiated with respect to allele frequency (Lingaas et al. 1996; Zajc et al. 1997; Zajc and Sampson 1999). In fact, recent analysis of 96 microsatellite loci in 85 dog breeds finds that nearly all breeds define discrete genetic groups that are further clustered into four genetic lineages. The most divergent lineage contains breeds from Asia, Africa, and the Arctic, suggesting that dogs existed in these regions from early in their history (Parker et al. 2004). Similarly, analysis of breed diagnostic nuclear markers has allowed phylogenetic reconstruction of breed history (Neff et al. 2004).

TABLE 19.1
Mitochondrial DNA Haplotypes Found in Dog Breeds and Their Distribution

	Clade I	Clade II	Clade III	Clade IV
Chow Chow (n = 3)	D1, D2, D3			
Border Collie (n = 3)	D1, D5			
Wirehaired Dachshund (n = 3)	D5			D10
Australian Dingo (n = 4)	D18			
Norwegian Elkhound (n = 9)	D3	D8		
German Shepherd (n = 8)	D4, D5		D7, D19	D6
Afghan Hound (n = 3)				D6
Siberian Husky (n = 3)	D3, D18		D7	
Jämthund (n = 3)		D8	D7	
Flat-coated Retriever (n = 3)	D4			D10
Golden Retriever (n = 6)	D4, D15			D6, D24
Labrador Retriever (n = 6)	D4, D12			
Samoyed (n = 3)	D1, D4, D5			
Giant Schnauzer (n =3)	D4		D7	
English Setter (n = 4)		D3, D5		
Irish Setter (n = 3)	D1, D9			
Mexican Hairless (n = 19)	D1, D3, D4, D26		D7, D21	D6

SOURCE: The table is adapted from Vilà et al. (1999b). Clades of dog haplotypes are from Figure 19.4.

The genealogical relationship among breeds is not apparent in the sequence tree, and sequences within breeds have often been classified in divergent sequence clades (see Mexican hairless haplotypes in Figure 19.4). This result is consistent with the recent origination of many breeds, mostly within the past few hundred years, such that unique breed-defining mutations have not occurred in the control region. Therefore, the relationship of sequences among breeds reflects divergence in the ancestral common gene pool of dogs, rather than specific ancestor-descendent relationships of recently diverged breeds. However, Parker et al. (2004) clearly detected a hierarchical grouping of breeds when a large number of markers were used to assess allele frequency differentiation among breeds. Such large-scale studies of multiple markers offer the best hope to resolve breed relationships.

In conclusion, despite the apparent phenotypic uniformity within breeds, the genetic diversity is high, reflecting a diverse genetic heritage. Most breeds have a recent origin (Dennis-Bryan and Clutton-Brock 1988), and the mitochondrial DNA data suggest that the founding stock is often drawn from a previously well-mixed and outbred pool of dogs. Even local dog gene pools were likely well-mixed, considering that dogs were transported with humans as companions or for trade (e.g., Schwartz 1997). However, nuclear data suggest that despite some mixing, allele frequencies have diverged such that breeds define unique genetic groups (Parker et al. 2004). Nuclear and mitochondrial markers provide different perspectives on isolation and drift, and the apparent disparity between them may reflect different rates of female and male geneflow, or effective population size. For example, a focus on the breeding of a limited number of popular sires, with less attention on females, might result in greater differentiation in biparentally inherited markers such as microsatellites, whereas maternally inherited markers would show less genetic drift.

Independent Origin for New World Dogs?

The oldest definitive dog remains from Central Europe and the Near East date from about 14,000 to 15,000 years ago (Clutton-Brock 1999; Olsen 1985; Dayan 1994; Reitz and Wing 1999; Sablin and Khlopachev 2002). Dogs have a long history in the New World as well, with the earliest securely dated dog remains coming from the Koster site in Illinois at about 8500 BP (Morey and Wiant 1992). The antiquity of dog remains in the New World, and the great distance separating these remains from the first Old World dogs, suggests an independent origination. Until recently, gray wolves were abundant in North America and had a geographic distribution stretching from Mexico to north of the Arctic Circle (Nowak 1999). The possibility of independent origination has been suggested by several authors (see, for example, Pferd III 1987). The hypothesis that American dogs derive from a separate American wolf domestication can be tested with genetic data. If Native American dogs have sequences very similar or identical to North American wolves, then a separate New World domestication would be supported. Alternatively, if North American dogs have sequences similar to those found in Old World dog breeds, then an ancient origination in common with Old World breeds of dog is supported. Further, if North American dogs are of Old World origin, then they have been isolated from them for at least about 11,000 years (Elias et al. 1996) and should have a level of sequence divergence consistent with the period of isolation.

The Xoloitzcuintle, an Ancient New World Dog Breed

To assess the possibility of an independent origination in the New World, we first surveyed an ancient extant breed, the Xoloitzcuintle (Mexican hairless) (Vilà et al. 1999b). The Xoloitzcuintle is a breed of hairless dog developed by Mesoamerican Indians several thousand years ago. In the sixteenth century, Francisco Hernández, a Spanish naturalist, described several unusual native dog breeds from Mexico (Valadez 1995). Among them was the Xoloitzcuintle, a medium-sized hairless dog that was used for food and companionship, and also to relieve pain associated with rheumatism (Cordy-Collins 1994). More recently, pottery from the Colima culture (250 BC to 450 AD) in western Mexico that depicts Xoloitzcuintle has been discovered (Cordy-Collins 1994). Living Xoloitzcuintle are usually missing several teeth, and old burials in western Mexico (700 to 1000 AD) have dog remains with incomplete dentitions that have been assumed to be Xoloitzcuintle.

Dogs of the Native Americans crossed with those brought by the Spanish conquistadors, and it is thought that native breeds were blended into nonexistence or were systematically eliminated as part of a program to replace the native traditions with Hispanic culture (Valadez and Mestre 1999). However, native people made special efforts to save the Xoloitzcuintle because of its religious value, and they hid the dogs in mountain villages in the western Mexican states of Guerrero, Michoacán, Colima, and Jalisco (Valadez 1995). In these retreats, breeding was carefully restricted to members of the breed. Moreover, due to their hairlessness the Xoloitzcuintle are very sensitive to extreme temperatures and excessive UV radiation, and were unlikely to survive in the wild. Hence, the breed was likely confined to a narrow geographic region.

The Xoloitzcuintle remains uncommon. Only a few breeders are active, mostly in Mexico and the United States. The breed, previously accepted by the American Kennel Club, was excluded in 1959 due to lack of registrants and show entries (Wilcox and Walkowicz 1995). The purported ancient origin of this breed, together with its presumed isolation from other dogs, makes the Xoloitzcuintle a good candidate breed to test for an independent wolf domestication in North America.

We obtained blood or hair samples from 19 individuals from seven breeders in the United States and Mexico. Dogs obtained from the same breeder were from different varieties (i.e., standard, coated, toy) or from different matrilines. We sequenced a 394-bp fragment of the control region and compared a 261-bp portion of this fragment to previously published sequences (Vilà et al. 1999b). We also included the Chinese Crested dog, a breed that the American Kennel Club considers may be ancestral to the Xoloitzcuintle (American Kennel Club 1997). However, this association seems based on superficial morphologic similarity and may not reflect a recent common ancestry.

We found a total of seven different haplotypes in Xoloitzcuintle dogs. These sequences were distributed among three clades of dog haplotypes (Clades I, III, and IV; Figure 19.4) but were not found in Clade II, which includes the dog haplotype D8 found only in two Scandinavian breeds (Norwegian Elkhound and Jämthund). No unique Xoloitzcuintle haplotypes were found. None of the haplotypes found was identical or similar to any of the sequences found in New World wolves, including Mexican wolves (Figure 19.4). The most common haplotype, D6, was found in five Xoloitzcuintle. This haplotype is the only one that has been found to be shared between dogs and wolves. However, the wolves presently found carrying this haplotype exist only in Romania and European Russia.

These results do not support a New World domestication for the Xoloitzcuintle. Instead, the sequences found in the Xoloitzcuintle are identical to sequences found in dog breeds originating in the Old World. Additionally, we found that none of the Xoloitzcuintle shared any of the haplotypes present in the Chinese Crested dog (D2 and D25, Figure 19.4). Consequently, our data do not support any ancestor-descendent relationship between the Xoloitzcuintle and the Chinese Crested dog. Our results also showed that the Xoloitzcuintle sequences were not highly divergent from those in the Old World. Four of the seven haplotypes were identical to those in Old World dog breeds, and the remaining ones differed by no more than one substitution (0.3%). Consequently, the low level of divergence between Xoloitzcuintle and Old World dogs does not support an ancient divergence for this breed.

The Xoloitzcuintle sampled had a surprisingly high diversity of control region sequences. We found seven distinct sequences in the 19 sampled Xoloitzcuintle, representing three distinct clades of dog control region sequences (Figure 19.4). This result implies that the population of dogs that founded the Mexican hairless was large and genetically diverse. Minimally, seven females contributed to the founding population. The number of founding females was likely much larger, however, considering that founding females may have had identical haplotypes and that many founder haplotypes were likely lost because of drift in small populations (see above). The phenotypic uniformity of the Xoloitzcuintle is surprising, given their diverse genetic heritage. The gene that determines hairlessness is dominant, but lethal when homozygous (Cordy-Collins 1994). Therefore, outcrossing can lead to mixed-breed dogs being hairless and subsequently identified as pure Xoloitzcuintle.

Thus, although there may have been more than one domestication event from gray wolves in the Old World, an

TABLE 19.2
Samples of Ancient Native American Dogs Studied by
Leonard et al. (2002)

Identification	mtDNA	Locality	Age
JAL 330	D27	Iwawi, Bolivia	>1000 BP
JAL 331	D28	Iwawi, Bolivia	>1000 BP
JAL 332	D29	Iwawi, Bolivia	>1000 BP
JAL 334	D28	Iwawi, Bolivia	>1000 BP
JAL 337	D28	Iwawi, Bolivia	>1000 BP
JAL 365	D26	Chiribaja Baja, Peru	1000 BP
PC 5	D30	Chiribaja Baja, Peru	1000 BP
PC 6	D31	Chiribaja Baja, Peru	1000 BP
PC 8	D25	Teotihuacan, Mexico	1300 BP
PC 10	D32	Texcoco, Mexico	800 BP
PC 12	D33	Tula (Hidalgo) Mexico	1400 BP
PC 13	D6	Tula (Hidalgo) Mexico	1400 BP
PC 14	D35	Tula (Hidalgo) Mexico	1400 BP
Perm597	D3	Fairbanks area, Alaska	430 ± 55 BP[a]
JAL 27	D36	Fairbanks area, Alaska	320 ± 50 BP[a]
JAL 42	D37	Fairbanks area, Alaska	228 ± 33 BP[a]
JAL 43	D38	Fairbanks area, Alaska	349 ± 37 BP[a]
JAL 44	D3	Fairbanks area, Alaska	222 ± 39 BP[a]
JAL 45	D18	Fairbanks area, Alaska	307 ± 40 BP[a]
JAL 46	D40	Fairbanks area, Alaska	265 ± 43 BP[a]
JAL 49	D1	Fairbanks area, Alaska	220 ± 43 BP[a]
JAL 53	D41	Fairbanks area, Alaska	278 ± 40 BP[a]
JAL 59	D42	Fairbanks area, Alaska	401 ± 40 BP[a]
JAL 62	D37	Fairbanks area, Alaska	442 ± 35 BP[a]

SOURCE: Haplotypes based on 257 bp are homologous to those analyzed in Vilà et al. (1997a). Haplotypes D1, D3, D6, D18, D25, and D26 have been identified in Old World dogs in Vilà et al. (1997a); all other haplotypes are novel.

[a] Uncorrected age in radiocarbon years.

analysis of one of the oldest New World dog breeds suggests that it was not derived by an independent origination from North American wolves. However, these conclusions are predicated on the assumption that our sample of the Xoloitzcuintle is representative of ancient North American breeds. Our sample could be questioned for at least two reasons. First, the sample of 19 individuals from 7 breeders may be a biased sample of the breed. We may have missed sequences that were derived from North American wolves or were highly divergent from Old World dogs. Second, although breeding of the Xoloitzcuintle was thought to be carefully controlled, extensive interbreeding in the recent and more distant past may have strongly affected the genetic composition of the breed. For example, the conquistadors made a special effort to eradicate local dogs and replace them with European dogs (Valadez and Mestre 1999), and there was a preference for the larger dogs that arrived with them (Schwartz 1997). Preferences for Old World breeds continue to the present, and crossing Xoloitzcuintle with these breeds may have obliterated unique sequences. Crosses between Xoloitzcuintle and other breeds can produce hairless offspring that could be misidentified as purebred. Thus an accurate reconstruction of the genome of ancient New World breeds requires the use of ancient DNA techniques.

DNA from Ancient Dogs

To better assess the genetic variability of native American dogs and further assess the New World origin hypothesis, we analyzed 425 bp of the mitochondrial DNA control region from dog remains predating contact by European colonists (Leonard et al. 2002). We sampled 13 dog specimens from archaeological sites in Mexico, Peru, and Bolivia that were dated to periods before the arrival of Columbus in the New World. Further, we analyzed sequences from 11 dog remains from Alaska that were deposited before the first arrival of European explorers.

We initially analyzed the dog remains from Latin America because these clearly were deposited well before European contact, and we found 12 different sequences in 13 samples. These sequences differed by 1 to 12 bp (0.2 to 3.1% divergence). The two identical sequences, JAL 332 and JAL 334, originated from the same locality: Bolivia (Table 19.2). To determine the relation of ancient sequences to modern ones, we compared the 257-bp region of the ancient sequences to previously published sequences from 165 dogs belonging to 67 diverse dog breeds, from Vilà et al. (1997a), and from several hundred other dogs available in public databases, as well as from 259 wolves obtained from 30 localities worldwide (Vilà et al. 1999a). Considering just this smaller fragment, the pre-Columbian sequences collapsed into 10 different haplotypes (Table 19.2). Almost all modern dogs were sampled in the Old World and corresponded to breeds originating there, with the exceptions of Eskimo dogs, the Mexican hairless, the Alaskan husky, the Newfoundland, and the Chesapeake Bay retriever. We also sampled two distinct dog populations from Oceania: the Australian dingo and the New Guinea singing dog. However, sequences from these dogs were also found in Eurasian breeds (Vilà et al. 1997a; but see Savolainen 2004).

To determine the evolutionary relationships of ancient and modern sequences, we made phylogenetic trees as above (Figure 19.5). The ancient American dog sequences clustered with two of the four previously defined dog clades (Vilà et al. 1997a; Leonard et al. 2002). All but one of the pre-Columbian American dog haplotypes grouped within the most diverse cluster of dog sequences, Clade I, which included the Australian dingo, the New Guinea singing dog, the African basenji, the greyhound, and other ancient breeds, as well as the majority of sequences found in modern breeds. Only the sequence from sample PC 13 was clustered inside Clade IV. American gray wolves (19.5, lu-28 to lu-33 in gray) did not cluster with the ancient dog sequences and differed from them by 3 to 13 bp. None of the ancient American dogs differed by more than 5 bp from a Eurasian dog sequence, and three of the 257-bp sequences found in ancient American

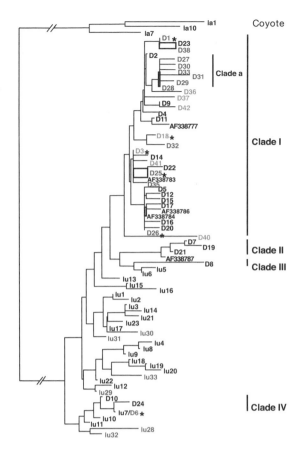

FIGURE 19.5 Neighbor-joining phylogeny of modern domestic dogs from throughout the world and ancient domestic dogs from the Americas, with the coyote as outgroup. Previously defined dog lineages (Clades I to IV) are marked (Vilà et al. 1997a), and modern dog haplotypes are in black, labeled D (from Vilà et al. 1997a) or with their GenBank accession number. Ancient American haplotypes are also labeled D, but are lettered in gray (Leonard et al. 2002). Clade *a* refers to a unique group of American haplotypes. Haplotypes shared between ancient American and modern dogs are marked with an asterisk. Wolf haplotypes are labeled with the prefix lu, Eurasian in black and American in gray (Vilà et al. 1999a). The figure is modified from Leonard et al. (2002).

dogs were identical to sequences observed in Eurasian dogs (Figure 19.5). However, when the sequence of the entire 425-bp segment from ancient samples was compared to over 350 modern dog sequences of different lengths, only the pre-Columbian sequence from sample PC13 was identical to one found in modern dogs (haplotype D6). These results, showing some divergence but a clear association with sequences from the Old World, strongly support the conclusion suggested by the analysis of Xoloitzcuintle that native New World dogs were not independently domesticated from North American wolves. Rather, New World dogs were derived from dogs domesticated in the Old World.

Gray wolves historically ranged in North America from the high Arctic south into Mexico. Therefore, to include specimens from the high latitudes, where gray wolves are abundant and a potential source for domestication and interbreeding, we obtained dog remains from the permafrost deposits of Alaska (Olsen 1985). These deposits often yield relatively well-preserved DNA from specimens as old as 50,000 years (Leonard et al. 2000; Barnes et al. 2002). However, the 11 dog remains that we tested date between 1450 and 1675 CE, and thus most postdate the first arrival of European colonists to America (Table 19.2). Nevertheless, all of these dog remains were deposited before the first sighting of Alaska by Europeans Vitus Bering and Aleksey Chirikov in 1741 (Hully 1958), and they likely represent native American dogs that did not breed extensively with dogs transported by European colonizers. Eight haplotypes were found in eleven samples of ancient Alaskan dogs; five of these were unique and three (D18, D3, D1) were shared with modern domestic dogs (Figure 19.5). All ancient Alaskan dog samples possessed Clade I haplotypes (Figure 19.6, dark gray fill) and hence were not most closely related to New World gray wolf sequences. Therefore, these results support the conclusion from the analysis of ancient Latin American dogs and the Xoloitzcuintle that New World dogs were not independently domesticated from New World wolves.

To better understand the sequence changes that occurred within New World dogs since their isolation from Eurasian dogs with the closing of the Bering land bridge, a minimum spanning network was constructed where haplotypes can occupy nodes and each branch represents a single nucleotide substitution or insertion/deletion (Figure 19.6). Such networks better display relationships in populations where ancestor and descendent sequences can co-occur. The minimum spanning network showed that haplotype D28, found in ancient samples from Bolivia, was possibly ancestral to a clade of unique New World haplotypes and differed from them by one or two substitutions. Haplotype D36, found in ancient Alaskan dogs, or the common haplotype D2 found in Old World dogs, were putatively ancestral to this endemic New World clade that we have designated as Clade *a* (Figure 19.6). The minimum spanning network suggests that the remaining ancient American sequences were derived from the common haplotypes D2, D3, D9, and D26 (Figure 19.6; Vilà et al. 1997a). Further, an additional lineage was derived from the Clade IV haplotype D6 (Figure 19.5). Consequently, our phylogenetic analysis suggested that a minimum of about five founding dog lineages (including the ancestor of Clade *a*) invaded North America with humans as they colonized the New World. The presence of the unique haplotype group (Clade *a*) derived from New World haplotype D28 suggests a substantial period of isolation and significant sequence evolution during the period of isolation.

No ancient dog sequence was similar to those from New World wolves. Therefore, we did not find any evidence of origin from or hybridization with New World gray wolves. These results differ from those of a previous study (Koop et al. 2000) that reported sequences closely related to New World wolves in five dog remains associated with Native Americans in British Columbia. This finding could represent localized interbreeding between domestic dogs and

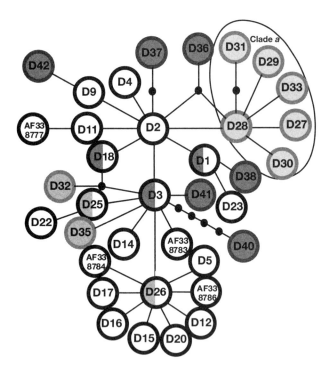

FIGURE 19.6 Statistical parsimony network of Clade I modern dogs (in white) from throughout the world and ancient dogs from America (dark gray from Alaska and light gray from Latin America). The figure is modified from Leonard et al. (2002).

North American wolves, as suggested by morphological (Clutton-Brock 1999; Valadez et al. 2002) and historical (Schwartz 1997; Valadez et al. 2002) data. However, our data suggest that widespread introgression of female wolf matrilines into the native dog population occurred infrequently, since all sequences from ancient native dogs were well-differentiated from those found in North American wolves. Similarly, extant North American gray wolves showed no genetic evidence of interbreeding with dogs, despite the high concentration of dogs in many areas occupied by gray wolves (Valadez et al. 2002). Hybridization has been occasionally observed in Europe, where dogs are common and gray wolves are rare (Andersone et al. 2002; Randi and Lucchini 2002; Vilà et al. 2003).

Our results also provide insight into the dynamics of human migration to the New World and the cultural interaction between Native Americans and European colonists. First, 6 of 12 ancient Latin American haplotypes were grouped in Clade *a* and include sequences found in dog remains from Bolivia, Peru, and Mexico (Figure 19.6). No sequences from Clade *a* have been found in samples from over 350 modern dogs (Leonard et al. 2002). The upper bound of a 95% confidence limit for the frequency that sequences from Clade *a* could have existed in modern dogs and be missed from our sample (observed frequency of zero) is 1.0% (Leonard et al. 2002). Consequently, the absence of Clade *a* sequences from modern dogs suggests an extensive replacement of Native American dogs by those introduced by Europeans. These lineages could be surviving in some unsurveyed modern Native American breeds or local dog populations (Valadez 1999). However, genetic analysis of a diverse sample of the Xoloitzcuintle (see above) only revealed mitochondrial DNA sequences previously observed in dogs of Eurasian origin (Vilà et al. 1999b). The absence of ancient North and South American dog haplotypes from a large diversity of modern breeds, including the Xoloitzcuintle, illustrates the considerable impact that invading Europeans likely had on native cultures. Native breeds may have been lost as a consequent of programs of cultural extermination practiced in Latin America. Additionally, larger European dogs may have been bred preferentially by Native Americans (Schwartz 1997).

Thus, our analysis of ancient specimens and the early fossil record of dogs in North America strongly support the hypothesis that ancient American and Eurasian domestic dogs share a common origin from Old World gray wolves. This implies that the earliest waves of human migrations that colonized America brought multiple lineages of fully domesticated dogs with them. The exclusion of ancient American dog haplotypes that were present in both North and South America from modern breeds illustrates the impact that European colonizers had on native cultures. Assuming that the Beringian land bridge was the primary source of canine migrants, our results suggest that by the time the land bridge closed, one element of human culture—the dog—had spread at least across Europe and Asia, and into the Americas. The large diversity of mtDNA lineages in the dogs that colonized the New World with upper Paleolithic people suggests that the number and diversity of dogs in Eurasia was already extensive at that time. These observations, together with the observation that remains of morphologically differentiated dogs have been found from the upper Paleolithic of central Europe (Sablin and Khlopachev 2002), suggest a date of domestication more ancient than 14,000 or 15,000 years ago, a conclusion consistent with previous genetic results (Vilà et al. 1997a). If dog domestication took place significantly earlier, humans first colonizing the subarctic mammoth-steppe of Siberia 26,000 to 19,000 years ago (Goebel 1999), and then invading the Americas, could have had dogs with them.

In contrast, a recent origin of domestic dogs, as suggested by the archaeological record and by the genetic analysis (discussed above) of Savolainen et al. (2002), places constraints on the rate of migration of people to the New World. If domestication occurred from 14,000 to 15,000 years ago, and if dogs first arrived in the New World at least by 10,000 years ago (or even earlier), then the dog as an element of culture would need to have been transmitted across Paleolithic societies on three continents in a few thousand years or less. This would imply a much more extensive intercultural exchange during the Paleolithic than previously assumed (Gamble 1999; Derev'anko 1998). Regardless, the common origin of New and Old World dogs demands a reconsideration of the relationship between humans and dogs in ancient societies.

Perspectives and Future Research

The Time of First Domestication

Our initial results suggested an ancient domestication event, perhaps 100,000 years ago or more (Vilà et al. 1997a). However, the more recent work on ancient New World samples (Leonard et al. 2002) indicates that a substantial amount of diversity may have been generated over at least 12,000 to 14,000 years since the New World dog population was founded (Figure 19.5). This may imply that the first date from Vilà et al. (1997a) could be an overestimation, although dates older than 15,000 years are still supported by the recent genetic and archaeological findings (see above). Further, recent control region analysis of a large sample of Asian dogs suggests an earliest date of about 15,000 years ago, although 41,000 years ago could be possible with different assumptions on the monophyly or polyphyly of Clade I (Savolainen et al. 2002).

Therefore, these results indicate that an ancient record of protodogs has yet to be discovered, although it may not extend as early as originally suggested. First occurrence data generally underestimate the time of divergence from wolves (Figure 19.3), and given the poor time sampling represented by early dog remains (fewer than a dozen remains greater than 10,000 years old; Clutton-Brock 1999; Savolainen et al. 2002), this underestimation is likely to be considerable (Figure 19.3). Additionally, early morphological signs of domestication may be difficult to determine if protodogs looked similar to wild gray wolves, as discussed earlier (Vilà et al. 1997a). Domestic species could show only subtle differences from their wild brethren, although at least partially reproductively isolated from them, especially if the conditions of selection are similar to those in the wild (see Chapter 13). Some domestic and wild cats are nearly indistinguishable, and it is difficult to differentiate primitive domestic horses from their wild progenitors based on cranioskeletal characteristics (Balharry and Daniels 1995; Clutton-Brock 1999; Chapter 17).

This suggests that a period of domestication during which humans and protodogs had a fundamentally different kind of association may have preceded the first evidence of morphologic changes in the archaeological record. We and others have speculated that humans may have originally interacted indirectly with the first domesticated wolves through sharing of food, resources, and shelter (Clutton-Brock 1995; Coppinger and Coppinger 2001). These protodogs would have followed nomadic humans, utilized food from refuse and carcasses, and in turn may have provided humans with early warning systems or defended occupation sites from other carnivores. The nomadic existence of these protodogs would have led to differences in reproductive timing or mate choice from their wild brethren.

Considering the uncertainties about the date of first domestication, remains from the early wolflike canids found in association with humans should be examined. Bones of wolves have been found in association with those of hominids from as early as the Middle Pleistocene, 400,000 years BP (Clutton-Brock 1995). Some authors have identified these bones as dog, suggesting very ancient origins as old as 125,000 years ago (Thurston 1996). These remains should be analyzed using modern morphometric techniques, and genetic characterization should also be attempted. In general, a fuller genetic study of a wide range of canid remains found in association with humans might provide a more definitive understanding of the origins of current diversity.

The Number of Domestication Events

The control region tree (Figure 19.4) suggests that following the first domestication event, a minimum of three origination/interbreeding events affected the genetic diversity of dogs (Vilà et al. 1997a; Savolainen et al. 2002). But more than just these four events likely occurred, as discussed above. Further, although interbreeding cannot easily be distinguished from origination, the placement of lineages in the sequence tree (Figure 19.4) and their distribution among dog breeds provide a weak indication of interbreeding in some cases, and separate origination in others. For example, only one dog haplotype from Clade IV is identical to one in wolves, and this haplotype is widely distributed among dog breeds. This is consistent with a recent hybridization event followed by backcrossing to a well-mixed gene pool. In contrast, the unique Clade II haplotype found at high frequency only in Scandinavian breeds suggests that they have a common origin from a distinct population of wolves. Y-chromosome typing and further sequencing of nuclear genes will provide a needed contrast to maternally derived mtDNA in this regard, and will allow further discrimination of hybridization from origination events.

The genetic results provide little geographic resolution for domestication events. The primary reason for the lack of resolution is that the genetic diversity in gray wolves is not partitioned geographically. Within a single population there is often as much diversity as there is between well-separated populations. Further, no subset of sequences characterizes regions in the Old World (Vilà et al. 1999a). Consequently, arguments for dog origin have to be based on differences in the amount of genetic variation; such arguments are not phylogenetic (Savolainen et al. 2002) and may be susceptible to sampling bias (see above). However, gray wolves of the New and Old World are clearly differentiated, and our results suggest that the former were not a source a genetic material in dogs of the recent or distant past. Humans that migrated to the New World appear to have had their own genetic stock of dogs alongside, and because of lack of intent, or differences in reproductive timing, these migrant dogs did not interbreed extensively with New World wolves (but see Koop et al. 2000).

Human and Dog Migrations and Associations

Our results suggest that the humans who colonized America at least 12,000 to 14,000 years ago brought multiple lineages of domesticated dogs with them. The large diversity of mtDNA lineages in the dogs that colonized the New World implies that the number and diversity of dogs in Eurasia was already extensive at that time. *A priori*, it does not seem likely that a single domestication process would have involved a large number of individuals. Consequently, the large diversity may be indicative of dispersion of domestic dogs from multiple origination centers (or sites where backcrossing took place) and exchange between localities. This idea is reinforced by the observation that domestic dogs existed in both Asia and Europe 14,000 to 15,000 years ago (Nobis 1979; Olsen 1985; Clutton-Brock 1999; Sablin and Khlopachev 2002), and soon thereafter also in the New World. This finding implies extensive intercultural exchange during the late Paleolithic, a time when human societies were essentially nomadic and presumably autonomous, and their movements are assumed to have been more spatially restricted (Gamble 1999). Further, the observation that human societies over at least three continents were living with dogs and exchanging them suggests that dogs offered valuable services to ancient societies, although we cannot be precise as to what kind of services those were.

In Figure 19.6, Clade *a* represents a group of mtDNA sequences found only in the Americas (Leonard et al. 2002). Interestingly, Clade *a* sequences were present in the samples from Mexico, Peru, and Bolivia, but were not detected in the Alaskan samples. Instead, several other haplotypes were detected, and one was found that was possibly ancestral to Clade *a*. These results are consistent with the hypothesis that the colonization of America occurred in different waves of human migration from northern North America to the south (e.g., Greenberg et al. 1986; Cavalli-Sforza et al. 1994), or that a small subset of northern dogs colonized areas to the south, most of which trace their ancestry to Alaskan haplotype D36 (Figure 19.6). These results indicate that the study of dog genetic diversity and that of other domestic animals or plants might be informative of human movements, and the history of the peopling of the New World (see, for example, Erickson et al. 2005).

Future Research

In order to further test hypotheses about the timing and number of origination events, more extensive sampling of dogs is needed. Asian dog populations are poorly sampled, although recent work has better characterized them (Savolainen et al. 2002). The dogs of the Middle East and of the New World that represent intermixed and potentially ancient populations need to be sampled as a first priority. Presumably, some ancient lineages might have been retained in these populations, giving further insight into the origin and diversification of dogs. Second, archaeological material of dogs from both the Old and the New World needs to be sampled for DNA sequencing. Our New World ancient sample (Leonard et al. 2002) represents a fraction of the dog remains in the New World, but it shows that a very large portion of the genetic diversity existing in the past may have been lost.

Presumably, the discovery of wolflike sequences or basal dog lineages in these ancient remains would provide a needed evolutionary perspective to sequence trees.

Further, very ancient remains of wolflike canids found in association with humans should be typed. If traces of dog ancestry are found in some of these specimens, then perhaps a careful comparative morphologic analysis might reveal the first hints of the effects of domestication on skeletal morphology and provide information as to the first human societies that kept dogs. The presence or absence of doglike sequences in very ancient wolflike canids found in association with humans would provide a direct test of the ancient domestication hypotheses, and would shed light on early human-dog associations. Additionally, an analysis of much longer mtDNA sequences would allow more accurate estimation of origination times, as well as a better assessment of the mutation rate (Cooper et al. 2001; Ingman et al. 2000). Finally, a genetic perspective provided by other molecular markers is urgently needed to counteract the biases associated with single-gene phylogenies and the maternal inheritance of mtDNA (Avise 1994, 2000; Vilà et al. 2001). These markers might included Y-chromosome sequences (Olivier and Lust 1998; Olivier et al. 1999; Sundqvist et al. 2001), microsatellite loci (Zajc and Sampson 1999; Breen et al. 2001; Richman et al. 2001), single nucleotide polymorphisms (Lindblad-Toh et al. 2000; Primmer et al. 2002; Parker et al. 2004; Vilà et al. 2005), or whole gene sequences (Ingman et al. 2000; Haworth et al. 2001). Until additional markers and samples are typed, hypotheses of origin and diversification based on genetic data remain tentative at best.

References

American Kennel Club. 1997. *The complete dog book: The histories and standards of breeds admitted to AKC registration, and the feeding, training, care, breeding, and health of pure-bred dogs*, 19th edition. Garden City, N.Y.: Doubleday.

Andersone, Z., V. Lucchini, E. Randi, and J. Ozolins. 2002. Hybridisation between wolves and dogs in Latvia as documented using mitochondrial and microsatellite DNA markers. *Mammalian Biology* 67: 79–90.

Ash, E. C. 1927. *Dogs: Their history and development*. London: E. Benn Ltd.

Avise, J. C. 1994. *Molecular markers, natural history and evolution*. New York: Chapman and Hall.

———. 2000. *Phylogeography. The history and formation of species*. Cambridge, Mass.: Harvard University Press.

Balharry, D. and M. J. Daniels. 1995. *Wild living cats in Scotland*. Perth, UK: Scottish Natural Heritage Research, Survey and Monitoring Report No. 23. Scottish Natural Heritage.

Barnes, I., P. Matheus, B. Shapiro, D. Jensen, and A. Cooper. 2002. Dynamics of Pleistocene population extinctions in Beringian brown bears. *Science* 295: 2267–2270.

Birky, C. W., Jr. 2001. The inheritance of genes in mitochondria and chloroplasts: Laws, mechanisms, and models. *Annual Review of Genetics* 35: 125–148.

Breen, M., S. Jouquand, C. Renier, C. S. Mellersh, and C. Hitte, et al. 2001. Chromosome-specific single-locus FISH probes allow anchorage of an 1800-marker integrated radiation-hybrid/linkage map of the domestic dog genome to all chromosomes. *Genome Research* 11: 1784–1795.

Bruford, M. W. and R. K. Wayne. 1993. Microsatellites and their application to population genetic studies. *Current Opinions in Genetics and Development* 3: 939–943.

Cavalli-Sforza, L. L., P. Menozzi, and A. Piazza. 1994. *The history and geography of human genes*. Princeton, N. J.: Princeton University Press.

Chase, K., D. R. Carrier, F. R. Adler, T. Jarvik, E. A. Ostrander, T. D. Lorentzen, and K. G. Lark. 2002. Genetic basis for systems of skeletal quantitative traits: Principal component analysis of the canid skeleton. *Proceedings of the National Academy of Sciences, USA* 99: 9930–9935.

Clutton-Brock, J. 1995. Origins of the dog: Domestication and early history. In *The domestic dog: Its evolution, behaviour and interactions with people*, J. Serpell (ed.), pp. 7–20. Cambridge: Cambridge University Press.

Clutton-Brock, J. 1999. *A natural history of domesticated mammals*, 2nd edition. Cambridge: Cambridge University Press.

Cooper, A., C. Lalueza-Fox, S. Anderson, A. Rambaut, J. Austin, and R. Ward. 2001. Complete mitochondrial genome sequences of two extinct moas clarify ratite evolution. *Nature* 409: 704–707.

Coppinger, R. and L. Coppinger. 2001. *Dogs: A new understanding of canine origin, behavior, and evolution*. Chicago: The University of Chicago Press.

Coppinger, R. and R. Schneider. 1995. Evolution of working dogs. In *The domestic dog: Its evolution, behaviour and interactions with people*, J. Serpell (ed.), pp. 21–47. Cambridge: Cambridge University Press.

Cordy-Collins, A. 1994. An unshaggy dog story. *Natural History* 2/94: 34–40.

Darwin, C. [1859] 1982. *The origin of species*. London: Penguin.

Dayan, T. 1994. Early domesticated dogs of the Near East. *Journal of Archaeological Science* 21: 663–640.

Dennis-Bryan, K. and J. Clutton-Brock. 1988. *Dogs of the last hundred years at the British Museum (Natural History)*. London: British Museum (Natural History).

Derev'anko, A. P. 1998. A short history of discoveries and the development of ideas in the Paleolithic of Siberia. In *The Paleolithic of Siberia: New discoveries and interpretations*, A. P. Derev'anko, D. B. Shimkin, and W. R. Powers (eds.), pp. 5–13. Champaign: The University of Illinois Press.

Elias, S. A., S. K. Short, C. H. Nelson and H. H. Birks. 1996. Life and times of the Bering land bridge. *Nature* 382: 60–63.

Erickson, D. L., B. D. Smith, A. C. Clark, D. H. Sanweiss, and N. Tuross. 2005. An Asian origin for a 10,000 year old domesticated plant in the Americas. *Proceedings of the National Academy of Sciences USA* 102: 18315–18320.

Ferrell, R. E., D. C. Morizot, J. Horn, and C. J. Carley. 1978. Biochemical markers in species endangered by introgression: The red wolf. *Biochemical Genetics* 18: 39–49.

Fondon, J.W. III and H. R. Garner. 2004. Molecular origins of rapid and continuous morphological evolution. *Proceedings of the National Academy of Sciences, USA* 101: 18058–18063.

Gamble, C. 1999. *The Paleolithic societies of Europe*. Cambridge: Cambridge World Archaeology, Cambridge University Press.

García-Moreno, J., M. D. Matocq, M. S. Roy, E. Geffen and R. K. Wayne. 1996. Relationship and genetic purity of the endangered Mexican wolf based on analysis of microsatellite loci. *Conservation Biology* 10: 376–389.

Goebel, T. 1999. Pleistocene human colonization of Siberia and peopling of the Americas: An ecological approach. *Evolutionary Anthropology* 8: 208–227.

Gottelli, D., C. Sillero-Zubiri, G. D. Applebaum, M. S. Roy, D. J. Girman, J. García-Moreno, E. A. Ostrander, and R. K. Wayne. 1994. Molecular genetics of the most endangered canid: The Ethiopian wolf, *Canis simensis*. *Molecular Ecology* 3: 301–312.

Greenberg, J. H., C. G. Turner II, and S. L. Zegura. 1986. The settlement of the Americas: A comparison of the linguistic, dental, and genetic evidence. *Current Anthropology* 27: 477–497.

Hart, B. L. 1995. Analysing breed and gender differences in behaviour. In *The domestic dog: Its evolution, behaviour and interactions with people*, J. Serpell (ed.), pp. 65–77. Cambridge: Cambridge University Press.

Haworth, K. E., I. Islam, M. Breen, W. Putt, E. Makrinou, M. Binns, D. Hopkinson, and Y. Edwards. 2001. Canine TCOF1: Cloning, chromosome assignment and genetic analysis in dogs with different head types. *Mammalian Genome* 12: 622–629.

Hofreiter, M., D. Serre, H. N. Poinar, M. Kuch, and S. Pääbo. 2001. Ancient DNA. *Nature Reviews Genetics* 2: 353–359.

Hully, C. C. 1958. *Alaska: Past and present*. Portland, Ore.: Binfords and Mort, Publishers.

Ingman, M., H. Kaessmann, S. Pääbo, and U. Gyllensten. 2000. Mitochondrial genome variation and the origin of modern humans. *Nature* 408: 708–713.

Koop, B. F., M. Burbidge, A. Byun, U. Rink, and S. J. Crockford. 2000. Ancient DNA evidence of a separate origin for North American indigenous dogs. In *Dogs through time: An archaeological perspective*, S. J. Crockford (ed.), pp. 271–285. British Archaeological Reports International Series 889. Oxford: BAR.

Lehman, N., A. Eisenhawer, K. Hansen, D. L. Mech, R. O. Peterson, P. J. P. Gogan, and R. K. Wayne. 1991. Introgression of coyote mitochondrial DNA into sympatric North American gray wolf populations. *Evolution* 45: 104–119.

Leonard, J. A., R. K. Wayne, and A. Cooper. 2000. Population genetics of Ice Age brown bears. *Proceedings of the Natural Academy of Sciences, USA* 97: 1651–1654.

Leonard, J. A., R. K. Wayne, J. Wheeler, R. Valadez, S. Guillen, and C. Vilà. 2002. Ancient DNA evidence for Old World origin of New World dogs. *Science* 298: 1613–1616.

Li, W.-H. 1997. *Molecular evolution*. Sunderland, Mass.: Sinauer Associates.

Lindblad-Toh, K., E. Winchester, M. J. Daly, D. G. Wang, J. N. Hirschhorn et al. 2000. Large-scale discovery and genotyping of single-nucleotide polymorphisms in the mouse. *Nature Genetics* 24: 381–386.

Lingaas, F., T. Aarskaug, A. Sorensen, L. Moe, and P.-E. Sundgreen. 1996. Estimates of genetic variation in dogs based on microsatellite polymorphism. *Animal Genetics* 27: 29.

Loftus, R. T., D. E. MacHugh, D. G. Bradley, P. M. Sharp, and P. Cunningham. 1994. Evidence for two independent domestications of cattle. *Proceedings of the National Academy of Sciences, USA* 91: 2757–2761.

Lorenz, K. 1954. *Man meets dog*. London: Methuen.

Luikart, G., L. Gielly, L. Excoffier, J.-D.Vigne, J. Bouvet, and P. Taberlet. 2001. Multiple maternal origins and weak phylogeographic structure in domestic goats. *Proceedings of the National Academy of Sciences, USA* 98: 5927–5932.

Morey, D. F. 1992. Size, shape, and development in the evolution of the domestic dog. *Journal of Archaeological Science* 19: 181–204.

Morey, D. 1994. The early evolution of the domestic dog. *American Scientist* 82: 336–347.

Morey, D. and M. D. Wiant. 1992. Early Holocene domestic dog burials from the North American Midwest. *Current Anthropology* 33: 224–229.

Neff, M. W., K. R. Robertson, A. K. Wong, N. Safra, K. W. Broman, M. Slatkin, K. L. Mealey, and N. C. Pedersen. 2004. Breed distribution and history of canine mdr1-1r, a pharmacogenetic mutation that marks the emergence of breeds from the collie lineage. *Proceedings of the National Academy of Sciences, USA* 101: 11725–11730.

Nobis, G. 1979. Der älteste Haushund lebte vor 14,000 Jahren. *Umschau* 19: 610.

Nowak, R. M. 1999. *Walker's mammals of the world*, 6th edition. Baltimore: The Johns Hopkins University Press.

Okumura, N., N. Ishiguro, M. Nakano, A. Matsui, and M. Sahara. 1996. Intra- and interbreed genetic variations of mitochondrial DNA major noncoding regions in Japanese native dog breeds *(Canis familiaris)*. *Animal Genetics* 27: 397–405.

Olivier, M. and G. Lust. 1998. Two DNA sequences specific for the canine Y chromosome. *Animal Genetics* 29: 146–149.

Olivier, M., M. Breen, M. M. Binns, and G. Lust. 1999. Localization and characterization of nucleotide sequences from the canine Y chromosome. *Chromosome Research* 7: 223–233.

Olsen, S. J. and J. W. Olsen. 1977. The Chinese wolf ancestor of New World dogs. *Science* 197: 533–535.

Olsen, S.J. 1985. *Origins of the domestic dog.* Tucson: University of Arizona Press.

Parker, G. H., L. V. Kim, N. B. Sutter, S. Carlson, T. D. Lorentzen, T. B. Malek, G. S. Johnson, H. B. DeFrance, E. A. Ostrander, and L. Kruglyak. 2004. Genetic structure of the purebred domestic dog. *Science* 304: 1160–1164.

Passos-Bueno, M. R., W. R. Wilcox, E. W. Jabs, A. L. Sertie, L. G. Alonso, and H. Kitoh. 1999. Clinical spectrum of fibroblast growth factor receptor mutations. *Human Mutation* 14: 115–125.

Pesole, G., C. Gissi, A. De Chirico, and C. Saccone. 1999. Nucleotide substitution rate of mammalian mitochondrial genomes. *Journal of Molecular Evolution* 48: 427–434.

Pferd, W. III. 1987. *Dogs of the American Indians*. Fairfax, Va.: Denlinger's Publishers.

Pilgrim, K. L., D. K. Boyd, and S. H. Forbes. 1998. Testing for wolf-coyote hybridization in the Rocky Mountains using mitochondrial DNA. *Journal of Wildlife Management* 62: 683–689.

Primmer, C. R., T. Borge, J. Lindell, and G. P. Saetre. 2002. Single-nucleotide polymorphism characterization in species with limited available sequence information: High nucleotide diversity revealed in the avian genome. *Molecular Ecology* 11: 603–612.

Rand, D. M. 2001. The units of selection on mitochondrial DNA. *Annual Reviews in Ecology and Systematics* 32: 415–448.

Randi, E. and V. Lucchini. 2002. Detecting rare introgression of domestic dog genes into wild wolf *(Canis lupus)* populations by Bayesian admixture analyses of microsatellite variation. *Conservation Genetics* 3: 31–45.

Randi, E., F. Francisci, and V. Lucchini. 1995. Mitochondrial DNA restriction-fragment-length monomorphism in the Italian wolf *(Canis lupus)*. *Journal of Zoological Systematics and Evolutionary Research* 33: 97–100.

Randi, E., V. Lucchini, M. F. Christensen, N. Mucci, S. M. Funk, G. Dolf, and V. Loeschcke. 2000. Mitochondrial DNA variability in Italian and East European wolves: Detecting the consequences of small population size and hybridization. *Conservation Biology* 14: 464–473.

Reitz, E. J. and E. S. Wing. 1999. *Zooarchaeology*. Cambridge: Cambridge University Press.

Richman, M., C. S. Mellersh, C. André, F. Galibert, and E. A. Ostrander. 2001. Characterization of a minimal screening set of 172 microsatellite markers for genome-wide screens of the canine genome. *Journal of Biochemical and Biophysical Methods* 47: 137–149.

Roy, M. S., E. Geffen, D. Smith, E. A. Ostrander, and R. K. Wayne. 1994. Patterns of differentiation and hybridization in North American wolf-like canids revealed by analysis of microsatellite loci. *Molecular Biology and Evolution* 11: 553–570.

Sablin, M. V. and G. A. Khlopachev. 2002. The earliest Ice Age dogs: Evidence from Eliseevichi. *Current Anthropology* 43: 795–799.

Sarich, V. M. 1977. Albumin phylogenetics. In *Albumin structure, function and uses*, V. M. Rosenoer, M. Oratz, and M. A. Rothschild (eds.), pp. 85–111. New York: Pergamon Press.

Savolainen, P., Y. P. Zhang, J. Luo, J. Lundeberg, and T. Leitner. 2002. Genetic evidence for an East Asian origin of domestic dogs. *Science* 298: 1610–1613.

Savolainen, P., T. Leitner, A. N. Wilton, E. Matisoo-Smith, and J. Lundeberg. 2004. A detailed picture of the origin of the Australian dingo, obtained from the study of mitochondrial DNA. *Proceedings of the National Academy of Sciences, USA* 101: 12387–12390.

Schwartz, J. 1997. *A history of dogs in the early Americas*. New Haven: Yale University Press.

Seal, U. S., N. I. Phillips, and A. W. Erickson. 1970. Carnivora systematics: Immunological relationships of bear albumins. *Comparative Biochemistry and Physiology* 32: 33–48.

Simonsen, V. 1976. Electrophoretic studies on blood proteins of domestic dogs and other Canidae. *Hereditas* 82: 7–18.

Stockard, C. R. 1941. *The genetic and endocrinic basis for differences in form and behavior, as elucidated by studies of contrasted pureline dog breeds and their hybrids*. Philadelphia: The Wistar Institute of Anatomy and Biology.

Sundqvist, A.-K., H. Ellegren, M. Olivier, and C. Vilà. 2001. Y-chromosome haplotyping in Scandinavian wolves *(Canis lupus)* based on microsatellite markers. *Molecular Ecology* 10: 1959–1966.

Tajima, F. 1993. Simple methods for testing the molecular evolutionary clock hypothesis. *Genetics* 135: 599–607.

Thurston, M. E. 1996. *The lost history of the canine race: Our 15,000-year love affair with dogs*. Kansas City: Andrews and McMeel.

Tsuda, K., Y. Kikkawa, H. Yonekawa, and Y. Tanabe. 1997. Extensive interbreeding occurred among multiple matriarchal ancestors during the domestication of dogs: Evidence from inter- and intraspecies polymorphisms in the D-loop region of mitochondrial DNA between dogs and wolves. *Genes and Genetic Systems* 72: 229–238.

Valadez, R. 1995. *El perro Mexicano*. México City: Universidad Nacional Autónoma de México.

Valadez, R. 1999. La patria del xoloitzcuintle. *Revista Asociación Mexicana de Médicos Veterinarios Especialistas en Pequeñas Especies (RevAMMVEPE)* 10: 76–81.

Valadez, R. and G. Mestre. 1999. *Historia del Xoloitzcuintle en México*. México City: Universidad Nacional Autónoma de México.

Valadez, R., B. Rodríguez, F. Viniegra, K. Olmos, A. Blanco, S. Tejeda, and M. Casas. 2002. Híbridos de lobos y perros en cuevas teotihuacanas. Crónica de un descubrimiento. *Revista Asociacion Mexicana de Médicos Veterinarios Especialistas en Pequeñas Especies* (RevAMMVEPE) 13: 6–23.

Vilà, C., P. Savolainen, J. E. Maldonado, I. R. Amorim, J. E. Rice, R. L. Honeycutt, K. A. Crandall, J. Lundeberg, and R. K. Wayne. 1997a. Multiple and ancient origins of the domestic dog. *Science* 276: 1687–1689.

Vilà, C., J. Maldonado, I. R, Amorim, R. K. Wayne, K. A. Crandall, and R. L. Honeycutt. 1997b. Man and his dog—Reply. *Science* 278: 206–207.

Vilà, C., I. R. Amorim, J. A. Leonard, D. Posada, J. Castroviejo, F. Petrucci-Fonseca, K. A. Crandall, H. Ellegren, and R. K. Wayne. 1999a. Mitochondrial DNA phylogeography and population history of the gray wolf, *Canis lupus*. *Molecular Ecology* 8: 2089–2103.

Vilà, C., J. Maldonado, and R. K. Wayne. 1999b. Phylogenetic relationships, evolution and genetic diversity of the domestic dog. *Journal of Heredity* 90: 71–77.

Vilà, C., J. A. Leonard, A. Götherström, S. Marklund, K. Sandberg, K. Lidén, R. K. Wayne, and H. Ellegren. 2001. Widespread origins of domestic horse lineages. *Science* 291: 474–477.

Vilà, C., C. Walker, A.-K. Sundqvist, Ø. Flagstad, A. Andersone, A. Casulli, I. Kojola, H. Valdmann, J. Halversone, and H. Ellegren. 2003. Combined use of maternal, paternal and bi-parental genetic markers for the identification of wolf-dog hybrids. *Heredity* 90:17–24.

Vilà, C., J. Seddon, and H. Ellegren. 2005. Genes of domestic mammals augmented by backcrossing with wild ancestors. *Trends in Genetics* 21: 214–218.

Wayne, R. K. 1986a. Cranial morphology of domestic and wild canids: The influence of development on morphological change. *Evolution* 4: 243–261.

———. 1986b. Limb morphology of domestic and wild canids: The influence of development on morphologic change. *Journal of Morphology* 187: 301–319.

Wayne, R. K. and S. J. O'Brien. 1987. Allozyme divergence within the Canidae. *Systematic Zoology* 36: 339–355.

Wayne, R. K., W. G. Nash, and S. J. O'Brien. 1987a. Chromosomal evolution of the Canidae: I. Species with high diploid numbers. *Cytogenetics and Cell Genetics* 44: 123–133.

———. 1987b. Chromosomal evolution of the Canidae: II. Divergence from the primitive carnivore karyotype. *Cytogenetics and Cell Genetics* 44: 134–141.

Wayne, R. K., R. E. Benveniste, and S. J. O'Brien. 1989. Molecular and biochemical evolution of the Carnivora. In *Carnivore behavior, ecology and evolution*, J. L. Gittleman (ed.), pp. 465–494. Ithaca, N.Y.: Cornell University Press.

Wayne, R. K., N. Lehman, M. W. Allard, and R. L. Honeycutt. 1992. Mitochondrial DNA variability of the gray wolf: Genetic consequences of population decline and habitat fragmentation. *Conservation Biology* 6: 559–569.

Wayne, R. K., E. Geffen, D. J. Girman, K. P. Koepfli, L. M. Lau, and C. Marshall. 1997. Molecular systematics of the Canidae. *Systematic Biology* 4: 622–653.

Wilcox, B. and C. Walkowicz. 1995. *The atlas of dog breeds of the world*, 5th edition. Neptune, N.J.: T.F.H. Publications.

Wilson, P. J., S. Grewal, I. D. Lawford, J. N. M. Heal, A. G. Granacki, D. Pennock, J. B. Theberge, M. T. Theberge, D. R. Voigt, W. Waddell, R. E. Chambers, P. C. Paquet, G. Goulet, D. Cluff, and B. N. White. 2000. DNA profiles of the eastern Canadian wolf and the red wolf provide evidence for a common evolutionary history independent of the gray wolf. *Canadian Journal of Zoology* 78: 2156–2166.

Wurster-Hill, D. H. and W. R. Centerwall. 1982. The interrelationships of chromosome banding patterns in canids, mustelids, hyena, and felids. *Cytogenetics and Cell Genetics* 34: 178–192.

Zajc, I. and J. Sampson. 1999. Utility of canine microsatellites in revealing the relationships of purebred dogs. *Journal of Heredity* 90: 104–107.

Zajc, I., C. S. Mellersh, and J. Sampson. 1997. Variability of canine microsatellites within and between different dog breeds. *Mammalian Genome* 8: 182–185.

CHAPTER 20

Origins and Diffusion of Domestic Goats Inferred from DNA Markers
Example Analyses of mtDNA, Y Chromosome, and Microsatellites

G. LUIKART, H. FERNÁNDEZ, M. MASHKOUR,
P. R. ENGLAND, AND P. TABERLET

Introduction

Goat domestication was an integral part of the rise of agriculture and the adoption of agricultural practices throughout much of the world. Insights into the evolution and spread of goats are likely to deepen our understanding of the origin and spread of agriculture and the rise of early human civilizations. As yet, the origins, diffusion patterns, and taxonomy of domestic goats are uncertain. For example, although many authors agree that domestic goats *(Capra hircus)* originated from the "wild goat" *Capra aegagrus*, some authors suggest the markhor *(C. falconeri)* gave rise to the cashmere breeds of eastern Asia (Hasnain 1985; Meadow 1996; Porter 1996). Genetics is proving to be one of the most powerful tools available for the investigation of these and other unresolved questions surrounding goat domestication. In this chapter we assess the origins and diffusion of domestic goats by synthesizing published and new genetic data from mtDNA (including ancient sequences), Y-chromosome, and microsatellite DNA markers for goats and their wild relatives *(Capra* spp.)

Goats were among the first farm animals domesticated (Clutton-Brock 1981; Zeder and Hesse 2000) and were important in the economies of societies as long as 10,000 years ago. Subsequently, goats were taken along during human migrations and colonizations, and in more recent times, on ships exploring new continents (Porter 1996). Domestic goat skin was apparently the main material used for writing biblical manuscripts, such as the Dead Sea Scrolls (Kahila Bar Gal et al. 1999). Goat skin was also used by the Roman military to construct hand-held battle shields and sandals. Goats and their skins were often used for tents, curtains, gift offerings, and sacrifices in Christian, Jewish, and Muslim societies. This is evident, for example, from many quotes in the Bible, including the following: "They made curtains of goat hair for the tent over the tabernacle—eleven altogether" (Exodus 36:14–19) and "The goats for the sin offering were brought before the king and the assembly, and they laid their hands on them" (Judges 13:19–24).

Goats are important to the economies of many contemporary societies, especially those in marginal areas and harsh climates. For the Nubian people in the hot, arid areas of northern Sudan, the Nubian goat (Figure 20.1a) represents the only source of milk. Goats also are of growing economic importance in developed countries, such as France and Switzerland, where they provide milk for infants and for specialty cheeses. In North America, goat farming is rapidly increasing thanks to government subsidies and the increased demand for goat meat. The Angora goat from Turkey is the world's main source of fine mohair (Figure 20.1b). The cashmere goat produces one of the finest fibers in the world. Cashmere fibers represent more than 10 percent of Mongolia's annual income. Finally, goats have been quite useful for controlling invasive weeds and plants (Porter 1996). However, they also have prevented reforestation and destroyed native vegetation, especially when introduced to predator-free islands such as Hawaii.

Goats are the most adaptive, hardy, and geographically widespread livestock species (Clutton-Brock 1981). For one example, goats (but not sheep or cattle) are able to survive the extreme heat (>50 °C) and infectious diseases in northern Sudan. Goats, unlike other livestock, have flexible feeding requirements and can digest woody material (browse) more efficiently than other livestock species. Some goats (e.g., those in North Africa) can even climb trees in search of leaves or fresh buds or twigs.

In this chapter we illustrate how DNA data can help address the following five objectives:

1. To identify the wild progenitor species of domestic goats
2. To assess the number of genetic origins
3. To determine the geographic locations of origins
4. To infer the relative date of population expansions for different genetic lineages (or for different continental populations)
5. To assess the amount of population structure and historical inter-continental movement of goats compared to humans and cattle

Sampling, DNA Markers, and Prior Research

In this section we discuss general considerations of sampling methods and DNA markers for studies of animal domestication. We also provide background information about prior genetic research on goat domestication, including the sampling and analyses (mainly from our laboratory).

Tissue Sampling and DNA Extraction

DNA can be obtained from any tissue material including liver, muscle, skin, blood, buccal cells, horn, bone, hair, milk,

FIGURE 20.1 Photos illustrating the morphological diversity and distinctiveness of two important goat breeds: (a) Nubian goat (top) from Sudan, which survives in hot and parasite-ridden areas where most other livestock cannot; (b) Angora breed (bottom), from near Ankara, Turkey, which is the world's only (or at least first and main) mohair-producing breed. (Photos by A. Beja-Pereira and G. Luikart, respectively).

cheese, feces, urine, and even fossil bones and ancient leather from clothing or writing parchment (Kahila Bar Gal et al. 1999; Taberlet et al. 1999). These tissues are listed roughly in decreasing order of DNA yield (quality and quantity). Standard commercial DNA extraction kits can be used to obtain DNA from any of these materials, for a cost of approximately 1 to 3 dollars per extraction.

Our sampling of contemporary domestic goats has relied mainly on a tiny (2 mm square) piece of ear or tail tissue stored in >6 volumes of alcohol (95% EtOH), at room temperature for up to 1 year before extraction. From some breeds we used fresh plucked hairs (stored in paper envelopes inside an airtight container with silica drying beads). DNA was extracted from fossil bone by first cleaning the exterior surface (3–5 mm square) and then pulverizing the bone into powder using liquid nitrogen and a mortar and pestle. DNA was extracted from all samples using standard commercial kits (Qiagen tissue and blood kits).

For our mtDNA control region studies, sampling included >400 individuals, 88 breeds, and 44 countries spanning most of the Old World distribution of goats from Nigeria to Iceland and Mongolia to Malaysia, including potential centers of domestication (e.g., Turkey, Egypt, Jordan, Iraq, and Pakistan) (for a list of samples, see Appendix 1 in Luikart et al. 2001, at http://www.pnas. org). However, our cytochrome *b* and Y-chromosome studies included fewer individuals (6 and ~50, respectively). Our microsatellite DNA studies included ~30 individuals per breed but only 9 breeds (from Mongolia, Malaysia, Nigeria, and England). It is important to note that we sampled only pure indigenous goats from small remote villages and excluded large cities to avoid sampling goats that might have recently been transported long distances.

DNA Markers and Sequences

MtDNA is the most widely used DNA marker because of its high variability and high abundance (copy number) in most cells, and because of its lack of recombination associated with strict maternal inheritance. This quality allows the identification of maternal lineages that have diverged through time only via accumulation of new mutations. Most mammalian cells contain hundreds of copies of the mtDNA genome, compared to only two copies (maternal and paternal) of the nuclear genome (and only one Y and X copy in males). The high copy number facilitates extraction of mtDNA from old or fossil material. The high variability facilitates study of closely related and recently evolved populations, such as domestic breeds.

Goat domestication issues were first addressed using mtDNA from several domestic goats and several captive wild goats *(Capra aegagrus)* (Takada et al. 1997). The mtDNA control region sequences suggested that the wild and domestic goats had similar mtDNA sequences. However, the sample sizes were small, and the wild samples were from zoos, where hybridization is possible and provenance can be dubious.

Our laboratory has sequenced the most variable part of the mtDNA molecule (hypervariable region 1; HVI), as have many early studies of human and livestock origins. We sequenced 481 base pairs of HVI from each of 406 goats using the following PCR primers:

CAP-FI: [5´-TCCATATAACGCGGACATAC-3´]
CAP-RI: [5´ATGGCCCTGAAGAAAGAAC-3´]

See Luikart et al. 2001 for more details.

For interspecific studies the mtDNA cytochrome *b* gene (cyt *b*) is more appropriate than the control region because it evolves less rapidly, being under functional constraint as a coding gene. Thus, it is easier to align between species (without sequence gaps) and to calibrate (via the molecular clock concept) using estimated divergence dates between taxa (e.g., sheep and goat diverged ~6 MYA). We sequenced 1,140 bp of cyt *b* for two goats (with different control region sequences) arbitrarily chosen from each divergent lineage (*C. hircus* A–C) identified by control region sequencing. The cyt *b* primers were: L14724V [3´-atgatatgaaaaaccatcgttg-5´] and H15915V: [3´-tctccttctctggtttacaagac-5´] (Luikart et al. 2001).

Y-chromosome DNA also provides excellent phylogenetic information because it is also transmitted without recombination. It is especially useful for reconstructing paternal lineages for a "male view" of evolutionary history. Information from Y chromosome is complementary to mtDNA data, although its DNA sequences are generally far less polymorphic, and therefore less informative. The Y chromosome is also less useful for inferring domestication events because of the confounding effects of paternal gene flow from wild males with domestic females. In contrast, maternal (mtDNA) gene flow is more likely to occur only during a domestication event where individuals are captured from the wild.

From the Y chromosome, about 1,200 base pairs (bp) were sequenced (950 bp of a ZFY gene intron and 280 bp of an amelogenin gene exon) from more than 50 domestic goats from across the Old World (Pidancier et al. n.d.). We also sequenced representatives of all the wild *Capra* morphotypes. Amelogenin is a protein that contributes to tooth enamel and the ZFY gene codes for a Zinc Finger protein. Primers for the fifth exon region of the amelogenin are

CAPY1F: 5´-CCCAGCAGACTCCCCAGAATC-3´
CAPY1R: 5´-CCAGAGGGAGGTCAGGAAGCA-3´

The ZFY (last intron) was amplified using these primers:

UEAZF2F: 5´-AAGACCTGATTCCAGGCAGTA-3´
UEAZFY: 5´-CTTCTTATTGGTAGTGTAGTAATC-

Microsatellites are the most useful DNA markers for measuring and comparing levels of diversity (e.g., number of alleles) within populations because these markers are highly variable, often having 5–10 alleles per locus. These markers can thus be used to compare levels of diversity within populations from different continental regions (e.g. inside and outside the centers of domestication). Microsatellites can also be useful for assessing genetic relationships among closely related populations (isolated for only several dozen generations). Here we analyzed 22 microsatellite loci in 9 domestic goat breeds from different continents, plus the wild *C. aegagrus* from Daghestan and *Capra ibex sibirica* from Mongolia (Luikart et al. 1999).

Statistical Analyses

For details on mtDNA and Y-chromosome DNA analyses see Luikart et al. 2001 and Pidancier et al. 2006. In brief, neighbor-joining, UPGMA clustering, and maximum likelihood trees were constructed using the PAUP* package version 0.64d (Swofford 1998). For the cyt *b* sequences, we estimated the time of the most recent ancestor (TMRCA) of the three observed maternal goat lineages. For this we assumed a molecular clock, using an estimated divergence time of 5 to 7 million years ago between sheep and goats (Savage and Russell 1983; Carroll 1988). Population structure was quantified for the mtDNA control region data set using the AMOVA

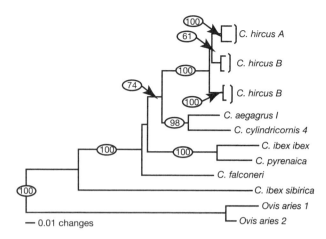

FIGURE 20.2 Phylogenetic tree of domestic and wild goat cytochrome *b* gene sequences (1,140 base pairs) constructed using neighbor-joining methods in the computer program PAUP* version 4.0b4 (Swofford 1998). Parsimony trees gave very similar topologies (not shown; G. Luikart, unpublished data). Note that *C. aegagrus* has the shortest branch length to the three domestic lineages. The *C. cylindricornis* also groups close to *C. aegagrus* (and the domestics), perhaps due to past mtDNA introgression (hybridization), which is possible because both were sampled from a region in the Caucasus Mountains.

framework, which used information both on allele divergence and allele frequency differentiation between populations. The expansion "dating" used the distribution of pairwise differences between mtDNA sequences (Schneider and Excoffier 1999). Microsatellite analyses were conducted using Genepop and Genetix software to test for Hardy Weinberg proportions and linkage disequilibrium and to compute heterozygosity and population differentiation (F_{ST}).

Results and Discussion

Here we will discuss how molecular genetic data can be used to address each of the five objectives stated earlier. We will finish with a brief discussion of perspectives and the promise of combining genetic and archaeological studies.

Wild Progenitors

To help identify the wild progenitor species of domestic goats, we sequenced the entire mitochondrial cytochrome *b* (cyt *b*) gene (1,040 base pairs) from six domestic goats and from representatives of each of the five wild *Capra* morphotypes. The shortest cyt *b* genetic distance (i.e., branch length) from domestic goats to a wild species was to the wild goat *C. aegagrus* (Figure 20.2). This is consistent with morphological studies, archaeological data, and inferred geographical distribution of wild *Capra* spp. (Smith 1998; Chapter 14), suggesting that *C. aegagrus* is the most probable progenitor. These mtDNA data cannot exclude other wild taxa (e.g., *C. cylindricornis*) from having been domesticated, however, because few wild populations were sampled and the sequence data do not group *C. aegagrus* substantially closer than *C. cylindricornis* to the domestic goat.

We have also sequenced 481 base pairs of the mtDNA control region (discussed below), but it is less useful for estimating interspecies relationships because of its high mutation rate and homoplasy (e.g., back mutation and multiple mutations per base pair). Nonetheless, our control region studies and other studies of mtDNA control region (using fewer individuals) have also found the *C. aegagrus* closely related to domestic goats (Takada et al. 1997).

The Y-chromosome data also suggest that *C. aegagrus* is the wild taxon most closely related to domestic goats. In fact, *C. aegagrus* shares two of three Y-chromosome haplotypes with domestic breeds (Figure 20.3). The second-closest taxon to domestic goats, based on the Y chromosome, is *C. falconeri*. This species is separated from both domestic goats and *C. aegagrus* by two to three mutations (substitutions), with a moderate bootstrap value of 64 (i.e., 64% of bootstrap-resampled trees grouped *C. falconeri* apart from *C. aegagrus*). Samples from additional wild populations of *C. falconeri* are needed to reject the possibility that *C. falconeri* gave rise to cashmere breeds or to other domestic goats from eastern Asia.

Our microsatellite data from nine domestic goat breeds and two wild taxa *(C. aegagrus* and *C. ibex sibirica)* are consistent with results from mtDNA and Y-chromosome sequences. The data suggest that *C. aegagrus* is more closely related to domestic goats than is the *C. ibex sibirica* from Mongolia (Figure 20.4). Nuclear DNA data (markers on autosomal chromosomes), such as microsatellites, are useful for giving a genome-wide view of relationships between breeds and species. Microsatellites are not the ideal marker for interspecific phylogenetic studies, however, because of their high mutation rates and back mutation (causing homoplasy). Nonetheless, microsatellites have been very useful in identifying the wild progenitor of domestic camelids (Chapter 23).

Our genetic data reinforce morphological and archaeological data that suggest *C. aegagrus* is the wild progenitor of domestic goats. Nonetheless, we cannot categorically exclude the possibility that other subspecies of *C. aegagrus*, or even *C. falconeri*, were also progenitors of domestication. This will require analysis of samples from areas such as Baluchistan and regions of the Indus Valley of Pakistan.

Number of Origins

At least three genetic origins of goats are supported by both mtDNA and Y-chromosome data. Three divergent mtDNA lineages were identified by the cytochrome *b* data (Figure 20.2). The same three maternal lineages were initially identified in a study of mtDNA control region sequences (Figure 20.5). Although the cytochrome *b* data include only 6 domestic goats, the control region data include extensive sampling of 406 goats from 88 breeds and 44 countries across the Old World (Luikart et al. 2001).

It is intriguing that three different Y-chromosome lineages have also been found among contemporary domestic

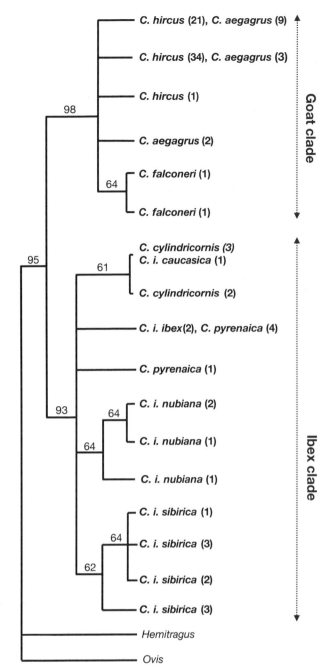

FIGURE 20.3 Phylogenetic tree constructed from two Y-chromosome gene fragments (265-bp amelogenin and 919 bp of the ZFY gene) obtained using maximum parsimony (2,000 replicates) and PAUP* version 4.0b4. The number in parentheses is the number of samples sequenced for both maximum likelihood and neighbor-joining methods. Numbers on branches are the percentage of 2000 bootstrap-resampled trees with the branch.

goats (Pidancier et al. n.d.). Finding three maternal and paternal lineages could be coincidental and not necessarily evidence of three origins or domestications. Interestingly, all individuals from a given divergent mtDNA lineage (e.g., lineage C) did not always come from the same distinctive Y-chromosome lineage. This is not surprising, however, in

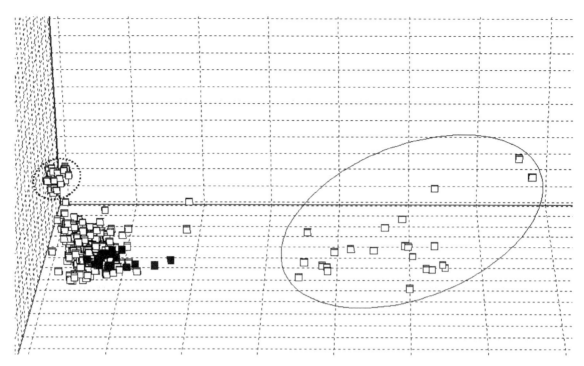

FIGURE 20.4 Relationships among nine domestic goat breeds and two wild *Capra* species assessed using 22 microsatellite DNA markers and frequency correspondence analysis (a multivariate clustering method in the Genetix software program). Each square represents one individual. The large, solid circle showing the Siberian ibex (*C. ibex sibirica* from Mongolia) is the most divergent. The filled squares are wild aegagrus goats (*Capra aegagrus* from the Caucasus Mountains), which group closely with the domestic breeds. The dashed circle is a highly inbred, bottlenecked breed of Keilder goats from England (Table 20.1 on page 301). These goats cluster apart because of recent genetic drift and loss of alleles associated with the population bottleneck.

light of the extensive gene flow among goat populations and regions, which is likely to mix maternal and paternal lineages within and among breeds (see the population structure section of this chapter). In domestic sheep *(Ovis aries)* three major mtDNA lineages have also been identified (Chapter 21). The combined sheep and goat data suggest that at least three major centers of domestication could have existed for sheep and goats. This hypothesis could be tested using ancient DNA from each of many different putative centers of domestication (see paragraphs on ancient DNA below).

A potential problem with inferring the number of domestications directly from the number of divergent DNA lineages is that recent hybridization with wild populations, not domestication, could explain the origin of a lineage found in contemporary domestic populations. For example, male wild goats *(C. aegagrus* and *C. ibex)* are known to occasionally appear in domestic flocks and breed with estrous females during the rut (D. Ozut, conversation with Luikart, 2001; Gauthier et al. 1991). The relatively large wild males would have no problem dominating and displacing the smaller domestic males to breed with the domestic females. Occasional historical gene flow from the wild is possible among all domestic ungulate species, and cross breeding with wild species is known or suspected, if not promoted (Loftus et al. 1994; Pulling and Van 1945; Yerxat 1995). Because a divergent genetic lineage (e.g., Y-chromosome lineage) could originate through hybridization, we must be cautious in inferring multiple origins from only the presence of multiple divergent lineages. Distinguishing between local domestication versus introgression might be aided by using ancient DNA samples from wild and domestic animals sampled through time (e.g., 12,000 to 5,000 years ago) in regions where domestication likely occurred (discussed later in this chapter).

Another approach we have used that strengthens the case for multiple domestication is to determine the age of divergence of the different genetic lineages found in domestic goats. The degree of sequence divergence between the lineages vastly predates the time of domestication (e.g., ~10,000 years ago). We estimate the time since divergence of the three mtDNA lineages to be from 201,000 to 282,000 years ago. This timeframe is much earlier than domestication, suggesting that the mtDNA in today's domestic goats does not originate from within one local population only 10,000 years ago. The three highly divergent lineages could not have evolved (i.e., accumulated so many mutations) in only 10,000 years. It is more likely that three genetic origins occurred from three different wild populations that already carried quite distinct mtDNA 10,000 years ago. Two assumptions are necessary to compute the divergence dates (201,000 to 282,000 years ago): first, a molecular clock with a constant rate of lineage evolution, and second, the divergence of sheep and goat taxa 5 to 7 million years ago, as estimated from the

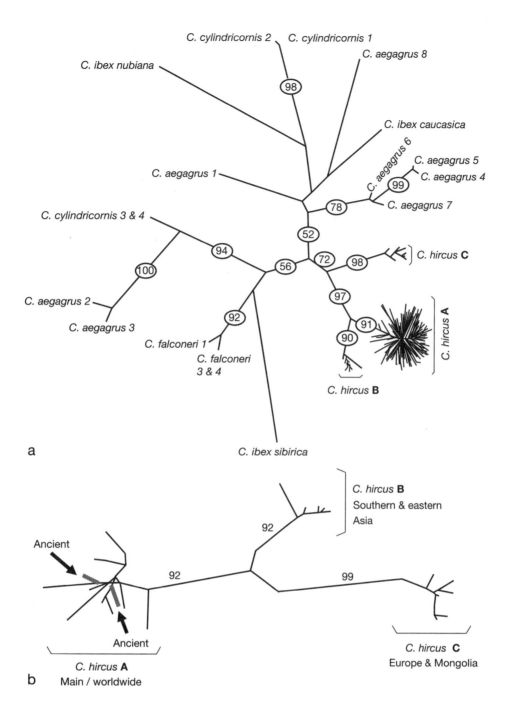

FIGURE 20.5 Phylogenetic trees from mitochondrial DNA (control region) sequences. (a) Neighbor-joining tree (PAUP* version 4.0b4) of mtDNA types from 406 domestic goats and 14 wild *Capra* (from Luikart et al. 2001). Trees constructed using other methods (e.g., UPGMA or neighbor-joining with alpha shape parameter = 0.20–0.40) were nearly identical in topology. The large star-shaped cluster (*C. hircus* A) contains 316 mtDNA types (found in 370 individuals and in all breeds). The two smaller lineages (*C. hircus* B and C) contain only eight and seven mtDNA types (found in 25 and 11 individuals, respectively). Numbers on branches are the percentages of 2,000 bootstrap trees with the same branch structure. The wild taxon with sequences most similar to domestic goats is *C. aegagrus* (61.3 substitutions, on average, using the gamma-corrected distance). The second most similar taxon is *C. cylindricornis*, which has 84.5 substitutions (on average) when compared to domestic goats. (b) Two ancient DNA sequences (130-bp control region; denoted by grey branches) along with ca. 10 arbitrarily chosen sequences from each of the three major mtDNA lineages (from Figure 20.5a) were used to construct this neighbor-joining tree using PAUP* version 4.0b4. Only bootstrap values >60 are given (after Fernández et al. 2005, Fig. 3).

fossil record (Carroll 1988; Savage and Russell 1983). Interestingly, two additional mtDNA lineages have been identified in domestic goats since this chapter was written (e.g., Joshi et al. 2004; Sultana et al. 2003; G. Luikart and A. Beja-Pereira, unpublished data).

Location of Origins

The geographic location of domestication events can be inferred from geographic patterns of distribution of genetic variation. For example, levels of molecular diversity in domestic breeds are expected to be highest near the center of origin, assuming that dispersal away from the center would lead to loss of genetic variation due to repeated founder effects (Loftus et al. 1999). In goats, however, mtDNA variation is apparently not higher in the Fertile Crescent region compared to most other continental regions. For example, the mtDNA diversity (mean number of base changes between two sequences) within the Near East, Europe, and Asia is approximately equal (10 base pair differences (i.e., substitutions)).

In contrast to mtDNA, Y-chromosome data does appear to suggest higher diversity in the Fertile Crescent region than in other regions. For example, we found two Y-chromosome lineages in contemporary goats from Jordan and Turkey, and not in other regions (Figure 20.6b). However, these data are preliminary since samples were small within geographic areas.

Microsatellite DNA diversity (heterozygosity) also appears to be highest in goats from the Fertile Crescent region of southern Turkey (Anatolian Black) (Table 20.1). This difference is only slight: H_e (expected heterozygosity) is 0.67, 0.60, and 0.56 in Turkey, Mongolia, and West Africa, respectively. This finding parallels those from microsatellite DNA data from cattle in the Fertile Crescent region (Loftus et al. 1999). The main drawback of inferring the center of origin from increased diversity is that we also expect increased diversity in a contact zone between two different centers of domestication. For example, we might expect increased diversity somewhere between the Indus River valley (Baluchistan) and the Fertile Crescent region if these two areas were domestication centers. Another problem might be introgression from local wild populations, which could increase diversity in the absence of a local domestication. Clearly, inferences about the location of origin from a single type of pattern of molecular data (e.g., diversity levels) should only be made with caution because they can be unsatisfactory or even potentially misleading. Rather, it is advisable to incorporate information from several analyses, such as the geographic distribution of lineages and also historical or temporal distributions (e.g., using ancient DNA, discussed below).

The geographic location of origin can be inferred from the geographic distribution of certain alleles or lineages as follows. In all livestock species, including goats, cattle, buffalo, pigs and sheep, a divergent mtDNA lineage occurs only in southern or eastern Asia (Luikart et al. 2001). This suggests a possible center of animal domestication in southern or eastern Asia. In goats, one mtDNA lineage (lineage B) was

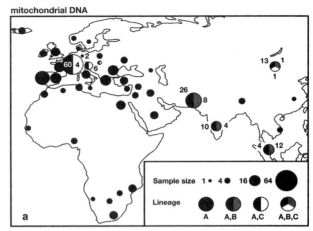

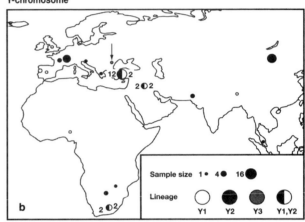

FIGURE 20.6 Map showing the distribution of maternal and paternal DNA lineages. (a) Geographic distribution of the three maternal (mtDNA) lineages (modified from Luikart et al. 2001). The size of each circle is proportional to the number of goats (1 to 64) sampled from each of 44 countries. The presence (not frequency) of each lineage in a country is represented by a different shade (lineage A = black, B = gray-stippled, and C = white). Thus, the Asian B lineage occurs in Pakistan, India, Malaysia, and Mongolia. The complete list of breeds and number of individuals sampled per breed and country is available from the supplementary material in Luikart et al. (2001).
(b) Geographic distribution of the three paternal (Y-chromosome DNA) lineages. The size of each circle is proportional to the sample size (1 to 12) from each of 15 countries.

detected only in the southern and eastern Asian countries of Pakistan, India, Malaysia, and Mongolia (Figures 20.5a and 20.6a).

Comparing the geographic distribution of goat mtDNA lineages and Y-chromosome lineages suggests a potential origin of goats in the Balkans or the Carpathian Mountains region of Romania. This is because we found a divergent mtDNA lineage (lineage C) mainly in Switzerland and Slovenia (Figure 20.6a), and a distinctive Y-chromosome lineage (lineage 3) only in the Carpathian Mountains (Figure 20.6b, see arrow). However, this interpretation is not convincing

TABLE 20.1
Average Heterozygosity and Allele Number for 22 Microsatellite Loci in Each of Nine Goat Breeds

Region	Breed	Heterozygosity[a]		Average Number of Alleles/Locus
		H_{obs}	H_e	
Turkey	Anatolian black	0.70	0.67	7.68
	Standard error	0.16	0.18	
Mongolia	Native cashmere	0.65	0.60	6.36
	Standard error	0.19	0.21	
Egypt	Zariabi	0.62	0.59	5.72
	Standard error	0.17	0.19	
Africa	West African dwarf	0.61	0.56	6.04
	Standard error	0.20	0.22	
Italy	Girgentana	0.56	0.55	4.13
	Standard error	0.16	0.21	
Spain	Murciana-Grenadina	0.61	0.54	5.81
	Standard error	0.19	0.17	
Switzerland	Saanen	0.56	0.51	5.00
	Standard error	0.18	0.19	
Malaysia	Malaysian native	0.52	0.42	5.04
	Standard error	0.20	0.20	
England	Keilder	0.33	0.29	2.31
	Standard error	0.23	0.26	

[a] H_{obs} is the actual observed heterozygosity; H_e is the expected heterozygosity (under random breeding at Hardy-Weinberg equilibrium).

because the sampling in the Y-chromosome study was limited to only about 50 domestic goats and because the mtDNA and Y-chromosome lineages could have actually originated elsewhere and subsequently been transported to the Balkans or Carpathian Mountains region. This hypothesis of movement after origins can be tested using historical or ancient DNA to elucidate patterns of temporal change in geographic distribution of DNA lineages.

Ancient DNA studies could be particularly helpful in pinpointing the site of domestication. For example, evidence for a local domestication would be provided by finding a DNA lineage in ancient domestic goats from one location (e.g., the Fertile Crescent) but not from other locations (e.g., East Asia). Even more compelling would be the discovery of a unique DNA lineage in both wild and domestic animals from one location, but a different DNA lineage in the wild and domestics from another location.

With this approach in mind, we sequenced a highly variable 130-bp subsection of control region (mentioned previously) in two ancient goat bones from Iran, dating ~4000–6000 years old (bones from Qabrestan, Qazvin Plain, Chalocolithic, 3782–3361 BC calibrated). We obtained two sequences that were most similar to the common A lineage from mtDNA of contemporary goats. This is consistent with the A lineage originating in the Fertile Crescent region. This interpretation is not surprising considering that the initial origin of goats is thought to be the Fertile Crescent region (Zeder and Hesse 2000; Chapter 14) and that the A lineage is by far the most common and widespread of the three mtDNA lineages in goats. We caution that these ancient DNA data are preliminary as we have not yet attempted to repeat the DNA extraction and sequencing, considered a necessary verification process in ancient DNA studies. In addition, more extensive sequencing of a diversity of even older material is required to comprehensively investigate the origins of domestication. Nonetheless, this example illustrates a potential usefulness of ancient DNA studies for inferring the origins and spread of domestic organisms.

Combining genetic and archaeological information provides the most fruitful approach for unlocking the secrets of the origins of domestic animals. Determining the time and location of origin of ancient genetic lineages can be best done using reliably dated fossils from archaeozoologists. Similarly, tracking the diffusion of these different genetics lineages requires well-dated fossil material from each of several archaeological sites.

Timing of Population Expansions

Geneticists have often used mtDNA sequences to detect historical population growth (expansions). One molecular signature of a population expansion is finding that most alleles (i.e., sequences) are of a similar age, that is, the degree of divergence between sequences is similar. This is because during rapid population growth, most new mutations are retained in the population (loss due to genetic drift is limited) and thus most mutations will date to the time of growth initiation in the past. For an expanding population, if we plot the distribution of divergence between sequence pairs, we obtain a smooth bell-shaped curve (Figure 20.7). The approximate time since expansion can be estimated from the mean of the bell-shaped curve (Rodgers and Harpending 1992). In stationary or declining populations, lineages are lost rapidly and at random, which generates a ragged or multiple-peaked distribution of genetic distances between pairs of DNA lineages (Rodgers and Harpending 1992).

Another signature of expansion is a star-shaped phylogeny (see Figure 20.5a, lineage A). An expansion from a small population size causes lineages to have similar branch lengths (and pairs of lineages to have a similar divergence). Again, this is because most new mutations are maintained during population expansion. Branch lengths represent the age of lineages (i.e., sequences). We computed the distribution of genetic distances between pairs of goat mtDNA sequences (mismatch distribution analysis) to test for population expansions and to compute the relative dates of expansion of the three goat lineages (from Figure 20.5).

Here we illustrate how we can infer the approximate relative dates of expansion of each goat mtDNA lineage by comparing the means of the bell-shaped curves. This is simply comparing the mean diversity (sequence variation) within each lineage. This comparison assumes that a similar (limited) amount of mtDNA diversity existed initially within each founder lineage. The relative dating of the population expansions is informative because the expansion dates should roughly correspond to the time that each goat lineage began to become numerous and thus to become important in historical human societies and economies. We emphasize that dates of expansion do not necessarily equal dates of domestication, unless rapid population growth began soon after the first domestication.

The mismatch analysis of goat mtDNA control region sequences suggested substantially different dates of expansion of the three major maternal lineages (Luikart et al. 2001). The mismatch distributions for each lineage had mean values of 10.9, 5.6, and 2.3. If we assume that the oldest domestication was ~10,000 years ago and that the very large A lineage represents the initial expansion 10,000 years ago, then the less abundant lineages can be estimated to have expanded about 6,000 years ago (C lineage), and 2,600 years ago (Asian B lineage). The less abundant lineages are widespread geographically (e.g., *C. hircus* B is found from Mongolia to Pakistan and Malaysia; see discussion later in this chapter), probably

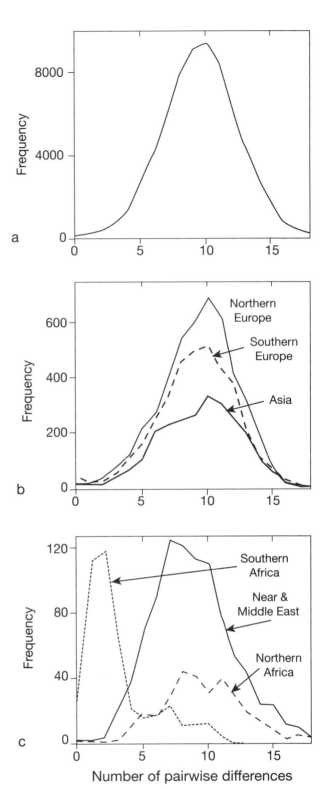

FIGURE 20.7 Mismatch distributions (i.e., distributions of all pairwise sequence differences) for mtDNA types from the major lineage of goat sequences (*C. hircus* A): (a) all 406 control region sequences; (b) sequences from Europe and Asia; (c) sequences from Africa and the Near/Middle East. The means of the distributions are different only for Southern Africa, suggesting more recent expansion dates for the goat population on the African continent south of the Sahara Desert.

resulting from a relatively recent spread of these mtDNA lineages. If the two smaller lineages had originated and spread 9,000–10,000 years ago, shortly after the estimated date of the initial domestication (Porter 1996; Zeder and Hesse 2000; Chapter 14), we would expect more divergent lineages to have evolved, as in the larger cluster of mtDNAs (*C. hircus* A). Timing of the expansion in the Africa population (Figure 20.7c) is recent (<3,000 years ago) as suggested by the lowered mtDNA diversity (in the "big" lineage A).

Computing the relative dates of population expansions is potentially useful because they should roughly correspond to the time that each goat lineage became important in historical human societies and economies. However, dating via the mismatch approach is speculative because it assumes low (and similar) levels of initial founder variability. If one population or lineage has far more founder variability, this could inflate the average pairwise diversity within the lineage and thus inflate estimates of the time since the expansion began. Similarly, introgression from wild populations could inflate diversity and time-since-expansion estimates. However, introgression of highly divergent lineages from the wild might also generate multiple (ragged) peaks in the distribution of pairwise differences, which would be detectable in the analysis. It is worth noting that another way to use mismatch distribution analysis is to estimate the absolute time since population expansion began using the molecular clock assumptions (e.g., Bradley et al. 1996), rather than the relative times since expansion as estimated here using the fossil record for goat domestication.

Population Structure and Gene Flow

Studies of genetic population structure can help infer the amount of movement or transportation of livestock between regions and continents. Any long distance movements must have been associated with human migration (or trade), because goats themselves could not have migrated between continents. Thus, if distant livestock populations are relatively similar genetically, we can infer that humans transported animals either during migration, colonization, or via trade. Differential movement of livestock species can help infer differential uses of livestock in historical societies.

We previously documented surprisingly low mtDNA population structure in domestic goats compared to cattle (Luikart et al. 2001). To quantify the amount of intercontinental population differentiation, we computed the percentage of the total mtDNA variation that is due to differences among continental populations using an approach analogous to ANOVA (Analysis of Variance), but called AMOVA (Analysis of Molecular Variance) (Excoffier et al. 1992). This analysis suggested that only about 10% of the total mtDNA variation in domestic goats (*C. hircus*) was due to differences among continents. This is far lower than the published estimates of 54–80% in cattle for the same mtDNA region (HVI). These results suggest that in the past goats were transported far more extensively than cattle. This finding seems reasonable considering cattle have only recently been moved around more than goats because of improved transportation and the greater economic importance of cattle. In addition, recent technological advances in cattle research have led to increased movement of gametes and embryos (through embryo transplants and artificial insemination). Thus genetic evidence of greater transportation of goats is not likely to be due to recent events, but rather to longer-term historical events.

Interestingly, the percentage of mtDNA variation (HVI) between continents for humans is surprisingly similar (10%) to that in goats (Table 20.2). This similarity might be coincidental, but it is intriguing to speculate that goats have had similar amounts of high intercontinental movement because they have been more important during human movements (migrations, colonizations) or more used in historical commerce and trade than cattle. However, it might not be appropriate to directly compare humans to goats because humans initially spread across the globe 50 thousand to 100 thousand years ago, which was much earlier than goats (10 thousand years ago). But the generation time in goats is four to six times shorter than in humans, making the number of generations since population expansion fairly similar in humans and goats. The relatively similar demographic history of goats and humans (compared to most vertebrate species) makes comparisons of the genetic structure between humans and goats interesting and not necessarily inappropriate.

It is important to assess population structuring using nuclear DNA, in addition to mtDNA, because the two DNA types could give different results due to their different evolutionary dynamics (see section on DNA Markers and Sequences). For example, if mainly males were transported via trade, we would expect lower genetic differentiation for nuclear DNA than for mtDNA data. Comparing our microsatellite data from nine populations (from Mongolia, Malaysia, the Middle/Near East, Europe, and Africa) with mtDNA data, we found the amount of variation explained by intercontinental differences was about 17% ($F_{ST} = 0.17$). If we exclude the highly inbred, bottlenecked Keilder breed from England (due to its very atypical history), F_{ST} is only 0.13. This means that the percentage of the total microsatellite variation in *Capra hircus* that is found within any given breed is approximately 83% (or 87% excluding Keilder). This agrees fairly well with the mtDNA data for which about 80% of the variation was distributed within local populations and 90% within continental regions. Thus there is little evidence of sex-biased transportation of domestic goats at the intercontinental level, implying the movement of domestic males and females, although further studies are needed, including more breed sampling for microsatellites (and the use of other nuclear DNA markers).

Summary and Perspectives

We have illustrated several ways that molecular data (including mitochondrial DNA, Y chromosome, and microsatellites) can help determine the origins and spread of domestic goats. All three types of DNA sequences suggest that the bezoar wild goat (*C. aegagrus*) is the most likely wild progenitor of domestic goats. But sampling of the wild

TABLE 20.2

Hierarchical Distribution of mtDNA (HVI) Diversity within and among Populations (and Continental Groups of Populations) for Goats, Cattle, and Humans

	Distribution of Sequence Variation (percentage of total)[a]		
Species[b]	Within Population	Within Groups	Among Continental Groups
Goats[1]	78.7	10.6	10.7 (N. & S. Europe/N. & S. Africa/Asia)
Goats[2]	74.0	12.1	13.9 (Europe/Africa/Asia)
Goats[3]	72.7	12.2	15.1 (Europe/Africa)
Cattle[2]	15.3	0.9	83.9 (Europe/Africa/Asia)
Cattle[3]	45.4	3.2	51.4 (Europe/Africa)
Humans[1]	82.1	9.4	8.5 (N. & S. Europe/N. & S. Africa/Asia)
Humans[2]	80.6	12.8	6.6 (Europe/Africa/Asia)
Humans[3]	77.0	16.1	6.9 (Europe/Africa)

[a] Computed under the AMOVA framework (Luikart et al 2001). Note the relatively low percentage of variation among continental groups for goats and humans, relative to cattle.

[b] "Populations" refer to breeds for goats and cattle. Populations for humans are ethnic groups (e.g., Basques, Turks, Swiss, Japanese; from Excoffier et al 1992). The six continental groups for Goats[1] and Humans[1] are northern Europe, southern Europe, northern Africa, southern Africa (south of the Sahara desert), Middle/Near East, and Asia. The continental groups for Goats[2], Goats[3], Humans[2], and Humans[3] were chosen for comparison with published AMOVA values from cattle having the same geographic population groupings; these consist of (1) Europe, Africa, and Asia and (2) Europe and Africa, respectively (Bradley et al 1996).

taxa is insufficient to exclude *C. falconeri* or subspecies of *C. aegagrus* as having been domesticated. We have shown how geographic patterns of molecular variation can help determine the number of times and location(s) where domestication might have occurred, assuming there has been no recent introgression from wild species. Ancient DNA sequencing combined with extensive sampling of contemporary local breeds provides exciting potential as a method of inferring the origins and diffusion of domestic taxa. Combining archaeozoological and ancient DNA studies to assess the pattern of diffusion of DNA lineages through time and space would be especially informative. Ideally this would include ancient DNA from domestics and their wild progenitors at each of several archaeological sites across the Fertile Crescent region and candidate locations of domestication.

Emerging ancient DNA technology and genomics technology will facilitate studies of animal domestication. For example, genome-wide studies in livestock are discovering single nucleotide polymorphisms (SNPs) that are diagnostic of breeds or morphological characteristics (e.g., color). These SNPs could be used to trace origins of breeds and characteristics. Importantly, SNPs can be studied using very short DNA fragments, an important requirement of ancient DNA work. It is exciting that future studies of ancient DNA, combining archaeozoology and genomics technologies, will likely yield new discoveries greatly advancing our understanding of the origin and spread of farm animals and their role in ancient human societies.

Acknowledgments

We gratefully thank V. Curri and M.-P. Biju-Duval for help with laboratory analyses. L. Excoffier helped with molecular dating and mismatch distributions analyses, and J.-D. Vigne with ancient material collection. Quotations were provided by L. Freesemann. We thank the following biologists and veterinarians who kindly provided samples for past research: S. Shahraki Khodabandeh, N. Hasima, A. Virk, A. Ghaffar, O. Hanotte (ILRI), E. Bedin, P. Weinberg, R. Soriguer, S. Dunner, M. K. Sanyasi, L. O. Ngere, D. Zygoyiannis, L. O. Eik, H. Larsen, A. Amcoff, M. Gough, P. Evans, N. Azzopardi, F. Pilla, Dr. Matassino, V. Fet, H. Amaturado, I. Coroiu, I. Moglan, T. M. Correic, E. Zimba, S. Breznik, E. Eytorsdottir, W. Hamdine, J. Honmode, J. M. Villemot, V. I. Glazko, E. Martyniuk, M. M. Shafie, Ferme du Pic Bois, C. Courturier, M-A. Dye, R. Del Olmo, T. Faure, and especially M. Abo-Shaheda, O. Ertugrul, M. Y. Zagdsuren, M. A. A. El-Barody, G. Obexer-Ruff, and G. Dolf. This work was funded by a grant from the European Commission (No. BIO4CT961189). H.F. was supported by a Swiss National Foundation grant.

References

Bradley D. G., D. E. MacHugh, P. Cunningham, and R. Loftus. 1996. Mitochondrial diversity and the origins of African and European cattle. *Proceedings of the National Academy of Sciences, USA* 93: 5131–5135.

Carroll, R. L. 1988. *Vertebrate paleontology and evolution*. New York: W. H. Freeman.

Clutton-Brock, J. 1981. *Domestic animals from early times*. London: Heinemann and British Museum of Natural History.

Excoffier, L., P. E. Smouse, and J. M. Quattro. 1992. Analysis of molecular variance inferred from metric distances among DNA haplotypes: Application to human mitochondrial DNA restriction data. *Genetics* 131: 479–491.

Fernández, H., P. Taberlet, M. Mashkour, J.-D. Vigne, and G. Luikart. 2005. Assessing the origin and diffusion of domestic goats using ancient DNA. In *The first steps of animal domestication*, J.-D. Vigne, J. Peters, and D. Helmer (eds.), pp. 50–54. Oxford: Oxbow Books.

Gauthier, D., J.-P. Martinot, J.-P. Choisy, J. Michallet, J.-C. Villaret, and E. Faure. 1991. Le bouquetin des Alpes. *Revue Ecologie (Terre Vie)* Suppl. 6: 233–275.

Hasnain, H. U. 1985. Sheep and goats in Pakistan. In *Food and agricultural organization*. Animal Health and Production Paper, 56. Rome: FAO.

Joshi, M. B., P. K. Rout, A. K. Mandal, C. Tyler-Smith, L. Singh, and K. Thangaraj. 2004. Phylogeography and origin of Indian domestic goats. *Molecular Biology and Evolution* 21: 454–462.

Kahila Bar Gal, G., C. Greenblatt, S. R. Woodward, M. Broshi, and P. Smith. 1999. The genetic signature of the Dead Sea Scrolls. In *Historical perspectives: From the Hasmoneans to Bar Kokhba in light of the Dead Sea Scrolls. Proceedings of the Fourth International Symposium of the Orion Center for the Study of the Dead Sea Scrolls and Associated Literature*, Volume 37, D. Goodblatt, A. Pinnick, and D. R. Schwartz (eds.). Leiden: Brill.

Loftus, R. T., D. E. MacHugh, D. E. Bradley, P. M. Sharp, and P. Cunningham. 1994. Evidence for two independent domestications of cattle. *Proceedings of the National Academy of Sciences, USA* 91: 2753–2761.

Loftus, R. T., O. Ertugrul, A. H. Harba, M. A. El-Barody, D. E. MacHugh, S. D. Park, and D. G. Bradley. 1999. A microsatellite survey of cattle from a centre of origin: The Near East. *Molecular Ecology* 8: 2015–2022.

Luikart, G., M.-P. Bidju-Duval, O. Ertugrul, Y. Zagdsuren, C. Maudet, and P. Taberlet. 1999. Power of 22 microsatellite markers in fluorescent multiplexes for semi-automated parentage testing in goats (*Capra hircus*). *Animal Genetics* 30: 31–38.

Luikart, G., L. Gielly, L. Excoffier, J.-D. Vigne, J. Bouvet, and P. Taberlet. 2001. Multiple maternal origins and weak phylogeographic structure in domestic goats. *Proceedings of the National Academy of Sciences, USA* 98: 5927–5930.

Meadow, R. H. 1996. The origins and spread of agriculture and pastoralism in northwestern South Asia. In *The origins and spread of agriculture and pastoralism in Eurasia*, D. R. Harris (ed.), pp. 390–412. Washington, D.C.: Smithsonian Institution Press.

Pidancier, N., G. Luikart, P. Weinberg, S. Jordan, and P. Taberlet. n.d. Evolution of the genus *Capra*: Discordance between mitochondrial and Y-chromosome sequences. Unpublished manuscript.

Porter, V. 1996. *Goats of the world*. Ipswich, UK: Farming Press.

Pulling, A. and S. Van. 1945. Hybridization of bighorn and domestic sheep. *Journal of Wildlife Management* 9: 82–83.

Rodgers A. R. and H. Harpending. 1992. Population growth makes waves in the distribution of pairwise genetic differences. *Molecular Biology and Evolution* 9: 552–569.

Savage, D. E. and D. E. Russell. 1983. *Mammalian paleofaunas of the world*. Reading, MA: Addison-Wesley.

Schneider, S. and L. Excoffier. 1999. Estimation of past demographic parameters from the distribution of pairwise differences when the mutation rates very among sites: application to humans. *Genetics* 152: 1079–1089.

Smith, B. D. 1998. *The emergence of agriculture*. New York: Scientific American Library.

Sultana, S, H. Mannen, and S. Tsuji. 2003. Mitochondrial DNA diversity of Pakistani goats. *Animal Genetics* 34: 417–421.

Swofford, D. L. 1998. *PAUP* phylogenetic analysis using parsimony and other methods*. Sunderland, MA: Sinauer.

Taberlet, P., L. Waits, and G. Luikart. 1999. Non-invasive genetic sampling: look before you leap. *Trends in Ecology and Evolution* 14: 323–327.

Takada, T., Y. Kikkawa, H. Yonekawa, S. Kawakami, and T. Amano. 1997. Bezoar (*Capra aegagrus*) is a matriarchal candidate for ancestor of domestic goat (*Capra hircus*): evidence from the mitochondrial diversity. *Biochemical Genetics* 35: 315–326.

Yerxat, M. Y. 1995. Application of wild goats in cashmere breeding. *Small Ruminant Research* 15: 287–291.

Zeder, M. A. and B. Hesse. 2000. The initial domestication of goats (*Capra hircus*) in the Zagros Mountains 10,000 years ago. *Science* 287: 2254–2257.

CHAPTER 21

Mitochondrial DNA Diversity in Modern Sheep
Implications for Domestication

MICHAEL W. BRUFORD AND SAFFRON J. TOWNSEND

Introduction

The domestic sheep (*Ovis aries*) is thought to descend from several candidate Eurasian species within the genus *Ovis* (e.g., Zeuner 1963). The genus *Ovis* is found within the tribe Caprini (sheep and goats), which is in turn found within the Caprinae, a subfamily of the Bovidae, most of whose modern members evolved during the Pleistocene. The divergence time of Caprinae from the subfamily Bovinae (cattle) is estimated at ~15–20 MYA, based on molecular studies of mitochondrial 12S and 16S rRNA genes (Allard et al. 1992). According to other molecular data (Randi et al. 1991; Irwin et al. 1991), *Capra* and *Ovis* diverged 5–7 MYA, with much of the subsequent diversification within these genera taking place during the Pleistocene (Geist 1971). The Pleistocene (1.8–0.01 MYA) saw pulses of rapid speciation among large herbivores (rates quadrupled over those in the Tertiary) resulting from extreme fluctuations in climate. Isolated areas of fertile, uncontested habitat became available, only to be re-covered with ice during glacial intervals (Hewitt 1996). During this time, populations were likely subject to constant and significant changes in selection pressures leading ultimately to high levels of diversification into post-Pleistocene genera (Geist 1983). Within the Caprini, sexual dimorphism is common, with most males displaying ornate horns specialized for frontal combat. A hierarchical social system exists for males, and there is usually spatial segregation of sexes outside the mating season. Most species inhabit upland and mountainous regions and flock throughout the year (Shackleton 1997).

The subfamily Caprinae remains in great need of taxonomic revision (Geist 1990; Shackleton 1997). The current disarray—within each genus, species, and subspecies, names often appear to be freely interchangeable—has come about primarily because the early taxonomy was written when age, sex, seasonal, ecotypic, allometric, and statistical variations were poorly understood, and taxonomic criteria were ill defined (Geist 1990). As a result, much of the taxon-delineating data are conflicting and names differ because of multiple (and often poor) historical identifications. *Ovis* taxonomy, in particular, is ill defined, and confused subspecies nomenclature is frequently encountered. However, at least five (but up to seven) wild *Ovis* species are recognized (Lydekker 1912; Geist 1990; Mason 1996; Shackleton 1997), of which three are potential domestic ancestors given that the abundance of archaeological and historical evidence pertaining to domestication falls only within their natural range.

O. orientalis (the Asiatic mouflon), *O. vignei* (the urial), and *O. ammon* (the argali) occupy the highlands extending from southwest to eastern Asia, and these are all candidate ancestors of modern sheep. Of these three, *O. ammon* and *O. vignei* are now thought to be the least likely domestic ancestors according to studies of ovine mitochondrial mtDNA (Hiendleder et al. 1998, 2002, and below; but see Zeuner 1963) and karyotype (Nadler et al. 1973a; Bunch et al. 1990; Lyapunova et al. 1997). However, these species are very often confused, named synonymously, and show extensive mitochondrial allele sharing (Hiendleder et al. 2002, below). The European mouflon, *O. musimon*, is no longer regarded as a potential ancestral source, but rather as a feral remnant of the first domestic populations to enter Europe, since *Ovis* is absent from the European fossil record immediately prior to the Neolithic (Clutton-Brock 1981; Hermans 1996).

At present, Eurasian domestic sheep are grouped into approximately 20 traditional breed types depending on commercial use and/or phenotypic trait and historic extraction (Mason 1996; Alderson 1994). Although these types are not definitive with regard to all Eurasian breeds, they allow an overview of the variety and partitioning among breeds. The characteristics most often used to describe breeds are those that have commercial relevance (meat, wool type, and milk), or those based on alternative phenotypes that provide simple visual cues (color, horns, tail shape and length).

The diploid chromosome number of domestic sheep (*Ovis aries*) and the seven *Ovis* species are *O. nivicola* (2n = 52); *O. dalli* (2n = 54); *O. aries* (2n = 54); *O. canadensis* (2n = 54); *O. musimon* (2n = 54); *O. ammon* (2n = 56); *O. orientalis* (2n = 54); and *O. vignei* (2n = 58). The karyotypes vary among species because of the number of biarmed metacentric chromosomes occurring as a result of Robertsonian translocations between acrocentric chromosomes (Bunch et al. 1990). The largest biarmed metacentric in all *Ovis* species is identical (Bunch et al. 1990). Therefore, this chromosome probably represents an ancestral *Ovis* translocation, while the other metacentrics in *Ovis* karyotypes evolved subsequently from the ancestral 2n = 58 karyotype (Bunch et al. 1976).

Crosses between species with different karyotypes have been observed and give rise to fertile offspring in the F1 generation: 2n = 54 × 2n = 58 can result in 2n = 55, 56, and 57 (Bunch et al. 1990; Hiendleder et al. 1998, 2002). Hybrid populations also exist (Valdez et al. 1978) in northern and southeastern Iran (Nadler et al. 1971; Valdez et al. 1978).

Bunch et al. (1976) found that 2n = 55 wild hybrid rams bred to 2n = 55 wild hybrid ewes resulted in a bias toward a 2n = 54 offspring karyotype, suggesting perhaps that forerunners of this karyotype could have arisen from hybridization between other wild sheep species. The 2n = 54 karyotype consists of three pairs of biarmed metacentric chromosomes, 23 acrocentric autosomes, and a sex chromosome pair, and has indistinguishable G-band patterns among those species that possess it (Nadler et al. 1973b). Thus, evidence at first glance implies common ancestry among *O. musimon*, *O. aries*, and *O. orientalis* (all 2n = 54). However, *O. canadensis* and *O. dalli* also have an identically G-banded 2n = 54, suggesting that this karyotype may have arisen twice independently within *Ovis*, since these species are found at opposite extremes of the *Ovis* geographic range, with all other karyotype variations in between. Thus, while the cytogenetics provides important data for uncovering domestic ancestry, the evidence needs to be augmented with other data before clear inferences can be made.

Relationships among wild and domestic sheep have also been studied using protein polymorphisms. Wang et al. (1991) studied hemoglobin polymorphisms in sheep and found that all wild individuals tested (from *O. dalli*, *O. musimon*, *O. canadensis*, *O. orientalis*, *O. vignei*, and *O. ammon* populations) were monomorphic for Hbβ^B. An Hbβ^A allele was also found and was exclusively confined to domestic sheep populations, albeit at lower frequency than Hbβ^B (subsequently, the Hbβ^A allele was also found in Corsican mouflon [Montelgard et al. 1994]). Wild sheep Hbβ^B was found to differ by five amino acids from domestic Hbβ^B, but from domestic Hbβ^A by only two positions (found at different sites). Assuming a constant rate of nonsynonymous substitution, it was thus hypothesized that Hbβ^A is a recent allele, which arose post-domestication (although no suggestion was made as to how the Hbβ^A might have evolved).

The system would appear to be less straightforward than this, however. Naitana et al. (1990) used isoelectric focusing to reveal two Hbβ^B alleles in Sardinian mouflon. One was found to be functionally related to domestic Hbβ^B (Hbβ^M), and the other to domestic Hbβ^A (Hbβ^B). This finding was reinforced by Rando et al. (1996), who reached the same conclusion by Southern blot analysis with probes specific for ancestral globin genes. Therefore, while Hbβ^A is an allele that may have arisen post-domestication from a wild Hbβ^B ancestral gene, there is no evidence to suggest that it may be less diverged from one wild ancestor than another. In addition, little comment can be made regarding presence/absence of Hb alleles in the wild European mouflon populations, since their recent history suggests that they have been subjected to severe population bottlenecking and founder effect (Naitana et al. 1990; Montelgard et al. 1994). Genealogical inference between European mouflon and domestics, therefore, requires multiple markers to address the lack of variability in mouflon.

Osfoori and Fesus (1996) used 10 polymorphic protein loci to show that the geographic distance separating the native locales of Iranian sheep breeds was in many cases negatively correlated to the genetic distances between them. They also found that these breeds segregated into three clades when dendrograms were constructed from the data, but that assortment of individuals to clade was not consistent between different distance measures applied. Manwell and Baker (1977) also studied 30 blood protein loci in two commercial breeds (Merino and British Poll Dorset) and found very different levels of variation between the breeds at individual loci. These markers proved fairly ineffective in the estimation of ancestral relationships, however. Nei's D (Nei 1972) was used to estimate a divergence time of 69,500 years between the breeds, which is over half an order of magnitude in excess of expected estimates (i.e., sometime after 10,000 BP). Thus, the degree of resolution/accuracy afforded by proteins is not particularly useful in domestic sheep, although it is possible that numerous markers might provide further evidence regarding the nature of ancestral relationships.

Studies of ovine evolutionary relationships using microsatellite markers have emerged sporadically, although over the coming years this is likely to be augmented in light of the many markers now available (de Gortari et al. 1998; Crawford 1999). Petit et al. (1997a, 1997b) used six microsatellite loci to investigate the social structure of introduced mouflon populations and found that sociospatial units correspond to genetically differentiated units. Studies of domestic sheep populations (Bancroft et al. 1995; Arranz et al. 1998; Farid et al. 2000; Diez-Tascon et al. 2000) also have shown these markers to be effective tools in identifying population subdivision both among and within breeds. Buchanan et al. (1994) used microsatellites to estimate times of divergence between Australian and British breeds of 1,094 years, a much more realistic estimate than that previously calculated using protein markers.

However, by far the most comprehensive and intriguing insights into the evolutionary origins of domestic sheep come from studies of mitochondrial mtDNA, and this will form the core of the remainder of this chapter. The first clue that mtDNA was going to reveal interesting patterns of diversity in modern sheep was uncovered by Wood and Phua (1996), who unexpectedly identified two major mtDNA lineages in New Zealand domestic sheep through sequence analysis of the control region (CR). However, since New Zealand populations were established from diverse European sources, this observation needed to be further investigated in Eurasia if its meaning was to be interpreted in context. In 1998, Hiendleder et al. followed up this work by examining whole mtDNA genome restriction fragment length polymorphisms in mtDNA lineages in *O. aries*, *O. ammon*, *O. vignei*, and *O. musimon*, but not in *O. orientalis*. The first phylogenetic reconstruction was carried out based on the genetic distances found between domestic sheep, European mouflon, argali, and urial. These were found to be between 0.54–0.87 percent, 1.78–2.45 percent, and 2.61–2.86 percent, respectively, based on restriction fragment data from the whole mitochondrial genome. Although no direct ancestor was identified as a result of this study, it was concluded, in agreement with the cytogenetic data, that it is not likely to be either the urial or the argali. The position of the European mouflon among domestic samples in the phylogeny was also given as support for the hypothesis that this taxon is derived from early European domesticates. Further CR sequence analysis of New Zealand sheep was carried out by Hiendleder et al. (1999), which reconfirmed the existence of two mitochondrial DNA "haplogroups" in sheep, whose mean estimated genetic distance was 0.716 percent (Hiendleder et al. 1998). These data were interpreted as identifying potentially separate "European" and "Asian" domestic genetic groups whose wild ancestors' common ancestry was estimated at between 750,000–375,000 BP (Hiendleder et al. 1998).

The most recent attempt to solve the origins of modern sheep was carried out by Hiendleder et al. (2002), who produced additional sequences for the hitherto somewhat limited domestic sheep dataset using seven additional samples from Kazakhstan, Tadjikistan, Syria, and Turkey. In addition, they produced new CR sequences for 11 new wild specimens, representing eight subspecies of different sheep species (following the nomenclature of Vorontsov et al. 1972), including two subspecies of *Ovis vignei* (*bochariensis* and *arkal* from Turkmenistan and Kazakhstan, respectively) and four subspecies of *Ovis ammon* (*ammon*, *darwini*, *nigrimontana*, and *collium* from Mongolia, Mongolia, Kazakhstan, and Kazakhstan, respectively). Again, *O. orientalis* sequences were not included. Mitochondrial NADH3, NADH4L, NADH4, and cytochrome *b* sequences were used to calibrate DNA divergence times, which are more reliable than CR sequences for this purpose. The full CR dataset comprised 61 unique sequences and also included three *Ovis musimon* and more than 40 sequences of European sheep, including Merino and Romney. The CR results were very clear: domestic sheep did not cluster with *O. ammon* or *O. vignei* sequences. Instead, once again, domestic CR haplotypes formed two haplogroups (named "A" and "B"), and the *O. musimon* sequences clustered within haplogroup "B." Two other clusters were found; one contained *Ovis ammon collium*, *O. a. nigrimontana*, and *O. vignei arkal*, and the other contained *O. a. ammon*, *O. a. darwini*, and *O. v. bochariensis*. Although the number of clusters recapitulated the existence of two major taxa, the clustering of *ammon* and *vignei* subspecies was mixed, further highlighting the taxonomic confusion within this group.

The authors also noted the existence of length heteroplasmy in sheep CR sequences because of a repeat sequence 75 bp in length that was repeated in variable numbers among the taxa. The number in nearly all domestic sheep was four, but five and six repeats were seen in wild Eurasian sheep. Protein-coding sequence-based divergence dates were estimated around 3.5 MY between the two urial/argali lineages and the two domestic sheep lineages, and 1.9 MY between the two domestic sheep lineages, with maximum divergence dates within the two domestic sheep lineages reaching less than one-third of this value. The results clearly point toward wild mouflon lineages as the most likely ancestors of modern sheep, although in the absence of *O. orientalis* wild samples, especially the wild populations from Turkey and western Iran (*O. o. anatolica* and *O. o. gmelini*, respectively) this could not be proved.

This study also left another major task remaining to be undertaken—a comprehensive assessment of mitochondrial DNA variation in Eurasian sheep and how the two domestic lineages are partitioned in space, since until now relatively few domestic sheep had been sampled. The importance of this study is illustrated by the work of Luikart et al. (2001) and Troy et al. (2001), who illustrated contrasting patterns of mtDNA diversity in goats and cattle across western Eurasia. Goats show three CR ancestral lineages coupled with little modern geographic structure, commensurate with their rapid expansion and high portability. Cattle show two ancestral CR lineages and a pattern of declining mtDNA diversity from southwestern Asia to northwestern Europe, commensurate with slower spread accompanying the Neolithic expansion of human pastoralists (see Bruford et al. 2003 for a fuller explanation).

The Present Study

To examine the distribution of domestic lineages in detail in western Eurasia, a large sample of European, Middle Eastern, and African sheep samples were analyzed either by sequencing or restriction fragment length analysis of the CR (Townsend 2000; Townsend et al. 2006).

A total of 836 samples were analyzed for domestic and wild sheep. Of these, 130 samples were sequenced for an 800-bp region of the CR, comprising 47 samples from Europe, representing 25 breeds spanning from the United Kingdom to Romania, 39 samples from 12 breeds of domestic sheep in the Middle East (defined here as including Turkey/Cyprus, Saudi Arabia, Israel, Jordan, Armenia, and Iraq), 8 samples from 8 African breeds, 2 samples from Asian domestic sheep (northwest India and Mongolia), and 19 samples from wild *Ovis* populations ranging from feral mouflon (France, Sardinia, and Cyprus, n = 5) and a small number of wild sheep from East Asia (Townsend et al. 2004). For all 836 samples, the same CR region was PCR-amplified, and haplogroup designations were made on the basis of restriction profile (see below). Table 21.1 shows the samples analyzed, and their distribution across Europe and Africa. In nearly all cases, the domestic animals came from purebred stocks of indigenous breeds. Exceptions were samples from Jodhpur, India, and Mongolia, whose breed origins are unknown. Samples from 21 European and Near East countries were included, comprising 60 breeds in total. Thirty-nine domestic samples were also obtained from 6 African countries.

DNA was extracted using either a standard phenol-chloroform protocol (Sambrook et al. 1989) or GFX™ DNA Purification Kit (Amersham Pharmacia Biotech). Extraction from hair follicles was carried out using a 5 percent chelex

TABLE 21.1
Samples Analyzed in This Study, Broken Down by Numbers per Breed and Region

Region	Breed	Abbreviation in Figure 21.1[a]	n	Total
Africa	Tekur (Ethiopia)	AF	4	40
	Karakul (Namibia)	AF	4	
	Tswana (Botswana)	AF	4	
	R. Maasa (Kenya)	AF	4	
	Pedi (South Africa)	AF	4	
	B. Persian (South Africa)	AF	4	
	Newala (Tanzania)	AF	4	
	Ugogo (Tanzania)	AF	4	
	Sukoma (Tanzania)	AF	4	
	Maasa (Tanzania)	AF	4	
Armenia	Caspian	CA	3	3
Cyprus	Cyprus Fat Tail	CFT	25	27
	Cyprus Mouflon	n/a	2	
Czech Republic	Sumavka	n/a	9	9
France	French Mouflon	n/a	8	22
	Thone-Marthod	n/a	14	
Germany	Skudde	SK	23	39
	Heidschnucke	HSN	8	
	Coburg	n/a	8	
Greece	Lesvos	LS	26	77
	Argos	AS	29	
	Chios	CS	22	
Hungary	Racka	RA	30	30
Iceland	Icelandic	IC	19	19
India	N/A	IND	1	1
Iraq	Awassi	AW	2	2
Israel	Awassi	AW	9	9
Italy	Comisana	CO	24	26
	Massese	n/a	2	
Jordan	Awassi	AW	6	6
Mongolia	N/A	MON	1	1
Netherlands	Zeeland	ZL	23	34
	Freesland	FL	11	
Romania	Tsigali	TS	27	52
	Turcana	TC	25	
Sardinia	Sardinian Mouflon	n/a	6	6
Saudi Arabia	Najdi	NJ	18	29
	Naemi	NA	10	
	Barqui	BQ	1	

TABLE 21.1 (continued)

Region	Breed	Abbreviation in Figure 21.1	n	Total
Spain	Churro	CH	15	61
	Merino	ME	27	
	Aragonese	n/a	19	
Turkey	Western Karaman	WK	2	26
	Karaman	KA	5	
	Kivircik	KI	4	
	Awassi	AW	5	
	Turkpeldi	n/a	5	
	Karxmerino	KM	5	
UK	Soay	SO	23	307
	Leicester Longw	LL	12	
	Swalendale	n/a	22	
	Wensleydale	n/a	23	
	Southdown	n/a	10	
	Welsh Mountain	WM	10	
	North Ron	NR	22	
	Hebridean	HE	21	
	Jacob	JC	20	
	Manx	MN	12	
	Shetland	SH	13	
	Boreray	BO	25	
	Scottish Blackface	n/a	19	
	Llanwenog	n/a	12	
	Herdwick	n/a	18	
	Wiltshire Horn	WH	14	
	Portland	PL	3	
	Cotswold	n/a	18	
	Suffolk	SF	10	
Yugoslavia	Istrian	n/a	11	11

[a] n/a = not applicable.

solution (BioRad Chelex 100) in 10mM Tris-EDTA buffer after overnight digestion with Proteinase K. The ovine mitochondrial control region has a modal length of 1,189 bp containing an 800-bp variable region 5' of conserved sequence block (CSB) 1 (Zardoya et al. 1994; Wood and Phua 1996). A hypervariable VNTR comprising 75-bp repeat motifs was found 150–375 bp upstream of tRNA PRO (Wood and Phua 1996). The variable region was PCR amplified and sequenced. This was achieved using PCR primers designed from bovine PRO and PHE tRNA flanking genes (Anderson et al. 1982) and two internal sequencing primers (Wood and Phua 1996). A further two internal sequencing primers (SOAA and SOAD) were designed, situated 209 bp and 778 bp, respectively, upstream from the Anderson et al. PRO primer site. Primer sequences for these were

SOAA 5' TGA ACG CTC ATG TCC G 3'
SOAD 5'TAG TCA ACA TGC GTA TCC TGT CC 3'

Final concentrations of reagents used for PCR were 16mM $(NH_4)_2 SO_4$; 67mM Tris-HCl (pH 8.8); 0.01% Tween-20; 1.25 mM each dNTP; 1.5 mM $MgCl_2$; 0.2 µM each primer; 0.5 U *Taq* polymerase (GibcoBRL). Amplification was carried out in a PE GeneAmp® 9700 thermal cycler, using a 5 min denaturation step at 95° C, followed by 30 cycles of 30 s at 95° C, 30 s at 57.5° C, 35 s at 72° C, and a final extension step at 72° C for 10 minutes. Sequencing was carried out using PE Applied Biosystems BigDye™ Terminator Cycle Sequencing Ready Reaction kit, according to manufacturers' instructions, followed by electrophoresis on an ABI Prism 377 automated sequencer.

Sequence analysis and alignment were performed using ABI Prism Sequencing Analysis and Sequencher™ version 3.1.1 for Power Macintosh (Gene Codes Corp.© 1991–1998). For the purposes of phylogenetic analyses, the hypervariable VNTR was excluded from all control region sequences after alignment. Kimura's 2-parameter (K2P) model of sequence evolution was used since equal base frequencies and unequal substitution rates across taxa were observed (data not shown). A neighbor-joining (NJ) tree was constructed to examine phylogenetic structure for the 130 produced sequences (Figure 21.1) and included an outgroup *(Pseudois nayaur)* sequence.

In order to examine the wide-scale geographic distribution of domestic haplogroups, an RFLP assay was designed for mass screening of samples. Polymorphic restriction sites in ovine control region sequences were identified using Sequencher™ version 3.1.1. A single restriction enzyme, *NsiI*, was found to differentiate among domestic haplogroups (Figure 21.2). RFLP screening was carried out on all 836 individuals (Table 21.1). Restriction digests were carried out overnight on whole control region PCR products. Samples were incubated at 37° C with 3 U of *NsiI* per reaction. Digest final concentrations were 3mM spermidine; 100mM NaCl; 10mM Tris-HCl (pH 8.4); 10mM $MgCl_2$; and 1mM dithiothreitol. Products were subsequently electrophoresed on 3% agarose gels. Pairwise F_{ST} values were generated using haplotype frequencies, and geographical correlations of the diversity uncovered were explored using Mantel matrix correlation tests (see Townsend 2000; Townsend et al. 2004 for further details).

Results

In 130 individuals for which control region sequence was obtained, 94 haplotypes were found, 80 of which were unique with only 14 found in more than one individual. Among *O. musimon* samples, there were 5 haplotypes present, of which 3 were unique. All other wild *Ovis* individuals had unique haplotypes. In contrast to previous results, 3 domestic haplogroups (HPGs) were recovered in the NJ tree (Figure 21.1), denoted here as A, B, and C. HPG A and B correspond to nomenclature used in previous studies (Wood and Phua 1996; Hiendleder et al. 1998; 1999; 2002), while HPG C was previously unidentified.

Average corrected pairwise control region genetic distances between domestic haplogroups were A/B 4.84%, A/C 5.07%, and B/C 5.75%. These results are in reasonable accordance with previous studies, which estimated a sequence divergence of 4.43% between HPG A and B, based on RFLP analysis of the whole mitochondrial genome (Hiendleder et al. 1998). Average corrected pairwise genetic distances between domestic haplogroups were A/B 0.42%, A/C 0.96%, and B/C 0.80%, which compares closely with the 0.716% estimate of Hiendleder et al. (2002).

The NJ tree shows clearly that HPG A is the most common and contains nearly all the sequences from European sheep, as well as sequences from Africa, a small number from the Middle East, and the sample from Mongolia. Importantly, all mouflon samples, regardless of geographic origin, group within HPG A. HPG B includes samples from the Middle East, from Europe (Romania, the Netherlands, and Germany), and the single wild mouflon (*O. orientalis anatolica*). The previously undiscovered HPG C comprises exclusively Middle Eastern samples plus the one sample from India and is the least represented of the three haplogroups. In agreement with previous studies, neither *O. ammon* nor *O. vignei* sequences group with those of modern sheep breeds.

Sequence checking of individuals that had been assigned to a specific haplogroup using the restriction enzyme *NsiI*, which shears DNA into diagnostic DNA fragments that can be separated on a gel, confirmed all HPG A and HPG B RFLP profiles but revealed the RFLP HPG C profile to be less than 100% diagnostic. This restriction profile was subsequently treated as A/C, since sequencing revealed 38.2% of profiles belonged to HPG C individuals while the remainder belonged to HPG A as the result of a G–A transition in a *NsiI* restriction site (Figure 21.2, Table 21.2). This analysis revealed that west of Greece 100% of A/C profiles belonged to HPG A, and east of Greece 100% belonged to HPG C. Individuals with A/C profiles remaining unassigned for HPGs A or C were not included in the final analyses. The geographic distribution of HPGs A, B, and C (Figure 21.3) shows a pattern of high variability in the Near East (all three haplogroups are present) and relative homogeneity in western Europe and Africa (HPG A only). The exceptions to this pattern were found in the Netherlands and Germany, in which low frequencies of HPG B were found (9.7% and 7.7%, respectively).

Geographic location and population pairwise F_{ST} were compared for European samples (see Townsend 2000 and Townsend et al. 2004 for details) using Mantel tests in order to examine patterns of outward expansion from the Near East. Testing (1,000 permutations) of positive correlations was carried out, but were nonsignificant between matrices including any combination of the Netherlands (NL), Hungarian (H), or German (GER) sample. This result may have resulted from the inclusion of HPG B individuals, so the analysis was re-run using European HPG A data only. Significant correlations were obtained only among the European HPG A sample (r = 0.435, P = 0.044) when the data from Germany, Hungary, and the Netherlands were removed from the analysis. Inclusion of these gave results of r = 0.0225, P = 0.200 (NL); r = 0.0237, P = 0.202 (GER); and r = 0.242, P = 0.156 (H). The correlation remained significant among the European HPG A sample upon addition of the African data (r = 0.495, P = 0.022), provided the NL/H/GER samples were excluded. No significant correlation was observed among the NL/H/GER HPG A sample when analyzed separately (Townsend 2000).

Discussion and Future Research

To augment previous studies, the analysis here attempted to describe the diversity of Eurasian and African domestic ovine mitochondrial lineages using a large and well-characterized sample. A previously undescribed domestic mtDNA haplogroup (HPG C) was unexpectedly identified. Further, analysis of CR sequences has shown that an individual from

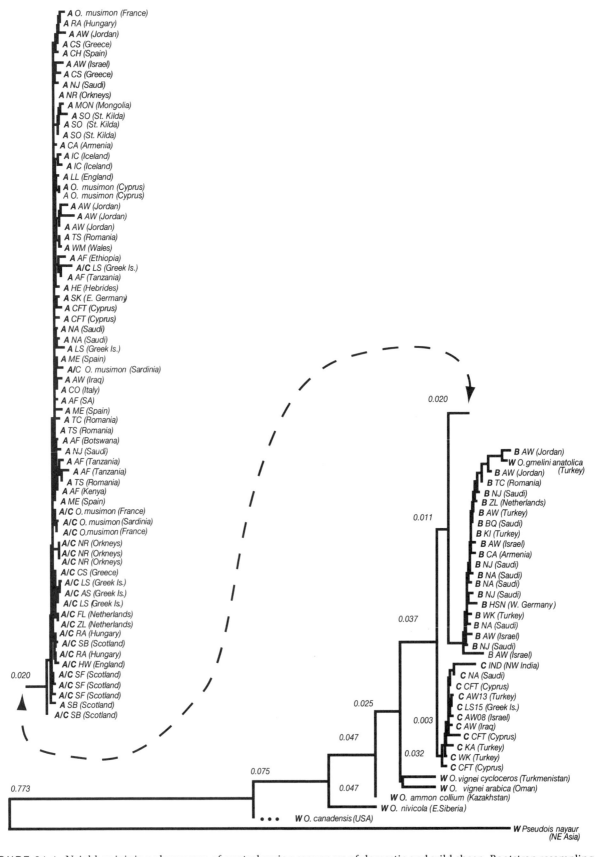

FIGURE 21.1 Neighbor-joining phenogram of control region sequences of domestic and wild sheep. Bootstrap resampling (not shown) of all species' boundaries and three haplogroups (A, B, and C) were greater than 95 percent. Breed origin uses two or three-letter descriptor (Townsend et al. 2000) and country of origin (see Table 21.1). W = wild; A, B, C = haplogroups A, B, or C.

TABLE 21.2
Frequency of *NsiI* Restriction Site Occurrence for All Three Major Restriction Profiles

NsiI Restriction Sites in Ovine Control Region	HPG A, Cut Frequency (%) (n = 736)	HPG B, Cut Frequency[a] (%) (n = 45)	A/C Restriction Profile, Cut Frequency (%) (n = 54)
SITE 1	100	4.44	100
SITE 2	100	4.44	100
SITE 3	100	6.66	100
SITE 4	100	2.22	100
SITE 5[b]	0	100	100
SITE 6	100	100	100

[a] The low levels of site occurrence 1–4 for HPG B (where none should exist) are the result of transition substitutions at these sites in different combinations of the same four individuals. None of these had a profile similar to A or C individuals, however, and were sequenced as a result.

[b] Denotes the site responsible for the difference between A and A/C profiles.

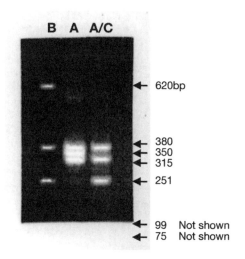

FIGURE 21.2 Restriction profiles of ovine control region showing banding patterns diagnostic for each haplogroup. The smallest fragments are not shown as they are frequently indistinct and not reliable as diagnostic tools at this level of resolution.

a previously unsampled wild population, *O. o. anatolica*, clusters with a domestic haplogroup (HPG B). The only other wild species to cluster within a domestic clade is the European mouflon, *O. musimon*, which other evidence has indicated is a feral remnant of the first domestic sheep to enter Europe.

In all analyses, domestic HPGs A, B, and C and the European mouflon are monophyletic relative to the other *Ovis* species analyzed. In a parallel analysis of cytochrome *b* diversity, Townsend et al. (2006) showed a domestic ancestral node (DAN) was estimated to coalesce at 245,100 BP, very close to that estimated by Hiendleder (1998). The most likely explanation of these data is that each domestic clade is derived from a different *O. orientalis* subspecies or divergent population. The earliest sites containing domestic ovine remains are found in elevated human settlements of the Taurus and Zagros mountains (Peters et al. 1999; Chapter 14), a region where Armenia, Turkey, and Iran intersect and the current range of *O. o. gmelini*. Therefore, it is possible that domestic sheep from these regions were first to travel west and south with farmers, while subsequent domestications of other local wild populations occurred afterwards such that HPGs B and C are derived from a slightly later domestication than HPG A, since they are not represented among the central European sample (even though B is found in a few locations in western Europe).

There are a number of hypotheses about the potential subspecific ancestor of each domestic clade compatible with the tree toplogy. For instance, HPG B individuals may be direct domestic descendants of *O. o. anatolica*, the most westerly wild *Ovis* population. Alternatively, *O. o. anatolica* may be a feral remnant of the first HPG B domestics, as is the case with *O. musimon* and HPG A. HPG C may be derived from either the *O. o. isphahanica* or the *O. o. laristanica* subspecies found to the south of Iran. Unlike the evolutionary relationships among domestic taxa, those between *O. ammon*/*O. vignei* and domestic clades remain unclear.

Domestic centers of origin frequently exhibit more genetic variation within domestic species relative to neighboring regions, and a decrease in diversity proportional to distance from the center of origin is not uncommon (Zohary 1996; Loftus et al 1999; Troy et al. 2001; but see Luikart et al. 2001 for an exception). The phylogeographic distribution of ovine haplogroups presented here is in accordance with this pattern. It is also in keeping with the Wave of Advance model of Neolithic human expansions (Ammerman and Cavalli-Sforza 1973), which describes proliferation into Europe as an extension from the outer edges of expanding Neolithic farming communities by means of gradual demic diffusion. The exceptions to the pattern of low diversity predicted for western Europe occur in Germany, the Netherlands, and Romania, where low levels of HPG B are seen. HPG A individuals from Germany, the Netherlands, and Hungary also appear to deviate from a European trend where degree of genetic similarity decreases in proportion to increasing distance. There are several processes by which these trends may have arisen.

First, recent introgression from imported eastern breeds may have taken place. A number of commercial antipodean breeds have been "improved" using Asian stock, as exemplified by the HPG B individuals (50 percent) found among a previous sample of 50 individuals whose founder stocks were originally imported from western Europe (Wood and Phua 1996). Second, it may reflect the proximity of these regions to Scandinavia, which could exhibit different patterns of haplogroup distribution when compared with the rest of Europe if it had received different founder stock from a separate Neolithic migration. Evidence suggests that this would

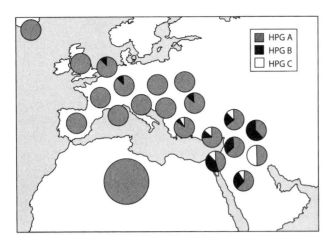

FIGURE 21.3 Geographic distribution of mitochondrial CR region haplogroups in domestic sheep [HPG] A, B, and C.

have taken place from the far north, however, since southern Scandinavia was subject to the same Neolithic migration as the rest of central Europe (Price 1996).

Similarly, it is possible that other separate routes of primary Neolithic expansion into Europe brought different founder stock with them. Archaeological and genetic evidence indicates that expansion out of the Near East occurred separately along northerly and westerly routes (Cavalli-Sforza et al. 1994; Chikhi et al. 1998), which may account for the absence of significant correlation of genetic divergence with distance between some European countries on a northeastern trajectory (i.e., the Netherlands, Germany, and Hungary) and others. The significant correlation among all other HPG A European and African ovine samples indicates, however, that this may not be an underlying cause of the pattern observed.

While considerable efforts have gone into documenting current patterns of sheep diversity in Europe and certain other regions of the world where sheep are used extensively, it is clear that we still have to find the origins of modern sheep. Work is under way to pursue this issue, including particularly a complete genetic study of the subspecific status and higher order taxonomic relationships of wild *Ovis* populations of Eurasia first started by Hiendleder et al. (2002). To date, however, a series of logistical problems in working with the potentially important wild, museum, and archaeological samples has prevented a definitive answer from being sought.

Clearly, a detailed study of *O. orientalis* would be of particular interest to the history of sheep domestication. It is possible that subspecies today are isolated, local remnants of a panmictic *O. orientalis* population that covered Southeast Asia at the time of domestication. If not, revealing the current geographic location of direct subspecific ancestors would add to existing knowledge regarding the source of different Neolithic human migrations. This will, in turn, be complemented by the further sampling of domestic populations in central Asia, Scandinavia, and Africa, while ongoing studies of Eurasian domestic ovines using Y-chromosome and autosomal loci are set to reveal further layers of genetic substructure among the domestic ovine populations sampled here.

Acknowledgments

We thank Juliet Clutton-Brock, Lounès Chikhi, and Michelle Bayes for advice and useful discussion, and Çan Biglin, Gordon Luikart, Osama Mohammed, Valery Fedosenko, Olivier Hanotte (ILRI) and the Harrison Museum for help with providing difficult samples. Thanks also go to the many other people who helped with the sampling process. This work was funded by the European Union Framework IV.

References

Alderson, G. L. H. 1994. *The chance to survive*. London: Cameron and Tayleur.

Allard, M. W., M. M. Miyamoto, L. Jarecki, F. Krans, and M. R. Tennant. 1992. DNA systematics and evolution of the artiodactyl family Bovidae. *Proceedings of the National Academy of Sciences, USA.* 89: 3972–3975

Ammerman, A. J. and L. L. Cavalli-Sforza. 1973. A population model for the diffusion of early farming in Europe. In *The explanation of culture change*, C. Renfrew (ed.), pp. 343–357. London: Duckworth.

Anderson, S., M. H. L. DeBruin, A. R. Coulson, I. C. Eperon, F. Sanger, and I. G. Young. 1982. Complete sequence of bovine mitochondrial DNA. Conserved features of the mammalian mitochondrial genome. *Journal of Molecular Biology* 156: 683–717.

Arranz, J. J., Y. Bayon, and F. S. Primitivo 1998. Genetic relationships among Spanish sheep using microsatellites. *Animal Genetics* 29: 435–440.

Bancroft, D. R., J. M. Pemberton, and P. King 1995. Extensive protein and microsatellite variability in an isolated, cyclic ungulate population. *Heredity* 74: 326–336.

Bruford, M. W., D. G. Bradley, and G. Luikart. 2003. Genetic analysis reveals complexity of livestock domestication. *Nature Reviews Genetics* 4: 900–910.

Buchanan, F. C., L. J. Adams, R. P. Littlejohn, J. F. Maddox, and A. M. Crawford. 1994. Determination of evolutionary relationships among sheep breeds using microsatellites. *Genomics* 22: 397–403.

Bunch, T. D., W. C. Foote, and J. J. Spillett. 1976. Translocations of acrocentric chromosomes and their implications in the evolution of sheep (*Ovis*). *Cytogenetics and Cell Genetics* 17: 122–136.

Bunch, T. D., R. M. Mitchell, and A. Maciulis. 1990. G-banded chromosomes of the *Gansu argali* (*Ovis ammon jubata*) and their implications in the evolution of the *Ovis* karyotype. *Journal of Heredity* 81: 227–230.

Cavalli-Sforza, L. L., P. Menozzi, and A. Piazza. 1994. *The history and geography of human genes*. Princeton, NJ: Princeton University Press.

Chikhi, L., G. Destro-Bisol, G. Bertorelle, V. Pascali, and G. Barbujani. 1998. Clines of nuclear DNA markers suggest a largely Neolithic ancestry of the European gene pool. *Proceedings of the National Academy of Sciences, USA.* 95: 9053–9058.

Clutton-Brock, J. 1981. *Domesticated animals from early times*. London: Heinemann/British Museum (Natural History).

Crawford, A. M. 1999. Recent advances in sheep genome mapping. Asian-Australasian. *Journal of Animal Science* 12: 1129–1134.

Diez-Tascon, C., R. P. Litlejohn, P. A. R. Aeida, and A. M. Crawford. 2000. Genetic variation within the Merino sheep breed: Analysis of closely related populations using microsatellites. *Animal Genetics* 31: 243–251.

Farid, A., E. O'Reilly, C. Dollard, and C. R. Kelsey. 2000. Genetic analysis of ten sheep breeds using microsatellite markers. *Canadian Journal of Animal Science* 80: 9–17.

Geist, V. 1971. *Mountain sheep: A study in behavior and evolution*. Chicago: University of Chicago Press.

———. 1983. On the evolution of ice age mammals and its significance to an understanding of speciations. ASB Bulletin 30: 109.

———. 1990. On the taxonomy of giant sheep. *Canadian Journal of Zoology* 69: 706–723.

de Gortari, M. J., B. A. Freking, R. P. Cuthbertson, S. M. Kappes, J. W. Keele, R. T. Stone, K. A. Leymaster, K. G. Dodds, A. M. Crawford, and C. W. Beattie. 1998. A second-generation linkage map of the sheep genome. *Mammalian Genome* 9: 204–209.

Hermans, W. A. 1996. The European mouflon, *Ovis musimon*. *Tijdschrift Voor Diergeneeskunde* 121: 515–517.

Hewitt, G. M. 1996. Some genetic consequences of ice ages, and their role in divergence and speciation. *Biological Journal of the Linnean Society* 58: 247–276.

Hiendleder, S., K. Mainz, Y. Plante, and H. Lewalski. 1998. Analysis of mitochondrial DNA indicates that domestic sheep are derived from two different maternal sources: No evidence for contributions from urial and argali sheep. *Journal of Heredity* 89: 113–120.

Hiendleder, S., S. H. Phua, and W. Hecht. 1999. A diagnostic assay discriminating between two major *Ovis aries* mitochondrial DNA haplogroups. *Animal Genetics* 30: 211–213.

Hiendleder, S., S. Kaupe, R. Wassmuth, and A. Janke. 2002. Molecular analysis of wild and domestic sheep questions current nomenclature and provides evidence for domestication from two different subspecies. *Proceedings of the Royal Society of London, Series B*. 269: 893–904.

Irwin, D. M., T. D. Kocher, and A. C. Wilson. 1991. Evolution of the cytochrome *b* gene of mammals. *Journal of Molecular Evolution* 32: 128–144.

Loftus, R. T., O. Ertugrul, A. H. Harba, M. A. El-Barody, D. E. MacHugh, S. D. Park, and D. G. Bradley. 1999. A microsatellite survey of cattle from a centre of origin: The Near East. *Molecular Ecology* 8: 2015–2022.

Luikart, G., L. Gielly, L. Excoffier, J.-D. Vigne, J. Bouvet, and P. Taberlet. 2001. Multiple maternal origins and weak phylogeographic structure in domestic goats. *Proceedings of the National Academy of Sciences, USA*. 98: 5927–5932.

Lyapunova, E. A., T. D. Bunch, N. N. Vorontsov, and R. S. Hoffman. 1997. Chromosomal complement and taxonomic position of Severtsov wild sheep *(Ovis ammon severtzovi)*. *Zoologichesky Zhurnal* 76: 1083–1093.

Lydekker, R. 1912. *The sheep and its cousins*. London: Allen.

Manwell, C. and C. M. A. Baker. 1977. Genetic distance between the Australian Merino and the Poll Dorset sheep. *Genetical Research* 29: 239–253.

Mason, I. L. 1996. *A world dictionary of livestock breeds, types and varieties*. Oxford: CAB International.

Montelgard, C., T. C. Nguyen, and D. Dubray. 1994. Genetic variability in French populations of the Corsican mouflon *(Ovis ammon musimon)*: Analysis of two blood proteins and red-cell blood groups. *Genetics Selection and Evolution*. 26: 303–315.

Nadler, C. F., D. M. Lay, and J. D. Hassinger. 1971. Cytogenetic analysis of wild sheep populations in northern Iran. *Cytogenetics* 10: 137–152.

Nadler, C. F., R. S. Hoffman, and A. Woolf. 1973a. G-band patterns as chromosomal markers and the interpretation of chromosomal evolution in wild sheep *(Ovis)*. *Experientia* 29: 117–119.

Nadler, C. F., K. V. Korobitsina, R. S. Hoffman, and N. N. Vorontsov. 1973b. Cytogenic differentiation, geographic distribution and domestication in Palaearctic sheep *(Ovis)*. *Zeitschrift für Saugetierkunde* 38: 109–125.

Naitana, S., S. Ledda, E. Cocco, L. Manca, and B. Masala. 1990. Haemoglobin phenotypes of the wild European mouflon sheep living on the island of Sardinia. *Animal Genetics* 21: 67–75.

Nei, M. 1972. Genetic distance between populations. *American Naturalist* 106: 283–292.

Osfoori, R. and L. Fesus 1996. Genetic relationships of Iranian sheep breeds using biochemical genetic markers. *Archiv Fur Tierzucht* Dummerstorf 39: 33–46.

Peters, J., D. Helmer, A. von den Driesch, and S. Segui. 1999. Animal husbandry in the northern Levant. *Paléorient* 25: 27–48.

Petit, E., S. Aulagnier, R. Bon, M. Dubois, and B. Crouau-roy. 1997a. Genetic structure of populations of the Mediterranean mouflon *(Ovis gmelini)*. *Journal of Mammology* 78: 459–467.

Petit, E., S. Aulagnier, D. Vaiman, C. Bouisseau, and B. Crouau-roy. 1997b. Microsatellite variation in an introduced mouflon population. *Journal of Heredity* 88: 517–520.

Price, T. D. 1996. The first farmers of southern Scandinavia., In *The origins and spread of agriculture and pastoralism in Eurasia*, D. R. Harris (ed.), pp. 346–362. London: University College of London Press.

Randi, E., G. Fusco, R. Lorenzini, S. Toso, and G. Tosi. 1991. Allozyme divergence and phylogenetic relationships among *Capra, Ovis* and *Rupicapra* (Artyodactyla, Bovidae). *Heredity* 67: 281–286.

Rando, A., P. DiGregorio, M. Capuano, C. Senese, L. Manca, S. Naitana, and B. Masala. 1996. A comparison between the beta-globin gene clusters of domestic sheep *(Ovis aries)* and Sardinian mouflon *(Ovis gmelini musimon)*. *Genetics Selection and Evolution* 28: 217–222.

Sambrook, J., E. F. Fritsch, and T. Maniatis 1989. *Molecular cloning: A laboratory manual*, 2nd edition. Cold Spring Harbor, NY: Cold Spring Harbor Laboratory Press.

Shackleton, D. M. 1997. *Wild sheep and goats and their relatives*. Gland, Switzerland: IUCN.

Townsend, S. J. 2000. *Patterns of genetic diversity in European sheep breeds*. Unpublished PhD thesis, University of East Anglia.

Townsend, S. J., K. Byrne, W. Jordan, G. Kierstein, O. Hanotte, G. M. Hewitt, and M. W. Bruford. 2006. Diversity, phylogeography and domestication of Eurasian and African sheep *(Ovis Aries)*. In review. *Proceedings of the Royal Society of London, Series B*.

Troy, C. S., D. E. MacHugh, J. F. Bailey, D. A. Magee, R. T. Loftus, P. Cunningham, A. T. Chamberlain, B. C. Sykes, and D. G.

Bradley. 2001. Genetic evidence for Near-Eastern origins of European cattle. *Nature* 410: 1088–1091.

Valdez, R., C. F. Nadler, and T. D. Bunch. 1978. Evolution of wild sheep in Iran. *Evolution* 32: 56–72.

Vorontsov, N. N., K. V. Korobitsyna, C. F. Nadler, R. Hofman, G. N. Sapozhnikov, and J. K. Gorelow. 1972. Chromosomi dikich baranow i proiskhzhdenie domashnikh ovets. *Priroda, Moscow* 3: 74–82.

Wang, S., W. C. Foote, and T. D. Bunch. 1991. Evolutionary implications of hemoglobin polymorphism in domesticated and wild sheep. *Small Ruminant Research* 4: 315–322.

Wood, N. J. and S. H. Phua. 1996. Variation in the control region sequence of the sheep mitochondrial genome. *Animal Genetics* 27: 25–33.

Zardoya, R., M. Villalta, M. J. Lopez-Perez, A. Garrido-Pertierra, J. Montoya, and J. M.Bautista. 1995. Nucleotide sequence of the sheep mitochondrial DNA D-loop and its flanking tRNA genes. *Current Genetics* 28: 94–96.

Zeuner, F. E. 1963. *A history of domesticated animals*. London: Hutchinson.

Zohary, D. 1996. The mode of domestication of the founder crops of southwest Asian agriculture. In *The origins and spread of agriculture and pastoralism in Eurasia*, D. R. Harris (ed.), pp. 142–158. London: University College of London Press.

CHAPTER 22

Genetics and the Origins of Domestic Cattle

DANIEL G. BRADLEY AND DAVID A. MAGEE

Introduction

Taxonomy places domesticated cattle within the mammalian Bovidae family, which belongs to the order Artiodactyla, or even-toed ungulates. The Bovidae first appeared in the Miocene approximately 20 MYA, and it is believed that their diversification coincided with the widespread emergence of savannah-based ecological niches during this epoch (Allard et al. 1992). Domesticated cattle are members of the Bovini tribe within the Bovidae. The ancestral species was *Bos primigenius*, and common usage gives two taxa for the domestic descendants, *Bos indicus* and *Bos taurus* (see Gentry et al. 2004). These are used here as shorthand terms (along with indicine and taurine, respectively) with an acknowledgment that their taxonomy is more complex.

Currently, there are approximately 1,300 million head of cattle distributed across the globe, constituting of approximately 800 separate extant cattle breeds, making this the most numerous and arguably most economically important of all the domestic livestock species (Bradley and Cunningham 1999). *Bos taurus* cattle predominate in the temperate lands of Europe, West Africa, and northern Asia, whereas *Bos indicus* cattle generally inhabit the hot-arid or semi-arid regions of South Asia and Africa. Indicine, or zebu, cattle differ from taurine cattle in a number of morphological and physiological aspects. Zebu animals are characterized externally by a prominent hump, a long face, usually steep upright horns, and a dewlap. They display a lower metabolic rate and nutrient requirement (food and water) in response to drought and food shortages. Indicine cattle are also more resistant to ticks, gastrointestinal parasites, and rinderpest than are taurine cattle, although indigenous *Bos taurus* animals exhibit comparative advantage in the tsetse fly–infected regions of Africa (Epstein and Mason 1984; Marshall 1989). However, both domestic taxa possess the same number of chromosomes ($2n = 60$) and freely interbreed.

Prior Research

The Wild Ancestor of Domesticated Cattle

It is now widely accepted that the progenitor of all domesticated cattle was the morphologically similar aurochs, *Bos primigenius* (Zeuner 1963). *Bos primigenius* is thought to have evolved from an Asiatic ancestor, *Bos acutifrons*, during the course of the Upper Pliocene, approximately 5–1.8 MYA (Zeuner 1963). The now-extinct aurochs dispersed throughout Asia, and into Northern Africa and Europe, reaching as far as Spain and Britain. Bulls were pronouncedly larger than modern domesticates, with a height of up to 6.5 feet at the shoulder, and were often equipped with longer horns (Zeuner 1963). Morphological differences in fossilized horn and body shapes have prompted some archaeozoologists to classify the aurochs into three major separate subspecies: a Eurasian subspecies (*Bos primigenius primigenius*), a South Asian subspecies (*Bos primigenius namadicus*), and a North African subspecies (*Bos primigenius opisthonomus*) (Clutton-Brock 1989).

The Domestic Origins of *Bos taurus* and *Bos indicus* Based on Archaeological Data

The earliest evidence for the domestication of cattle, during the eighth millennium BC, points toward the marshlands and forests of the Middle Euphrates Basin. Peters et al. (1999) suggest an early date, within the Mid Pre-Pottery Neolithic B (ca. 8500 BP uncalibrated), when size reduction is clear in *Bos* remains. Later evidence in the region for elevated female:male ratios and the absence of older individuals within slaughter remains (ca. 7000 BC) suggests a culling strategy and domestication. Other data suggesting early cattle domestication has emerged from Cyprus, where Pre-Pottery Neolithic cattle remains must be those of introduced and thus domesticated animals (Vigne and Buitenhaus 1999). Interestingly, the early cattle on Cyprus are quite large, suggesting that herding may have been present for several centuries before detectable change from the wild-type morphology emerged.

It has been postulated that both *Bos taurus* and *Bos indicus* cattle were derived from the western Eurasian aurochs (*B. primigenius*) within Southwest Asia during the early Neolithic development of pastoralism. Within this scenario, *Bos indicus* cattle would have been derived from initial *Bos taurus* herds by means of artificial selection during early pastoral movements across the Iranian Plateau into the Indian subcontinent ca. 4000 BC (Epstein 1971; Epstein and Mason 1984; Payne 1991).

However, a more widely held view is that *Bos indicus* cattle were domesticated independently or subsequently to *Bos taurus* cattle from a biologically distinct wild progenitor within the Indian subcontinent (Zeuner 1963). This assertion is supported by detailed craniometric and osteological measurements that reveal similarity between *Bos indicus* and the South Asiatic aurochs, *Bos primigenius namadicus*, and *Bos taurus* and the Eurasian aurochs, *Bos primigenius primigenius* (Grigson 1978; Grigson 1980). These similarities point to

B. p. namadicus as the ancestor of zebu cattle and *B. p. primigenius* as the ancestor of taurine cattle. Indeed, archaeological surveys located within Baluchistan have provided plausible evidence for the local domestication of *Bos indicus* cattle, most probably from local *Bos primigenius namadicus* populations, ca. fifth millennium BC (Meadow 1993, 1996; Bökönyi 1997). Some also argue for the domestication of *Bos indicus* within peninsular India (Allchin and Allchin 1968; Naik 1978). Divergent ancestries for taurine and zebu cattle are also supported by biological and genetic data, which are discussed later.

The Dispersal of Domestic Cattle from Initial Centers and Local Contribution

The first European farmers were settled in Greece before 6500 BC (Renfrew, 1987), but the limited ranges of the wild progenitors of many of the primary domestic species, such as barley, emmer wheat, einkorn, sheep, and goat, point to their origins farther east, in Anatolia or the Fertile Crescent (Zohary 1996; Renfrew 1987). Similarly, it is held that European *Bos taurus* cattle were introduced into Europe as a result of early pastoralist migrations from the Near East (Epstein and Mason 1984; Medjugorac et al. 1994). It has been suggested that domestic taurine animals, along with other Near Eastern domesticates, entered Europe via two routes: (1) a Danubian route, in which domesticates first entered Greece ca. 6000 BC and later followed the inland course of the River Danube into Southeast Europe, finally reaching the northwest coast of mainland Europe ca. 4000–3000 BC, and (2) a more rapid maritime Mediterranean coastal route, via the Dalmatian coast, Sicily, and southern France, reaching Spain and Portugal ca. 5400 BC (Waterbolk 1968).

Despite a prevailing view pointing toward a derivative Near Eastern origin of European *Bos taurus* cattle, some authors contend that wild cattle may have been domesticated within Europe, as a feature of local contributions to the farming economy (Bökönyi 1974). Indeed, archaeological data indicate that the aurochs ranged throughout much of continental Europe and Britain and was available for domestication. Similarly, some authors have argued, albeit controversially, that the founding *Bos taurus* domesticates of the African continent were derived from a separate domestication of an indigenous aurochsen strain within the eastern Sahara ca. 9000–8000 BP (Wendorf and Schild 1994).

Biochemical analysis of bovine genetic variation has been an active field for decades, with perhaps only studies in humans having produced more data. An example is the collation of protein polymorphism data from about a thousand research papers describing 216 cattle breeds by Manwell and Baker (1980). These authors identified east-west gradients of gene frequencies and noted the pronounced divergence between *Bos indicus* and *Bos taurus* populations. However, they acknowledged that it was difficult to isolate selective effects from domestication events in the genesis of these patterns. This is a particular issue with classical polymorphisms because the accessible proteins primarily used (blood and milk proteins) are often strong candidates for the effects of both natural and artificial selection, which can be confounding factors in the elucidation of domestic history.

Genetic Signatures of Animal Domestication

It is sometimes contended (e.g., Ingold 1996) that the domestic state is not a definite category, but rather part of a continuum of human-animal relationships that stretch from hunting, through loose association, to the highly managed herding of animals. Although it may be true that a strict dichotomy between hunting and herding is an oversimplification, the process on which genetic analysis may inform is a defined one: the managed breeding of domestic animals by herders. As genetic data are transmitted to succeeding generations, modern (and ancient) populations carry the genetic signatures of the past demographic processes, which have molded them over time. Incontrovertibly, in the case of livestock domestication, all modern domesticates are descended from the wild animals that were incorporated into a finite genetic domestic pool at some stage (or stages) within domestic history. Therefore, modern populations are the endpoint of a temporal genetic stream that flows from specific past interactions between wild populations and herders. It is these events to which genetic study may add information, not less influential interactions that have left no genetic legacy.

Importantly, alternate genetic marker systems, because of different modes of inheritance, information content, and mutation rates, may give information on different aspects of past demography. The marker systems employed beneficially in domesticated cattle include mitochondrial DNA (mtDNA) sequence, Y-chromosome markers, autosomal microsatellite markers, single-nucleotide polymorphisms (SNPs), and classical markers (blood groups and protein polymorphisms). To date, the two most informative and extensively used marker systems in the investigation of domestic cattle origins have been mtDNA sequence diversity and allelic variability at autosomal microsatellite loci.

MITOCHONDRIAL DNA

Surveys of mitochondrial DNA (mtDNA) variation continue to propel molecular phylogeographic surveys in a wide range of species. This is due to several properties. First, the high copy number of mitochondrial chromosomes within each eukaryotic cell facilitates the retrieval of mtDNA from difficult substrate such as archaeological bone. Second, mtDNA is highly polymorphic because of a rate of evolution that is 5–10 times that of nuclear DNA, permitting phylogenetic analysis within and between populations involving shallow evolutionary time depths, such as that associated with domestic history (Brown et al. 1979). Third, mammalian mitochondrial chromosomes seem to display exclusively maternal inheritance. Thus, mtDNA lineages are clonal, information-rich, singly inherited units (or haplotypes), lacking the complication of bi-parental recombination (although this has been questioned recently; see reviews by Eyre-Walker and Awadalla

(2001) and Innan and Nordberg (2002)). Finally, because mtDNA is haploid and contributed to the next generation by only one parent, it has an effective population size one-quarter that of nuclear DNA. Consequently, mtDNA is a more sensitive indicator of population dynamics than is nuclear DNA. For example, population bottlenecks can dramatically reduce mtDNA diversity within a population, while having little effect on nuclear variability (Jorde et al. 1998).

Phylogenetic reconstruction using mtDNA sequence data is not without complication. Human data have revealed considerable variation in the rate of evolution between sites situated within the control-region, which, along with other factors, can hinder mitochondrial mutation rate (or molecular clock) calibration (Gibbons 1998). Hypervariability also implies that certain sites may have mutated more than once in the genealogy of a species, and selection of the most parsimonious topology consequently becomes difficult (Richards and Macaulay 2001). Lastly, it must be emphasized that phylogenetic inference involving mtDNA is based effectively on one segregating locus, albeit a whole chromosome—one that is subject to atypical population dynamics. Thus it is important to complement the rich information that can be gleaned from mtDNA with analysis of nuclear diversity and Y-chromosomal polymorphisms.

MICROSATELLITE LOCI

Microsatellite genetic markers consist of tandemly arranged reiterated units of noncoding DNA sequence, typically ranging between 2–5 base pairs (bp) in length. Also referred to as simple tandem repeat loci, they are ubiquitously distributed throughout eukaryotic genomes. These genetic elements are often highly polymorphic because of variation in the number of repeat units, which account for a multitude of distinguishable alleles that are inherited in a Mendelian fashion. Estimates of microsatellite evolutionary rates range between 10^{-5} and 10^{-2} mutations per generation (Rosenbaum and Deinard 1998). Microsatellite marker variation is presumed to be neutral; that is, it is without the complication of selection unless closely linked to a selected coding region.

Y-CHROMOSOME MARKERS

The phylogeography of bovine Y chromosome has not yet been studied in depth, with the published level of definition limited to distinction between *Bos indicus* and *Bos taurus* (Bradley et al. 1994; Hanotte et al. 2000). However, this system holds much promise, as more markers are now available and current human studies of Y-chromosome variation are proving especially fruitful.

Research Design: Geographic and Temporal Contexts

Samples Modern and Ancient

Modern samples discussed here have been detailed elsewhere. They include, for mtDNA analysis, 18 European, 9 African, 9 Near Eastern, and 6 Indian breed samples of approximately 12 animals each (Troy et al. 2001; Magee 2002). Microsatellite data comprised 10 European, 7 Near Eastern, and 2 Indian samples (Loftus et al. 1999; Schmid et al. 1999; Edwards et al. 2000). These sample sizes typically were 35–40 per breed.

Six ancient samples of *Bos primigenius* have yielded reliable genetic data. These were from skeletal remains excavated from chronologically distinct sites dispersed across England (Troy et al. 2001).

Analytical Approaches

There are two main approaches to the kinds of genetic data presented here: population-based and phylogeographic methods. Both are employed here and contribute to an overall understanding of the domestication and diffusion of domestic cattle. The former has the advantage that it considers the data in groups of breeding individuals—ultimately the focus of inference. However, it has the drawback that we may not accurately suppose that present populations follow trajectories backward through time without complicating such events as hybridizations, which are difficult to model. A phylogeographic perspective examines the evolutionary trajectories of specific loci, such as mtDNA, that can often afford high resolution and powerful inference. This approach has one major pitfall: single loci are themselves a sample of whole genomic diversity and may not represent the only, or even the major, evolutionary history bound up within a population. It is clear that multiple lines of evidence are required to piece together the different strands of ancestry in a subject such as domestic cattle.

Presentation of Results

Population-Based Analyses

An overview of genetic variation may be conveniently generated using summary statistics, such as genetic distance between populations and conventional visualization methods. Here, genetic distances among cattle populations were calculated for both mtDNA (pairwise FST values; Slatkin 1991) and microsatellites (D_A genetic distances; Nei 1987) and are summarized in two dimensions following multidimensional scaling (MDS; Kruskal 1964) of the genetic distance data (Figure 22.1). Conveniently, each dimension revealed by MDS is orthogonal and may summarize separate genetic trends within the data.

The primary feature of the MDS plot for both mtDNA and microsatellite genetic systems is the marked divergence of *Bos taurus* and *Bos indicus* populations, which is clearly defined along the major axis of variation (i.e., dimension 1). Also, in both systems, some Near Eastern cattle populations occupy intermediate positions between the European *Bos taurus* and Indian *Bos indicus* groups, which undoubtedly reflects degrees of hybrid *Bos taurus–Bos indicus* ancestry within these populations (Loftus et al. 1999; Troy et al. 2001; Magee 2002). The

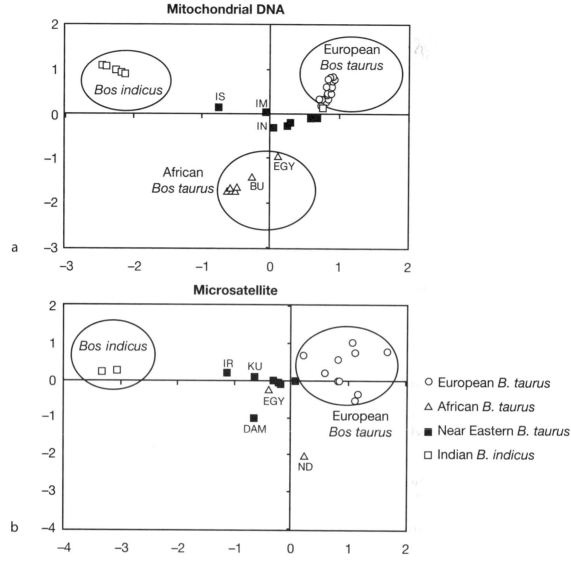

FIGURE 22.1 Scaled two-dimensional representation of genetic distances between cattle populations: (a) mitochondrial variation; (b) microsatellite variation. Symbols are given for regional groups. Specific breed codes, where given, are (a) IS, southern Iraq; IM, middle Iraq; IN northern Iraq; EGY, Egyptian Baladi; BU, Sudanese Butana; and (b) IR, Iraq; KU, Iraqi Kurdish; DAM, Syrian Damascus; and ND, Guinean Ndama.

minor axis of variation in both systems distinguishes African populations from other regional groups, although it is interesting that the Egyptian sample groups more closely with the Near Eastern breeds in the microsatellite plot and is more similar to the African breeds in the mtDNA analysis. This is presumably indicative of substantial, but predominantly male-driven, geneflow from the Near East into Northeast Africa, which has not perturbed mtDNA diversity appreciably.

This broad pattern of variation is similar to that produced by previous examinations of classical polymorphisms (Manwell and Baker 1980). However, an advantage of mtDNA-based investigations is the potential for highly informative haplotypes, which enable the phylogenetic and geographic (or phylogeographic) analysis of single loci as well as populations—an approach that can bring great insight.

Phylogeographic Analyses

A more detailed examination of the mtDNA diversity that underlies the indicine-taurine divergence reveals major phylogeographic distinctions between the mtDNA variations that occur within each domestic group. Whereas Near Eastern and other hybrid populations display intermediate mtDNA haplotype frequencies between European and Indian breeds, no hybrid indicine-taurine mtDNA haplotypes are detected. This feature is illustrated on the map in Figure 22.2, where mtDNA sequence phylogenies taken from four hybrid Near Eastern breeds are superimposed upon their approximate sampling locations. Populations may have intermediate frequencies of zebu and taurine mitochondrial variants, but the quantitative divergence between the variants themselves remains.

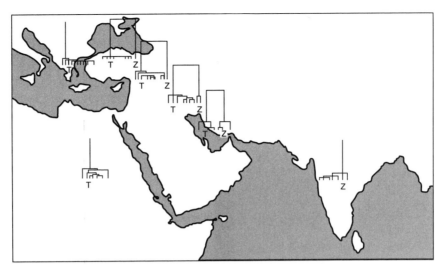

FIGURE 22.2 Phylogenies of seven cattle mtDNA breed samples superimposed on approximate sample origins. In each case *Bos taurus*–type mtDNA sequences (T) are found in the left-hand phylogenetic clade and indicine sequences (Z) on the right-hand side. Two aspects are notable: whereas there are intermediate populations consisting of both types, there are no intermediate mtDNA sequences; and there is a gradation of indicine mtDNA frequency within the samples, which peaks in peninsular India.

Thus, it has emerged that there is a strict dichotomy in modern bovine mtDNA sequence haplotypes. All of over 500 known mtDNA partial control region sequences sampled widely from four continents fall firmly into either a *Bos indicus* or *Bos taurus* family of variants (Loftus et al. 1994a; Bradley et al. 1996; Mannen et al. 1998; Cymbron et al. 1999; Troy et al. 2001). The mtDNA sequences within each of these mitochondrial clades are similar to each other, differing by only a few base-pair substitutions at most (on average, 0.8%). However, the divergence between the two sequence families is considerable, at 8.3% in the control region, with a whole mtDNA genome estimate of 2.3% (Loftus et al. 1994a; Loftus et al. 1994b). Previous studies have estimated the divergence between the *Bos taurus* and *Bos indicus* mtDNA sequence families, which may be calibrated by several methods, including the use of bison mtDNA sequences as an outgroup and an estimated time depth of one million years since the *Bos-Bison* divergence (Loftus et al. 1994a). Although such calibrations are complicated by the complexity of modeling control region mutation patterns, this divergence has consistently yielded estimates of hundreds of thousands of years (Bradley et al. 1996). It is therefore clear that the diversity between these clades could not have arisen within the purported 10,000-year history of animal herding.

The divergence between the two taxa is also clear from close examination of autosomal microsatellite loci. A common pattern of these markers is that alternate collections of alleles from several loci display high frequencies in either indicine or taurine populations, respectively, providing another indication of the pronounced time depth in their genetic separation (MacHugh et al. 1997). Y chromosomes also show a strict dichotomy, both at a cytogenetic level, where there are two morphologies of chromosome that can be easily identified by microscopy, and at the molecular level, where both Y-chromosome-specific microsatellites and other analyses have identified indicine and taurine variants, which have been used to trace male lineage origins in hybrid populations (Bradley et al. 1994; Hanotte et al. 2000).

Notably, the phylogenetic motif of distinct mtDNA haplotypic clades separated by long internal branches is a recurring one in domestic ungulate mtDNA diversity studies (MacHugh and Bradley 2001). These clades (usually two or three per species) have a tendency to be geographically distributed, often along an east-west axis. In pig, for example, one type is derived from East Asian breeds, while another is exclusive to domesticates of European origin (Giuffra et al. 2000; see also Chapter 20 for a similar pattern in goat). In cattle, one mtDNA clade is clearly of South Asian origin and is encountered in the *Bos indicus* breeds of India and surrounding regions. The other clade may be securely identified as *Bos taurus*, and its presence in all assayed African taurine and indicine samples is explained by the hybrid origins of African indicine populations (Bradley et al. 1996). Furthermore, this finding highlights the pronounced geographic inertia of mtDNA lineages, particularly with respect to secondary introgressions, in which most genetic exchange between locales under managed breeding schemes is predominantly male-mediated (MacHugh et al. 1997; Troy et al. 2001; Hanotte et al. 2002).

Genetic diversity within domestic ungulate mtDNA clade groupings is often markedly smaller than that between them. For example, in cattle, one finds approximately 10 times less divergence within the indicine and taurine groups than between the groups. Because there is an expectancy among

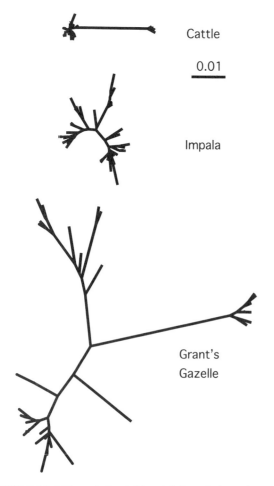

FIGURE 22.3 Unrooted neighbor-joining phylogenies constructed using control region sequences from cattle, impala, and Grant's gazelle, drawn to the same scale. Comparison of two arbitrarily chosen wild data sets with cattle illustrates the restriction of genetic diversity resulting from the domestication process.

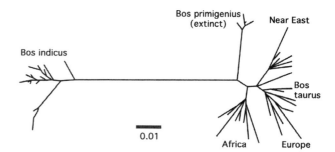

FIGURE 22.4 Neighbor-joining phylogeny of *Bos indicus*, *Bos taurus*, and extinct British *Bos primigenius* mtDNA control region haplotypes. Sequences compared were 201 bp in length and included positions 16042–16158 and 16179–16262. Gaps were excluded, and no distance correction was employed. The four ancient sequences cluster tightly (bootstrap value of 94%; 1,000 replications) and form an outgroup to the modern *Bos taurus* haplotypes. This is despite their origin in animals dispersed geographically and chronologically. These ancient sequences all share eight transitions that separate them from the modern *Bos taurus* root sequence (after Troy et al. 2001, Fig. 1).

natural populations to display geographic and genetic correlation (i.e., animals of similar location tend to have related genetic variation), each ungulate clade is credibly the result of a domestication-induced attenuated sampling of wild variation incorporated into the domestic pool (i.e., capture limited in space and time). This is illustrated by the unrooted phylogenies in Figure 22.3. In these, mtDNA sequences from two arbitrarily chosen wild African ungulate data collections, namely impala and Grant's gazelle, are compared to data from cattle sampled over three continents (Arctander et al. 1996; Nersting and Arctander 2001; Mannen et al. 1998). Each wild ungulate phylogeny shows multiple clades and a complex branching structure resulting from a wide geographic sampling of wild variability. The cattle phylogeny, in contrast, is simple, with only two major distinct clades. Presumably these correspond to two regional populations of the wild ox, namely the Near Eastern and South Asian landraces, which were sampled during the domestication process. Furthermore, the extant diversity within both *Bos taurus* and *Bos indicus* clades comprises the limited amount of diversity accrued via mutation within the historic time frame of domestication. Interestingly when ancient DNA data from six British aurochsen bone samples are added, these form a third, distinct clade, the only indication thus far of the genetic character of the western European wild ox (Figure 22.4; Bailey et al. 1996; Troy et al. 2001).

The phylogeographic patterns of diversity within the *Bos taurus* mtDNA clade require some closer examination. In analogy with Vavilov's principle, it is predicted that centers of domestic origin display the highest levels of genetic diversity. For example, if domestication of the *Bos taurus* progenitor was limited in time and space to the Near East, one might expect the greatest retention of the captured wild genetic diversity within the modern breeds of this region. Correspondingly, as the formation of the taurine populations of Europe would have presumably involved a subsampling of this Near Eastern mtDNA diversity, one should observe a reduction in diversity in populations situated away from the primary domestication center. This phenomenon is illustrated in Figure 22.5, which shows one measure of mtDNA diversity—mean number of pairwise differences or nucleotide diversity—calculated for 30 breeds sampled from six different geographic regions. A simple pattern is evident. There is higher diversity in Anatolia and the Near East, with the lowest levels of diversity in the most peripheral regional sample, western Europe, which comprises breeds from Scandinavia and Ireland.

Other useful information about diversity levels can be gained by an examination of the data in Tables 22.1 and 22.2. These summarize genetic sequence diversity in terms of the total number of haplotypes per regional group and the number of haplotypes unique to each (i.e., present in one group but not in the other groups). A total of 193 mtDNA

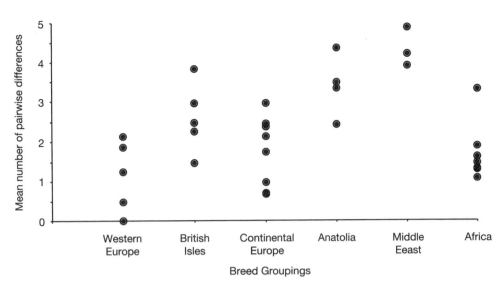

FIGURE 22.5 A graph of the mean pairwise difference for each breed plotted by breed groupings. The Middle Eastern breed group displays the highest levels of diversity, followed by the Anatolian breeds. Continental Europe and African groups show similar levels of diversity. Levels of mtDNA diversity in the British Isles are slightly higher, and levels in western Europe slightly lower than in continental Europe.

haplotypes were detected in 504 sequences from five regional samples of approximately equal size (Britain/western fringe Europe, continental Europe, the Near East, Africa, and India; see Table 22.1). In accord with the mtDNA diversity estimates, the Near Eastern sample displays the highest number of haplotypes (66) and unique mtDNA haplotypes (44) of all regional groups surveyed. It is also instructive that the African sample has a high level of unique haplotypes (34); similar to that found within Indian *Bos indicus* (34).

A similar pattern of diversity is observed for nuclear microsatellite data in which a total of 214 alleles were observed across 18 loci analyzed in a total of 806 samples from four regional groups (Table 22.2). The Near Eastern regional group exhibited both the highest total number of alleles (199) and the greatest number of unique alleles (27) of all regional groups surveyed, and it was also the only region with a substantial display of private variants.

The puzzlingly high amount of mtDNA variation unique to Africa (Table 22.1) is explained when the phylogenetic substructure of the *Bos taurus* clade is examined. When reduced median phylogenetic networks (Bandelt et al. 1995) of Near Eastern, European, and African haplotypes are constructed, they show a striking recurring pattern of sequence ancestry (Figure 22.6). Almost all sequences root back to the phylogeny through one of four main haplotypes that are topologically and numerically predominant. We have termed these central haplotypes T1, T2, T3, and T, a nomenclature that also may be applied to the four subfamilies of variants that surround these (Figure 22.6b; Troy et al. 2001). A fifth *Bos taurus* subfamily (J5 or T4) has also been described in Japanese native cattle (Figure 22.6c; Mannen et al. 1998). Figure 22.6b shows a skeleton network that estimates the phylogenetic relationship between the central haplotypes, each of which is separated by only a few substitutions in the 240-bp region analyzed.

The starlike pattern of derived variants surrounding each of these four central haplotypes suggests that they are ancestral and is consistent with a history of population expansion—an assertion that is supported by other statistical treatments of these data (Troy et al. 2001). Interestingly, when the diversity around each haplogroup is calibrated, expansion dates are consistent with the time depth of cattle domestication, suggesting that these represent phylogenetically distinct, yet closely related, mtDNA subfamilies that were incorporated into the domestic gene pool because of the domestication of wild female animals. It must be added, however, that the molecular clock may not be sufficiently accurate to exclude a Holocene time depth for these expansions.

A striking finding about these subfamilies of taurine variants is that they are geographically distributed (Figure 22.6a). In Anatolia and the Middle East, three predominate: T, T2, and T3. Typically, European samples are composed almost completely of just one of these Near Eastern subfamilies, T3. In contrast, African samples are almost exclusively composed of T1, a variant found only sporadically in the Near East. Interestingly, when T1 mtDNA is found elsewhere, there is a possibility of African influence, such as in Portugal, where T1 variants in southern breeds are undoubtedly a legacy of African migration, perhaps during the Moorish occupation of the Iberian peninsula (Cymbron et al. 1999). T1 haplotypes have also been discovered in Caribbean Creole cattle, where they may represent the genetic signatures of secondary introgressions of West African domesticates to the region (Magee et al. 2002). In symmetry, when non-T1 variants are encountered in Africa, they are most often found in samples from Egypt, Sudan, and Morocco, which would have been likely

TABLE 22.1
Intra-regional mtDNA Genetic Diversity

Regional Group	Sample Size	Total Number of Haplotypes (B. taurus / B. indicus)	Number of Unique Haplotypes (B. taurus / B. indicus)
Britain/western fringe Europe	116	43 (43 / 0)	34 (34 / 0)
Continental Europe	91	32 (32 / 0)	18 (18 / 0)
Near East	107	66 (62 / 4)	44 (43 / 1)
Africa	95	43 (43 / 0)	39 (39 / 0)
India	95	37 (3 / 34)	34 (3 / 31)
Overall sample	504	193 (158 / 35)	169 (137 / 32)

TABLE 22.2
Intra-regional Microsatellite Genetic Diversity

Regional Group	Sample Size	Total Number of Alleles	Mean Number of Alleles per Region[a]	Total Number of Unique Alleles
Europe *B. taurus*	383	161	7.28	10
Near East *B. taurus*	290	199	9.25	27
Africa *B. taurus*	76	143	7.36	1
India *B. indicus*	57	111	5.87	1
Overall sample	806	214		39

[a] These means were calculated using 1,000 resamplings of genotypes for a standardized sample size.

to have encountered introgression because of geographic proximity to Europe and the Near East.

Thus, within *Bos taurus* mtDNA subfamilies, the relationship between Europe and the Near East seems to be a derivative one, with the former possessing a subset of the diversity of the latter. However, the relationship between African haplotypic variation and that of the Near East seems to be qualitatively different, with the former possessing a legacy of a domestication-induced expansion that is absent from other regions. This suggests that the expansion took place outside the Near Eastern complex and provides some support for archaeological assertions of a separate African domestication of cattle (Wendorf and Schild 1994).

Although mtDNA sequence diversity has proved to be the most informative genomic element in the investigation of domestic cattle origins, it must be emphasized that phylogenetic inferences drawn from mtDNA data are based solely on one segregating locus and that alternative evolutionary histories may be deduced by analyzing variation at different genetic loci. In this regard, microsatellites give a more general picture of the ancestry of modern populations, as they allow the sampling of genetic variation from around the total genome. One finding from these analyses is that many populations are composites of several disparate ancestral strands.

African breeds are a case in point. Figure 22.1 shows the phylogenetic distinction of African cattle through a cluster of breeds sampled for mtDNA (Figure 22.1a) and in the distinct position of Ndama from Guinea, which have been typed for 18 microsatellites (Figure 22.1b). This sub-Saharan West African *Bos taurus* population is often regarded as the modern descendants of the founding African *Bos taurus* cattle, which have been least affected by Indian *Bos indicus* introgression to the continent (MacHugh et al. 1997). However, analysis of microsatellite data reveals that Egyptian cattle, despite showing predominantly African mtDNA variation, are more Near Eastern in genetic character over the bulk of their genomes. Recently, a wide sampling of African cattle microsatellite diversity vividly illustrated the geography of the different demographic influences that have molded the genetic variation on the African continent (Hanotte et al. 2002). These include a component of variation that is restricted to Africa and possibly represents the initial spread of taurine cattle within the continent; a major component that corresponds to the secondary influx of indicine and centers around the coast of East Africa; and a strand of Near Eastern or European influence that is most prevalent in Morocco, Egypt, and some southern African breeds. It seems that an interwoven ancestry of possibly indigenous African, Near Eastern, and South Asian ancestry is the nature of most African cattle, with the only likely exceptions being the relatively purebred African *Bos taurus* breeds of coastal West Africa.

Admixture between taurine and indicine components is also a feature of the modern cattle of the Near East. Loftus et al. (1999) and Kumar et al. (2003) show that the *Bos taurus* populations of Anatolia and the Middle East have incurred

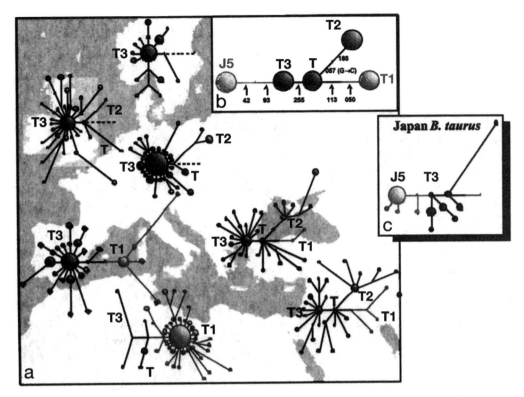

FIGURE 22.6 Skeleton and reduced media networks for *Bos taurus* cattle haplotypes. (a) Samples grouped by geographical origin for the Middle East, Anatolia, Mainland Europe, Britain and western fringe Europe, Japan, and Africa; reduced median phylogenetic networks were constructed manually. Haplotypes encountered in each region are represented by filled circles, and small empty circles denote unsampled intermediate nodes or unsampled primary haplotypes. Circle areas are proportional to haplotype frequencies, and shading patterns indicate which of the four skeleton network haplotypes they coalesce to. The major haplogroups within each regional sample are labeled. It is of interest that the major European haplogroup, T3, is encountered at reasonable frequency in the Near East but the African haplogroup T1 is encountered elsewhere at frequency only within the Iberian peninsula. (b) Network defining the relationships of the five central, primary *Bos taurus* mtDNA haplotypes—T, T1, T2, T3, and J5/T4; positions of sequence differences that distinguish these and their derivatives are indicated. These haplotypes are deduced to be ancestral to the other taurine sequences. The spatial arrangement of the skeleton network and the shading codes are preserved in the full networks beneath. (c) Network showing diversity within a sample of native Japanese cattle; a dominant feature is that of an additional haplogroup centered around J5 (alternatively denoted T4) (after Troy et al. 2001, Fig. 2).

zebu introgression to the extent of roughly 30 percent, and, with some symmetry, the *Bos indicus* breeds of northern India possess a minority component of taurine ancestry.

Archaeological Inference from Patterns of Genetic Diversity

The phylogenetic coherence within both the indicine and taurine mtDNA clades, together with the large divergence between them, clearly implies that two geographically and biologically separate landraces of *Bos primigenius* were domesticated. The obvious inference is that this supports a domestication of *Bos taurus* in the traditionally identified region of origin in the Near East and a different center for the domestication of *Bos indicus*, possibly in Baluchistan, where the site of Mehrgarh has yielded archaeological evidence pertaining to the domestication of indicine cattle, possibly as early as 7000 BC (Meadow 1993). Earlier ideas that all cattle had ancestry stretching back to a single domestic origin are clearly not consistent with the hundreds of thousands of years' divergence between the domestic cattle types. However, it should be stated that the observed biological separation of the strains that were domesticated from the eastern and western ranges of the aurochs does not necessarily prove that the domestication processes were culturally independent. Indeed, some authors have proposed that Baluchistan was greatly influenced by the technological developments from regions situated farther west (Meadow 1993, 1996; Diamond 2002).

As previously stated, the starlike phylogenies that become visible once the taurine mtDNA clade is examined in detail are indicative of past population expansions that may be dated. These dates are imprecise in archaeological terms, however, and typically give a plausible range between 5,000 and 20,000 years ago. Domestication itself is an excellent candidate process for causing these expansions, involving, as it did, the selecting of limited numbers from the wild and their promulgation to eventually arrive at the estimate of 1.3 billion head of domestic cattle that we have today. Clearly, the time estimates are consistent with domestication, but they may also be consistent with other processes, for example a postglacial expansion in the wild ox. Interestingly, preliminary estimates for expansion times within the indicine clade (Bradley and Magee unpublished data) are somewhat later than those within *Bos taurus*, and they may concur with the eastern domestication being a later, perhaps culturally derivative, process.

The aurochs ranged throughout much of Europe, and an interesting question is whether it was domesticated locally as a contribution to the farming economy, which was expanding from the Near East (Harris 1996; Bökönyi 1974). Two aspects of cattle genetic variation argue against this. First, *Bos taurus* variation is greater in Anatolia and the Middle East than in European populations. This is true of both microsatellite and mtDNA data, where analysis of each implies that European variation is a subset of that found in the Near East (Loftus et al. 1999; Troy et al. 2001). Second, the one published regional sample of *Bos primigenius*, from Britain, implies that European wild ox mtDNA variation was markedly and consistently different from that in modern cattle. Thus, the data to date do not suggest any detectable maternal contribution to the European domestic cattle gene pool from locally domesticated wild cattle (but see Götherström et al. 2004 for Y-chromosome data showing hybridization with European aurochs bulls).

In contrast, the *Bos taurus* heritage within African cattle that derives from the pre-indicine layer of ancestry in that continent does not seem to be a simple subset of that in the Near East. The genetic relationship between these two regions is qualitatively different from that between Europe and the latter and lends some support for assertions that aurochs may have been separately domesticated in North Africa and possibly at an early stage (Wendorf and Schild 1994). This assertion is strengthened by the discovery of a significant component of microsatellite variation within African cattle that seems to be indigenous in origin (Hanotte et al. 2002). However, it remains a (perhaps less plausible) possibility that the discontinuity in genetic variation between the Near East and Africa is the result of a severe demographic process, such as a foundation bottleneck, that was not encountered in the passage of cattle to Europe.

Traces of secondary migration processes are clearly visible in the genetic record of Africa. In particular, a major center for the influx of indicine genetics, which has introgressed into modern cattle herds in most regions, may be identified in coastal East Africa (Hanotte et al. 2000; Hanotte et al. 2002). This influx is interesting in that it has parallels with other exchanges of domesticates between Asia and Africa, such as chicken and camel. An entry via the Horn of Africa, rather than through the Isthmus of Suez is definitely implied from genetic data, and one factor favoring a major influence of sea-borne communication is the identification of cattle populations from the island of Madagascar as possessing the most indicine character of African breeds (Hanotte et al. 2002).

Partial zebu ancestry in the taurine breeds of the Near East is also interesting; it forms part of a gradient that stretches back to northern India. There is some evidence that indicine cattle may have been in Jordan as early as 3400 BC (Zeuner 1963; Clason 1978; Bökönyi 1997). There were extensive trading links between the civilizations of the Near East and those farther East, and it is likely that these links facilitated genetic exchange. Climate warming and increase in aridity may have provided an impetus for the directional expansion of *Bos indicus* from its initial center westward and into Africa, resulting in its extensive distribution today. It is tempting to speculate that the presence of a minor layer of taurine variants in Indian cattle may be a legacy of initial movements from the Near East, but as yet there is no time depth from genetic analysis of this ancestral influence.

Discussion and Future Research

Although the analysis of genetic variability in extant domestic cattle populations has provided some novel insights pertaining to the origins and demographic history of *Bos taurus* and *Bos indicus* cattle, several key issues cannot be answered securely via the analysis of modern genetic diversity. In several cases the analysis of DNA from fossil remains is required, and the preliminary data from these investigations, although limited, are promising.

Extensive analysis of modern genetic variation in cattle has made them the best-studied domestic species and has revealed new sources of inference about the prehistory of cattle herding. Genetic data have the power to illustrate interesting aspects of the domestication process and may also inform regarding secondary migrations. However, it is notable that the genetic analysis of extant variability has raised a number of questions that are likely to be resolved only by increasing the database of ancient DNA results. In particular, the dating of demographic events from mtDNA presently lacks the resolution offered by archaeological investigations. The direct analysis of fossil DNA will undoubtedly help by adding a temporal dimension to the geographic description of the genetic diversity observed in modern domestic populations.

Unlike the case of other large domesticates, the wild ancestor of cattle is extinct. Several British aurochsen have been sequenced, but there is a clear need for a wider description of the genetic diversity in the domestic cattle progenitor; indeed, work toward this description is continuing in the authors' and other laboratories. This may help to illustrate the natural geographic distribution of the aurochs landraces from which domestic cattle were derived, and it could also substantiate genetic continuity between wild and domestic African *Bos taurus* cattle, or perhaps in other regions, through male introgression (see Götherström 2005). Also, it is not clear whether the British aurochs sequences are typical of European variants and where the (presumably) Near Eastern aurochsen with domestic-like variation ranged. This temporal dimension will be important for definitively discriminating among Paleolithic demographic restrictions, domestication bottlenecks, and subsequent expansions as candidate processes for causing the diversity patterns in modern DNA.

Finally, the studies outlined in this chapter are focused on neutral, or noncoding, aspects of genetic variation. Thus it is assumed that selection, artificial or natural, has no major role in shaping the observed patterns. An interesting future focus is bound to be the study of the genes that have been selected through time as a result of domestication, a process that affected the important and dramatic changes distinguishing modern cattle breeds from each other in physiology and productivity and from their remote and formidable ancestors, the aurochsen of three continents.

References

Allard, M. W., M. M. Miyamoto, L. Jarecki, F. Kraus, and M. R. Tennant. 1992. DNA systematics and evolution of the artiodactyl family Bovine. *Proceedings of the National Academy of Sciences, USA* 89: 3972–3976.

Allchin, B. and R. A. Allchin. 1968. *The birth of Indian civilization: India and Pakistan before 500 BC*. Harmondsworth, UK: Penguin.

Arctander, P., P. W. Kat, R. A. Aman, and H. R. Siegismund. 1996. Extreme genetic differences among populations of *Gazella granti*, Grant's gazelle in Kenya. *Heredity* 76: 465–475.

Bandelt, H. J., P. Forster, B. C. Sykes, and M. B. Richards. 1995. Mitochondrial portraits of human populations using median networks. *Genetics* 141: 743–753.

Bökönyi, S. 1974. Development of early stock rearing in the Near East. *Nature* 264: 19–23.

Bökönyi, S. 1997. Zebus and Indian wild cattle. *Anthropozoologica* 25/26: 647–654.

Bailey, J. F., M. B. Richards, V. A. Macaulay, I. B. Colson, I. T. James, D. G. Bradley, R. E. Hedges, and B. C. Sykes. 1996. Ancient DNA suggests a recent expansion of European cattle from a diverse wild progenitor species. *Proceedings of the Royal Society of London B Biological Sciences* 263: 1467–1473.

Bradley, D. G., D. E. MacHugh, R. T. Loftus, R. S. Sow, C. H. Hoste, and E. P. Cunningham. 1994. Zebu-taurine variation in Y chromosomal DNA: A sensitive assay for genetic introgression in west African trypanotolerant cattle populations. *Animal Genetics* 25: 7–12.

Bradley, D. G. and P. Cunningham. 1999. Genetic aspects of domestication, common breeds and their origins. In *The genetics of cattle*, R. Fries and A. Ruvinsky (eds.), pp. 15–31. Wallingford, UK: CAB International.

Bradley, D. G., D. E. MacHugh, P. Cunningham, and R. T. Loftus. 1996. Mitochondrial diversity and the origins of African and European cattle. *Proceedings of the National Academy of Sciences, USA* 93: 5131–5135.

Brown, W. M., M. George, Jr., and A. C. Wilson. 1979. Rapid evolution of animal mitochondrial DNA. *Proceedings of the National Academy of Sciences, USA* 76: 1967–1971.

Clason, A. T. 1978. Late Bronze-Iron Age zebu in Jordan? *Journal of Archaeological Science* 5: 91–93.

Clutton-Brock, J. 1989. Cattle in ancient North Africa. In *The walking larder: Patterns of domestication, pastoralism and predation*, J. Clutton-Brock (ed.), pp. 201–206. London, UK.: Unwin Hyman.

Cymbron, T., R. T. Loftus, M. I. Malheiro, and D. G. Bradley. 1999. Mitochondrial sequence variation suggests an African influence in Portuguese cattle. *Proceedings of the Royal Society of London, series B* 266: 597–604.

Diamond, J. 2002. Evolution, consequences and future of plant and animal domestication. *Nature* 418: 700–707.

Edwards, C. J., G. Dolf, C. Looft, R. T. Loftus, and D. G. Bradley. 2000. Relationships between the endangered Pustertaler Sprinzen and three related European cattle breeds as analyzed with 20 microsatellite loci. *Animal Genetics* 31: 329–332.

Epstein, H. 1971. *The origin of the domestic cattle of Africa*. New York: Africana.

Epstein, H. and I. L. Mason. 1984. Cattle. In *Evolution of domesticated animals*, I. L. Mason (ed.), pp. 6–27. London: Longman.

Eyre-Walker, A. and P. Awadalla. 2001. Does human mtDNA recombine? *Journal of Molecular Evolution* 53: 430–435.

Gentry, A., J. Clutton-Brock, and C. P. Groves. 2004. The naming of wild animal species and their domestic derivatives. *Journal of Archaeological Science* 31: 645–651.

Gibbons, A. 1998. Calibrating the mitochondrial clock. *Science* 279: 28–29.

Giuffra, E., J. H. M Kijas, V. Amarger, Ö. Carlborg, J. T. Jeon, and L. Andersson. 2000. The origins of the domestic pig: Independent domestication and subsequent introgression. *Genetics* 154: 1785–1791.

Götherström, A. 2005. Cattle domestication in the Near East was followed by hybridization with aurochs bulls in Europe. *Proceedings of the Royal Society, Series B* 22: 2345–2350.

Grigson, C. 1978. The craniology and relationships of four species of *Bos*. 4. The relationship between *Bos primigenius* Boj. and *Bos taurus* L. and its implications for the phylogeny of the domestic breeds. *Journal of Archaeological Sciences* 5: 123–152.

———. 1980. The craniology and relationships of four species of *Bos*. 5. *Bos indicus* L. *Journal of Archaeological Science* 7: 3–32.

Hanotte, O., C. L. Tawah, D. G. Bradley, M. Okomo, Y. Verjee, J. Ochieng, and J. E. Rege. 2000. Geographic distribution and frequency of a taurine *Bos taurus* and an indicine *Bos indicus* Y specific allele amongst sub-Saharan African cattle breeds. *Molecular Ecology* 9: 387–396.

Hanotte, O., D. G. Bradley, J. W. Ochieng, Y. Verjee, E. W. Hill, and J. E. Rege. 2002. African pastoralism: Genetic imprints of origins and migrations. *Science* 296: 336–339.

Harris, D. R. 1996. Introduction: themes and concepts in the study of early agriculture. In *The origins and spread of agriculture and pastoralism in Eurasia*, D. R. Harris (ed.), pp. 1–9. London: University College London Press.

Ingold, T. 1996. Growing plants and raising animals: An anthropological perspective on domestication. In *The origins and spread of agriculture and pastoralism in Eurasia*, D. R. Harris (ed.), pp. 12–24. London: University College London Press.

Innan, H. and M. Nordberg. 2002. Recombination or mutational hot spots in human mtDNA? *Molecular Biology and Evolution* 19: 1122–1127.

Jorde, L. B., M. Bamshad, and A. R. Rogers. 1998. Using mitochondrial and nuclear DNA markers to reconstruct human evolution. *Bioessays* 20: 126–136.

Kruskal, J. 1964. Multidimensional scaling by optimising goodness of fit to a non-metric hypothesis. *Psychometrika* 29: 1–27.

Kumar, P., A. R. Freeman, R. T. Loftus, C. Gaillard, D. Q. Fuller, and D. G. Bradley. 2003. Admixture analysis of South Asian cattle. *Heredity* 91:43–50.

Loftus, R. T., D. E. MacHugh, D. G. Bradley, P. M. Sharp, and P. Cunningham. 1994a. Evidence for two independent domestications of cattle. *Proceedings of the National Academy of Sciences, USA* 91: 2757–2761.

Loftus, R. T., D. E. MacHugh, L. O. Ngere, D. S. Balain, A. M. Badi, D. G. Bradley, and E. P. Cunningham. 1994b. Mitochondrial genetic variation in European, African and Indian cattle populations. *Animal Genetics* 25: 265–271.

Loftus, R. T., O. Ertugrul, A. H. Harba, M. A. El-Barody, D. E. MacHugh, S. D. Park, and D. G. Bradley. 1999. A microsatellite survey of cattle from a centre of origin: The Near East. *Molecular Ecology* 8: 2015–2022.

MacHugh, D. E. and D. G. Bradley. 2001. Livestock genetic origins: Goats buck the trend. *Proceedings of the National Academy of Sciences, USA* 98: 5382–5384.

MacHugh, D. E., M. D. Shriver, R. T. Loftus, P. Cunningham, and D. G. Bradley. 1997. Microsatellite DNA variation and the evolution, domestication and phylogeography of taurine and zebu cattle (*Bos taurus* and *Bos indicus*). *Genetics* 146: 1071–1086.

Magee, D. A. 2002. Molecular genetic investigations of the diversity and origins of Old and New World cattle populations. PhD dissertation, Genetics Department, University of Dublin, Trinity College, Ireland.

Magee, D. A., C. Meghan, S. Harrison, C. S. Troy, T. Cymbron, C. Gaillard, A. Morrow, J. C. Maillard, and D. G. Bradley. 2002. A partial African ancestry for the Creole cattle populations of the Caribbean. *Journal of Heredity* 93: 429–432.

Mannen, H., S. Tsuji, R. T. Loftus, and D. G. Bradley. 1998. Mitochondrial DNA variation and evolution of Japanese black cattle (*Bos taurus*). *Genetics* 150: 1169–1175.

Manwell C. and C. M. A. Baker. 1980. Chemical classification of cattle. 2. Phylogenetic tree and specific status of the Zebu. *Animal Blood Groups and Biochemical Genetics* 11: 151–162.

Marshall, F. 1989. Rethinking the role of *Bos indicus* in sub-Saharan Africa. *Current Anthropology* 30: 235–240.

Meadow, R. H. 1993. Animal domestication in the Middle East: A revised view from the Eastern Margin. In *Harappan civilization*, 2nd edition, G. Possehl (ed.), pp. 295–320. New Delhi: Oxford and IBH.

———. 1996. The origins and spread of pastoralism in northwestern South Asia. In *The origins and spread of agriculture and pastoralism in Eurasia*, D. R. Harris (ed.), pp. 390–412. London: University College London Press.

Medjugorac, I., W. Kustermann, P. Lazar, I. Russ, and F. Pirchner. 1994. Marker-derived phylogeny of European cattle supports demic expansion of agriculture. *Animal Genetics* 25, suppl. 1: 19–27.

Naik, S. 1978. Origin and domestication of zebu cattle (*Bos indicus*). *Journal of Human Evolution* 7: 23–30.

Nei, M. 1987. *Molecular evolutionary genetics*. New York: Columbia University Press.

Nersting, L. G. and P. Arctander. 2001. Phylogeography and conservation of impala and greater kudu. *Molecular Ecology* 10: 711–719.

Payne, W. J. A. 1991. Domestication: A step forward in civilization. In *Cattle genetic resources*, 1st edition, C. G. Hickman (ed.), pp. 51–72. World Animal Science, Volume B. Amsterdam: Elsevier.

Peters, J., D. Helmer, A. von den Driesch, and S. Segui. 1999. Animal husbandry in the northern Levant. *Paléorient* 25: 27–48.

Renfrew, C. 1987. *Archaeology and language*. Harmondsworth, UK: Penguin.

Richards, M. and V. Macaulay. 2001. The mitochondrial gene tree comes of age. *American Journal of Human Genetics* 68: 1315–1320.

Rosenbaum, H. C. and A. S. Deinard. 1998. Caution before claim: An overview of microsatellite analysis in ecology and evolutionary biology. In *Molecular approaches to ecology and evolution*, R. DeSalle and B. Schierwater (eds.), pp. 87–106. Boston/Berlin: Birkhäuser.

Schmid, M., N. Saitbekova, C. Gaillard, and G. Dolf. 1999. Genetic diversity in Swiss cattle breeds. *Journal of Animal Breeding and Genetics* 116: 1–8.

Slatkin, M. 1991. Inbreeding coefficients and coalescence times. *Genetical Research* 58: 167–175.

Troy, C. S., D. E. MacHugh, J. F. Bailey, D. A. Magee, R. T. Loftus, P. Cunningham, A. T. Chamberlain, B. C. Sykes, and D. G. Bradley. 2001. Genetic evidence for Near-Eastern origins of European cattle. *Nature* 410: 1088–1091.

Vigne, J.-D. and H Buitenhuis. 1999. Lest premiers pas de la domestication animale a l'Ouest de l'Euphrate: Chypre et l'Anatolie Centrale. *Paléorient* 25: 49–62.

Waterbolk, H. T. 1968. Food production in prehistoric Europe. *Science* 162: 1093–1102.

Wendorf, F. and R. Schild, 1994. Are the Early Holocene cattle in the Eastern Saharan domestic or wild? *Evolutionary Anthropology* 3: 118–128.

Zeuner, F. E. 1963. *A history of domesticated animals*. London: Hutchison.

Zohary, D. 1996. The mode of domestication of the founder crops of Southwest Asian agriculture. In *The origins and spread of agriculture and pastoralism in Eurasia*, D. R. Harris (ed.), pp. 142–158. London: University College London Press.

CHAPTER 23

Genetic Analysis of the Origins of Domestic South American Camelids

JANE C. WHEELER, LOUNÈS CHIKHI, AND
MICHAEL W. BRUFORD

Introduction

Ancestors of the family Camelidae originated in North America during the Eocene, 40–45 MYA, with the division between Lamini and Camelini (the tribes of New and Old World camelids, respectively) dating to 11 MYA (Webb 1974; Harrison 1979). Their subsequent migration to South America and Asia occurred 3 MYA (Webb 1974), with representatives of the extant New World genera *Lama* and *Vicugna* appearing 2 MYA (Hoffstetter 1986) in South America.

Two branches of the Lamini evolved from the ancestral North American *Pliauchenia* (11–9 MYA). The first exclusively North American branch contains *Alforjas* (10–4.5 MYA) and *Camelops* (4.5–0.1 MYA), while the second includes *Hemiauchenia* (10–0.1 MYA), *Palaeolama* (2–0.1 MYA), *Lama* (2 MYA–present), and *Vicugna* (2 MYA–present), all of which are found in South America. Although a recent article suggests that *Hemiauchenia* should be classified within *Palaeolama* (Guerin and Faure 1999), it remains clear that *Lama* and *Vicugna* evolved from *Hemiauchenia*. By the end of the Pleistocene, the only surviving members of the Lamini were the South American *Lama* and *Vicugna*.

The Lamini are classified within the order Artiodactyla, suborder Tylopoda and family Camelidae. Some taxonomists have favored classification into two genera and four species: *Lama guanicoe* (guanaco), *L. pacos* (alpaca), *L. glama* (llama) and *Vicugna pacos* (vicuña) (Cabrera and Yepes 1960), while many others have favored placing the four species within the genus *Lama*. Recent genetic studies, however, have shown that the Camelidae are most likely composed of two genera, each containing two species: *L. guanicoe* (guanaco) and *L. glama* (llama), and *V. vicugna* (vicuña) and *V. pacos* (alpaca) (Kadwell et al. 2001).

Four subspecies of guanaco (*L. guanicoe guanicoe* in Patagonia, Tierra del Fuego, and Argentina south of 35° S; *L. g. huanacus* in Chile; *L. g. cacsilensis* in the high Andes of Peru, Bolivia, and northeast Chile; and *L. g. voglii* on the eastern slope of the Andes in Argentina between 21–32° S) and two of vicuña (*V. vicugna mensalis* from 9°30'–18° S and *V. v. vicugna* from 18–29° S) have been described. In the case of the guanaco, virtually nothing is known about the northernmost *L. g. cacsilensis*, whose relict populations, perhaps 3,500 animals in total, are highly endangered. Research has been carried out on the behavior and ecology of *L. g. guanicoe* and *V. v. mensalis* (Franklin 1983), and recently information on the genetic diversity of extant wild populations has become available (Palma et al. 2001; Wheeler et al. 2001). These studies indicate that the two northernmost forms, *L. g. cacsilensis* and *V. vicugna mensalis*, are the ancestors of the domestic llama and alpaca respectively.

The alpaca has variously been described as descending from the guanaco, the vicuña, and as a llama × vicuña hybrid, while the llama is thought to originate from the guanaco. These contradictory hypotheses have been developed primarily from the study of morphological and behavioral variations among living animals, while archaeozoological evidence has pointed toward domestication of the alpaca from the vicuña in the wet puna of Peru's central Andes 6,000–7,000 years ago and toward possible multiple domestications of the llama from the guanaco in the dry punas of southern Peru, Chile, and Argentina (Wheeler 1995; see Chapter 16).

In 1775, Frisch (1775) attributed the origin of the llama to the guanaco and the alpaca to the vicuña, an opinion subsequently supported by Ledger (1860), Darwin (1868), Antonius (1922), Faige (1929), Krumbiegel (1944, 1952), Steinbacher (1953), Frechkop (1955), Capurro and Silva (1960), Akimushkin (1971), and Semorile et al. (1994). Other authors have concluded that both domestic camelids descend from the guanaco, and that the vicuña was never domesticated (Thomas 1891; Peterson 1904; Hilzheimer 1913; Lönnberg 1913; Brehm 1916; Cook 1925; Weber 1928; Herre 1952, 1953, 1976, 1982; Röhrs 1957; Fallet 1961; Zeuner 1963; Herre and Thiede 1965; Herre and Röhrs 1973; Bates 1975; Pires-Ferreira 1981/82; Kleinschmidt et al. 1986; Kruska 1982; Jürgens et al. 1988; and Piccinini et al. 1990). In the 1930s, López Aranguren (1930) and Cabrera (1932) suggested that the llama and the alpaca evolved from presently extinct wild precursors, based on the discovery of 2 Myr Plio-Pleistocene *L. glama*, *L. pacos*, *L. guanicoe* and *V. vicugna* fossils in Argentina, and that the guanaco and vicuña were never domesticated. This position is no longer considered a possible alternative. Finally, Hemmer (1975, 1983, 1990) attributes llama ancestry to the guanaco, but has deduced on the basis of shared morphological and behavioral traits that the alpaca originated from hybridization between the llama and vicuña.

Conclusions about llama and alpaca ancestry have, in large part, been based upon morphological changes produced by the domestication process. During the 1950s, Herre and Röhrs (Herre 1952, 1953, 1976; Röhrs 1957; Herre and Röhrs 1973) examined alterations in the mesotympanal area of the skull related to a decrease in llama and alpaca hearing acuity and reported an overall reduction in cranial capacity of both domestic species relative to the guanaco. In contrast, they found the vicuña cranium to be the smallest of all living Lamini, and, based on the premise that domestic animals are smaller than their ancestors, concluded that this species was never brought under human control. Herre and Röhrs consider the llama and alpaca to be "races of the same domestic species bred for different purposes" (Herre 1976: 26). Research on the relationship of brain size relative to body size by Kruska (1982) also found the vicuña to be smaller than the alpaca and the llama, which in turn were smaller than the guanaco, suggesting that the latter is the only ancestral form. Nonetheless, papers by Jerison (1971) and Hemmer (1990) report the ratio of alpaca brain size to body size to be smaller than in the vicuña, permitting a different conclusion about origins of the domestic forms. These contradictory data on size reduction are almost certainly a product of sampling as neither subspecific variation in the wild forms nor the possibility of hybridization between the domestic animals were considered in any of the studies (see Chapter 16).

Based on the study of pelage characteristics (skin thickness, follicle structure, secondary/primary ratio, fiber length and diameter, coloration) in living camelids, Fallet (1961) found the llama to be an intermediate evolutionary stage between the wild guanaco and the specialized, fiber-producing alpaca. Fallett concluded that the absence of transitional characteristics between vicuña and alpaca fleeces eliminates the former from consideration as an ancestral form. This deduction is, in part, based on the assumption that llamas have been selected exclusively for use as pack animals, whereas alpacas have been bred for fiber production. Nonetheless, new data on preconquest llama and alpaca breeds in Peru have revealed the prior existence of a fine-fiber-producing llama, as well as an extra-fine-fiber alpaca that is transitional between the vicuña and a second, prehispanic, fine-fiber alpaca breed (Wheeler, Russel, and Redden 1995).

Research on camelid behavior has produced contradictory hypotheses concerning llama and alpaca origins. Krumbiegel (1944, 1952) and Steinbacher (1953) argue that the alpaca is the domestic vicuña based on unique, shared behavioral traits that are said to differ from those observed in the guanaco and llama. Hemmer, on the other hand, concludes that although some alpaca behavior patterns match those of the vicuña, others are intermediate between those of vicuña and guanaco, suggesting that "the alpaca is a mixture of both lines, [produced] by crossbreeding of captured vicunas with the only initially available domestic animal, the llama" (1990: 63). It has also been suggested that the vicuña was never domesticated because it is more territorial than the guanaco (Franklin 1974). Nonetheless, this assumption is open to question because it is based on a study of guanacos located at the southernmost extreme of their range where seasonal migration in response to severe climatic changes is essential for survival (Franklin 1982, 1983). Further to the north, where vicuña and guanaco ranges overlap and where llama and alpaca domestication occurred (Wheeler 1995), a more benign climate and a constant food supply permit the characteristic sedentary social organization of the vicuña (Franklin 1982, 1983). Although data concerning behavior of the guanaco in this region are lacking, it is possible that the limited sedentary territorial organization observed in some Patagonian groups plays a more important role in these less-extreme climatic conditions.

Analysis of hemoglobin amino acid sequences in vicuña, alpaca, llama, and guanaco from Hannover Zoo, Germany, led Kleinschmidt et al. (1986), Jürgens et al. (1988), and Piccinini et al. (1990) to the conclusion that the vicuña was never domesticated. However, earlier research on blood and muscle samples with descending bidimensional chromatography (circular and descending) for hydrolyzed muscle samples and horizontal electrophoresis for blood serum samples from llama, alpaca, vicuña, guanaco, and alpaca × vicuña hybrids at Santiago Zoo (Cappuro and Silva 1960) indicated a llama-guanaco and alpaca-vicuña subdivision, as have more recent data from ribosomal genes (Semorile et al. 1994). Other researchers using immunological, electrophoretic analysis and protein sequencing have found it impossible to draw conclusions about llama and alpaca ancestry (Miller et al. 1985; Penedo et al. 1988). Cytogenetic studies (Capanna and Civitelli 1965; Taylor et al. 1968; Larramendy et al. 1984; Gentz and Yates 1986) indicate that all four species of the South American Camelidae (SAC) have the same $2n = 74$ karyotype. Analysis of satellite DNA, mitochondrial cytochrome b gene, and nuclear microsatellites in large sample sets has documented extensive hybridization among the domestic SAC (Vidal Rioja et al. 1987; Saluda-Gorgul et al. 1990; Stanley et al. 1994; Kadwell et al. 2001). Recent studies of the fiber from mummified ninth- and tenth-century llamas and alpacas provide additional evidence that postconquest hybridization has modified the genetic makeup of living populations (Wheeler et al. 1995), a fact that may well explain the diversity of conclusions about their ancestry.

In an attempt to solve the question once and for all, the first South American camelid mitochondrial DNA sequences were analyzed by Stanley, Kadwell, and Wheeler in 1994. In this study, sequence data from a short (158 bp) but highly informative region of the cytochrome b gene were used to examine the phylogenetic affiliations of alpaca and llama. Unfortunately, although the results confirmed that *Lama* and *Vicugna* are valid genera, which separated 3–2 MYA, the origin of the domestic forms remained unclear since there was evidence for considerable bidirectional hybridization.

From the mid–late 1990s, nuclear microsatellite DNA markers began to be isolated from a number of South American camelids (Lang et al. 1996; McPartlan et al. 1998; Penedo et al. 1998; Obreque et al. 1998; Obreque et al. 1999; Penedo et al. 1999a, 1999b; Sarno et al. 2000), and now in excess of 70 such markers are available. Because the strict maternal inheritance of mitochondrial DNA in most mammals restricts its use in studies of hybridization, especially in domestic livestock (e.g., MacHugh et al. 1997), the most recent work on SAC domestication has also included analysis of microsatellites. Recently, Kadwell et al. (2001) used a large sample set collected throughout the geographic range of the

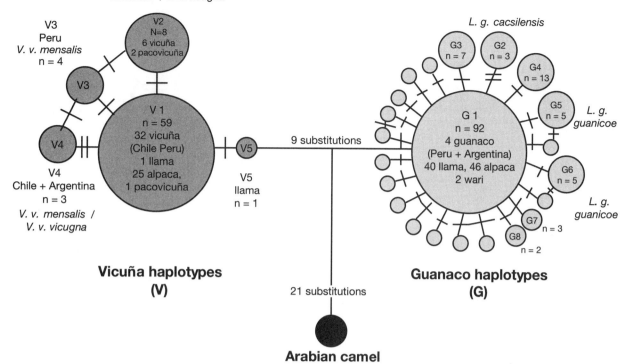

FIGURE 23.1 Minimum spanning network representing the relationships among cytochrome *b* mitochondrial haplotypes. Halpotypes are shown as circular nodes, and the number of substitutions connecting each sequence is represented by numbers or dashes on each connecting line. The relative frequency of each haplotype is represented by the area of each circle. Dark gray circles indicate vicuña haplotypes, and light gray circles represent guanaco haplotypes. Wild samples are specifically referred to where present. Phylogenetic analysis (maximum parsimony and neighbor joining based on uncorrected p, JC, and K2P sequence distances) recovered an equivalent pattern to the network, and high (>90%) bootstrap support was always found for the major split between V and G haplotypes (not shown).

four species (771 samples) and analyzed it for four microsatellites. Cytochrome *b* sequences were also analyzed from a subset comprising 211 samples. The results of this study with some subsequent statistical analysis of the results and their implications for camelid evolution form the basis of the material in the remainder of this chapter.

The Present Study

As far as possible, sample collection sites spanned the geographic range of both wild species and included alpaca samples (including "suri" and "huacaya" fleece types) from Peru, Argentina, Chile, and Bolivia (n ≤ 141); llama samples (a range of morphological types) from the same countries (n ≤ 60); guanaco *(L. g. guanicoe* and *L. g. cacsiliensis)* from Peru and Argentina (n ≤ 122); and vicuña *(V. v. vicugna* and *V. v. mensalis)* from Peru, Argentina, and Chile (n ≤ 440). Samples were taken only from those individuals whose phenotype conformed to accepted morphological criteria for domestic forms.

The phylogenetic relationships of the llama and alpaca were first established by sequencing the same region of the cytochrome *b* gene as in Stanley et al. (1994) (deposited in Genbank accessions U06425–30). Two hundred and eleven individuals were sequenced (21 guanaco, 42 vicuña, 54 llama, 84 alpaca, and 10 hybrids including alpaca/vicuña and llama/alpaca crosses). DNA was extracted from blood or skin using standard Proteinase K digestion followed by organic extraction using phenol and phenol/chloroform, and total DNA was precipitated in 100% ethanol (Stanley et al. 1994; Bruford et al. 1998). DNA samples were stored in TE buffer (10 mM Tris-HCl, 1 mM EDTA, pH 8.0). The cytochrome *b* primers L14724 and H14900 were used for PCR, and amplifications were carried as in Stanley et al. (1994). PCR products were purified and DNA sequencing was carried out as in Stanley et al. 1994. Sequences were aligned and the unique sequences (from here on called "haplotypes") were also deposited in Genbank under accession numbers AF373809–373833. Among-haplotype divergence and haplotype frequencies were calculated for the guanaco, vicuña, llama, and alpaca samples (Stanley et al. 1994), and a minimum spanning network (Kruskal 1965; Bandelt et al. 1999) was generated using the program MINSPNET (Excoffier 1993; Figure 23.1). The distribution patterns of domestic SAC haplotypes were then compared with those of the wild SAC sample.

The maternal inheritance of mitochondrial DNA means that hybridization studies will only inform us on female introgression, and this can be very misleading, especially in domestic livestock (e.g., MacHugh et al. 1997). Here, we also

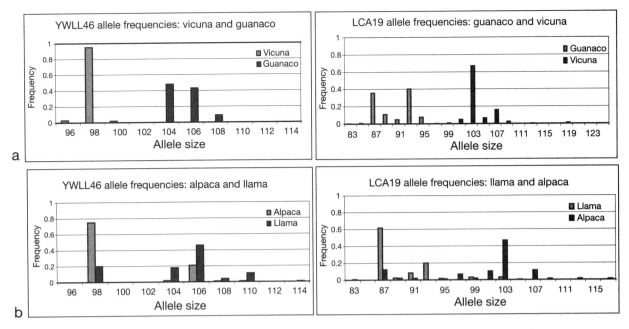

FIGURE 23.2 Allele size and frequency histograms for LCA 19 and YWLL 46, respectively: (a) vicuña and guanaco; (b) llama and alpaca. The histograms represent the frequencies of each allele for both loci in the four SAC "species." The domestic species (alpaca and llama) are plotted together, as are the wild species (vicuña and guanaco). Distributions were generated using guanaco n = 104 [LCA 19] and 177 [YWLL 46]; llama, n = 56 [LCA 19] and 58 [YWLL 46]; alpaca, n = 80 [LCA19] and 82 [YWLL 46]; vicuña, n = 227 [LCA19] and 231 [YWLL 46].

applied nuclear DNA markers that are biparentally inherited and thus equally represent female and male lineages. Four microsatellite loci (YWLL 38, YWLL 43, YWLL 46, and LCA 19; Lang et al. 1996; Penedo et al. 1998) were typed for 669–771 individuals, including the 211 individuals sequenced above (Kadwell et al. 2001; Figure 23.2). Three measures of genetic distances were used. First, an allele-sharing distance was estimated as 1 p(s) (where p(s) is the proportion of shared alleles between two individuals). This measure is very useful as it allows calculation of genetic distances between individuals when a number of loci are available. The second measure of genetic distance used was Reynolds' distance (Reynolds et al. 1983), a measure commonly used in livestock analysis where genetic drift has a major impact on allele frequencies. The third measure used in this study, (delta mu)^2 (Goldstein et al. 1995), estimates population differences using allele size differences under a stepwise mutation model. This distance was developed especially to analyze microsatellite data, as they are thought to evolve under a similar mutational model where the gain, or loss, of repeated DNA units is commonly observed. All genetic distances were estimated using the program MICROSAT v1.5d (Eric Minch, Stanford University 1999).

A form of ordination, known as factorial correspondence analysis, was then performed on allele frequency data. Here, the genetic diversity among populations (in this case vicuña, guanaco, llama, and alpaca) is expressed as factors that explain the inertia of the set of points (representing an individual) in a multidimensional space defined by the presence and absence of alleles within samples. In Figure 23.3, we show the dimensions that explain the highest proportion of the inertia (Benzécri 1973). This ordination can be thought of as analogous to displaying the two principle components in a Principle Components Analysis (PCA), although the approach is slightly different. The relationships between populations can be judged by examining how individuals from the sample cluster in two, three, or more dimensions. Because of the large number of alleles, we commonly find that the axis can very clearly identify clusters while providing relatively low proportions of the total inertia. Correspondence analysis is used as an exploratory tool. The analysis was performed using the Genetix software (Belkhir 1999).

Combined Analysis

To further assess introgression in llama and alpaca populations, two approaches were used. First, we applied methods that allow us to estimate admixture between two so-called parental populations in a hybrid or admixed population. This means that, in the three cases, we assumed for simplicity that llama and alpaca were the result of an admixture event between vicuña and guanaco some time in the past and that gene flow was limited afterwards. To estimate the proportion of vicuña (or guanaco) genes within llama and alpaca, we used three methods. The first two estimators, mC and mY, were developed by Chakraborty et al. 1992 (based on a previous method developed by Long 1991), Bertorelle (1998), and Bertorelle and Excoffier (1998). The first estimator uses only information on allele frequencies while the second also uses molecular information such as the number of substitutions for mtDNA or the allele size differences for microsatellites. These two estimators are implemented in the

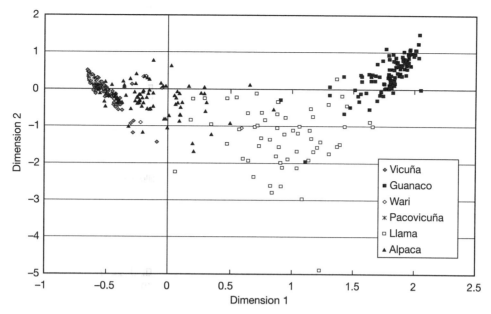

FIGURE 23.3 Two-dimensional factorial correspondence plot for allele frequencies at four microsatellite loci in all SACs. Almost half of the explained correspondence (15%) is found in factors 1 and 2, represented on the horizontal and vertical axes respectively. Clusters in two-dimensional FCA plots represent samples with similar allele frequencies. As expected, vicuña and guanaco clusters are far apart in the plot and form tight groupings. In contrast, alpaca (which group most closely to vicuña) are more disperse and llama are the most dispersed group (and are found in the same half of factor 1 in the plot as guanaco).

program ADMIX1_0 (Bertorelle 1998). The third method allows us to estimate (and account for) both drift and admixture (Chikhi et al. 2001) and produces posterior distributions for the parameters of interest, namely, p1, the contribution of one parental population (it was arbitrarily chosen to be vicuña) and three parameters measuring drift in the two parental and the hybrid population. For details, see Chikhi et al. (2001) and Langella et al (2001) for descriptions of the corresponding software. All three methods are model-dependent and assume that there has been a major admixture event that was relatively limited in time. There is good reason to hypothesize that the admixture between llama and alpaca occurred primarily on two occasions—at the time of the Spanish conquest and during the past 25 years. These methods are therefore expected to work best when the model is a reasonable approximation of reality. As should be clear from previous sections, the admixture processes that have taken place in llama and alpaca are more complex than assumed by any of the three methods. However, previous work has shown that even if admixture is more complex, such methods are expected to estimate overall levels of admixture, even when they are not as discrete as assumed by the models (see review by Chakraborty 1986).

The second approach was made possible by the fact that for mtDNA and for two of the four microsatellites, differences between vicuña and guanaco were extremely clear. As we were able to identify "typical" guanaco and vicuña alleles, concordance between mitochondrial and microsatellite data in terms of introgression patterns could be assessed in more detail. We thus re-analyzed the 211 individuals typed for both mtDNA and microsatellites. Genotypes were coded "V" (vicuña) or "G" (guanaco) for mtDNA and "V," "G," or "H" (hybrid) for YWLL 46 and LCA 19 depending on their allele sizes with reference to the guanaco and vicuña ranges, and we examined the data for each locus separately and combined. This type of approach is not always possible as the "typical" status of an allele, in practice, is impossible to state with certainty (drift might have eliminated an allele from one population, making it look typical of another one). This type of approach, therefore, cannot be used to make quantitative statements but can be powerful as a qualitative or exploratory approach (see Goldstein and Chikhi 2002).

Results

Mitochondrial DNA

Twenty-six unique haplotypes found in the 211 SACs were analyzed. Uncorrected distances within SACs ranged between 0.006 (one substitution) and 0.089 (14 substitutions). The minimum spanning network (Figure 23.1) revealed two major groups, which represent the same reciprocally monophyletic clades found previously by Stanley et al. (1994). The first group contained all vicuña ("V") and the second contained all guanaco ("G"). The branch leading to the Arabian camel exhibited 21 substitutions. The domestic camelids were found within both groups, but 81% (including 73% of alpaca) were found within the "G" (guanaco) group. A minority (19%) comprising alpaca, pacovicuña, and just two llamas were found within the "V" group. These results add support to the findings of Stanley et al. (1994), who found no evidence

TABLE 23.1
Three Locus Genotypes[a] for Samples Where All Three Types of Data Are Available

	Vicuña (n = 42)	Alpaca (n = 84)	Guanaco (n = 21)	Llama (n = 54)	Wari (n = 7)	Alpaca/Vicuña (n = 3)
GGG	—	—	21	32	—	—
GGH	—	—	—	15	1	—
GGV	—	1	—	—	—	—
GHH/GHX/GXH	—	2/1/1	—	2	1	—
GXV	—	1	—	—	—	—
GVH/GHV	—	18	—	2	1	—
GVG	—	2	—	1	—	—
GVV/GXV	—	34/1	—	—	4	—
VVV	42	17	—	—	—	3
VVH/VHV	—	3	—	—	—	—
VGG	—	—	—	2	—	—
VGV	—	1	—	—	—	—
VHH	—	2	—	—	—	—

[a] Loci are ordered mtDNA, LCA 19, and YWLL 46; for example, GVH indicates guanaco mtDNA, vicuña genotype at LCA 19, and a hybrid genotype at YWLL 46 (X signifies that the sample could not be typed).

for consistent segregation of mtDNA alleles with taxa defined in the analysis. Wild vicuña and guanaco mtDNA were reciprocally monophyletic with 5.8–8.9% sequence divergence being found between the two lineages, recapitulating the suggestion in Stanley et al. (1994) that these species diverged from a common ancestor 2–3 MYA. Furthermore, Stanley's finding that nearly all modern llamas possess a "guanaco" haplotype was also supported in this analysis (Table 23.1), where all except 2 llamas from a sample of 54 individuals possessed guanaco mtDNA. However, the much-expanded alpaca data (n = 84) revealed a different picture, with only 27% of individuals possessing vicuña mtDNA (Table 23.1), in contrast to the 50% described previously.

Microsatellites

Unusually in such studies, the microsatellite markers feature a large number of alleles exclusive to either the vicuña or the guanaco. These alleles, which range in frequency between 33% and 100% at different loci, also occupy predominantly different allele size ranges. The graphs in Figure 23.2a show allele frequency histograms for the loci YWLL 46 and LCA 19 for wild vicuña and guanaco: the allele sizes do not overlap. Such loci provide powerful tools for the discrimination of ancestral genomes in modern domestic stock. The graphs in Figure 23.2b show equivalent histograms for the llama and alpaca, which displayed similar patterns. However, the patterns of genetic similarity were in direct contrast to those revealed by mtDNA. Strong similarities between the allele size distributions of vicuña and alpaca and between guanaco and llama were observed. For YWLL 46, the 98-bp allele had a frequency of 0.95 in the vicuña and 0.75 in alpaca, while the 104- and 106-bp alleles had a combined frequency of 0.91 in the guanaco and 0.64 in llama. Analysis of the four microsatellites showed that the genetic distances between vicuña and alpaca and between guanaco and llama (Table 23.2) were almost always much lower than those between vicuña and guanaco, vicuña and llama, or guanaco and alpaca. Distances between alpaca and llama were mostly intermediate.

However, a second feature of microsatellite frequencies was the presence, at a low frequency, of "vicuña" alleles in the llama sample and of "guanaco" alleles in the alpaca sample, suggesting bidirectional introgression in both domestic forms. Notwithstanding, a striking pattern emerged from the factorial correspondence analysis (Figure 23.3), where guanaco and vicuña formed two tightly clustered and distinct groups. Additionally, alpaca formed a cohesive group, clustering strongly with the vicuña. In contrast, the llamas and hybrids formed a much more diffuse group. The llama samples, although tending to cluster with guanaco on axis 1, were more intermediate with respect to the wild species when compared with the alpaca sample and were also the most genetically diffuse group on axis 2. The most likely explanation for the separation between llama and guanaco in our sample is that the guanaco samples are from the austral form, *L.g. guanicoe*, and not the highland *L.g. cacsilensis*. The two samples of *L.g. cacsilensis*, the most likely ancestral subspecies of the llama, fall in the middle of the llama samples.

Combined Data and Admixture Analysis

The admixture proportions were nonconcordant between the mitochondrial and microsatellite analyses in the alpaca, where the estimated microsatellite proportion of vicuña genome was much higher than the mtDNA estimate (0.310 mtDNA; 0.903 ± 0.108 mC and 0.823 mY for microsatellites). In the llama, although both estimates were low, the microsatellite admixture proportions were an order of magnitude higher. It is nonetheless evident using all markers and

TABLE 23.2
Pairwise Genetic Distances between the Four SACs[a]

	Vicuña	Guanaco	Alpaca	Llama
(delta mu)²				
Vicuña	—			
Guanaco	28.928	—		
Alpaca	1.089	19.781	—	
Llama	17.162	10.784	9.892	—
Reynolds' distance				
Vicuña	—			
Guanaco	0.729	—		
Alpaca	0.173	0.433	—	
Llama	0.627	0.174	0.267	—
1 p(s)				
Vicuña	—			
Guanaco	0.963	—		
Alpaca	0.337	0.841	—	
Llama	0.825	0.522	0.616	—

[a] (delta mu)² is the genetic distance based on mean-squared difference in allele size; Reynolds' distance is a genetic distance based on the difference in allele frequencies, assuming an Infinite Alleles Model; 1 p(s) is a proportion of shared alleles.

estimates that the proportion of vicuña DNA is much lower in llama than in alpaca.

In order to investigate the reason for these results, we applied a computer-intensive method that we had developed (Chikhi et al. 2001). Apparent differences of admixture estimates between loci can be the result of both admixture and drift (or even selection). As simulations have shown, if drift has been important, for instance, different loci may appear to indicate very different admixture levels, even though they have been submitted to the same process (Chikhi et al. 2001). This is crucial as the two previous methods do not allow a separation of the effects of admixture and drift. Figure 23.4 shows the results of the admixture analysis for the four microsatellite loci. For comparison, we have also plotted the intervals as vertical bars (estimate ± one SD) obtained with mC (admixture estimator based on allele frequencies in the parental and hybrid populations, shown as solid lines) and mY (admixture estimator based on allele size differences in the parental and hybrid populations, as dashed lines). This figure clearly shows that (1) the four loci exhibit much less variability among themselves in the alpaca than they do in the llama, suggesting a very strong and consistent admixture signal in the alpaca in contrast to the llama; (2) the alpaca and llama distributions appear very different, especially highlighting the limited introgression of vicuña into llama; and (3) the same pattern is observed for the three different estimators, even though they can be rather different in some cases (in particular for locus YWLL 43 in llama). Note also that both mC and mY can give estimates of admixture that are outside the expected [0,1] range (twice in llama for mY and once for mC in alpaca). When all four loci are used together, the difference in admixture levels between alpaca (Figure 23.5b) and llama (Figure 23.5a) becomes even more obvious, with the posterior distribution becoming much thinner (and therefore more precise) and with the modal (i.e., the most probable) value for the vicuña contribution being 84.1% for alpaca and 14.4% for llama.

The combined three-locus analysis of the 211 individuals produced striking results (Table 23.1). Of the 54 llamas, 96% possessed a "G" mtDNA haplotype, with 90% and 61% possessing a pure "G" genotype for LCA19 and YWLL 46, respectively. Of the 84 alpacas, only 27% possessed a "V" mtDNA haplotype, while 70% and 79% possessed pure "V" LCA19 and YWLL 46 genotype respectively. Of the llamas tested in this study, 60% exhibited a "GGG" type, but only 20% of alpacas exhibited a "VVV" type. Extensive nuclear introgression was detected in the llama, with 37% showing one or more "vicuña" alleles at LCA19 and/or YWLL46. In contrast, much of the introgression in the alpaca was mitochondrial, with 40% of samples showing a "GVV" type.

Discussion and Future Research

In isolation, the finding that a large proportion of modern-day alpacas possess guanaco mtDNA is in accordance with hypotheses that alpacas, in common with llamas, descend from the guanaco. However, as in Stanley et al. (1994), the presence of substantial numbers of alpaca possessing vicuña mtDNA also raises the possibility that the alpaca is of mixed origin or has undergone substantial hybridization. The limitations of mtDNA in the context of gene flow and evolution in livestock are obvious since historical and modern-day agricultural practice has often used desirable males to sire large numbers of females.

The microsatellites provided a stark contrast to the mtDNA, and the existence of two loci with nonoverlapping allele size ranges in the wild ancestors allowed us an unusual opportunity to compare patterns of divergence in relatively large numbers of domestic animals. Inspection of allele frequency distributions, genetic distances, and factorial correspondence all revealed a striking similarity between the alpaca and the vicuña. Each genetic distance estimate was lowest for the alpaca-vicuña comparison, and factorial correspondence showed that alpaca and vicuña overlap almost completely. These data point toward a very close genetic affinity between alpaca and vicuña, a finding in direct conflict with the mtDNA data.

The microsatellites also supported a close relationship between llama and guanaco. Of the genetic distance estimates, both Reynolds's and the allele-sharing distances were second lowest for the guanaco-llama comparison, and the Reynolds's distance estimate was almost identical to that between alpaca and vicuña. Other data, however, were more equivocal, with factorial correspondence revealing a dispersed pattern for the llama intermediate between vicuña

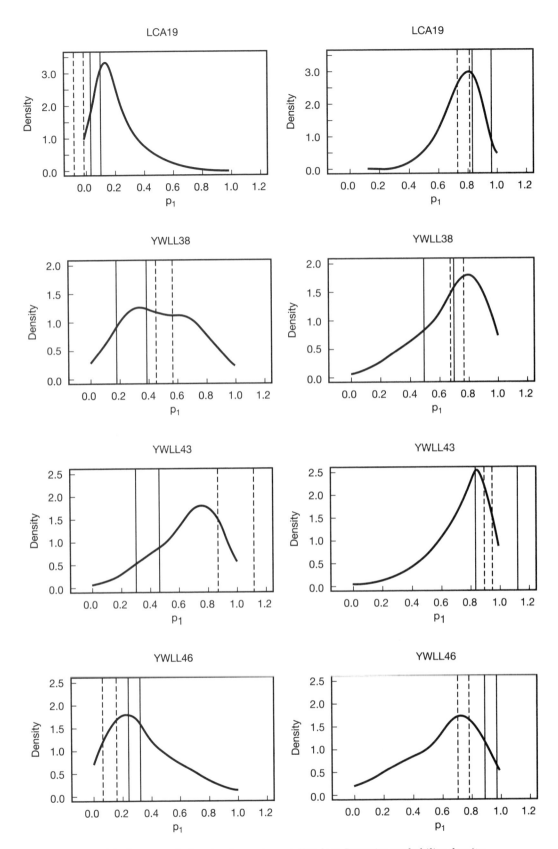

FIGURE 23.4 Admixture analysis using four microsatellite loci. Posterior probability density distributions for vicuña contribution in llama (p_1, left-hand graphs) and vicuña contribution in alpaca (p_1, right-hand graphs) are given for each locus. For comparison, mC and mY estimates (± one SD) are given as solid and dashed lines, respectively.

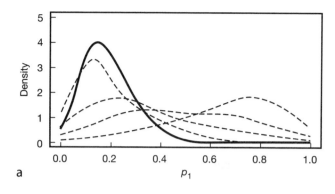

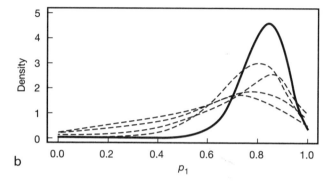

FIGURE 23.5 Posterior density distributions for admixture contributions (p_1) calculated from a combined analysis of four loci (solid line). For comparison, dashed lines show distributions for the individual loci. (a) Vicuña contribution in llama; (b) vicuña contribution in alpaca.

and guanaco and with (delta mu)² distances slightly lower for the llama-alpaca comparison than for llama-guanaco. Although none of the above is indicative of a close relationship between llama and vicuña, they suggest nuclear geneflow between llama and vicuña, or, more likely, between llama and alpaca. Another possibility is that the ancestral llama was extremely genetically diverse. The guanaco and llama have much greater geographic ranges than do the vicuña and alpaca, which may have led to greater intraspecific differentiation historically, reflected in the greater diversity in nuclear and also mitochondrial DNA (21 guanaco haplotypes as opposed to 5 in vicuña). In fact, the most likely explanation lies in the geographic distribution of our samples. It was very difficult to obtain samples of *L.g. cacsilensis*, the northern, high-altitude guanaco subspecies, and the few samples available fall in with the llama distribution. Archaeozoological data suggest that the llama could have derived from this subspecies, and recent mtDNA data published by Palma et al. (2001) indicate an ancestral relationship between *L.g. cacsilensis* and *L. glama*. Additionally, the archaeozoological data point to the possibility of at least three independent domestication events, one each in Peru, Chile, and Argentina (Wheeler 1995; see also Mengoni Goñalons and Yacobaccio, Chapter 16, this volume).

The results presented here show significant differences between mitochondrial and microsatellite levels of introgression. These differences can occur for a number of reasons.

The most obvious is that mtDNA is transmitted by females: the results presented here are simply the reflection of differential contributions from males and females of vicuña and guanaco origin. This seems to be a reasonable explanation, as we have noted above, given practices used in domesticated camelids. It may be worth noting that this hypothesis could be tested if data were available on the Y chromosome, as we would then expect a higher vicuña contribution in both llama and alpaca than observed here. Another possibility is suggested by our previous work on the estimation of admixture on simulated data sets (Kadwell et al. 2001). We found that loci simulated under the same conditions of drift and admixture could generate different posterior distributions, pointing sometimes to very different point estimates. This had led us to suggest that point estimates should be used with care, and the whole distribution should always be checked. Here, we found that the four loci generated very similar posterior distributions for alpaca, whereas for llama significant differences could be observed. For instance, the posterior distribution is rather flat for YWLL 38, whereas for locus YWLL43 the modal value is 0.76 (i.e., 76% of vicuña genes instead of the 14.4% figure obtained for the four loci together). This pattern could be the result of a much larger level of genetic drift in llama as compared to alpaca, at least in the samples considered here. As our method allows us to estimate drift, we checked what the estimated drift was for alpaca and for llama. We do find, indeed, that drift appears larger for llama than for alpaca with modal values being 0.07 vs. 0.041 (measured in units of generations/population size), respectively. One has to be cautious here, as we have already stated that the admixture methods used assume very simple models, which are unlikely to be correct. For instance, the vicuña contribution in llama might actually be the result of introgression with alpaca rather than with vicuña. The effect of drift, which we have found, is the compound effect of drift and of sampling. We wish to note that the increased drift observed in llama could be the result of different factors, which include sampling, and could be investigated in the future.

The suggestion of substantial mitochondrial introgression in the alpaca and nuclear introgression in the llama was reinforced when admixture was measured for both markers. The low estimated admixture proportion of vicuña mtDNA present in alpaca (0.31) was in contrast with the high proportion estimated for the microsatellites (0.82–0.90). Further, the extremely low admixture proportion of vicuña mtDNA in the llama (0.02) contrasted equally strongly with microsatellite estimates (0.22–0.39), which also suggest substantial nuclear admixture in the llama.

Inspection of the three-locus genotypes confirmed many of the above findings. Only 27% of alpacas were mitochondrially "vicuña," although 40% of alpaca possessed vicuña microsatellite alleles with guanaco mtDNA haplotypes. Such a pattern suggests that introgression of guanaco (or more likely llama) mtDNA may have occurred recurrently within alpaca populations but may have been accompanied more recently by a reversion to line or stock breeding within local alpaca populations.

The lack of written records, both preconquest documents and present-day breed registries, in the Andean region means that any such inference is speculative. Table 23.1 suggests that mitochondrial introgression has occurred much less frequently in the llama. However, although nuclear introgression is similar in alpaca and llama, in the expanded microsatellite dataset it seems to have occurred at a higher level in llamas (two to three times higher than in alpaca from the admixture analysis), which may partly account for the more dispersed factorial correspondence pattern, but this warrants further investigation.

The implications of these data are potentially important for the way in which these genetic resources are managed in the future. In our sample, only 35% of domestic animals have not undergone any detectable hybridization. In particular, there are very large numbers of detectable hybrids in the alpaca (80%)—accentuated when using mitochondrial DNA. Forty percent of llama show detectable signs of hybridization, with mitochondrial introgression virtually absent.

During the last 20–25 years, large-scale hybridization between llamas and alpacas has been carried out in the Andes (Bustinza 1989). Specifically, male alpacas have been bred to female llamas to increase the population of animals producing higher-priced "alpaca" fiber, and male llamas have been bred to female alpacas to obtain greater fleece weights and, thus, increased income. With sale price traditionally determined by weight, and no consideration given to fineness, the quality of alpaca fiber has decreased markedly over the past 25 years. Indigenous Quechua- and Aymara-speaking herders subdivide the hybrids into llamawari or waritu (llama-like) and pacowari or wayki (alpaca-like) respectively, depending upon physical appearance (Flores Ochoa 1977; Dransart 1991). The F_1 offspring are fertile, tend to be intermediate in size, and can be back-crossed to either parental type. Further, recent intensive selection for white fleece in modern alpaca may also have involved bidirectional hybridization. A combination of these practices and our results could explain the taxonomic confusion surrounding the domestic forms in the recent past, as it is likely that many specimens used in previous taxonomic studies were hybrids.

Given the extreme hybridization in present-day alpacas, DNA analysis has been critical in resolving the origin of this domestic form. Since our results suggest the vicuña as the ancestor of the alpaca, we propose that the classification of the alpaca should be changed from *Lama pacos* L. to *Vicugna pacos* L. The degeneration of quality and value in fleece of present-day alpacas and llamas has therefore probably been the result of extensive hybridization, probably beginning with the conquest and continuing to the present day. While it was believed that these crosses were between different forms of a single domestic animal descended from the guanaco, there was little concern about the economic impact of such introgression. However, given that the alpaca is likely to be descended from the vicuña, the negative impact on fleece quality of such crosses is now evident. The use of DNA analysis to identify and eliminate hybrid animals from the breeding pool is essential, since the antiquity of the ongoing hybridization process makes it impossible to accurately identify all hybrids on the basis of phenotypic characteristics. Additionally, the knowledge that the alpaca descends from the vicuña opens new routes for the improvement of alpaca fiber production, not only through the identification of hybrids and their elimination from purebred elite herds, but through the back crossing of purebred alpacas to their vicuña ancestor in order to possibly improve fiber fineness.

Although 90% of the alpaca fiber produced in Peru today has a diameter greater than 25 µm and fetches low prices on the world market ($3–$30/kg, 1980–1995), preconquest animals produced fiber of 17–22 µm (Wheeler et al. 1995), similar to cashmere (15–17 µm: $60–$120/kg, 1980–1995). It is possible, therefore, that identification of the remaining pure alpacas may aid in recovery of the fine-fiber characteristics of preconquest animals. In 2002, CONOPA, Coordinadora de Investigación y Desarrollo de Camélidos Sudamericanos, began a survey of alpaca populations in the central Peruvian Andes, designed to identify genetically pure alpacas and to determine the relationship between fiber fineness and purity. In the future, a core herd will be established and accelerated reproductive technology will be applied to rapidly increase selected purebred animals in order to ensure survival of the species and promote repopulation programs in the Andes.

The knowledge that the alpaca is the domestic vicuña also necessitates a reevaluation of vicuña conservation policy. Although the vicuña has been listed as endangered under the Convention on International Trade in Endangered Species of Wild Fauna and Flora CITES (Appendix I) since its inception in 1975, all Peruvian vicuñas, and large segments of the Chilean, Argentine, and Bolivian populations, have been reclassified as threatened (Appendix II), permitting controlled commercialization of live shorn fiber. With unprocessed fiber currently valued at approximately $500/kg, vicuña fleece is the most expensive natural fiber in the world and represents an important potential source of income for the extremely poor rural populations on whose lands the animals live. To date, Peru's rational use policy has produced an important increase in vicuña numbers, but demands for greater control over the species through construction of fences, intensive rearing, and selection are growing. Judging by the impact of such measures on the alpaca, such interventions in the long run will lead to a deterioration of fiber quality and fineness (which, at 12–14 mm, is the basis of its value), and increased limitation on movement, especially of the nonterritorial male bachelor bands, represents a significant new threat to this species.

Acknowledgments

We would like to acknowledge the following people who generously helped by providing information, samples or permits for this study. Argentina: M. Knobel, Argentine Guanaco Products, Esquel; S. Poncet; Eduardo Frank, Universidad de Cordoba; Freddy Sossa, EU Supreme Project,

Jujuy; Gustavo Rebuffi, INTA Abra Pampa; Daniel Almeida, INTA Bariloche; Hector Guillermo Villanueva F., Director, Recursos Naturales, Salta; Teresa Raquel Chalabe, Universidad de Salta; G. Moseley, AFRC-IGER, Aberystwyth, UK. Chile: Hernan Torres, Grupo de Especialistas Camélidos Sudamericanos, UICN; Eduardo Nuñez and Rafael Fernández, CONAF; Calogero Santoro, Universidad de Tarapacá, Arica. Peru: Alfonso Martinez, Domingo Hoces, Jorge Herrera, and Marco Antonio Zuñiga, CONACS, Lima; Alex Montufar and Marco Antonio Escobar, CONACS Puno; Roberto Bombilla, CONACS Junín; Carlos Ponce del Prado, Conservación Internacional, Lima; Maximo Gamarra and Santiago Baudilio, SAIS Tupac Amaru, Pachacayo; Rosa Perales, Jose Alva, and Néstor Falcón, Facultad de Medicina Veterinaria, Universidad Nacional Mayor de San Marcos, Lima; Clive Woodham, Labvetsur, Arequipa; Felipe San Martín, Facultad de Medicina Veterinaria, UNMSM, Lima. Nicola Anthony, Peter Arctander, Mark Beaumont, Jon Bridle, Kate Byrne, Gordon Luikart, and Benoît Goossens provided invaluable comments on the manuscript, and Georgio Bertorelle gave valuable advice on the admixture analysis. This study was supported by the Institute of Zoology, Instituto Veterinario de Investigaciones Tropicales y de Altura (IVITA), UNMSM, Lima; NERC grant GST/02/828 to H. F. Stanley and J. C. W.; and Darwin Initiative grant 162/06/126 to H. F. Stanley, J. C. W, and M. W. B. L. C. is now supported by the CNRS.

References

Akimushkin, J. 1971. *Migraciones*. Buenos Aires: Editorial Cartaga.
Antonius, O. 1922. *Stammesgeschichte der haustiere*. Jena. Verlag Gustav Fischer, 305–310.
Bahn, P. G. 1994. Archaeozoology—Time for a change. *Nature* 367: 511–512.
Bandelt, H. J., P. Forster, and A. Röhl. 1999. Median joining networks for inferring intraspecific phylogenies. *Molecular Biology and Evolution* 16: 37–48.
Bates, M. 1975. *Südamerika-Flora und Fauna*. Reinbek: Rowehlt.
Belkhir, L. 1999. GENETIX v4.0. Belkhir Biosoft, Laboratoire des Genomes et Populations, Université Montpellier II.
Benzécri, J. P. 1973. *L'Analyse des Données: T. 2, I' Analyse des correspondances*. Paris: Dunod.
Bertorelle, G. 1998. ADMIX1_0. www.unife.it/genetica/Giorgio/giorgio_soft.html#ADMIX
Bertorelle, G. and L Excoffier. 1998. Inferring admixture proportions from molecular data. *Molecular Biology and Evolution* 15: 1298–1311.
Brehm A. 1916. *Die Säugetiere*. Leipzig: Bibiographilches Insitut.
Bruford, M. W., O. Hanotte, J. F. Y. Brookfield, and T. Burke. 1998. Single and multilocus DNA fingerprinting. In *Molecular genetic analysis of populations: A practical approach*, 2nd edition, A. R. Hoelzel (ed.), pp. 287–336. Oxford: Oxford University Press.
Bustinza, V. 1989. Algunas consecuenciias de la agresion cultural en la ganadeía Andina. In Crianza de llamas y alpacas en los Andes. PAL/ PRATEC, pp. 115–130. Lima: Talleres Graficos de ART Lautrec, S.R.L.
Cabrera, A. 1932. Sobre los camélidos fosiles y actuales de la América austral. *Revista del Museo de la Plata* 33: 89–117.
Cabrera, A. and J. Yepes. 1960. *Mamiferos SudAmericanos*. Buenos Aires: Ediar.
Capanna, E. and M. V. Civitelli. 1965. The chromosomes of three species of Neotropical Camelidae. *Mammalian Chromosomes Newsletter* 17: 75.
Capurro, L. F. and F. Silva. 1960. Estudios cromatagráficos y electroforéticos en Camélidos Sudamericanos. *Investigaciones Zoológicas Chilenas* 6: 49–64.
Chakraborty, R. 1986. Gene admixture in human populations: Models and predictions. *Yearbook of Physical Anthropology*, 29: 1–43.
Chakraborty, R., M. I. Kamboh, M. Nwankwo, and R. E. Ferrell. 1992. Caucasian genes in American blacks—New data. *American Journal of Human Genetics*. 50: 145–155.
Chikhi, L., M. W. Bruford, and M. Beaumont. 2001. Estimation of admixture proportions: A likelihood-based approach using Markov Chain Monte Carlo. *Genetics* 158: 1347–1362.
Cook, O. F. 1925. Peru as a center of domestication. *Journal of Heredity* 16: 1–41.
Darwin, C. 1868. *The variation of animals and plants under domestication*. London: Murray.
Dransart, P. Z. 1991. *Fiber to fabric: The role of fiber in camelid economies in prehispanic and contemporary Chile*. Unpublished D.Phil thesis, Linacre College, Oxford University.
Excoffier, L. 1993. MINSPNET. Available from http://lgb.unige.ch/software/win/min-span-net/
Faige, F. 1929. *Haustierkunde u haustiereucht*. Leipzig.
Fallet, M. 1961. Vergleichende untersuchungen zur wollbildung südamerikanischer tylopoden. *Zeitschrift fur Tierzüchtung und Züchtngsbiologie* 75: 34–56.
Flores Ochoa, J. A. 1977. Patores de alpacas de los Andes. In *Pastores de puna*, J. A. Flores Ochoa (ed.), pp. 15–52. Lima: Instituto de Estudios Peruanos.
Franklin, W. L. 1974. The social behaviour of the vicuña. In *The behaviour of ungulates and its relation to management*, V. Geist and F. Walther (eds.), pp. 477–487. International Union for Conservation of Nature, I.U.C.N., Morges, Switzerland.
———. 1982. Biology, ecology, and relationship to man of the South American camelids. In *Mammalian biology in South America*, M. A. Mares and H. H. Genoways (eds.), pp. 457–489. Linesville, Pymatuning Laboratory of Ecology Special Publication 6. Pittsburgh: University of Pittsburgh.
———. 1983. Contrasting socioecologies of South America's wild camelids: The vicuña and the guanaco. In *Advances in the study of mammalian behavior*, J. F. Eisenberg and D. K. Kleinman (eds.), pp. 573–629. Special Publication of the American Society of Mammalogists 7.
Frechkop, S. 1955. Ruminants digitigrades ou superfamille des Tylopodes. In *Traite de Zoologie* 17(1), P. P. Grassé (ed.), pp. 5180–5189. Paris: P. P. Masson, Editeur.
Frisch, J. L. 1775. *Das natur-system der vierfüssigen thiere*, Glogau. E. J. Gentz and T. L. Yates. 1986. Genetic identification of camelids. *Zoo Biology* 5: 349–354.
Goldstein, D. B. and L. Chikhi. 2002. Human migrations and population structure: What we know and why it matters. *Annual Review of Genomics and Human Genetics* 3: 129–152.
Goldstein, D. B., A. R. Linares, L. L. Cavalli-Sforza, and M. W. Feldman. 1995. An evaluation of genetic distances for use with microsatellite Loci. *Genetics* 139: 463–471.
Guerin, C. and M. Faure. 1999. Palaeolama (Hemiauchenia) niedae nov. sp., nouveau camelidae du nordeste brésilien,

et sa place parmi le Lamini d'Amérique du Sud. *Geobios* 32: 629–659.

Harrison, J. A.1979. Revision of the Camelinae (Artiodactyla Tylopoda) and description of the new genus *Alforjas*. *Paleontological Contributions, University of Kansas* 95: 1–20.

Hemmer, H. 1975. Zer herkunft des alpakas. *Zeitschrift des Kölner Zoo* 18 (2): 59–66.

———. 1983. *Domestikation verarmung der merkwelt*. Braunschweig/Wiesbaden: Friedr. Vieweg and Sohn.

———. 1990. *Domestication: The decline of environmental appreciation*. Cambridge: Cambridge University Press.

Herre, W. 1952. Studien über die wilden und domestizierten Tylopoden Südamerikas. *Der Zoologische Garten*, Leipzig, N.F. 19 (2–4): 70–98.

———. 1953. Studien am skelet des mittelhores wilder und domestizierter formen der Gattung *Lama* Frisch. *Basel Acta Anatomica* 19: 271–289.

———. 1976. Die herkunft des alpaka. *Zeitschrift des Kölner Zoo* 19 (1): 22–26.

———. 1982. Zur stammesgeschichte der Tylopoden. *Verhandlugen der Deutschen Zoologischen Gesellschaft* 75: 159–171.

Herre, W. and M. Röhrs. 1973. *Haustiere—Zoologisch gesehen*. Stuttgart: Gustav Fischer Verlag.

Herre, W. and U. Thiede. 1965. Studien an ghirnen Südamerikanischer Tylopoden. *Zoologisches Jahrbuch Anatomie* 81: 155–176.

Hilzheimer, M. 1913. Überblick über die geschichte der haustierforschung, besonders der letzen 30 Jahre. *Zoologischer Anzeiger* 5: 233–254.

Hoffstetter, R. 1986. High Andean mammalian faunas during the Plio-Pleistocene. In *High altitude tropical biogeography*, F. Vulleumier and M. Monasterio (eds.), pp. 218–245. Oxford: Oxford University Press.

Jerison, H. J. 1971. Quantitative analysis of the evolution of the camelid brain. *American Naturalist* 105: 227–239.

Jürgens, K. D., M. Pietschmann, K. Yamaguchi, and T. Kleinschmidt. 1988. Oxygen binding properties, capillary densities and heart weights in high altitude camelids. *Journal of Comparative Physiology B* 158: 469–477.

Kadwell, M., M. Fernández, H. F. Stanley, R. Baldi, J. C. Wheeler, R. Rosadio, and M. W. Bruford. 2001. Genetic analysis reveals the wild ancestors of the llama and alpaca. *Proceedings of the Royal Society of London, B*. 268: 2575–2584.

Kleinschmidt, T., J. Marz, K. D. Jürgens, and G. Braunitzer. 1986. The primary structure of two Tylopoda hemoglobins with high oxygen affinity: Vicuña *(Lama vicugna)* and alpaca *(Lama pacos)*. *Biological Chemistry Hoppe-Seyler* 367: 153–160.

Krumbiegel, I. 1944. Die neuveltlichen tylopoden. *Zoologischer Anzeiger* 145: 45–70.

———. 1952. *Lamas*. Leipzig: Neue Brehmbücherei.

Kruska, D. 1982. Hirngrößenänderungen bei Tylopoden während der stammesgeschichte und in der domestikation. *Verhandlugen der Deutschen Zoologischen Gesellschaft* 75: 173–183.

Kruskal, J. B. 1956. On the shortest spanning subtree of the graph and the travelling salesman problem. *Proceedings of the American Mathematical Society*. 7: 48–57.

Lang, K. D., Y. Wang, and Y. Plante. 1996. Fifteen polymorphic dinucleotide microsatellites in llamas and alpacas. *Animal Genetics* 27: 293.

Langella, O., L. Chikhi, and M. A. Beaumont. 2001. LEA (Likelihood-based estimation of admixture): A program to simultaneously estimate admixture and the time since admixture. *Molecular Ecology Notes* 1: 357–358.

Larramendy, M., L. Vidal-Rioja, L. Bianchi, and M. Bianchi. 1984. Camélidos Sudamericanos: Estudios genéticos. *Boletín de Lima* 6: 92–96.

Ledger, C. 1860. Sur un tropeau d'Alpacas introduit en Australie. *Bulletin Société Impérial Zoologique d'Acclimatation 7*. Paris.

Long, J. 1991. The genetic structure of admixed populations. *Genetics* 127: 417–428

Lönnberg, E. 1913. Notes on guanacos. *Arkiv für Zoologi* (Stockholm) 8: 1–8.

López Aranguren, D. J. 1930. Camélidos fosiles argentinos. *Anales de la Sociedad Cientifica Argentina* (Buenos Aires) 109: 15–35, 97–126.

MacHugh, D. E., M. D. Shriver, R. T. Loftus, P. Cunningham, and D. G. Bradley. 1997. Microsatellite DNA variation and the evolution, domestication and phylogeography of taurine and zebu cattle *(Bos taurus* and *Bos indicus)*. *Genetics* 146: 1071–1086.

McPartlan, H. C., M. E. Matthews, and N. A. Robinson. 1998. Alpaca microsatellites at the VIAS A1 and VIAS A2 loci. *Animal Genetics* 29: 158–159.

Miller, W. J., J. P. Hollander, and W. L. Franklin. 1985. Blood typing South American camelids. *Journal of Heredity* 76: 369–371.

Obreque, V., L. Coogle, P. J. Henney, E. Bailey, R. Mancilla, J. Garcia-Huidobro, P. Hinrichsen, and E. G. Cothran. 1998. Characterization of 10 polymorphic alpaca dinucleotide microsatellites. *Animal Genetics* 29: 461–462.

Obreque, V., R. Mancilla, J. Garcia-Huidobro, E. G. Cothran, and P. Hinrichsen. 1999. Thirteen new dinucleotide microsatellites in alpaca. *Animal Genetics* 30: 397–398.

Palma, R. E., J. C. Marín, A. E. Spotorno, and J. L. Galaz. 2001. Phylogenetic relationships among South American subspecies of camelids based on sequences of mitochondrial genes. In *Progress in South American camelid research*, M. Gerken and C. Renieri (eds.), pp. 44–52. *Proceedings of the III Symposium on South American Camelids, University of Goettingen*. Germany: Wageningen Press.

Penedo, M. C. T., M. E. Fowler, A. T. Bowling, D. L. Anderson, and L. Gordon. 1988. Genetic variation in the blood of llamas, *Lama glama*, and alpacas, *Lama pacos*. *Animal Genetics* 19: 267–276.

Penedo, M. C., A. R. Caetano, and K. I. Cordova. 1998. Microsatellite markers for South American camelids. *Animal Genetics* 29: 411–412.

Penedo, M. C., A. R. Caetano, and K. I. Cordova. 1999a. Eight microsatellite markers for South American camelids. *Animal Genetics* 30: 166–167.

Penedo, M. C., A. R. Caetano, and K. I. Cordova. 1999b. Six microsatellite markers for South American camelids. *Animal Genetics* 30: 399.

Peterson, O. A. 1904. Osteology of Oxydactylus. *Annals of the Carnegie Museum* 2: 434–476.

Piccinini, M., T. Kleinschmidt, K. D. Jürgens, and G. Baunitzer. 1990. Primary structure and oxygen-binding properties of the hemoglobin from guanaco *(Lama Guanacoë [sic], Tylopoda)*. *Biological Chemistry Hoppe-Seyler* 371: 641–648.

Pires-Ferreira, E. 1981/1982. Nomenclatura y nueva classificacion de los camélidos Sudamericanos. *Revista do Museu Paulista* 28: 203–219.

Reynolds, J., B. S. Weir, and C. C. Cockerham. 1983. Estimation of the co-ancestry coefficient—Basis for a short-term genetic-distance. *Genetics* 105: 767–779.

Röhrs, M. 1957. Ökologische Beobachtungen an wildlebenden Tylopoden Südamerikas. *Velhandlung der Deutchen Zoologischen Gesellschaft* 538–554.

Saluda-Gorgul, A., J. Jaworski, and J. Greger. 1990. Nucleotide sequence of satellite I and II DNA from Alpaca *(Lama pacos)* genome. *Acta Biochimica Polonica* 37: 283–297.

Sarno, R. J., V. A. David, W. L. Franklin, S. J. O'Brien, and W. E. Johnson. 2000. Development of microsatellite markers in the guanaco, *Lama guanicoe*: Utility for South American camelids. *Molecular Ecology* 9: 1922–1924.

Semorile, L. C., J. V. Crisci, and L. Vidal-Rioja. 1994. Restriction sites variation of the ribosomal DNA in Camelidae. *Genetica* 92: 115–122.

Stanley, H. F., M. Kadwell, and J. C. Wheeler. 1994. Molecular evolution of the family Camelidae—A mitochondrial DNA study. *Proceedings of the Royal Society of London, B.* 256: 1–6.

Steinbacher, G. 1953. Zur abstammung des alpakka, *Lama pacos* (Linne, 1758). *Säugetierkundliche Mitteilungen* 1: 78–79.

Taylor, K. M., D. A. Hungerford, R. L. Snyder, and F. A. Ulmer, Jr. 1968. Uniformity of karyotypes in the Camelidae. *Cytogenetics* 7: 8–15.

Thomas, O. 1891. Notes on some ungulate mammals. *Proceedings of the Zoological Society of London*: 384–389.

Vidal Rioja, L., L. Semorile, N. O. Bianchi, and J. Padron. 1987. DNA composition in South American camelids I. Characterization and *in situ* hybridization of satellite DNA fractions. *Genetica* 72: 137–146.

Webb, S. D. 1974. Pleistocene llamas of Florida, with a brief review of the Lamini. In *Pleistocene mammals of Florida*, S. D. Webb (ed.), pp. 170–213. Gainesville: University Presses of Florida.

Weber, M. 1928. *Die Säugetiere*. Teil: Jane Jena.

Wheeler, J. C. 1995. Evolution and present situation of the South American Camelidae. *Biological Journal of the Linnean Society* 54: 271–295.

Wheeler, J. C., A. J. F. Russel, and H. Redden. 1995. Llamas and alpacas: Pre-conquest breeds and post-conquest hybrids. *Journal of Archaeological Science*, 22: 833–840.

Wheeler, J. C., M. Fernández, R. Rosadio, D. Hoces, M. Kadwell, and M. W. Bruford. 2001. Diversidad genética y manejo de poblaciones de vicuñas en el Perú. *RIVEP Revista de Investigaccciones veterinarias del Perú* Suplemento 1: 170–183.

Zeuner, F. E. 1963. *A History of the domesticated animals*. London: Hutchinson.

CHAPTER 24

Genetic Documentation of Horse and Donkey Domestication

CARLES VILÀ, JENNIFER A. LEONARD, AND
ALBANO BEJA-PEREIRA

Introduction

Humans and horses have had a close relationship for many thousands of years. During the Glacial Maxima at the end of the Pleistocene, grasslands dominated extensive areas of Eurasia, and large herds of horses roamed them. Horse remains are common in many archaeological assemblages, and they were clearly a very important part of the diet in some human groups (Olsen 1989, 1995). The horse was also the most frequently depicted species in the art of the final Upper Paleolithic of Europe (Levine 1999a).

After the Ice Age, however, wild horses disappeared from extensive areas that became covered with forests and were only abundant in the Eurasian steppes or along the steppe borderlands—a region that spans a vast area running east-west across Eurasia, between the Black Sea and Mongolia (Olsen 1996; Clutton-Brock 1999). Consequently, it is these areas where the permanent association between humans and horses may have been most likely to develop (Anthony 1996). As a result, active, ongoing efforts to document the oldest evidence for domestic horse have focused on the Eurasian steppe, especially the area from the Ukraine, across the steppe lands of central Russia, and into Kazakhstan (Anthony and Brown 1991; Clutton-Brock 1992; Brown and Anthony 1998; Olsen 2003 and Chapter 17). From this heartland of initial domestication, horses spread progressively across the continents (Clutton-Brock 1992, 1996)

Documenting the initial domestication of the horse on the Eurasian steppe has proven exceptionally difficult using traditional archaeological markers of domestication (see Chapter 17). Although the dramatic increase in the abundance of horse remains in Eurasia archaeofaunal assemblages at about 6,000 years ago might be a signal of the horse's initial domestication, an argument can also be made that this increase was caused by intensification in the hunting of wild horse populations (Levine 1999b). Since wild and domestic horses cannot be indisputably differentiated using traditional osteological markers (Levine 1990; Chapter 17), researchers have used a wide range of other morphological and nonmorphological evidence of domestication in attempts to document the process of horse domestication. Some authors suggested that the notching and anterior beveling of ancient horse incisors were the result of crib biting and thus are an indication of human control (Bahn 1980). According to these authors, crib biting, a result of boredom and prolonged inactive confinement, creates a characteristic wear pattern not observed in wild individuals. However, other authors claim that similar forms of wear can be observed in horse remains from Early and Middle Pleistocene deposits in North America and thus cannot be related to human control (Rogers and Rogers 1988; White 1989). Similarly, it has been suggested that tooth wear patterns can be indicative of horseback riding and that this could have originated about 6,000 years ago (Anthony and Brown 1991; Anthony et al. 1991; Brown and Anthony 1998; Anthony 1996). These results imply that horses may have been ridden since the moment the herds started to be under human control. Indeed, a number of researchers believe that steppe herding may have demanded horseback riding (Barclay 1982; Levine 1999b; Chapter 17). Nevertheless, the association between tooth wear and horseback riding is still controversial (Levine 1999b), and wear patterns that mimic those caused by bit wear can be found in Pleistocene-age horses (Chapter 17).

The geographic origin of domestication for the donkey, the second domesticated equid, is also a controversial issue. The oldest archaeological evidence of domestic donkeys is reported in Sudan at 5000 BP (Peters 1986). However, there is clear evidence for the presence of domestic donkeys in Syria at about 4300 BP (Clutton-Brock and Davis 1993) and even earlier in highland Iran at about 4800 BP (Zeder 1986). Domestic donkeys arrived in Europe during the second millennium BC, brought by the Greeks to their colonies along the north coast of the Mediterranean. Later, the Romans continued to disperse donkeys through Europe to the limits of their empire (Bodson 1985; Bökönyi 1991).

In addition to the rapid expansion and sparse archaeological data, other factors make it difficult to identify the time and place of donkey domestication. As with horses, the large morphological diversity of domestic forms and the lack of adequate modern or archaeological osteological collections make it exceedingly difficult, if not impossible, to distinguish the remains of domestic asses from those of wild asses (Davis 1987). The most probable ancestor of the donkey is the African wild ass *(Equus africanus)* (Clutton-Brock 1992), an animal that inhabited a wide territory that in the Late Pleistocene and Early Holocene extended across North and northeast Africa, and likely into Arabia and the Levant (Groves 1986; Uerpmann 1987), areas that, with the exception of the Levant, are not well studied. Moreover, the range of the African wild ass and the areas into which the domestic ass was introduced overlap with the geographic range of its close cousin, the wild half-ass, or onager *(Equus hemionus)*, an equid that likely was never domesticated, and whose remains are also extremely difficult to distinguish both from the wild ass and from the domestic ass.

Equids were the last of the common Old World livestock animals to be domesticated (Clutton-Brock 1999). All the other major livestock species and a majority of the major crops were domesticated many millennia earlier (Cowan and Watson 1992; Smith 1998). Despite this wide array of already domesticated species that could provide food resources, domestic equids—especially horses—were readily adopted and quickly spread all over Eurasia. The rapid spread of domestic equids likely stems from the fact that, compared to the other domesticates, horses and donkeys permitted a significant change in the way of life in some ancient societies. Besides providing meat, milk, and leather like the other domestic species, equids represented an improvement in transportation capabilities. They allowed people to move faster, farther, and with heavier loads (Clutton-Brock 1992). Associated with the ability to move people and goods more quickly over large territories came the enhanced ability to wage war, a tactical advantage that clearly played a substantial role in the rise and collapse of civilizations across the Near East and the Eurasian steppe (Diamond 1991). Horse domestication has also been argued to have played a key role in the spread of Indo-European languages (ancestors of English, German, Spanish, French, Italian, Swedish, Greek, and many other modern languages) from the Iberian Peninsula to India (Anthony 1986; Diamond 1991). Recent research contradicts this view, however, and suggests that the expansion of Indo-European languages was slightly earlier and therefore not related to the domestication of horses but to the spread of domestic crops (Gray and Atkinson 2003).

The cultural responses associated with the acquisition of horses have been recorded in the New World, where wild horses became extinct and domestic horses were introduced by European colonists. The impact of the introduction of the horse to Native American groups was profound, causing "a revolution in virtually every aspect of life of the Plains tribes" (Anthony et al. 1991: 98) and providing a model for what may have happened when horses were first domesticated in the Old World (Anthony 1986; Anthony et al. 1991).

Donkey domestication was also important to human societies because of their usefulness in traction (agriculture) and long distance mobility (trade and movement of people). Since the ecological requirements of the donkey are different from those of the horse, their role was especially important in the southern and more arid regions of Eurasia and northern Africa. Another important consequence of the domestication of donkeys was the possibility of creating hybrids with horses (e.g. mules, hinnies). These large and sturdy animals also had considerable effect in augmenting the possibilities of traction and mobility.

Recently, genetic techniques have been added to the suite of archaeological approaches to documenting the domestication of horses and donkeys. These new techniques have illuminated many aspects of the domestication of these important species and have provided an additional view on their possible role in ancient human societies. In this chapter, we summarize how molecular genetic techniques can complement archaeological approaches to obtain a better understanding of the domestication of equids, their subsequent spread, and their more recent division into breeds.

Genetic Markers

Molecular genetics has provided a new point of view from which to study the process of domestication. Many questions related to the origin and complexity of this process have been clarified through the use of molecular genetics markers. The different modes of inheritance and rates of mutation make particular markers appropriate to address different issues. Most markers that are used for studying the history or relationship of a population or species are neutral; that is, whatever allele (variant of a genetic marker) an individual has does not affect its fitness. For example, mitochondrial DNA (mtDNA) sequences, microsatellite markers either in the autosomal DNA or on the Y chromosome, and sequences of other regions of noncoding DNA are considered neutral.

Mitochondrial DNA has been very popular for addressing many issues surrounding domestication of animals (Fumihito et al. 1994; Vilà et al. 1997; Vilà et al. 2001; Giuffra et al. 2000; Kadwell et al. 2001; Luikart et al. 2001; Leonard et al. 2002). This DNA molecule is located inside mitochondria, the organelles responsible for the production of energy in the cells. Mitochondrial DNA evolves fairly rapidly, so variation can accumulate within species, although it has also been useful for examining the relationship between closely related taxa. Mitochondrial DNA is only inherited from the mother, so only the female lineage can be investigated with this marker. Additionally, the mtDNA sequence (or mtDNA haplotype) is not subject to recombination, which simplifies the analysis of data from this marker. Since mitochondria are very abundant in most cells, there are many copies of each mtDNA per cell, which greatly facilitates its study. However, males and females may have very different natural histories, so it may not be reasonable to generalize across the sexes in some cases. Recently, markers on the Y chromosome have started to be applied to a variety of organisms (Hellborg and Ellegren 2003). The Y chromosome is inherited only from father to son and thus can trace paternal lineages. Other genetic markers are located on non-sex chromosomes (autosomes) and are inherited from both parents. Most nuclear microsatellites belong to this category of markers. They are often used to characterize populations or to determine paternity. Combining markers with different modes of inheritance can allow tracking paternal and maternal lineages separately, and thus a more complete view of the domestication process can be obtained.

Recent technological advances make it possible to examine DNA from some ancient remains. This kind of data opens a wide new field of opportunities, since the entire period of domestication lies within the time period DNA can theoretically survive (Wayne et al. 1999; Hofreiter et al. 2001). In addition to using ancient DNA to look at archaeological remains of domestic animals (i.e., Leonard et al. 2002), it can be used to look at wild populations of the progenitor species, which may have since become extinct or changed dramatically (i.e., Vilà et al. 2001). Although all markers discussed above are available for ancient DNA analysis, mitochondrial DNA sequences have been the most commonly used (Wayne et al. 1999).

The Ancestor of the Horse

Phylogenetic reconstruction based on morphological (Forstén 1992) and genetic data (George and Ryder 1986; Oakenfull and Clegg 1998; Oakenfull et al. 2000) indicate that all the extant equids can be divided in two lineages—stenoid (including zebras and asses) and caballoid (horses)—that may have diverged between 2 and 3.9 million years ago (however, see Bennett 1980; Jansen et al. 2002). This implies that the wild ancestor of the domestic horse was a caballoid horse, and it could be the Przewalski's horse (the only extant wild caballoid horse) or an extinct species (see Chapter 17). Representations of horses in Upper Paleolithic art often have an appearance much like that of a Przewalski's horse, with a massive head, erect mane, dun coat, pale belly, and stripes on the shoulders and legs. Consequently, it has been suggested that domestic horses could have derived from Przewalski's horses. This species became extinct in the wild in the 1960s after a handful of individuals were captured to initiate a captive breeding program (Bouman and Bouman 1994). All recognized individuals existing in zoos and breeding facilities descend from 13 individuals, one of which was a domestic horse and another of which was a hybrid (Oakenfull and Ryder 1998). According to the studbook, four female lineages are represented in the living population (neither the domestic horse nor the hybrid individual left any female descendents). These four lineages are represented by two mitochondrial DNA haplotypes that are practically indistinguishable from those in domestic horses (George and Ryder 1986; Ishida et al. 1995; Oakenfull and Ryder 1998; Oakenfull et al. 2000). Accordingly, this Asian wild horse could be very closely related to the ancestor of modern horses.

Despite the similarity of mtDNA sequences, other lines of evidence suggest that Przewalski's horse may not be the direct progenitor. While Przewalski's horses have 2n = 66 chromosomes, domestic horses have 64 (Benirschke et al. 1965), and although the hybrids are fertile (Allen and Short 1997), the difference in the number of chromosomes has been interpreted as having an ancient origin. Additionally, all Przewalski's horses that have been typed have the same Y-chromosome sequence (Wallner et al. 2003), and this sequence is different from the only one identified in domestic horses (Wallner et al. 2003; Lindgren et al. 2004). Consequently, although Przewalski's horse is very closely related to the ancestor of our domestic horse, it may not be the direct progenitor.

As a result of the small founding population, the current population of Przewalski's horses has little genetic diversity (Ryder 1994; Hedrick et al. 1999). In addition to recent founding effects, a large amount of genetic diversity could have been lost previously. It is clear that there was a severe historic population decline resulting from anthropogenic causes, which led to the depletion of the natural populations, but it is likely that climate and environmental change began reducing their numbers thousands of years ago. For these reasons, the current genetic diversity of Przewalski's horse is not representative of the amount of genetic diversity that could have been found in a wild population of horses around the time that horses were domesticated.

Molecular Genetic Analysis of Horse Domestication

Surveys of the mitochondrial DNA variation in domestic species have been used to estimate the number, location, and date of several domestication events (for cattle: Loftus 1994; Troy 2001; Chapter 22; for dogs: Vilà et al. 1997; Savolainen et al. 2002; Chapter 19; for pigs: Giuffra 2000; Chapter 15; for goats: Luikart 2001; Chapter 20; for sheep: Hiendleder et al. 2002; Chapter 21). This approach has also been used to study the domestication of horses (Lister et al. 1998; Vilà et al. 2001; Jansen et al. 2002). Mitochondrial DNA surveys in domestic horses have uncovered a large amount of genetic variation within the species (George and Ryder 1986; Ishida et al. 1994, 1995; Marklund et al. 1995; Bowling et al. 2000; Vilà et al. 2001; Jansen et al. 2002) and even inside a single population (Wang et al. 1994). The average sequence (haplotype) divergence for the mitochondrial control region was 2.6% across a large number of horses (range: 0.2 to 5.0%; Vilà et al. 2001). Similar or higher divergence values across mtDNA sequences have also been observed in other domestic mammals. However, in these other species, all sequences are partitioned into a limited number of clades (or groups of sequences with a single origin) that indicate a small number of domestication events (see review in Bruford et al. 2003; Chapter 18). This is not the case for the horse. When the relationships between mtDNA sequences are reconstructed in a phylogenetic tree (Figure 24.1), they do not form a small number of well-supported clades that could indicate different domestication events.

It could be speculated that all modern horses derive from a single domestication, and that all the sequences have evolved since a single taming and domestication of horses. However, the comparison of the horse haplotypes with those in donkeys allowed Vilà et al. (2001) to estimate the mutation rate for this region of the equid mtDNA and led them to conclude that the different lineages in modern horses originated at least 300,000 years ago, thus excluding their possible origination after domestication. The diversity found in horses, together with the absence of a reduced number of clades that could be identified as different domestication events and the lack of geographic patterns in the distribution of modern domestic horse haplotypes (see below), has been used by Lister et al. (1998) to suggest that domestic horses could have derived "from wild stock distributed over

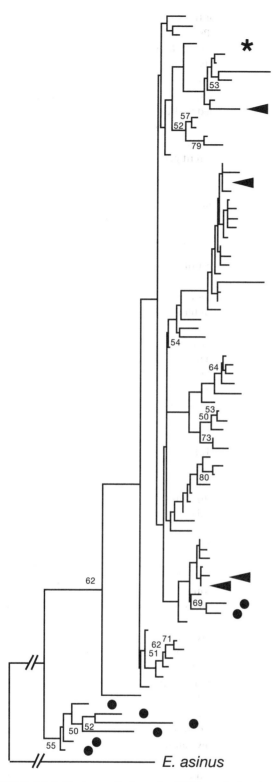

FIGURE 24.1 Phylogenetic tree of mtDNA control region sequences (length: 355 bp) in horses. Support for internal nodes is indicated with bootstrap values when higher than 50%. Sequences corresponding to Alaskan horses dated 12,000 to 28,000 years old are indicated with solid black circles. Sequences found in Viking horses (1,000–2,000 years old) are marked with arrows. The asterisk marks a sequence observed in Przewalski's horses. (Based on Vilà et al. 2001.)

a moderately extensive geographic region, large enough to have contained within it considerable pre-existing haplotype diversity" (Lister et al. 1998: 276).

To confirm the existence of multiple lineages leading to modern horses and to infer their geographic origin without making assumptions regarding mutation rates, some knowledge about the levels and distribution of genetic diversity in the ancestral species is necessary. Unfortunately, as we have seen above, this is not possible to obtain using modern samples because the progenitor species of modern horses has either gone through a major bottleneck or has become extinct. Nevertheless, this may be addressed through DNA analysis of horse remains from ancient populations. Using ancient DNA techniques, Vilà et al. (2001) sequenced a fragment of the mtDNA control region from horse bones from a single population in Alaska dating to between 12,000 and 28,000 BP. These bones had been frozen in the permafrost soil, allowing the preservation of short DNA fragments for a very long time. Horses were extinct in North America after the terminal Pleistocene extinction event, but they were abundant throughout the Pleistocene (Simpson 1951). Although these Alaskan animals were not involved in the domestication process, the numerous remains can be used to determine how much genetic diversity could be contained within one wild horse population.

Each of the eight wild Late Pleistocene horses that were analyzed had a different haplotype. These data suggested that a relatively large amount of genetic diversity may have been present in a single population of the wild progenitor species. None of these sequences was the same as that found in Przewalski's horse or the same as the sequences in modern domestic horses, although they were very closely related. The Alaskan sequences clustered in two clades (Figure 24.1). One of them was inside the tree of modern horses, implying that all modern horses did not form a single monophyletic (with one single origin) clade. This confirms that modern horses do not derive from a single maternal lineage and that multiple breeding females had to be involved in the domestication process. Supporting this view, the sequence from Przewalski's horses, a lineage close to modern horses but clearly differentiated (see above), was also inside the diversity of modern horses, confirming that modern horses could not derive from one single lineage. On the other hand, although the ancient American horses were very closely related to the domestic horses, several haplotypes formed a clade that was not present in modern animals, indicating some amount of geographic partitioning of the genetic diversity. Also, the presence of just two groups of sequences in a single locality during a period of several thousand years suggested that the genetic diversity per locality was likely to be limited in the progenitor of modern horses, and thus the large diversity observed in modern horses might imply the contribution of many different populations. Nevertheless, a final assessment of the geographic distribution of the wild horses that contributed to the diversity of modern horses will not be reached until a larger number of populations of the wild ancestor are studied using ancient DNA techniques.

Considering the mutation rate for horse mtDNA, Jansen et al. (2002) tried to quantify the number of maternal lineages present in domestic horses. These authors estimated that a minimum of 77 successfully breeding mares (each one founding one lineage) had been recruited from the wild. The extensive genetic diversity led them to conclude that several distinct populations were involved in the domestication of the horse. These results contrast dramatically with the patterns observed in all other domesticated mammals, where fewer than 4 lineages are commonly observed (Bruford et al. 2003; Chapter 18; but see Larsen et al. 2005 for pigs).

On the other hand, a recent study (Lindgren et al. 2004) has suggested that horses have very reduced diversity on the Y chromosome. Only one haplotype was found in a very large diversity of modern horses. This could indicate that the ancestral species was depleted of variability in this chromosome, but since different sequences were found in Przewalski's horses (Wallner et al. 2003; Lindgren et al. 2004), the authors presumed that some diversity could have existed in the ancestral species and that the lack of diversity in modern horses may indicate that very few stallions were involved in the domestication process. This could indicate that the domestication of horses may have had a localized origin involving just a few individuals and that domestic populations had later been supplemented with large numbers of wild mares. However, this explanation will remain tentative until the diversity present in the Y chromosome in natural populations is better known.

What could account for such a disparate pattern in horses as compared with the other domestic mammals? When horses were domesticated, several other domestic mammal species (cattle, sheep, goats, pigs, and dogs) were already widespread in the same region. We can speculate that there was no need for a new domesticate, unless it provided something that the other species could not. Horses revolutionized transportation, and therefore trade and warfare (Diamond 1991). Perhaps the need of some human groups to cope with their horseback-riding neighbors forced them to augment their own equine stocks by capturing more mares. If horses were as important to ancient societies as suggested above, this importance may be reflected in the large number of mares recruited from the wild over an extensive area. Genetic analyses may thus support archaeological evidence pointing to the profound and widespread impact of the domestic horse on ancient societies.

Breeding Practices and Origin of Horse Breeds

Over 170 horse and pony breeds exist today (Bongianni 1987). These breeds are highly diverse in morphology and behavior, from the very small Falabella to the gigantic Shire and the graceful Arabian. Since the domestication process may have occurred over a very large area, it is possible that the different horse types and/or local breeds derived from the

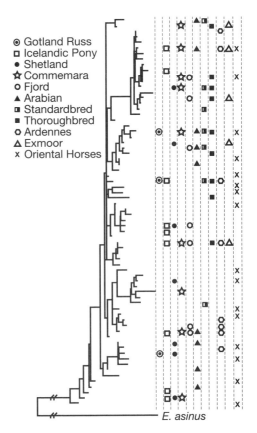

FIGURE 24.2 Genetic diversity in horse breeds. Distribution of mtDNA sequences found in several breeds of domestic horses, superimposed on a tree of sequences in modern horses. (Based on Vilà et al. 2001.)

recruitment of local horses (Bennett 1992). It was thought that whereas males were exchanged across regions, the mares usually had a local origin, and subsequently "the 'native broodmare' (not the stallion) has been the foundation of European and Asian horse breeding for millennia ... the bloodline of importance was the broodmare's" (Bennett 1992: 55–56). Accordingly, since the mitochondrial DNA is maternally inherited, it would be expected that different breeds, from different geographic areas, would be characterized by different mtDNA lineages and that the diversity within breeds would be quite reduced. However, when mtDNA sequences have been studied in multiple individuals from the same breed, the levels of genetic diversity found have been surprisingly high (Wang et al. 1994; Kavar et al. 1999; Bowling et al. 2000; Yang et al. 2002). Very divergent breeds share mtDNA lineages, and these lineages do not tend to form monophyletic clades that would imply a single origin (Lister et al. 1998; Kim et al. 1999; Jansen et al. 2002; Vilà et al. 2001; Yang et al. 2002). The study of about 20 individuals from each of 10 horse breeds (Vilà et al. 2001) showed that the sequences found in each breed were distributed across the entire phylogenetic tree without showing clades of similar sequences within a breed (Figure 24.2). Breeds as morphologically divergent as the Arabian and the Shetland or the Swedish Ardennes and the Exmoor share some mtDNA haplotypes (Figure 24.2).

TABLE 24.1
Assignment Test[a]

	Arabian	Thoroughbred	Standardbred	Ardennes	Fjord	Connemara	Exmoor	Gotland Russ	Icelandic	Shetland
Arabian	19 (95)	0	1	0	0	0	0	0	0	0
Thoroughbred	0	19 (100)	0	0	0	0	0	0	0	0
Standardbred	1	1	17 (85)	0	0	0	0	1	0	0
Ardennes	0	0	0	19 (100)	0	0	0	0	0	0
Fjord	0	0	0	0	13 (93)	1	0	0	0	0
Connemara	0	0	0	0	0	13 (87)	1	0	0	1
Exmoor	0	0	0	0	0	0	19 (100)	0	0	0
Gotland Russ	0	0	0	0	0	0	0	20 (100)	0	0
Icelandic	0	0	0	0	0	0	0	0	20 (100)	0
Shetland	0	0	0	0	2	0	0	0	0	8 (90)

[a] Number of individuals from the breeds in the rows that are assigned to the breeds are shown in the columns. Percentages of correct assignments are indicated in parentheses.

The lack of mtDNA differentiation between breeds in separate geographic areas indicates that, contrary to previous assumptions, breeds have not resulted solely from recruitment of local wild mares for the creation of each breed. This admixture of mtDNA lineages instead suggests extensive translocation of horses from one region to another in recent times. However, mtDNA sequences from Viking-age horse remains (1,000 to 2,000 years old) from around the Baltic Sea indicated that the diversity present at that time within a single region was already very large (Figure 24.1; Vilà et al. 2001). This implies that the admixture of horse lineages has not been only recent, as could be suspected from the role of horses in warfare and their use for extended transportation from the earliest times (Clutton-Brock 1992, 1996).

A more extensive study (Jansen et al. 2002) supports the general lack of exclusive groups of haplotypes corresponding to single breeds or groups of breeds, but the authors observe some degree of geographic structure. They observe that a small number of clades are present at higher frequencies in horses deriving from certain regions. These results may indicate that the admixture of mtDNA lineages may have occurred largely across limited geographic areas, at least in some regions.

Despite the high degree of admixture suggested by the mtDNA analyses, breeds are very well defined from the morphological point of view. Similarly, studies of horse breeds based on autosomal markers (i.e., microsatellites) are able to differentiate the breeds (Bjørnstad and Røed 2001, 2002; Kelly et al. 2002). The same horses used to compare mtDNA variability across breeds (Figure 24.2) were typed for 15 microsatellite markers (Vilà et al. 2001), and the degree of differentiation between breeds was assessed by an assignment test. For each individual horse, the probability of obtaining its genotype (genetic constitution at all microsatellites studied) was calculated for each one of the breeds according to the breed's allele frequencies. The genotype was then assigned to the breed for which it had the highest probability. If breeds were well separated from each other, a large number of horses should assign to their nominal breed as opposed to being assigned to another breed. In this case, the frequency of correct assignments ranged between 85% and 100% for each breed (Table 24.1), demonstrating their clear separation. Using similar methods, Bjørnstad and Røed (2001) observed that even among closely related breeds, the use of 26 microsatellites allowed them to correctly assign 96% of the individuals to their source.

How is it possible to have such a great degree of interbreed differentiation in nuclear markers while mtDNA displays such intense admixture? Because mtDNA reflects only maternal inheritance, the strong separation between breeds implied by autosomal markers could indicate strong differentiation in paternal lines. The study of paternal lineages through the analysis of Y-chromosome sequences could provide a direct measure of exchange of males across breeds. Unfortunately, the studies available so far show a complete lack of sequence variation on the Y chromosome of domestic horses (Wallner et al. 2003; Lindgren et al. 2004) that prevents the tracing of paternal lines. However, we know that current breeding practices involve the use of a limited number of stallions that are used to define the characteristics of the breed. The breeding potential of males is much higher than that of females, allowing a more efficient selection for specific traits. A clear example of this is the case of the thoroughbred horses. Although the breed numbers more than 300,000 individuals worldwide, an exhaustive pedigree and microsatellite analysis revealed that one single founder stallion was responsible for 95% of the paternal lineages. On the female side, 10 founder mares accounted for 72% of maternal lineages (Cunningham et al. 2001).

The relative role of females and males in the formation of breeds can be roughly visualized by comparing the average degree of genetic exchange between breeds estimated as the

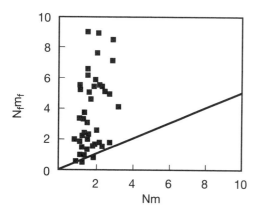

FIGURE 24.3 Pair-wise number of female migrants per generation $(N_f m_f)$ and absolute number of migrants per generation (Nm) between horse breeds. The thick, continuous line marks an equal number of migrant males and females (i.e., the number of female migrants is half of the total number of migrants). Because almost all pairwise comparisons are above this line, it can be concluded that females are responsible for most of the exchange between breeds.

number of migrants per generation. Since the breeds today are highly isolated, this is a measure of co-ancestry (or common genetic legacy) among breeds more than current genetic exchange. The number of migrants per generation can be estimated based on mtDNA sequences (ϕ_{ST}) and provides information regarding only the exchange of females $(N_f m_f,$ where N_f is the effective number of females and m_f is the female migration rate per generation). Estimates based on microsatellites (θ) can provide information regarding the frequency of exchange in both sexes combined (Nm). Plotting these two estimates together (Figure 24.3) demonstrates that exchange between breeds may have been more intense for females than for males, so males could have been more subject to selection within each breed.

The degree of genetic similarity at autosomal markers has also been used to infer the relationships between breeds in an effort to reconstruct their history. For example, such data have been used to estimate the relationship between Spanish Celtic horse breeds (Cañon et al. 2000), to indicate that Japanese horses originated from Mongolian horses migrating through the Korean Peninsula (Tozaki et al. 2003), and to support the hypothesis that horses from Central Asia may have contributed to northern European populations (Bjørnstad et al. 2003).

Molecular Genetic Analysis of Donkey Domestication

In arid areas, the domestic donkey has played a role similar to that of the horse in other regions. The capacity to carry wood and water and the ability to ride over short distances have been crucial for survival in desert areas, and donkeys may have offered a more efficient transportation system than horses in these dry habitats. In spite of the important role played by the donkey in small rural communities all over the world, their use has been regarded as synonymous with backwardness, underdevelopment, and low socioeconomic status. Today, the domestic donkey is widely distributed across the world (85 million), although its numbers are decreasing in many regions. In the last five decades, populations have declined largely in Asia and Europe while they have increased in sub-Saharan Africa, northern India, and the tropical highlands of Latin America (FAO 2000).

Preliminary surveys of the relationships between *Equus* species based on mtDNA sequences indicate that domestic donkeys and the African ass, *E. africanus*, group together, and their genetic divergence is comparable to that within the other equid species (Oakenfull et al. 2000). These results suggest a common origin for both species. However, this is based on very few sequences, and the authors indicate that the evidence is insufficient to fully understand the relationship between the species. A more extensive study analyzed a 479-bp sequence fragment of the hypervariable region I of the mtDNA from Asian wild half-asses *(E. hemionus, E. kiang)*, and from each of the two extant wild African ass *(E. africanus)* subspecies: *E. a. africanus*, or the Nubian ass, and *E. a. somaliensis*, or the Somali wild ass, and 150 domestic donkeys (Beja-Pereira et al. 2004). These data were used to assess the relationship between domestic donkeys and their wild relatives and the number of maternal genetic lineages involved in domestication, and to determine the geographic origin of this species. Samples of domestic donkeys were collected across the Old World (32 countries), paying special attention to the hypothesized centers of domestication (northeast Africa, Near East, and Arabic Peninsula).

Phylogenetic analysis of domestic donkeys and all of the living asses and half-asses clearly shows that domestic donkeys were domesticated from African wild asses (Figure 24.4). It has been suggested that the Asiatic half-ass (onager, *E. hemionus*) was also either domesticated in western Asia or used to hybridize with the domestic donkeys there (Clutton-Brock 1999). The genetic data reveal no evidence to support this. Although the genetic data do not exclude the possibility of hybridization in the ancient world between Asiatic half-asses and domestic donkeys, the offspring of such a pairing would be sterile (Clutton-Brock 1999) and therefore a genetic dead end. Therefore, it is most likely that the African wild asses are the sole progenitors of the domestic donkey.

Two distinct groups of domestic donkey mtDNA haplotypes were revealed by the phylogenetic analysis (Figure 24.5). The average sequence divergence between these two clades was 15.73 ± 0.61 substitutions (3.29 ± 0.01%) with a maximum divergence of 27 substitutions (5.64 ± 0.01%) (Beja-Pereira et al. 2004). Because these levels of divergences are similar to the levels of divergence found between subspecies of African wild ass (16.33 ± 0.21; 3.41 ± 0.01%), the authors suggest at least two distinct maternal origins for the modern domestic donkeys that may indicate there were two distinct centers of domestication (Bruford et al. 2003).

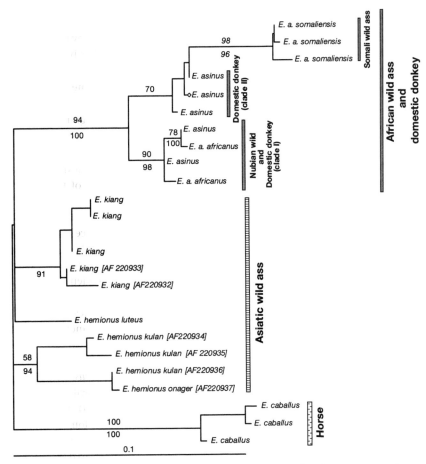

FIGURE 24.4 Phylogenetic tree based on mtDNA control region sequences from donkeys *(Equus asinus)* and their kin. The grouping of domestic donkeys and African wild asses *(E. africanus)* suggests that this is the ancestor species and excludes the Asiatic half-asses *(E. hemionus* and *E. kiang)* as progenitors. Bootstrap support for internal branches is indicated when higher than 50%. (After Beja-Pereira et al. 2004, Fig. 1a).

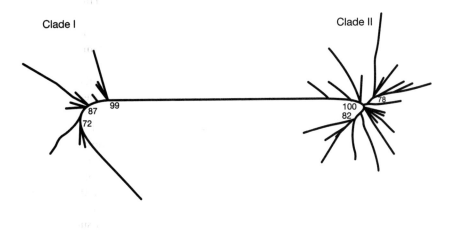

FIGURE 24.5 Unrooted tree of all domestic donkey mtDNA haplotypes. Two clades (I and II) of haplotypes are indicated, suggesting two domestication events. Branch lengths are drawn to scale. Bootstrap support for internal branches is indicated when higher than 50%. (Based on Beja-Pereira et al. 2004.)

TABLE 24.2
Nucleotide Diversity (and Standard Deviation) for Each mtDNA Donkey Clade, for Each Continent and for Each Hypothetical Domestication Center

	Region[a]	Nucleotide Diversity	Standard Deviation
Clade I	Africa	0.0101	0.0006
	Asia	0.0067	0.0020
	Europe	0.0033	0.0017
	Near East	0.0076	0.0028
	Arabian Peninsula	0.0008	0.0004
	Northeast Africa	0.0177	0.0008
Clade II	Africa	0.0118	0.0027
	Asia	0.0075	0.0015
	Europe	0.0082	0.0034
	Near East	0.0140	0.0049
	Arabian Peninsula	0.0028	0.0009
	Northeast Africa	0.0976	0.0033

[a] Northeast Africa: Egypt, Sudan, Ethiopia, and Eritrea; Arabian Peninsula: Saudi Arabia, Yemen, and Oman; Near East: Turkey, Syria, Jordan, Lebanon, Israel, and Iraq.

It has been suggested that centers of domestication might retain higher genetic diversity within each clade of mtDNA haplotypes (Troy et al. 2001; Savolainen et al. 2002). Consequently, analyses of the diversity of mtDNA sequences belonging to each one of the clades could assist in locating where donkeys were first domesticated. Three regions have been suggested as possible centers of domestication for the donkey: the Near East, the Arabian Peninsula, and northeast Africa. Overall, Africa is the continent in which the highest nucleotide diversity (Nei 1987) has been observed. Among the three hypothesized centers of domestication, northeast Africa (Egypt, Sudan, Eritrea, and Ethiopia) displayed significantly higher values for both clades (Table 24.2). This could imply that the two domestic lineages could have derived from the same region. It is important, however, to keep in mind that the genetic patterns within domestic animals may have been shaped largely by the social history and customs of the people caring for them (Leonard et al. 2002).

To test the hypotheses that (1) there were two centers of domestication for the donkey and that (2) both of those centers were in northern Africa, it is important to examine the relationships between genetic diversity present in domestic donkeys and the progenitor species, the African wild ass, across their ranges. Until the Holocene, the African wild ass was present in southwest Asia (southern Israel, along the Red Sea hill shores of the Arabian Peninsula), in northeast Africa (Egypt to Somalia), and in the Atlas ridge (northwest Africa) (Cattani and Bökönyi 2002). Presently, only two subspecies remain: the *E. a. africanus*, or Nubian ass, and the *E. a. somaliensis*, or Somali wild ass. Currently, the Nubian ass is circumscribed to the northeastern part of Sudan close to the Red Sea and the Eritrea border, whereas the Somali ass is mainly present in Ethiopia, Somalia, and Eritrea. Both subspecies are considered critically endangered and are included in the red list of the IUCN (Moehlman 2002). The reduced range of the African wild ass, and the possibility that a lot of genetic variation has been lost because of its endangered status, may hinder the identification of the precise location(s) of domestication, but it is still an important exercise. A phylogenetic analysis of the wild and domestic asses shows that the Nubian wild ass cluster is within one of the two domestic clades. This is strong evidence that one of the centers of domestication was around the northeast region of Sudan around the Red Sea. The second clade of domestic donkeys is more closely related to the Somali wild ass but not obviously within it. This might be expected if the actual progenitor population is now extinct but was closely related to the Somali ass. This may suggest that the second center of domestication was further southeast along the coast of the Red Sea or Gulf of Aden.

Although the precise reasons why the donkey was domesticated remain unknown, two main events that precede the domestication of the donkey may have influenced it. First was the desertification of the Sahara after the end of the last glacial maximum (ca. 9000–12,000 BP). The increased aridity would have necessitated more frequent relocations, greater traveling distances between camps, and permanent oases for early pastoralists. Domestication of the donkey, therefore, may represent the response of pastoralists and other societies in northeastern Africa to these changes. Second was the appearance of the first organized nation in that region, the Pharaonic kingdom, which inevitably brought new patterns of life and movement. Better communication was necessary to enforce the power of the pharaoh across the country and to promote economic growth through trade.

Our genetic data added two main findings to the complex history of the donkey. First, that the donkey was likely domesticated from two divergent African wild-ass lineages and, second, that this likely took place in northeast Africa. Surveys of the remaining wild populations as well as of ancient remains are needed to confirm that such divergent clades could have existed inside the same region. These results strengthen the archaeological data that suggest that northeast Africa is the most probable origin of domestic donkeys (Clutton-Brock 1992; Zeuner 1963). If this is true, the donkey is the only livestock ungulate species to have been domesticated solely in Africa.

Future Prospects

There is still a lot that can be learned from the continued genetic analysis of equids. Further work on larger numbers of nuclear markers may allow us to reconstruct the history of particular breeds and the relationships between breeds. Unraveling the story of donkey domestication, in particular, will require a great deal of additional work. Much more data is necessary on the diversity, or lack of diversity, in Y-chromosome markers of donkeys. From this, we may be able to track paternal lineages and obtain a complementary view on the domestication to that offered by mtDNA as has been done with horses. Most importantly, a survey of wild asses across their range could provide a lot of necessary information regarding location of domestication.

Ancient DNA may be useful in the study of both horses and donkeys since the wild progenitor species of both have become highly endangered or extinct in the wild. The analysis of ancient bone remains may offer a more unbiased view of the distribution of the genetic diversity across the range of the ancestral species. Overall, future molecular genetic studies of equid domestication promise to provide important perspectives to complement ongoing archaeological and biogeographic studies of these important domesticates.

Acknowledgments

A. B.-P. is supported by a research grant from Fundação para a Ciência e Tecnologia (SFRH/BD/2746/2000) through the graduate program in areas of basic and applied biology from University of Porto (GABBA), Portugal. Funding to J. A. L. was provided by a Smithsonian Fellowship and by NSF (OPP 0352634) and to J. A. L. and C. V. by the Swedish Research Council.

References

Allen, W. R. and R. V. Short. 1997. Interspecific and extraspecific pregnancies in equids: Anything goes. *Journal of Heredity* 88: 384–392.

Anthony, D. W. 1986. The "Kurgan culture," Indo-European origins, and the domestication of the horse: A reconsideration. *Current Anthropology* 27: 291–313.

———. 1996. Bridling horse power. In *Horses through time*, S. L. Olsen (ed.), pp. 57–82. Boulder, CO: Roberts Rinehart for Carnegie Museum of Natural History.

Anthony, D. W. and D. R. Brown. 1991. The origin of horseback riding. *Antiquity* 65: 22–38.

Anthony, D. W., D. Y. Telegin, and D. Brown. 1991. The origin of horseback riding. *Scientific American* 265: 94–100.

Bahn, P. G. 1980. Crib-biting: Tethered horses in the Palaeolithic? *World Archaeology* 12: 212–217.

Barclay, H. B. 1982. Another look at the origins of horse riding. *Anthropos* 77: 244–249.

Beja-Pereira, A., P. R. England, N. Ferrand, S. Jordan, A. O. Bakhiet, M. A. Abdalla, M. Mashkour, J. Jordana, P. Taberlet, and G. Luikart. 2004. African origins of the domestic donkey. *Science* 304: 1781.

Benirschke, K., N. Malouf, R. J. Low, and H. Heck. 1965. Chromosome complement differences between *Equus caballus* and *Equus przewalskii*, Poliakoff. *Science* 148: 382–383.

Bennett, D. K. 1980. Stripes do not a zebra make, part I: A cladistic analysis of *Equus*. *Systematic Zoology* 29: 272–287.

Bennett, D. 1992. Origin and distribution of living breeds of the domestic horse. In *Horse breeding and management*, J. W. Evans (ed.), pp. 41–61. World Animal Science, C7, Amsterdam: Elsevier Science Ltd.

Bjørnstad, G. and K. H. Røed. 2001. Breed demarcation and potential for breed allocation of horses assessed by microsatellite markers. *Animal Genetics* 32: 59–65.

———. 2002. Evaluation of factors affecting individual assignment precision using microsatellite data from horse breeds and simulated breed crosses. *Animal Genetics* 33: 264–270.

Bjørnstad, G., N. ø. Nilsen, and K. H. Røed. 2003. Genetic relationship between Mongolian and Norwegian horses? *Animal Genetics* 34: 55–58.

Bodson, L. 1985. L'utilisation de l'ane dans l'antiquité Greco-Romaine. *Ethnozootechnie* 56: 27–41.

Bökönyi, S. 1991. The earliest occurrence of domestic asses in Italy. In *Equids in the ancient world*, Volume 2, R. Meadow and H. Uerpmann (eds.), pp. 178–216. Wiesbaden: Ludwig Reichert Verlag.

Bongianni, M. (ed.). 1987. *Simon & Schuster's guide to horses & ponies of the World*. New York: Simon & Schuster Inc.

Bouman, I. and J. Bouman. 1994. The history of Przewalski's horse. In *Przewalski's horse: The history and biology of an endangered species*, L. Boyd and K. A. Houpt (eds.), pp. 5–39. Albany: State University of New York Press.

Bowling, A. T., A. Del Valle, and M. Bowling. 2000. A pedigree-based study of mitochondrial D-loop DNA sequence variation among Arabian horses. *Animal Genetics* 31: 1–7.

Brown, D. and D. Anthony. 1998. Bit wear, horseback riding and the Botai site in Kazakhstan. *Journal of Archaeological Science* 25: 331–347.

Bruford, M. W., D. G. Bradley, and G. Luikart. 2003. DNA markers reveal the complexity of livestock domestication. *Nature Reviews Genetics* 4: 900–910.

Cañon, J., M. L. Checa, C. Carleos, J. L. Vega-Pla, M. Vallejo, and S. Dunner. 2000. The genetic structure of Spanish Celtic horse breeds inferred from microsatellite data. *Animal Genetics* 31: 39–48.

Cattani, M. and S. Bökönyi. 2002. Ash-Shumah an early Holocene settlement of desert hunters and mangrove foragers in the Yemeni Tihamah. In *Essays in the late prehistory of Arabian Peninsula*, S. Cleuziou, M. Tosi, and J. Zarins (eds.), pp. 31–53. Rome: Instituto Italiano Per l'Africa e l'Oriente.

Clutton-Brock, J. 1992. *Horse power. A history of the horse and the donkey in human societies*. Cambridge, MA: Harvard, University Press.

———. 1996. Horses in history. In *Horses through time*, S. L. Olsen (ed.), pp. 83–102. Boulder, Co: Roberts Rinehart for Carnegie Museum of Natural History.

———. 1999. *A natural history of domesticated mammals*, 2nd edition. Cambridge: Cambridge University Press.

Clutton-Brock, J. and S. Davis. 1993. More donkeys from Tell Brak. *Iraq* 55: 209–221.

Cowan, C. W. and P. J. Watson (eds.). 1992. *The origins of agriculture*. Washington, D.C.: Smithsonian Institution Press.

Cunningham, E. P., J. J. Dooley, R. K. Span, and D. G. Bradley. 2001. Microsatellite diversity, pedigree relatedness and the contributions of founder lineages to thoroughbred horses. *Animal Genetics* 32: 360–364.

Davis, S. 1987. *The archaeology of animals*. London: Yale University Press.

Denzau, G. and H. Denzau. 1999. *Wildesel*. Stuttgart: Jan Thorbecke.

Diamond, J. M. 1991. The earliest horsemen. *Nature* 350: 275–276.

FAO. 2000. *FAO Statistical Database Website*. Rome: Food and Agriculture Organization.

Forstén, A. 1992. Mitochondrial-DNA time-table and the evolution of *Equus*: Comparison of molecular and paleontological evidence. *Annales Zoologici Fennici* 28: 301–309.

Fumihito, A., T. Miyake, S.-I. Sumi, M. Takada, S. Ohno, and N. Kondo. 1994. One subspecies of the red junglefowl *(Gallus gallus gallus)* suffices as the matriarchic ancestor of all domestic breeds. *Proceedings of the National Academy of Sciences, USA.* 91: 12505–12509.

George, M., Jr., and O. A. Ryder. 1986. Mitochondrial DNA evolution in the genus *Equus*. *Molecular Biology and Evolution* 3: 535–546.

Giuffra, E., J. M. H. Kijas, Ö. Carlborg, J.-T. Jeon, and L. Andersson. 2000. The origin of the domestic pig: Independent domestication and subsequent introgression. *Genetics* 154: 1785–1791.

Gray, R. D. and Q. D. Atkinson. 2003. Language-tree divergence times support the Anatolian theory of Indo-European origin. *Nature* 426: 435–439.

Groves, C. 1986. The taxonomy, distribution and adaptations of recent equids. In *Equids in the ancient world*, Volume 1, R. Meadow and H.-P. Uerpmann (eds.), pp. 11–65. Weisbaden: Ludwig Reichert..

Hedrick, P. W., K. M. Parker, E. L. Miller, and P. S. Miller. 1999. Major histocompatibility complex variation in the endangered Przewalskii's horse. *Genetics* 152: 1701–1710.

Hellborg, L. and H. Ellegren. 2003. Y chromosome conserved anchored tagged sequences (YCATS) for the analysis of mammalian male-specific DNA. *Molecular Ecology* 12: 283–291.

Hiendleder, S., B. Kaupe, R. Wassmuth, and A. Janke. 2002. Molecular analysis of wild and domestic sheep questions current nomenclature and provides evidence for domestication from two different subspecies. *Proceedings of the Royal Society of London, Series B* 269: 893–904.

Hofreiter, M., D. Sere, H. N. Poinar, M. Kuch, and S. Pääbo. 2001. Ancient DNA. *Nature Reviews* 2: 353–329.

Ishida, N., T. Hasegawa, K. Takeda, M. Sakagami, A. Onishi, S. Inumaru, M. Komatsu, and H. Mukoyama. 1994. Polymorphic sequence in the D-loop region of equine mitochondrial DNA. *Animal Genetics* 25: 215–221.

Ishida, N., T. Oyunsuren, S. Mashima, H. Mukoyama, and N. Saitou. 1995. Mitochondrial DNA sequences of various species of the genus *Equus* with special reference to the phylogenetic relationship between Przewalskii's wild horse and domestic horse. *Journal of Molecular Evolution* 41: 180–188.

Jansen, T., P. Forster, M.A. Levine, H. Oelke, M. Hurles, C. Renfrew, J. Weber, and K. Olek. 2002. Mitochondrial DNA and the origins of the domestic horse. *Proceedings of the National Academy of Sciences, USA* 99: 10905–10910.

Kadwell, M., M. Fernandez, H. F. Stanley, R. Baldi, J. C. Wheeler, R. Rosadio, and M. W. Bruford. 2001. Genetic analysis reveals the wild ancestors of the llama and the alpaca. *Proceedings of the Royal Society of London, Series B* 268: 2575–2584.

Kavar, T., F. Habe, G. Brem, and P. Dovc. 1999. Mitochondrial D-loop sequence variation among the 16 maternal lines of the Lipizzan horse breed. *Animal Genetics* 30: 423–430.

Kelly, L., A. Postiglioni, D. F. De Andrés, J. L. Vega-Plá, R. Gagliardi, R. Biagetti, and J. Franco. 2002. Genetic characterisation of the Uruguayan Creole horse and analysis of relationships among horse breeds. *Research in Veterinary Science* 72: 69–73.

Kim, K.-I., Y.-H. Yang, S.-S. Lee, C. Park, R. Ma, J. L. Bouzat, and H. A. Lewin. 1999. Phylogenetic relationships of Cheju horses to other horse breeds as determined by mtDNA D-loop sequence polymorphism. *Animal Genetics* 30: 102–108.

Larsen, G., K. M. Dobney, U. Albarella, M. Fang, E. Matisoo-Smith, J. Robins, S. Lowden, H. Finlayson, T. Brand, E. Willerslev, P. Rowley-Conwy, L. Andersson, and A. Cooper. Worldwide phylogeography of wild boar reveals multiple centers of domestication. *Science* 307: 1618–1621.

Leonard, J. A., R. K. Wayne, J. Wheeler, R. Valadez, S. Guillén, and C. Vilà. 2002. Ancient DNA evidence for Old World origin of New World dogs. *Science* 298: 1613–1616.

Levine, M.A. 1990. Dereivka and the problem of horse domestication. *Antiquity* 64: 727–740.

———. 1999a. The origins of horse husbandry on the Eurasian steppe. In *Late prehistoric exploitation of the Eurasian steppe*, M. Levine, Y. Rassamakin, A. Kislenko, and N. Tatarintseva (eds.), pp. 5–58. Cambridge: McDonald Institute Monographs, University of Cambridge.

———. 1999b. Botai and the origins of horse domestication. *Journal of Anthropological Archeology* 18: 29–78.

Lindgren, G., N. Backstrom, J. Swinburne, L. Hellborg, A. Einarsson, K. Sandberg, G. Cothran, C. Vilà, M. Binns, and H. Ellegren. 2004. Limited number of patrilines in horse domestication. *Nature Genetics* 36: 335–336.

Lister, A. M., M. Kadwell, L. M. Kaagan, W. C. Jordan, M. B. Richards, and H. F. Stanley. 1998. Ancient and modern DNA from a variety of sources in a study of horse domestication. *Ancient Biomolecules* 2: 267–280.

Loftus, R. T., D. E. MacHugh, D. G. Bradley, P. M. Sharp, and P. Cunningham. 1994. Evidence for two independent domestications of cattle. *Proceedings of the National Academy of Sciences, USA.* 91: 2757–2761.

Luikart, G., L. Gielly, L. Excoffier, J.-D. Vigne, J. Bouvet, and P. Taberlet. 2001. Multiple maternal origins and weak phylogeographic structure in domestic goats. *Proceedings of the National Academy of Sciences, USA.* 98: 5927–5932.

Marklund, S., R. Chaudhary, L. Marklund, K. Sandberg, and L. Andersson. 1995. Extensive mtDNA diversity in horses revealed by PCR–SSCP analysis. *Animal Genetics* 26: 193–196.

Moehlman P. D. 2002. Status and action plan for the African wild ass *(Equus africanus)*. In *Equids: Zebras, asses and horses. Status survey and conservation action plan. IUCN/SSC Equid Specialist Group*, P. D. Moehlman (ed.), pp. 2–10. Gland, Switzerland: IUCN Publications.

Nei, M. 1987. *Molecular Evolutionary Genetics*. New York: Columbia University Press

Oakenfull, E. A. and J. B. Clegg. 1998. Phylogenetic relationships within the genus *Equus* and the evolution of α and θ globin genes. *Journal of Molecular Evolution* 47: 772–783.

Oakenfull, E. A. and O. A. Ryder. 1998. Mitochondrial control region and 12S rRNA variation in Przewalski's horse *(Equus przewalskii)*. *Animal Genetics* 29: 456–459.

Oakenfull, E. A., H. N. Lim, and O. A. Ryder. 2000. A survey of equid mitochondrial DNA: Implications for the evolution, genetic diversity and conservation of *Equus*. *Conservation Genetics* 1: 341–355.

Olsen, S. L. 1989. Solutré: A theoretical approach to the reconstruction of Upper Palaeolithic hunting strategies. *Journal of Human Evolution* 18: 295–327.

———. 1995. Pleistocene horse-hunting at Solutré: Why bison jump analogies fail. In *Ancient peoples and landscapes*, E. Johnson (ed.), pp. 65–75. Lubbock: Texas Tech University Press.

———. 1996. Horse hunters of the ice age. In *Horses through time*, S. L. Olsen (ed.), pp. 35–56. Boulder, CO: Roberts Rinehart for Carnegie Museum of Natural History.

———. 2003. The exploitation of horses at Botai, Kazakhstan. In *Prehistoric steppe adaptation and the horse*, M. Levine, C. Renfrew, and K. Boyle (eds.), pp. 83–104. Cambridge: McDonald Institute Monographs.

Peters, J. 1986. *Bijdrage tot de archeozoologie van Soeden en Egypte* Doctoral thesis, Rijksuniversiteit.

Rogers, R. A. and L. A. Rogers. 1988. Notching and anterior beveling on fossil horse incisors: Indicators of domestication? *Quaternary Research* 29: 72–74.

Ryder, O.E. 1994. Genetic studies of Przewalski's horse and their impact on conservation. In *Przewalski's horse: The history and biology of an endangered species*, L. Boyd and K. A. Houpt (eds.), pp. 75–92. Albany: State University of New York Press.

Savolainen, P., Y. Zhang, J. Luo, J. Lundeberg, and T. Leitner. 2002. Genetic evidence for an East Asian origin of domestic dogs. *Science* 298: 1610–1613.

Simpson, G. G. 1951. *Horses*. New York: Oxford University Press.

Smith, B. D. 1998. *The emergence of agriculture*. New York: W. H. Freeman.

Tozaki, T., N. Takezaki, T. Hasegawa, N. Ishida, M. Kurosawa, M. Tomita, N. Saitou, and H. Mukoyama. 2003. Microsatellite variation in Japanese and Asian horses and their phylogenetic relationship using a European horse outgroup. *Journal of Heredity* 94: 374–380.

Troy, C. S., D. E. MacHugh, J. F. Bailey, D. A. Magee, R. T. Loftus, P. Cunningham, A. T. Chamberlain, B. C. Sykes, and D. G. Bradley. 2001. Genetic evidence for Near-Eastern origins of European cattle. *Nature* 410: 1088–1091.

Uerpmann, H.-P. 1987. *The ancient distribution of ungulate mammals in the Near East*. Wiesbaden: Dr. Ludwig Reichert.

Vilà, C., P. Savolainen, J. E. Maldonado, I. R. Amorim, J. E. Rice, R. L. Honeycutt, K. A. Crandall, J. Lundeberg, and R. K. Wayne. 1997. Multiple and ancient origins of the domestic dog. *Science* 276: 1687–1689.

Vilà, C., J. A. Leonard, A. Götherström, S. Marklund, K. Sandberg, K. Lidén, R. K. Wayne, and H. Ellegren. 2001. Widespread origins of domestic horse lineages. *Science* 291: 474–477.

Wallner, B., G. Brem, M. Müller, and R. Achmann. 2003. Fixed nucleotide differences on the Y chromosome indicate clear divergence between *Equus przewalskii* and *Equus caballus*. *Animal Genetics* 34: 453–456.

Wang, W., A.-H. Liu, S.-Y. Lin, H. Lan, B. Su, D.-W. Xie, and L.-M. Shi. 1994. Multiple genotypes of mitochondrial DNA within a horse population from a small region in Yunnan Province of China. *Biochemical Genetics* 32: 371–378.

Wayne, R. K., J. A. Leonard, and A. Cooper. 1999. Full of sound and fury: The recent history of ancient DNA. *Annual Review of Ecology and Systematics* 30: 457–477.

White, R. 1989. Husbandry and herd control in the Upper Paleolithic. *Current Anthropology* 30: 609–616.

Yang, Y. H., K. I. Kim, E. G. Cothran, and A. R. Flannery. 2002. Genetic diversity of Cheju horses *(Equus caballus)* determined by using mitochondrial DNA D-loop polymorphism. *Biochemical Genetics* 40: 175–186.

Zeder, M. A. 1986. The equid remains from Tal-e Malyan. In *Equids of the Old World*, Volume 1, R. Meadow and H.-P. Uerpmann (eds.), pp. 367–412. Beihefte zum Tubinger Atlas des Vorderern Orients. Tubingen, Germany: University of Tubingen.

Zeuner, F. E. 1963. *A history of domesticated animals*. London: Hutchinson.

INDEX

Absolute dating, 131
Adaptive syndrome of domestication, 3, 25, 99, 106
　defined, 18–19
Admixture analysis, 332–333, 334–335, 336, 337
Adventive crops, 100
Aggressive nature, as attribute of animal domestication, 4, 171
Agriculture, origins of, 1
Albarella, Umberto, 209
Allele sharing, 129
Allium tuberosum. *See* Chinese chive
Allopolyploids, 153, 158
Alpaca, 329. *See also* Camelids, South American
　fiber, 338
　hybridization between llama and, 338
　mitochondrial introgression in, 334, 337
　phylogenetic relationship with llama, 331
Amplified fragment length polymorphism (AFLP), 7, 136
Ancient DNA (aDNA), 343–344
　promise of, 10
　to study crop domestication, 115–116
　to study domestication of dogs, 286–289
　to trace evolutionary history of domesticated animals, 7, 276–277
Animal domestication. *See also individual animals by name*
　attractive candidates for, 4
　body size, changes in, 172–173
　breeding partners, human selection of, 172
　causality, 181
　demographic profiling and, 174–175, 188
　dietary shifts and, 174
　documenting
　　anchoring genetic data to archaeological narrative, 275–276
　　ancient DNA, uses of, 7, 276–277
　　archaeogenetics, 273
　　archaeogenomics, 277–278
　　archaeological approaches to, 171–180
　　geographic focus on, 176–177
　　phylogeographic discontinuities, 273–275
　environmental appreciation, decline of, 171
　facial structure, changes in, 171–172
　genetically driven markers of, 4, 171–173
　genetic documentation of, 5–7
　and genetic isolation from wild progenitor populations, 173

　horns, changes in size and shape of, 172, 181–182, 183
　human intervention, evidence of, 171, 172, 176
　juvenile behavioral traits, 171
　microsatellite loci, 319
　mitochondrial DNA, 318–319
　moment vs. process, 178
　morphological markers of, 4, 171–174
　multiple markers of, 177–178
　nonmorphological markers of, 174–176
　origin of, 1
　paedomorphic behavioral traits, 171
　penning as evidence of, 4, 174, 176
　phylogeographic discontinuity, 273–275
　plastic responses in, 173–174
　progenitor population, distribution of, 176
　random genetic drift, 173
　selection for less aggressive behavior, 171
　tooth placement and, 171–172
　y-chromosome markers, 319, 343, 346
　zoogeography and abundance, 175–176, 210–212
Annual seed plants, human intervention in, 22
Arbitrarily primed PCR (AP-PCR), 103, 136
Archaeogenetics, 273
Archaeogenomics, 277–278
Arrowroot (*Maranta arundinacea*), 46
Atomic mass spectrometer (AMS) dating, 5
Autoallopolyploid, 161
Autopolyploid, 153, 161

Banana, domestication of in Africa, 68–81
　categories, 68
　complexes, timing and diffusion of, 68, 78
　Nkang, South Cameroon archaeological case study
　　analysis of phytoliths from, 74–77
　　animal taxa represented at, 72
　　archaeological and paleo-environmental context, 69–72
　　chronology of, 70
　　early food-producing complex in Southern Cameroon, 76–78
　　environmental context, reconstruction of, 70
　　faunal data from, 70–72
　　location of, 69
　　musaceae phytoliths from, 72–76, 77
　　plant taxa recovered from, 71
　　radiocarbon dates recovered from, 70
　　soil condition, 77

　　phytoliths, 68, 72–77
　　progenitors, 68
Behavioral change, as evidence of domestication, 16, 171
Beadle, George, 104
Beja-Pereira, Albano, 342
Berville, André, 143
Besnard, Guillaume, 143
Bitter manioc, 47, 49
Blattner, Frank R., 134
"Blue-rim" phenomenon, 183–184
Body size
　changes in, as evidence of domestication, 172–173, 215
　factors affecting, 184–185
　of goats and gazelles compared, 197, 199
Bottleneck of domestication, 107–108. *See also* Founder effect phenomenon
Bovine genetic variation, biochemical analysis of, 319
Bradley, Dan, 1, 2, 3, 273, 317
Braidwood, Robert, 1
Breton, Catherine, 143
Bruford, Michael W., 306, 329
Bruno, Maria, C., 32
Budares, 49
Burning, 20

Caballoid horse, 344
Camelids, South American. *See also* Alpaca; Guanaco; Llama; Vicuña
　admixture analysis, 332, 334–335, 336
　ancestors of, 329
　classification of, 228, 329
　domestication of, 228–244
　　archaeological sites for study of, 236
　　bone morphology, 232, 238
　　contextual indicators, 234, 239
　　dental morphology, 230, 231, 238
　　exploitation, 239–240
　　fleece, analysis of, 232–233, 338, 239
　　human control over reproduction, 230
　　indicators of, 230–234
　　intensification of use, 233, 235
　　mortality patterns, 234, 238–239, 240
　　origins of, 229
　　osteometry, 230–232, 237–238
　　prior research on, 229–230
　　in south-central Andes, recent research on, 234–240
　　species diversity and temporal trends, 233–234, 235

Camelids, South American. *See also* Alpaca; Guanaco; Llama; Vicuña (*continued*)
  future research, suggestions for, 240–241, 335, 337–338
  genetic analysis of, 329–341
  geographic variation in size of, 231
  hybridization between llamas and alpacas, 338
  morphological changes, 329–331
  mtDNA diversity in, 274–275
  skeletal differences among, 232
  uses of, 228, 235
Candidate gene approach, 107
Cassava (*Manihot esculenta* Crantz ssp. *esculenta*)
  caveats regarding research on, 130–131
  compilospecies hypothesis re:, 123–124, 131
  domestication of, 113
  genetic diversity in, 127
  geographic location of, 126
  global importance as food source, 123
  hybrid origin vs. single progenitor, 124
  molecular markers of domestication, 124–126
    DNA sequences, 125
    microsatellites, 125–126
  origins of, 123, 130
  prior research on, 123–124
  single wild progenitor, 130
  wild populations most closely related to, 129
Cattle, domestic
  archaeological inference from patterns of genetic diversity, 325–326
  dispersal from initial centers and local contribution, 318
  future research, suggestions for, 326
  genetic variation, 318
  global distribution of, 317
  mtDNA diversity in, 274, 324
  origins of
    Bos taurus and *Bos indicus*, 317–318
    genetics, 317–328
  phylogeographic analyses, 320–325
  population-based analysis, 319–320
  and population expansion, 323
  research design, 319
  secondary migration processes, 326
  study results, 319–325
  taxonomy, 317
  wild ancestor of, 317, 326
Causal chain theory of domestication, 16–17
Centers of origin concept, 100, 101
Chemical composition, changes in, 99
Chenopods, Andean (*Chenopodium*)
  crop/weed complexes, 33–34
  domestication of
    morphological approach to, 34–36
    taxonomic and genetic approaches to, 32–34
  future research, suggestions for, 43
  human interaction with, 32
  margin configuration, 42
  pericarp patterning, 41, 42
  samples, 37, 38
  seed(s)
    coat texture, 41
    diameter, 37, 39
    morphological analysis of, 32, 36, 37–39, 41–43
    size, 34, 35, 38–39
    testa thickness, 32, 34–36, 38, 42
  of the Southern Lake Titicaca Basin, Bolivia, case study of, 36–37, 40–41
  wild and domesticated compared, 34
Chikhi, Lounès, 329
Childe V. Gordon, 1
Chinese chive (*Allium tuberosum*)
  domestication of, 134
    evidence for, 135–137
    genomic markers for, 135–137
    morphological and physiological markers for, 135
    via a single event, 135
  geographical distribution of, 140
  intertaxon hybridization, potential for, 134
  origins of, 114
  prior research on, 134–135
  study
    materials and methods of, 137–138
    results, 138–140
  uses of, 134
  wild
    and cultivated forms compared, 134
    progenitor of, 140
Chinese leek. *See* Chinese chive
Chipped tool residues, analysis of, 64
Chloroplast genome of plants (cpDNA), 7
"Chromosome painting," 102
Climate change
  in domestication, 11
  in emergence of agriculture, 1
Clonal propagation, vs. sexual reproduction, 5–6
Co-adapted complexes, 112
Cob phytoliths, identification of, 62–63
Cob types, 88
Compilospecies hypothesis, 123–124, 131
Convention on International Trade in Endangered Species of Wild Fauna and Flora (CITES), 338
Convergent domestication, hypothesis of, 115
Copia-SSR, 103
Crop domestication. *See also* Plant domestication; *individual plants by name*
  centers of origin concept, 100, 101
  comparisons, 115
  defined, 99
  future prospects for study of, 114–116
  genetic data used to study, 101–104
    case studies illustrating, 113–114
  use of ancient DNA to study, 115–116
Crop evolution, effects of selection on, 5
Crop origins, 99–100, 101
Crop-wild-weed geneflow, 111–112, 141
Cross-breeding, 1
Cucurbita pepo. *See* Squash
Cytogenetic studies, 101–102

Darwin, Charles, 1
Day length sensitivity, 99
Decker-Walters, Deena, 2
De Langhe, E., 68
De Maret, P., 68

Demographic profiling, and animal domestication, 174–175, 188
Denaturing gradient gel electrophoresis (DGGE), 103
Dereivka cult stallion, 245, 247, 251
Diploid, defined, 153
DNA sequencing, 102–104
  evolution, rate of, 6
  to study origins of cassava, 125, 127–129
  to study origins of oca, 114
Dobney, Keith, 209
Doebley, John, 2
Dog(s)
  breeds
    diverse genetic heritage, 284
    genealogical relationship among, 284
    origin of, 283–284
    sequence diversity, 283
  domestication of
    DNA from ancient dogs, 286–289
    events, number of, 289–290
    future research, suggestions for, 290
    gray wolf as progenitor, 279, 282, 287, 288
    human migration and, 289–290
    mitochondrial control region sequencing, 280–281, 283, 285
    molecular genetic analysis of, 281–284
    molecules vs. morphology, 281
    mtDNA diversity in, 275
    new world dogs, independent origin of, 284–289
    oldest definitive remains, 284
    origin and timing of, 279, 281–283
    origin/interbreeding events, 283, 289
    previous research, 279–280
    protodogs, record of, 289
    time of first, 289
    Xoloitzcuintle, 285–286
Domestic ass, mtDNA diversity in, 275
Domestication. *See also* Animal domestication; Crop domestication; Plant domestication
  adaptive syndrome of, 3, 18–19, 25, 99, 106
  advantages of using multiple molecular markets to study, 150
  aspects of, studied with genetic data, 99–101
  behavioral change as evidence of, 16, 171
  captivity as requirement for, 210
  causal chain theory of, 16–17
  common features underlying 1–2
  convergent, 115
  defined, 15, 99, 209–210, 230
  dispersal of domesticates, tracing, 9
  documenting
    archaeological approaches to, 3–5
    contributions of archaeology and genetics to, 7–10
    and location of domestication events, 8–9
    temporal sequence of domestication, 9–10
  evidence for, three general categories of, 16
  evolution under, 100–101
  founder effect phenomenon, 100–101, 107–108, 113

full, 16
genetic basis of, 112–113
genetic changes as evidence of, 16
genetic diversity, loss of, 113
genetic documentation of, 5–7
human control over reproduction, 230
human intervention and, 2, 16, 18, 210
incipient, 205
independent, 100
indicators of, 230–234
interdisciplinary approaches to study of, 1
moment vs. process, 178, 205
morphological change as evidence of, 4–5, 15, 16
multiple, 100, 109–111, 218, 298
origin of, 1
plant and animal, differences in, 3
and polyploidy, 157–158
population genetic techniques, application of, 9
possible evolutionary relationships of crop species and their wild relatives, 136
via a single event, 13
technological advances influencing study of, 1
temporal sequence of
  documenting, 9–10
  molecular-clock approach to, 10, 275
universalist approaches to explaining, 10–11
using molecular data to address questions of, 124–125
Domestication bottleneck, 100, 113
Domestication events
  documenting number and location of, 8–9
  in dogs, number of, 289–290
  in goats, geographic location of, 300–301
Domestication genes, 6, 101
  clusters, 112
  defined, 106
  maize, 106–107
Dominance hierarchy, 4
Donkey, domestication of
  from African wild asses, 348
  future research, suggestions for, 350–351
  geographic origin of, 342
  molecular genetic analysis of, 348–350
  time and place of, 342
Doutrelepont, H., 68
"Dry ashing," 83
Durham Pig Project, 210

Edgeground cobble groundstone tool, 59, 61
Emshwiller, Eve, 1, 2, 99, 153
England, P. R., 294
Ensete phytoliths, 73, 74, 77
Environmental appreciation, decline of in animals, 171
Equid domestication, timing of, 343. *See also* Donkey, domestication of; Horse(s), domestication of
Equus ferus, 245, 247
Evolution
  under domestication, 100–101
  genetic basis of, fundamental processes in, 101
  phylogeny, 273
Extra virgin olive oil, 143

Factorial correspondence analysis, 332
Fermented mare's milk, 250
Fernández, H., 294
Fixed heterozygosity, 158, 161
Flow cytometry, 162
  defined, 160
Fluorescent *in situ* hybridization (FISH), 102
Food residue phytolith analysis, 21, 82–83
Footprint of selection, 99, 108–109
Founder effect phenomenon, 100–101, 107–108, 113
Fragment analysis, 102
Free-living population(s), 100
  introgression between domesticates and, 111–112
Friesen, Nikolai, 134
Full domestication, 16

Gene exchange, 101
Geneflow
  as a complication for determining crop origin, 131
  crop-wild-weed, 111–112, 141
  from teosinte to maize, 106
Generation time, 5
Genetic bottlenecks, 99, 135
Genetic changes, as evidence of domestication, 16
Genetic diversity
  in cassava crop, 127
  geographic trends within, 276
  interpretations of, 273
  Przewalski horse, 344
  reduction in, 108, 113
Genetic drift, random, 173
Genetic hitchhiking, 108
Genetic isolation, 173, 202, 205
Genetic plasticity, 2
Genetic variability, causes of, 141
Gene trees, 125, 128
Genome analysis, 101
Genome typing, 277
Genomic *in situ* hybridization (GISH), 102
Genotyping, 277
Geometric Impressed Pottery Province (GIPP), 256
Germination dormancy, 25
Glume induration, 86
Goat(s) (*Capra hircus*)
  adaptive abilities of, 294
  domestic, origins and spread of, 303–304
  domestication of
    age, analysis of, 190–191
    animal-oriented markers of, 181–184
    archaeological case study, 192–202
    "blue-rim" phenomenon, 183–184
    body size, as marker of, 184–185, 190–192, 193, 196–197, 202–203
    bone structure, changes in, 183–184
    demographic markers of, 187–189, 191–192, 197–198, 201–202, 203–204
    DNA markers and sequences, 295–296
    domestic events, geographic location of, 300–301
    genetic lineages, divergence of, 300
    harvest strategies and, 185–186, 187, 189, 204
    head-butting behavior and, 182
    herd management and, 201–205
    horns, changes in, 181–183
    human intervention and, 181, 185–189
    loss of habitat, 186
    markers of, 181–189
    molecular variation, geographic patterns of, 304
    morphological diversity and distinctiveness, 295
    origins of, 297–298
    overhunting and, 186
    population expansions, timing of, 303–303
    population structure and gene flow, 303
    restraints, use of, 189
    sex-specific harvest profiles, 204
    statistical analyses, 296
    study results, 296–303
    tissue sampling and DNA extraction, 294–295
    using mtDNA, 295
    y-chromosome DNA, 296
    in the Zagros, 202
    zoogeographic markers of, 186–187
  importance of, to economies, 294
  modern skeletal collections, analysis of, 190–192
  wild
    distribution of, 182
    horns, 181
    progenitor, 181, 286–297
Grater chips, for processing root crops, 49
Grater flakes, 49, 64
Gray wolves, 279, 282, 287, 288
Greaves' Effect, 255
Grinding stones, 59, 60–62
Guanaco, 329, 332. *See also* Camelids, South American
  bidirectional introgression in, 334
  and vicuña compared, 334

Habitat, loss of, 186
Haplotype(s), 125, 273, 285
Haplotypic variation, 129, 273
Heliconia phytoliths, 78
Herding and farming, transition to, 1
Heterozygosity, 300, 301
  fixed, 158, 161
Hidden beneficial alleles, 115
Homology assessment, 104
Horns, changes in as marker of domestication, 172, 181–182, 183
Horse(s)
  Acquisition of, 343
  ancestor of, 245–246, 344–346
  bloodline, 346
  breeds, breeding practices and origins of, 346–348
  classification difficulties in, 246
  cultural responses associated with acquisition of, 343
  Dereivka cult stallion, 245, 247, 251
  domestication of, 245–269, 342

domestication of (*continued*)
adult males, role of, 249
artifactual and architectural evidence of, 253–254
bit wear, 254, 255, 256
body size, 247–248
breeding practices, 348–350
cranial morphology, 246–247
culling strategies of, 249, 256
detection of, in archaeological records, 256
early, documenting, 245
forests, gradual succession of, 252
future research, suggestions for, 252
Geaves' Effect, 255, 261
genetic markers, 343–344
geographic distribution and, 251–253
humans and, 342
increase of frequencies in horse remains, 253
livestock culling strategies, 249
mare's milk, importance of, 250
molecular genetic analysis of, 344–346
molecular studies, 245
mortality patterns, 248–251
multiple lineages, 345
pathologies associated with, 254–255
possible wild progenitors, 245–246
prior research, 246–255
process of, 255–256
remains, increased frequencies of, 253
riding as indication of, 254, 255–256
saddle fitting and padding, 254
skeleton as proof of, 256
tooth size, 247, 249
traction and, 254
translocation and, 347
Equus ferus, 245, 247
mixed use of horses for food and riding, 265
Przewalski horse, 246, 247, 344
research at Botai culture settlements
bit wear, 260, 261
body-part distribution, 261–262
Botai settlement patterns, 257, 263–264
distribution of horses by sex and age, 258
horse manure, presence of, 264
horse rituals, 263
juvenile and adult horses compared, 259
lithic transport, evidence of, 262–263
milking, evidence of, 264–265
mortality patterns, 257–259
pathologies, 260–261
proportions of adult female/male horses, 258
slaughter methods, 259–260
teeth, changes in, 258
thong-smoothers, presence of, 261, 262
use of horses for traction or as beasts of burden, 260
ritual horse sacrifice, 249, 255
role of females and males in formation of breeds, 347–348
wild
disappearance of, 342

progenitor species, genetic diversity and, 345
Human intervention
in animal reproduction, 230
in domestication
animals, 171, 172, 176
general, 2, 16
plants, 18, 22
Hunter-gatherers, social and economic complexity among, 235
Hunting and gathering, transition from, 1, 15, 22
Hybridization, 9, 10, 111, 123
fluorescent *in situ* (FISH), 102
genomic *in situ* (GISH), 102
interspecies, 211
interspecific, 6, 100
intertaxon, 134
introgressive, 101, 111
between llamas and alpacas, 338

Incipient domestication, 205
Independent domestication, 100
Interbreeding, 100
Inter-MITE polymorphisms (IMP), 103
Inter-retrotransposon amplified polymorphism (IRAP), 103
Inter-simple sequence repeats (ISSRs), 103, 136
Interspecies hybridization, 211
Interspecific hybridization, 6, 100
Intertaxon hybridization, 134
Intraspecific phylogenies, 111
Introgression
between domesticates and free-living populations, 111–112
mitochondrial, in alpacas, 334, 337
Introgressive hybridization, 101
Isozymes, 102
ITS sequencing, 160

Jaccard's Coefficient of Genetic Similarity, 110

Koumiss, 250

Leonard, Jennifer, A., 279, 342
Linear enamel hypoplasia (LEH), as evidence of domestication, 221
Lirén (*Calathea allouia*), 46
Llama. *See also* Camelids, South American
hybridization between alpaca and, 338
nuclear introgression in, 337
origin of, 329
phylogenetic relationship with alpaca, 331
Luikart, G., 294

MacNeish, Richard, 1
Magee, David, 317
Maize (*Zea mays*)
analysis of food residues to document presence of in Central/South America, 82–95
case study, 88–93
collection sites, map of, 90
phytolith taxonomy, 84

residue samples, recovery of, 84, 91
silica body assemblages from different maize types, 85–87
statistical analysis of assemblage data, 84–85
bottleneck of domestication, 107–108
cob phytolith assemblages, identification of, 62–63, 93
diffusion, rates and routes of, 82
domestication genes, 6, 106–107
domestication of
research on, aspects of, 108–109
study of using genetic data, 104–109
early diversification of, models for, 83
evolution of, as a continuous process, 90
geneflow between maize and teosinte, 106
glume development, 86, 93
importance of, in the ceramic-phase subsistence economy, 62
initial introduction of, into South America, 83–84
Northern Flint Maize Complex, 88
origins of, 104–106
presence of cob chaff in organic residue, 93
rondel phytolith assemblage profiles, 84, 85, 88
signature of selection, testing, 108–109
single domestication of and its subsequent dispersal, 106
starch grains, identification of, 62, 63–64
use, social implications of, 93–94
Mangelsdorf, Paul, 105
Manioc (*Manihot esculenta* Crantz). *See also* Cassava
archaeological evidence from Aguadulce Rock Shelter site, 58–64
briefly described, 58–60
starch grains from, illustration of, 61
starches from other domesticated and wild plants at, 62–64
stone tools and sediments in, 60–62
bitter forms of, 47, 49
domesticated and wild species compared, 49, 57
domestication of, overview, 46, 47
evidence of, in Mesoamerica, 49
future research, suggestions for, 64
geographic locations of, 126
origins and dispersals, prior research on, 47–49
processing of, 49
reference collections and identification of, 51–58, 64
standard starch markers for, 58
starch grain analysis, case study in Panama, 46–67
starch grain size, table of, 57
use and spread of, 47–49
Mare's milk, 250
Mashkur, M., 294
Mbida, Ch., 68
Medical/veterinary genetics, 273
Mendel, Gregor, 1, 104
Mengoni Goñalons, Guillermo L., 228
Methodical selection
in annual seed crops, 19
in perennial root crops, 20–22

358   INDEX

Microsatellites, 7
    defined, 103
    distribution of, 319
    to measure and compare diversity, 296, 298, 299
    to study camelid origins, 334, 335–336, 337
    to study origins of cassava, 125–126, 127, 129
Milk genes, 277
Minisatellites, defined, 103
Mitochondrial DNA (mtDNA), 343
    advantages of using, 295
    control region sequencing, 280–281, 283, 285
    diversity, 274, 306–316, 324
    horse, mutation rate for, 346
    molecular evolution of, in animals, 6
    neighbor-joining networks linking, 274
    of plants, 7
    phylogenic reconstruction using, 318–319
    in study of camelid origins, 333–334, 337
Molecular clock, 9–10, 275
Molecular data, for study of crop domestication, 102–104
Molecular ecology, 273
Morphological markers, 99
    of animal domestication, 4, 171–174
    choosing appropriate, 124–126
    of plant domestication, 4–5, 15, 16, 17–19, 25–27, 99
mtDNA. See Mitochondrial DNA (mtDNA)
Multiple domestication, 100, 298
    in pigs, 218
    studies of, 109–111
Musa phytoliths, 72–73, 74, 77, 78

Native broodmare, 346
ncpGS sequencing, 160–162
Non-molecular data, for study of crop domestication, 101–102
Northern Flint Maize Complex, 88
Nuclear DNA (nDNA), 6, 7, 277

Oca (*Oxalis tuberosa*)
    cytological work concerning origins of, 156
    domestication, inferences from data concerning, 162–163
    future studies, suggestions for, 164–165
    habitat of domestication, speculation concerning, 154
    octoploid level, 163
    origins of
        comparisons with prior hypotheses, 163–165
        use of DNA sequence data to study, 114
    Oxalis tuberosa alliance, 114, 156
    preserved remains of, 154–155
    taxonomy, 155
    wild
        distribution of, 156
        tuber-bearing populations, 155–156
Oleaster, 143
    ancestor(s) of, 148
    Corsican, 149–150
    cultivars, 150

diversity, 148–149
molecular markers, distribution of, 147
and olive compared, 148
populations, analysis of, 145
study of, results, 147–148
Olive
    cultivars, 143, 147–148, 150, 151
    domestication
        and dispersal of, 114, 143–144, 147
        timing of, 151
    future research, suggestions for, 151
    history of, 143–144
    molecular markers, distribution of, 147
    new tools for studying origins of, 144
    and oleaster compared,148
    oil, 143
    production, 143
    study of
        cultivars, list of, 146
        data analyses, 145, 147
        molecular data, 145
        plant material, 144–145
        results, 147–148
    types of, 143
Olsen, Kenneth M., 123
Olsen, Sandra L., 245
The Origin of the Species (Darwin), 1
Overhunting, 186
Oxalis tuberosa. See Oca

Panama. See also Manioc
    flora, study of, 46
    map of archaeological sites, 47
Penning, as a marker of domestication, 4, 174, 176
Perennial root crops
    dietary importance of, 23
    domestication in, 20
    human intervention and, 22
    methodical selection in, 20–22
Phenology, 99
Phylogenetic reconstruction, 319
Phylogenetic relationship, between alpacas and llamas, 331
Phylogeny
    defined, 273
    intraspecific, 111
Phylogeographic discontinuities, 273–275
Phytolith(s)
    analysis of, as evidence of domestication, 21–22
    assemblages, food residue in, 21, 82–83
    banana, 68, 72–77
    defined, 21, 68
    Ensete, 73, 74, 77
    Heliconia, 78
    Musa, 72–73, 74, 77, 78
    rondels, 62–63, 84, 85, 86
Pig(s) (*Sus scrofa*)
    as adaptable and generalized omnivores, 209
    domestication
        aging, 217–218
        biometrical markers of, 212–217
        body size reduction and, 215
        dietary changes, 219–222
        future research, suggestions for, 223

        geographic distribution, 209, 211, 213–214
        hunting, seasonality of, 217
        isotope analysis, 221–222
        kill-off pattern, variation of, 217
        linear enamel hypoplasia (LEH), 221
        morphological similarity between different species, 211
        multiple domestication hypothesis, 218
        origins of, 218
        peak and tail pattern, 213, 215
        population genetics, 218–219
        prior research on, 209
        sex bias and, 214
        sexual compositions of assemblages, 214
        size variation as indication of, 212, 215–216
        snout, shortening of, 217
        temporal context for, 222–223
        tooth size, 215, 216–217
        tooth wear and calculus, evidence of, 220–221
        zoogeographic markers for, 210–212
    interspecies hybridization, 211
    prehistoric European, 209
    size index scaling technique, 212
    types of, 210
    wild, distribution of, 209
Piperno, Dolores R., 46
Plantains, 76–77
Plant domestication
    archaeological indications of, 15–17
    attractive candidates for, 3
    chemical composition, changes in, 99
    crop origins, 99–100
    crop-wild-weed geneflow, 111–112
    documenting, challenges to, 3–4
    evolution under domestication, 100–101
    founder effect phenomenon, 100–101, 107–108, 113
    future prospects for, 114–116
    generation time, 5
    genetic basis of, 112–113
    genetic data and, 99–122
    genetic documentation of, 5–7
    human intervention in, 18, 22–23
    morphological markers of, 4–5, 15, 16, 17–19, 25–27, 99
    multiple domestication, 109–111
    origin of, 1
    and polyploids, 111, 157–158
    and seed dispersal mechanisms, 3, 19
    signature of selection, testing, 108–109
Pole-axing, 259–260
Polymerase chain reaction (PCR), 99, 102, 103
Polyploidy, 6
    accessions sampled, 159
    choosing sequences for study, 158–159
    classification of, 153
    defined, 153
    domestication and, 111, 157–158
    flow cytometry data, 160
    genome complements, 153
    molecular data and the study of, 157–158
    oca as example of, 154–156, 158–165
    origins of, 100, 153–168

Polyploidy (*continued*)
   prior research on, 153–156
   sequencing and analyses, 159–160
   study results, 160–162
   terminology, 153
Population abundance, as marker of domestication, 176, 187
Population bottleneck, 6–7, 100
Population expansion(s), 275
   computing the relative dates of, 303
   indicators of, 302–303
   role of, in domestication, 10, 323
   timing of, 302–303
Population genetics, 277
Pressure facets, 51
Principal coordinate analysis (PCoA), 138
Principle Components Analysis (PCA), 332
Progenitor species
   distribution of, 176
   geneflow distinguished from, 130
   identification of, 100
Protodogs, 289
Protodomesticated population, 205
Przewalski horse, 246, 247, 344
Pumpelly, Raphael, 1

Quantitative trait, defined, 106
Quantitative trait loci (QTL) mapping, 6, 106–107, 112

Random amplified microsatellite polymorphism (RAMP), 103, 136
Random amplified polymorphic DNA (RAPD), 7, 103, 113–114, 134, 136–137, 144
Random genetic drift, 173
Reproduction, human control over, 230
Reserve starch, 50
Resource optimization, role in domestication, 10
Restriction fragment length polymorphism (RFLP), 102, 103, 144, 311
Restriction site, 102
Retrotransposon-based insertion polymorphism (RBIP), 103
Retrotransposon-microsatellite amplified polymorphism (REMAP), 103
Retrotransposons, 103
Ritual horse sacrifice, 249, 255
Robertsonian translocation, 246
Rondel phytolith, 62–63, 84, 85, 86
Root crops. *See also* Manioc; Perennial root crops
   domestication, identifying, 49
   problems of preservation in, 20
Rowley-Conwy, Peter, 209

Sampling, 131
   residues, procedure for, 84
   sample size, 19
Saturated genetic map, 107
Scanning electron microscopy (SEM), 1
Schaal, Barbara A., 123

Schlepp Effect, 261
SDS (sodium dodecyl sulfate) polyacrylamide gel electrophoresis (SDS-PAGE), 102
Seed(s)
   coat thickness, 15, 18, 25, 32, 38
   diameter, 37
   dispersal mechanisms, 3, 19
   germination dormancy, 25
   morphology, 32
   plants, morphological markers of domestication in, 19
   size, 15, 25, 26–27
   stock, storage and protection of, 18
Selection
   for breeding, 9, 10
   footprint (signature) of, 99, 108–109
   for less aggressive behavior, 171
   methodical
      in annual seed crops, 19
      in perennial root crops, 20–22
Selective breeding, 9, 10
Selectively amplified microsatellite polymorphic loci (SAMPL), 103
Selective sweep, 108
Sequence-related amplified polymorphism (SRAP), 103
Sexual precocity, as attribute of animal domestication, 4
Sexual reproduction, vs. clonal propagation, 5
Sheep
   breed types, 306
   crosses between species with different karyotypes, 306–307
   diploid chromosome number, 306
   diversity of, in Europe, 314
   domestic, origins of, 306, 307
   domestication
      future research, suggestions for, 311, 313–314
      horns as evidence of, 183
      mitochondrial DNA diversity and, 306–316
      overview, 306–308
      relationships between wild and domestic, 307
      study of, 308–311
      research techniques for, 308, 310
      results, 311
      samples analyzed, 309
      sequence analysis and alignment, 310–311
Signature of selection, 99, 108–109
Simple sequence repeats (SSRs), 144
Single-strand conformation polymorphism (SSCP), 103
Smith, Bruce, 1, 2, 15, 25
Social tensions, role of in domestication, 10
Spontaneous populations, 100
Squared chord distances, defined, 84

Squash (*Curcurbita pepo*)
   future research, suggestions for, 29–30
   lineage, 27
   morphological markers of domestication, 25–27
   multiple domestication, 109–110
   seed and peduncle assemblage
      from Guilá Naquitz Cave, 27–28
      from the Phillips Spring Site, 28–29
   seed size as marker of domestication in, 25, 26–27
SSRs. *See* Simple sequence repeats (SSRs)
Starch grains
   bell-shaped, 51, 53, 54–55, 57
   compound, 53
   defined, 50
   identification, basic aspects of, 50–51
   illustrations of, 52, 53, 56, 58, 61, 63, 64
   pressure facets, 51
   reference collections, 51
   research issues, 50
   size of, in modern wild and domesticated *Manihot*, 57
   types of, 50
Sweet potato (*Ipomoea batatas*), 46
Swennen, Ro., 68
Synteny, among taxa, 113

Taberlet, P., 294
Taq error, 160
Tarpan, 246–247
Teeth
   changes in, 258
   linear enamel hypoplasia (LEH), 221
   placement, 171–172
   size of, 212, 215–216, 247, 249
   wear and calculus, 220–221
Teosinte, geneflow from to maize, 106
Testa thickness. *See* Seed(s), coat thickness
Theoretical population genetics, 273
Thompson, Robert G., 82
Thong-smoothers, presence of, 261, 262
Threatened species, 338
Townsend, Saffron J., 306
Traits, inheritance of, 1, 104
Transitory starch, 50
Transposon tagging, 107
Tropical forests, seasonal classification of, 47

Unconscious selection, 18

VanNeer, W., 68
Vicuña, 332. *See also* Camelids, South American
   bidirectional introgression in, 333
   and guanaco compared, 334
Vilà, Carles, 279, 342
Vrydaghs, L., 68

Wayne, Robert K., 279
Wheeler, Jane C., 329

Wild Asiatic horse. *See* Przewalski horse
Wild goats
    geographic distribution of, 182
    horns in, 181–182
Wild populations, extinction of, 141
Wild progenitors, identification of, 1, 7–8, 100, 106. *See also* Progenitor species

World core germplasm collection, 127

Xoloitzcuintle, 285–286

Yacobaccio, Hugo D., 228
Yams (*Dioscorea trifida*), 46, 62, 64
Yautia (*Xanthosoma sagittifolium*), 46, 62

Y-chromosome DNA, 296, 297, 303
Y-chromosome markers, 319, 343, 346

Zea mays. *See* Maize
Zeder, Melinda, 1, 2, 3, 171, 181
Zoogeographic data, as evidence of domestication, 175–176, 210–212